Biosurfactants for a Sustainable Future

Biosurfactants for a Sustainable Future

Production and Applications in the Environment and Biomedicine

Edited by

Hemen Sarma
Department of Botany
Nanda Nath Saika College
Titabar, Assam, India

Majeti Narasimha Vara Prasad
School of Life Sciences
University of Hyderabad (an Institution of Eminence)
Hyderabad, Telangana, India

This edition first published 2021
© 2021 by John Wiley & Sons Ltd

All rights reserved. No part of this publication may be reproduced, stored in a retrieval system, or transmitted, in any form or by any means, electronic, mechanical, photocopying, recording or otherwise, except as permitted by law. Advice on how to obtain permission to reuse material from this title is available at http://www.wiley.com/go/permissions.

The right of Hemen Sarma and Majeti Narasimha Vara Prasad to be identified as the authors of this editorial work has been asserted in accordance with law.

Registered Offices
John Wiley & Sons Ltd, The Atrium, Southern Gate, Chichester, West Sussex, PO19 8SQ, UK
John Wiley & Sons, Inc., 111 River Street, Hoboken, NJ 07030, USA

Editorial Office
The Atrium, Southern Gate, Chichester, West Sussex, PO19 8SQ, UK

For details of our global editorial offices, customer services, and more information about Wiley products visit us at www.wiley.com.

Wiley also publishes its books in a variety of electronic formats and by print-on-demand. Some content that appears in standard print versions of this book may not be available in other formats.

Limit of Liability/Disclaimer of Warranty
In view of ongoing research, equipment modifications, changes in governmental regulations, and the constant flow of information relating to the use of experimental reagents, equipment, and devices, the reader is urged to review and evaluate the information provided in the package insert or instructions for each chemical, piece of equipment, reagent, or device for, among other things, any changes in the instructions or indication of usage and for added warnings and precautions. While the publisher and authors have used their best efforts in preparing this book, they make no representations or warranties with respect to the accuracy or completeness of the contents of this work and specifically disclaim all warranties, including without limitation any implied warranties of merchantability or fitness for a particular purpose. No warranty may be created or extended by sales representatives, written sales materials or promotional statements for this work. This work is sold with the understanding that the publisher is not engaged in rendering professional services. The advice and strategies contained herein may not be suitable for your situation. You should consult with a professional where appropriate. Neither the publisher nor author shall be liable for any loss of profit or any other commercial damages, including but not limited to special, incidental, consequential, or other damages. The fact that an organization, website, or product is referred to in this work as a citation and/or potential source of further information does not mean that the publisher and authors endorse the information or services the organization, website, or product may provide or recommendations it may make. Further, readers should be aware that websites listed in this work may have changed or disappeared between when this work was written and when it is read. Neither the publisher nor authors shall be liable for any loss of profit or any other commercial damages, including but not limited to special, incidental, consequential, or other damages.

Library of Congress Cataloging-in-Publication Data

Names: Sarma, Hemen, editor. | Prasad, Majeti Narasimha Vara editor.
Title: Biosurfactants for a sustainable future : production and
applications in the environment and biomedicine / Hemen Sarma,
Nanda Nath Saika College, Department of Botany, 785630, Titabar, India;
Majeti Narasimha Vara Prasad, University of Hyderabad, School of Life Sciences,
500046 Hyderabad, India.
Description: Hoboken, NJ : Wiley, 2021. | Includes bibliographical
references and index.
Identifiers: LCCN 2020051310 (print) | LCCN 2020051311 (ebook) | ISBN
9781119671008 (cloth) | ISBN 9781119671039 (adobe pdf) | ISBN
9781119671053 (epub)
Subjects: LCSH: Biosurfactants.
Classification: LCC TP248.B57 B565 2021 (print) | LCC TP248.B57 (ebook) |
DDC 668/.1–dc23
LC record available at https://lccn.loc.gov/2020051310
LC ebook record available at https://lccn.loc.gov/2020051311

Cover Design: Wiley
Cover Image: @Kennia Barrantes

Set in 9.5/12.5pt STIXTwoText by SPi Global, Pondicherry, India

C9781119671008_300321
Printed and bound by CPI Group (UK) Ltd, Croydon CR0 4YY

Contents

List of Contributors

Shashi Bhushan Agrawal
Department of Botany
Institute of Science
Banaras Hindu University
Varanasi
Uttar Pradesh
India

Akil Ahmad
School of Industrial Technology
Universiti Sains Malaysia
Gelugor
Penang
Malaysia

Subramania Angaiah
Electro-Materials Research Lab
Centre for Nanoscience and Technology
Pondicherry University
Puducherry
India

Juan José Araya
Escuela de Química
Centro de Investigaciónen Electroquímica y
Energía Química (CELEQ)
Universidad de Costa Rica
San José
Costa Rica

Zarith Asyikin Abdul Aziz
School of Chemical and Energy Engineering
Faculty of Engineering
University Teknologi Malaysia
Johor Bahru
Johor Malaysia

Kenia Barrantes
Nutrition and Infection Section
Health Research Institute
University of Costa Rica
San Jose
Costa Rica

Káren Gercyane Oliveira Bezerra
Northeastern Network of Biotechnology
Federal Rural University of Pernambuco
Recife
Pernambuco
Brazil

Advanced Institute of Technology and
Innovation (IATI)
Recife
Pernambuco
Brazil

Catholic University of Pernambuco
Recife
Pernambuco
Brazil

Siddhartha Narayan Borah
Royal School of Biosciences
Royal Global University
Guwahati
Assam, India

Luz Chacón
Nutrition and Infection Section
Health Research Institute
University of Costa Rica
San Jose, Costa Rica

Liang Cheng
School of Environment and Safety Engineering
Jiangsu University
Zhengjiang
China

Jéssica Correia
CEB – Centre of Biological Engineering
University of Minho
Braga
Portugal

Suresh Deka
Environmental Biotechnology Laboratory
Resource Management and Environment Section
Life Sciences Division
Institute of Advanced Study in Science and Technology (IASST)
Guwahati
Assam
India

Santiago de Frutos
Departamento de Química Física
Facultad de Ciencias
Universidad de Santiago de Compostela
Lugo
Spain

Maribel Farfán
Department of Biology
Healthcare and the Environment
Section of Microbiology
University of Barcelona
Barcelona
Spain

Francisco Fraga
Departamento de Física Aplicada
Facultad de Ciencias
Universidad de Santiago de Compostela
Lugo
Spain

S. J. Geetha
Department of Biology
College of Science
Sultan Qaboos University
Muscat
Oman

Madhurankhi Goswami
Environmental Biotechnology Laboratory
Resource Management and Environment Section
Life Sciences Division
Institute of Advanced Study in Science and Technology (IASST)
Guwahati
Assam
India

Eduardo J. Gudiña
CEB – Centre of Biological Engineering
University of Minho
Braga
Portugal

Sonam Gupta
Department of Biotechnology
National Institute of Technology, Raipur
Chhattisgarh
India

Gabriel Ibarra
Department of Public Health Sciences
College of Health Sciences
University of Texas at El Paso
El Paso
TX
USA

Nazim F. Islam
Department of Botany
N N Saikia College
Assam
India

Sanket J. Joshi
Oil & Gas Research Center
Central Analytical and Applied Research Unit
Sultan Qaboos University
Muscat
Oman

Aida Jover
Departamento de Química Física
Facultad de Ciencias
Universidad de Santiago de Compostela
Lugo
Spain

Arun Karnwal
Department of Microbiology
School of Bioengineering and Biosciences
Lovely Professional University
Phagwara
Punjab
India

Asma Khatoon
Centre of Lipids Engineering and Applied Research (CLEAR)
Universiti Teknologi Malaysia
Johor Bahru
Johor
Malaysia

Anand K. Kondapi
Laboratory for Molecular Therapeutics
Department of Biotechnology and Bioinformatics
School of Life Sciences, University of Hyderabad
Hyderabad
India

Current address: Department of Microbiology
Immunology and Pathology
Colorado State University
Fort Collins
CO
USA

Fernanda Lugo
Department of Public Health Sciences
College of Health Sciences
University of Texas at El Paso
El Paso
TX, USA

Chandana Malakar
Institute of Advanced Study in Science and Technology (IASST)
Garchuk
Assam
India

Ana María Marqués
Department of Biology
Healthcare and the Environment
Section of Microbiology
University of Barcelona
Barcelona
Spain

Francisco Meijide
Departamento de Química Física
Facultad de Ciencias
Universidad de Santiago de Compostela
Lugo
Spain

Monoj Kumar Mondal
Department of Chemical Engineering and Technology
Indian Institute of Technology
(Banaras Hindu University)
Varanasi
Uttar Pradesh
India

Amy R. Nava
Department of Interdisciplinary Health Sciences
College of Health Sciences
University of Texas
El Paso
TX, USA

Kannan Pakshirajan
Department of Biosciences and Bioengineering
Indian Institute of Technology Guwahati
Guwahati
Assam
India

Punniyakotti Parthipan
Electro-Materials Research Lab
Centre for Nanoscience and Technology
Pondicherry University
Puducherry
India

Yogesh Patil
Symbiosis Centre for Research and Innovation
Symbiosis International University
Pune
Maharashtra
India

Lourdes Pérez
Department of Surfactant and Nanobiotechnology
IQAC, CSIC
Barcelona
Spain

Aurora Pinazo
Department of Surfactant and Nanobiotechnology
IQAC, CSIC
Barcelona
Spain

Rolando Procupez-Schtirbu
General Chemistry
Department of Chemistry
University of Costa Rica
San Jose
Costa Rica

Vikas Pruthi
Department of Biotechnology
Indian Institute of Technology Roorkee
Roorkee
Uttarakhand
India

Aruliah Rajasekar
Environmental Molecular Microbiology Research Laboratory
Department of Biotechnology
Thiruvalluvar University
Vellore
Tamilnadu
India

Ashwini N. Rane
Department of Environmental Science
Savitribai Phule Pune University
Pune
Maharashtra
India

Selvan Ravindran
Symbiosis School of Biological Sciences
Symbiosis International University
Pune
Maharashtra
India

Ana I. Rodrigues
CEB – Centre of Biological Engineering
University of Minho
Braga
Portugal

Lígia R. Rodrigues
CEB – Centre of Biological Engineering
University of Minho
Braga
Portugal

Rashmi Rekha Saikia
Department of Zoology
Jagannath Barooah College
Jorhat
Assam
India

Hemen Sarma
Department of Botany
N N Saikia College
Titabar
Assam
India

Leonie Asfora Sarubbo
Advanced Institute of Technology and Innovation (AITI)
Recife
Pernambuco
Brazil

Catholic University of Pernambuco
Recife
Pernambuco
Brazil

Julio A. Seijas
Departamento de Química Orgánica
Facultad de Ciencias
Universidad de Santiago de Compostela
Lugo
Spain

Suparna Sen
Environmental Biotechnology Laboratory
Resource Management and Environment Section
Life Sciences Division
Institute of Advanced Study in Science and Technology
Guwahati
Assam
India

Siti Hamidah Mohd Setapar
School of Chemical and Energy Engineering
Faculty of Engineering, Universiti Teknologi Malaysia
Johor Bahru
Johor
Malaysia;

Department of Chemical Processes
Malaysia-Japan
International Institute of Technology
University Teknologi Malaysia
Skudai
Johor
Malaysia

SHE Empire Sdn., Jalan Pulai Ria
Bandar Baru Kangkar Pulai
Skudai
Johor
Malaysia

Pooja Singh
Symbiosis School of Biological Sciences
Symbiosis International University
Pune
Maharashtra
India

Victor H. Soto
School of Chemistry
Research Center in Electrochemistry and Chemical Energy (CELEQ)
University of Costa Rica
Costa Rica

Shalini Srivastava
Department of Botany
Institute of Science
Banaras Hindu University
Varanasi
Uttar Pradesh
India

José A. Teixeira
CEB – Centre of Biological Engineering
University of Minho
Braga
Portugal

José Vázquez-Tato
Departamento de Química Física
Facultad de Ciencias
Universidad de Santiago de Compostela
Lugo
Spain

M. Pilar Vázquez-Tato
Departamento de Química Orgánica
Facultad de Ciencias
Universidad de Santiago de Compostela
Lugo
Spain

Preface

This book is useful for the petrochemical industry (enhanced oil recovery from sludge), the pharmaceutical industry (developed technology for controlling multidrug-resistant pathogens), and the agro-industry (using byproducts), as well as environmental scientists and engineers (developing sustainable remediation technologies). As bioremediation is becoming green and a sustainable approach to environmental pollution control, the articles in this book will be relevant for future research that could benefit our stakeholders. The chapters in this reference book may be a unique collection that has been covered by most of the recent studies and provides systematic material produced by contemporary experts in the field. Focusing on research and development over the last 10 years, the study highlights relevant developments in the field. We hope that this book will support researchers by adding a new dimension to environmental studies and the remediation of emerging pollutants. A further benefit would be the understanding of the processes involved from the production to the sustainable use of biosurfactants in the environment and biomedicine.

- This book explains how various methods can be used to recognize and classify microorganism-producing biosurfactants in the environment. In addition, the various aspects of biosurfactants, including structural characteristics, developments, production, bioeconomics and their sustainable use in the environment, and biomedicine, are addressed. It presents metagenomic strategies to facilitate the discovery of novel biosurfactants (mechanistic understanding and future prospects) for the sustainable remediation of emerging pollutants.
- The use of microbes for human well-being is a prospective challenge, as they have developed novel chemicals and their metabolic pathway could be altered through omics approaches to the production of high-value chemicals (HVCs), including biosurfactants. These chemicals may be used in sustainable remediation techniques such as the regulation of the antibiotic resistance gene (AGR) and microbe-enhanced oil recovery (MEOR). We continue to face new and difficult challenges in the restoration of the environment, because current methods of remediation require so many chemicals that have again polluted the environment. There is a need to turn to more efficient alternative approaches and to find environmentally friendly chemicals for sustainability. As a result, the microbial world has the option of offering a replacement for green high-value chemicals to replace certain hazardous compounds already used in environmental reclamation.

This book opens a window on the rapid development of microbiology sciences by explaining how microbes and their products are used in advanced medical technology and in the sustainable remediation of emerging environmental contaminants. The authors concentrate on the environment as well as the biomedical field and highlight the role of microbes in the real world. This book will be updated to reflect current knowledge, the latest developments in the field of biosurfactants, sustainable remediation applications, and applied medical sciences, and the biotechnological strategies being developed to improve production processes. The most important goal of writing this book will be to communicate current advances and challenges in biosurfactant research. This will allow the reader to understand the dynamics of applied science that underlie microbially derived surfactants, called biosurfactants, and their use in sustainable remediation technology. The basic aim is to include updated content throughout in order to keep pace with this advancing field.

Key features:

- Addresses the applications of biosurfactants in sustainable remediation technology, for example, as agents to form emulsions and biofilm formation for desorption of hydrophobic pollutants.
- Discusses the current state of understanding of the different microbial surfactants, their classifications, properties, how to achieve higher yields, and new applications.
- There is a substantial research result on biosurfactants that envisages our capacity to build a consolidated framework for further development of applications. Biosurfactants for sustainable remediation technology should fill this need, covering the latest trend on biosurfactant research and their applications.

The book was contributed by 56 authors from leading surfactants research groups from Brazil, Costa Rica, China, India, Malaysia, Oman, Portugal, Spain, and the United States, comprising 22 chapters.

1) Introduction to Biosurfactants
2) Metagenomics Approach for Selection of Biosurfactant Producing Bacteria from Oil Contaminated Soils: An Insight into Its Technology
3) Biosurfactant Production Using Bioreactors from Industrial Byproducts
4) Biosurfactants for Heavy Metal Remediation and Bioeconomics
5) Application of Biosurfactants for Microbial Enhanced Oil Recovery (MEOR)
6) Biosurfactant Enhanced Sustainable Remediation of Petroleum Contaminated Soil
7) Microbial Surfactants Are Next-Generation Biomolecules for Sustainable Remediation of Polyaromatic Hydrocarbons
8) Biosurfactants for Enhanced Bioavailability of Micronutrients in Soil: A Sustainable Approach
9) Biosurfactants: Production and Role in Synthesis of Nanoparticles for Environmental Applications
10) Green Surfactants: Production, Properties, and Application in Advanced Medical Technologies
11) Antiviral, Antimicrobial, and Antibiofilm Properties of Biosurfactants: Sustainable Use in Food and Pharmaceuticals
12) Biosurfactant-Based Antibiofilm Nano Materials
13) Biosurfactants from Bacteria and Fungi: Perspectives on Advanced Biomedical Applications
14) Biosurfactant-Inspired Control of Methicillin-Resistant *Staphylococcus aureus* (MRSA)
15) Exploiting the Significance of Biosurfactant for the Treatment of Multidrug-Resistant Pathogenic Infections
16) Biosurfactants Against Drug-Resistant Human and Plant Pathogens: Recent Advances
17) Surfactant- and Biosurfactant-based Therapeutics: Structures, Properties, and Recent Developments in Drug Delivery and Therapeutic Applications
18) The Potential Use of Biosurfactants in Cosmetics and Dermatological Products: Current Trends and Future Prospects

19) Cosmeceutical Applications of Biosurfactants: Challenges and Perspectives
20) Biotechnologically Derived Bioactive Molecules for Skin and Hair-Care Application
21) Biosurfactants as Biocontrol Agents Against Mycotoxigenic Fungi
22) Biosurfactant-Mediated Biocontrol of Pathogenic Microbes of Crop Plants

The book explores how these twenty-first century multifunctional biomolecules improve or replace chemically synthesized surface-active agents with the aid of the industrial application of biosurfactant production based on renewable resources. This book is also useful for scholars, academicians in bioengineering and biomedical sciences, undergraduate and graduate students in microbiology, environmental biotechnology, health, clinical, and pharmaceutical sciences.

1

Introduction to Biosurfactants

José Vázquez Tato[1], Julio A. Seijas[2], M. Pilar Vázquez-Tato[2], Francisco Meijide[1], Santiago de Frutos[1], Aida Jover[1], Francisco Fraga[3], and Victor H. Soto[4]

[1] *Departamento de Química Física, Facultad de Ciencias, Universidad de Santiago de Compostela, Avda, Lugo, Spain*
[2] *Departamento de Química Orgánica, Facultad de Ciencias, Universidad de Santiago de Compostela, Avda, Lugo, Spain*
[3] *Departamento de Física Aplicada, Facultad de Ciencias, Universidad de Santiago de Compostela, Avda, Lugo, Spain*
[4] *Escuela de Química, Centro de Investigación en Electroquímica y Energía Química (CELEQ), Universidad de Costa Rica, San José, Costa Rica*

CHAPTER MENU

1.1 Introduction and Historical Perspective

Surface tension is a property that involves the common frontier (boundary surface) between two media or phases. Strictly speaking, the surface tension of a liquid should mean the surface tension of the liquid in contact and equilibrium with its own vapor. However, as the gas phase has normally a small influence on the surface, the term is generally applied to the liquid–air boundary. The phases can also be two liquids (interfacial tension) or a liquid and solid. According to IUPAC, the surface tension is the work required to increase a surface area divided by that area [1]. This is the reversible work required to carry the molecules or ions from the bulk phase into the surface implying its enlargement and corresponds to the increase in Gibbs free energy (G) of the system per unit surface area (A),

$$\left(\frac{\partial G}{\partial A}\right)_{T,P} = \gamma \tag{1.1}$$

Biosurfactants for a Sustainable Future: Production and Applications in the Environment and Biomedicine, First Edition. Edited by Hemen Sarma and Majeti Narasimha Vara Prasad.
© 2021 John Wiley & Sons Ltd. Published 2021 by John Wiley & Sons Ltd.

where γ is the interfacial tension. Therefore, the units of γ are J/m^2 or N/m, but it is normally recorded in mN/m (because it coincides with the value in dyn/cm of the cgs system). In 1944, Taylor and Alexander [2] collected some representative published (1885–1939) values for the surface tension of water at 20 °C. Their own value was 72.70 ± 0.07 mN/m (calculated by extrapolation) in agreement with more recent determinations, the accepted value being 71.99 ± 0.36 mN/m at 25 °C [3]. This is a rather high value when it is compared with those of other common solvents as ethanol (22.39 ± 0.06 mN/m), acetic acid (27.59 ± 0.09 mN/m), or acetone (29.26 ± 0.05 mN/m) (values from [4]) at 20 °C.

The decrease in the surface tension of water has been traditionally achieved by using soaps or soap-like compounds. According to IUPAC a "soap is a salt of a fatty acid, saturated or unsaturated, containing at least eight carbon atoms or a mixture of such salts. A neat soap is a lamellar structure containing much (e.g. 75%) soap and little (e.g. 25%) water. Soaps have the property of reducing the surface tension of water when they are dissolved in soap-like compounds in water." This reduction facilitates personal care, washing of clothes and other fabrics, etc. The early documents with descriptions of soaps and their uses are typically related with medicinal aspects, and nowadays there is almost a specific type of soap for each requirement. Levey [5] has reviewed the early history of "soaps" used in medicine, cleansing, and personal care. For instance, he mentions that "in a prescription of the seventh century BC, soap made from castor oil (source of ricinoleic [12-hydroxy-9-*cis*-octadecenoic] acid) and horned alkali is used. . . as a mouth cleanser, in enemata, and also to wash the head." However, Levey concludes that a true soap using caustic alkali was probably not produced in antiquity but "evidence has been adduced to indicate that salting out was in use in early Sumerian times." In his *Naturalis Historia*, Pliny the Elder [6] refers to soap (sapo) as *prodest et sapo, Galliarum hoc inventum rutilandis capillis. fit ex sebo et cinere, optimus fagino et caprino, duobus modis, spissus ac liquidus, uterque apud Germanos maiore in usu viris quam feminis*, which may be translated as "There is also soap, an invention of the Gauls for making their hair shiny (or glossy). It is made from suet and ashes, the best from beechwood ash and goat suet, and exists in two forms, thick and liquid, both being used among the Germans, more by men than by women."

Hunt [7] indicates that centers of soap production by the end of the first millennium were in Marseilles (France) and Savona (Italy), while in Britain some references appear in the literature around 1000 AD. For instance, in 1192 the monk Richard of Devizes referred to the number of "soap makers in Bristol and the unpleasant smells which their activities produced." Hunt also resumed other aspects as the chemistry of soap, the British alkali industry, the expansion of soap production, soap manufacturers, and manufacturing methods. As early as 1858, Campbell presented a USA patent [8] for the production of soaps. He described the process as consisting in "the use of powdered carbonate of soda for saponifying the fatty acids generally, and more particularly the red oil or 'red (oleic) acid oil' and converting them, by direct combination, into soap in open pans or kettles, at temperatures between 32 and 500 °F." Mitchell [9] revised the *Jabón de Castilla* or Castile soap (named from the central region of Spain), probably the first white hard soap. It was an olive oil-based soap and soaps with this name can still be bought today. Traditional recipes and videos can be easily found on the Internet. In the paper "Literature of Soaps and Synthetic Detergents", Schulze [10] recorded the literature (including books, periodicals, abstracts, indexes, information services, patent publications, association publications, conference proceedings) on soaps, surfactants, and synthetic detergents up to 1966.

Nowadays descriptions for soap-making from fats and oils are frequent for teaching purposes. For instance, Phanstiel et al. [11] have described the saponification process (basic hydrolysis of fats). It involves heating either animal fat or vegetable oil in an alkaline solution. The alkaline solution hydrolyses the triglyceride into glycerol and salts of the long-chain carboxylic acids (Scheme 1.1).

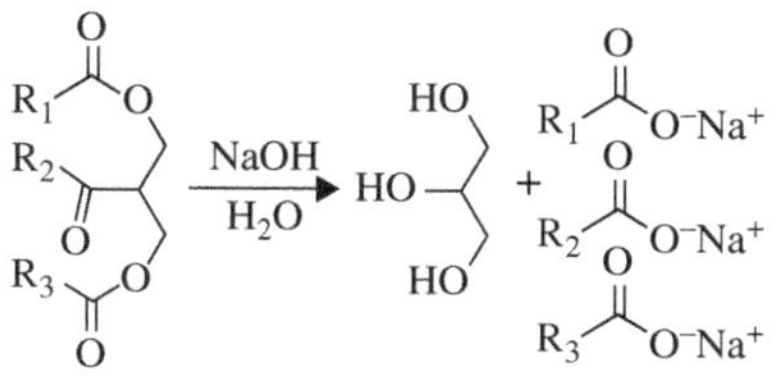

Scheme 1.1 Alkaline hydrolysis of a triglyceride to obtain soaps.

To overcome the shortcomings of the carboxylic group of soaps, during the first decades of the twentieth century, new surface-active agents were obtained in chemistry laboratories. Kastens and Ayo [12] and Kosswig [13] reviewed the main achievements of these decades. The first result of this search was Nekal, an alkyl naphthalene sulfonate, although it probably was a mixture of various homologs [14]. Other pioneer compounds were Avirol series (sulfuric acid esters of butyl ricinoleic acid), Igepon A series (fatty acid esters of hydroxyethanesulfonic acid), Igepon T series (amide-derivatives of taurine). All these products represented different approaches to the elimination of the carboxylic group of soaps. IUPAC defines a surfactant as a substance that lowers the surface tension of the medium in which it is dissolved and/or the interfacial tension with other phases, and, accordingly, is positively adsorbed at the liquid/vapor and/or at other interfaces. By detergent, IUPAC refers to a surfactant (or a mixture containing one or more surfactants) having cleaning properties in dilute solutions. Thus, soaps are surfactants and detergents.

It is not easy to whom the use of the word surfactant should be ascribed for the first time. A search in SciFinder® suggests that the word was first used by Bellon and LeTellier in a French patent (1943) [15]. The SciFinder abstract of this patent indicates that "Surfactants such as wetting agents, detergents, emulsifiers, and stickers are prepared by treating by-product materials containing starches, cellulose, amino acids, and smaller quantities of inedible fats with NaOH and neutralizing the reaction product."

Because of their physicochemical properties, surfactants have found applications in almost any kind of industry. A list of the relevant ISO and DIN regulations for a utility evaluation of surfactants has been provided by Kosswig [13]. For instance, in 1950 Lucas and Brown [16] measured the wetting power of 13 surfactants to find a wetting agent that would enable sulfuric acid to wet peaches quickly and uniformly so as to permit acid peeling. Anionic, cationic, and neutral surfactants were tested. In the *Application Guide* appendix of the book *Chemistry and Technology of Surfactants* [17] there is a list that illustrates the variety of surfactants and their versatility in a wide range of applications. Among others the following are mentioned: Agrochemical formulations, Civil engineering, Cosmetics and toiletries, Detergents, Household products, Miscellaneous industrial applications, Leather, Metal and engineering, Paints, inks, coatings, and adhesives, Paper and pulp, Petroleum and oil, Plastics, rubber, and resins, and Textiles and fibers. For instance, their wetting properties have been early used in food technology. We have already mentioned the early connection of soap and medicine and correspondingly the use of surfactants in pharmacy in the formulation (as emulsifying agents, solubilizers, dispersants, for suspensions) and as wetting agents, which cannot be a surprise [18]. Nursing care makes a continuous use of surface-active agents.

The soaps of Scheme 1.1 show the most important structural characteristic of surfactants: the coexistence of one lyophilic group (alkyl chain) and one lyophobic group (carboxylate ion). In aqueous solutions, it is more frequent to use the terms hydrophilic and hydrophobic. A graphical representation head–tail (hydrophobic group–hydrophilic group) is widely used, the alkyl chain being the tail and the carboxylate group the head (Figure 1.1). This structure gives the amphiphile character to surfactant compounds.

More generally, the head can be any polar group and the tail any apolar group, leading to a wide range of structures and types of surfactants. Among anionic heads, typical groups are carboxylate, sulfate, sulfonate, and phosphate, while the most frequent counterions are monovalent and divalent cations. Polycharged heads are also common, EDTA derivatives being well-known examples [19]. Cyclopeptides constitute another important group [20]. Among cationic heads, typical groups are tetralkylammonium, *N,N*-dialkylimidazolinium and *N*-alkylpyridinium ions, while chloride and bromide are the most common counterions. Among neutral heads, polyethylene glycol ethers, polyglycol ethers, and carbohydrates can be mentioned. Zwitterionic heads are very important as phospholipids belong to this group, as well as sulfobetaines and trialkylamine oxides. Many examples can be found elsewhere [13].

However, the structures of surfactants may be more complex than the head–tail model suggests. For instance, the number of polar and non-polar groups can be higher than one, the phospholipid phosphatidylcholine with two alkyl–allyl chains and a zwitterion as the head being an example. Gemini surfactants are dimeric surfactants [21] carrying two charged groups and two alkyl groups. The two amphiphilic moieties are connected at the level of the head groups, which are separated by a spacer group. They are characterized by critical micelle concentrations that are one to two orders of magnitude lower than those corresponding to conventional (monomeric) surfactants [22].

Bolaamphiphilic molecules contain a hydrophobic skeleton (e.g. one, two, or three alkyl chains, a steroid, or a porphyrin) and two water-soluble groups on both ends [23]. They can be symmetric or asymmetric [24, 25]. Recent examples of bolaamphiphilic, Y-shaped and divalent surfactants have been published by Baccile et al. [26] (Figure 1.1).

Some surfactants, instead of the mentioned head–tail structure, present a bifacial polarity with the hydrophilic and hydrophobic characteristics at two opposite sides of the molecule. The best-known examples are bile salts (see Figure 1.2) [27, 28]. Many membrane-active compounds are facial amphiphiles including cationic peptide antibiotics [29]. The facial amphiphilic conformation adopted by these peptides is a consequence of their secondary and tertiary structures, allowing

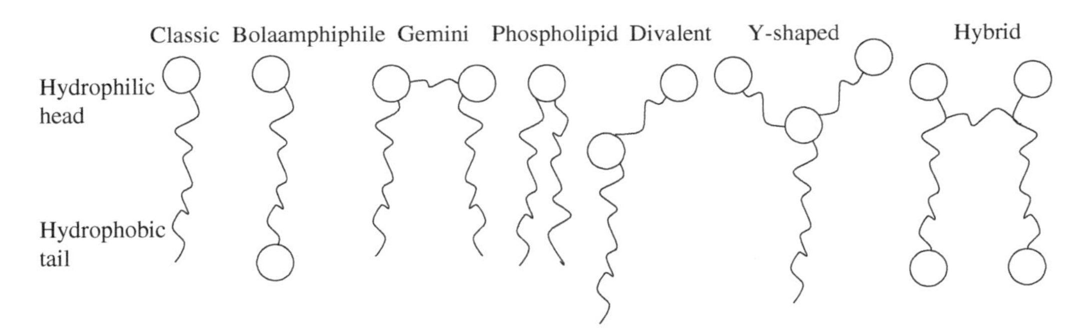

Figure 1.1 Schematic representation of the structure of some surfactants.

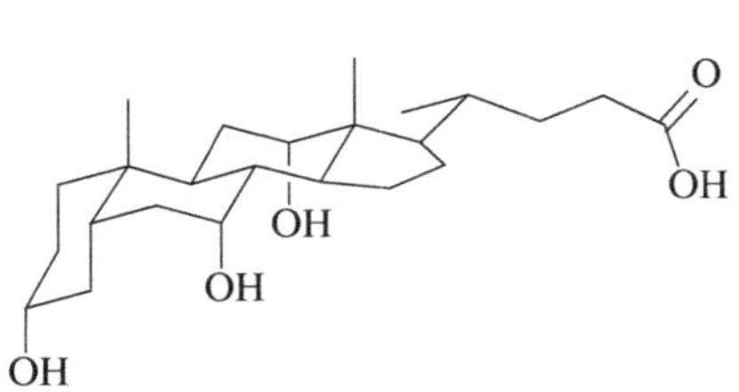

Figure 1.2 Bifacial structure of cholic acid.

that one face of the molecule presents cationic groups (protonated amines or guanidines) and the other face contains hydrophobic groups. An example may be magainin I [30]. Among other surfactant structures, diblock copolymers and polymeric surfactants, fluorosurfactants and silicone-based surfactants can be mentioned [13].

1.2 Micelle Formation

The necessity of a quantitative measurement of the surface tension of soap solutions was soon evident. By the time that I. Traube published his earliest paper in 1884, significant theories of capillarity from La Place, Poisson, or Gauss were known [31]. Early measurements of the surface tension only imply inorganic salts, acids, and bases. In 1864 Guthrie [32, 33] measured some organic liquids. At the same time, Musculus [34] studied the capillarity of aqueous solution of alcohol observing that "the capillarity of the water decreases considerably with the addition of the least amount of alcohol, in the beginning, much faster than in the presence of more alcohol." He also noticed that "all derivatives of ethyl alcohol which are soluble in water (as acetic acid) behave like this, and probably this is also the case with the other alcohols," but substances such as "sugars, and salts if they are not present in a great amount, almost do not influence the capillarity of water." He proposed the use of capillarity for measuring the concentration of alcohol and acetic acid in water, among other reasons, because "it offers the advantage that one needs only very little fluid for analysis, one drop being enough." He continued that, as "the animal fluids, such as blood serum, urine, have a capillarity which is equal to that of water, it is possible to detect and quantify substances in the urine," making reference, for instance, to bile.

Traube started the measurement of the influence of many organic substances on the surface tension of water in the period 1884–1885 [31] and observed that "the surface tension of capillary-active compounds belonging to one homologous series decreased with each additional CH_2 group in a constant ratio which is approximately 3:1," leading him to propose *Traube's Rule*.

A nice historical paper was published by Traube [31] in 1940, in which he mentioned previous works related to the investigation of aqueous solutions of inorganic salts, acids, and bases, employing the method of capillary tubes, and, particularly, the dropping method applied by Quinke. Traube developed this method and designed a simple instrument, the stalagmometer – together with the stagonometer – which found general application in science and industry. In the mentioned paper, Traube refers mainly to his publications that appeared in the period 1886–1887. By 1906, the measurement of the surface tension by the capillary rise was so important that it was included in the book *Practical Physical Chemistry* by A. Findlay. The use of Traube's stalagmometer for such a purpose was proposed in the 3rd edition of the book, published in 1915. The experiment is still proposed in recent textbooks on practical Physical Chemistry [35].

Seventeen of the more important methods of measuring surface tension were described in 1926 by Dorsey [36]. According to his own words, "The list of references does not pretend to be complete but is intended merely to direct the reader to one or more of the sources from which the required information can be obtained most satisfactorily." Even so, the number of cited papers was greater than 110, while the number of citations corresponding to the nineteenth century was 63 (56%). Eminent scientists such as Bohr, Rayleigh, Thomson, Kelvin, Maxwell, Laplace, and Poisson were among them. Tate [37] published his famous law in 1864 and Wilhelmy in 1863.

Even at low concentrations, surfactants reduce the surface tension of water due to its tendency to migrate toward the air–water interface, forming a monolayer. This was first suggested in 1907 by Milner [38] and, previously, Marangoni in 1871 "suggested that this capability [local variation in the tension of its surface] is due to the presence on the surface of the film of a pellicle, composed of matter having a smaller capillary tension than that of water." Milner clearly established that "in several organic solutions the surface tension is less than that of water, and there is consequently an excess of solute in the surface." Later, Langmuir [39] indicated that "the -COOH, -CO, and –OH groups have more affinity for water than for hydrocarbons. . . [and] when an oil is placed on water, the –COO– groups combine with the water, while the hydrocarbon chains remain combined with each other." In other words, the tail of a surfactant (the hydrocarbon chain) must be located at the air interface, with the tail upwards oriented and the head (hydrophilic groups) at the water interface.

Rising the surfactant concentration, the surface concentration increases as well until the full coverage of the interface by the molecules or ions. If the interface is completely covered, further increment of the surfactant concentration does not (almost) modify the surface tension. Furthermore, the additional surfactant molecules (or ions) have to remain in the bulk solution, and following Langmuir "hydrocarbon chains remain combined with each other, thus forming micelles" (or other aggregates).

The term micelle was commonly used by the first years of the twentieth century [40, 41] in relation to colloid solutions (frequently inorganic gels). In 1920 McBain and Salmon [42] (see also [43]) described a brief *résumé* of previous work, citing, for instance, Krafft's work. From the summary of this paper we extract the following sentences:

- 3. These colloidal electrolytes are salts in which one of the ions has been replaced by an ionic micelle.
- 5. This is exemplified by any one of the higher soaps simply on change of concentration. Thus, in concentrated solution there is little else present than colloid plus cation, whereas in dilute solution both undissociated and dissociated soap are crystalloids of simple molecular weight.
- 8. The ionic micelle in the case of soaps exhibits an equivalent conductivity quite equal to that of potassium ion. Its formula may correspond to $P_n^{n-}.mH_2O$ but more probably it is $\left(NaP\right)_x P_n^{n-}.mH_2O$, where P^- is the anion of the fatty acid in question.

Therefore, the essential definition of the present concept of a micelle was established. IUPAC indicates that "Surfactants in solution are often association colloids, that is, they tend to form aggregates of colloidal dimensions, which exist in equilibrium with the molecules or ions from which they are formed. Such aggregates are termed micelles."

In 1922, McBain and Jenkins [44] studied solutions of sodium oleate and potassium laurate by ultrafiltration, using this technique for separating the ionic micelle from the neutral colloid. For both surfactants they showed that the proportion (simple potassium laurate or sodium oleate)/(ionic micelle) increases fast at low concentrations and reached a plateau at high concentrations (see graphs of the paper). They also concluded that the diameter of the ionic micelle is only a few times the length of the molecule and "the particles of sodium oleate are about ten times larger than those of potassium laurate."

By the end of the twenties and the beginning of thirties of the twentieth century, the research activity on micelle-forming substances experienced an extraordinary blooming spring. The paper by Grindley and Bury [45] is a landmark on the subject, being particularly illustrative for the purposes of this review. They represented the formation of micelles by butyric acid in solution by the equation

$$nC_3H_7CO_2H \rightleftarrows \left(C_3H_7CO_2H\right)_n \tag{1.2}$$

where n is "the number of simple molecules in a micelle" or aggregation number (which is a relatively large number) and write the equilibrium constant as

$$K^n = s^n / m_n \tag{1.3}$$

where s and m are the concentrations of butyric acid as monomers and as micelles, respectively. The previous equation can be written as

$$m_n = (s/K)^n \tag{1.4}$$

from which they deduced that if s/K is appreciably smaller than unity, the concentration of micelles will be negligible. Only when s approaches the value K does the concentration of micelles become appreciable, and "will rapidly increase as the total concentration increases." From this analysis they conclude that "if any physical property of aqueous butyric acid solutions be plotted against the concentration, the slope of the curve will change abruptly near this point." A few months later, Davies and Bury [46] named that concentration as the *critical concentration for micelles.*

Previous analysis constitutes the basis of all experimental techniques so far used for determining the critical concentration for micelles (from here *cmc*). For instance, the association of monomers in micelles reduces the number of particles in the solution and, consequently, colligative properties (freezing point, vapor pressure. . .) also drastically change at this concentration. Other properties such as solubilization of solutes as dyes or the conductivity of the solution also change significantly. As an example, we shall mention the paper by Powney and Addison [47] who measured the surface tension of aqueous solutions of sodium dodecyl, tetradecyl, hexadecyl, and octadecyl sulfates and plotted the results in the form of vs log (concentration), as we do nowadays. The curves showed breaks at critical concentrations, which correspond to transitions from single ions to micelles, these single ions constituting the surface-active species. Figure 1.3 shows a typical plot for an unspecified surfactant. Powney and Addison noticed that the magnitude of the surface activity and the critical concentration for micelles were governed by chain length, temperature, and the valency of the added cation.

In 1895, Krafft and Wiglow [48] observed the formation of crystals at 60°, 45°, 31.5°, 11°, 35°, and 0° with hot aqueous solutions (1%) of stearate, palmitate, myristate, laurate, and elaidate sodium salts, respectively. Each of these temperatures is now known as the Krafft point (T_k). IUPAC defines it as the temperature (more precisely, narrow temperature range) above which the solubility of a surfactant rises sharply. At this temperature the solubility of the surfactant becomes equal to the *cmc*.

In 1955, Hutchinson et al. [49] published the paper "A new interpretation of the properties of colloidal electrolyte solutions" in which "the formation of micelles was treated as a phase separation rather than as an association governed by the law of mass action." Seven years later, Shinoda and Hutchinson [50] used this model to interpret the Krafft point, associated to the micellization process. These authors proposed micellization as a "similar phase separation, with the important distinction that micellization does not lead to an effectively infinite aggregation number, such as corresponds to true phase separation." If correct, the model requires that the activity of

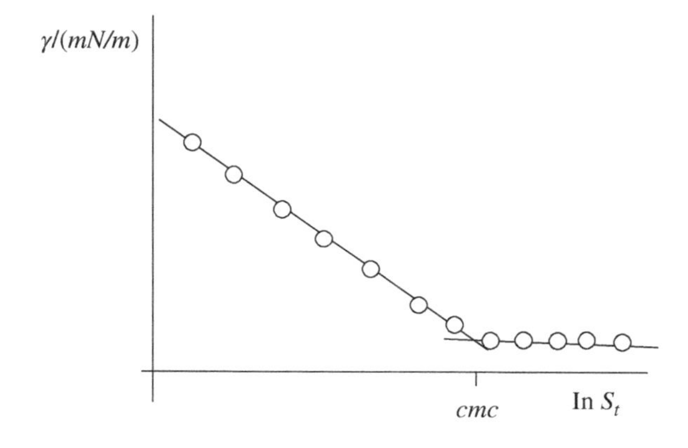

Figure 1.3 Typical surface tension vs ln (surfactant) plot showing the break point corresponding to *cmc*.

micelle-forming compounds should be practically constant above the *cmc*. Among others, the authors invoke Nilsson results [51] with radiotracers as evidence for their proposition. In a frequently reproduced graph, Shinoda and Hutchinson [50] plotted the concentration vs temperature for sodium decyl sulfonate near the Krafft point. The plot resembles the phase diagram of water near its triple point. If micelles are considered as a phase, by the phase rule, the system should become invariant at constant temperature and pressure [49, 50]. In other words, "the equilibrium hydrated solid monomers micelles is univariant, so that at a given pressure the point is fixed." As temperature increases, the solubility also increases until T_k where the *cmc* is reached. Above this temperature, the surfactant is dissolved in the form of micelles.

Let us go back to 1915. In this year, Allen [52] published a paper in which he showed the use of the surface tension measurement for the determination of bile salts in urine. In the introduction of his paper, Allen refers to Hay's method of testing the presence of bile salts in the urine. That method consists in "shaking flowers of sulphur upon the surface of the urine... When the surface tension of the urine is lowered [by bile salts] the powdered sulphur sinks to the bottom, and the lower the surface tension the more rapidly this takes place. [But] The method is very unsatisfactory... and if possible, a quantitative method, would be very desirable." Thus Allen proposed a very accurate measurement of the surface tension of a solution by the stalagmometric method "to determine the feasibility of estimating the amount of bile salts present in pathological urines from measurements of the surface tension taken with a portable Traube stalagmometer." He computed the surface tension value of a solution in per cent of that of distilled water according to the formula

$$\frac{\textit{Number of drops of distillided water}}{\textit{Number of drops of solution}} \times \textit{specific gravity of the solution} \tag{1.5}$$

The method relies on the fact that bile salts possess the property of lowering the surface tension of a solution very markedly, even when present in small concentrations. Table I of his paper shows some results for sodium glycocholate (NaGC) in distilled water. In a reanalysis of these data by plotting the surface tension vs ln(concentration), it is possible to determine a value of 0.0117 M for the *cmc* of NaGC, a value in perfect agreement with recent measurements. From the Reis et al. [53] compilation, an average value of $(1.04 \pm 0.29) \times 10^{-2}$ M may be estimated for the *cmc* of this bile salt.

The "excess of solute [surfactant] in the surface" indicated by Milner has traditionally been analyzed through the Gibbs equation [54]

$$\Gamma = -\frac{1}{nRT}\left(\frac{\partial \gamma}{\partial \ln c}\right) \tag{1.6}$$

where Γ is the surface excess, $(\partial\gamma/\partial \ln c)$ is the slope of the dependence of γ with the logarithm of the concentration of the surfactant (frequently being linear), R is the ideal gas constant, T the temperature, and n a factor that depends on the nature of the surfactant. The equation allows the determination of the area occupied per molecule at the interface, which is the inverse of the surface excess, i.e.

$$a = \frac{1}{N_A \Gamma} \tag{1.7}$$

where N_A is Avogadro's number.

Recently Menger et al. [55] have questioned the validity of the Gibbs equation on the basis that in the region of concentration where the equation is applied the adsorption at the interface does not generally reach saturation. This criticism has been supported by measurements from a radioactive surfactant [56], results that suggest that the γ-ln c linearity is not indicative of surface saturation, a hypothesis required for the deduction of the Gibbs equation. Neutron reflection measurements also support the fact that there are serious limitations in applying the Gibbs equation accurately to surface tension data [57, 58].

In the late nineteen thirties, other important papers were published. Wright and co-workers [59–61] measured the conductivity, density, viscosity, and solubility of several sodium alkyl (decyl, dodecyl, and hexadecyl) sulfonates at several temperatures. In all cases breaks at the curves or linear dependences of the property with the sulfonate concentration were observed. They also reported that the addition of sodium chloride to solutions of sodium dodecyl sulfonate lowered the *cmc* and that the lowering becomes less marked with a rise in temperature. Hartley [62] demonstrated that paraffin chain salts behave as strong electrolytes at low concentrations. For cetane sulfonic acid, a value of about 0.008 N in water at 60 °C was given for *cmc* and that it increased by about 2% per degree. This is an important question since the formation (or not) of premicellar aggregates is still under debate.

By the end of this decade Hartley [63] reviewed (36 references) the subject, the title of the paper being *Ion aggregation in solutions of salts with long paraffin chains*. In the abstracts we can read about the structure of micelles which are "aggregates of paraffin-chain ions with some adsorbed opposite ions," micelles are spherical with a radius equal to the length of a completely stretched paraffin-chain and have a liquid interior and the strong dependence of *cmc* with the length of the hydrocarbon chain and nature of the ionized terminal groups and opposite ions (counterions), and with temperature (in less extension). He also affirmed that "the spherical micelle is more stable than ion pairs."

Thus, by this time, the essential parameters that define a micelle were introduced or established: change of properties at the *cmc*, variables that influence the *cmc*, shape and size, internal and peripheral structures, and the essential thermodynamics (mass action law).

In the period 1946–1947, immediately after the Second World War, the activity on surfactant research experiences an important enhancement.

Corrin and Harkins [64] proposed the equation $\log(cmc) = -A \times \log(counterion^{\pm}) - B$ to relate the dependence of the *cmc* with the concentration of added salts (the sign at the superscript of the counterion is opposite to that of the surfactant ion). Table 1.1 resumes the values for the constants A and B for several surfactants. They also noticed that urea has a negligible effect in lowering the *cmc*.

Three years later, Lange [65] applied the mass action law to ionic micelles and wrote the equilibrium of formation of the micelle as

$$pK + qA \rightleftarrows K_pA_q \left(q > p\right) \tag{1.8}$$

where K is the counterion, A the surfactant ion, and p and q the stoichiometric coefficients. Although Lange considered the activity coefficients of the different species, for simplicity we will ignore them and write the equilibrium constant as

$$\left[K\right]^p \left[A\right]^q = K_{eq}\left[K_pA_q\right] \tag{1.9}$$

Table 1.1 Parameters A and B of the Corrin–Harkins equation. The number of figures on the values of *A* and *B* has been reduced.

Surfactant	*A*	*B*
Potassium laurate	0.570	2.62
Sodium dodecyl sulfate	0.458	3.25
Dodecyl ammonium chloride	0.562	2.86
Decyltrimethylammonium bromide	0.343	1.58

Source: Corrin and Harkins [64], p. 683.

Writing $[A] = c_k$ and $[K] = c_k + N$, where N is the equivalent concentration of added salt, it is finally found that

$$\log c_k = -\frac{p}{q}\log\left(c_k + N\right) + \frac{1}{q}\log L \tag{1.10}$$

where $L = [K_p A_q]$. Thus with $\log c_k$ as ordinate and $\log(c_k + N)$ as abscissa, this is the equation of a straight line with the slope $-p/q$, which corresponds to the empirical one found by Corrin and Harkins. This point has been discussed in detail by Hall [66] in his theory for dilute solutions of polyelectrolytes and of ionic surfactants.

The effects of solvents (alkyl alcohols $C_nH_{2n+1}OH$, $n = 1–4$; $HOCH_2CH_2OH$, glycerol, 1,4-dioxane, and heptanol) on the critical concentration for micelle formation of cationic soaps was studied by Corrin and Harkins [67], Herzfeld et al. [68], and Reichenberg [69]. Klevens [70] found that increasing the temperature causes an apparent decrease in the *cmc*, as determined by spectral changes in various dyes. However, this same author found the opposite effect when the micelles formation was determined by refraction [71].

Simultaneously, other experimental techniques, mainly spectroscopic ones, were introduced for the determination of the *cmc*. After a paper published by Sheppard and Geddes [72], in which the authors reported that by the addition of cetyl pyridinium chloride, the absorption spectrum of aqueous pinacyanol chloride was shifted from that exhibited in aqueous solutions to that in non-polar solvents, Corrin et al. [73] used this property to determine the *cmc* of laurate and myristate potassium salts, giving values of 6×10^{-3} M and 0.023–0.024 M, respectively. The concentration of soap at which this spectral change occurs was taken as the *cmc*, proposing that the dye is solubilized in a non-polar environment within the micelle. Klevens [74] performed a similar work by studying the changes in the spectrum of pinacyanol chloride in solutions of myristate, laurate, caprate and caprylate potassium salts, and sodium lauryl sulfate. These studies were extended to other surfactants [75] and other dyes as p-dimethylaminoazobenzene [76]. By using suitable dyes (Rhodamine 6G, Fluorescein, Acridine Orange, Acridine Yellow, Acriflavine, and Dichlorofluorescein) fluorescence spectroscopy was soon adopted [77, 78].

In 1950, Klevens [79] studied the solubility of some polycyclic hydrocarbons in water and in solutions of potassium laurate (at 25 °C). For all the polycyclic hydrocarbons, he showed that by increasing the concentration of the surfactant, their solubility also increased. Particularly, for pyrene he measured solubilities of 0.77×10^{-6} and 2.24×10^{-3} M in water and potassium laurate (0.50 M), respectively.

One year later, Ekwall [80] studied the sodium cholate association by measuring the fluorescence intensity, and determined that the lowest concentration at which polycyclic hydrocarbons (3,4-benzopyrene included) are solubilized is 0.018 M. This corresponds to the beginning of the micelle formation, although "at first relatively small amounts of cholate ion aggregates and the actual micelle formation occurs at about 0.040 to 0.044 M." Foerster and Selinger [81] observed that in micelles of cetyldimethylbenzylammonium chloride, pyrene forms dimers in excited states (excimers).

In the period 1971–1980, the number of papers on solubilized pyrene in micelle solutions increased very quickly. The fluorescence decay of the excited state of pyrene received an important attention. The aggregation number and microviscosities of the micellar interior [82], the permeability of these micelles with respect to nonionic and ionic quenchers [83], oxygen penetration of micelles [84], or the environmental effects on the vibronic band intensities in pyrene monomer fluorescence in micellar systems [85, 86] were published. Kalyanasundaram and Thomas carefully analyzed the lifetime of the monomer fluorescence and the ratio I_3/I_1 of the third and first vibronic band intensities of pyrene in sodium lauryl sulfate as a function of its concentration. Both curves have a sigmoidal shape (see Figure 1.3 of the paper). A value of 8×10^{-3}M for the *cmc* of the surfactant was given.

However, Nakajima [86] plotted the ratio I_1/I_3 and accepted the *cmc* as the concentration at which the first break is observed (point A in Figure 1.4). At low concentrations of the surfactants the values of the I_1/I_3 ratio are high, typical of a hydrophilic environment for pyrene, the value in water being 1.96 [87] while at high surfactant concentrations the I_1/I_3 ratio tend to typical values of non-polar solvents. For instance, at high surfactant concentrations of sodium cholate and sodium deoxycholate, the I_1/I_3 ratio is around 0.7 [88] while the value in cyclohexane is 0.61 [89]. This suggests that the polarity of the microenvironment of pyrene is a lipophilic one. Andersson and Olofsson [90], when performing a calorimetric study of nonionic surfactants, also made use of Nakajima's approach.

Other authors have proposed the inflection point of the curve (point B) as *cmc* [91]. As such it fulfills the condition

$$\left(\frac{d\varphi}{dS_t}\right)_{S_t=cmc}=0 \tag{1.11}$$

where φ would be the I_1/I_3 ratio. The expression is also valid for any other property that exhibits a sigmoidal behavior as the obtained enthalpograms from isothermal titration calorimetry (ITC) [92]. The plot of $(d\varphi/dS_t)$ vs S_t is shown in Figure 1.4 (right) and the *cmc* is easily obtained from the peak.

Aguiar et al. [93] have analyzed both points (A and B) for several surfactants and proposed an approach for choosing between one or the other point. Occasionally, both A and C points have been accepted as an indication that the system has two *cmc* values. We consider that this is not correct. These different points of view introduce an important question related to the determination of the *cmc* from sigmoidal curves, which are frequently found when using some experimental techniques.

By now, some different approaches to determine the *cmc* have already been introduced. Rusanov [94] has reviewed the definitions of *cmc* based on the application of the mass action law to the aggregation process in surfactant solutions. Among them, we must mention the definition given by the equation

$$\left(\frac{d^3\varphi}{dS_t^3}\right)_{S_t=cmc}=0 \tag{1.12}$$

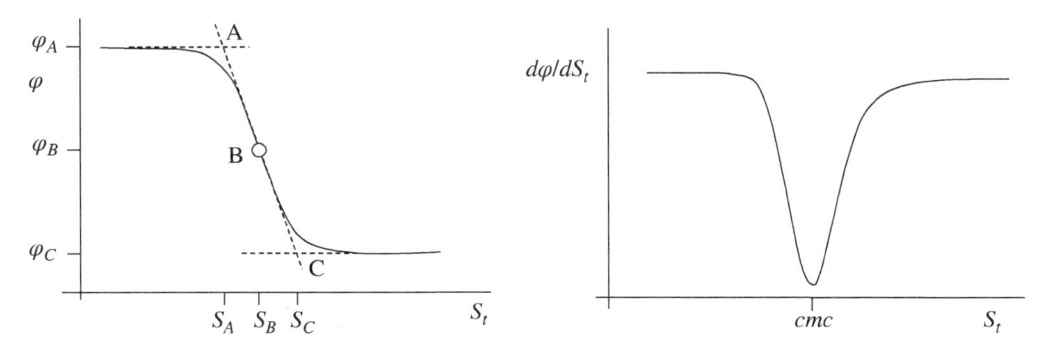

Figure 1.4 Typical plot of a sigmoidal curve. Example $\varphi = I_1/I_3$ (ratio of the intensities of the first and third vibronic peaks of pyrene) vs increasing concentration of a surfactant (left) and its first derivative (right). The shape of curves from isothermal titration calorimetry are similar in shape (see below for a description).

which was proposed by Phillips [95] in 1955 for determining the *cmc* for an ideal measured property (φ)-concentration (S_t) relationship. Phillips pretended that Eq. (1.12) corresponds to the point of maximum curvature, but this is not the case. Nakajima's approach fulfills this condition as well as the methodology proposed by Olesen et al. [96] for determining the aggregation number of aggregates from ITC curves. However, the definition of *cmc* as corresponding to the inflection point in the φ vs S_t curve has been recommended [97] for the determination of the *cmc* from ITC curves. For large absolute values of the slope at the inflection point ($= (\varphi_C - \varphi_A)/(S_C - S_A)$) in previous sigmoidal curves, the difference in the values obtained from any of the two previous equations may be considered negligible for practical purposes. If the *cmc* is fairly sharp, Hall [98] has proposed that it can be regarded approximately as a second order phase transition.

Among the other definitions for *cmc* analyzed by Rusanov we would like to remark on the following one. The focus is on a system in which micelles are composed of a single sort of particle. For further details and the analysis of more complex systems, the two papers by Rusanov [94] are recommended.

Let us redefine the Grindley and Bury [45] equilibrium constant as

$$K = K_o^{n-1} = m_n/s^n \tag{1.13}$$

The equilibrium constant K_o would correspond to a hypothetical single step in which a virtual aggregate m_j is formed by the binding of an additional monomer to a virtual aggregate of size $m_{j\text{-}1}$ according to

$$s + m_{j-1} \rightleftarrows m_j \tag{1.14}$$

its equilibrium constant being

$$K_j = m_j/\left(m_{j-1}s\right) \tag{1.15}$$

The isodesmic model accepts that all K_j constants are equal to K_o. The difference with the Grindley and Bury equilibrium constant comes from the fact that only $(n - 1)$ steps are required to form a micelle with n monomers. Interestingly, in 1935 Goodeve [99] have pointed out that forming micelles of, say, about 20 molecules must pass through all the intermediate stages of association. The formation of the micelles from the monomer in one stage is, of course, highly improbable as it requires "a collision of 20 molecules at one time." Goodeve presented Eq. (1.14) as representing the equilibrium according to this point of view.

Equation (1.13) is better understood in the form

$$\frac{m_n}{s} = \left(K_o \times s\right)^{n-1} \tag{1.16}$$

where K_o and s are both positive, n is usually large, and, independently of the value of n, for $K_o \times s = 1$, the concentration of micelles and monomers are the same. Deviations of the product $K_o \times s$ from that value lead to either $m_n < s$ or $m_n > s$. For instance, for $n = 50$, the ratio m_n/s changes by a factor 1.86×10^4 when $K_o \times s$ varies from 0.9 to 1.1. This is in fact the analysis by Grindley and Bury [45].

This suggests a definition of *cmc* by the condition

$$s_{cmc} = K_o^{-1} \tag{1.17}$$

and from the conservation of material ($S_t = s + n \times m_n$) it follows that at *cmc* $S_{t,cmc} = (n+1)s_{cmc} = (n+1)m_n$.

From Eq. (1.16) it also follows that

$$\frac{m_n}{s} = \frac{1}{n}\left(\frac{\partial m_n}{\partial s}\right) \tag{1.18}$$

or

$$n = \left(\frac{\partial \ln m_n}{\partial \ln s}\right) \tag{1.19}$$

In a monodisperse system, this equation may be simplified to

$$n = \left(\frac{\partial m_n}{\partial s}\right)_{cmc} \tag{1.20}$$

since at *cmc*, $m_n = s$ and n is constant for the whole range of surfactant concentrations. Thus, the larger the rate of change of the micelle concentration is with respect to the change in the monomer concentration, the higher the aggregation number will be.

Once the equilibrium constant and the aggregation number are known, all the thermodynamic functions may be obtained. These thermodynamic quantities have traditionally been determined from the measurement of *cmc* at different temperatures, the range of temperatures being around 40 °C (or less). The problem is that the dependence of the *cmc* with temperature is usually low for most of the surfactants and, as the dependence of ΔG^o with the concentration is logarithmic, the range of experimental values is even shorter. This introduces an error in the determination of the thermodynamic amounts, which is necessarily rather high.

The commercial introduction of high quality isothermal titration calorimeters has provided a routine way for the determination of previous amounts, which have a much higher precision. In a typical measurement a sample cell is filled with water (or any other appropriate solvent). A surfactant solution is placed in a syringe, which allows the injection of small aliquots into the sample cell at different intervals of time. The solvent of this solution and the one filling the sample cell must be identical to prevent some effects as the dilution heat of inert salts or buffers. The concentration of the surfactant ranges from 10 to 30 times the *cmc* value. Each injection increases the surfactant concentration in the sample cell from zero to a concentration clearly above the *cmc*. The heat involved in the process (the concentration in the syringe is always higher than in the sample cell) after each injection is measured and plotted vs the increasing concentration in the sample cell. Figure 1.4 imitates a typical enthalpogram and its derivative. In this case, $\varphi = \Delta H$ (in kJ/mol of injectant) is the involved heat after each injection.

An ideal enthalpogram can be subdivided into three concentration ranges. In region I (first injections, till point A in Figure 1.4) the increasing concentration in the cell is still below the *cmc*. Here, the large enthalpic effects observed are mainly due to breaking micelles into monomers (demicellization process) and dilution of monomers [92]. In region III (final injections, after point C in Figure 1.4), the increasing concentration in the cell is above the *cmc*. Here, the low enthalpic effects observed are mainly due to dilution of micelles. In the central region (between A and C in Figure 1.4), a sharp decrease is observed and corresponds to a transition from reaching the *cmc* and exceeding it. Therefore, the *cmc* corresponds to the inflection point of the curve, which can easily be determined as the first derivative of the curve (Eq. (1.11), Figure 1.4, right). The heat of demicellization ΔH_{demic} is equal to the enthalpy difference between the two extrapolated lines in Figure 1.4. Thus, the *cmc* and the enthalpy of micellization are simultaneously determined, but independently to each other.

Repetition of the ITC experiment at other temperatures allows the determination of the change in the heat capacity of the demicellization process, ΔC_P^o. The interval of temperatures used in these studies is rarely larger than 30–40 °C. Within this interval, the dependence of ΔH_{demic} for most of the surfactants is linear with T, meaning that ΔC_P^o is constant ([100] and references therein). While ΔH_{demic} may be either positive (endothermic) or negative (exothermic), ΔC_P^o for the demicellization process is always positive. This means that the hydrophobic surface of monomers, being exposed to water, increases upon demicellization. For this reason, it is frequently observed that ΔH_{demic} is negative at low temperatures and positive at high ones. The formation of a micelle requires that some water molecules surrounding each monomer must be lost in the aggregation process to form the final aggregate. The process also contributes to a favorable entropy term for micellization. Thus, the transfer of surfactant monomers from an aggregate to the bulk water has many facts in common with the dissolution of liquid alkanes into water [101]. Gill et al. [102] have noticed that the experimental heat capacity difference between gaseous and dissolved non-polar molecules in water is correlated with the number of water molecules in the first solvation shell. They concluded that a two-state model, in which each water molecule in the solvation shell behaves independently, provides a satisfactory basis to quantitatively describe the heat capacity properties of the solvation shell. For a series of solutes (most of them being hydrocarbon compounds), an average value of ~13.3 J/mol K (see the theoretical line shown in Figure 1.1 at 25 °C of that paper) was estimated for the contribution of each water molecule to ΔC_P^o.

Calorimetric measurements of vapor equilibrium of the system cyclohexane-heptane were performed almost 70 years ago by Crutzen et al. [103]. These authors observed that between 40 and 60 °C, the increase in the molar free Gibbs energy becomes small because of the partial compensation of the heat of mixing and the entropy of mixing. Since then, many papers have been published in which the concept enthalpy–entropy compensation (EEC) has been taken into consideration. Arguments for or against EEC have been published and, for surfactant systems, EEC has been reviewed several times [100, 104–106]. For demicellization (or equivalently micellization), the relationship is written linearly as

$$\Delta H = \Delta H_c + T_c \Delta S \tag{1.21}$$

where $T_c = (\partial \Delta H / \partial \Delta S)_P$ is known as the compensation temperature.

Recently, Vázquez-Tato et al. [100] have shown that "it is possible to obtain as many compensation temperature values as the number of temperature intervals in which the dependencies of enthalpy and entropy changes with temperature are analyzed." Furthermore, "the value of each T_c will agree with the central value T_o of each temperature interval." These authors concluded that "T_c is simply such experimental T_o" without any physical meaning and concluded that it "does not provide any additional information about the systems." In other words, any physical interpretation derived from T_c (and by extension from ΔH_c) is meaningless.

1.3 Average Aggregation Numbers

Recently Olesen et al. [96] have published a method for analyzing the ITC curves that allows the determination of the aggregation number. As in previous cases, we will limit the presentation to neutral surfactants. The mass-action model and monodispersity of micelles are assumed.

Other significant papers for determining the aggregation number of micelles are those by Debye and by Turro and Yekta. After his landmark paper published in 1947 for the molecular weight determination by light scattering, Debye [107] immediately published the first determinations of micellar molecular weights by this technique in 1949 [108], the surfactants being alkyl quaternary ammonium salts and amine hydrochlorides. A few years later, Tartar and Lelong [109] determined the micellar molecular weights of some paraffin chain salts by the same technique. Nowadays, the technique is almost routinely applied in laboratories for determination of molecular weights of polymers, micelles, and so on.

In 1978, Turro and Yekta [110] presented a simple procedure for determination of the mean aggregation number of micelles by measuring the steady-state fluorescence quenching of a luminescent probe by a hydrophobic quencher. The Poisson statistics to describe the distribution of the luminescent molecule D and the quencher Q in a solution that contains a well-defined but unknown micelle concentration $[M]$ was accepted. Both D and Q are selected in such a way that they reside exclusively in the micellar phase. D will partition itself both among micelles containing Q and among "empty" micelles. They also assumed that only excited micelles of D, D^*, emit in the micelles containing no Q, i.e. D^* is completely quenched when it occupies a micelle containing at least one Q. Under these conditions a "very simple expression" for obtaining the aggregation number is deduced.

In 1899 Biltz [111] published the book *Practical Methods for Determining Molecular Weights*. The book is a summation of the practical methods for determining molecular weights by vapordensity and other methods based on colligative properties, mainly from the measurement of the increase of the boiling point of a solution with respect to a pure solvent, and the freezing-point method. Both methods, together with the Nernst method (based upon the principle of lowering of solubility) are the only ones that "find practical application in the laboratory." When the freezing-point method is applied to some salts, Biltz accepts the Arrhenius dissociation theory of electrolyte solutions. Otherwise "the electrolytic dissociation in aqueous solutions can lead to smaller molecular weights than would be expected from the formula of the substance." Earlier in 1896, Krafft [112] noticed that the sodium salts of the shorter fatty acids exist in the "molecular state" (meaning that they are in the state of single molecules, and not in that of molecular aggregates) in aqueous solution and that they give twice the normal rise of boiling point which in fact it would correspond with a hydrolytic decomposition into sodium (hydroxide, in the original) and the fatty acid. Following Biltz, the reverse can take place if ". . . by condensation several simple molecules form a more complex molecule. . .. The term *association* has recently been proposed for this kind of condensation." Similarly, Kahlenberg and Schreiner [113] observed that the reduction of the freezing point of solutions of sodium oleate resulted in a molecular weight, which is nearly twice as large, like the theoretical formula. Botazzi and d'Errico [114] investigated glycogen of different concentrations by viscosity, freezing point, and electrical conductivity and observed that when the concentration of glycogen solutions reaches a certain maximum it appears that the colloidal particles combine with one another to form micelles. McBain et al. [115] measured the freezing point and the conductivity of sodium and potassium salts of saturated fatty acids that remain liquid at 0°. From this paper we must remark the comment that "free ions of charge equal and opposite to that of the charged colloid are present in the sol or gel." In 1935 McBain and Betz [116] measured the freezing point of undecyl and lauryl sulfonic acids, expressing the results in terms of the osmotic coefficient. They concluded that in dilute solutions they behave as simple moderately weak electrolytes but with increased concentration molecules and ions associate into neutral and ionic micelles. They also considered that micelles "owing to the wide spacing of their charges, have the ionic strength similar to uni-univalent electrolytes." McBain and Betz [116] and Johnston and McBain [117] carried out careful freezing point measurements on colloidal electrolytes (potassium and sodium oleate, sodium decyl and dodecyl sulphate, sodium decyl sulfonate and sodium deoxycholate) and, in 1947, Gonick and McBain [118] showed a "cryoscopic" evidence of a micellar association in aqueous solution of nonionic detergents. Following their own words, "since the depression of the freezing point is determined primarily by the number of solute particles per unit weight of solvent, the osmotic coefficient expresses directly, to a first approximation, the ratio of the true number of solute particles to that obtaining at complete dispersion of the solute and thus gives a measure of the average degree of association into micelles or other aggregates." The critical concentration of micelles formation was evident from an abrupt drop in the coefficient. Furthermore, they noticed that the addition of potassium chloride to a solution of nonaethylene glycol (mono) laurate caused no significant change in the degree of association of the detergent. Herrington and Sahi [119] studied the nonionic surfactants sucrose monolaurate and sucrose monooleate in aqueous solution by the freezing point and vapor pressure methods. For the analysis of the results they accepted that micelles are monodisperse and observed that the aggregation numbers do not increase significantly with temperature.

Starting from a model published by Burchfield and Woolley [120], Coello et al. [121, 122] combined freezing point and sodium ion activities measurements to obtain both the aggregation number and the fraction of bound counterions of aqueous bile salt solutions. The mass action model for an ionic micelle and the fraction of bound counterions to the micelles were considered. The theory of the freezing point depression is very well known [54] and does not need further attention.

While in a monodisperse system (as we have analyzed above) the aggregation number is a single-value variable [123], in a polydisperse self-aggregation system the aggregation number can assume all possible values from 2 to ∞. For these systems only the average aggregation number of the aggregates is obtained. However, different experimental techniques lead to aggregation numbers that are not the same. For instance, the average aggregation number measured from colligative properties is the number average aggregation number, $\overline{n}$, given by

$$\overline{n} = \frac{\Sigma iX_i}{\Sigma X_i} \tag{1.22}$$

where X_i is the concentration of the ith species. Similarly, the average aggregation numbers obtained from static light scattering and viscosity techniques are the weight average and the z-average aggregation numbers, $\overline{n}_w$ and $\overline{n}_z$, respectively, given by

$$\overline{n}_w = \frac{\Sigma i^2 X_i}{\Sigma iX_i} \tag{1.23}$$

$$\overline{n}_z = \frac{\Sigma i^3 X_i}{\Sigma i^2 X_i} \tag{1.24}$$

For analyzing the experimental results from the experimental techniques, the hypothesis of a constant average aggregation number with concentration is frequently used. Let us examine this hypothesis by following the landmark paper by Israelachvili et al. [124]. In this presentation we are following the notation of this original paper.

Each aggregate in the solution is characterized by its own free energy. They also accepted that the dilute solution theory holds. Equilibrium thermodynamics requires that in a system of molecules that form aggregated structures in solution, the chemical potential of all identical molecules in different aggregates is the same [125]. For an aggregate containing i monomer molecules, this statement is expressed as

$$\mu_1^o + kT \ln X_1 = \mu_i^o + \frac{kT}{i} \ln \frac{X_i}{i} \tag{1.25}$$

where μ_i^o is the standard part of the chemical potential (the mean interaction free energy per molecule) in aggregates of aggregation number i and X_i is the concentration of the ith aggregate. The subscript "1" corresponds to monomers in solution. That is to say, the left-side term corresponds to the chemical potential of the free monomer and the right-side term is the chemical potential per monomer incorporated in a micelle.

From the previous equation and the material balance in a polydispersed micellar solution, it follows that the number average aggregation number is [124]

$$n = \left(\frac{\partial \ln\left(S_t - X_1 \right)}{\partial \ln X_1} \right) \tag{1.26}$$

which relates the rate of change of micelle concentration with monomer concentration to the mean micelle aggregation number. It should be noticed that this equation reduces to Eq. (1.19) for a monodisperse system. In such a case, the various average aggregation numbers are identical to n [123].

The standard deviation is given by

$$\sigma^2 = \left(\frac{\partial \ln \bar{n}}{\partial \ln\left(S_t - X_1 \right)} \right) \tag{1.27}$$

which relates it to the rate of change of the mean aggregation number with respect to micelle concentration and means that if the standard deviation is close to zero, the aggregation number must almost be independent of the total surfactant concentration.

Well above *cmc*, Eq. (1.27) may be approximated by [125]

$$\sigma^2 = \left(\frac{\partial \ln \bar{n}}{\partial \ln(S_t)} \right) \tag{1.28}$$

which shows that if the system is highly polydisperse the average aggregation number will be very sensitive to the total surfactant concentration. Therefore, a rapid change of concentration is evidence of a large distribution of polydispersity in micelle size. Nagarajan [123] remarks that "this is a general thermodynamic result applicable to any self-assembling system" and that "interpreting any experimental data, one must ensure that this equation is not violated." Similar comments have been provided by Rusanov [126].

A well-known approach to study a fast process is to rapidly perturb a system in equilibrium and measure the relaxation time required by the system to adjust to the new equilibrium. Typical techniques are temperature jump, pressure jump, and ultrasonics [127]. The time scale depends on the relaxation technique [128]. The mathematical analysis to study simple A↔B and A+B↔C systems by temperature jump has been didactically published by Finholt [129]. Kresheck et al. [130] applied the temperature-jump (jump 5.2 °C) technique to study the dissociation of the dodecylpyridinium iodide micelle. Pressure-jump studies have also been carried out by several authors [131–133].

In the ultrasonic methods (Ultrasonic Absorption Spectrometry) the perturbation of equilibrium is due to the periodic variation of pressure and temperature due to the passage of the sound wave through the system. Relaxation of the equilibrium gives rise to changes in the quantity α/f^2 (where

α is the sound absorption coefficient at frequency f) and from its dependence with frequency the relaxation time is obtained [128].

Platz [134] has presented a simplified analysis of the Aniansson and Wall [135] isodesmic model for analyzing the kinetics of micelle association and dissociation in surfactant solutions. The theory is nowadays commonly known as the Teubner–Kahlweit–Aniansson–Wall theory [136, 137], the key equation being Eq. (1.14), characterized by forward (k_i^f) and back (k_i^b) kinetic constants, which correspond to the exchange of monomers between micelles. The ratio between the kinetic constants is directly the equilibrium constant. Further developments of the theory are due to Lang et al. [133] and Telgmann and Kaatze [138, 139]. In relaxation experiments two characteristic times are observed. The so-called "fast process" corresponds to the kinetic analysis of Aniansson and Wall and the "slow process" is assumed to be a change of the total number of micelles.

After several assumptions (micellar distribution is Gaussian-like, reactions between aggregates do not have to be considered, and the change in the number of the micelles is much slower than the interchange of the monomer) Eq. (1.29) is deduced [140]:

$$\frac{1}{\tau_1} = k_b\left(\frac{1}{\sigma^2} + \frac{1}{\overline{n}}\frac{S_t - \overline{s}}{\overline{s}}\right) \tag{1.29}$$

where τ_1 is the relaxation time (associated to k_b), σ^2 is the variance of the micellar distribution, $k_b = k_{\overline{n}}^b$ is the backward rate constant at micelle sizes around the mean micelle size $\overline{n}$, and $\overline{s}$ is the average monomer concentration. In addition, $\overline{s}$ is identified with the concentration at *cmc*, s_{cmc}, at concentrations above the *cmc*. When $\overline{s}$, k_b, the aggregation number and the standard deviation (see above) do not appreciably change with the concentration, and the previous equation suggests a linear relationship between τ_1^{-1} and the concentration. This fact has been verified, for instance, for alkyl sulfates [128].

The backward kinetic constant depends on the length of the alkyl chain of the surfactant. For instance, values in the interval 10–0.8 × 10^9/s have been measured by Kaatze [136] for ammonium chloride surfactants $CH_3C_{x-1}H_{2(x-1)}NH_3^+Cl^-$ ($x = 5$–8).

1.4 Packing Properties of Amphiphiles

In 1964 Reiss-Husson and Luzzati [141] used small-angle X-ray scattering methods to study micellar solutions of several amphiphiles in water, without added electrolytes. The amphiphiles were sodium salts of lauryl sulfate (SLS), laurate (NaC_{12}), myristate, palmitate, stearate and oleate, and cetyltrimethylammonium chloride (CTACl) and bromide. The studies were performed at different concentrations and temperatures. SLS (67, 25 °C), NaC_{12} (25, 70 °C), and CTACl (84, 27 °C) form spherical micelles with the aggregation numbers given in parenthesis. However, at high surfactant concentrations the spheres become rods and the concentration at which the sphere–rod transition takes place was also determined.

The sphere–rod transition was observed by Hayashi and Ikeda [142] for SLS when the concentration of NaCl is increased. The transition is accompanied by a large increment of the aggregation number. For instance, at NaCl 0.01 M the apparent aggregation number is 70 while at NaCl 0.80 M the value is 1630 and the length of the rod 597 Å. A simple version of this experiment was provided by Coello et al. [143].

The monomer packing of an alkyl surfactant in a given geometrical shape, for instance, a sphere, will depend on parameters such as the micellar radius, R, chain monomer volume, v, maximum length that the chain can assume (critical chain length, l_c), and the interfacial area per monomer, a_o. For spherical micelles, simple geometry gives [125],

$$n = \frac{4\pi R^3}{3v} \tag{1.30}$$

(from the volume) and

$$n = \frac{4\pi R^2}{a_o} \tag{1.31}$$

(from the surface). It follows that

$$R = \frac{3v}{a_o} \tag{1.32}$$

According to Tanford [144], for an alkyl chain with n_c carbon atoms the maximum chain length is given by

$$l_{\max} = 1.5 + 1.265 n_c \ \text{Å} \tag{1.33}$$

and the volume of the alkyl chain as

$$v = 27.4 + 26.9 n_c \ \text{Å}^3 \tag{1.34}$$

Only with the condition (l_c critical length)

$$\frac{v}{a_o l_c} < \frac{1}{3} \tag{1.35}$$

will the amphiphiles be able to fill the whole volume of the spherical micelle without leaving an empty space in the micelle core. For two surfactants having identical a_o values, previous equations predict that both the radius ($R_2 = (v_2/v_1)^2 R_1$) and the aggregation number ($n_2 = (v_2/v_1)^2 n_1$) will increase with the alkyl chain length. Figure 1.5 shows the plot of ln n vs ln v for sodium alkyl sulphates at 25 °C [145] and sodium alkyl sulfonates at 23–60 °C [76], the values of the slopes being 2.06 ± 0.05 and 2.20 ± 0.11, respectively, in good agreement with the theoretical expected one equal to 2. For the original measurements of Debye the value of the slope is 3.53. The value of a_o depends not only on the structure of the hydrophilic head group but also on the environmental properties (presence of inert salts, pH, additives, temperature).

The ratio ($v/a_o l_c$) of Eq. (1.35) is named as packing parameter P.

Surfactants can self-aggregate in other well-defined structures, different from spherical, ellipsoidal or rod-like micelles. Flat lamellar or disk-like structures are also common. Jung et al. [146] have studied the origins of stability of spontaneous vesicles. The cryo-TEM images provided by these authors, unequivocally show the formation of vesicles with a well-defined limiting membrane [147]. Similarly, Terech and Talmon [148] have demonstrated the formation of long single-walled tubes, and mechanisms for the formation of these tubes have been proposed for other bile acid derivatives [149, 150].

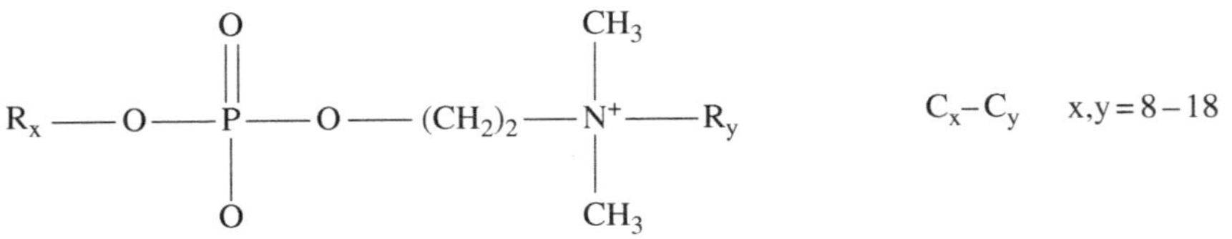

$$R_x\text{—O—}\overset{\overset{\displaystyle O}{\|}}{\underset{\underset{\displaystyle O}{|}}{P}}\text{—O—}(CH_2)_2\text{—}\overset{\overset{\displaystyle CH_3}{|}}{\underset{\underset{\displaystyle CH_3}{|}}{N^+}}\text{—}R_y \qquad C_x\text{–}C_y \quad x,y = 8-18$$

Scheme 1.2 Structure of synthetized gemini surfactants. *Source:* Peresypkin and Menger [153].

The determining factor of the surfactant molecules to self-organize in one or another structure is the packing parameter P. Israelachvili et al. [124] have indicated that the critical conditions for cylindrical micelles and planar bilayers are

$$\frac{v}{a_o l_c} = \frac{1}{2}\text{(cylindrical micelles)} \tag{1.36}$$

$$\frac{v}{a_o l_c} = 1\text{(planar bilayers)} \tag{1.37}$$

The structure of gemini surfactants [151, 152] provides different ways of changing their physicochemical properties. The ionic groups may be positive, negative, or zwitterionic, the spacer may be rigid or flexible allowing for different separation lengths between the hydrophilic groups, and the hydrophobic chains may be identical or not. In this last case, the gemini is clearly asymmetric. Thus, different values can be obtained for the packing parameter. A particularly instructive example is the family of zwitterionic geminis synthesized by Peresypkin and Menger [153] in which the negative group is a phosphodiester and the other is a cationic quaternary ammonium salt (Scheme 1.2).

Dynamic light scattering experiments suggest that only the C8–C8 derivative originates micelles exclusively, while others (for example, C8–C12, C12–C8, C14–C8, C10–C14, and C10–C16) self-assemble in structures such as tubules or vesicles of different sizes, forming also coacervate droplets (C8–C10).

1.5 Biosurfactants

Although phospholipids or bile acids are of a biological origin, the term "biosurfactant" concerns those amphiphilic derivatives that are produced by microorganisms. In general, they have excellent surface-active properties.

Different classifications have been purposed for biosurfactants [154–156]. For instance, Otzen indicates that, based on their overall structure, biosurfactants fall into four classes: glycolipids, lipopeptides, saponins, and all the rest. Glycolipids can be subdivided into rhamnolipids, sophorolipids, trehalolipids . . ., in which the head group are different saccharides (rhamnose, sophorose, trehalose, . . .). Similarly, lipopeptides can be divided into several families as surfactin, iturin or fengycin [157, 158], some of them having a peptide-cycle structure. Gemini type biosurfactants are also common [159]. The hydrophobic part is commonly one or more unsaturated or saturated hydrocarbon chains. The saponin Aescin or glycyrrhic acid are examples of biosurfactant complex structures [159, 160].

The number of biosurfactants can enormously be enlarged by chemical modifications, which can be carried out in laboratories. Sugar surfactants are well-known examples in which glucose, galactose, xylose, sucrose, or lactose are common hydrophilic heads [161]. They can easily be obtained chemically [162] or enzymatically [163]. Baccile et al. [26] have obtained and characterized the self-assembly properties of a broad family of amino derivatives of sophorolipid biosurfactants, including asymmetric (sophorose–ammonium and sophorose–amine oxide) and symmetric (sophorose–sophorose bearing three and four hydrophilic centers) bolaamphiphiles.

Previous naturally found biosurfactants and similar derivatives will all together show a broad range of physicochemical properties. Let us analyze some significant examples.

The surface tension vs concentration curves of biosurfactants are similar to those of classical surfactants. The break points in these plots, corresponding to *cmc* values, are normally well defined, and the surface tension values at *cmc*, γ_{cmc} are frequently found to be <30 mN/m. However, differences in reported *cmc* values are not uncommon, probably due to the presence of impurities or the use of mixtures instead of pure samples. For instance, Saini et al. [164] reported a value of 54 mg/l for the *cmc* of viscosin (Figure 1.6), a cyclic oligopeptide lipid which contains nine amino acids: two (leu-glu) are linked to the fatty acid tail, and the remaining seven [val-leu-ser-leu-ser-ile-allo(thr)] form a cycle. However, other values have also been reported for *cmc*, as Saini et al. have noticed, with values ranging from 4 to 9 mg/l [165] to 150 mg/l [166].

General rules observed in classical surfactants are also observed for biosurfactants. Figure 1.5 shows a linear relationship between the aggregation number and the alkyl chain volume in classical surfactants. This is just an example of linear relationships for different properties in a series of homologous surfactants. For instance, Garofalakis et al. [161] have observed a reduction in the critical aggregation concentration (*cac*) of the surfactants when increasing the carbon chain length for a series of monoesters of xylose, galactose, sucrose, and lactose with different hydrophobic chain lengths (C12–C16). These authors also observed that the more hydrophilic head groups, the higher is *cac*, though this trend was moderated by the alkyl chain length. Some observed differences between maltose and glucose derivatives have been ascribed to a higher degree of hydration of maltose compared to that of the glucose head group as molar heat capacities for these two sugars suggest [167]. Another example corresponds to the standard free energy change. For n_c-alkyl-D-maltosides (n_c, number of carbon atoms of the alkyl chain, = 8, 10, 12, and 14), Varga et al. [168] have shown that the dependence of the standard free energy change (and from here the *cmc* as well) with the length of the alkyl chain for these sugar surfactants is parallel to the one for alkyltrimethylammonium bromides and sodium alkylsulfates.

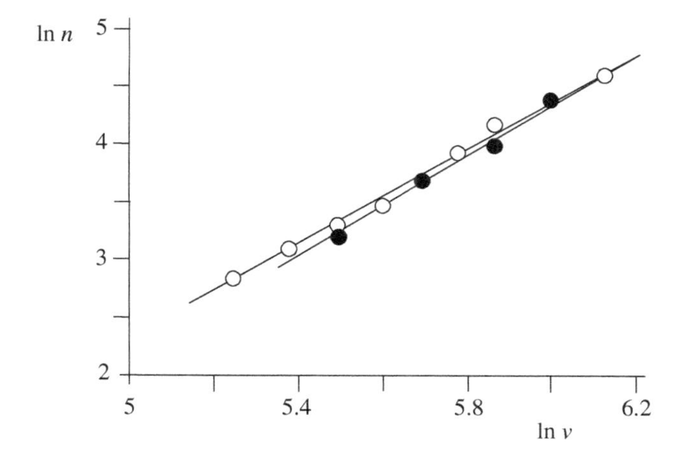

Figure 1.5 Plot of logarithm (aggregation number) vs logarithm (alkyl chain volume) of sodium alkyl sulfates (o) and sodium alkyl sulfonates (•). *Source:* Original data from Aniansson et al. [145] and Tartar and Lelong [109], respectively.

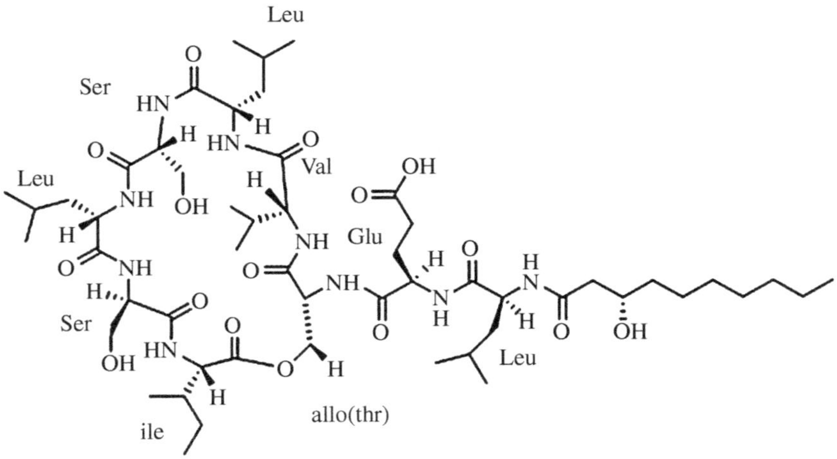

Figure 1.6 Structure of viscosin. *Source:* Saini et al. [164]. Reproduced by permission of the American Chemical Society.

Similarly, Ribeiro et al. [169] have obtained diacetylated lactonic sophorolipids with different hydrophobic chains (C18:0, C18:1, C18:2, and C18:3). The *cmc* values (in mg/l) increase linearly with the number of double bonds from 29.2 (C18:0) to 39.1 (C18:3), the slope being 3.35 ($r^2 = 0.979$). The surface tension at *cmc* (γ_{cmc}) also increases with that number (from 35.7 to 38.8 mN/m).

Sugar alkyl surfactants also display additivities with the number of carbon atoms of the alkyl chain. For instance, Angarten and Loh [170] studied two series of surfactants Cn_cG_x using ITC, where $x = 1,2$ is the number of glucoside units (G) of the hydrophilic head and $n_c = 7$–12 (the limits depending on x). Both families exhibit a linear ΔC_p vs n_c plot, the slope representing the contribution of each methylene group to the heat capacity for micellization. The obtained values for the slope are 63 ± 2 J/(K mol) and 59 ± 2 J/(K mol) for $x = 1$ and $x = 2$, respectively. Therefore, within experimental error, there are not significant differences between them, suggesting that the contributions of the mono or diglucoside head groups are essentially the same. Let us remember that a value of 33 J/(K mol) was found for the contribution of each hydrogen atom of each methylene group for the removal of alkanes from water to air [171]. This also has been related to the number of water molecules in the first solvation shell that contribute to the thermodynamics of hydrophobic solvation [102]. Other ITC measurements for alkyl sugar surfactants have been carried out by Blume et al. [92, 172]. Similar studies [173] were carried out for monorhamnose and dirhamnose rhamnolipids (R1, R2) (and their mixtures) (Figure 1.7). The *cmc* values are shown at Table 1.2.

Another example of the general behavior of biosurfactants corresponds to the kinetics of the micelle formation (see above). For instance, Haller and Kaatze [140] have studied the kinetics of micelle formation in aqueous solution of sugar surfactants as hexyl-, heptyl-, octyl-, nonyl-, and decyl-β-D-maltopyranoside (C_xG_2, $x = 6$, 8–10) as well as of decyl-β-D-maltopyranoside $C_{10}G_2$. As for other alkyl surfactants, there is a general tendency in the backward rate constant to increase with increasing *cmc* and with decreasing length of the alkyl chain.

From small angle neutron scattering (SANS) experiments, Chen et al. [174] found that, in dilute solution (<20 mM), R1 and R2 form small globular micelles (aggregation numbers being about 50 and 30, respectively), while at higher concentrations, R1 aggregates have a predominantly planar structure (unilamellar or bilamellar vesicles) whereas R2 remains globular, with an aggregation number that increases with increasing surfactant concentration. Over the concentration range

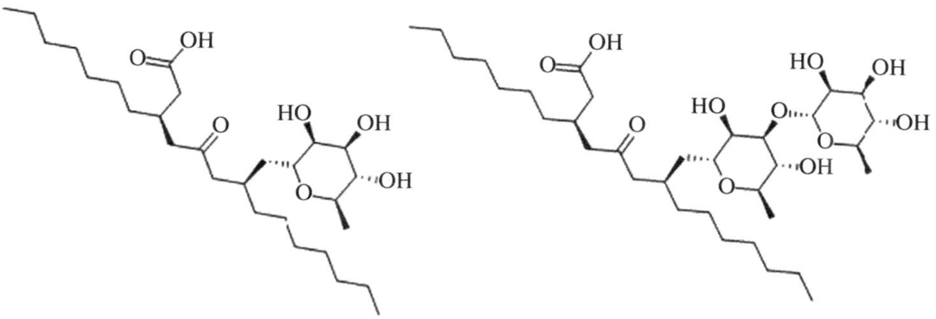

Figure 1.7 Structure of biosurfactants monorhamnose and dirhamnose rhamnolipids (R1 -left- and R2 right). *Source:* Chen et al. [174], p. 18281.

measured, for both surfactants the mean thickness of the adsorbed monolayer is about 21 ± 1 Å. From neutron reflectivity (NR) data and the Langmuir isotherm, the molecular areas at the surface are 62 ± 2 Å^2 (R1) and 77 ± 2 Å^2 (R2).

Gouzy et al. [24] have obtained two series of asymmetric bipolar surfactants with lactose as one of the hydrophilic groups. Their structures resemble those of asymmetric gemini surfactants but without a second hydrophobic moiety (Figure 1.1). These surfactants are known as divalent [26]. They have a long hydrocarbon chain, a nonionic polar head (lactose), a hydrocarbon spacer (of length n_c), and a second polar head (of length m_c) at the end of the spacer. At constant n_c, the results evidence a linear variation of log (*cmc*) with m_c but with a positive slope, i.e. the largest the hydrophobic alkyl side chain, the larger the *cmc*, which is the opposite trend observed for classical alkyl surfactants. These results are in line with those observed for *gemini* surfactants. Menger and Littau affirm that they are "counter to all previously reported trends in surfactant chemistry" [151], are anomalous as "in the protected duchies of academia, it is taught that a longer hydrocarbon tail always lowers the *cmc*" [152], while Rosen et al. [181] described this behavior as "aberrant." These authors accept that this unconventional behavior is indicative of substantial premicellar aggregation.

As indicated, low critical aggregation values have been measured and they favorably compare with both ionic and nonionic surfactants. The values obtained for the sucrose hexadecyl and dodecyl derivatives by Garofalakis et al. [161] (see Table 1.2) canbe compared with the *cmc* values for sodium hexadecyl sulfate (4.5×10^{-4} M) and sodium dodecyl sulfate (8.2×10^{-3} M) [145], as well as with those for polyoxyethylenated nonionic surfactants of structure $C_{nc}H_{nc+1}(OC_2H_4)_xOH$ (C_nE_x). For instance, for the series $n_c = 12$, $x = 2$, Rosen et al. [182] have measured values in the interval 3.3×10^{-5} ($x = 2$) to 1.09×10^{-4} M ($x = 12$) (all data at 25 °C), where it is obvious that the larger the hydrophilic head, the higher the *cmc*. Similar values for other members of this type of surfactant have been published elsewhere [183–185].

Some observed differences between maltose and glucose derivatives [161] have been ascribed to a higher degree of hydration of maltose compared to that of the glucose head group, as molar heat capacities for these two sugars suggest [167]. Significative differences in the hydration shell of the uncharged head groups were observed by Tyrode et al. [186, 187]. These authors have studied the OH stretching region of water molecules in the vicinity of nonionic surfactant monolayers by using vibrational sum frequency spectroscopy, thus allowing the study of the hydration shell around the head groups. Compounds such as dodecanol, sugar surfactants (n-decyl-β-D-glucopyranoside and n-decyl-β-D-maltopyranoside), and polyoxyethylene surfactants ($C_{12}E_4$ and

Table 1.2 Examples of *cmc* values of biosurfactants.

Compound	Cmc/M	γ/mN/m	References
Sucrose hexadecyl	4.1×10^{-6}	31.0–43.0	Garofalakis et al. [161]
Sucrose dodecyl	2.1×10^{-4}		
R1 monorhamnose rhamnolipid L-rhamnosyl-β-hydroxydecanoyl-β-hydroxydecanoate ($RhaC_{10}C_{10}$).	$(3.6 \pm 0.2) \times 10^{-4}$ 30 °C	31.2 ± 0.2	Chen et al. [173]
R2 L-rhamnosyl-L-rhamnosyl-β-hydroxydecanoyl-β-hydroxydecanoate ($Rha2C_{10}C_{10}$); The surface tension, NR, and SANS measurements were all made at pH 9 (buffer consisted of 0.023 M borax and 0.008 M HCl).	$(1.8 \pm 0.2) \times 10^{-4}$ 30 °C	37.4 ± 0.2	
Rhamnolipid A	6.22×10^{-5} pH 7.35		Ishigami et al. [175]
Rhamnolipid B	1.50×10^{-4} pH 7.35		
Rhamnolipid	50 mg/l		Whang et al. [176]
Surfactin	4.72×10^{-5} pH 8.0 20 mM phosphate buffer		Onaizi et al. [177]
Surfactin	45 mg/l	<30	Whang et al. [176]
Sophorolipid (lactonic form) C18:1 LS	2.8×10^{-5} potassium phosphate buffer (0.1 M, pH 7.4) at 25 °C	36.1	Otto et al. [178]
Diacetyl LS (Sophorolipid)	6×10^{-5}	36	Chen et al. [179]
Diacetyl AS (Sophorolipid)	6.7×10^{-4}	38.5	
Nonacetyl AS (Sophorolipid)	6.2×10^{-4}	39	
LAS (Sophorolipid)	1.6×10^{-3}		
SL-p (palmitic)	>200 mg/l	35	Ashby et al. [180]
SL-s (stearic)	35 mg/l	35	
SL-o (oleic)	140 mg/l	36	
SL-l (linoleic)	250 mg/l	36	
Glycyrrhizic acid	2.9×10^{-3} pH 5 5.3×10^{-3} pH 6 No clear *cmc* at pH 7	55.2 56.8	Matsuoka et al. [160]

$C_{12}E_8$) were studied. For all the surfactants, it was detected that the water molecules located in proximity to the surfactant hydrocarbon tail phase have both hydrogen atoms free from forming hydrogen bonds. For the two sugar surfactants, the strength of the hydrogen bonds in the hydration shell was found to be similar to those observed for tetrahedrally coordinated water molecules in ice. Despite being itself disordered, the polyoxyethylene head group induces a significant ordering and structuring of water at the surface. The orientation of the $C_{12}E_5$ molecules changes with concentration from lying on the surface with their hydrocarbon tails close to the surface plane (the observed results being consistent with the formation of disk-like "surface micelles" with a flat orientation of the amphiphiles at low surface concentrations) to a more upright configuration as the surface covering liquid layer is formed [187].

The formation of surface micelles was discussed in another paper [188] in which surface tension measurements were used to study the adsorption isotherms for sugar surfactants (n-decyl-β-D-glucopyranoside (Glu), n-decyl β-D-maltopyranoside (Mal), and n-decyl-β-D-thiomaltopyranoside (S-Mal)). A gradual change in molecular areas is observed when the surfactant concentration is increased. As the area/molecule is comparatively large, the resulting surface phase cannot be a coherent hydrocarbon film and should include a large portion of unperturbed air–water interface. The formation of surface micelles can account for this observation. A hard-disk simulation allowed the calculation of the number of molecules per micelle as a function of bulk surfactant concentration for Mal (values in the interval 9–12) and Glu (values in the interval 10–14), the surfactant molecules strongly favoring an orientation in the plane of the surface.

1.6 Sophorolipids

The head hydrophilic group of sophorolipids is the disaccharide sophorose (2-*O*-β-D-glucopyranosyl-β-D-glucopyranose). It is linked to a hydroxy fatty acyl moiety by a glycosidic bond between the 1′-hydroxy group of the sophorose-sugar and the ω or (ω−1) carbon atom of the fatty acid. Sophorolipids were first isolated by Gorin et al. [189] from the oil formed during fermentation by a strain of *Torulopsis magnoliae*. It has been demonstrated that they have taste-sensory properties [190]. Sophorolipids may have a free carboxylic acid (AS) structure or a lactone (LS) one. Figure 1.8 shows both structures for the derivative with a C18:1 (oleic) chain.

Ashby et al. [180] have obtained other derivatives by fed-batch fermentation of *Candida bombicola* on glucose and several fatty acids as palmitic acid (SL-p), stearic acid (SL-s), oleic acid (SL-o), and linoleic acid (SL-l). The *cmc* values obtained by these authors are shown in Table 1.2. The exact composition can vary with the type of hydrocarbon substrate used in the sophorolipid production and the production conditions [178], and correspondingly different *cmc* values have been published. For a pure diacetylated C18:1 LS, Otto et al. [178] have reported a *cmc* value of 2.8×10^{-5} M (Table 1.2). Higher values have been published by Chen et al. [179] for diacetyl LS, diacetyl AS, and nonacetyl AS.

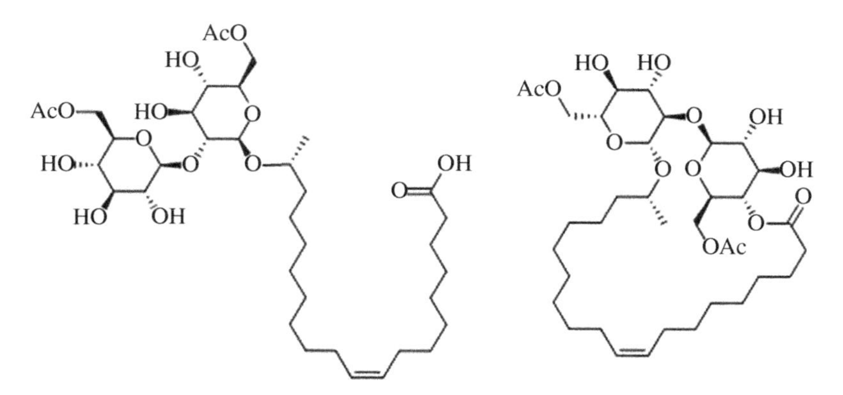

Figure 1.8 Chemical structures of the acidic (AS) and lactonic (LS) C18:1 sophorolipids.

Penfold et al. [191] have studied sophorolipids by SANS. At low surfactant concentrations (0.2–3 mM), data for LS are consistent with the formation of small unilamellar vesicles, with inner and outer radii increasing with concentration. The shell thickness also increases from about 15 to 24 Å. At high concentrations (30 mM), dynamic light scattering measurements are consistent with large aggregates (~300 nm). The solutions of AS with one and two acetyl groups have a hazy appearance, indicating the presence of large aggregates, while the solution of AS with no acetyl groups are consistent with small micellar structures, the aggregation number increasing steadily from 28 to 40 at the concentration range 5–50 mM. These results are consistent with predictions from the packing parameter.

At the air–solution interface, NR measurements were also carried out by Chen et al. for the deuterated surfactants (d-LS and d-AS) [179]. This technique provides different parameters as adsorbed amounts, composition, thickness of the adsorbed layer, and structure at the surface. The adsorption obtained values are consistent with a Langmuir isotherm (Eq. (1.38))

$$\Gamma = \frac{\Gamma_{max} C}{C + k} \tag{1.38}$$

where Γ and Γ_{max} are the adsorbed amounts and the maximum adsorption, C is the surfactant concentration, and k is the adsorption coefficient. AS and LS have similar k values (2.2×10^{-6}), suggesting that both sophorolipids have similar affinities for the air–water interface. Above *cmc*, the thickness is around 23 Å while the area/molecule is around 74 Å^2. For the less hydrophobic AS, the authors obtained a value of 85 Å^2. These results for the adsorbed amount are in good agreement with the values obtained from surface tension data.

Studies by Manet et al. [192] have shown that the micellar morphology of no acetylated C18:1 AS is a prolate ellipsoid. Depending on experimental conditions (the salts cause an increase of the aggregation number and an elongation of the micellar aggregates), the equatorial radius of the ellipsoid varies between 6.1 and 8.0 Å, the axial core ratio varies between 4.7 and 9.4, and the aggregation number between 24 and 73. The fraction of CH_2 groups inserted in the dry core of the micelle is in the interval 0.5–0.7, meaning that the core/shell interface is located far from the sugar head group. However, the equatorial shell thickness is almost constant (12.0 ± 0.5 Å). The shell thickness that best describes the sophorolipid micelles is a variable one from the equatorial value given above to zero, i.e. the hydrophilic shell has a nonhomogeneous distribution of matter containing carboxylic groups, sophorose, salt, water, and part of the aliphatic chain. This is an atypical result since most of surfactant systems are described by a homogeneous shell thickness. The area per sophorolipid between the alkyl chain and the sugar/carboxylate head group has been estimated between 102 and 141 Å^2 for the most ionized micelles. For nonacetyl AS, Chen et al. [179] reported a value of 104 ± 8 Å^2 for the area at the air—water interface.

Previously, Cecutti et al. [193] had noticed that the sugar rings represent a major part of the molecular volume for glycolipids and, consequently, they differentiate between the micelle-solvent interface and the hydrophobic core-sugar head group interface. The best result for interpreting neutron and X-ray small-angle scattering intensity curves for -dodecyl maltoside in water (6% w/v, 310 K) is by a short ellipsoid model with an ellipticity of 1.2. The difference between the total short radius of the micelle (24 Å), and the short radius of the apolar hydrophobic core (18 Å) allows enough space for the sugar head groups. Other obtained parameters are the areas per surfactant head at the water-micelles interface (87 Å^2), and at the chain-head group interface (50 Å^2), the aggregation number (82) and the number of water molecules per surfactant molecules (10).

AS has a bolaamphiphile asymmetrical structure with cis-9-octadecenoic chain linking the sophorose disaccharide and the carboxylic acid groups. Zhou et al. [194] have observed that, at a concentration of AS of 2.0 mg/ml and pH 2.0, ribbon formation may be noticed by a light microscope. The length and the width of the ribbons grow with time, and after two hours, twisted ribbons with lengths of a few hundred micrometers and a width of ~5 m had formed. Ribbon formation is slowed by increasing the pH. After a time, which depends on pH, precipitates were observed. After 28 hours, at pH 4.1 dynamic light scattering measurements of the solution showed that the large aggregates coexisted with small micelles. Small-angle X-ray scattering (SAXS) and wide-angle X-ray scattering (WAXS) measurements were carried out in aqueous solutions and dried solid ribbons formed at pH 4.1. All WAXS diffractograms indicate high crystallinity of the hydrocarbon chains and the disaccharide head groups inside the ribbons. At pH 5.1 individual ribbons were rarely found. At 0.02 mg/ml and pH 7.8 (at which the carboxylic groups are in their negative carboxylate form), small aggregates are formed, the hydrodynamic radius R_h being 37 nm (measured at a scattering angle of 15°). At concentrations of 0.97–1.78 mg/ml, nearly monodisperse micellar aggregates were formed with R_h about 100 nm. An apparent radius of gyration R_g of 175 nm was estimated for the radius of gyration by measuring the scattering intensity at the angle range of 15–35° at a surfactin concentration of 1.40 mg/ml. The R_g/R_h ratio is about 1.75, a rather large value which indicates that the large micellar aggregates have a very anisotropic geometry.

As described in the previous paragraph, the carboxylic acid group of C18:1 AS makes that its aggregation behavior is sensitive to pH. Baccile et al. [195] have studied the system by SANS at different amounts of NaOH and other bases (NH_3, KOH, and $Ca(OH)_2$). A core–shell prolate micelle structure with an interaction potential (which combines hard-sphere and screened Coulomb potentials) was used to fit SANS data. The total effective cross-radius of the micelle (core radius + shell thickness) decreases (from 21.8 to 18.2 Å) with increasing NaOH, the core radius contributing the most to this reduction as the shell is practically constant (~8 Å). The core size and the length of the oleic acid chain suggest that the chain is partially folded and that the formation of more carboxylate/Na^+ pairs favors such a bending. The eccentricity of the prolate also reduces (from 3.3 to 2.6) and micelles become more negatively charged as the effective micellar charge varies from about −0.5 to −5.3 with increasing NaOH concentration. Baccile et al. have also noticed that in the presence of $Ca(OH)_2$, the system evolves toward a better surface charge screening, which has the effect of reducing the repulsive potential (between micelles) and the effective surface charge. The micellar length is also elongated.

In a nice piece of work, Baccile et al. [26] prepared a broad range (38 new compounds) of amino derivatives of sophorolipid biosurfactant, comprising quaternary ammonium salts, amine oxides, and symmetric and asymmetric bolaamphiphiles with three or four hydrophilic centers, as well as divalent and Y-shaped derivatives. The compounds are constituted by at least two hydrophilic head groups, different in nature and charge since sophorose is neutral and the nitrogen atom is charged, being separated by a spacer. SAXS experiments were used for the estimation of the aggregation number.

Bolaform derivatives with two (sophorose and ammonium) hydrophilic groups either do not form aggregated or exhibit a poor tendency to self-aggregate. Correspondingly, the aggregation numbers are small (<20) and the aggregates are highly hydrated. The behavior of compounds with three hydrophilic groups (sophorose–ammonium–sophorose) depends on the nature of the ammonium group. If this is small, the compounds behave as the bolaform derivatives with two groups. If the group is charged and bulky, small hydrated aggregates are observed, coexisting with fibrilary systems. Tetra-center compounds (sophorose–ammonium–ammonium–sophorose) with small and/or charged ammonium groups, have a poor tendency for self-aggregation, and again small aggregation numbers and highly hydrated micelles are observed. However, divalent and Y-shaped surfactants tend to form larger micellar aggregates as a function of the size of the hydrophobic chain and spacers between sophorose and nitrogen, the aggregation numbers usually being >50. From these experiments Baccile et al. concluded that modification of acidic sophorolipids does not necessarily improve their aggregation behavior in water.

1.7 Surfactin

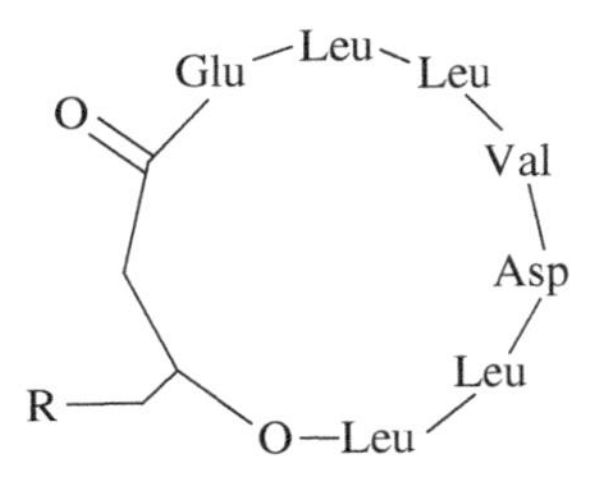

Figure 1.9 Schematic structure of a surfactin.

In 1968, Arima et al. [196] reported the isolation and characterization of a potent clotting inhibitor produced in the culture fluids of several strains of *Bacillus subtilis*. As its surface activity was much stronger than that of sodium lauryl sulfate they named the product *surfactin*. Figure 1.9 shows the structure of a typical surfactin. Other structures are also known [20].

According to Arima et al., surfactin is a peptide lipid composed of L-aspartic acid, L-glutamic acid, L-valine, L-leucine, D-leucine (1 : 1 : 1 : 2 : 2) and unidentified fatty acids. The complete elucidation of the structure was published by Kakinuma et al. in a series of papers one year later [197–200]. Bonmatin et al. [201] studied surfactin by two-dimensional ^{1}H-NMR in DMSO and observed two conformations characterized by a saddle-shape topology with polar Glu and Asp side chains oppositely oriented to that of the C11–13 aliphatic chain. The conformation of surfactin was reinvestigated by FTIR spectroscopy by Vass et al. [202]. Circular dichroism (CD) and FTIR spectroscopic data in different solvents with or without Ca^{2+} ions indicate that surfactin has a unique ability of adopting strongly different conformations depending on the conditions. The carboxyl groups of Glu^1 and Asp^5 are responsible for Ca^{2+} binding at low concentration (ratio of calcium/lipopeptide < 1). The NMR structure of surfactin has been determined in sodium dodecyl sulfate and dodecylphosphocholine micellar solutions [203]. p*Ka* values of Asp and Glu are around 4.3 and 4.5 [204, 203]. When comparing results for surfactins from different authors it is important to notice that different hydrophobic alkyl chains may be involved. Razafindralambo et al. [205] have isolated a homologous series of surfactins containing β-hydroxy fatty acids having different alkyl chain lengths (13, 14, or 15 carbon atoms). These authors have investigated their dynamic surface properties and found a dependence with both bulk concentration and hydrophobic character of the alkyl chain. The *cmc* values depend on the length of the alkyl chain and the tendency is the same as that observed for classical alkyl chain surfactants, i.e. the values are lower the larger is the alkyl chain. The surface tension at *cmc* changes in the same direction. These authors also observed that at low concentrations, the longer the alkyl chain, the faster is the decrease of the surface with time, but at high concentrations the maximum rate was observed for $n_c = 14$. Surfactin reduces to 27 mN/m at a concentration as low as 2×10^{-5} M. For surfactin, several authors have found two break points in the plot surface tension vs log (concentration), both being dependent on the experimental conditions [177]. The first one of 1.89×10^{-6} M (20 mM phosphate buffer) has been ascribed to a premicellization phenomenon, while the second one was assigned to the *cmc* (Table 1.2).

The effect of monovalent and divalent cations on surfactin was studied by Thimon et al. [206]. The experiments were performed at pH 9.5 (5 mM Tris), and Mg^{2+}, Ca^{2+}, and Mn^{2+} chloride salts were added at 0.5 mM concentration and monovalent ions (Li^+, Na^+, K^+, and Rb^+) were added at 0.1 M. Both type of cations, as well their concentrations, had a strong effect on *cmc*. For instance, values of 2.0×10^{-5} M (Ca^{2+}), 5.7×10^{-5} M (Mn^{2+}), 6.0×10^{-7} M (Li^+) and 1.47×10^{-6} M (Rb^+) were found. Molecular areas (determined from the analysis of the premicellar surface tension-log (concentration) plot) are also affected and the authors concluded that the micellization of surfactin is highly favored in the presence of divalent cations. Later Thimon et al. [207] calculated the association constants of some cations, the obtained values being K = 1.5×10^5/M and 1.9×10^4/M for Ca^{2+} and Mg^{2+}, respectively. The stoichiometry of complexes is 1 : 1 (Ca^{2+}) or 2 : 1 (Rb^+).

Li et al. [208] studied the influence of Na^+ ions on surfactin-C_{16} by fluorescence using pyrene as the fluorescent probe. From the plot of I_1/I_3 vs log S_t a value of 2.47×10^{-5} M (0.05 M Tris buffer, pH 8.5–8.6, 293 K) was obtained. These authors also observed that the micropolarity surrounding the pyrene molecules decreases with the addition of enough Na^+. The authors observed a decrease of the aggregation number (determined by the steady-state fluorescence quenching method) with increasing Na^+, which is contrary to what it should be expected. A similar study was conducted by these authors [209] to analyze the effect of other ions as Li^+, K^+, Mg^{2+}, and Ca^{2+}. As previously, monovalent ions reduced the micropolarity, and tend to originate small spherical micelle particles. The effect of Mg^{2+} concentration on micropolarity (expressed by the I_1/I_3 ratio) is less obvious than for monovalent ions while the behavior in the presence of Ca^{2+} is different as the ratio strongly increased to reach a maximum and then "was almost unchanged at other concentrations." The morphology of surfactin-C_{16} micelles with different counterions was observed by Freeze-Fracture Transmission Electron Microscopy (FFTEM). The smallest micelles were observed in the Li^+ solution. In the presence of divalent cations, large aggregates about 200 nm wide and more than 500 nm length were observed.

Surfactin was studied by ITC in phosphate buffer of pH 7.4 at 30 °C [210]. Values of 11.09 kJ/mol and 3.8×10^{-5} M were obtained for the enthalpy of micellization and the *cmc*, respectively. As the process is endothermic at this temperature, the process has to be entropy-driven. The distribution of hydrodynamic radius (dynamic light scattering experiments), in terms of the relative number, shows one peak at 4–6 nm at both 0.1 and 0.3 mM surfactin concentrations. In terms of intensity, large aggregates were also observed (~85 nm at 0.1 mM surfactin and ~108 nm at 0.3 mM surfactin). TEM images at 0.3 mM also show the coexistence of small micelles and large aggregates. CD spectra are concentration dependent, showing that the secondary structure of surfactin adopts a β-turn at low concentrations (0.1–0.3 mM) and begins to adopt a β-sheet conformation at a relatively high concentration (0.5 mM).

At pH 8.7 (0.1 M $NaHCO_3$) Ishigami et al. [211] measured a value of 9.4×10^{-6} M for the *cmc* of surfactin (3-hydroxymyristic acid) at which the surface tension was 30 mN/m (25 °C). In these conditions, CD measurements confirmed that the secondary structure of surfactin was a -sheet conformation, the molar ellipticity being stable, while the lactone ring was confirmed by FTIR. The micellar aggregation number was 173 and, assuming a cylindrical shape, from static light scattering measurements, the dimensions would be 231 nm (length) and 5.8 nm (diameter). The authors concluded thati"surfactin formed a large elongated rod-shaped micelle in spite of its bulky molecular structure."

The surface pressure increases from areas A_o = 1.84, 1.82, and 2.02 nm^2 and reaches breaking points at areas of A_t = 0.89, 0.81, and 0.79 nm^2, the values being obtained at pH values of 4.2, 4.8, and 5.4, respectively (20 °C, pK_a = 5.8). The A_t values are close to the molecular area (0.75 nm^2) estimated for a surfactin model with alkyl chains and the peptide ring vertically and supine oriented, respectively, to the plane surface. It was concluded that the compression from A_o to A_t leads to a reorientation of the alkyl chains (from flat on the surface to vertical with respect to the surface plane) but the peptide rings remain with a supine orientation. Different values were obtained by Maget-Dana and Ptak [212].

Knoblich et al. [213] studied the aggregates of surfactin by cryo-TEM in water at pH 7, 9.5, and 12. Spherical, ellipsoidal, and/or cylindrical micelles were observed as a function of pH. At pH 7, spherical (diameters between 5 and 9 nm) and ellipsoidal micelles (length 19 nm and width 11 nm) were measured. At pH 9.5 the dimension of cylindrical micelles had a length of 40–160 nm and a width of 10–14 nm. At this pH, the addition of 0.1 M NaC1 and 0.2 M $CaC1_2$ transforms the cylindrical micelles into spherical and/or ellipsoidal micelles of small size. However, at pH 12 the micrographs show mostly spherical (average diameter ~ 8 nm) and ellipsoidal micelles (size ~9 and 6 nm). At this pH the characteristic FTIR band for the lactone group disappeared, meaning that this group is opened. Consequently, the authors suggest that the formation of the surfactin micelles at pH 12 is different from that at low pH values.

For surfactin-C15, Zou et al. [214] obtained a *cmc* value of 1.54×10^{-5} M and $\gamma_{cmc} = 27.7$ mN/m (0.01 M phosphate buffer at pH 7.4; 25 °C). From the Gibbs isotherm, a value of 107.8 Å^2 was calculated for the molecular area at the interface. The experimental SANS data, fitting the curves for sphere-like aggregates, show that the radius of gyration of surfactin aggregates increases from 16 ± 0.4 Å to 20.1 ± 0.6 Å when the concentration increases from 4.0×10^{-5} M to 2.4×10^{-4} M. The pressure-area isotherm at the air-water interface shows that the pressure starts to increase at 231 Å^2/molecule and reaches a breaking point at 123 Å^2/molecule (pH 7.4, 25 °C). These values are higher than those published by Ishigami et al. [211] at lower pH values.

Osman et al. [215] have studied the effects of pH, Ca^{2+} ions, and the nonionic surfactant $C_{12}E_7$ on the conformation of surfactin in aqueous solutions using CD. Gradual alterations in the CD spectra of surfactin were observed that were related to the aggregational behavior of surfactin. The aggregation number (static light scattering measurements) raised to 144 at the surfactin/$C_{12}E_7$ molar ratio of 25 : 75, indicating enhancement of micellization. The molar ellipticity suggests that $C_{12}E_7$ enhanced the formation of micelles by promoting the assembly of surfactin molecules in sheets, even at very low surfactin concentrations. This could be due to an intercalation of $C_{12}E_7$ surfactin molecules in the micelle. The formation of sheets is enhanced by a temperature increase. Below the *cmc*, at pH > 8.5, CD spectra suggest an unordered conformation, but at neutral pH the conformation changed to sheets and the surfactin monomers have a helical conformation. Above the *cmc*, the pH effect was different. At pH 9, helices were formed, and below this pH, until a value of 6 occurred, sheets were formed. At pH > 9, the cyclic lactone ring of surfactin may be cleaved to form linear surfactin in solutions stored for long periods of time, and consequently above this pH value the observations could be related to the linear surfactin derivative, in concordance with Knoblich et al. [213]. The transitions induced in the monomers were dependent on the concentration of Ca^{2+} (α-helices–unordered structure–β-sheet conformation), a phenomenon that could be due to the binding of surfactin to Ca^{2+} and formation of surfactin clusters. These transitions were also observed above the *cmc*. Thus Ca^{2+} affects the surfactin conformation and also induces concentration-dependent transitions. These results suggest that β-sheets is the preferred bioactive conformation of surfactin.

Shen et al. [216] have produced from the *B. subtilis* strain three different deuterated surfactins (one perdeuterated, one with the four leucines perdeuterated, and one with everything except the four leucines perdeuterated) and used them in NR and SANS studies at pH 7.5. As expected, the largest signal in the reflectivity profiles is from the perdeuterated sample. Fitting the layer as a Gaussian distribution normal to the surface, in all three cases the area per molecule at the surface is 147 ± 10 Å^2 and the overall thickness of the layer is 14 ± 2 Å. The results also suggest an unusually close-packed surface layer and that the alkyl chain must be folded back into the leucines of the heptapeptide ring in order to give the observed compactness and the low extent of immersion in the aqueous subphase. Therefore, surfactin adopts a compact globular structure at the surface. The best fit of data was obtained by a core-shell model in which the core contains the alkyl chain and the four leucines, and the remaining head group, water, and the counterions are at the shell. The aggregation number was 20 ± 4, the overall micelle diameter 50 ± 5 Å, and the radius of the hydrophobic core 22 ± 2 Å.

The structure of other cyclolipopeptides different from surfactin have been reviewed by Kaspar et al. [20]. Among them families of iturins (heptapeptides), fengycin (decapeptides), or locillomycins (nonapeptides) can be mentioned, as well as linear lipopeptides. In comparison to surfactin, much less is known about their physicochemical properties.

1.8 Final Comments

Most of the isolated biosurfactants have not been characterized as deeply as sophorolipids or surfactins reviewed above. In many cases, the tiny amounts obtained prevent more careful studies and, for many of them, the *cmc* is the only physicochemical variable so far provided. Even the published values for *cmc* must be handled cautiously as the purity degree of the biosurfactant is low. However, if the biosurfactant belongs to a given family of derivatives, many of their properties can be estimated with different degrees of accuracy. Aggregation number or the change in heat capacity are provided examples in this chapter for classical and sugar surfactants. Conversely, a new measured physical quantity may be checked and compared with published values for similar compounds, although, occasionally, results (such as those obtained for some gemini surfactants) can break an accepted rule. The absence of physicochemical characterization is also related to the origin of research groups involved in their discovery, as they are more interested in biological properties and applications, as following chapters of this book will review. As classical surfactants, biosurfactants are affected by revisions of old theories or by new proposals. Although we have presented the theories and models in their simplest versions, we have also illustrated that concepts largely accepted during decades are nowadays under scrutiny.

Also, the knowledge of the structure/property relationship allows a possible improvement of some properties of a new biosurfactant by enlargement of the alkyl chain, introduction of a hydrophobic or hydrophilic residue, or the synthesis of new structures (gemini, Y-shaped, bolaamphiphile. . .). All of them are nowadays well-known options. The cooperation between chemistry and biology research groups, with a wide range of capacities, can be decisive in the improvement of desired properties and applications. Comments by Menger [217] about host–guest systems, typical examples of supramolecular entities, are valid for stimulating such a cooperation. Emulating his comments, it might be easy to design on paper a new surfactant bearing a wish-list of optimally oriented capacities and properties. Of course, the risk will be that after spending hours (and money) in the synthesis laboratory, such a molecule does not fulfill our expectations. We have unpublished experience on several negative projects. The task may be facilitated by the ability of researchers to further develop molecules from a biological origin that are the result of evolution. Bile salts (which have been our focus of interest for the last three decades) might be good examples of previous assertions.

The steroid nucleus with some specific organic functions located at certain positions and different orientations and, mainly, its enormous transcendence in living organisms (including human beings) is the result of a billion years of evolution of nature. Although all the human scientific knowledge would not probably be able to design it from zero, we (and others) have been able to modify their hydrophobic/hydrophilic balance by attaching specific residues into the structure. As a result, the formation of initially unexpected supramolecular structures is now well documented. Publications of the potential use of these new derivatives for the formation of gels, resolution of enantiomers, complexing a single water molecule, and synthesis of new antibiotics and antidotes can be found elsewhere [218–222]. These are just examples that pretend to encourage chemical modifications on new biosurfactants that microorganisms provided us.

Acknowledgement

The authors thank the Ministerio de Economía, Industria y Competitividad (Spain) (Project MAT2017-86109-P) for financial support.

References

1 IUPAC (1997). *Compendium of Chemical Terminology. (the "Gold Book")*, 2e (eds. A.D. McNaught and A. Wilkinson). Oxford: Blackwell. Online version created by Chalk. S. J. ISBN: 0-9678550-9-8. (2019).
2 Taylor, H.J. and Alexander, J. (1944). The measurement of surface tension by means of sessile drops. *Proc. Indian Acad. Sci., Math. Sci.* 19: 149–158.
3 Vargaftik, N.B., Volkov, B.N., and Voljak, L.D. (1983). International tables of the surface tension of water. *J. Phys. Chem. Ref. Data Monogr.* 12: 817–820.
4 Jasper, J.J. (1972). The surface tension of pure liquid compounds. *J. Phys. Chem. Ref. Data Monogr.* 1: 841–1009.
5 Levey, M. (1954). The early history of detergent substances: A chapter in Babylonian chemistry. *J. Chem. Educ.* 31: 521–524.
6 Mayhoff, K.F.T. (ed.) (1906). *Pliny the Elders, (AD 23–79) Naturalis Historia*. Lipsiae: Teubner.
7 Hunt, J.A. (1999). A short history of soap. *Pharm. J.* (1 Dec.).
8 Campbell, M. 1858. Improved process of making soap. US Patent Office, Patent no. 19667.
9 Mitchell, R.W. (1927). *Castile Soap-a Monograph Covering the Origin, History and Significance*. Boston: Brackett & Co.
10 Schulze, E.L. (1966). Literature of soaps and synthetic detergents. *Lit. Chem. Technol.*: 231–248.
11 Phanstiel, O. IV, Dueno, E., and Wang, Q.X. (1998). Synthesis of exotic soaps in the chemistry laboratory. *J. Chem. Educ.* 75: 612–614.
12 Kastens, M.L. and Ayo, J.J. Jr. (1950). Pioneer surfactant. *Ind. Eng. Chem.*, 42: 1626–1638.
13 Kosswig, K. (2012). *Surfactants. Ullmann's Encyclopedia of Industrial Chemistry*. Weinheim: Wiley VCH.
14 Kanno, S., Suzuki, A., Baba, H., and Hanzawa, Y. (1977). Structure of a Nekal-type surfactant a - commercial Twitchell reagent "Idrapidspalter". *Yukagaku* 26: 789–791.
15 Bellon, J.L.M. & LeTellier, P.A. Surfactants. P.A. FR 881893 19430511 (1943).
16 Lucas, F.H. and Brown, A.H. (1950). Activity of wetting agents-temperature effects. *Food Technol.* 4: 121–124.
17 Farn, R.J. (ed.) (2006). *Chemistry and Technology of Surfactants*. Oxford: Blackwell Publishing Ltd.
18 Attwood, D. and Florence, A.T. (2013). *Surfactant Systems, Their Chemistry, Pharmacy and Biology*. London: Chapman and Hall.
19 Meijide, F., Trillo, J.V., de Frutos, S. et al. (2013). Symbiotic and synergic effects in amide and ester derivatives of EDTA. In: *EDTA: Synthesis, Uses and Environmental Concerns* (ed. A. Molnar). Nova Publishers: New York.
20 Kaspar, F., Neubauer, P., and Gimpel, M. (2019). Bioactive secondary metabolites from *Bacillus subtilis*: a comprehensive review. *J. Nat. Prod.* 82: 2038–2053.

21 Zana, R. (2002). Dimeric and oligomeric surfactants. Behavior at interfaces and in aqueous solution: a review. *Adv. Colloid Interface Sci.* 97: 205–253.

22 Zana, R. and Xia, J. (eds.) (2004). *Gemini Surfactants. Synthesis, Interfacial and Solution -Phase Behavior, and Applications, Surfactants Science Series*. New York: Dekker.

23 Fuhrhop, J.-H. and Wang, T. (2004). Bolaamphiphiles. *Chem. Rev.* 104: 2901–2937.

24 Gouzy, M.-F., Guidetti, B., Andre-Barres, C. et al. (2001). Aggregation behavior in aqueous solutions of a new class of asymmetric bipolar Amphiphiles investigated by surface tension measurements. *J. Colloid Interface Sci.* 239: 517–521.

25 Guilbot, J., Benvegnu, T., Legros, N. et al. (2001). Efficient synthesis of unsymmetrical bolaamphiphiles for spontaneous formation of vesicles and disks with a transmembrane organization. *Langmuir* 17: 613–618.

26 Baccile, N., Delbeke, E.I.P., Brennich, M. et al. (2019). Asymmetrical, symmetrical, divalent, and Y-shaped (bola)amphiphiles: the relationship between the molecular structure and self-assembly in amino derivatives of Sophorolipid biosurfactants. *J. Phys. Chem. B* 123: 3841–3858.

27 Hofmann, A.F. and Mysels, K.J. (1988). Bile salts as biological surfactants. *Colloids Surf.* 30: 145–173.

28 Small, D.M. (1971). The physical chemistry of Cholanic acids. In: *The Bile Acids, Chemistry, Physiology, and Metabolism* (eds. P.P. Nair and D. Kritchevski). Plenum Press: New York.

29 Savage, P.B., Li, C., Taotafa, U. et al. (2002). Antibacterial properties of cationic steroid antibiotics. *FEMS Microbiol. Lett.* 217: 1–7.

30 Savage, P.B. (2002). Design, synthesis and characterization of cationic peptide and steroid antibiotics. *Eur. J. Org. Chem.* 759-768.

31 Traube, I. (1940). The earliest history of capillary chemistry. *J. Chem. Educ.* 17: 324–329.

32 Guthrie, F. (1864). II. On drops. *Proc. R. Soc. London, Ser. B* 13: 444–457.

33 Guthrie, F. (1864). On drops. Part II. *Proc. R. Soc. London, Ser. B* 13: 457–483.

34 Musculus, C. (1864). Ueber die Veränderungen der Molecularcohäsion des Wassers (about the changes of the molecular cohesion of water). *Chem. Zentralbl.* 922.

35 Yadav, J.B. (2010). *Advanced Practical Physical Chemistry*. India: Krishna Prakashan Media.

36 Dorsey, N.E. (1926). Measurement of the surface tension. *Sci. Paper* 21: 563–595.

37 Tate, T. (1864). On the magnitude of a drop of liquid formed under different circumstances. *Philos. Mag.* 27: 176–180.

38 Milner, S.R. (1907). IV. On surface concentration, and the formation of liquid films. *London, Edinburgh Dublin Philos. Mag. J. Sci.* 13: 96–110.

39 Langmuir, I. (1917). The shapes of group molecules forming the surfaces of liquids. *Proc. Natl. Acad. Sci. USA* 3: 251–257.

40 Malfitano, G. (1909). On the properties of colloidal particles called micellae. *Compt. Rend.* 148: 1045.

41 Wyrouboff, G. (1901). Some remarks over the colloids. *Bull. Soc. Chim. Fr.* 25: 1016–1022.

42 McBain, J.W. and Salmon, C.S. (1920). Colloidal electrolytes. Soap solutions and their constitution. *J. Am. Chem. Soc.* 42: 426–460.

43 Laing, M.E. and McBain, J.W. (1920). Investigations of sodium oleate solutions in the three physical states of curd, gel and sol. *J. Chem. Soc. Trans.* 117: 1508–1528.

44 McBain, J.W. and Jenkins, W.J. (1922). Ultrafiltration of soap solutions. Sodium oleate and potassium laurate. *J. Chem. Soc., Trans.* 121: 2325–2344.

45 Grindley, J. and Bury, C.R. (1929). The densities of butyric acid–water mixtures. *J. Chem. Soc.*: 679–684.

46 Davies, D.G. and Bury, C.R. (1930). The partial specific volume of potassium octoate in aqueous solution. *J. Chem. Soc.*: 2263–2267.

47 Powney, J. and Addison, C.C. (1937). The properties of detergent solutions. II. The surface and interfacial tensions of aqueous solutions of alkyl sodium sulfates. *Trans. Faraday Soc.* 33: 1243–1253.

48 Krafft, F. and Wiglow, H. (1895). Behaviour of the alkali salts of the fatty acids and of soaps in presence of water. *Ber. Dtsch. Chem. Ges.* 28: 2566–2573, 2573–2582.

49 Hutchinson, E., Inaba, A., and Baley, L.G. (1955). The properties of colloidal electrolyte solutions. *Z. Phys. Chem.* 5: 344–371.

50 Shinoda, K. and Hutchinson, E. (1962). Pseudo-phase separation model for thermodynamic calculations on micellar solutions. *J. Phys. Chem.* 66: 577–582.

51 Nilsson, G. (1957). The adsorption of Tritiated sodium dodecyl sulfate at the solution surface measured with a windowless, high humidity gas flow proportional counter. *J. Phys. Chem.* 57: 1135–1142.

52 Allen, G.D. (1915). The determination of the bile salts in urine by means of the surface tension method. *J. Biol. Chem.* 22: 505–524.

53 Reis, S., Guimaraes Moutinho, C., Matos, C. et al. (2004). Noninvasive methods to determine the critical micelle concentration of some bile acid salts. *Anal. Biochem.* 334: 117–126.

54 Atkins, P.W. and de Paula, J. (2002). *Physical Chemistry*, 7e. Oxford: Oxford University.

55 Menger, F.M., Shi, L., and Rizvi, S.A.A. (2009). Re-evaluating the Gibbs analysis of surface tension at the air/water Interface. *J. Am. Chem. Soc.* 131: 10380–10381.

56 Menger, F.M. and Rizvi, S.A.A. (2011). Relationship between surface tension and surface coverage. *Langmuir* 27: 13975–13977.

57 Li, P.X., Thomas, R.K., and Penfold, J. (2014). Limitations in the use of surface tension and the Gibbs equation to determine surface excesses of cationic surfactants. *Langmuir* 30: 6739–6747.

58 Xu, H., Li, P.X., Ma, K. et al. (2013). Limitations in the application of the Gibbs equation to anionic surfactants at the air/water surface: Sodium dodecylsulfate and sodium dodecylmonooxyethylenesulfate above and nelow the CMC. *Langmuir* 29: 9335–9351.

59 Tartar, H.V. and Wright, K.A. (1939). Sulfonates. III. Solubilities, micelle formation and hydrates of the sodium salts of the higher alkyl sulfonates. *J. Am. Chem. Soc.* 61: 539–544.

60 Wright, K.A. and Tartar, H.V. (1939). Studies of sulfonates. IV. Densities and viscosities of sodium dodecyl sulfonate solutions in relation to micelle formation. *J. Am. Chem. Soc.* 61: 544–549.

61 Wright, K.A., Abbott, A.D., Sivertz, V., and Tartar, H.V. (1939). Sulfonates. V. Electrical conductance of sodium decyl-, dodecyl- and hexadecyl-sulfonate solutions at 40°, 60° and 80°. Micelle formation. *J. Am. Chem. Soc.* 61: 549–551.

62 Hartley, G.S. (1936). Critical concentration for micelles in solutions of cetanesulfonic acid. *J. Am. Chem. Soc.* 58: 2347–2354.

63 Hartley, G.S. (1939). Ion aggregation in solutions of salts with long paraffin chains. *Kolloidn. Zh.* 88: 22–40.

64 Corrin, M.L. and Harkins, W.D. (1947b). The effect of salts on the critical concentration for the formation of micelles in colloidal electrolytes. *J. Am. Chem. Soc.* 67: 683–688.

65 Lange, H. (1950). Application of the law of mass action to micelle formation in colloidal electrolytes. *Kolloidn. Zh.* 117: 48–51.

66 Hall, D.G. (1981). Thermodynamics of solutions of polyelectrolytes, ionic surfactants, and other charged colloidal system. *J. Chem. Soc. Faraday Trans.* 1 (77): 1121–1156.

67 Corrin, M.L. and Harkins, W.D. (1946a). The effect of solvents on the critical concentration for micelle formation of cationic soaps. *J. Chem. Phys.* 14: 640–641.

68 Herzfeld, S.H., Corrin, M.L., and Harkins, W.D. (1950). The the effect of alcohols and of alcohols and salts on the critical micelle concentration of dodecylammonium chloride. *J. Phys. Colloid Chem.* 54: 271–283.

69 Reichenberg, D. (1947). Colloidal crystallites and micelles. I. The micelle in solution. Apparent anomalies in the surface- and interfacial-tension-concentration curves of aqueous solutions of paraffin-chain salts. *Trans. Faraday Soc.* 43: 467–479.

70 Klevens, H.B. (1947a). Effects of temperature on the critical concentrations of anionic and cationic detergents. *J. Phys. Chem.* 51: 1143–1154.

71 Klevens, H.B. (1947b). Effect of temperature on micelle formation as determined by refraction. *J. Colloid Sci.* 2: 301–303.

72 Sheppard, S.E. and Geddes, A.L. (1945). Amphipathic character of proteins and certain lyophile colloids as indicated by absorption spectra of dyes. *J. Chem. Phys.* 13: 63.

73 Corrin, M.L., Klevens, H.B., and Harkins, W.D. (1946). Critical concentration for the formation of micelles as indicated by the absorption spectrum of a cyanine dye. *J. Chem. Phys.* 14: 216–217.

74 Klevens, H.B. (1946). The critical micelle concentration of anionic soap mixtures. *J. Chem. Phys.* 14: 742.

75 Corrin, M.L., Klevens, H.B., and Harkins, W.D. (1946.a). The determination of critical concentrations for the formation of soap micelles by the spectral behavior of pinacyanol chloride. *J. Chem. Phys.* 14: 480–486.

76 Kolthoff, I.M. and Johnson, W.F. (1946). Solubilization of p-dimethylaminoazobenzene in soap solutions. *J. Phys. Chem.* 50: 440–442.

77 Corrin, M.L. and Harkins, W.D. (1946). Determination of critical concentrations for micelle formation in solutions of cationic soaps by changes in the color and fluorescence of dyes. *J. Chem. Phys.* 14: 641.

78 Corrin, M.L. and Harkins, W.D. (1947). Determination of the critical concentration for micelle formation in solutions of colloidal electrolytes by the spectral change of a dye. *J. Am. Chem. Soc.* 69: 679–683.

79 Klevens, H.B. (1950). Solubilization of polycyclic hydrocarbons. *J. Phys. Colloid Chem.* 54: 283–298.

80 Ekwall, P. (1951). Micelle formation in sodium cholate solutions. *Acta Acad. Abo., Ser. B* 17: 1–10.

81 Foerster, T. and Selinger, B. (1964). Concentration change of the fluorescence of aromatic hydrocarbons in micellar colloidal solution. *Z. Naturforsch.* 19a: 38–41.

82 Dorrance, R.C. and Hunter, T.F. (1974). Absorption and emission studies of solubilization in micelles. 2. Determination of aggregation numbers and solubilizate diffusion in cationic micelles. *J. Chem. Soc. Faraday Trans.* 1 (70): 1572–1580.

83 Chen, M. and Graetzel, J.K. (1974). Thomas, photochemical reactions in micelles of biological importance. *Chem. Phys. Lett.* 24.

84 Geiger, M.W. and Turro, N.J. (1975). Pyrene fluorescence lifetime as a probe for oxygen penetration of micelles. *Photochem. Photobiol.* 22: 273–276.

85 Kalyanasundaram, K. and Thomas, J.K. (1977). Environmental effects on vibronic band intensities in pyrene monomer fluorescence and their application in studies of micellar systems. *J. Am. Chem. Soc.* 99: 2039–2044.

86 Nakajima, A. (1977). Variations in the vibrational structures of fluorescence spectra of naphthalene and pyrene in water and in aqueous surfactant solutions. *Bull. Chem. Soc. Jpn.* 50: 2473–2474.

87 Acharya, D.P., Kunieda, H., Shiba, Y., and Aratani, K. (2004). Phase and rheological behavior of novel gemini-type surfactant systems. *J. Phys. Chem. B* 108: 1790–1797.

88 Jover, A., Meijide, F., Rodríguez Núñez, E. et al. (1996). Unusual pyrene excimer formation during sodium deoxycholate gelation. *Langmuir* 12: 1789–1793.

89 Hashimoto, S. and Thomas, J.K. (1984). Photophysical studies of pyrene in micellar sodium taurocholate at high salt concentrations. *J. Colloid Interface Sci.* 102: 152–163.

90 Andersson, B. and Olofsson, G. (1988). Calorimetric study of nonionic surfactants: enthalpies and heat-capacity changes for micelle formation in water of C8E4 and Triton X-100 and micelle size of C_8E_4. *J. Chem. Soc. Faraday Trans.* 1 (84): 4087–4095.

91 Chung, H.S. and Heilweil, I.J. (1970). Statistical treatment of micellar solutions. *J. Phys. Chem.* 74: 488–494.

92 Paula, S., Sues, W., Tuchtenhagen, J., and Blume, A. (1995). Thermodynamics of micelle formation as a function of temperature: A high sensitivity titration calorimetry study. *J. Phys. Chem.* 99: 11742–11751.

93 Aguiar, J., Carpena, P., Molina-Bolivar, J.A., and Carnero Ruiz, C. (2003). On the determination of the critical micelle concentration by the pyrene 1:3 ratio method. *J. Colloid Interface Sci.* 258: 116–122.

94 Rusanov, A.I. (1993). The mass action law theory of micellar solutions. *Adv. Colloid Interface Sci.* 45: 1–78.

95 Phillips, J.N. (1955). Energetics of micelle formation. *Trans. Faraday Soc.* 51: 561–569.

96 Olesen, N.E., Holm, R., and Westh, P. (2015). Determination of the aggregation number for micelles by isothermal titration calorimetry. *Thermochim. Acta* 588: 28–37.

97 Olofsson, G. and Loh, W. (2009). The use of titration calorimetry to study the association of surfactants in aqueous solutions. *J. Braz. Chem. Soc.* 20: 577–593.

98 Hall, D.G. (1972). Exact phenomenological interpretation of the micelle point in multicomponent systems. *J. Chem. Soc. Faraday Trans.* 2 (68): 668–679.

99 Goodeve, C.F. (1935). General discussion on "equilibrium between micelles and simple ions, with particular reference to the solubility of long-chain salts. Discussion on equilibrium between micelles and simple ions, with particular reference to the solubility of long-chain salts". *Trans. Faraday Soc.* 31: 197–198.

100 Vázquez-Tato, M.P., Meijide, F., Seijas, J.A. et al. (2018). Analysis of an old controversy: The compensation temperature for micellization of surfactants. *Adv. Colloid Interface Sci.* 254: 94–98.

101 Gill, S.J., Nichols, N.F., and Wadsö, I. (1976). Calorimetric determination of enthalpies of solution of slightly soluble liquids. II. Enthalpy of solution of some hydrocarbons in water and their use in establishing the temperature dependence of their solubilities. *J. Chem. Thermodyn.* 8: 445–452.

102 Gill, S.J., Dec, S.F., Olofsson, G., and Wadsö, I. (1985). Anomalous heat capacity of hydrophobic solvation. *J. Phys. Chem.* 89: 3758–3761.

103 Crutzen, J.L., Hasse, R., and Sieg, L. (1950). Vapor equilibrium and heat of mixing in the systems cyclohexane-heptane and methylcyclohexane-heptane. *Z. Naturforsch., B: J. Chem. Sci.* 5a: 600–604.

104 Jolicoeur, C. and Philip, P.R. (1974). Enthalpy–entropy compensation for micellization and other hydrophobic interactions in aqueous solutions. *Can. J. Chem.* 52: 1834–1839.

105 Pan, A., Kar, T., Rakshit, A.K., and Moulik, S.P. (2016). Enthalpy–entropy compensation (EEC) effect: decisive role of free energy. *J. Phys. Chem. B* 120: 10531–10539.

106 Sugihara, G., Nakano, T.-Y., Sulthana, S.B., and Rakshit, A.K. (2001). Enthalpy–entropy compensation rule and the compensation temperature observed in micelle formation of different surfactants in water. What is the so-called compensation temperature? *J. Oleo Sci.* 50: 29–39.

107 Debye, P. (1947). Molecular weight determination by light scattering. *J. Phys. Chem.* 51: 18–32.

108 Debye, P. (1949). Light scattering in soap solutions. *J. Phys. Colloid Chem.* 53: 1–8.

109 Tartar, H.V. and Lelong, A.L.M. (1955). Micellar molecular weights of some paraffin-chain salts by light scattering. *J. Phys. Chem.* 59: 1185–1190.

110 Turro, N.J. and Yekta, A. (1978). Luminescent probes for detergent solutions. A simple procedure for determination of the mean aggregation number of micelles. *J. Am. Chem. Soc.* 100: 5951–5952.

111 Biltz, H. (1899). *Practical Methods for Determining Molecular Weights*. Easton: The Chemical Publishing Company.

112 Krafft, F. (1896). A theory of colloidal solutions. *Ber. Dtsch. Chem. Ges.* 29: 1334–1344.

113 Kahlenberg, L. and Schreiner, O. (1898). The aqueous solutions of the soaps. *Z. Phys. Chem.* 27: 552–566.

114 Botazzi, F. and d'Errico, G. (1906). Physico-chemical investigations of glycogen. *Pfluegers Arch. Gesamte Physiol. Menschen Tiere* 115: 359–386.

115 McBain, J.W., Laing, M.E., and Titley, A.F. (1919). Colloidal electrolytes. Soap solutions as a type. *J. Chem. Soc., Trans.* 115: 1279–1300.

116 McBain, J.W. and Betz, M.D. (1935). The predominant role of association in the dissociation of simple straight-chain sulfonic acids in water. II. Freezing point. *J. Am. Chem. Soc.* 57: 1909–1912.

117 Johnston, S.A. and McBain, J.W. (1942). Freezing-points of solutions of typical colloidal electrolytes; soaps, sulphonates, sulphates and bile salt. *Proc. R. Soc. London, Ser. A* 181 (985): 119–133.

118 Gonick, E. and McBain, J.W. (1947). Cryoscopic evidence for micellar association in aqueous solutions of nonionic detergents. *J. Am. Chem. Soc.* 69: 334–336.

119 Herrington, T.M. and Sahi, S.S. (1986). Temperature dependence of the micellar aggregation number of aqueous solutions of sucrose monolaurate and sucrose monooleate. *Colloids Surf.* 17: 103–113.

120 Burchfield, T.E. and Woolley, E.M. (1984). Model for thermodynamics of ionic surfactant solutions. 1. Osmotic and activity coefficients. *J. Phys. Chem.* 88: 2149–2155.

121 Coello, A., Meijide, F., Rodríguez Núñez, E., and Vázquez Tato, J. (1993). Aggregation behavior of sodium cholate in aqueous solution. *J. Phys. Chem.* 97: 10186–10191.

122 Coello, A., Meijide, F., Rodríguez Núñez, E., and Vázquez Tato, J. (1996). Aggregation behavior of bile salts in aqueous solution. *J. Pharm. Sci.* 85: 9–15.

123 Nagarajan, R. (1994). On interpreting fluorescence measurements: what does thermodynamics have to say about change in Micellar aggregation number versus change in size distribution induced by increasing concentration of the surfactant in solution? *Langmuir* 10: 2028–2034.

124 Israelachvili, J.N., Mitchell, D.J., and Ninham, B.W. (1976). Theory of self-assembly of hydrocarbon amphiphile into micelles and bilayers. *J. Chem. Soc. Faraday Trans.* 2 (72): 1525–1568.

125 Israelachvili, J. (2011). *Intermolecular and Surface Forces*, 3e. Santa Barbara, CA: Academic Press.

126 Rusanov, A.I. (2014). The mass-action-law theory of micellization revisited. *Langmuir* 30: 14443–14451.

127 Hoffmann, H. (2012). Structure formation in surfactant solutions. A personal view of 35 years of research in surfactant science. *Adv. Colloid Interface Sci.* 178: 21–33.

128 Hall, D.G. and Wyn-Jones, E. (1986). Chemical relaxation spectrometry in aqueous surfactant solutions. *J. Mol. Liq.* 32: 63–82.

129 Finholt, J.E. (1968). The temperature-jump method for the study of fast reactions. *J. Chem. Educ.* 45: 394.

130 Kresheck, G.C., Hamori, E., Davenport, G., and Scheraga, H.A. (1966). Determination of the dissociation rate of dodecylpyridinium iodide micelles by a temperature-jump technique. *J. Am. Chem. Soc.* 88: 246–253.

131 Folger, R., Hoffmann, H., and Ulbricht, W. (1974). Mechanism of the formation of micelles in sodium dodecyl sulfate (SDS) solutions. *Ber. Bunsenges.* 78: 986–997.

132 Inoue, T., Tashlro, R., Shlbuya, Y., and Shimozawa, R. (1978). Chemical relaxation studies in micellar solutions of dodecylpyridinium halides. *J. Phys. Chem.* 82: 2037.

133 Lang, J., Tondre, C., Zana, R. et al. (1975). Chemical relaxation studies of micellar equilibria. *J. Phys. Chem.* 79: 276–283.

134 Platz, G. (1979). The kinetics of micelle formation. *NATO Adv. Study Inst. Ser., Ser. C* C50: 239–248.

135 Aniansson, E.A.G. and Wall, S.N. (1974). Kinetics of step-wise micelle association. *J. Phys. Chem.* 78: 1024–1030.

136 Kaatze, U. (2011). Kinetics of micelle formation and concentration fluctuations in solutions of short-chain surfactants. *J. Phys. Chem. B* 115: 10470–10477.

137 Teubner, M. (1979). Theory of ultrasonic absorption in micellar solutions. *J. Phys. Chem.* 83: 2917–2920.

138 Telgmann, T. and Kaatze, U. (1997). On the kinetics of the formation of small micelles. 1. Broadband ultrasonic absorption spectrometry. *J. Phys. Chem. B* 101: 7758–7765.

139 Telgmann, T. and Kaatze, U. (1997). On the kinetics of the formation of small micelles. 2. Extension of the model of stepwise association. *J. Phys. Chem. B* 101: 7766–7772.

140 Haller, J. and Kaatze, U. (2009). Ultrasonic spectrometry of aqueous solutions of alkyl maltosides: kinetics of micelle formation and head-group isomerization. *ChemPhysChem* 10: 2703–2710.

141 Reiss-Husson, F. and Luzzati, V. (1964). The structure of the micellar solutions of some amphiphilic compounds in pure water as determined by absolute small-angle X-ray scattering techniques. *J. Phys. Chem.* 68: 3504–3511.

142 Hayashi, S. and Ikeda, S. (1980). Micelle size and shape of sodium dodecyl sulfate in concentrated sodium chloride solutions. *J. Phys. Chem.* 84: 744–751.

143 Coello, A., Meijide, F., Mougan, M.A. et al. (1995). Spherical and rod SDS micelles. *J. Chem. Educ.* 72: 73–75.

144 Tanford, C. (1972). Micelle shape and size. *J. Phys. Chem.* 76: 3020–3024.

145 Aniansson, E.A.G., Wall, S.N., Almgren, M. et al. (1976). Theory of the kinetics of micellar equilibria and quantitative interpretation of chemical relaxation studies of micellar solutions of ionic surfactants. *J. Phys. Chem.* 80: 905–922.

146 Jung, H.T., Coldren, B., Zasadzinski, J.A. et al. (2001). The origins of stability of spontaneous vesicles. *Proc. Natl. Acad. Sci. U. S. A.* 98: 1353–1357.

147 Coldren, B., Van Zanten, R., Mackel, M.J. et al. (2003). From vesicle size distributions to bilayer elasticity via cryo-transmission and freeze-fracture electron microscopy. *Langmuir* 19: 5632–5639.

148 Terech, P. and Talmon, Y. (2002). Aqueous suspensions of steroid nanotubules: structural and rheological characterizations. *Langmuir* 18: 7240–7244.

149 Meijide, F., Trillo, J.V., de Frutos, S. et al. (2012). Formation of tubules by p-tert-butylphenylamide derivatives of chenodeoxycholic and ursodeoxycholic acids in aqueous solution. *Steroids* 77: 1205–1211.

150 Soto, V.H., Jover, A., Meijide, F. et al. (2007). Supramolecular structures generated by a *p-tert*-butylphenyl-amide derivative of cholic acid. From vesicles to molecular tubes. *Adv. Mater.* 19: 1752–1756.

151 Menger, F.M. and Littau, C.A. (1991). Gemini-surfactants: Synthesis and properties. *J. Am. Chem. Soc.* 113: 1451–1452.

152 Menger, F.M. and Littau, C.A. (1993). Gemini surfactants: A new class of self-assembling molecules. *J. Am. Chem. Soc.* 115: 10083–10090.

153 Peresypkin, A.V. and Menger, F.M. (1999). Zwitterionic Geminis. Coacervate formation from a single organic compound. *Org. Lett.* 1: 1347–1350.

154 Nitschke, M. and Pastore, G.M. (2002). Biosurfactants: Properties and applications. *Quim. Nova* 25: 772–776.

155 Otzen, D.E. (2017). Biosurfactants and surfactants interacting with membranes and proteins: Same but different? *Biochim. Biophys. Acta* 1859: 639–649.

156 Rosenberg, E. and Ron, E.Z. (1999). High- and low-molecular-mass microbial surfactants. *Appl. Microbiol. Biotechnol.* 52: 154–162.

157 Mnif, I. and Dhouha, G. (2015). Lipopeptide surfactants: Production, recovery and pore forming capacity. *Peptides* 71: 100–112.

158 Sałek, K. and Euston, S.R. (2019). Sustainable microbial biosurfactants and bioemulsifiers for commercial exploitation. *Process Biochem.* 85: 143–155.

159 Ishigami, Y. and Suzuki, S. (1997). Development of biochemicals-functionalization of biosurfactants and natural dyes. *Prog. Org. Coat.* 31: 51–61.

160 Matsuoka, K., Miyajima, R., Ishida, Y. et al. (2016). Aggregate formation of glycyrrhizic acid. *Colloids Surf., A* 500: 112–117.

161 Garofalakis, G., Murray, B.S., and Sarney, D.B. (2000). Surface activity and critical aggregation concentration of pure sugar esters with different sugar head groups. *J. Colloid Interface Sci.* 229: 391–398.

162 Goueth, P.Y., Gogalis, P., Bikanga, R. et al. (1994). Synthesis of monoesters as surfactants and drugs from D-glucose. *J. Carbohydr. Chem.* 13: 249–272.

163 Sarney, D.B. and Vulfson, E.N. (1995). Application of enzymes to the synthesis of surfactants. *Trends Biotechnol.* 13: 164–172.

164 Saini, H.S., Barragan-Huerta, B.E., Lebron-Paler, A. et al. (2008). Efficient purification of the biosurfactant viscosin from *Pseudomonas libanensis* strain M9-3 and its physicochemical and biological properties. *J. Nat. Prod.* 71: 1011–1015.

165 Laycock, M.V., Hildebrand, P.D., Thibault, P. et al. (1991). Viscosin, a potent peptidolipid biosurfactant and phytopathogenic mediator produced by a pectolytic strain of *Pseudomonas fluorescens*. *J. Agric. Food Chem.* 39: 483–489.

166 Neu, T.R., Haertner, T., and Poralla, K. (1990). Surface active properties of viscosin: A peptidolipid antibiotic. *Appl. Microbiol. Biotechnol.* 32: 518–520.

167 Banipal, P.K., Banipal, T.S., Lark, B.S., and Ahluwalia, J.C. (1997). Partial molar heat capacities and volumes of some mono-, di- and tri-saccharides in water at 298.15, 308.15 and 318.15 K. *J. Chem. Soc. Faraday Trans.* 93: 81–87.

168 Varga, I., Mészáros, R., Stubenrauch, C., and Gilányi, T. (2012). Adsorption of sugar surfactants at the air/water interface. *J. Colloid Interface Sci.* 379: 78–83.

169 Ribeiro, I.A.C., Faustino, C.M.C., Guerreiro, P.S. et al. (2015). Development of novel sophorolipids with improved cytotoxic activity toward MDA-MB-231 breast cancer cells. *J. Mol. Recognit.* 28: 155–165.

170 Angarten, R.G. and Loh, W. (2014). Thermodynamics of micellization of homologous series of alkyl mono and di-glucosides in water and in heavy water. *J. Chem. Thermodyn.* 73: 218–223.

171 Gill, S.J. and Wadsö, I. (1976). An equation of state describing hydrophobic interactions. *Proc. Natl. Acad. Sci. U. S. A.* 73: 2955–2958.

172 Majhi, P.R. and Blume, A. (2001). Thermodynamic vharacterization of temperature-induced micellization and demicellization of detergents studied by differential scanning calorimetry. *Langmuir* 17: 3844–3851.

173 Chen, M., Penfold, J., Thomas, R.K. et al. (2010). Mixing behavior of the biosurfactant, rhamnolipid, with a conventional anionic surfactant, sodium dodecyl benzene sulfonate. *Langmuir* 26: 17958–17968.

174 Chen, M., Penfold, J., Thomas, R.K. et al. (2010). Solution self-assembly and adsorption at the air–water interface of the monorhamnose and dirhamnose rhamnolipids and their mixtures. *Langmuir* 26: 18281–18292.

175 Ishigami, Y., Gama, Y., Nagahora, H. et al. (1987). The pH-sensitive conversion of molecular aggregates of rhamnolipid biosurfactant. *Chem. Lett.*: 16(5):763–16(5):766.

176 Whang, L.-M., Liu, P.-W.G., Ma, C.-C., and Cheng, S.-S. (2008). Application of biosurfactants, rhamnolipid, and surfactin, for enhanced biodegradation of diesel-contaminated water and soil. *J. Hazard. Mater.* 151: 155–163.

177 Onaizi, S.A., Nasser, M.S., and Twaiq, F.A. (2012). Micellization and interfacial behavior of a synthetic surfactant-biosurfactant mixture. *Colloids Surf., A* 415: 388–393.

178 Otto, R.T., Daniel, H.-J., Pekin, G. et al. (1999). Production of sophorolipids from whey. II. Product composition, surface active properties, cytotoxicity and stability against hydrolases by enzymatic treatment. *Appl. Microbiol. Biotechnol.* 52: 495–501.

179 Chen, M., Dong, C., Penfold, J. et al. (2011). Adsorption of sophorolipid biosurfactants on their own and mixed with sodium dodecyl benzene sulfonate, at the air/water interface. *Langmuir* 27: 8854–8866.

180 Ashby, R.D., Solaiman, D.K.Y., and Foglia, T.A. (2008). Property control of sophorolipids: Influence of fatty acid substrate and blending. *Biotechnol. Lett.* 30: 1093–1100.

181 Rosen, M.J., Mathias, J.H., and Davenport, L. (1999). Aberrant aggregation behavior in cationic gemini surfactants investigated by surface tension, interfacial tension, and fluorescence methods. *Langmuir* 15: 7340–7346.

182 Rosen, M.J., Cohen, A.W., Dahanayake, M., and Hua, X.Y. (1982). Relationship of structure to properties in surfactants. 10. Surface and thermodynamic properties of 2-dodecyloxypoly(ethenox yethanol)s, $C_{12}H_{25}(OC_2H_4)_xOH$, in aqueous solution. *J. Phys. Chem.* 86: 541–545.

183 Bakshi, M.S., Singh, K., Kaur, G. et al. (2006). Spectroscopic investigation on the hydrophobicity in the mixtures of nonionic plus twin tail alkylammonium bromide surfactants. *Colloids Surf., A* 278: 129–139.

184 Chen, L.-J., Lin, S.-Y., Huang, C.-C., and Chen, E.-M. (1998). Temperature dependence of critical micelle concentration of polyoxyethylenated non-ionic surfactants. *Colloids Surf., A* 135: 175–181.

185 Sulthana, S.B., Bhat, S.G.T., and Rakshit, A.K. (1997). Studies of the effect of additives on the surface and thermodynamic properties of poly(oxyethylene(10)) lauryl ether in aqueous solution. *Langmuir* 13: 4562–4568.

186 Tyrode, E., Johnson, C.M., Kumpulainen, A. et al. (2005). Hydration state of nonionic surfactant monolayers at the liquid/vapor interface: Structure determination by vibrational sum frequency spectroscopy. *J. Am. Chem. Soc.* 127: 16848–16859.

187 Tyrode, E., Johnson, C.M., Rutland, M.W., and Claesson, P.M. (2007). Structure and hydration of poly(ethylene oxide) surfactants at the air /liquid interface. A vibrational sum frequency spectroscopy study. *J. Phys. Chem. C* 111: 11642–11652.

188 Kumpulainen, A.J., Persson, C.M., Eriksson, J.C. et al. (2005). Soluble monolayers of n-decyl glucopyranoside and n-decyl maltopyranoside. Phase changes in the gaseous to the liquid-expanded range. *Langmuir* 21: 305–315.

189 Gorin, P.A.J., Spencer, J.F.T., and Tulloch, A.P. (1961). Hydroxy fatty acid glycosides of sophorose from Torulopsis magnoliae. *Can. J. Chem.* 39: 846–855.

190 Ozdener, M.H., Ashby, R.D., Jyotaki, M. et al. (2019). Sophorolipid biosurfactants activate taste receptor type 1 member 3-mediated taste responses and block responses to bitter taste in vitro; and in vivo. *J. Surfactant Deterg.* 22: 441–449.

191 Penfold, J., Chen, M., Thomas, R.K. et al. (2011). Solution self-assembly of the sophorolipid biosurfactant and its mixture with anionic surfactant sodium dodecyl benzene sulfonate. *Langmuir* 27: 8867–8877.

192 Manet, S., Cuvier, A.-S., Valotteau, C. et al. (2015). Structure of bolaamphiphile sophorolipid micelles characterized with SAXS, SANS, and MD simulations. *J. Phys. Chem. B* 119: 13113–13133.

193 Cecutti, C., Focher, B., Perly, B., and Zemb, T. (1991). Glycolipid self-assembly: Micellar structure. *Langmuir* 7: 2580–2585.

194 Zhou, S., Xu, C., Wang, J. et al. (2004). Supramolecular assemblies of a naturally derived sophorolipid. *Langmuir* 20: 7926–7932.

195 Baccile, N., Pedersen, J.S., Pehau-Arnaudete, G., and Van Bogaertf, I.N.A. (2013). Surface charge of acidic sophorolipid micelles: Effect of base and time. *Soft Matter* 9: 4911–4922.

196 Arima, K., Kakinuma, A., and Tamura, G. (1968). Surfactin, a crystalline peptidelipid surfactant produced by *Bacillus subtilis*: Isolation, characterization, and its inhibition of fibrin clot formation. *Biochem. Biophys. Res. Commun.* 31: 488–494.

197 Kakinuma, A., Hori, M., Sugino, H. et al. (1969). Determination of the location of the lactone ring in surfactin. *Agric. Biol. Chem.* 33: 1523–1524.

198 Kakinuma, A., Ouchida, A., Shima, T. et al. (1969). Confirmation of the structure of surfactin by mass spectrometry. *Agric. Biol. Chem.* 33: 1669–1671.

199 Kakinuma, A., Sugino, H., Isono, M. et al. (1969). Determination of fatty acids in surfactin and elucidation of the total structure of surfactin. *Agric. Biol. Chem.* 33: 973–976.

200 Kakinuma, A., Hori, M., Isono, M. et al. (1969d). Determination of amino acid sequence of surfactin, a crystalline peptide-lipid surfactant produced by *Bacillus subtilis*. *Agric. Biol. Chem.* 33: 971–972.

201 Bonmatin, J.M., Genest, M., Labbe, H., and Ptak, M. (1994). Solution three-dimensional structure of surfactin: A cyclic lipopeptide studied by ^{1}H-NMR, distance geometry, and molecular dynamics. *Biopolymers* 34: 975–986.

202 Vass, E., Besson, F., Majer, Z. et al. (2001). Ca^{2+}-induced changes of surfactin conformation: An FTIR and circular dichroism study. *Biochem. Biophys. Res. Commun.* 282: 361–367.

203 Tsan, P., Volpon, L., Besson, F., and Lancelin, J.-M. (2007). Structure and dynamics of surfactin studied by NMR in micellar media. *J. Am. Chem. Soc.* 129: 1968–1977.

204 Zou, A., Liu, J., Garamus, V.M. et al. (2010). Micellization activity of the natural lipopeptide [Glu1, Asp5] surfactin-C15 in aqueous solution. *J. Phys. Chem. B* 114: 2712–2718.

205 Razafindralambo, H., Thonart, P., and Paquot, M. (2004). Dynamic and equilibrium surface tensions of surfactin aqueous solutions. *J. Surfactant Deterg.* 7: 41–46.

206 Thimon, L., Peypoux, F., and Michel, G. (1992). Interactions of surfactin, a biosurfactant from *Bacillus subtilis*, with inorganic cations. *Biotechnol. Lett.* 14: 713–718.

207 Thimon, L., Peypoux, F., Wallach, J., and Michel, G. (1993). Ionophorous and sequestering properties of surfactin, a biosurfactant from *Bacillus subtilis*. *Colloids Surf. B. Biointerfaces* 1: 57–62.

208 Li, Y., Ye, R.-Q., and Mu, B.-Z. (2009). Influence of sodium ions on micelles of surfactin-C16 in solution. *J. Surfactant Deterg.* 12: 31–36.

209 Li, Y., Zou, A.-H., Ye, R.-Q., and Mu, B.-Z. (2009). Counterion-induced changes to the micellization of surfactin-C16 aqueous solution. *J. Phys. Chem. B* 113: 15272–15277.

210 Han, Y., Huang, X., Cao, M., and Wang, Y. (2008). Micellization of surfactin and its effect on the aggregate conformation of amyloid β(1-40). *J. Phys. Chem. B* 112: 15195–15201.

211 Ishigami, Y., Osman, M., Nakahara, H. et al. (1995). Significance of β-sheet formation for micellization and surface adsorption of surfactin. *Colloids Surf. B. Biointerfaces* 4: 341–348.

212 Maget-Dana, R. and Ptak, M. (1992). Interfacial properties of surfactin. *J. Colloid Interface Sci.* 153: 285–291.

213 Knoblich, A., Matsumoto, M., Ishiguro, R. et al. (1995). Electron cryo-microscopic studies on micellar shape and size of surfactin, an anionic lipopeptide. *Colloids Surf. B. Biointerfaces* 5: 43–48.

214 Zou, A., Liu, J., Garamus, V.M. et al. (2010). Interaction between the natural lipopeptide [Glu1, Asp5] surfactin-C15 and hemoglobin in aqueous solution. *Biomacromolecules* 11: 593–599.

215 Osman, M., Hoiland, H., Holmsen, H., and Ishigami, Y. (1998). Tuning micelles of a bioactive heptapeptide biosurfactant via extrinsically induced conformational transition of surfactin assembly. *J. Pept. Sci.* 4: 449–458.

216 Shen, H.-H., Thomas, R.K., Chen, C.-Y. et al. (2009). Aggregation of the naturally occurring lipopeptide, surfactin, at interfaces and in solution: an unusual type of surfactant? *Langmuir* 25: 4211–4218.

217 Menger, F.M. (2002). Supramolecular chemistry and self-assembly. *Proc. Natl. Acad. Sci. U. S. A.* 99: 4819–4822.

218 Bhattacharya, S., Maitra, U., Mukhopadhyay, S., and Srivastava, A. (2006). *Advances in molecular hydrogels* (eds. G. Weiss and P. Terech). Springer: *Molecular Gels*. Dordrecht.

219 Galantini, L., di Gregorio, M.C., Gubitosi, M. et al. (2015). Bile salts and derivatives: rigid unconventional amphiphiles as dispersants, carriers and superstructure building blocks. *Curr. Opin. Colloid Interface Sci.* 20: 170–182.

220 Savage, P.B. (2002). Cationic steroid antibiotics. *Curr. Med. Chem.: Anti-Infect. Agents* 1: 293–304.

221 Svobodova, H., Noponen, V., Kolehmainen, E., and Sievaenen, E. (2012). Recent advances in steroidal supramolecular gels. *RSC Adv.* 2: 4985–5007.

222 Vázquez Tato, J. (2014). *Molecular biomimicry*. Santiago: Servicio de Publicaciones, USC. ISBN 978-84-16183-11-1.

2

Metagenomics Approach for Selection of Biosurfactant Producing Bacteria from Oil Contaminated Soil

An Insight Into Its Technology

Nazim F. Islam and Hemen Sarma

Department of Botany, N N Saikia College, Titabar, Assam, India

CHAPTER MENU

2.1 Introduction

Humans have been using soap and soap-like substances for thousands of years. The development of soap-like materials is evident in ancient Babylon about 2800 BCE [1]. In India, reetha or soapnut (*Sapindus mukorossi* Gaertn.) has traditionally been used as a hair cleanser, among other things. Today, soap comprises a half of the total global production of surfactants, which is projected to be 15 mt/y. Linear alkylbenzene sulfonate (LABS) (1700 kt/y), lignin sulfonate (600 kt/y), fatty alcohol ethoxylates (700 kt/y), and alkylphenol ethoxylates (500 kt/y) are particularly widely produced surfactants [2]. Surfactants are hydrophobic and hydrophilic molecules capable of altering surface/interface tension and improving the solubility of polar compounds in non-polar solvents [3]. Similarly, biosurfactants are amphiphilic metabolites derived from microorganisms. These are interesting alternatives to conventional chemical surfactants because they are easily degradable in the environment. Hydrophilic moieties of biosurfactant molecules are mainly carbohydrates, carboxylic acids, phosphates, amino acids, cyclic peptides, or alcohols, whereas hydrophobic moieties are fatty acids, hydroxy fatty acids, and alkyl and β-hydroxy fatty acids [4]. Natural as well as synthetic surfactants, which are widely used in industrial processes, have various

Biosurfactants for a Sustainable Future: Production and Applications in the Environment and Biomedicine, First Edition. Edited by Hemen Sarma and Majeti Narasimha Vara Prasad.
© 2021 John Wiley & Sons Ltd. Published 2021 by John Wiley & Sons Ltd.

properties; therefore, they are used as solvents, stabilizers, lubricants, and foaming agents. Biosurfactants are produced by different strains of bacteria, fungi, and yeast in natural environments. These multifunctional microorganism-derived biomolecules have been extensively studied and reported in scientific literature [3, 5].

Biosurfactants are generally classified into a low or a high molecular weight group based on their chemical nature. Low molecular weight surfactants are widely used to lower surface-to-surface stress, while high molecular weight surfactants are generally used as emulsifiers and stabilizers [6]. For details of the composition, classification, critical concentration of micellization (CMC) values and properties of biosurfactants, see Chapter 1.

Microbe-derived surfactants appear to display a performance similar to synthetic surfactants [7]. While synthetic surfactants are commercially preferred during industrial applications, their use leads to the development of undesirable environmental pollutants [5, 8]. The majority of synthetic surfactants such as linear alkylbenzene sulfonate (LABS) are non-biodegradable with adverse environmental effects. Contrary to these, biosurfactants are less persistent and biodegradable in the environment owing to their biological origins [9]. In addition, most biosurfactants are active in a wide range of temperatures, pH and other environmental conditions [10].

Microbiologically derived surfactants have been widely used in industries such as emulsifiers, dispersants, foaming agents, and wetting agents [11], with a lower CMC value, which improve their performance over synthetic surfactants [12]. Some common sustainable applications of biosurfactants in the environment and in biomedicine (bioremediation, medical technology, food processing and pharmaceutical formulations, and cosmetics) are discussed in more detail in Chapters 5, 10, 11, and 19.

Many microorganisms, which are potential producers of biosurfactants, inhabit oil-contaminated soil in and around oil fields. One of the major hindrances for the discovery of novel biosurfactant-producing strains is the isolation and cultivation of biosurfactant-producing microbes. The metagenomics approach allows for the extraction of DNA (eDNA) from the environmental DNA pool and the screening for biosurfactant-producing genes [13]. The aim of this review is therefore to discuss the possibility of oil field soil being a repository for bacteria-producing biosurfactants that help in the desorption of oil during microbial degradation, their isolation, and screening techniques. This could be of enormous scope for industrial application and bioremediation. In addition, this article discusses current developments in the research on molecular techniques such as metagenomics combined with a stable isotope probe (SIP) for the discovery of new microbial strains that produce biosurfactants.

2.2 Metagenomics Application: A State-of-the-Art Technique

Most metagenomic studies have focused mainly on screening environmental DNA samples that produce novel biomolecules and on the diversity of microbes in different environments [14]. Metagenomics offers multiple uses and does not have an exhaustive list of applications. This technique has been used successfully to discover novel genes and microbes (for biodegradation and bioremediation), investigate microbial diversity, discover medicines, identify enzymes, monitor pollutions, and so on (Figure 2.1).

Techniques of metagenomics may also be used to classify genes/microbes from environmental samples that produce biomolecules. These multifunctional, fascinating biomolecules with diverse structural complexities can be used in a number of advanced environmental and bio-science applications [15].

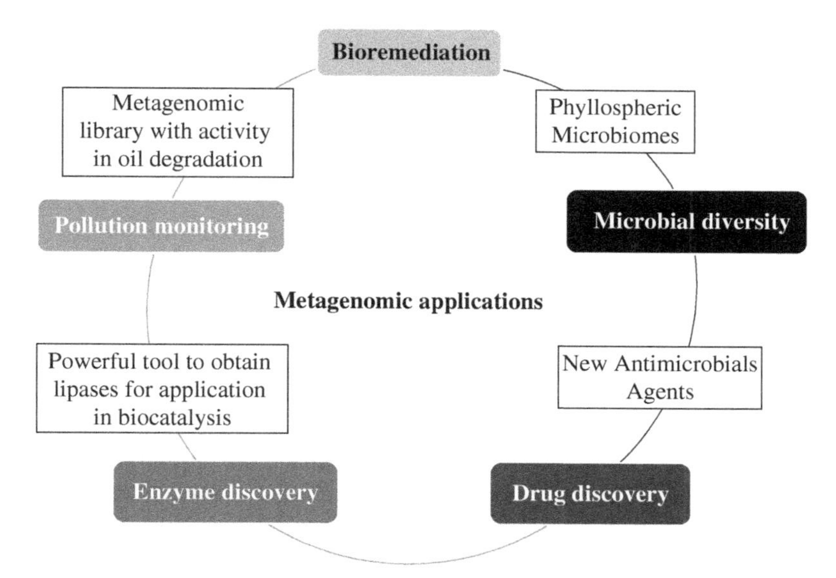

Figure 2.1 Application of the metagenomic technique for environmental management of advanced biomedical applications.

In environmental research, metagenomics is primarily used to monitor and evaluate microbes in environmental samples and to assist in bioremediation techniques [14]. Microbes can be screened through metagenomics for potential genes that can be used as biomarkers for pollution [16] or for the production of novel biomolecules for environmental management.

Biosurfactants are key agents in the remediation of persistent heavy metals in soil. They have been found to be effective in the remediation of heavy metals contaminated soil by the formation of a surfactant-associated complex which has already been established [3]. Metagenomics figure centrally in this context as well. For instance, metallothionein (MT) genes have been discovered from soil microbiomes using this approach. MT genes from novel metal-tolerant bacterial strains that confer Cu/Cd resistance and biosorption have been used for the development of metal bioremediation tools [13]. Similarly, metagenomics can help isolate genes that specifically target the degradation of PAHs in contaminated soil. This is crucial in the context of producing biosurfactants since they play a significant role in the biodegradation of polyaromatic hydrocarbons (PAHs) from soil. Biosurfactants increase the mobility of PAHs by reducing surface and interface stresses [17] and reduce the half-life of three- and five-ring PAHs by accelerating the degradation process of contaminated soil PAHs [4]. In addition to PAH degradation, biosurfactants are used to clean oil sludge from storage tanks, enhance oil recovery from refinery sludge and reservoirs, and mobilize oil flow through pipelines [4, 5, 18].

Metagenomics is used to identify genes from an environmental sample that have the potential to produce enzymes of industrial importance. Several enzymes, such as lipase, β-glucosidase, amylase, proteases, and esterases, have been described to have been investigated using metagenomics approaches [19]. They have also been used successfully in the creation of novel products for drug molecules [15]. Recent research has expanded our knowledge in the field of biomedicine by showing that both terrestrial and aquatic environments harbor microorganisms involved in the production of drug molecules [20]. Several microbial metabolites with antibacterial, antiviral, and antifungal properties have been screened through functional metagenomics [21]. In the field of health care, an individual's clinical diagnosis for pathogenic organisms can be traced with the help of metagenomics. Since most pathogens can be grown in selective media and are difficult to cultivate, metagenomics offers the screening of the individuals' microbiome as a whole [14].

Metagenomics is further employed in a range of other diverse sectors. In the food sector, it can be used in the detection of potential microbes for biosurfactant production for use in food industries. While the use of biosurfactants has not been very common in the food industry, they are being used to stabilize the agglomeration of fat globules and to improve the quality of foods based on fat [3]. In the field of agriculture, they are mainly used to monitor plant pathogens. Early detection of pathogens may help prevent plant diseases and minimize the loss of economically important crop plants [14]. They can also be used to successfully track viral pathogens, which often pose difficulties in screening using other conventional methods.

From this, it becomes evident that metagenomics has become an integral technique with multiple uses, from the detection of new molecules to advanced medical technology. Enzymes and other industrially important bioactive compounds, such as biosurfactants, have led to sustainable industrial growth through metagenomics. This technology also greatly contributes to the environmental monitoring of microbes and toxins, leading to the identification, treatment, and prevention of many diseases, and to the prevention of epidemics. Early identification with microbial metagenomics contributes to the reduction and elimination of health threats. Bioremediation and innovation of new drugs also help to improve the quality of life. Given the novelty of this technique, however, many other aspects of metagenomics are yet to be explored.

2.3 Hydrocarbon-Degrading Bacteria and Genes

Exploration of crude oil often results in accidental spillage and environmental contamination due to its various toxic components [22, 23]. Crude oil exploration fields are also home to oil-degrading microbes that are capable of using spilled oil as their carbon source and can remove crude oil from contaminated sites [24]. Numerous oil-degrading bacteria strains have been isolated from both cold [9] and hot [25] environments. Microbe-enhanced oil recovery tests performed using biosurfactant-producing microorganisms are briefly described in Chapter 5.

In the last few years, attempts have been made to identify possible biosurfactant-producing microorganisms [4]. Some of the key genera that make biosurfactants are *Acinetobacter*, *Bacillus*, *Azotobacter*, *Candida* [18], *Enterobacter*, *Micrococcus*, *Oceanobacillus*, *Pseudomonas*, *Rhodococcus*, *Serratia* and *Stenotrophomonas* [18, 26]. *Rhodococcus* sp. HL-6 reported from petroleum-contaminated soil produces glycolipid biosurfactants and has been successfully exploited for the remediation of crude oil contaminated sites [27]. *Pseudomonas* is one of the most widely described genera for the production of biosurfactants [28]. *Bacillus subtilis* and *Pseudomonas aeruginosa* have also been reported from oil-contaminated soils and have been shown to be a potential candidate for the degradation of petroleum hydrocarbons [18, 28, 29]. Similarly, biosurfactants derived from the consortium of *P. aeruginosa* and *Rhodococcus* strains have been reported to degrade more than 90% of oil sludge [30]. *P. aeruginosa* RS29 isolated from crude oil-contaminated sites has been reported to produce potent biosurfactants with enhanced foaming and emulsifying properties [28]. The thermophilic hydrocarbon-degrading bacteria *P. aeruginosa* AP02-1 are known to produce biosurfactants using hydrocarbon as the sole source of carbon [31]. Biosurfactant BSW10 derived from *P. aeruginosa* W10 has been successful in phenanthrene and fluoranthene biodegradation from oil-contaminated sites [32].

The marine microbiome is a global collective of all microorganisms. This is a good source of useful microbes for use in bioproducts. Specific microbial communities living in marine and coral reefs have been reported to be beneficial to humans. *Acinetobacter*, *Alteromonas*, *Azotobacter*, *Corynebacteria*, and *Myroids* are some marine microorganisms that have been reported to produce biosurfactants [12], while *Alcanivorax* and *Halomonas* have also been reported to produce biosurfactants in marine environments and to degrade hydrocarbons [33]. *Alcanivorax* uses n-alkanes from oil-contaminated sites producing glycolipid biosurfactants [34]. Species of *Halobacterium* viz. *Haloferax*, *Halovivax*, and *Haloarcula* have been described as biosurfactant producers and are known to utilize different hydrocarbons [35]. *Marinobacter*, *Methylophagia*, *Roseovarius*, *Thalassospria*, *Rheinheimera*, and *Sphingomonas* are the other genera known to produce biosurfactants and have been able to degrade aliphatic and aromatic hydrocarbons [36].

Many potential genes that are responsible for biosurfactant production have now been identified and described. Leite et al. [37] describe the presence of *rhlAB* and *alkB* genes from bacterial genomes isolated from soil contaminated with crude oil. The main pollutants of crude oil are alkanes and aromatic hydrocarbons. These bacterial genomes contain *RhlAB* genes and are responsible for the production of rhamnolipid biosurfactants and the *alkB* gene (alkane mono-oxygenase) mediates the degradation of petroleum hydrocarbons by the alkane mono-oxygenase enzyme system. Similarly, another gene reported from oil-contaminated soil is the naphthalene dioxygenase (*Nah*) gene, which contributes to the degradation of both alkanes and aromatic hydrocarbons [38]. Likewise, a lipopeptide biosurfactant surfactin is produced by three genes, *srfA*, *srfB*, and *srfC*, present in *srfA operon* [35, 39].

All the results reported in this section provide scientific evidence that oil-contaminated soil and marine environment are good sources of potential microbes that produce biosurfactants and degrade hydrocarbons. Microbial genes from contaminated environments could be used to produce environmentally safe biosurfactants that help with bioremediation and at the same time reduce production costs involved in the process.

2.4 Metagenomic Approaches in the Selection of Biosurfactant-Producing Microbes

Metagenomic approaches for the selection of biosurfactants producing genes are scanty, and few studies have been conducted to detect commercially important biosurfactants using metagenomic tools [40, 41]. In view of the enormous diversity of microbes in oil-contaminated soil, the search for novel biosurfactant molecules should be stepped up [42, 43].

Standard cultivation techniques do not encourage the isolation and screening of novel biosurfactant producers due to the diverse culture requirements of the microbial population. Moreover, due to structural and molecular diversity, the vast majority of them remain uncultured [43–45]. Metagenomics is an excellent way to bypass traditional cultivation techniques and to explore the undiscovered microbial population from oil-contaminated soil [44, 46, 47]. Metagenomic tools make it easier to isolate DNA and diverse microbial populations from environmental samples for potential uses [48–50].

Isolated metagenomic DNA is subjected to DNA sequencing or screening for functional activity [51, 52]. Metagenomic DNA sequencing involves next generation sequencing (NGS) and polymerase chain reaction (PCR) amplification. NGS allows the identification of coding sequences based on the homology of known genes [53]. The screening process is based on previously designed probes and primers based on sequences of known gene coding for an enzyme or bioactive compound [14].

Shotgun metagenomic sequencing is normally used to acquire a gene pool from a sample of environmental DNA [13]. Metagenomic sequence homology to reference sequence database is performed through similarity search tools such as BLAST (basic local alignment search tool), KEGG (Kyoto Encyclopedia of Genes and Genomes), COG (clusters of orthologous droups of proteins), etc. [46]. A software tool such as antiSMASH (antibiotic and secondary metabolite analysis shell) enables the identification of gene clusters linked to important biosynthetic pathways such as biosurfactant production [52]. Furthermore, screening for other related genes parallel to known genes for biosynthetic pathways may help to explore new structural compounds with improved biosurfactant properties from oil-contaminated soil.

Functional metagenomics involves the cloning of environmental DNA into a suitable vector and the heterologous expression of genes of interest. This technique allows the detection and testing of heterologous biomolecules and bioactivity using high-throughput monitoring systems [46, 51]. Functional screening aids in the identification of genes or new biomolecules from the Environmental Clone Library without prior knowledge of sequence isolation [14, 54, 55].

The construction of a metagenomics library for the desired gene expression is based on the precise selection of DNA fragments from the DNA pool extracted. The ideal vectors and hosts are then selected for target gene expression. Various vectors, such as plasmids, cosmids, fosmids, and bacterial artificial chromosomes (BAC), are used based on the size of the DNA fragment. Both single-host and multiple-host expression systems are used for gene expression systems [19].

Although functional metagenomics has made significant progress in the last few years, it has some limitations. Environmental DNA expression depends on the heterologous expression system of choice. Expressions of foreign genes in heterologous hosts may be hindered by host transcription machinery, which leads to low targeted gene expression. In contrast, the screening method may not be sufficiently sensitive to detect gene expression [56]. In addition, the proteins expressed may have a toxic effect on the host and the desired number of new biomolecules may not be achieved. [46]. Another caveat for functional metagenomics is searching for the targeted novel genes or its functional products from the large community DNA pool. To overcome this limitation, recent research has come to focus on ecological enhancement, i.e. enhancement of *in situ* environmental conditions by addition of specific substrates or altering the microhabitat for targeted microbial communities so that the desired functions from the extracted metagenome is achieved [56].

Thus, at present, metagenomics could be considered to be a powerful molecular technique for the detection of both bacterial and gene-producing biosurfactants. However, the efficacy of metagenomics can be improved by the use of the stable isotope probe (SIP) discussed in the next section.

2.5 Metagenomics with Stable Isotope Probe (SIP) Techniques

The metagenomic approach in mapping microbial population is one of the preferred strategies in environmental samples, and the use of this tool has increased considerably. Metagenomics is a genetic strategy that allows for the study of entire genomic microbial communities covering all genes, catabolic genes, and whole operons in environmental samples. The major advantage of shotgun sequencing is the ability to reconstruct the entire genome from identified library clone fragments to determine a biosynthetic pathway [57]. Although we have been led by metagenomics to explore non-cultivable microbes from environmental samples, it has certain limitations too.

Conventional metagenomics may not be a viable option for determining the functional aspects of low-abundance microbial populations [58]. The extraction of entire DNA from environmental cells is also a challenging task. Some of the DNA may be degraded or left behind erroneously due to different extraction procedures. As a result, genes of significant metabolic function or novel biomolecules may remain unknown [58]. Owing to the above-mentioned limitations, metagenomics might be used in combination with an SIP (Figure 2.2) to detect specific microbial communities

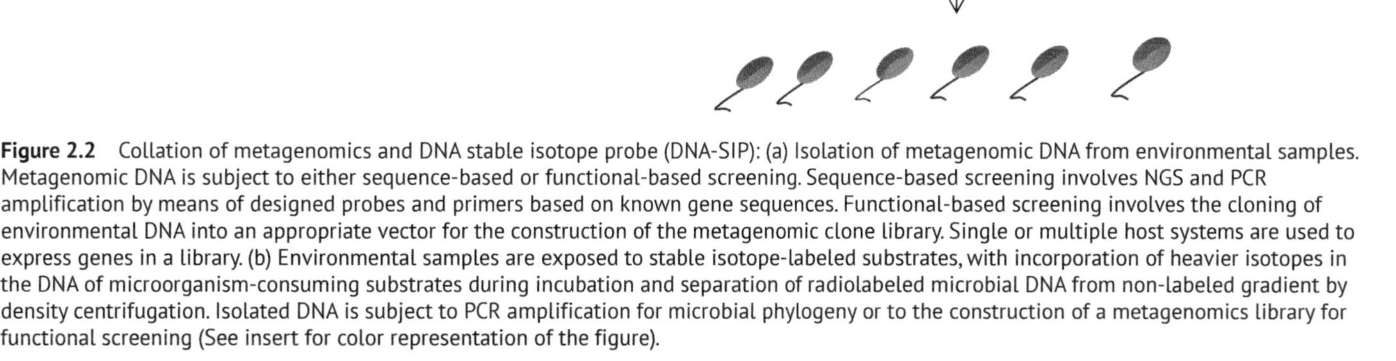

Figure 2.2 Collation of metagenomics and DNA stable isotope probe (DNA-SIP): (a) Isolation of metagenomic DNA from environmental samples. Metagenomic DNA is subject to either sequence-based or functional-based screening. Sequence-based screening involves NGS and PCR amplification by means of designed probes and primers based on known gene sequences. Functional-based screening involves the cloning of environmental DNA into an appropriate vector for the construction of the metagenomic clone library. Single or multiple host systems are used to express genes in a library. (b) Environmental samples are exposed to stable isotope-labeled substrates, with incorporation of heavier isotopes in the DNA of microorganism-consuming substrates during incubation and separation of radiolabeled microbial DNA from non-labeled gradient by density centrifugation. Isolated DNA is subject to PCR amplification for microbial phylogeny or to the construction of a metagenomics library for functional screening (See insert for color representation of the figure).

degrading hydrocarbons. The technique employs the incorporation of stable isotopes (^{13}C or ^{15}N) from the labeled substrate in the environmental sample into the microbial DNA to determine the functional aspects of the microbes. DNA stable isotope probing (DNA-SIP) enables the identification of biosynthetic pathways of the metabolically active microbes through the built-in radiolabeled substrate along with the identity of the microbe of interest [58]. In other words, the expression of specific functions of the targeted microbial community can be determined using a specific labeled substrate. Stable-isotope labeled DNA, also known as heavy DNA from active microorganisms, may be separated from the unmarked microbial community by ultracentrifugal gradient density for further analysis [59]. Isotope-labeled DNA is tested through isotope-ratio mass spectrometry (IRMS). Targeted DNA is then amplified by PCR or subjected to a multiple displacement amplification reaction (MDA) for the construction of a metagenomics library [60]. One of the deficiencies of DNA-SIP is that a high concentration of radiolabeled substrate is required during incubation to be incorporated into DNA. In addition, a high concentration of the substrate may hinder the growth of some microorganisms. This technique has the ability to isolate the microorganism that produces biosurfactants in oil and gas exploration sites.

2.6 Screening Methods to Identify Features of Biosurfactants

The diverse properties of biosurfactants do not make it possible to identify the molecules directly. Moreover, the biological roles of surfactant producers and the conditions of culture have a major impact on the production of biosurfactants. Instead, the physical shift in liquid media is the only general indicator for biosurfactant production. A number of screening techniques for biosurfactants have been developed and every technique has its own advantages and limitations.

For example, in the drop collapse method an aliquot of bacterial culture or culture supernatant is placed over the surface of the oil. The drops that collapse are considered positive for biosurfactants, while the drops that remain beaded are considered negative. Improved methods, such as the atomized spray method, have several advantages over the commonly used drop collapse method. The method helps to identify strains that have bypassed other detection methods and also helps to detect surfactants that are present in very low concentrations [61]. Another common method used for biosurfactant screening is the hemolytic method. In the hemolytic method, biosurfactant producing bacterial strains are tested on the basis of hemolytic activity. However, there are some bacterial strains that are capable of producing biosurfactant-like substances but are non-hemolytic. For example, glucose lipids produced by *Alcaligenes* sp. have properties similar to biosurfactants but are non-hemolytic [62]. Although it is a rapid test method, the breakdown and destruction of blood cells in the blood agar plate may be mediated by factors other than biosurfactants, resulting in false positive results. In addition, the quantity of biosurfactants may not be sufficient to lyse the blood cells in a given experimental condition. The other rapid screening method is the emulsification assay. The production of biosurfactants in the emulsification assay is determined by the emulsification index. The higher the emulsification index the more will be the emulsifying capacity of the biosurfactants in emulsifying oil hydrocarbons [63]. Detailed features of each screening method are presented in Table 2.1. A combination of screening methods is warranted for the high-throughput screening of biosurfactant producers. The combined assay will facilitate the detection of surfactants produced at a much lower concentration. Biosurfactants produced may be subjected to an analysis of Fourier transform infrared spectroscopy (FTIR) for the determination of chemical bonds and functional groups [70]. Compounds present in biosurfactants may be analyzed using gas chromatography–mass spectrometry (GCMS) [71].

Table 2.1 Comparison of screening techniques for determination of biosurfactants production.

Screening techniques	Key features	Limitations	References
CTAB-methylene blue agar	• Allows the identification of biological anionic glycolipid biosurfactants. • Biological surfactants form insoluble ion pairs with cationic surfactant cetyl trimethyl ammonium bromide and base dye methylene blue, indicated by the formation of dark blue halo around the culture colonies. • Ideal method for detecting extracellular glycolipids (rhamnolipids).	• CTAB is toxic to certain bacterial strains, e.g. *E. coli*. • The method is only appropriate for anionic biosurfactants.	[64]
Drop-collapse test	• Suitable for the detection of large-scale metagenomic clones. • Sensitive method for the determination of surfactant activity using a small volume of cell-free broth culture. • It can be used for both qualitative and quantitative tests. • Cell-free broth droplets are transferred to an oil-coated surface. Surfactant containing droplets collapses, whereas those lacking remain beaded. • In quantitative test diameter of droplet is measured. Droplet diameter of test broth larger than control indicates positive for biosurfactants.	• Bacterial strains producing low levels of surfactant cannot be detected. • May show false positive due to the hydrophobicity of certain bacterial cells acting as biosurfactants themselves.	[61, 65]
Atomized oil assay	• The oil droplet/liquid paraffin mist is sprayed over the culture plates. The formation of a bright zone or halo around the bacterial colonies indicates the production of biosurfactants. • Facilitates the simultaneous assessment of the number of colonies and is ideal for the library of metagenomic clones. • The production of surfactants can be detected even at very low concentrations.	• Method uses the detection of surfactant producers on the basis of the formation of bright halos around the bacterial colonies. Some synthetic surfactants may imitate the formation of bright halos. The distinction between "bright" and "dark" halo is arbitrary. • The assay is limited to cultivable microbes only. • The assay shows positive results on a solid medium.	[61, 66]
Oil-spreading technique	• Crude oil is added to the surface of the distilled water taken on a petri dish. • An aliquot of bacterial culture is placed on the surface of the oil. • Biosurfactant production is indicated by the formation of a dispersion zone.	• The method is suitable for primary screening and qualitative testing.	[37]

(*Continued*)

Table 2.1 (Continued)

Screening techniques	Key features	Limitations	References
Haemolytic method (Blood plate method)	• Rapid biosurfactant detection test is indicated by the formation of halo around the spot-inoculated bacterial colony on the blood agar plate. • Ideal method for the detection of rhamnolipids or surfactins.	• Extracellular metabolites other than biosurfactants can provide false-positive results.	[37, 67]
Emulsification assay	• The production of biosurfactants is determined by the ability of cell-free broth to emulsify crude oil in the test solution. • The activity of emulsification is calculated on an emulsification index basis.	• The activity of emulsification may not be correlated with reduction in surface tension. • The method only indicates for biosurfactant presence.	[68]
Bacterial adhesion to hydrocarbons (BATH) assay	• Measures the hydrophobicity of bacterial cells to hydrocarbons. • An indirect method to screen biosurfactants producing bacteria. • Increase in cell adhesion to liquid hydrocarbons indicates the production of biosurfactants.	• Affinity for hydrocarbon may vary between different bacterial strains. • Cell adherence may also be due to other bacterial cellular components.	[69]

2.7 Functional Metagenomics: Challenge and Opportunities

Sequence/homology-based screening is routinely used to screen metagenomic DNA using designed PCR primers or through NGS. A sequence-based approach is primarily based on shotgun sequencing of target genes from a library of clones to look at the important metabolic pathways [72–74]. The main advantage of shotgun sequencing is that the entire genome can be reconstructed from identified fragments of library clones to determine the biosynthetic pathway [57, 73, 75]. However, shotgun sequencing may not be a viable option for determining the functional aspects of a complex microbial population or those present in low abundance [58]. Functional metagenomics-based screening has several advantages over sequencing-based screening. The main advantage is that novel genes or their functional products can be traced without prior knowledge of gene sequences [15]. Heterologous gene expression is one of the challenges faced by functional metagenomics. Studies suggest that sizeable fractions of the target genes are insufficiently expressed in the expression host [15]. This may be due to a lack of optimal codon usages by host transcriptive machinery, discrepancy during protein synthesis and processing, lack of a suitable substrate required for a biosynthetic pathway, the toxic nature of the gene products, or other unknown associated factors [56]. In order to evaluate a complete biosynthetic pathway, a single metagenomic clone containing all the genes for the pathway is needed. Moreover, in order to represent the entire metagenome, the library should have a sufficient number of clones, taking into account the diversity of the community.

2.7.1 Single vs Multiple Host Expression System

The expression of genes in a metagenomics library by a single host expression system is a widely used strategy in recent times. The potentiality of an expression host is determined by its ability to replicate the vectors containing inserted DNA fragments, impeding recombination, and conferring resistance to background gene products and lytic phages. *Escherichia coli* is the most favored host system for the expression of foreign genes. However, only 40% of the genes with functional activity have been reported to express in an *E. coli* host system [15]. The metabolic potential of most functional genes of remotely related microbes may not be sufficiently expressed [56]. This may also be attributed to the lack of an appropriate biosynthetic pathway substrate in a single host or to the fact that the host transcription machines may not recognize the sequence of promoters or favor the expression of foreign genes by limiting the essential factors [19]. However, some of the shortcomings of the *E. coli* system are rectified by augmenting it with plasmids equipped with an additional tRNA gene, simultaneous expression of chaperonin genes, etc. [76, 77].

In order to mitigate the limitations of a single host, multiple hosts are used to increase the likelihood of expression of targeted genes. *Bacillus*, *Burkholderia*, *Sphingomonas*, *Streptomyces*, and *Pseudomonas* [78] are the alternate host systems used for the functional screening of metagenomic libraries. Parallel screening using multiple hosts allows the successful expression of gene products from the Metagenomic Clone Library by providing a diverse host cell environment. Furthermore, a multiple host system permits the screening of metagenomic libraries for biosynthetic pathways that may remain undiscovered through a single host expression system. Despite all these limitations and prospects, functional screening using multiple hosts is one of the viable options in discovery of novel biosurfactants from oil contaminated environments.

2.7.2 Metagenomic Clone Libraries

In addition to the transcriptive machinery of the host, the expression of genes in a metagenomic library host also depends on the DNA insert size [56]. The selection of an appropriate vector based on the nature of the host expression is another hurdle in functional metagenomics. Small insert sizes may limit the detection of important biosynthetic pathways required for novel biomolecule synthesis. Fosmid and Cosmid libraries could accommodate insert sizes of ~15–40 kb, which may be limiting for cloning large size DNA fragments. Large insert libraries like BACs are preferred as they permit the cloning of genes for entire biosynthetic pathways of targeted biomolecules [79]. BACs (Bacterial Artificial Chromosome) can accommodate 100–200 kb of DNA fragments, rendering them suitable for metagenomic libraries. However, maintaining large size metagenomic DNA with a high molecular weight is the other challenge in functional screening. Owing to difficulties such as this in screening large clones, newer technologies like fluorescence-based assay are gaining importance for the rapid detection of an enzymatic activity. Furthermore, use of a robotic assay simplifies the screening process in high-throughput screening within a short time period [19].

2.8 Conclusion

Although the metagenomic approach for detecting microbes that produce novel biosurfactants is still at a growing stage, there is an ample opportunity to use these molecular tools in the future. As of now, these techniques have been used in the collection of environmental DNA samples for research purposes. In the above sections, we discussed metagenomic approaches to microorganism screening, which produce biosurfactants with add-on techniques to overcome barriers faced by conventional metagenomics. Emphasis has been placed on the synergy of techniques and methodologies with a functional or sequence-based approach for more favorable outcomes. We present how DNA-SIP coupled with metagenomic strategies can help identify specific microbes of interest. The technique could be exploited in screening microbes or genes producing novel biosurfactants from oil-contaminated soil. In addition to the above, we emphasize the characteristics and limitations of biosurfactant screening methods. The challenges and opportunities for functional metagenomics have been presented in detail.

Acknowledgements

The opinions, ideas, conceptions, and design presented in this chapter are of the authors themselves. The authors are grateful to NNS College for providing research and logistics facilities. The authors are grateful to Professor M.N.V. Prasad for his critical suggestions and inputs during the preparation of this chapter.

References

1 Willcox, M. (2000). Soap. In: *Poucher's Perfumes, Cosmetics and Soaps*, 10e (ed. H. Butler), 453. Dordrecht: Kluwer Academic Publishers. ISBN: 978-0-7514-0479-1. Archived from the original on 2016-08-20. The earliest recorded evidence of the production of soap-like materials dates back to around 2800 BCE in ancient Babylon.

2 Kosswig, K. (2005). *"Surfactants" in Ullmann's Encyclopedia of Industrial Chemistry*. Weinheim: Wiley VCH https://doi.org/10.1002/14356007.a25747.

3 Santos, D.K.F., Rufino, R.D., Luna, J.M. et al. (2016). Biosurfactants: multifunctional biomolecules of the 21st century. *IJMS* 17 (3): 401–431.

4 Sarma, H., Bustamante, K.L.T., and Prasad, M.N.V. (2019). Biosurfactants for oil recovery from refinery sludge: magnetic nanoparticles assisted purification. In: *Industrial and Municipal Sludge* (eds. M.N.V. Prasad, P.J. de Campos Favas, M. Vithanage and S. Venkata Mohan), 107–132. Butterworth-Heinemann, UK https://doi.org/10.1016/B978-0-12-815907-1.00006-4.

5 Fenibo, E.O., Ijoma, G.N., Selvarajan, R., and Chikere, C.B. (2019). Microbial surfactants: the next generation multifunctional biomolecules for applications in the petroleum industry and its associated environmental remediation. *Microorganisms* 7: 581.

6 Shekhar, S., Sundaramanickam, A., and Balasubramanian, T. (2015). Biosurfactant producing microbes and their potential applications: A review. *Crit. Rev. Environ. Sci. Technol.* 45: 1522–1554.

7 Muller, M.M., Kügler, J.H., Henkel, M. et al. (2012). Rhamnolipids – Next generation surfactants? *J. Biotechnol.* 161: 366–380.

8 Myers, D. (2010). *Surfactant Science and Technology*, 3e. Hoboken: Wiley.

9 Chaudhary, D.K. and Kim, J. (2019). New insights into bioremediation strategies for oil-contaminated soil in cold environments. *Int. Biodeter. Biodegr.* 142: 58–72.

10 Malavenda, R., Rizzo, C., Michaud, L. et al. (2015). Biosurfactant production by Arctic and Antarctic bacteria growing on hydrocarbons. *Polar Biol.* 38: 1565–1574.

11 Geys, R., Soetaert, W., and Van Bogaert, I. (2014). Biotechnological opportunities in biosurfactant production. *Curr. Opin. Biotechnol.* 30: 66–72.

12 Jahan, R., Bodratti, A.M., Tsianou, M., and Alexandridis, P. (2020). Biosurfactants, natural alternatives to synthetic surfactants: Physicochemical properties and applications. *Adv. Colloid Interface Sci.* https://doi.org/10.1016/j.cis.2019.102061.

13 Li, X., Islam, M.M., Chen, L. et al. (2020). Metagenomics-guided discovery of potential metallothionein genes from the soil microbiome that confer Cu and/or Cd resistance. *Appl. Environ. Microbiol.* 86 (9): e02907–e02919.

14 Dutta, S., Rajnish, K.N., Samuel, M.S. et al. (2020). Metagenomic applications in microbial diversity, bioremediation, pollution monitoring, enzyme and drug discovery. A review. *Environ. Chem. Lett.* https://doi.org/10.1007/s10311-020-01010-z.

15 Trindade, M., van Zyl, L.J., Navarro-Fernandez, J., and Elrazak, A.A. (2015). Targeted metagenomics as a tool to tap into marine natural product diversity for the discovery and production of drug candidates. *Front. Microbiol.* 6: 1–14.

16 Kisand, V., Valente, A., Lahm, A. et al. (2012). Phylogenetic and functional metagenomic profiling for assessing microbial biodiversity in environmental monitoring. *PLoS One* 7 (8): e43630. https://doi.org/10.1371/journal.pone.0043630.

17 Sarma, H., Nava, A.R., and Prasad, M.N.V. (2019). Mechanistic understanding and future prospect of microbe-enhanced phytoremediation of polycyclic aromatic hydrocarbons in soil. *Environ. Technol. Innov.* 13: 318–330.

18 Sarma, H. and Prasad, M.N.V. (2018). Metabolic engineering of rhizobacteria associated with plants for remediation of toxic metals and metalloids. In: *Transgenic Plant Technology* (ed. M.N.V. Prasad). Elsevier, Netherlands, 299–318. eBook ISBN: 9780128143902, Paperback ISBN: 9780128143896.

19 Kennedy, J., Flemer, B., Jackson, S.A. et al. (2010). Marine metagenomics: new tools for the study and exploitation of marine microbial metabolism. *Mar. Drugs* 8: 608–628.

20 Montaser, R. and Luesch, H. (2011). Marine natural products: a wave of new drugs? *Future Med. Chem.* 3: 1475–1489. https://doi.org/10.4155/fmc.11.118.

21 Rocha-Martin, J., Harrington, C., Dobson, A., and O'Gara, F. (2014). Emerging strategies and integrated systems microbiology technologies for biodiscovery of marine bioactive compounds. *Mar. Drugs* 12: 3516–3559. https://doi.org/10.3390/md12063516.

22 Sarma, H., Islam, N.F., Borgohain, P. et al. (2016). Localization of polycyclic aromatic hydrocarbons and heavy metals in surface soil of Asia's oldest oil and gas drilling site in Assam, Northeast India: Implications for the bio economy. *Emerging Contam.* 2 (3): 119–127.

23 Sharma, D., Sarma, H., Hazarika, S. et al. (2018). Agro-ecosystem diversity in petroleum and natural gas explored sites in Assam state, north-eastern India: Socio-economic perspectives. In: *Sustainable Agriculture Reviews 27* (ed. E. Lichtfouse). Springer, Cham, 37–60.

24 Sarma, H. and Prasad, M.N.V. (2016). Phytomanagement of polycyclic aromatic hydrocarbons and heavy metals-contaminated sites in Assam, north eastern state of India, for boosting bioeconomy. In: *Bioremediation and Bioeconomy* (ed. M.N.V. Prasad), 609–626. Elsevier, USA, Chapter 24. doi:https://doi.org/10.1016/B978-0-12-802830-8.00024-1. ISBN: 978-0-12-802830-8.

25 Sharma, N., Lavania, M., Kukreti, V., and Lal, B. (2020). Instigation of indigenous thermophilic bacterial consortia for enhanced oil recovery from high temperature oil reservoirs. *PLoS One* 15 (5): e0229889. https://doi.org/10.1371/journal.pone.0229889.

26 Varjani, S.J. (2017). Microbial degradation of petroleum hydrocarbons. *Bioresour. Technol.* 223: 277–286.

27 Tian, Z.-J., Chen, L.-Y., Li, D.-H. et al. (2016). Characterization of a biosurfactant-producing strain *Rhodococcus* sp. hl-6. *Rom. Biotechnol. Lett.* 21 (4): 11650–11659.

28 Saikia, R.R., Deka, S., Deka, M., and Sarma, H. (2012). Optimization of environmental factors for improved production of rhamnolipid biosurfactants by *Pseudomonas aeruginosa* RS29 on glycerol. *J. Basic Microbiol.* 52: 446–457.

29 Sarma, H. and Prasad, M.N.V. (2015). Plant-microbe association-assisted removal of heavy metals and degradation of polycyclic aromatic hydrocarbons. In: *In: S Mukherjee (ed.), Petroleum Geosciences: Indian Contexts*, Switzerland, 219–Switzerland, 236. Springer International Publishing https://doi.org/10.1007/978-3-319-03119-4_10. ISBN: 978-3-319-03118-7.

30 Cameotra, S.S. and Singh, P. (2008). Bioremediation of oil sludge using crude biosurfactants. *Int. Biodeter. Biodegr.* 62: 274–280.

31 Perfumo, A., Banat, I.M., Canganella, F., and Marchant, R. (2006). Rhamnolipid production by a novel hydrocarbon-degrading *Pseudomonas aeruginosa* AP02-1. *Appl. Microbiol. Biotechnol.* 72: 132–138.

32 Ma, Z., Liu, J., Dick, R.P. et al. (2018). Rhamnolipid influences biosorption and biodegradation of Phenanthrene by Phenanthrene- degrading strain *Pseudomonas* sp. pH6. *Environ. Pollut.* 240: 359–367.

33 Chen, W., Wilkes, G., Khan, I.U. et al. (2018). Aquatic bacterial communities associated with land use and environmental factors in agricultural landscapes using a metabarcoding approach. *Front. Microbiol.* 9: 2301.

34 Nisenbaum, M., Corti-Monzon, G., Villegas-Plazas, M. et al. (2020). Enrichment and key features of a robust and consistent indigenous marine-cognate microbial consortium growing on oily bilge wastewaters. *Biodegradation* https://doi.org/10.1007/s10532-020-09896.

35 De Silva Araujo, S.C., Silva-Portela, R.C.B., de Lima, D.C. et al. (2020). MBSP1: A biosurfactants protein derived from a metagenomic library with activity in oil degradation. *Sci. Rep.* 10: 1340.

36 Vigor, S., Joessar, M., Soares-Castro, P. et al. (2020). Microbial metabolic potential of phenol degradation in wastewater treatment plant of crude oil refinery: Analysis of metagenomes and characterization of isolates. *Microorganisms* 8: 652.

37 Leite, G.G.F., Figueirora, J.V., Almedia, T.C.M. et al. (2016). Production of rhamnolipids and diesel oil degradation by bacteria isolated from soil contaminated by petroleum. *Am. Inst. Chem. Eng.* 32: 262–270.

38 Qinglong, L., Tang, J., Bai, Z. et al. (2015). Distribution of petroleum degrading genes and factor analysis of petroleum contaminated soil from the Dagang oilfield, China. *Sci. Rep.* 5: 11068.

39 Sachdev, D.P. and Cameotra, S.S. (2013). Biosurfactants in agriculture, mini-review. *Appl. Microbiol. Biotechnol.* 97: 1005–1016.

40 Kebbouche-Gana, S., Gana, M.L., Ferrioune, I. et al. (2013). Production of biosurfactant on crude date syrup under saline conditions by entrapped cells of *Natrialba* sp. strain E21, an extremely halophilic bacterium isolated from a solar saltern (Ain Salah, Algeria). *Extremophiles* 17: 981–993.

41 Rizzo, C., Michaud, L., Hörmann, B. et al. (2013). Bacteria associated with sabellids (Polychaeta: Annelida) as a novel source of surface active compounds. *Mar. Pollut. Bull.* 70: 125–133.

42 Handelsman, J., Rondon, M.R., Brady, S.F. et al. (1998). Molecular biological access to the chemistry of unknown soil microbes: a new frontier for natural products. *Chem. Biol.* 5: R245–R249.

43 Rappe, M.S. and Giovannoni, S.J. (2003). The uncultured microbial majority. *Annu. Rev. Microbiol.* 57: 369–394.

44 Handelsman, J., Liles, M., Mann, D., and Riesenfeld, C. (2002). Cloning the metagenome: Culture-independent access to the diversity and functions of the uncultivated microbial world. *Methods Microbiol.* 33: 241–255.

45 Rondon, M.R., August, P.R., Bettermann, A.D. et al. (2000). Cloning the soil metagenome: a strategy for accessing the genetic and functional diversity of uncultured microorganisms. *Appl. Environ. 66(6):2541–2547.*

46 Jackson, S.A., Borchert, E., O'Gara, F., and Dobson, A.D. (2015). Metagenomics for the discovery of novel biosurfactants of environmental interest from marine ecosystems. *Curr. Opin. Biotechnol.* 33: 176–182.

47 Kennedy, J., O'Leary, N.D., Kiran, G.S. et al. (2011). Functional metagenomic strategies for the discovery of novel enzymes and biosurfactants with biotechnological applications from marine ecosystems. *J. Appl. Microbiol.* 111: 787–799.

48 Gloux, K., Leclerc, M., Iliozer, H. et al. (2007). Development of high-throughput phenotyping of metagenomic clones from the human gut microbiome for modulation of eukaryotic cell growth. *Appl. Environ. Microbiol.* 73: 3734–3737.

49 Gurgui, C. and Piel, J. (2010). Metagenomic approaches to identify and isolate bioactive natural products from microbiota of marine sponges. In: *Metagenomics: Methods and Protocols, Methods in Molecular Biology* (eds. W.R. Streit and R. Daniel), 247–263. Berlin: Springer Science + Business Media.

50 Zhou, J., Bruns, M., and Tiedje, J.M. (1996). DNA recovery from soils of diverse composition. *Appl. Environ. Microbiol.* 62: 316–322.

51 Walter, V., Syldatk, C., and Hausmann, R. (2010). Screening concepts for the isolation of biosurfactant producing microorganisms. In: *Biosurfactants* (ed. R. Sen), 1–13. New York: Landes Bioscience and Springer Science.

52 Weber, T., Blin, K., Duddela, S. et al. (2015). AntiSMASH 3.0 – a comprehensive resource for the genome mining of biosynthetic gene clusters. *Nucleic Acids Res.* 43: 1–7.

53 Altschul, S.F., Gish, W., Miller, W. et al. (1990). Basic local alignment search tool. *J. Mol. Biol.* 215: 403–410.

54 Suenaga, H. (2012). Targeted metagenomics: A high-resolution metagenomics approach for specific gene clusters in complex microbial communities. *Environ. Microbiol.* 14: 13–22. https://doi.org/10.1111/j.1462-2920.2011.02438.x.

55 Tuffin, M., Anderson, D., Heath, C., and Cowan, D. (2009). Metagenomic gene discovery: How far have we moved into novel sequence space? *Biotechnol. J.* 4: 1671–1683.

56 Ekkers, D.M., Cretoiu, M.S., Kielak, A.M., and Elsas, J.D. (2012). The great screen anomaly — a new frontier in product discovery through functional metagenomics. *Appl. Microbiol. Biotechnol.* 93: 1005–1020.

57 Montiel, D., Kang, H.-S., Chang, F.-Y. et al. (2015). Yeast homologous recombination-based promoterengineering for the activation of silent natural product biosynthetic gene clusters. *Proc. Natl. Acad. Sci. USA.* 112 (29): 8953–8958. https://doi.org/10.1073/pnas.1507606112.

58 Chen, Y. and Murrell, J.C. (2010). When metagenomics meets stable-isotopes probing: Progress and perspectives. *Trends Microbiol.* 18: 4.

59 Dumont, M.G. and Murrell, J.C. (2005). Stable isotope probing – Linking microbial identity to function. *Nat. Rev. Microbiol.* 3: 499–504.

60 Binga, E.K., Lasken, R.S., and Neufeld, J.D. (2008). Something from (almost) nothing: The impact of multiple displacement amplification on microbial ecology. *ISME J.* 2: 233–241.

61 Burch, A.Y., Browne, P.J., Dunlap, C.A. et al. (2011). Comparison of biosurfactant detection methods reveals hydrophobic surfactants and contact-regulated production. *Environ. Microbiol.* 13: 2681–2691.

62 He, S., Ni, Y., Lu, L. et al. (2020). Simultaneous degradation of n-hexane and production of biosurfactants by *Pseudomonas* sp. strain NEE2 isolated from oil-contaminated soils. *Chemosphere* 242: 125237.

63 Lenchi, N., Kebbouche-Gana, S., Servais, P. et al. (2020). Disel biodegradation capacities and biosurfactants production in saline-alkaline conditions by *Delftia* sp. NL1, isolated from an Algerian oilfield. *Geomicrobiol. J.* https://doi.org/10.1080/01490451.2020.1722769.

64 Siegmund, I. and Wagner, F. (1991). New method for detecting rhamnolipids excreted by *Pseudomonas* species during growth on mineral agar. *Biotechnol. Tech.* 5: 265–268.

65 Bodour, A.A. and Maier, R.M. (1998). Application of a modified drop collapse technique for surfactant quantification and screening of biosurfactant-producing microorganisms. *J. Microbiol. Methods* 32: 273–280.

66 Burch, A.Y., Shimada, B.K., Browne, P.J., and Lindow, S.E. (2010). Novel high-throughput detection method to assess bacterial surfactant production. *Appl. Environ. Microbiol.* 76: 5363–5372.

67 Thavasi, R., Sharma, S., and Jayalakshmi, S. (2011). Evaluation of screening methods for the isolation of biosurfactant producing marine bacteria. *J. Pet. Environ. Biotechnol.* S1: 001.

68 Batista, S.B., Mounteer, A.H., Amorim, F.R., and Totola, M.R. (2006). Isolation and characterization of biosurfactant/bioemulsifier-producing bacteria from petroleum contaminated sites. *Bioresour. Technol.* 97: 868–875.

69 Rosenberg, M., Gutnick, D., and Rosenberg, E. (1980). Adherence to bacteria to hydrocarbons: A simple method for measuring cell-surface hydrophobicity. *FEMS Microbiol. Lett.* 9: 29–33.

70 Gidudu, B., Mudenda, E., and Chirwa, E.M.N. (2020). Biosurfactant produced by *Serrati* sp. and its application in bioremediation enhancement of oil sludge. *Chem. Eng. Trans.* 79: 433–438.

71 Ashitha, A., Radhakrishnan, E.K., and Jyothis, M. (2020). Characterization of biosurfactant produced by the endophyte *Burkholderia* sp. WYAT7 and evaluation of its antibacterial and antibiofilm potentials. *J. Biotechnol.* https://doi.org/10.1016/j.jbiotec.2020.03.005.

72 Charlop-Powers, Z., Milshteyn, A., and Brady, S.F. (2014). Metagenomic small molecule discovery methods. *Curr. Opin. Microbiol.* 19C: 70–75.

73 Kim, J.H., Feng, Z., Bauer, J.D. et al. (2010). Cloning large natural product gene clusters from the environment: Piecing environmental DNA gene clusters back together with TAR. *Biopolymers* 93: 833–844.

74 Owen, J.G., Reddy, B.V.B., Ternei, M.A. et al. (2013). Mapping gene clusters within arrayed metagenomic libraries to expand the structural diversity of biomedically relevant natural products. *Proc. Natl. Acad. Sci. USA.* 110: 11797–11802.

75 Loeschcke, A., Markert, A., Wilhelm, S. et al. (2013). TREX: a universal tool for the transfer and expression of biosynthetic pathways in bacteria. *ACS Synth. Biol.* 2: 22–33.

76 Ferrer, M., Chernikova, T.N., Yakimov, M.M. et al. (2003). Chaperonins govern growth of *Escherichia coli* at low temperatures. *Nat. Biotechnol.* 21: 1266–1267.

77 Makrides, S.C. (1996). Strategies for achieving high-level expression of genes in *Escherichia coli. Microbiol. Rev.* 60: 512–538.

78 Van Elsas, J.D., Speksnijder, A.J., and van Overbeek, L.S. (2008). A procedure for the metagenomics exploration of disease-suppressive soils. *J. Microbiol. Methods* 75: 515–522.

79 Kakirde, K.S., Parsley, L.C., and Liles, M.R. (2010). Size does matter: Application-driven approaches for soil metagenomics. *Soil Biol. Biochem.* 42: 1911–1923.

3

Biosurfactant Production Using Bioreactors from Industrial Byproducts

Arun Karnwal

Department of Microbiology, School of Bioengineering and Biosciences, Lovely Professional University, Phagwara, Punjab, India

CHAPTER MENU

3.1 Introduction

Surfactants are amphiphilic molecules with two contrary components, one component being hydrophobic and other hydrophilic in nature [1, 2]. Hydrophobic and hydrophilic components depend on the polar charge, which may be anionic, cationic, neutral, or amphoteric. Biosurfactants are emerging as a promising alternative for synthetic surfactants in the industrial sector, as companies develop environmentally safe biosurfactants using various renewable and organic materials. Usually, the surfactants derived from organic substances comprise both hydrophilic and hydrophobic components. Natural surfactants are a group of secondary metabolites that are widely present in many plants, microorganisms and several sea animals [3]. Surfactants are surface-active

Biosurfactants for a Sustainable Future: Production and Applications in the Environment and Biomedicine,
First Edition. Edited by Hemen Sarma and Majeti Narasimha Vara Prasad.
© 2021 John Wiley & Sons Ltd. Published 2021 by John Wiley & Sons Ltd.

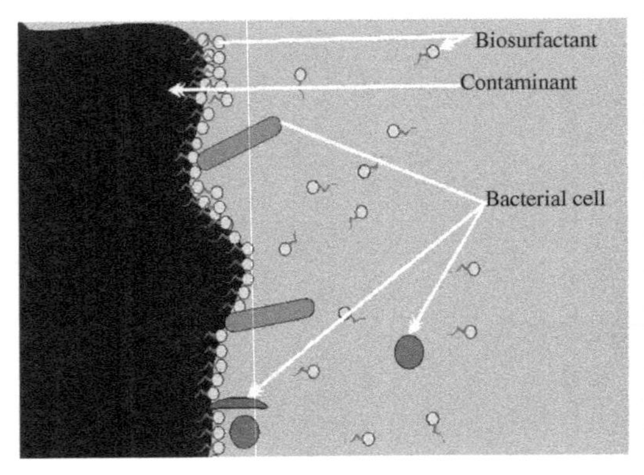

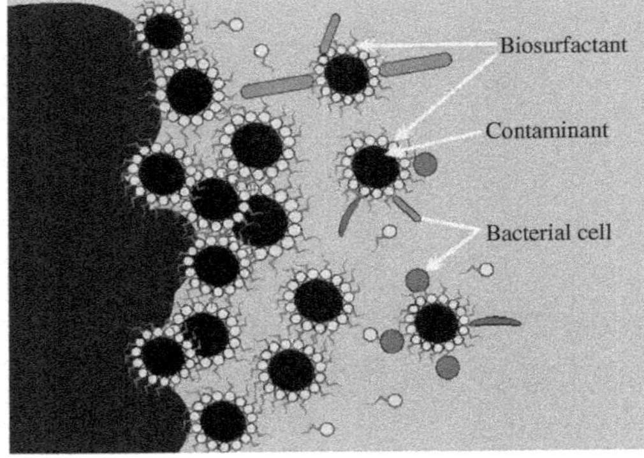

Figure 3.1 Schematic representation of the adhesion of bio surfactant molecules to the containment in which bacterial cell is associated (See insert for color representation of the figure) (Guerra-Santos et al., 1984).

chemicals used in detergents and soaps for reducing surface tension. Biosurfactants can be produced from various low-cost industrial waste materials (Figure 3.1).

They have many benefits over chemical-derived surfactants, including minimal toxicity, biologically available, biologically degradable, high foaming, and environmentally safe [4, 5]. Therefore, they are safer substitutes for synthetic surfactants, notably in food, medicine, cosmetics, and edible oils [6, 7]. Biosurfactants have an extensive range of applications in different domains such as cosmetics, pharmaceuticals, milk, energy, irrigation, forestry, textiles, painting, and several other sectors. These molecules are commonly known as multifunctional compounds, like stabilizers, wetting agents, antimicrobials, moisturizers, emulsifiers, and antiadhesives [8–11].

3.2 Significance of the Production of Biosurfactants from Industrial Products

The major barrier in the production of biosurfactants is the cost of the production process. It was documented that the main obstacle to large-scale application of biosurfactants is related to their higher production cost (10–30%) than chemical surfactants. In particular, carbon and energy sources used during the fermentation process cost 50% of the total costs of production of biosurfactants [12–14]. However, the use of alternative nutrient sources that are readily available and cheap may drastically reduce this cost [15]. The usage of industrial waste or byproducts as a source of energy for biosurfactant production may be an effective way to reduce production cost and sustainability of the production process for industries. It was reported earlier [16] that agro-industrial waste, with a higher protein, fat, and carbohydrate component, is desirable as a production medium component for biosurfactant production.

Similarly, industrial byproducts, i.e. glycerol, petroleum sludge, sugar cane bagasse, and fish waste, could be used as a carbon source for the fermentation process and microbial growth [17]. Several microorganisms, like *Bacillus*, *Corynebacterium*, *Pseudomonas*, and *Rhodococcus* are capable of producing biosurfactants from various industrial byproducts. Aguiar et al. [18] reported the use of *Corynebacterium aquaticum* as emulsifying biosurfactant producers by using agro-waste. Waste discharged from oil-based industries is severely harmful to the environment and is assumed to be primarily responsible for worldwide pollution. Many of them are neurotoxic, carcinogenic, and poisonous and, therefore, can impact human and animal health [19–21]. Similarly, the solvents used for the removal of paint adversely affect the environment. Searching for new techniques that effectively reduce pollutants is therefore extremely important in order to reduce the adverse impact of industrial wastes.

3.3 Factors Affect Biosurfactant Production in Bioreactor

Various factors influence the biosurfactant production efficiency of microbes. These factors are divided into various categories, but primarily belong to three major groups:

1) The first group of factors includes medium components, i.e. carbon, nitrogen source.
2) The second group includes physicochemical parameters of the microbial community and growth conditions in the bioreactor, i.e. upstream processing [22, 23].
3) The third group includes the product isolation parameters, i.e. downstream processing.

The components of all of these groups have a direct impact on production cost and product quality.

3.4 Microorganisms

Different groups of microorganisms, i.e. bacteria, fungi, and yeasts, are known as potential sources for the production of biosurfactants. Their production potential relies on various environmental and physiological conditions. Each microbe does have optimal conditions for its growth. However, the optimum growth conditions for cells are not always very suitable for the production of the desired product [14, 24]. The quantity of biosurfactants produced depends primarily on the type of microorganism and nutrient type. Most of the microorganisms that have a biosurfactant production capability have been screened from various industrial waste sites, i.e. contaminated soils, effluents, and wastewater discharge points [25]. Therefore, these organisms can grow on industrial byproducts and could be utilized for biosurfactant production under controlled conditions by using industrial waste as a carbon source. It was earlier reported that the nature, quantity, and quality of the carbon substrates and other media components (i.e. nitrogen, phosphorus, magnesium, iron, and manganese) mostly determine the type of microbial biosurfactant [16, 26–28]. Similarly, microbial growth parameters, i.e. pH, temperature, agitation, and dilution rate, also influence the nature of biosurfactant produced in fermentation [29].

3.4.1 Bacteria

Bacteria plays an essential role in the biosynthesis of biosurfactants on the industrial scale. *Pseudomonas* has been reported to be the leading genus, followed by others for biosurfactant production [30]. *Pseudomonas nautica,* isolated from a Mediterranean coastal area, was reported to produce extracellular biosurfactants with excellent emulsifying behavior [31]. Based on the various kinds of carbon as well as hydrocarbons, microorganisms can yield different types of emulsifiers [32]. A group of researchers [33] proved this practically by providing various hydrocarbons as a carbon source to hydrocarbon-degrading *Pseudomonas fluorescens* for the biosynthesis of trehalose lipid-o-dialkyl monoglycerides-protein emulsifier [34].

Bacillus sp. are mostly recognized for the biosynthesis of lipopeptides (lipid connected to a peptide) [35], lichenysin (anionic cyclic lipoheptapeptide biosurfactant), surfactin (bacterial cyclic lipopeptide) [36], lipid–protein complex [37] and subtilisin (a protein-digesting enzyme) [38]. In 1983, Jenneman et al. [39] documented the application of thermotolerant and halotolerant *Bacillus licheniformis* JF2 in microbial-enhanced oil recovery. Similarly, *Bacillus brevis* and *Bacillus polymyxa* were documented to produce a large number of cyclic lipopeptides by using agro wastes [40]. Horowitz and Griffin [41] reported that biosurfactant BL-86, which has been produced by *B. licheniformis*, is useful for various applications and is capable of remediating heavy metal contaminated soil. BL-86 is capable of reducing surface tension and emulsification of hydrocarbons.

There have already been 160 biosurfactants producing bacterial strains identified in soils contaminated with petroleum [12, 42, 43]. Abouseoud et al. [44] reported the potential of *P. fluorescens* Migula 1895-DSMZ for biosurfactant production from olive oil. In their study, it was concluded that *P. fluorescens* Migula 1895-DSMZ is the primary organism responsible for the production of biosurfactants from various industrial wastes.

Acinetobacter sp. are readily available in nature and are among the most frequently available marine microbes in ocean habitats [45]. Due to their separate existence, *Acinetobacter* sp. have received significant attention from many scientists in the last few years. Hydrocarbon degrading *Acinetobacter* sp. perform a significant role in the bioremediation processes of various hydrocarbons in the natural environment [46]. Choi et al. [47] isolated *Acinetobacter calcoaceticus* RAG-1, which produces a commercially important biosurfactant called "emulsan," from the Mediterranean Sea, by using oil industry waste.

Rhodococcus sp. are prominent for producing glycolipids like surface-active molecules [48]. Peng et al. [49] reported the abundance of *Rhodococcus erythropolis* strain 3C-9 in Xiamen Island coastal area soil, and they can remediate the oil-contaminated soil in the area. A few scientists [13, 50] documented the production of various types of biosurfactants (glycolipid, polysaccharides, free fatty acids, and trehalose dicorynomycolate) by *R. erythropolis* [51] and *Rhodococcus* sp. [52].

3.4.2 Fungi and Yeast

Many scholars [53–57] reported the ability of different fungi to produce surfactants by using diverse nutrient sources. Earlier, it was documented that *Candida* sp. were the most common fungal species used for biosurfactant production compared to other fungi. Dubey et al. [58] outlined the application of *Candida bombicola* in sophorolipid production by using various carbon sources from industrial byproducts. *Yarrowia lipolytica* is a well-known fungus used for the synthesis of bioemulsifiers based on lipid, carbohydrate, and proteins [59]. This polysaccharides bioemulsifier improved the cellular hydrophobicity throughout the developmental stages.

Yeast-based biosurfactant research (*Candida* sp. *Pseudozyma* sp. and *Yarrowia* sp.) has gained the growing interest of researchers [60]. The significant advantage of using yeast for biosurfactant production is its GRAS (generally recognized as safe) status, which includes *Y. lipolytica*, *Saccharomyces cerevisiae*, and *Kluyveromyces lactis* [61]. GRAS designated organisms are not toxic or pathogenic so their products can be used in the FMCG and pharma companies. Zinjarde et al. [62] demonstrated that extracellular bioemulsifier production occurs once cells reach a stagnant growth phase. It has been shown that *Y. lipolytica* NCIM 3589 screened from the ocean sites produces a cell wall-associated emulsifier (a mixture of carbohydrate, lipid, and protein) using alkanes or crude oil. Fontes et al. [63] described another bioemulsifier type Yansan, produced from the *Y. lipolytica* wild strain, IMUFRJ 50682, in glucose-enriched fermentation media of molasses.

Pseudozyma and *Torulopis* are the second most explored yeasts preceded by *Candida* for biosurfactant production. Stüwer et al. [64] reported the production of sophorolipids derived from *Torulopis apicola* on agricultural waste under lab conditions. Similarly, Vacheron et al. [65] also reported sophorolipid production from *Torulopis petrophilum* under different growth parameters by using waste materials. Morita et al. (2007) reported *Candida antarctica* and *Pseudozyma rugulosa* as Glycolipid (Mannosylerythritol lipids) producing yeasts, which have superior vesicle-producing and surface-active properties. Sarubbo et al. [66] reported biosurfactant production by using *Y. lipolytica* strain in the fermentation medium enriched with protein (47%), carbohydrate (45%), and lipids (5%), where glucose was the primary carbon source utilized for biosurfactant production.

3.5 Bacterial Growth Conditions

Biosurfactant production in a bioreactor is a complicated process involving the close monitoring of the conditions involved in it. First of all, it is crucial to choose a suitable culture method (i.e. Continuous, Batch, or Fed-batch), which depends on the type of organism, product, and type of bioreactor [67–69]. Atlić et al. [70] reported that every culture method has specific biomass kinetics, substrates, or final product or byproduct concentrations (Figure 3.2).

For example, researchers [71] described an advanced method for surfactin production in the stirred-tank bioreactor with the help of *Aureobasidium pullulans* LB 83. Their group used a centrally ordered facial model intending to evaluate the impact of aeration levels (0.1–1.1 per min) and sucrose concentrations (20–80 g/l) on total biosurfactant productivity. Their results reported an increase in tensoactivity of 8.05 cm in an oil spreading test and productiveness of 0.0838 cm/h when all parameters were used at high levels.

3.5.1 Continuous Cultures

The continuous cultivation of microbes is one of the methodology of growing significance [72]. This methodology is primarily characterized by constant microbial growth in continuous culture, i.e. a constant rate of growth in a constant environment. In continuous culture, parameters such

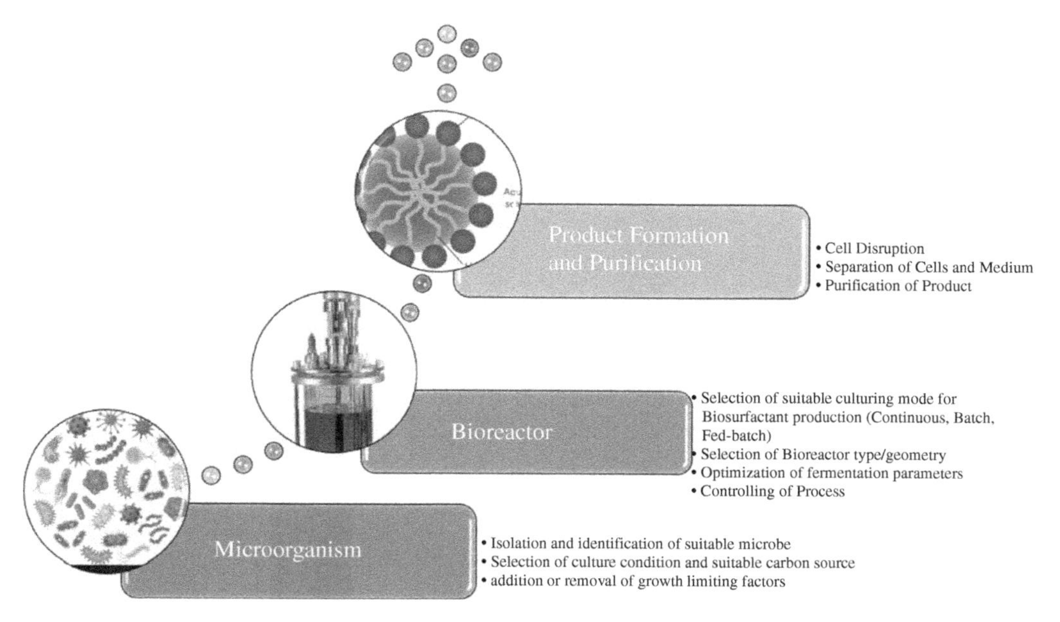

Figure 3.2 Key stages used for bioproduct formation in bioreactor.

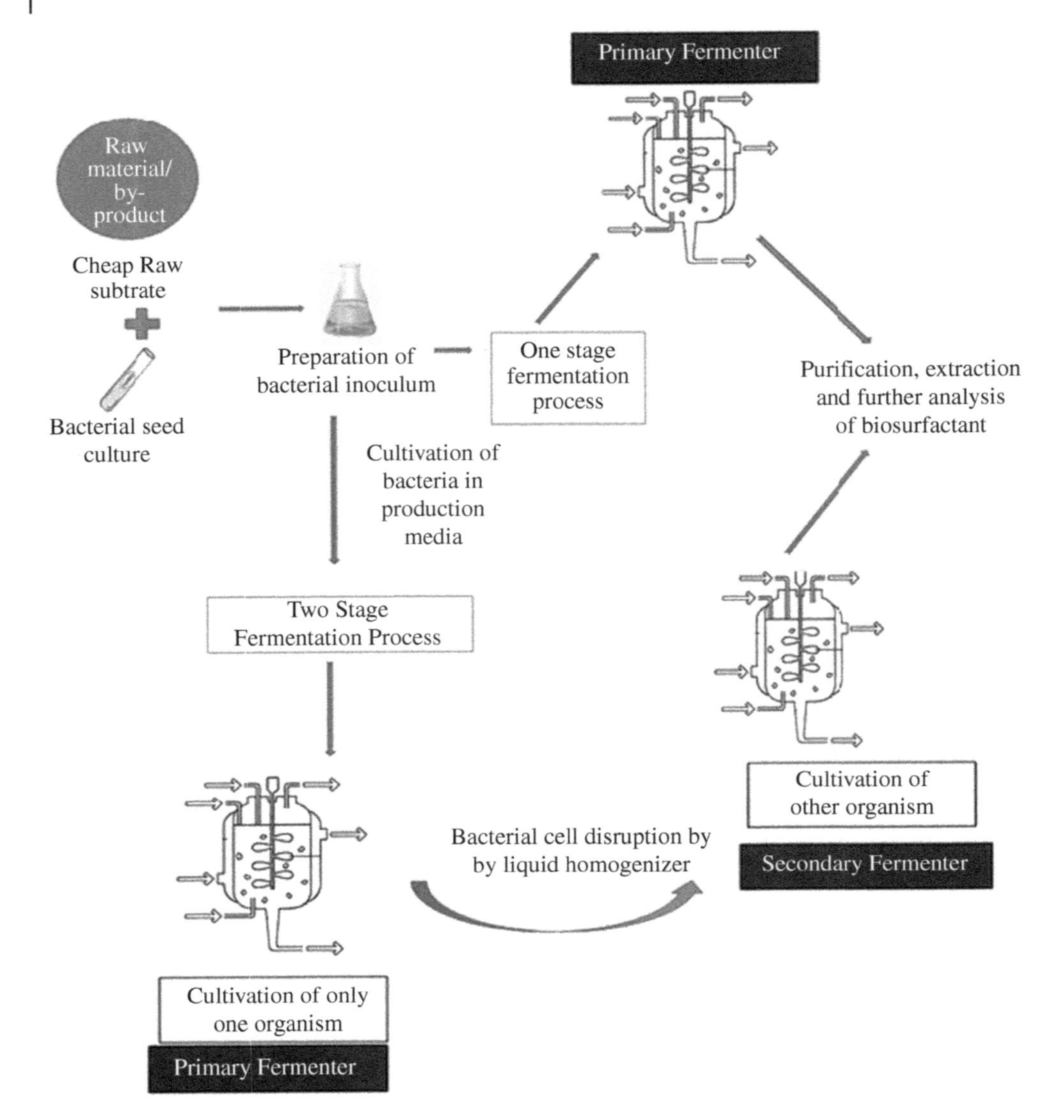

Figure 3.3 Diagrammatic presentation of one stage and two stage fermentation process for biosurfactant production.

as pH, substrate concentrations, metabolic products, and oxygen that eventually shift during the "batch cultivation" growth cycle are all constant; however, the experimenter may individually monitor and control them [73–75]. These attributes of the continuous culture method make it a desirable option for research while offering the industrial microbiologist several benefits in terms of more affordable production techniques (Figure 3.3).

The continuous culture of microbes is performed in bioreactors called chemostats [76]. Chemostat is a type of bioreactor to which freshly prepared substrate is constantly supplied, while culture liquid comprising remaining nutrients, microbial end products, and microbe culture is continually withdrawn simultaneously at the same rate to maintain a constant culture volume. By adjusting the rate at which freshly prepared culture medium is supplied to the bioreactor, the specific growth rate of microorganisms can be effectively managed under limits [77, 78].

There may be several types of the continuous bioreactor, for example:

a) **Auxostat:** An auxostat is a method that uses inputs from a microbial growth chamber analysis to monitor the constant media flowrate, keeping the calculation at a constant level [79].
b) **Retentostat:** Retentostat is a modified form of chemostat in which a filtration assembly is linked to the effluent line and thus recycles the biomass to the bioreactor [19].
c) **Turbidostat:** A turbidostat is a continuous cultivation system with input on the optical density and dilution rate of the culture vessel [80].

3.5.2 Batch Processes

Batch fermentation is widely used in fermentation industries for the production of various microbial products, including vitamins, hormones, drugs, and secondary metabolites [81]. In the batch process, microorganisms and substrates are supplied into a bioreactor on a batch-wise basis for product synthesis [82]. The batch approach is a simple way of conducting the fermentation and ensuring controlled environments inside a bioreactor. However, during the fermentation process, competitive changes may occur in microbial biomass, acid concentration, and byproduct concentration. The batch bioreactor consists of a mechanically agitated container with many other fittings, including gas sprayer, insulation jacket for regulating temperature shifts, pH meter, and air spargers [83, 84]. Despite the easiness of the batch process, batch fermentation incurs huge expenses and consumes a large amount of time spent on preliminary and post-run operations, including bioreactors emptying, filling, and cleaning. In certain situations, such operations can take a much longer time than growing the microbial biomass [85].

3.5.3 Fed-Batch Process

The batch process is a customized form of batch fermentation. It is the most popular mode of operation of fermentation in the bioprocessing sector. Microorganisms are inoculated and cultivated under the batch system for a period of time after the introduction of nutrients to the fermenter in order to feed them. Fermentation is interrupted only when the fermentation broth volume approaches 75% of the bioreactor volume [86]. In the fed-batch systems, the steady feed flow of the media substrate enables the target secondary metabolites to achieve very high concentrations/levels. The benefit of this method of culturing is that the fed substrate level can be managed at the target level (very often relatively low) [87]. This will allow the prevention of undesirable changes, like changes responsible for substrate inhibition or changes in cellular metabolism at a high concentration of substrate. Also, fed-batch systems can be applied when large amounts of biomass are required [88, 89].

3.6 Substrate for Biosurfactant Production

The application of different industrial byproducts/waste from agro-industrial or industrial processes used for the production of biosurfactant agents is a cost-effective solution for waste management [33, 84, 90]. Industrial production of biosurfactants, such as the petroleum industry, the sugarcane/syrup industry's starch, the sugar industry's byproduct residues, fruit and vegetable processing, distilleries, and slaughterhouse animal fat [91], are shown in Table 3.1.

Table 3.1 Summary of various agricultural and industrial byproducts used for biosurfactant production and respective producing microorganisms.

Microorganism	Byproducts/Carbon sources
Pseudomonas sp.	D-glucose/Molasses/Paneer -whey
Pseudomonas sp.	Vegetable oil/Rice water/Petroleum product/Milk whey
Bacillus subtilis	Glucose/Sunflower oil amended with unrefined petroleum oil
Bordetella hinizi strain *DAFI*	Sucrose/Molasses amended with unrefined petroleum oil
Trichosporon asahii	Diesel engine/Motor oil
Pseudomonas aeruginosa strain *LBI, Acinetobacter calcoaceticus*	Soapstock
Serratia marcescens	Glycerin
Candida sp. strain *SY-16*	Soyabean oil and D-glucose
Pseudomonas aeruginosa strain *SP4*	Palm oil
Rhodococcus sp.	Sucrose/Petroleum product/hydrocarbons
Bacillus subtilis, P. aeruginosa	Edible oil/Refined petroleum product
Pseudomonas aeruginosa strain J4	D-glucose/Refined petroleum product/Glycerin/Olive oil/ Sunflower oil
Pseudomonas aeruginosa strain EM1	D-glucose/Glycerin/Sucrose/n-Hexane/Soyabean oil
Pseudomonas aeruginosa strain SR17	Cheese whey
Bacillus licheniformis strain KC710973	Orange-peel
Pseudomonas sp. strain NAF1	Solid-waste from dates and condensed fermented corn extractives
Pseudomonas cepacia strain CCT6659	Waste frying rapeseed oil and condensed fermented corn extractives
Bacillus subtilis strain LAMI005	Purified CAJ (cashew-apple juice)
Candida lipolytica strain UCP0988	Animal fat and condensed fermented corn extractives
Candida sphaerica strain UCP0995	Refinery residue of soyabean oil/condensed fermented corn extractives
Pseudomonas aeruginosa, B. subtilis, Pseudomonas aeruginosa strain GS3	Molasses
Bacillus subtilis strain ATCC 21332, *Bacillus subtilis* strain LB5	Cassava flour, wastewater
Bacillus subtilis	Sweet potatoes
Bacillus subtilis	Potato waste
Bacillus sp.	Engine oil
Bacillus subtilis strain ATCC 21332	Potato waste
Candida antarctica, Candida apicola	Oil refinery waste
Candida bombicola, Candida lipolytica	Rapeseed oil
Candida lipolytica	Industrial residue
Candida sp. strain SY16 95 45, *Pseudomonas aeruginosa* strain AT10	Soyabean oil and waste
Pseudomonas cepacia	Sunflower oil
Cladosporium resinae	Jet fuel JP8

Source: Modified based on Bhardwaj et al. [92] and Banat et al. [93].

Biosurfactant production by using industrial waste is used:

- to achieve lower operating costs,
- to achieve higher affordability of different low-cost sustainable substrates,
- to achieve large quantities of substrates universally available for production purposes,
- to retain the natural features of the final product,
- to create products that are non-toxic for microbial growth,
- to ensure that the product components are environmentally friendly and safe [94].

The next sections discuss the research carried out on the development of biosurfactants by utilizing various waste by-products or agricultural by-products.

3.6.1 Production of Biosurfactant with Food and Vegetable Oil Waste

Vegetable oil processing units produce significant quantities of garbage and byproducts (like soap products, oilseed cakes, lipid residues, semi-solid effluents, and water-soluble effluents) that are rich in fats, oils, and other compounds [91]. Such waste products have become a major source of pollution in the hydrosphere and lithosphere because of their low biodegradable lipid content. This being so, it is crucial to deploy waste material as a substrate for product formation since it puts such waste to more constructive use.

Mercade et al. [95] presented the application of olive-oil mill effluent as a substrate for rhamnolipid biosynthesis using *Pseudomonas* sp. JAMM with 0.058 g/g biosurfactant production by using 100 g/l of olive oil mill effluents and $NaNO_3$ (2.5 g/l). The glycolipids produced in this process reduced the media surface tension from almost 40 to around 30 mN/m. In another study conducted by Abalos et al. [96], the application of *Pseudomonas aeruginosa* AT10 was reported for rhamnolipid production using soybean-oil refinery waste in the fermentation medium. They reported that *P. aeruginosa* AT10 produced 9.5 g/l of glycolipids with 26.8 mN/m surface tension and 122 mg/l critical micelle concentration (CMC) value. Benincasa et al. [97] documented that water-immiscible waste (soapstock) from refining vegetable oil processing unit can be used as a substrate for rhamnolipid production of 15.8 g/l using *P. aeruginosa* LBI. Meanwhile De Faria et al. [98] supplemented raw glycerol as substrate produced from a biodiesel fermentation unit for surfactin (C14/Leu7) synthesis as sole carbon sources with 1.36 g/l final product. In another study, George and Jayachandran [99] communicated the use of waste coconut oil for rhamnolipid production (1.97 g/l) using *P. aeruginosa*. Similarly, Moya Ramírez et al. [100] evaluated the role of olive mill waste (OMW) as a substrate for rhamnolipid production using *P. aeruginosa* and reported production of 29.5 mg/l rhamnolipids. They found the rhamnolipids and surfactins production reached up to 299 and 26.5 mg/l under optimum fermentation conditions. Most of the above research findings used various types of lipid-rich waste as a carbon source and *Pseudomonas* as a primary organism for rhamnolipid synthesis. They concluded that water-immiscible substrate produced a better amount of biosurfactant than water-miscible substrate, like fructose and glucose. The byproducts of food and vegetable oil used as raw materials for the production of biosurfactants are shown in Table 3.2.

In 2004, Bednarski et al. [101] described the use of two *Candida* yeast strains (*C. antarctica* strain ATCC 20509 and *C. apicola* strain ATTC 96134) for the biosynthesis of glycolipids from waste residues isolated from two oil refineries. They supplemented the fermentation media with 5–12% v/v soap stock and found that yeast strains produced ~7–13 g/l glycolipid content, respectively, whereas 6.6 and 10.5 g/l glycolipid were produced from the use of post-refinery trans-fatty acids at 2–5% v/v, respectively. The researchers [101] concluded that adding soap stock seemed to have a beneficial impact on glycolipid biosynthesis.

Nitschke et al. [102] experimented with soybean, cottonseed, babassu, and maize seed oil waste from oil refineries for rhamnolipid synthesis by applying *P. aeruginosa* LBI strain. Their findings

Table 3.2 The byproduct of foodstuffs and vegetable oil used in the production of biosurfactants.

Raw material/byproduct	Type of biosurfactant	Microbial strain
Molasses	Rhamnolipids	*Pseudomonas putida* strain B17
Canola oil	Rhamnolipids	*Pseudomonas* sp. strain DSM 2874
Cusi oil	Sophorolipids	*Candida lipolytica* strain IA 1055
Turkish maize oil	Sophorolipids	*Candida bombicola* strain ATCC 22214
Sunflower and Soybean oil	Rhamnolipids	*Pseudomonas aeruginosa* strain DS10–129
Sunflower oil	Lipopeptide	*Serratia marcescens*
Soybean oil	Mannosylerythritol lipid	*Candida* sp. strain SY16
Whey and Liquor industry waste	Rhamnolipids	*Pseudomonas aeruginosa* strain BS2

Source: Modified based on Kaur et al. [21].

revealed that out of four kinds of oil refinery waste, 2% w/v soybean soap stock could be utilized as the most preferred raw material for 11.7 g/l rhamnolipid production with 26.9 mN/m surface tension and 51.5 mg/l CMC. A sequential factorial method was introduced by Rufino et al. [103] to maximize the development of *C. lipolytica* in a fermentation medium supplemented with soybean oil as a substrate supplemented with refinery waste, glutamic acid, and yeast extract for biosurfactant production. They reported biosurfactant production with 25.29 mN/m surface tension when a combination of oil residue (6%) and glutamic acid (1%) was used in the fermentation medium. The biosurfactants produced had stability over a more comprehensive pH (2–12), temperature (0–120 °C) and salinity (2–10% sodium chloride) levels. Rufino et al. [103] investigated the cell surface characteristics, including its relation to the development of biosurfactants for diverse uses. This research group [103] cultivated six strains of Candida in fermentation medium supplemented with glucose, corn-steep liquor, soybean refinery oil residues, peanut oil refinery residues, and n-hexadecane. Their studies indicated that the biosurfactants produced with yeast could be used to extract hydrophobic substances with 90% capability to eliminate hydrophobic pollutants from soil.

3.6.2 Development of Biosurfactants Using Waste Frying Oil

Several crops are cultivated primarily for the production of foodstuffs for industries, which are then used in industrial processes that generate byproducts [76, 104, 105]. Go et al. [17] reported that increasing population and growing living standards have contributed to increased demand for edible oils, as they supply the necessary nutritional components and energy for physical activities. Edible vegetable oils are usually composed of triacylglycerols (more than 95%) and various fatty acids [106]. The waste cooking oil/frying oil contains several hazardous chemicals that cause health risks when consumers use it or process it. The composition of the used frying oil depends on the food fried in it and also the number of times it is reused; usually, though, recycled oil has 30% higher polar-hydrocarbons compared to fresh frying oil [107]. Waste frying oil represents a renewable energy source for the production of new industrial products and alternative feedstock in place of pure and expensive chemicals. Haba et al. [108] studied sunflower oil and olive oil waste for rhamnolipid production by using *P. aeruginosa* 47T2 (2.7 g/l). They reported 0.34 g/g rhamnolipid production with $NaNO_3$ (5 g/l) and waste frying oil (40 g/l).

Researchers [109] developed a modified approach of biosurfactant production under submerged culture conditions through *Bacillus subtilis* MTCC2423 strain using sunflower bran, paddy bran, waste frying oil (50 g/l), yeast extract (5 g/l) and rock salt (1.7 g/l). They observed that the surface tension decreased in glucose + sunflower bran, paddy bran, and frying oils waste by approximately

29.0, 32.0, and 34.5 mN/m, respectively, while glucose produced the best output (2.1 g/l) of surfactin. It was concluded from their results that the surfactin production process was safe for waste disposal and low-cost biosurfactant development. Pan et al. [110] had also reported rhamnolipids production using *P. aeruginosa* strain DG30 in fermentation medium enriched with 5% w/v discarded vegetable oil (w/v) and reported 15.6 g/l of biosurfactant during the fermentation process.

Moreover, the coconut fried oil waste (2%) was used by George and Jayachandran [99] for the synthesis of rhamnolipids using *P. aeruginosa* strain D, with a recorded emulsification index (EI) of 71% and a yield of 3.55 g/l. It was reported that *Mucor circinelloides*, grown in culture media with 5% waste frying oil, produced ~12.4 g/l of glycolipids [111]. The obtained glycolipids lowered the surface tension by up to 26 mN/m, produced a consistent 129 mm diameter hollow area in the oil dispersal assay, and proved the emulsification capability of up to 65% of crude oil in marine water. All the results reported in this section provide scientific evidence that waste oil has been a good source of carbon to support microbial growth and production of biosurfactants. The waste substrates used for the production of biosurfactants themselves can reduce the cost of production and can be considered environmentally safe by reducing the pollution problem.

3.6.3 Fruit and Vegetable Industry Byproducts for Biosurfactant Processing

The commercial manufacturing units use vegetable and fruit items (cassava, apple, banana juice and peels, pineapple, mango, carrot, and lime) for the production of various consumer products, but also in this process enormous quantities of residual waste is produced that is rich in carbon content and can be used to produce biosurfactants [112, 113]. The production of cashew nut generates a massive amount of cashew apples as waste, with only 12% being used as a fruit or for commercial processing, whereas more than 70% of cashew apples remain as waste in the soil and cause pollution [114]. Rocha et al. [115] reported that the cashew apple is an invaluable raw material for varied practical applications due to its abundant carbohydrate, vitamin, and mineral content. They evaluated the *A. calcoaceticus* strain RAG-1's ability to produce emulsions by utilizing cashew apple juice, which lowered the kerosene surface tension by ~17 and 59% of EI value. They also assessed the potential of ATCC-10145 strain of *P. aeruginosa* in the nutrient media enriched with cashew apple juice as having 90–97 g/l of carbohydrate for rhamnolipid production. The maximum surface tension reduction was 29.5 mN/m, whereas the maximum rhamnolipid synthesis was 3.8 g/l, which was obtained by adding peptone (5 g/l) to cashew apple juice. During the research, they analyzed the surfactin synthesis using *B. subtilis* LAMI008 in nutrient media supplemented with 86.1 g/l carbon content with cashew apple juice.

Subsequently, a related study was conducted by Giro et al. [116] using *B. subtilis* LAMI005, where they reported that after 48 hours of fermentation, the highest amount of surfactin was 123 mg/l with clarified cashew apple juice. The production levels were, however, two times smaller than those of mineral media enriched with glucose 10 g/l and fructose 8.7 g/l. The CMC value of biosurfactant from cashew apple juice was 2.5 times lower than that of the CMC value of biosurfactant derived from glucose and fructose media, indicating an increase in efficiency of biosurfactant. These results suggest that cashew apple juice can be used as an appropriate substrate for the production of biosurfactants using *B. subtilis* LAMI005 and may be used as a rich source of carbon for large-scale industrial production.

3.6.4 Starch-Rich Byproduct from the Industry for Biosurfactant Production

High volumes of effluents, extremely rich in starch and cellulose, are generated during the commercial extraction of starch using various staple crops like maize, rice, cassava, wheat, and potato, which could be utilized as a growth medium to produce different products like surfactins [76, 117, 118].

Potato waste, for example, comprises 16–20% of starchy material, 2–2.5% of proteins, 1–1.8% of fibers, and 0.15% of fatty acids. It was earlier reported [119] that potato with surface skin contains elevated potassium, B complex vitamins and vitamin C, and minerals like P, Mg, and Fe.

Thompson et al. [120] examined potato waste as a possible source of carbon in shake flask culture to produce biosurfactants using *B. subtilis* ATCC-21332. They evaluated different potato-based fermentation media for biosurfactant production, which includes defined potato media, liquefied and solid potato waste media, synthetically made starchy medium by the addition of pure starch in mineral media. In a solid medium, the surface tension decreased from 71.3 to 28.3 mN/m, and CMC 100 mg/l was reported when only 60 g/l of potato substrate was used for microbial cultivation, without adding any other nutrient in the fermentation medium. They also examined the surfactin synthesis by using *B. subtilis* 21332 strain in a medium containing potato industrial effluent with 16.2 and 6.5 g/l of potato solid components. The potato effluent was diluted at 1:10 by adding minerals and corn steep liquor to the modified and unmodified media. Surfactin produced using small potato solids showed a better production of biosurfactants with a production concentration of 0.44 g/l than that of large potato solids. Thompson et al. [120] and Noah et al. [76] demonstrated the usage of corn steep liquor for surfactin production. Noah et al. [76] subsequently produced surfactin with a low-solid potato effluent with the same microbial strain in batch-mode operated chemostat and recorded ~0.8–0.9 g/l production after 52 h of fermentation. Another study conducted by Das and Mukherjee [121] documented the production of lipopeptides using *B. subtilis* DM03 and DM04 strain with 5 g potato peel waste under solid-state fermentation and 2% w/v substratum in submerged fermentation. During fermentation, the production of lipopeptide by *B. subtilis* DM-03 was reported with 80 and 67 mg/g in submerged and solid-state fermentation, respectively.

Wang et al. [122] used *B. subtilis* B6–1 for fengycin and poly-β-glutamic acid(α-PGA) production by incorporating 5 g/l of soy curd and 5 g/l of sweet potato residue in solid-state fermentation. The quantity of lipopeptide was reached at the maximum level after 54 hours of incubation; however, the highest amount of γ-PGA (3.63%) was achieved after 42 hours of incubation. The researchers also emphasized the potential use of these lipopeptides as a biocontrol agent and fertilizer synergists.

Cassava wastewater is another extremely rich carbohydrate waste used for biosurfactant production [123, 124]. The *Bacillus* sp. strain LB5a produced biosurfactants from cassava wastewater [125]. The results of a Nitschke and Pastore [126] study showed that bacteria were able to grow and yield biosurfactants in both solid and liquid medium, but the best results were reported in broth medium with the surface tension of 26.6 mN/m. They also examined the efficiency of *B. subtilis* ATCC 21332 and *B. subtilis* LB5a for biosurfactant production using cassava wastewater in another study. *B. subtilis* LB5a lowered the medium surface tension up to 26 mN/m with 3.0 g/l of biosurfactant, while the strain ATCC-21332 produced crude biosurfactant (2.2 g/l) and changed medium surface tension up to 25.9 mN/m [127, 128]. The above studies emphasized the potential use of starchy byproducts and associated carbon sources for synthesis of biosurfactants. The potential of starch-rich waste as a carbon source for the production of biosurfactants and some other useful products is promising; however, multidisciplinary collaborative research is needed to meet the industrial needs in terms of product quantity and quality.

3.6.5 Biosurfactant Synthesis from Lignocellulosic Industrial Byproducts

Lignocellulosic substances comprise cellulose, lignin, and hemicellulose. The paper industry primarily utilizes cellulose from lignocellulose and applies a variety of techniques to eliminate certain fiber elements (such as hemicellulose and lignin), as a result of which these other elements remain, more often than not, unused, thus becoming waste products. Furthermore, in bioethanol research, various pretreatment techniques are used to extract cellulose for fermentation, with very little attention paid to hemicellulose and lignin. Such pretreatment technologies are, therefore, only midway to becoming primary methods of refining lignocellulosic materials, and only realize the use of one or two key elements, whereas other contents are wasted [35, 129, 130]. Portilla-Rivera et al. [131] examined *Lactobacillus pentosus*'s ability for biosurfactant and lactic acid production in fermentation medium composed with 10.0 g/l of the byproduct of corn wet-milling, 10 g/l of yeast extract, and hydrolyzed pomace (10.8% cellulose, 11.2% hemicellulose, and 51% lignin). They obtained 5.5 g/l of lactic acid and 4.8 mg/l of cell-bound biosurfactant. In a subsequent study, the sustainability and emulsification of biosurfactants obtained from *L. pentosus* from distilled pomace hydrolysates, walnut, and hazelnut shell-based fermentation media were also examined by Portilla-Rivera et al. [131]. The emulsion density reported was 14.1% for gasoline and 27.2% for kerosene, which was higher than those produced for industrial surfactin. Cortés-Camargo et al. [132] used the ZSB10 strain of *Bacillus tequilensis*, separated from Mexican brine, for intracellular and extracellular biosurfactant production. Their findings revealed that the extracellular biosurfactant production using lignocellulosic waste was 1.52 g/l with a surface tension reduction of 38.6 mN/m and 177.0 mg/l CMC value. The intracellular biosurfactant produced was only 0.078 g/l in quantity and exhibited a lesser EI (41%) than an extracellular EI (47%). Jokari and group [133] reported the impact of aeration levels on *B. subtilis* ATCC 6633 growth and the amount of biosurfactant produced in a miniaturized bioreactor ventilated flask. In the ideal conditions where the filling and shaking rates were 15 ml and 300 rpm, the maximum biosurfactant content (0.0485 g/l/h) was achieved. Their findings indicated the noticeable increase in surfactin productivity under environments that were not oxygen-limiting.

These studies have shown that lignocellulosic compounds and waste products could be considered as raw materials for the production of biosurfactants due to their low processing costs and higher nutrient quality. The results of these studies also show that lignocellulosic waste can be a potential carbon source for fermentation. The biosurfactant synthesis of these industrial byproducts offers a promising financial advantage that can be used to achieve cost savings over the production of synthetic surfactants.

3.7 Conclusions

The large-scale production of valuable products in the bioreactor is a complex process, requiring an assessment of the different parameters affecting its efficiency. The skills of a biotechnologist and a chemical engineer need to be combined in order to achieve a practical approach to the production of biosurfactants. Several research efforts have been made to evaluate the potential of different microbes for the production of industrial byproduct biosurfactants as substrates.

The use of low-cost industrial waste and renewable materials may significantly reduce the operating costs of biosurfactant production (by almost 50%). The use of industrial byproducts/waste for the production of biosurfactant is therefore a sustainable option.

Acknowledgement

The editor (Hemen Sarma) has extensively revised the readability and carried out editing on the basis of the original text of the authors, without altering the meaning of the text in this chapter. However, any competing interest arises from any statement, and the author is liable.

References

1 Ochsner, U.A. and Reiser, J. (1995). Autoinducer-mediated regulation of rhamnolipid biosurfactant synthesis in *Pseudomonas aeruginosa*. *Proc. Natl. Acad. Sci. USA* 92 (14): 6424–6428.

2 Das Neves, L.C.M., De Oliveira, K.S., Kobayashi, M.J. et al. (2007). Biosurfactant production by cultivation of *Bacillus atrophaeus* ATCC 9372 in semidefined glucose/casein-based media. *Appl. Biochem. Biotecnol.* 137: 539–554.

3 Batista, B.D., Taniguti, L.M., Almeida, J.R. et al. (2016). Draft genome sequence of multitrait plant growth-promoting *Bacillus* sp. strain RZ2MS9. *Genome Announc.* 4 (6): e01402–e01416.

4 Nguyen, T.T. and Sabatini, D.A. (2011). Characterization and emulsification properties of rhamnolipid and sophorolipid biosurfactants and their applications. *Int. J. Mol. Sci.* 12: 1232–1244.

5 Sarma, H., Bustamante, K.L.T., and Prasad, M.N.V. (2018). Biosurfactants for oil recovery from refinery sludge: magnetic nanoparticles assisted purification. In: Industrial and Municipal Sludge (eds. M.N.V. Prasad, P.J. de Campos, F. Meththika and V.S. Venkata Mohan (eds.)). Elsevier. ISBN: 9780128159071.

6 Mata-Sandoval, J.C., Karns, J., and Torrents, A. (2002). Influence of rhamnolipids and triton X-100 on the desorption of pesticides from soils. *Environ. Sci. Technol.* 36: 4669–4675.

7 Das, P., Mukherjee, S., and Sen, R. (2009). Substrate dependent production of extracellular biosurfactant by a marine bacterium. *Bioresour. Technol.* 100 (2): 1015–1019.

8 Geetha, S.J., Banat, I.M., and Joshi, S.J. (2018). Biosurfactants: Production and potential applications in microbial enhanced oil recovery (MEOR). *Biocatal. Agric. Biotechnol.* 14: 23–32.

9 Geissler, M., Oellig, C., Moss, K. et al. (2017). High-performance thin-layer chromatography (HPTLC) for the simultaneous quantification of the cyclic lipopeptides surfactin, iturin A and fengycin in culture samples of *Bacillus* species. *J. Chromatogr. B* 1044: 214–224.

10 Guzik, M.W., Kenny, S.T., Duane, G.F. et al. (2014). Conversion of post consumer polyethylene to the biodegradable polymer polyhydroxyalkanoate. *Appl. Microbiol. Biotechnol.* 98: 4223–4232.

11 Haeri, S.A. (2016). Bio-sorption based dispersive liquid-liquid microextraction for the highly efficient enrichment of trace-level bisphenol A from water samples prior to its determination by HPLC. *J. Chromatogr. B Anal. Technol. Biomed. Life Sci.* 1028: 186–191.

12 GhayyomiJazeh, M.G., Forghani, F., and Deog-Hwan, O. (2012). Biosurfactan production by *Bacillus* sp. isolated from petroleum contaminated soils of Sirri Island. *Am. J. Appl. Sci.* 9: 1–6.

13 Grosso-Becerra, M.V., Gonzalez-Valdez, A., Granados-Martinez, M.J. et al. (2016). *Pseudomonas aeruginosa* ATCC 9027 is a non-virulent strain suitable for mono-rhamnolipids production. *Appl. Microbiol. Biotechnol.* 100: 9995–10004.

14 He, S., Ni, Y., Lu, L. et al. (2020). Simultaneous degradation of n-hexane and production of biosurfactants by *Pseudomonas* sp. strain NEE2 isolated from oil-contaminated soils. *Chemosphere* 242: 125237.

15 Vecino, X., Rodríguez-López, L., Gudiña, E.J. et al. (2017). Vineyard pruning waste as an alternative carbon source to produce novel biosurfactants by *Lactobacillus paracasei*. *J. Ind. Eng. Chem.* 55: 40–49.

16 Cavalcanti, M.H.C., Magalhaes, V.M., Farias, C.B.B. et al. (2020). Maximization of biosurfactant production by *Bacillus invictae* using agroindustrial residues for application in the removal of hydrophobic pollutants. *Chem. Eng. Trans.* 79: 55–60.

17 Go, A.W., Conag, A.T., Igdon, R.M.B. et al. (2019). Potentials of agricultural and agro-industrial crop residues for the displacement of fossil fuels: A Philippine context. *Energ. Strat. Rev.* 23: 100–113.

18 Aguiar, G.P.S., Limberger, G.M., and Silveira, E.L. (2014). Alternativas tecnológicas para o aproveitamento de resíduos provenientes da industrialização de pescados. *Rev. Eletrônica Interdiscip.* 1 (11): 229–225.

19 Jørgensen, T.R., Nitsche, B.M., Lamers, G.E. et al. (2010). Transcriptomic insights into the physiology of *Aspergillus niger* approaching a specific growth rate of zero. *Appl. Environ. Microbiol.* 76 (16): 5344–5355.

20 Karnwal, A. (2018). Use of bio-chemical surfactant producing endophytic bacteria isolated from rice root for heavy metal bioremediation. *Pertanika J. Trop. Agric. Sci.* 41 (2): 699–713.

21 Kaur, H.P., Prasad, B., and Kaur, S. (2015). A review on application of biosurfactants produced from unconventional inexpensive wastes in food and agriculture industry. *World J. Pharm. Res.* 4 (8): 827–842.

22 Kertesz, M.A. and Thai, M. (2018). Compost bacteria and fungi that influence growth and development of *Agaricus bisporus* and other commercial mushrooms. *Appl. Microbiol. Biotechnol.* 102 (4): 1639–1650.

23 Lima, F.A., Santos, O.S., Pomella, A.W.V. et al. (2020). Culture medium evaluation using low-cost substrate for biosurfactants lipopeptides production by *Bacillus amyloliquefaciens* in pilot bioreactor. *J. Surfactant Deterg.* 23 (1): 91–98.

24 Satpute, S.K., Bhuyan, S.S., Pardesi, K.R. et al. (2010). Molecular genetics of biosurfactant synthesis in microorganisms. *Adv. Exp. Med. Biol.* 672: 14–41.

25 Kiran, G.S., Ninawe, A.S., Lipton, A.N. et al. (2016). Rhamnolipid biosurfactants: evolutionary implications, applications and future prospects from untapped marine resource. *Crit. Rev. Biotechnol.* 36: 399–415.

26 Maheshwari, D.K. (2012). Bacteria in Agrobiology: Stress Management. Heidelberg, New York: Springer.

27 Schiano, C.A., Bellows, L.E., and Lathem, W.W. (2010). The small RNA chaperone Hfq is required for the virulence of *Yersinia pseudotuberculosis*. *Infect. Immun.* 78: 2034–2044.

28 Whang, L.M., Liu, P.W., Ma, C.C., and Cheng, S.S. (2008). Application of biosurfactants, rhamnolipid, and surfactin, for enhanced biodegradation of diesel-contaminated water and soil. *J. Hazard. Mater.* 151: 155–163.

29 Karnwal, A., Bhardwaj, V., Dohroo, A. et al. (2018). Effect of microbial surfactants on heavy metal polluted wastewater. *Pollut. Res.* 37: 39–46.

30 Mishra, S. and Singh, S.N. (2012). Microbial degradation of n-hexadecane in mineral salt medium as mediated by degradative enzymes. *Bioresour. Technol.* 111: 148–154.

31 Husain, D.R., Goutx, M., Bezac, C. et al. (1997). Morphological adaptation of *Pseudomonas nautica* strain 617 to growth on eicosane and modes of eicosane uptake. *Lett. Appl. Microbiol.* 24 (1): 55–58.

32 Das, P., Mukherjee, S., Sivapathasekaran, C., and Sen, R. (2010). Microbial surfactants of marine origin: Potentials and prospects. In: Biosurfactants. Advances in Experimental Medicine and Biology, vol. 672 (ed. R. Sen), 88–101. New York, NY: Springer.

33 Zhang, J., Lin, X.G., Liu, W.W., and Yin, R. (2012). Response of soil microbial community to the bioremediation of soil contaminated with PAHs. *Huan Jing Ke Xue* 33: 2825–2831.

34 Cazals, F., Huguenot, D., Crampon, M. et al. (2020). Production of biosurfactant using the endemic bacterial community of a PAHs contaminated soil, and its potential use for PAHs remobilization. *Sci. Total Environ.* 709: 136143.

35 Satyanarayana, T., Johri, B.N., and Prakash, A. (2012). Microorganisms in Sustainable Agriculture and Biotechnology. New York: Springer, Dordrecht.

36 de Almeida Couto, C.R., Alvarez, V.M., Marques, J.M. et al. (2015). Exploiting the aerobic endospore-forming bacterial diversity in saline and hypersaline environments for biosurfactant production. *BMC Microbiol.* 15: 240.

37 Silva, M.A., Silva, A.F., Rufino, R.D. et al. (2017). Production of biosurfactants by *Pseudomonas* species for application in the petroleum industry. *Water Environ. Res.* 89: 117–126.

38 Nguyen, T.T., Quyen, T.D., and Le, H.T. (2013). Cloning and enhancing production of a detergent- and organic-solvent-resistant nattokinase from *Bacillus subtilis* VTCC-DVN-12-01 by using an eight-protease-gene-deficient *Bacillus subtilis* WB800. *Microb. Cell Fact.* 12 (1): 79.

39 Jenneman, G.E., McInerney, M.J., Knapp, R.M., Clark, J.B., Feero, J.M., Revus, D.E. and Menzie, D.E., (1983). Halotolerant, biosurfactant-producing *Bacillus* species potentially useful for enhanced oil recovery. *Dev. Ind. Microbiol.* (United States), 24(CONF-8208164-).

40 Almeida, P.F.D., Moreira, R.S., Almeida, R.C.D.C. et al. (2004). Selection and application of microorganisms to improve oil recovery. *Eng. Life Sci.* 4 (4): 319–325.

41 Horowitz, S. and Griffin, W.M. (1991). Structural analysis of *Bacillus licheniformis* 86 surfactant. *J. Ind. Microbiol.* 7 (1): 45–52.

42 Coronel-Leon, J., Pinazo, A., Perez, L. et al. (2017). Lichenysin-geminal amino acid-based surfactants: Synergistic action of an unconventional antimicrobial mixture. *Colloids Surf. B Biointerfaces* 149: 38–47.

43 Makkar, R.S. and Cameotra, S.S. (1997). Utilization of molasses for biosurfactant production by two *Bacillus* strains at thermophilic conditions. *J. Am. Oil Chem. Soc.* 74 (7): 887–889.

44 Abouseoud, M., Maachi, R., Amrane, A. et al. (2008). Evaluation of different carbon and nitrogen sources in production of biosurfactant by *Pseudomonas fluorescens*. *Desalination* 223 (1–3): 143–151.

45 Mohanram, R., Jagtap, C., and Kumar, P. (2016). Isolation, screening, and characterization of surface-active agent-producing, oil-degrading marine bacteria of Mumbai Harbor. *Mar. Pollut. Bull.* 105: 131–138.

46 Khan, A.H.A., Tanveer, S., Alia, S. et al. (2017). Role of nutrients in bacterial biosurfactant production and effect of biosurfactant production on petroleum hydrocarbon biodegradation. *Ecol. Eng.* 104: 158–164.

47 Choi, J.W., Choi, H.G., and Lee, W.H. (1996). Effects of ethanol and phosphate on emulsan production by *Acinetobacter calcoaceticus* RAG-1. *J. Biotechnol.* 45 (3): 217–225.

48 Hassanshahian, M., Emtiazi, G., and Cappello, S. (2012). Isolation and characterization of crude-oil-degrading bacteria from the Persian Gulf and the Caspian Sea. *Mar. Pollut. Bull.* 64: 7–12.

49 Peng, F., Liu, Z., Wang, L., and Shao, Z. (2007). An oil-degrading bacterium: *Rhodococcus erythropolis* strain 3C-9 and its biosurfactants. *J. Appl. Microbiol.* 102: 1603–1611.

50 Wojciechowski, K., Orczyk, M., Gutberlet, T., and Geue, T. (2016). Complexation of phospholipids and cholesterol by triterpenic saponins in bulk and in monolayers. *Biochim. Biophys. Acta* 1858: 363–373.

51 Ma, T., Li, G., Li, J. et al. (2006). Desulfurization of dibenzothiophene by *Bacillus subtilis* recombinants carrying dszABC and dszD genes. *Biotechnol. Lett.* 28: 1095–1100.

52 Mishra, S., Singh, S.N., and Pande, V. (2014). Bacteria induced degradation of fluoranthene in minimal salt medium mediated by catabolic enzymes in vitro condition. *Bioresour. Technol.* 164: 299–308.

53 Miao, S., Dashtbozorg, S.S., Callow, N.V., and Ju, L.K. (2015). Rhamnolipids as platform molecules for production of potential anti-zoospore agrochemicals. *J. Agric. Food Chem.* 63: 3367–3376.

54 Mouillon, J.M. and Persson, B.L. (2006). New aspects on phosphate sensing and signalling in *Saccharomyces cerevisiae*. *FEMS Yeast Res.* 6 (2): 171–176.

55 Rivera, O.M.P., Moldes, A.B., Torrado, A.M., and Domínguez, J.M. (2007). Lactic acid and biosurfactants production from hydrolyzed distilled grape marc. *Process Biochem.* 42 (6): 1010–1020.

56 Sachdev, D.P. and Cameotra, S.S. (2013). Biosurfactants in agriculture. *Appl. Microbiol. Biotechnol.* 97 (3): 1005–1016.

57 Sriram, M.I., Kalishwaralal, K., Deepak, V. et al. (2011). Biofilm inhibition and antimicrobial action of lipopeptide biosurfactant produced by heavy metal tolerant strain *Bacillus cereus* NK1. *Colloids Surf. B Biointerfaces* 85 (2): 174–181.

58 Dubey, P., Kumar, S., Aswal, V.K. et al. (2016). Silk fibroin-sophorolipid gelation: Deciphering the underlying mechanism. *Biomacromolecules* 17: 3318–3327.

59 Yilmaz, F., Ergene, A., Yalcin, E., and Tan, S. (2009). Production and characterization of biosurfactants produced by microorganisms isolated from milk factory wastewaters. *Environ. Technol.* 30: 1397–1404.

60 Basak, G. and Das, N. (2014). Characterization of sophorolipid biosurfactant produced by *Cryptococcus* sp. VITGBN2 and its application on Zn (II) removal from electroplating wastewater. *J. Environ. Biol.* 35 (6): 1087.

61 Falode, O.A., Adeleke, M.A., and Ogunshe, A.A. (2017). Evaluation of indigenous biosurfactant-producing bacteria for de-emulsification of crude oil emulsions. *Microbiol. Res. J. Int.* 18: 1–9.

62 Zinjarde, S., Chinnathambi, S., Lachke, A.H., and Pant, A. (1997). Isolation of an emulsifier from *Yarrowia lipolytica* NCIM 3589 using a modified mini isoelectric focusing unit. *Lett. Appl. Microbiol.* 24 (2): 117–121.

63 Fontes, G.C., Fonseca Amaral, P.F., Nele, M., and Zarur Coelho, M.A. (2010). Factorial design to optimize biosurfactant production by *Yarrowia lipolytica*. *Biomed. Res. Int.* 2010: 821306.

64 Stüwer, O., Hommel, R., Haferburg, D., and Kleber, H.P. (1987). Production of crystalline surface-active glycolipids by a strain of *Torulopsis apicola*. *J. Biotechnol.* 6 (4): 259–269.

65 Vacheron, J., Desbrosses, G., Bouffaud, M.L. et al. (2013). Plant growth-promoting rhizobacteria and root system functioning. *Front. Plant Sci.* 4: 356.

66 Sarubbo, L.A., do Carmo Marçal, M., Neves, M.L.C. et al. (2001). Bioemulsifier production in batch culture using glucose as carbon source by *Candida lipolytica*. *Appl. Biochem. Biotechnol.* 95 (1): 59–67.

67 Bernard, A. and Payton, M. (1995). Fermentation and growth of Escherichia coli for optimal protein production. *Curr. Protoc. Protein Sci.* 1: 5–3.

68 Blank, L.L., Grosso, L.J., and Benson, J.J. (1984). A survey of clinical skills evaluation practices in internal medicine residency programs. *J. Med. Educ.* 59 (5): 401–406.

69 Brück, H., Coutte, F., Delvigne, F., Dhulster, P. and Jacques, P., (2020). Optimization of biosurfactant production in a trickle-bed biofilm reactor with genetically improved bacteria. Poster presented at the 25th National Symposium for Applied Biological Science. Available at: http://hdl.handle.net/2268/247270.

70 Atlić, A., Koller, M., Scherzer, D. et al. (2011). Continuous production of poly ([R]-3-hydroxybutyrate) by *Cupriavidus necator* in a multistage bioreactor cascade. *Appl. Microbiol. Biotechnol.* 91 (2): 295–304.

71 Brumano, L.P., Antunes, F.A.F., Souto, S.G. et al. (2017). Biosurfactant production by *Aureobasidium pullulans* in stirred tank bioreactor: new approach to understand the influence of important variables in the process. *Bioresour. Technol.* 243: 264–272.

72 Amutha, R. and Gunasekaran, P. (2001). Production of ethanol from liquefied cassava starch using co-immobilized cells of *Zymomonas mobilis* and *Saccharomyces diastaticus*. *J. Biosci. Bioeng.* 92 (6): 560–564.

73 Rebroš, M., Rosenberg, M., Grosová, Z. et al. (2009). Ethanol production from starch hydrolyzates using *Zymomonas mobilis* and glucoamylase entrapped in polyvinylalcohol hydrogel. *Appl. Biochem. Biotechnol.* 158 (3): 561–570.

74 Saikia, R.R., Deka, S., Deka, M., and Sarma, H. (2012). Optimization of environmental factors for improved production of rhamnolipid biosurfactant by *Pseudomonas aeruginosa* RS29 on glycerol. *J. Basic Microbiol.* 52 (4): 446–457.

75 Santos, D.K., Rufino, R.D., Luna, J.M. et al. (2016). Biosurfactants: multifunctional biomolecules of the 21st century. *Int. J. Mol. Sci.* 17 (3): 401. https://doi.org/10.3390/ijms17030401.

76 Noah, K.S., Bruhn, D.F., and Bala, G.A. (2005). Surfactin production from potato process effluent by *Bacillus subtilis* in a chemostat. *Appl. Biochem. Biotechnol.* 121–124: 465–473.

77 Kiran, G.S., Sabu, A., and Selvin, J. (2010). Synthesis of silver nanoparticles by glycolipid biosurfactant produced from marine *Brevibacterium casei* MSA19. *J. Biotechnol.* 148 (4): 221–225.

78 Samad, A., Zhang, J., Chen, D., and Liang, Y. (2015). Sophorolipid production from biomass hydrolysates. *Appl. Biochem. Biotechnol.* 175: 2246–2257.

79 Adamberg, K., Kask, S., Laht, T.M., and Paalme, T. (2003). The effect of temperature and pH on the growth of lactic acid bacteria: a pH-auxostat study. *Int. J. Food Microbiol.* 85 (1–2): 171–183.

80 Klok, A.J., Verbaanderd, J.A., Lamers, P.P. et al. (2013). A model for customising biomass composition in continuous microalgae production. *Bioresour. Technol.* 146: 89–100.

81 Kebbouche-Gana, S., Gana, M.L., Ferrioune, I. et al. (2013). Production of biosurfactant on crude date syrup under saline conditions by entrapped cells of *Natrialba* sp. strain E21, an extremely halophilic bacterium isolated from a solar saltern (Ain Salah, Algeria). *Extremophiles* 17: 981–993.

82 Vanavil, B., Perumalsamy, M., and Rao, A.S. (2013). Biosurfactant production from novel air isolate NITT6L: screening, characterization and optimization of media. *J. Microbiol. Biotechnol.* 23: 1229–1243.

83 Behrens, B., Helmer, P.O., Tiso, T. et al. (2016). Rhamnolipid biosurfactant analysis using online turbulent flow chromatography-liquid chromatography-tandem mass spectrometry. *J. Chromatogr. A* 1465: 90–97.

84 Zhang, Q., Li, Y., and Xia, L. (2014). An oleaginous endophyte *Bacillus subtilis* HB1310 isolated from thin-shelled walnut and its utilization of cotton stalk hydrolysate for lipid production. *Biotechnol. Biofuels* 7 (1): 152.

85 Probert, H.M. and Gibson, G.R. (2002). Investigating the prebiotic and gas-generating effects of selected carbohydrates on the human colonic microflora. *Lett. Appl. Microbiol.* 35 (6): 473–480.

86 Rodriguez-Contreras, A., Koller, M., de Sousa Dias, M.M. et al. (2013). Novel poly [(R)-3-hydroxybutyrate]-producing bacterium isolated from a Bolivian hypersaline lake. *Food Technol. Biotechnol.* 51 (1): 123–130.

87 Sarilmiser, H.K., Ates, O., Ozdemir, G. et al. (2015). Effective stimulating factors for microbial levan production by *Halomonas smyrnensis* AAD6T. *J. Biosci. Bioeng.* 119 (4): 455–463.

88 Xu, N., Liu, S., Xu, L. et al. (2020). Enhanced rhamnolipids production using a novel bioreactor system based on integrated foam-control and repeated fed-batch fermentation strategy. *Biotechnol. Biofuels* 13: 1–10.

89 Yao, S., Zhao, S., Lu, Z. et al. (2015). Control of agitation and aeration rates in the production of surfactin in foam overflowing fed-batch culture with industrial fermentation. *Rev. Argent. Microbiol.* 47: 344–349.

90 Zhu, Y., Gan, J.J., Zhang, G.L. et al. (2007). Reuse of waste frying oil for production of rhamnolipids using *Pseudomonas aeruginosa* zju. u1M. *J. Zhejiang Univ. Sci. A* 8 (9): 1514–1520.

91 Aguilera-Segura, S.M., Vélez, V.N., Achenie, L. et al. (2016). Peptides design based on transmembrane *Escherichia coli*'s OmpA protein through molecular dynamics simulations in water–dodecane interfaces. *J. Mol. Graph. Model.* 68: 216–223.

92 Bhardwaj, G., Cameotra, S.S., and Chopra, H.K. (2013). Utilization of oleo-chemical industry by-products for biosurfactant production. *AMB Express* 3 (1): 68.

93 Banat, I.M., Satpute, S.K., Cameotra, S.S. et al. (2014). Cost effective technologies and renewable substrates for biosurfactants' production. *Front. Microbiol.* 5: 697.

94 Thavasi, R., Jayalakshmi, S., Balasubramanian, T., and Banat, I.M. (2008). Production and characterization of a glycolipid biosurfactant from *Bacillus megaterium* using economically cheaper sources. *World J. Microbiol. Biotechnol.* 24 (7): 917–925.

95 Mercade, M.E., Manresa, M.A., Robert, M. et al. (1993). Olive oil mill effluent (OOME). New substrate for biosurfactant production. *Bioresour. Technol.* 43 (1): 1–6.

96 Abalos, A., Pinazo, A., Infante, M.R. et al. (2001). Physicochemical and antimicrobial properties of new rhamnolipids produced by *Pseudomonas aeruginosa* AT10 from soybean oil refinery wastes. *Langmuir* 17 (5): 1367–1371.

97 Benincasa, M., Abalos, A., Oliveira, I., and Manresa, A. (2004). Chemical structure, surface properties and biological activities of the biosurfactant produced by *Pseudomonas aeruginosa* LBI from soapstock. *Antonie Van Leeuwenhoek* 85: 1–8.

98 De Faria, A.F., Teodoro-Martinez, D.S., De Oliveira Barbosa, G.N. et al. (2011). Production and structural characterization of surfactin (C14/Leu7) produced by *Bacillus subtilis* isolate LSFM-05 grown on raw glycerol from the biodiesel industry. *Process Biochem.* 46: 1951–1957.

99 George, S. and Jayachandran, K. (2013). Production and characterization of rhamnolipid biosurfactant from waste frying coconut oil using a novel *Pseudomonas aeruginosa* D. *J. Appl. Microbiol.* 114: 373–383.

100 Moya Ramírez, I., Altmajer Vaz, D., Banat, I.M. et al. (2016). Hydrolysis of olive mill waste to enhance rhamnolipids and surfactin production. *Bioresour. Technol.* 205: 1–6.

101 Bednarski, W., Adamczak, M., Tomasik, J., and Płaszczyk, M. (2004). Application of oil refinery waste in the biosynthesis of glycolipids by yeast. *Bioresour. Technol.* 95 (1): 15–18.

102 Nitschke, M., Costa, S.G., and Contiero, J. (2005). Rhamnolipid surfactants: an update on the general aspects of these remarkable biomolecules. *Biotechnol. Prog.* 21: 1593–1600.

103 Rufino, R.D., Sarubbo, L.A., Neto, B.B., and Campos-Takaki, G.M. (2008). Experimental design for the production of tensio-active agent by *Candida lipolytica*. *J. Ind. Microbiol. Biotechnol.* 35: 907–914.

104 Jang, J.Y., Yang, S.Y., Kim, Y.C. et al. (2013). Identification of orfamide A as an insecticidal metabolite produced by *Pseudomonas protegens* F6. *J. Agric. Food Chem.* 61: 6786–6791.

105 Menon, V., Prakash, G., Prabhune, A., and Rao, M. (2010). Biocatalytic approach for the utilization of hemicellulose for ethanol production from agricultural residue using thermostable xylanase and thermotolerant yeast. *Bioresour. Technol.* 101: 5366–5373.

106 Di Martino, C., Catone, M.V., Lopez, N.I., and Raiger Iustman, L.J. (2014). Polyhydroxyalkanoate synthesis affects biosurfactant production and cell attachment to hydrocarbons in *Pseudomonas* sp. KA-08. *Curr. Microbiol.* 68: 735–742.

107 Marmesat, S., Rodrigues, E., Velasco, J., and Dobarganes, C. (2007). Quality of used frying fats and oils: comparison of rapid tests based on chemical and physical oil properties. *Int. J. Food Sci. Technol.* 42 (5): 601–608.

108 Haba, E., Espuny, M.J., Busquets, M., and Manresa, A. (2000). Screening and production of rhamnolipids by *Pseudomonas aeruginosa* 47T2 NCIB 40044 from waste frying oils. *J. Appl. Microbiol.* 88: 379–387.

109 Vedaraman, N. and Venkatesh, N. (2011). Production of surfactin by *Bacillus subtilis* MTCC 2423 from waste frying oils. *Braz. J. Chem. Eng.* 28 (2): 175–180.

110 Pan, L.S., Xu, N., Tian, Z. et al. (2011). Preparation and characterization of poly(propylene carbonate)/alkali lignin composite sheets by calendering process. In: Advanced Materials Research, vol. 233–235, 1786–1789. Trans Tech Publications Ltd.

111 Hasanizadeh, P., Moghimi, H., and Hamedi, J. (2018). Biosurfactant production by *Mucor circinelloides*: Environmental applications and surface-active properties. *Eng. Life Sci.* 18 (5): 317–325.

112 Banasik, A., Kanellopoulos, A., Claassen, G.D.H. et al. (2017). Closing loops in agricultural supply chains using multi-objective optimization: A case study of an industrial mushroom supply chain. *Int. J. Prod. Econ.* 183: 409–420.

113 Garg, V.K., Suthar, S., and Yadav, A. (2012). Management of food industry waste employing vermicomposting technology. *Bioresour. Technol.* 126: 437–443.

114 Ponte Rocha, M.V., Gomes Barreto, R.V., Melo, V.M., and Barros Goncalves, L.R. (2009). Evaluation of cashew apple juice for surfactin production by *Bacillus subtilis* LAMI008. *Appl. Biochem. Biotechnol.* 155: 366–378.

115 Rocha, M.V., Souza, M.C., Benedicto, S.C. et al. (2007). Production of biosurfactant by *Pseudomonas aeruginosa* grown on cashew apple juice. *Appl. Biochem. Biotechnol.* 137–140: 185–194.

116 Giro, M.E., Martins, J.J., Rocha, M.V. et al. (2009). Clarified cashew apple juice as alternative raw material for biosurfactant production by *Bacillus subtilis* in a batch bioreactor. *Biotechnol. J.* 4: 738–747.

117 Liu, X., Ren, B., Chen, M. et al. (2010). Production and characterization of a group of bioemulsifiers from the marine *Bacillus velezensis* strain H3. *Appl. Microbiol. Biotechnol.* 87: 1881–1893.

118 Verma, S., Prasanna, R., Saxena, J. et al. (2012). Deciphering the metabolic capabilities of a lipase producing *Pseudomonas aeruginosa* SL-72 strain. *Folia Microbiol. (Praha)* 57: 525–531.

119 FAO (2008). International Year of the Potato 2008 New Light on a Hidden Treasure. FAO.

120 Thompson, D.N., Fox, S.L. and Bala, G.A., (2000). Biosurfactants from potato process effluents. In: M. Finkelstein and B.H. Davison (eds), *Twenty-First Symposium on Biotechnology for Fuels and Chemicals. Applied Biochemistry and Biotechnology*, pp. 917–930. Humana Press, Totowa, NJ.

121 Das, K. and Mukherjee, A.K. (2007). Comparison of lipopeptide biosurfactants production by *Bacillus subtilis* strains in submerged and solid state fermentation systems using a cheap carbon source: Some industrial applications of biosurfactants. *Process Biochem.* 42 (8): 1191–1199.

122 Wang, Q., Chen, S., Zhang, J. et al. (2008). Co-producing lipopeptides and poly-γ-glutamic acid by solid-state fermentation of *Bacillus subtilis* using soybean and sweet potato residues and its biocontrol and fertilizer synergistic effects. *Bioresour. Technol.* 99 (8): 3318–3323.

123 Araújo, H.W., Andrade, R.F., Montero-Rodríguez, D. et al. (2019). Sustainable biosurfactant produced by Serratia marcescens UCP 1549 and its suitability for agricultural and marine bioremediation applications. *Microb. Cell Fact.* 18 (1): 1–13.

124 Barros, F.F.C., Ponezi, A.N., and Pastore, G.M. (2008). Production of biosurfactant by *Bacillus subtilis* LB5a on a pilot scale using cassava wastewater as substrate. *J. Ind. Microbiol. Biotechnol.* 35 (9): 1071–1078.

125 Nitschke, M. and Pastore, G. (2003). Cassava flour wastewater as a substrate for biosurfactant production. *Appl. Biochem. Biotechnol.* 105–108: 295–301.

126 Nitschke, M. and Pastore, G.M. (2006). Production and properties of a surfactant obtained from *Bacillus subtilis* grown on cassava wastewater. *Bioresour. Technol.* 97: 336–341.

127 Makkar, R.S., Cameotra, S.S., and Banat, I.M. (2011). Advances in utilization of renewable substrates for biosurfactant production. *AMB Express* 1 (1): 5.

128 Nitschke, M., Ferraz, C., and Pastore, G.M. (2004). Selection of microorganisms for biosurfactant production using agroindustrial wastes. *Braz. J. Microbiol.* 35: 81–85.

129 Marcelino, P.R.F., Gonçalves, F., Jimenez, I.M. et al. (2020). Sustainable production of biosurfactants and their applications. In: A.P. Ingle, A.K. Chandel, and S.S. Silva (eds),. Lignocellulosic Biorefining Technologies: 159–183. Available at: https://doi.org/10.1002/9781119568858.ch8.

130 Rinaldi, R., Jastrzebski, R., Clough, M.T. et al. (2016). Paving the way for lignin valorisation: recent advances in bioengineering, biorefining and catalysis. *Angew. Chem. Int. Ed.* 55 (29): 8164–8215.

131 Portilla-Rivera, O., Torrado, A., Domínguez, J.M., and Moldes, A.B. (2008). Stability and emulsifying capacity of biosurfactants obtained from lignocellulosic sources using *Lactobacillus pentosus*. *J. Agric. Food Chem.* 56 (17): 8074–8080.

132 Cortés-Camargo, S., Pérez-Rodríguez, N., de Souza Oliveira, R.P. et al. (2016). Production of biosurfactants from vine-trimming shoots using the halotolerant strain *Bacillus tequilensis* ZSB10. *Ind. Crop Prod.* 79: 258–266.

133 Jokari, S., Rashedi, H., Amoabediny, G.H. et al. (2012). Effect of aeration rate on biosurfactin production in a miniaturized bioreactor. *Int. J. Environ. Res.* 6 (3): 627–634.

134 Morita, T., Fukuoka, T., Konishi, M. et al. (2009). Production of a novel glycolipid biosurfactant, mannosylmannitol lipid, by *Pseudozyma parantarctica* and its interfacial properties. *Appl. Microbiol. Biotechnol.* 83 (6): 1017–1025.

4

Biosurfactants for Heavy Metal Remediation and Bioeconomics

Shalini Srivastava[1], Monoj Kumar Mondal[2], and Shashi Bhushan Agrawal[1]

[1] *Department of Botany, Institute of Science, Banaras Hindu University, Varanasi, Uttar Pradesh, India*
[2] *Department of Chemical Engineering and Technology, Indian Institute of Technology (Banaras Hindu University), Varanasi, Uttar Pradesh, India*

CHAPTER MENU

Biosurfactants for a Sustainable Future: Production and Applications in the Environment and Biomedicine, First Edition. Edited by Hemen Sarma and Majeti Narasimha Vara Prasad.
© 2021 John Wiley & Sons Ltd. Published 2021 by John Wiley & Sons Ltd.

4.1 Introduction

In the present era, irresponsible and irrational actions of innumerable industrial units such as steel manufacturing, glass manufacturing, electroplating, leather tanning, ceramics, wood preservations, and chemical processing, along with applications of huge amounts of chemical fertilizers, release too much toxic metal ions in the surrounding atmosphere and becomes a major problem for environmental pollution [1–5]. In the current scenario, a major environmental problem is the pollution of heavy metals due to their non-degradable and bioaccumulative nature in the environment. The toxicity and bioaccumulation tendency of heavy metals in living organisms is a serious health hazard. Environmental contamination due to heavy metals has greatly increased the recommended limit by various concerned agencies [6–10]. With the chemical or biological processes, one cannot break heavy metals into non-toxic form but can only transform them into less toxic forms [11]. Even at very low concentrations, heavy metals are toxic and also have the potential to contaminate the food chain, where they accumulate and impose damage to living organisms. The metal ion toxicity depends on the exposure quantity to the organism, the absorbed dose and its type, the route, and the duration of exposure [12]. Liver and kidney damage, certain learning disabilities, and in extreme cases even birth defects are some common ailments that have a direct connection with metal toxicity [13]. Therefore, it has become an extremely important responsibility of scientists to find an eco-friendly approach for metal ion remediation from the environment and consequently to preserve the health of the living [14].

In this effort, numerous attempts have been made, fixing specifically on toxic metal remediation from a ruined environment [15]. Treatment of contaminated soils with water, inorganic and organic acids, chemical and metal chelating agents such as EDTA are some of the most common techniques used for metal ion remediation from soils [16]. Techniques like thermal treatment, stabilization, excavation, and hazardous waste sites transportation of contaminated soil for landfill have several fundamental drawbacks since they cannot totally remediate metals, but only restrain them in the polluted soil and also cover huge land spaces [17]. A good metal complexing agent is required that possesses the properties of solubility, environmental stability, and a good complexation potential for the efficient removal of heavy metal ions from contaminated environments [18]. Biosurfactant application in both aqueous solutions and soils for metal ion removal have gained much attention in several studies [19, 20]. Biosurfactants that are surface-active metabolite possess metal complexing properties, reported to be effective in remediation of heavy metal contaminated sites [13]. There are several reasons to assign biosurfactants as capable substitute agents for remediation purposes [4, 21]. These reasons are their better environmental compatibility and biodegradability, less toxic nature, and most importantly utilization of inexpensive agro-based raw materials and organic wastes for their production [22]. The biosurfactants have a specific property to retain their activity even at extreme condition of high or low temperature, pH, and salt concentration. To date, a majority of informed biosurfactants are of microbial origin.

In recent years, scientific communities have paid attention to the economic aspects of biological processes and from here the concept of a "circular economy" has been introduced for biosurfactant production and its application. The circular economy strategy is focused on waste prevention or where it is generated it can be utilized in bioeconomical ways. In a circular economy, biorefining, which is the sustainable dispensation of waste biomass into a spectrum of marketable products and energy, play a crucial role. The biosurfactant synthesis and its bioeconomical utilization help in upliftment of a deteriorating environment condition, especially by heavy metals. Therefore, the aim of the present chapter is to provide information about biosurfactant production and their potential application in a bioeconomic way as environment-friendly products in metal remediation.

4.2 Concept of Surfactant and Biosurfactant for Heavy Metal Remediation

The wide, extensive, and significant role of surfactants has been introduced in various sectors, such as environmental pollution mitigation, the petroleum industry, the detergent industry and food production facilities [23]. Due to the amphiphilic nature of surfactants, the surface tension in an oil–water interface decreased and hence water-immiscible substance solubility increased. The surfactant environmental impacts and their easy disposal after utilization is of principal concern due to sundry and widespread utilization in the environment. This demanded screening of an environmentally friendly alternative to synthetic surfactants with equal efficiency. These classes of novel surfactants, having a biological origin, stated as "biosurfactant," shows significant variation in terms of their chemical composition, structure, and mode of action. In present environmental remediation techniques, biosurfactants are gaining more attention due to their biodegradability and eco-friendly attributes [24].

When focusing on biosurfactants of microbial origins, the extremophiles have gained more attention in the last few years as they can also show efficient remediation capacity in extreme punitive conditions. Their chemical structure involves hydrophilic polar moiety as oligo- or monosaccharide and proteins as well as polysaccharides or peptides and the hydrophobic moiety has unsaturated or saturated fatty alcohols or hydroxylated fatty acids [25]. The balance between the hydrophilic and lipophilic ends of biosurfactants grounds the hydrophilic and hydrophobic ends to be determined in substances that are surface active. The amphiphilic behavior of biosurfactants enables them to enhance the hydrophobic substance surface area as well as to give them modification ability in the microbial cell surface property.

The real breakthrough in the biosurfactant production and application research comes only after the knowledge of genetically engineered microorganisms because they have the capability of giving high yields. The detailed knowledge of genetics of the microorganisms plays a crucial role in this esteem. Previous studies on the molecular genetics and biochemistry of several biosurfactants revealed the operons, the enzymes, and the metabolic pathways for their extracellular production. One of the examples is surfactin – a cyclic lipopeptide biosurfactant. Its production occurred as a result of catalysis of a large multienzyme peptide synthase complex, i.e. surfactin synthase via a non-ribosomal biosynthesis pathway. Similar enzyme complexes are responsible for the synthesis of other lipopeptides such as iturin, lichenysin, and arthrofactin. A very high level of compositional similarity has been shown by various lipopeptide synthesizing non-ribosomal peptide synthetases (NRPSs). *Psuedomonas* species required plasmid-encoded-*rhl* A, B, R, and I genes of an *rhl* quorum-sensing system for production of glycolipid biosurfactants. The molecular genetics of biosynthesis of alasan and emulsan synthesized by *Acinetobacter* species, along with some other biosurfactants of fungal origin including mannosylerythritol lipids (MEL) and hydrophobins, have also been researched by scientists.

Microbes involved in biosurfactant production show their excellent efficiency in metal ion bioremediation from both aqueous and terrestrial environments. Some of the key characteristics of these biological compounds that play a crucial and target-specific role during the remediation process are their higher surface activity with high tolerance to various environmental factors. The biosurfactants also can withstand from mean to extreme conditions, such as ionic strength, temperature, acidity or basicity of medium salt concentrations, biodegradable nature, demulsifying–emulsifying ability, anti-inflammatory potential and antimicrobial activity.

The scientists and researchers have identified and characterized different types of biosurfactants produced from various biological sources [26–28]. The main criteria behind biosurfactant classification are their chemical structure, source of origin, antimicrobial activity, efficiency of pollutant removal from the environment, and surface tension reduction ability [29]. A variety of materials have been used by the microbial community as carbon and energy sources for their production. Microorganisms release tensio-active substances as biosurfactants in the medium during degradation of hydrocarbons [30].

4.3 Mechanisms of Biosurfactant–Metal Interactions

Two main pathways have been identified for the desorption of metal ions from contaminated land using biosurfactants [31]. In the first pathway, there is a complex formation between the free, nonionic form of metal and biosurfactant molecules. In this interaction, using the principle of Le Chatelier, the solution phase activity of the metal ions is reduced and thus its desorption from the medium increases. In the second pathway, it is proposed that there is an accumulation of biosurfactants at the solid–solution interface and absorption of metal ions occurs as the interfacial tension reduces between the two.

According to Rufino et al. [32], ion exchange, precipitation dissolution, counter-ion association, and electrostatic interaction are some of the chief mechanisms that govern metal–biosurfactant binding in the contaminated environment. Studies reveal that the complex formation capability of the biosurfactant with the metal ions is the chief cause for their usefulness in metal ion remediation. Precisely, ionic bonds formed in between metal ions and anionic biosurfactants lead to a generation of stronger stabilizing forces as nonionic complexes form, which, of course, are stronger as compared to metal and soil interaction. Because of the neutral charge of the complex with a subsequent amalgamation of the metal into micelles, the complex form of metal–biosurfactants desorb from the soil matrix and move into the soil solution. A detailed study of the proposed mechanism reveals that either an outer-sphere surface complex formation occurs in between negatively charged surfaces and metals due to strong electrostatic attraction or an inner-sphere surface complex formation is established between the metal ions and biosurfactant molecules due to chemical bonding in which hydroxide groups serve as ligands. The mechanism of metal binding through both pathways is smoothed in the presence of water molecules and easy protonation and deprotonation of oxide functional groups. The mechanism of metal–biosurfactant interaction is represented in Figure 4.1.

4.4 Substrates Used for Biosurfactant Production

Microorganisms are identified as the most important source for production of biosurfactants. Willumsen and Karlson [33] in their study found that many of the biosurfactant-producing microorganisms are hydrocarbon degraders. The proficiency of microbial biosurfactant in the bioremediation as well as in the enhanced oil recovery have been researched extensively [34]. The verities of substrate used for common biosurfactant production is represented in Figure 4.2.

4.4.1 Biosurfactants of Bacterial Origin

In the growth medium, the hydrocarbons are emulsified by ionic surfactants excreted by some of the bacteria and yeast. *Pseudomonas* sp. that produce rhamnolipids (RLs) and *Torulopsis* sp. that are mainly involved in the production of sophorolipids are some examples of these groups of biosurfactants [35, 36].

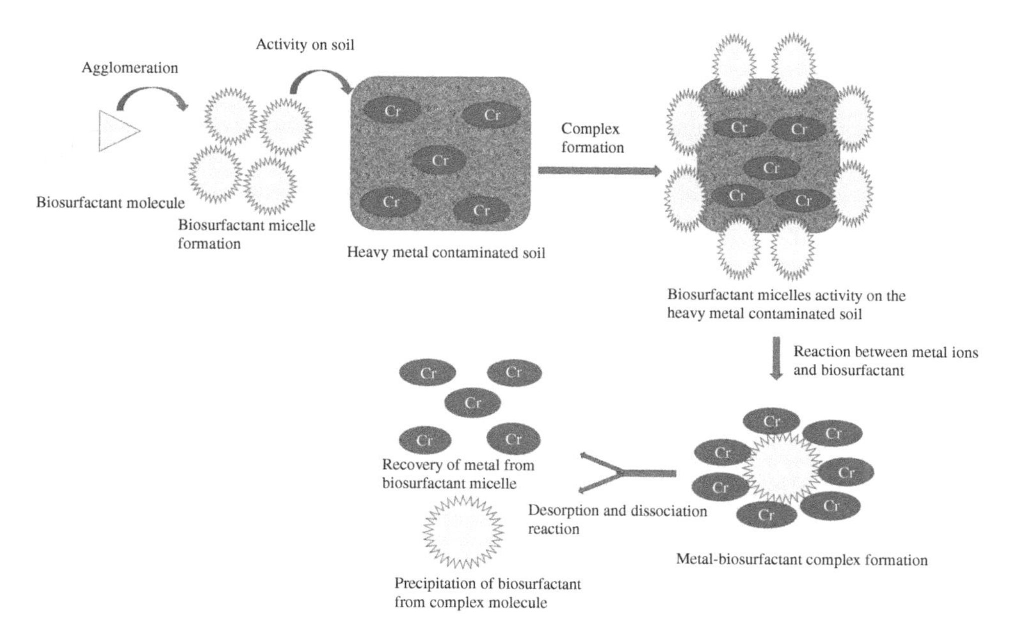

Figure 4.1 Biosurfactant mediated heavy metal remediation (See insert for color representation of the figure).

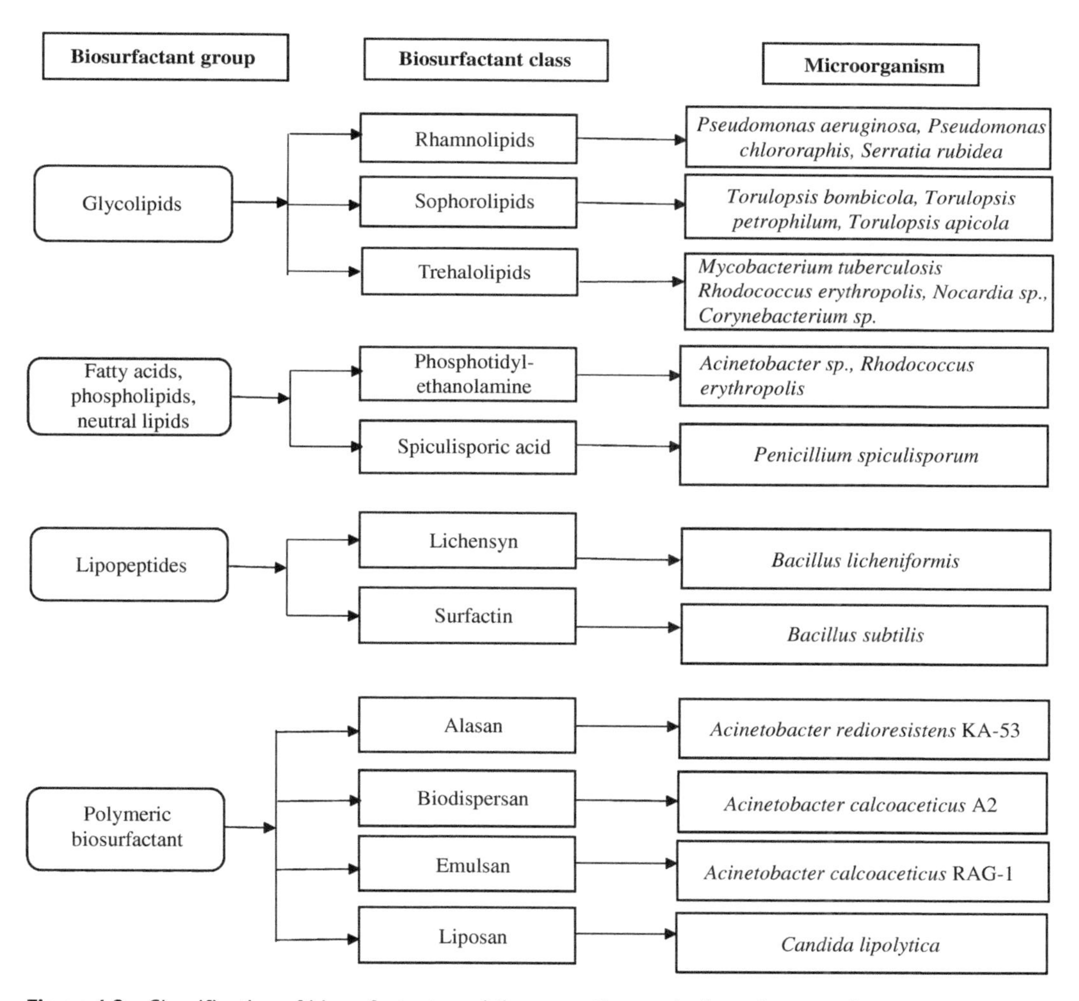

Figure 4.2 Classification of biosurfactants and the respective producing microorganisms.

Some bacterial species have the ability to alter their cell membrane structure by producing some nonionic or lipopolysaccharide biosurfactants. Examples of some nonionic trehalose corynomycolates producing bacterial strains are: *Rhodococcus erythropolis, Arthrobacter* sp., and various *Mycobacterium* sp. [37]. *Acinetobacter* sp. produce lipopolysaccharides, such as emulsan, and *Bacillus subtilis* produces extensive quantities of lipoproteins, such as surfactin and subtilisin [38, 39]. Table 4.1 depicts biosurfactants produced by various strains of bacteria.

4.4.2 Biosurfactanats of Fungal Origin

Only a few species of fungi are identified for the production of biosurfactants in comparison to bacterial species. Some of the typical fungal strains explored for the production of biosurfactants, as investigated by researchers are, *Candida bombicola* [40], *Candida ishiwadae* [41], *Candida lipolytica* [42], *Candida batistae* [43], *Aspergillus ustus* [44], and *Trichosporon ashii* [45]. The best part of these fungal strains is that they have produced biosurfactants using low-cost raw materials as their growth substrate. Glycolipids and sophorolipids are one of the most important class of biosurfactants produced by these fungal strains. The various biofactants produced by fungi are shown in Table 4.2.

Table 4.1 Biosurfactants derived from bacteria.

Bacteria	Biosurfactant
Serratia marcescens	Serrawettin
Rhodotorula glutinis, Rhodotorula graminis	Polyol lipids
Rhodococcus erythropolis, Corynebacterium sp. *Mycobacterium* sp., *Arhtrobacter* sp., *Nocardia erythropolis*	Trehalose lipids
Pseudomonas sp., *Thiobacillus thiooxidans, Agrobacterium* sp.	Ornithine lipids
Pseudomonas fluorescens, Leuconostoc mesenteriods	Viscosin
Pseudomonas aeruginosa, Pseudomonas chlororaphis, Serratia rubidea	Rhamnolipids
Pseudomonas fluorescens, Debaryomyces polmorphus	Carbohydrate-lipid
Pseudomonas aeruginosa	Protein PA
Lactobacillus fermentum	Diglycosyl diglycerides

Table 4.2 Biosurfactants derived from fungi.

Fungi	Biosurfactant
Torulopsis bombicola	Sophorose lipid
Candida bombicola	Sophoro lipids
Candida lipolytica	Protein-lipidpolysaccharide complex
Candida lipolytica	Protein-lipidcarbohydrate complex
Candida ishiwadae	glycolipid
Candida batistae	sophorolipids
Aspergillus ustus	Glycolipoprotein
Tichosporon ashii	sophorolipids

4.5 Classification of Biosurfactants

Origin and composition are the two main factors on the basis of which the classification of biosurfactants has been performed. According to Rosenberg and Ron [46], based on the molecular weight, biosurfactants are categorized into two types. The first one is comprised of those compounds that have low molecular weight compounds with lower surface and interfacial tensions and the second one is comprised of high molecular weight compounds with strong surface binding capacity. The majority of low molecular weight biosurfactants comes under the glycolipids, lipopeptides, and phospholipids category while high molecular weight ones are mainly particulate and polymeric surfactants [47]. Another basis for biosurfactant classification is the presence and type of charge on individual polar moiety. The negatively charged surfactants, i.e. anionic usually have a sulphonate or sulfur group as the chief functional group on their cell surface while positively charged or cationic surfactants mainly possess an ammonium and hydroxyl group. Also, surfactants with a neutral or non-ionic nature are identified and are the products of a 1, 2-epoxyethane polymerization reaction. When both positively and negatively charged functional groups are present on the same surfactant molecules, they are identified as amphoteric surfactants [48].

4.6 Types of Biosurfactants

There have been so many forms of biosurfactants and each have a common microbial origin. Some of the broad categories of biosurfactant are now discussed.

4.6.1 Glycolipids

Glycolipids are the most common type of biosurfactants and consist of mono-, di-, tri-, and tetrasaccharides. The saccharides include glucose, mannose, galactose, glucuronic acid, rhamnose, and galactose sulphate. Some of the microorganisms usually have the same fatty acids and phospholipid composition [49, 50]. Carbohydrates in combination with long-chain aliphatic acids or hydroxyaliphatic acids are the key component of glycolipids [26]. According to Karanth et al. [51], rhamnolipids, trehalolipids, and sophorolipids are the best-known glycolipids.

4.6.2 Rhamnolipids

Rhamnose and 3-hydroxy fatty acids containing glycolipid surfactant have been produced by *Pseudomonas* sp. [52]. As shown in Figure 4.3, one or two molecules of rhamnose are linked to one or two molecules of hydroxyl decanoic acid and represent the basic structure of rhamnolipids. *Pseudomonas aeruginosa* and *Burkholderia* sp. play a key role in the production of rhamnolipids and are identified as one of the most effective surfactants in the remediation of hydrodrophobic compounds from contaminated soils [53].

4.6.3 Sophorolipids

Torulopsis sp., mainly *T. bombicola* and *T. apicola*, are the main strains involved in the production of sophorolipids. Asmer et al. [54] proposed the chemical composition of sophorolipids as the dimeric carbohydrate sophorose and long chain hydroxyl fatty acids, linked by a β-glycosidic bond (Figure 4.4).

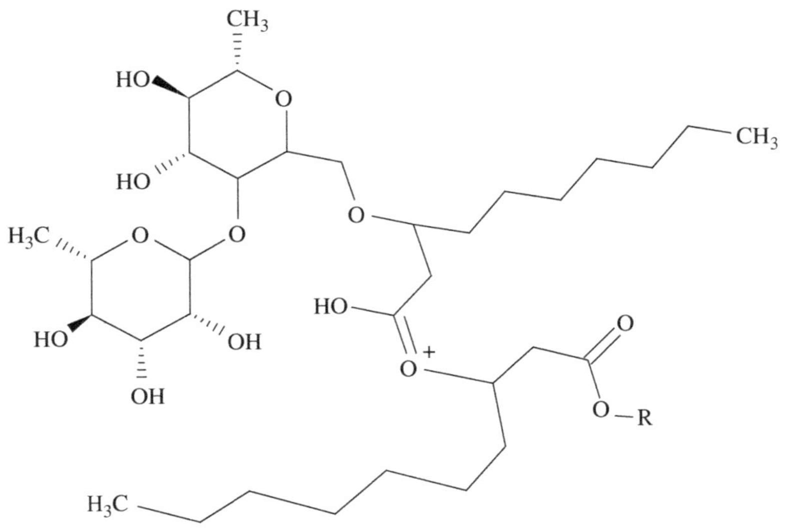

Figure 4.3 Structure of dirhamnolipid.

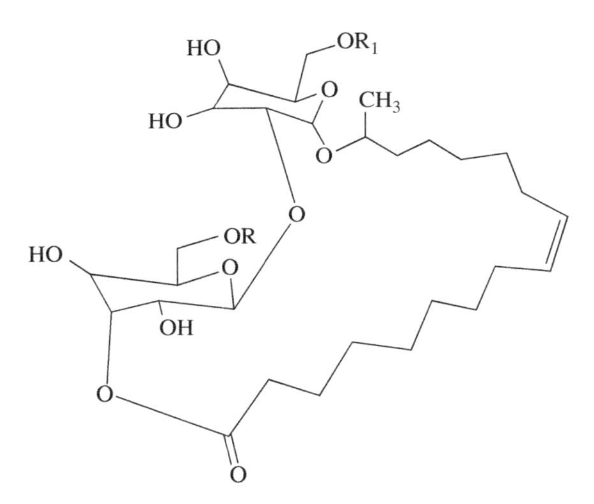

Figure 4.4 Structure of sophorolipids.

4.6.4 Trehalolipids

Trehalolipids is another important biosurfactant of glycolipid nature. Many members of the genus *Mycobacterium* show the presence of the serpentine group because of the presence of trehalose esters on their cell surface. Disaccharide trehalose linked at C-6 and C-6 to mycolic acid is associated with most species of *Mycobacterium*, *Norcadia*, and *Corynebacterium*. The trehalolipids have a complex chemical composition and possess the ability to alter the size and structure of their mycolic corrosive ends. They often occur in the form of a complex mixture whose composition varies depending on the strain physiology and growth condition.

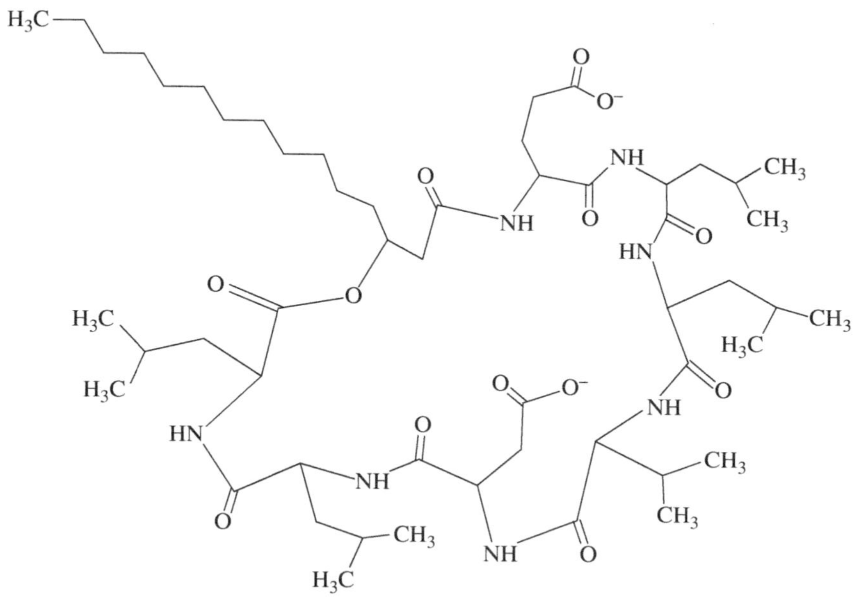

Figure 4.5 Structure of surfactin.

4.6.5 Surfactin

Bacillus subtilis produces the most potential biosurfactant, i.e. surfactin is a complex mixture of different amino acids and fatty acid chain that bind with each other through a lactone linkage (Figure 4.5). The reduction in surface tension and interfacial tension of the water molecules occurs in the presence of surfactin. Surfactin has the potential to inactivate herpes and retroviruses.

4.6.6 Lipopeptides and Lipoproteins

According to Rosenberg and Ron [46], these structures consist of lipid molecules in attachment with a polypeptide chain. Antimicrobial action has been shown by many of the biosurfactants against various bacteria, algae, fungi, and viruses. The antifungal and antibacterial property of the lipopeptide, iturin, produced by *B. subtilis* have been reported by Besson et al. [55] and Singh and Cameotra [13], respectively. A study conducted by Nitschke and Pastore [56] shows lipopeptide activity even after autoclaving between pH 5 and 11 and with a shelf life of six months at −18 °C.

4.6.7 Fatty Acids, Phospholipids, and Neutral Lipids

Huge amounts of unsaturated fats and phospholipid surfactants have been produced by microbes and yeast. For therapeutic applications, these classes of biosurfactants have the utmost importance. The major cause for the respiration failure in prematurely born children is chiefly due to the deficiency of phospholipid protein complex [57].

4.6.8 Polymeric Biosurfactant

The best-studied polymeric biosurfactants are emulsan, liposan, mannoprotein, and polysaccharide–protein complexes [26]. For hydrocarbons in water, even at a concentration as low as 0.001–0.01%, emulsan works as a very effective emulsifying agent. With a composition of 83% carbohydrate and 17% protein, liposan behaves as an extracellular water-soluble emulsifier. Mannoproteins, which contain 44% amnnose and 17% protein, are produced in large quantity by *Saccharomyces cerevisiae*. For oil spills, organic solvents, and alkanes, these mannoproteins show excellent emulsifying activity.

4.6.9 Particulate Biosurfactants

An essential fragment for the remediation of alkanes by microbes is microemulsion, consisting of extracellular film vesicle segments of hydrocarbon. Protein, phospholipids, and lipopolysaccharides are the key components of vesicles formed by *Acinetobacter* sp. that have a thickness of 1.158 cg/cm^3 and 20–50 nm diameter.

4.7 Factors Influencing Biosurfactants Production

Biosurfactant production and their chemical compositions are influenced by a number of factors. Environmental conditions (temperature, pH, air circulation, divalent cation, and saltiness) as well as the nature of the available energy source in the form of carbon and nitrogen, reaction media composition, and limited nutrient supply have a powerful impact on synthesized biosurfactants.

4.7.1 Environmental Factors

The biosurfactants production is either enhanced or inhibited by the reaction conditions. Therefore, for the large-scale production of the desired biosurfactants, it is important to constantly upgrade the bioprocess as the item might be influenced by changes in reaction conditions, i.e. temperature, pH, air circulation, or unsettling speed. Most of the synthesized biosurfactants provide their best performance in a temperature range of 25–300 °C. A change in the biosurfactant composition can occur during temperature variations. Zinjarde and Pant [58] in their study on *Yarrowia lipolytica* illustrated the best biosurfactant production at pH 8.0, i.e. at regular ocean water pH, which is also the natural surrounding pH of *Y. lipolytica*. Rhamnolipid production by *Pseudomonas* sp. occurred at its most extreme pH (6–6.5) and decreases above pH 7.

4.7.2 Carbon and Nitrogen Sources for Biosurfactant Production

Scientists have used various carbon compounds as fundamental substrates and energy sources for the production of biosurfactants. The substrate composition, i.e. carbon and nitrogen source, influences the biosurfactant type, quality, and quantity [59]. Ilori et al. [60] in their study identify glucose, sucrose, glycerol, diesel, and raw petroleum as good sources of carbon for biosurfactant production. A nitrogen source is also very crucial for biosurfactant production. Medium containing nitrogen sources fulfills the basic requirement of multiplication and development of biosurfactant-producing microorganisms for protein and chemical blends. Some of the distinctive sources of nitrogen used for biosurfactant production are ammonium sulfate, ammonium nitrate, sodium nitrate, meat concentrates, and malt extricate, etc.

4.8 Strategies for Commercial Biosurfactant Production

The high production cost, low yield, and sophisticated product recovery restricts widespread application of biosurfactants as compared to chemical surfactants. The inefficient bioprocess engineering, usage of a cost–credit substrate, and poor strain improvement are some of the major drawbacks behind high production costs of biosurfactants. For commercial applications of biosurfactants in various fields, these stated drawbacks should be solved. Therefore, the net economic gain between the production cost and application benefit will govern the future of biosurfactants. For large-scale industrial production and applications, the links between the production parameters of these molecules, their structure and their functions, need to be optimized. The following proposed strategy can smooth cost-effective biosurfactant production and application.

4.8.1 Raw Material: Low Cost from Renewable Resources

In their natural environment, the microbial population produces surface-active agent in extremely minute quantities. Keeping in mind these microbial behaviors, researchers try to maximize the biomolecule yield and extract more and more concentrations of highly efficient biosurfactants.

In lowering the overall production cost of biosurfactants, a selection of suitable and efficient low-cost raw materials is important. Raw materials with higher concentrations of carbohydrate, nitrogen, and lipids highlight the necessity for biosurfactants in commercial production. Utilization of agricultural wastes and byproduct materials that are available in abundant quantity along with the benefit of reduced environmental pollution chances serve as the best raw material for biosurfactants. In a study conducted by Ashby et al. [61] on the effect of raw materials on biosurfactant cost, the authors found that approximately 75% of the total operating cost accounted for 90.7 million kg of sophorolipid production biosurfactant was due to glucose and oleic acid as the raw materials. The sophorolipid production costs vary depending on the raw material used; for example, when glucose and high oleic sunflower oil were used, the cost was estimated to be $2.95/kg and when glucose and oleic acid were used, it was reported to be $2.54/kg. This estimated high cost of sophorolipids production can be reduced after replacing the costly substrate with a low-cost industrial and agro-based byproduct. In another study by Rodrigues et al. [62], authors utilized low-cost materials for production of biosurfactants and the yields were increased by 1.5 times to that of the original cost and a 60–80% reduction in the medium cost was observed.

4.8.2 Production Process: Engineered for Low Capital and Operating Costs

Currently, only very few biosurfactants have been used in metal ion remediation processes on a commercial scale due to lack of cost-effective production processes. Due to the high costs of producing biosurfactants, their industrial application has been hindered.

Biosurfactant production on a large scale seemed to be very effective, but there is an urgent need to overcome competitiveness with their synthetic counterparts. Scientists made few attempts to develop large-scale biosurfactants. The *Bacillus subtilis* FE-2 strain has been used by Veenanadig et al. [50] to produce biosurfactants in a packed column bioreactor with a volumetric capacity of

30 l. In another study conducted by Daniels et al. [63] for large-scale production of rhamnose and 3-hydroxydecanoic acid from *Pseudomonas* sp., they claimed, in their patent, to produce in a defined culture medium that contained corn oil as a carbon source a high level of rhamnolipids at a concentration from about 30 g/l to about 50 g/l.

4.8.3 Improved Bioprocess Engineering

Process optimization plays a crucial role in cost reduction of large-scale biosurfactant production. Synthesis of biosurfactant can be categorized into four foremost types:

1) Biosurfactant production associated with the growth medium and substrate utilization.
2) Under a growth-limiting condition, biosurfactant production, e.g. *P. aeruginosa* shows an overproduction of biosurfactants when nitrogen and iron are limited.
3) Resting or immobilized cell utilization in biosurfactant production. This type of biosurfactant production shows high efficiency because microbial cells keep using a carbon source only for biosurfactant synthesis and not for multiplication, thus helping to reduce the production cost.
4) Precursor addition for biosurfactant production. The quality and quantity of biosurfactant polymer influenced chemical and physical parameters, and the type of carbon and nitrogen source along with their ratios in the culture media [64].

4.8.4 Strain Improvement: Engineered for Higher Yield

In a study, Kosaric et al. [65] suggested four factors to reduce the cost of biosurfactants production. The first one was the type of microbes (selected, adapted, or engineered for higher yields). The second one was the nature of reaction condition (selected, adapted, or engineered for low capital and operating costs). The third one was the growth media composition and raw material nature and the fourth one was the process byproducts (minimum or managed as saleable products rather than as waste). In order to make commercially viable biosurfactants, it is important to improve and optimize the reaction condition using bioprocess engineering along with the use of hyperproducing microbial strain. To economize the production process and to obtain products with better commercial characteristics, the availability of hyperproducer strains and recombinants is important.

4.8.5 Enzymatic Synthesis of Biosurfactants

Enhanced enzyme productivity in microbes after genetic modification for enhanced biosurfactants production has been used by scientists to improve the productivity-to-cost ratio. The effectiveness and efficiency of enzymes have been maximized through the use of biotechnological techniques. The specificity of microbial enzymes, their catalytic properties, and mode of action can be altered and modified into more effective forms using these techniques.

4.9 Application of Biosurfactant for Heavy Metal Remediation

In the last few decades, many studies have been conducted and published by the scientific communities on biosurfactant production and remediation application. Due to the vast variation in chemical composition, their eco-friendly behavior, and wide range of applications in various processes, biosurfactants are utilized extensively in many sectors, including hydrophobic organic compounds and heavy metal ions remediation, enhanced oil recovery, and cosmetics and pharmaceutical sectors, etc. In the present review article, authors pay attention to the biosurfactant application for remediation of heavy metals (Table 4.3).

Mekwichai et al. [72] conducted a case study conducted in the Moe Sot District of Thailand, which had been reported to be contaminated over more than a 600-hectare (ha) area of paddy field. Researchers in their study utilized the potential of biosurfactants (rhamnolipid (RL) and saponin (SP)) for Cd remediation.

Table 4.3 Heavy metal removal efficiency of different biosurfactants.

Organism	Biosurfactant type	Contaminated environment	pH	Temperature (°C)	Metals	Efficiency	References
Commercial	Rhamnolipid	Soil	6.5	25	Cu	37	Dahrazma and Mulligan [16]
					Ni	33.2	
					zn	7.5	
Torulopsis bombicola	Sophorolipid	Soil	5.4	—	Cu	25	Mulligan et al. [17]
					Zn	60	
Bacillus subtilis	Surfactin				Cu	15	
					Zn	6	
Pseudomonas aeruginosa	Rhanmolipid				Cu	65	
					Zn	18	
Candida sphaerica	Anionic	Soil/water	—	—	Fe	95	Luna et al. [66]
					Zn	90	
					Pb	79	
Bacillus subtilis	Surfactin	Soil	—	—	Cd	15	Mulligan et al. [67]
					Cu	70	
					Zn	25	
Bacillus subtilis	Lipopeptide	Soil	9	25	Cd	44.2	Singh and Cameotra [68]
					Co	35.4	
					Cu	26.2	
					Ni	32.2	
					Pb	40.3	
					Zn	32.07	
Bacillus circulans	Crude surfactant	Soil	—	—	Cd	97.66	Das et al. [69]
					Pb	100	
Candida lipolytica UCP 0988	Lipoprotein	Soil	—	—	Cd	50	Rufino et al. [70]
					Cu	96	
					Fe	16.5	
					Pb	15.4	
					Zn	96	
Pseudomonas aeruginosa CVCM 411	Rhamnolipid	Soil	8	25	Fe	19	Diaz et al. [71]
					Zn	52	

Researchers utilized corn, which is a low cost agro waste for reduction of toxic Cd, for the contaminated site and corn biomass utilization. The Cd phytoextraction was enhanced by adding RL and SP., where 4 mmol/kg was set as the optimum dose for both RL and SP. To execute the phytoextraction experiment on corn, the optimum dose of biosurfactants had been applied at different growth stages of corn, i.e. the 7th, 45th, and 80th day from the sowing. On the 45th day of the experiment, the highest Cd uptake levels were recorded and RL showed a maximum Cd uptake capacity of 39.06 mg/kg. The recorded data was also validated by bioaccumulation factors, which reflected that RL increased soil Cd uptake by corn plants to the highest extent. Also, Cd leaching after biosurfactant addition was studied and results confirmed a lower level of leaching compared to EDTA applications.

The performance of an anionic biosurfactant from *Candida sphaerica* for the removal of heavy metal ions collected from soil of an automotive battery industry have been evaluated by Luna et al. [66]. They also evaluated metal remediation performance of biosurfactant from an aqueous solution. Multiple combinations of biosurfactant solution, sodium hydroxide, and hydrogen chloride were tested. Biosurfactant showed a very efficient removal rate with values of 95, 90, and 79% for Fe, Zn, and Pb, respectively. Treatment of biosurfactant solution with 0.1 and 0.25% HCl solution increased the metal removal rate. The recycled biosurfactant also showed 70, 62, and 45% of Fe, Zn, and Pb removal efficiency, respectively. In another study, Rufino et al. [32] extracted lipopeptide biosurfactant from *C. lipolytica* (UCP 0988). Both Zn and Cu metal ions were reduced by up to 96% of their initial concentration, and also there was significant reduction in the concentrations of Pb, Cd, and Fe.

Scientific communities for the production of biosurfactants have also utilized many species of *Bacillus*. In one study, surfactin extracted from *B. subtilis* have been tested for the removal of heavy metals from a contaminated soil (890 mg/kg Zn, 420 mg/kg Cu, 12.6% oil and grease) and sediments (110 mg/kg Cu and 3300 mg/kg Zn). Results showed that 25 and 70% of the Cu, 6 and 25% of the Zn, and 5 and 15% of the Cd could be removed by 0.1% surfactin with 1% NaOH, respectively, after one and five batch washings of the soil. Also, 15% of the Cu and 6% of the Zn could be removed after a single washing with 0.25% surfactin/1% NaOH from the sediment [67]. In their subsequent study, a batch study was performed by Mulligan et al. [17] to evaluate the feasibility of biosurfactants extracted from different strains for the removal of metal ions from sediments. Surfactin, rhamnolipids, and sophorolipid extracted from *B. subtilis*, *P. aeruginosa* and *T. bombicola*, respectively, were evaluated using sediment polluted with metals (110 mg/kg Cu and 3300 mg/kg Zn); 65% of the Cu and 18% of the Zn were removed by studied biosurfactant after a single washing with a concentration of 0.5% rhamnolipid, whereas 25% of the Cu and 60% of the Zn were removed by 4% sophorolipids. Compared to rhamnolipid and sophorolipids, surfactin was less effective, removing 15% of the Cu and only 6% of the Zn. Singh and Cameotra [68] utilized *B. subtilis* A21 species to synthesize lipopeptide biosurfactant, consisting of surfactin and fengycin, for the removal of petroleum hydrocarbons and heavy metals from contaminated soil. Soil washing with lipopeptide biosurfactant solution removed significant amounts of petroleum hydrocarbons (64.5%) and metals, namely Cd (44.2%), Co (35.4%), Pb (40.3%), Ni (32.2%), Cu (26.2%), and Zn (32.07%).

To evaluate the efficiency of environmentally compatible rhamnolipid biosurfactant produced by *P. aeruginosa* BS2 for the remediation of Cd and Pb from the artificially contaminated soil, Juwarkar et al. [18] focused their research on column experiments. Results revealed that extracted biosurfactant removes not only the leachable or available fraction of heavy metals but also the bound metals as compared to tap water, which removed the mobile fraction of the metal ions only. Contaminated soil washing with tap water shows only 2.75% of Cd and 9.8% of Pb removal whereas washing with rhamnolipids removed 92% of Cd and 88% of Pb after 36 hours of leaching.

A study of Diaz et al. [71] on biosurfactants application shows their ability to change the surface of many metal ions and their aggregation on interfaces favoring the metal separation from contaminated environments. The authors evaluated the metal removal efficiency of rhamnolipids and bioleaching with a mixed bacterial culture of *Acidithiobacillus thiooxidans* and *Acidithiobacillus ferrooxidans* from mineral waste/contaminated soils using alternate cycles of treatment. Results reflect that bioleaching alone is effective in Zn removal with a value of 50% but for Fe it was not very effective and removed only 19%. When rhamnolipids were used at low concentration (0.4 mg/ml), 11% of Fe and 25% of Zn were removed, while at 1 mg/ml concentration, 19% of Fe and 52% of Zn removal occurred. A combination of bioleaching and biosurfactant in the cycling treatment process enhanced metal removal efficiency and reached up to 36% for Fe and 63–70% for Zn.

Dahrazma and Mulligan [16] conducted their experiment with the objective to estimate the Cu, Zn, and Ni removal efficiency of rhamnolipid in a continuous flow configuration. The effect of process parameters such as concentration of rhamnolipids and the additives, time, and solution flowrate on the column performance have been analyzed. The removal of metal ions was up to 37% of Cu, 13% of Zn, and 27% of Ni when rhamnolipid without additives was applied. Addition of 1% of NaOH to 0.5% of rhamnolipid enhanced the Cu removal up to four times as compared to 0.5% rhamnolipid solution alone.

4.10 Bioeconomics of Metal Remediation Using Biosurfactants

In recent times, the developments of the "bioeconomy" have been promoted by most of the world's developing and developed countries. The bioeconomy policies and strategies have been formulated by the scientific communities to reach this sustainable goal. A strong support for this thoughtful concept was provided by the Global Bioeconomy Summit organized in Berlin in the year 2015, which gave the chance to the bioeconomic experts and stakeholders from more than 50 countries to come together and discuss their critical views on a stated smoldering topic. In the summit, it was declared that nothing has a unified definition of the bioeconomy. However, experts agreed on a common understanding of the bioeconomy as the "knowledge-based production and utilization of biological resources, innovative biological processes and principles to sustainably provide goods and services across all economic sectors."

In the present review article, authors thoroughly studied the bioeconomy of biosurfactants production and utilization. With more and more stringent regulations on greener processes and catering to the huge demand, biosurfactants form a major share of the surfactant market. The global biosurfactant market in 2013 was 344 068.40 tons and had been expected to reach 461 991.67 tons by 2020, growing at a CAGR of 4.3% from 2014 to 2020 [73]. Revenue generation of the biosurfactant market was found to be over $1.8 billion in 2016 and is expected to reach $2.6 billion by 2023 (540 ktons by 2024), with the rhamnolipid market set to witness a gain of over 8% [74]. Some other market research reports showed the global biosurfactant market at over 5.52 billion by 2022, at a CAGR of 5.6% from 2017 to 2022 [75]. At the present time, Europe is rising and is projected to prolong to rise as the biggest market (nearly 53%) followed by the United States, mainly due to stricter regulatory guidelines in the region. In the meanwhile, the growing wakefulness and infrastructures in Asian countries is making them a rising consumer of biosurfactants. With the detergent industry leading the product application sector, sophorolipids among all different type of biosurfactants were found to have the largest global market share. The Germany-based company BASF and Belgium Company Ecover have emerged as the top two biosurfactants producers in the surfactant market. Some other prominent companies involved in biosurfactant production and supplies are MG Intobio, Urumqui Unite, Saraya, Sun Products Corporation, Akzo Noble, Croda International PLC, Evonik Industries (Germany), Mitsubishi Chemical Corporation, and Jeneil Biosurfactant [74, 76]. On the other hand, in spite of the huge market demand, biosurfactant production is not as competitive as its synthetic counterparts. Therefore, economizing the biosurfactant production process assumes significance in order to sustain the market for these compounds in the current environmentally fragile scenario and long-term sustainable development.

4.11 Conclusion

This chapter reflects detailed information on the utilization of biosurfactants as a potential substitute in the heavy metal ions bioremediation process from the polluted environment. Designing the new strategies and technologies is the need of society in order to minimize the biosurfactant production cost at a commercial scale and make the production process economically competitive. The biosurfactants formation costs can be decreased by lowering the number of agro-industrial mistakes and the waste that treatment uses up. Agro-industrial waste-based biosurfactant generation renders another option for cheap and commercially viable biosurfactant production. For various applications, biosurfactants moved to generally regarded as safe (GRAS) microorganisms like lactobacilli and yeasts, which have astounding promise. However, significantly more research is now required in this field.

The production and commercial scale application of biosurfactant molecules remains a testing subject, as the planning of an irrefutable item is influenced by various key factors. For the utilization of biosurfactants in various areas, the rules and directions should be detailed. It is very much expected that in the near future such types of microbial strains have been synthesized using genetic engineering techniques that large-scale production using crude materials as well as their industrial scale application will be possible. In terms of lowering the cost of biosurfactant production, these discussed strategies could prove to be the most economical ones. A stricter positioning of these structures is, therefore, an authoritative step toward production enhancement, production procedure economization, and establishing an economically competitive and successful biosurfactant market, as well as addressing the solid waste disposal issue by efficient conversion of low-cost solid industrial and agricultural waste into revenue generating value-added products.

References

1 Khan, M.S., Zaidi, A., Wani, P.A., and Oves, M. (2009). Role of plant growth promoting rhizobacteria in the remediation of metal contaminated soils. *Environmental Chemistry Letters* 7 (1): 1–19.

2 Oliveira, S., Pessenda, L.C., Gouveia, S.E., and Favaro, D.I. (2011). Heavy metal concentrations in soils from a remote oceanic island, Fernando de Noronha, Brazil. *Anais da Academia Brasileira de Ciências* 83 (4): 1193–1206.

3 Sarma, H., Islam, N.F., Borgohain, P. et al. (2016). Localization of polycyclic aromatic hydrocarbons and heavy metals in surface soil of Asia's oldest oil and gas drilling site in Assam, northeast India: implications for the bio economy. *Emerging Contaminants* 2 (3): 119–127.

4 Sarma, H., Sonowa, S., and Prasad, M.N.V. (2019). Plant-microbiome assisted and biochar-amended remediation of heavy metals and polyaromatic compounds – A microcosmic study. *Ecotoxicology and Environmental Safety* 176: 288–299.

5 Tian, H.Z., Lu, L., Cheng, K. et al. (2012). Anthropogenic atmospheric nickel emissions and its distribution characteristics in China. *Science of the Total Environment* 417: 148–157.
6 Dixit, R., Malaviya, D., Pandiyan, K. et al. (2015). Bioremediation of heavy metals from soil and aquatic environment: an overview of principles and criteria of fundamental processes. *Sustainability* 7 (2): 2189–2212.
7 Gaur, N., Flora, G., Yadav, M., and Tiwari, A. (2014). A review with recent advancements on bioremediation-based abolition of heavy metals. *Environmental Science: Processes and Impacts* 16 (2): 180–193.
8 Sarma, H. and Prasad, M.N.V. (2015). Plant-microbe association-assisted removal of heavy metals and degradation of polycyclic aromatic hydrocarbons. In: *Petroleum Geosciences: Indian Contexts*, 219–236. Springer https://doi.org/10.1007/978-3-319-03119-4_10. ISBN: 978-3-319-03118-7.
9 Sarma, H. and Prasad, M.N.V. (2016). Chapter 24 – Phytomanagement of polycyclic aromatic hydrocarbons and heavy metals-contaminated sites in Assam, North Eastern State of India, for boosting bioeconomy. In: *Bioremediation and Bioeconomy* (ed. M.N.V. Prasad), 609–626. Elsevier https://doi.org/10.1016/B978-0-12-802830-8.00024-1.
10 Tak, H.I., Ahmad, F., and Babalola, O.O. (2013). Advances in the application of plant growth-promoting rhizobacteria in phytoremediation of heavy metals. In: *Reviews of Environmental Contamination and Toxicology*, vol. 223, 33–52. New York, NY: Springer.
11 Sarma, H., Islam, N.F., and Prasad, M.N.V. (2017). Plant-microbial association in petroleum and gas exploration sites in the state of Assam, north-east India – Significance for bioremediation. *Environmental Science and Pollution Research* 24 (9): 8744–8758.
12 Mani, D. and Kumar, C. (2014). Biotechnological advances in bioremediation of heavy metals contaminated ecosystems: an overview with special reference to phytoremediation. *International Journal of Environmental Science and Technology* 11 (3): 843–872.
13 Singh, P. and Cameotra, S.S. (2004). Enhancement of metal bioremediation by use of microbial surfactants. *Biochemical and Biophysical Research Communications* 319 (2): 291–297.
14 Sekhar, K.C., Kamala, C.T., Chary, N.S. et al. (2004). Removal of lead from aqueous solutions using an immobilized biomaterial derived from a plant biomass. *Journal of Hazardous Materials* 108 (1-2): 111–117.
15 Sarma, H. and Prasad, M.N.V. (2018). Metabolic engineering of rhizobacteria associated with plants for remediation of toxic metals and metalloids. In: *Transgenic Plant Technology* (ed. M.N.V. Prasad). Elsevier. eBook ISBN: 9780128143902, Paperback ISBN: 9780128143896.
16 Dahrazma, B. and Mulligan, C.N. (2007). Investigation of the removal of heavy metals from sediments using rhamnolipid in a continuous flow configuration. *Chemosphere* 69 (5): 705–711.
17 Mulligan, C.N., Yong, R.N., and Gibbs, B.F. (2001). Heavy metal removal from sediments by biosurfactants. *Journal of Hazardous Materials* 85 (1–2): 111–125.
18 Juwarkar, A.A., Nair, A., Dubey, K.V. et al. (2007). Biosurfactant technology for remediation of cadmium and lead contaminated soils. *Chemosphere* 68 (10): 1996–2002.
19 Herman, D.C., Artiola, J.F., and Miller, R.M. (1995). Removal of cadmium, lead, and zinc from soil by a rhamnolipid biosurfactant. *Environmental Science and Technology* 29 (9): 2280–2285.
20 Tan, H., Champion, J.T., Artiola, J.F. et al. (1994). Complexation of cadmium by a rhamnolipid biosurfactant. *Environmental Science and Technology* 28 (13): 2402–2406.
21 Saikia, R.R., Deka, S., Deka, M., and Sarma, H. (2012). Optimization of environmental factors for improved production of rhamnolipid biosurfactant by *Pseudomonas aeruginosa* RS29 on glycerol. *Journal of Basic Microbiology* 52 (4): 446–457.
22 Mukherjee, S., Das, P., and Sen, R. (2006). Towards commercial production of microbial surfactants. *Trends in Biotechnology* 24 (11): 509–515.

23 Cameotra, S.S., Makkar, R.S., Kaur, J., and Mehta, S.K. (2010). Synthesis of biosurfactants and their advantages to microorganisms and mankind. In: *Biosurfactants. Advances in Experimental Medicine and Biology*, vol. 672 (ed. R. Sen), 261–280. New York, NY: Springer.

24 Shekhar, S., Sundaramanickam, A., and Balasubramanian, T. (2015). Biosurfactant producing microbes and their potential applications: A review. *Critical Reviews in Environmental Science and Technology* 45 (14): 1522–1554.

25 Rodrigues, L.R. (2015). Microbial surfactants: fundamentals and applicability in the formulation of nano-sized drug delivery vectors. *Journal of Colloid and Interface Science* 449: 304–316.

26 Desai, J.D. and Banat, I.M. (1997). Microbial production of surfactants and their commercial potential. *Microbiology and Molecular Biology Review* 61 (1): 47–64.

27 Lin, S.C. (1996). Biosurfactants: Recent advances. *Journal of Chemical Technology and Biotechnology: International Research in Process, Environmental and Clean Technology* 66 (2): 109–120.

28 Parkinson, M. (1985). Bio-surfactants. *Biotechnology Advances* 3 (1): 65–83.

29 Tabatabaei, M. (2015) Design and fabrication of integrated plasmonic platforms for ultra-sensitive molecular and biomolecular detections. Doctorate Thesis.

30 Leuchtle, B., Xie, W., Zambanini, T. et al. (2015). Critical factors for microbial contamination of domestic heating oil. *Energy & Fuels* 29 (10): 6394–6403.

31 Miller, R.M. (1995). Biosurfactant-facilitated remediation of metal-contaminated soils. *Environmental Health Perspectives* 103: 59–62.

32 Rufino, R.D., Luna, J.M., Campos-Takaki, G.M. et al. (2012). Application of the biosurfactant produced by *Candida lipolytica* in the remediation of heavy metals. *Chemical Engineering* 27: 61–66.

33 Willumsen, P.A. and Karlson, U. (1996). Screening of bacteria, isolated from PAH-contaminated soils, for production of biosurfactants and bioemulsifiers. *Biodegradation* 7 (5): 415–423.

34 Tabatabaei, A., Nouhi, A.A., Sajadian, V., and Mazaheri, A.M. (2005). Isolation of biosurfactant producing bacteria from oil reservoirs. *Iranian Journal of Environmental Health Science and Engineering* 2 (1): 6–12.

35 Burger, M.M., Glaser, L., and Burton, R.M. (1963). The enzymatic synthesis of a rhamnose-containing glycolipid by extracts of *Pseudomonas aeruginosa*. *Journal of Biological Chemistry* 238 (8): 2595–2602.

36 Cooper, D.G. and Paddock, D.A. (1984). Production of a biosurfactant from *Torulopsis bombicola*. *Applied Environmental Microbiology* 47 (1): 73–176.

37 Ristau, E. and Wagner, F. (1983). Formation of novel anionic trehalosetetraesters from *Rhodococcus erythropolis* under growth limiting conditions. *Biotechnology Letters* 5 (2): 95–100.

38 Cooper, D.G., Macdonald, C.R., Duff, S.J.B., and Kosaric, N. (1981). Enhanced production of surfactin from *Bacillus subtilis* by continuous product removal and metal cation additions. *Applied Environmental Microbiology* 42 (3): 408–412.

39 Kretschmer, A., Bock, H., and Wagner, F. (1982). Chemical and physical characterization of interfacial-active lipids from *Rhodococcus erythropolis* grown on n-alkanes. *Applied Environmental. Microbiology* 44 (4): 864–870.

40 Casas, J.A., de Lara, S.G., and Garcia-Ochoa, F. (1997). Optimization of a synthetic medium for *Candida bombicola* growth using factorial design of experiments. *Enzyme and Microbial Technology* 21 (3): 21–229.

41 Thanomsub, B., Watcharachaipong, T., Chotelersak, K. et al. (2004). Monoacylglycerols: glycolipid biosurfactants produced by a thermotolerant yeast, *Candida ishiwadae*. *Journal of Applied Microbiology* 96 (3): 588–592.

42 Sarubbo, L.A., Farias, C.B., and Campos-Takaki, G.M. (2007). Co-utilization of canola oil and glucose on the production of a surfactant by *Candida lipolytica*. *Current Microbiology* 54 (1): 68–73.

43 Konishi, M., Fukuoka, T., Morita, T. et al. (2008). Production of new types of sophorolipids by *Candida batistae*. *Journal of Oleo Science* 57 (6): 359–369.

44 Alejandro, C.S., Humberto, H.S., and Maria, J.F. (2011). Production of glycolipids with antimicrobial activity by *Ustilago maydis* FBD12 in submerged culture. *African Journal of Microbiol Research* 5: 2512–2523.

45 Chandran, P. and Das, N. (2010). Biosurfactant production and diesel oil degradation by yeast species *Trichosporon asahii* isolated from petroleum hydrocarbon contaminated soil. *International Journal of Engineering Science and Technology* 2 (12): 6942–6953.

46 Rosenberg, E. and Ron, E.Z. (1999). High-and low-molecular-mass microbial surfactants. *Applied Microbiology and Biotechnology* 52 (2): 154–162.

47 Saenz-Marta, C.I., de Lourdes Ballinas-Casarrubias, M., Rivera-Chavira, B.E., and Nevarez-o, G.V. (2015). Biosurfactants as useful tools in bioremediation. In: *Advances in Bioremediation of Wastewater and Polluted Soil*, 2e, 94–109. Rijeka, Crotia (InTechOpen) https://doi.org/10.5772/60751.

48 Van Ginkel, C.G. (1996). Complete degradation of xenobiotic surfactants by consortia of aerobic microorganisms. *Biodegradation* 7 (2): 151–164.

49 Chen, S.Y., Wei, Y.H., and Chang, J.S. (2007). Repeated pH-stat fed-batch fermentation for rhamnolipid production with indigenous *Pseudomonas aeruginosa* S2. *Applied Microbiology and Biotechnology* 76 (1): 67–74.

50 Veenanadig, N.K., Gowthaman, M.K., and Karanth, N.G.K. (2000). Scale up studies for the production of biosurfactant in packed column bioreactor. *Bioprocess Engineering* 22 (2): 95–99.

51 Karanth, N.G.K., Deo, P.G., and Veenanadig, N.K. (1999). Microbial production of biosurfactants and their importance. *Current Science* 77: 116–126.

52 Rahman, K.S.M., Rahman, T.J., McClean, S. et al. (2002). Rhamnolipid biosurfactant production by strains of *Pseudomonas aeruginosa* using low-cost raw materials. *Biotechnology Progress* 18 (6): 1277–1281.

53 Rahman, K.S.M., Rahman, T.J., Banat, I.M. et al. (2007). Bioremediation of petroleum sludge using bacterial consortium with biosurfactant. In: *Environmental Bioremediation Technologies* (eds. S.N. Singh and R.D. Tripathi), 391–408. Berlin, Heidelberg: Springer.

54 Asmer, H.J., Lang, S., Wagner, F., and Wray, V. (1988). Microbial production, structure elucidation and bioconversion of sophorose lipids. *Journal of the American Oil Chemists Society* 65 (9): 1460–1466.

55 Besson, F., Peypoux, F., Michel, G., and Delcambe, L. (1976). Characterization of iturin A in antibiotics from various strains of *Bacillus subtilis*. *The Journal of Antibiotics* 29 (10): 1043–1049.

56 Nitschke, M. and Pastore, G.M. (2006). Production and properties of a surfactant obtained from *Bacillus subtilis* grown on cassava wastewater. *Bioresource Technology* 97 (2): 336–341.

57 Gautam, K.K. and Tyagi, V.K. (2006). Microbial surfactants: A review. *Journal of Oleo Science* 55 (4): 155–166.

58 Zinjarde, S.S. and Pant, A. (2002). Emulsifier from a tropical marine yeast, *Yarrowia lipolytica* NCIM 3589. *Journal of Basic Microbiology: An International Journal on Biochemistry, Physiology, Genetics, Morphology, and Ecology of Microorganisms* 42 (1): 67–73.

59 Raza, Z.A., Rehman, A., Khan, M.S., and Khalid, Z.M. (2007). Improved production of biosurfactant by a *Pseudomonas aeruginosa* mutant using vegetable oil refinery wastes. *Biodegradation* 18 (1): 115–121.

60 Ilori, M.O., Amobi, C.J., and Odocha, A.C. (2005). Factors affecting biosurfactant production by oil degrading *Aeromonas* spp. isolated from a tropical environment. *Chemosphere* 61 (7): 985–992.

61 Ashby, R.D., McAloon, A.J., Solaiman, D.K. et al. (2013). A process model for approximating the production costs of the fermentative synthesis of sophorolipids. *Journal of Surfactants and Detergents* 16 (5): 683–691.

62 Rodrigues, L.R., Teixeira, J.A., and Oliveira, R. (2006). Low-cost fermentative medium for biosurfactant production by probiotic bacteria. *Biochemical Engineering Journal* 32 (3): 135–142.

63 Daniels, L., Linhardt, R.J., Bryan, B.A., Mayerl, F. and Pickenhagen, W., Method for producing rhamnose. University of Iowa Research Foundation (UIRF) (1990) US Patent 4,933,281.

64 Calvo, C., Toledo, F.L., Pozo, C. et al. (2004). Biotechnology of bioemulsifiers produced by micro-organisms. *Journal of Food Agriculture and Environment* 2 (3): 238–243.

65 Kosaric, N., Cairns, W.L., Gray, N.C.C. et al. (1984). The role of nitrogen in multi organism strategies for biosurfactant production. *Journal of the American Oil Chemists' Society* 61 (11): 1735–1743.

66 Luna, J.M., Rufino, R.D., and Sarubbo, L.A. (2016). Biosurfactant from *Candida sphaerica* UCP0995 exhibiting heavy metal remediation properties. *Process Safety and Environmental Protection* 102: 558–566.

67 Mulligan, C.N., Yong, R.N., and Gibbs, B.F. (1999). Removal of heavy metals from contaminated soil and sediments using the biosurfactant surfactin. *Journal of Soil Contamination* 8 (2): 231–254.

68 Singh, A.K. and Cameotra, S.S. (2013). Efficiency of lipopeptide biosurfactants in removal of petroleum hydrocarbons and heavy metals from contaminated soil. *Environmental Science and Pollution Research* 20 (10): 7367–7376.

69 Das, P., Mukherjee, S., and Sen, R. (2009). Biosurfactant of marine origin exhibiting heavy metal remediation properties. *Bioresource Technology* 100 (20): 4887–4890.

70 Rufino, R.D., Rodrigues, G.I.B., Campos-Takaki, G.M. et al. (2011). Application of a yeast biosurfactant in the removal of heavy metals and hydrophobic contaminant in a soil used as slurry barrier. *Applied and Environmental Soil Science* 2011: 1–7.

71 Diaz, M.A., De Ranson, I.U., Dorta, B. et al. (2015). Metal removal from contaminated soils through bioleaching with oxidizing bacteria and rhamnolipid biosurfactants. *Soil and Sediment Contamination: An International Journal* 24 (1): 16–29.

72 Mekwichai, P., Tongcumpou, C., Kittipongvises, S., and Tuntiwiwattanapun, N. (2020). Simultaneous biosurfactant-assisted remediation and corn cultivation on cadmium-contaminated soil. *Ecotoxicology and Environmental Safety* 192: 110–298.

73 Grand View Research (2015). Available at: https://www.grandviewresearch.com/industry-analysis/biosurfactants-industry. Accessed March 8, 2018.

74 Global Market Insights (2018). Available at: https://www.gminsights.com/industry-analysis/biosurfactantsmarket-report. Accessed March 8, 2018.

75 Markets and Markets (2017). Available at: https://www.marketsandmarkets.com/Market-Reports/biosurfactant-market-163644922.html. Accessed March 8, 2018.

76 Research and Markets (2017). Available at: https://www.researchandmarkets.com/reports/4437552/biosurfactants-marketby-type-glycolipids#pos-1. Accessed March 8, 2018.

77 Zouboulis, A.I., Loukidou, M.X., and Matis, K.A. (2004). Biosorption of toxic metals from aqueous solutions by bacteria strains isolated from metal-polluted soils. *Process Biochemistry* 39 (8): 909–916.

5

Application of Biosurfactants for Microbial Enhanced Oil Recovery (MEOR)

Jéssica Correia, Lígia R. Rodrigues, José A. Teixeira, and Eduardo J. Gudiña

CEB – Centre of Biological Engineering, University of Minho, Braga, Portugal

CHAPTER MENU

5.1 Energy Demand and Fossil Fuels

World population is expected to surpass 9000 million by 2040. At the same time, the economic growth is projected to continue in the upcoming decades, mainly due to the development of emerging economies, which will contribute to increase the demand for energy. In 2017, fossil fuels provided 81% of the total primary energy supply worldwide, distributed as follows: crude oil 32%; coal 27%; natural gas 22% [1]. Despite the increasing concerns regarding climate change and global warming, and the development and implementation of renewable energies, fossil fuels will remain the main energy source in the short and medium term. Total energy demand is expected to increase at an annual rate of 1.3% between 2018 and 2040, and crude oil production is expected to rise from 95 million barrels per day (bpd) in 2018 up to 106 million bpd in 2040 [1].

To fulfill the increasing energy demand in the upcoming years, it is necessary to have an efficient exploitation of the existing crude oil reserves. The estimated proven oil reserves worldwide (conventional oil that can be recovered from oil reservoirs using existing technologies) is 1.67×10^{12} barrels, and there is not expected to be the discovery of new oilfields in the future that could significantly increase these reserves. Furthermore, most of the already known oil reservoirs are approaching or will achieve their economic limit of exploitation in the near future [2]. According to the oil consumption profiles estimated for the next few years, this crude oil reserves will be exhausted in 50 years. Consequently, the development of technologies that allow the recovery of

Biosurfactants for a Sustainable Future: Production and Applications in the Environment and Biomedicine, First Edition. Edited by Hemen Sarma and Majeti Narasimha Vara Prasad.
© 2021 John Wiley & Sons Ltd. Published 2021 by John Wiley & Sons Ltd.

most of the oil present in the reservoirs is necessary, as well as the exploitation of unconventional crude oil reserves (oils that cannot be produced using conventional recovery technologies). Conventional oils represent about 30% of the total oil reserves, whereas unconventional oils represent nowadays around 5% of total oil production worldwide. However, in the future, due to the depletion of conventional, easy to recover crude oil reserves, it is expected that unconventional oil reserves will significantly contribute to the global energy supply [3].

Oil recovery operations involve three phases. During primary recovery, the natural stored energy of the reservoir, resulting mainly from the expansion of fluids (gas, oil, and water), drives crude oil to the extraction wells. As the reservoir pressure decreases, it is necessary to use pumps to assist in oil recovery. Primary recovery is the least expensive phase of oil recovery, as it does not require special infrastructures. It can produce between 5 and 20% of the original oil in place (OOIP), depending on the properties of the reservoir and crude oil [4]. When primary recovery becomes inefficient, it is necessary to inject water or gases to maintain the reservoir pressure and displace the remaining oil to the surface (secondary recovery). It is estimated that 50% of the oil produced worldwide is recovered through water flooding. This phase does not require great investment in new infrastructures (preexisting oil production wells can be used as injection wells). These techniques can extract up to 60% of the OOIP [5]. When secondary recovery becomes uneconomical, between 30 and 60% of the OOIP still remains entrapped in the oil reservoir, it will be necessary to apply tertiary oil recovery techniques, known as Enhanced Oil Recovery (EOR). The application of these techniques is also necessary to recover unconventional oil reserves (e.g. extra-heavy crude oils, bitumen, oil sands, oil shale). EOR techniques can be divided into gas injection, chemical flooding, and thermal recovery [6].

The chemical processes consist of the addition of chemical compounds to the injected water to increase oil recovery. The compounds used include polymers, surfactants, acids, and solvents. Polymers are added to the injected water in order to increase its viscosity and reduce its mobility, improving the water–oil mobility ratio. This results in a more uniform oil displacement front and a more efficient displacement of the entrapped oil. Furthermore, polymers block the high-permeability channels of the oil reservoir, reducing its permeability and directing the injected water to oil-rich channels [7]. Surfactants are amphiphilic compounds, containing both hydrophilic and hydrophobic moieties. Due to their structure, surfactants accumulate at the interface between the crude oil and water, reducing the interfacial tension (IFT) and promoting the displacement of the entrapped oil by the injected water. Furthermore, surfactants emulsify crude oil, allowing the mobilization of oil droplets by the injected water. Additionally, the adsorption of surfactants to the reservoir rock alters its wettability, changing the distribution of oil and water in the rock/fluid system [8]. Surfactants are commonly used in EOR in combination with alkalis, which improve their activity by providing the appropriate environment to reduce the IFT. Furthermore, alkalis react with acidic components of crude oil to form surface-active compounds *in situ,* which contribute to decrease the IFT, reducing the amount of surfactant required to mobilize the entrapped oil. Surfactants can also be used in combination with polymers (surfactant–polymer flooding or alkali–surfactant–polymer flooding) to achieve high oil recoveries through the synergistic effect of combining different oil recovery mechanisms in a single treatment [9]. Acids and solvents alter the porosity and permeability of oil reservoirs as they can dissolve the reservoir rocks, releasing part of the entrapped oil. They can also act as emulsifiers and reduce crude oil viscosity [10].

Gases (natural gas, flue gas, CO_2, N_2) are used in tertiary oil recovery, not to repressurize the oil reservoir as in secondary oil recovery, but to reduce the oil density and viscosity, and to reduce the IFT between the displacing (injected water) and displaced fluid (crude oil). Among the different gases, CO_2 is the most widely used [5].

Regarding the thermal processes, the objective is to increase the reservoir temperature, in order to reduce the viscosity of crude oil and consequently improve its mobility through the porous network. This can be achieved through the injection of hot water or steam (cyclic or continue steam injection) into the reservoir; another approach is *in situ* combustion of crude oil [3].

The main disadvantage of chemical EOR is the negative environmental impact, as most of the polymers and surfactants used exhibit low biodegradability and are toxic to microorganisms, plants, and aquatic ecosystems. Regarding the thermal techniques, they consume high amounts of energy [4, 6]. Consequently, in order to ensure secure global energy supplies and minimize the environmental impact, it is necessary to develop alternative technologies to conventional EOR.

5.2 Microbial Enhanced Oil Recovery (MEOR)

Microbial Enhanced Oil Recovery (MEOR) is a cost-effective and environmentally friendly alternative to conventional EOR. The idea of using microorganisms to increase the productivity of oil reservoirs was first proposed almost 100 years ago. The basis of this technology is the use of microorganisms to produce metabolites and promote activities that contribute to increase the oil recovery [11]. MEOR can be applied *in situ* or *ex situ*. In *ex situ* approaches, selected microorganisms are used to produce metabolites (mainly biosurfactants and biopolymers) out of the reservoir. Subsequently, those metabolites (whole cultures, cell-free supernatants (CFS), or partially purified metabolites) are injected into the oil reservoir as an alternative to the chemical surfactants and polymers used in conventional EOR. In the *in situ* approaches, selected microorganisms are injected into the oil reservoir together with appropriate nutrients to allow their growth. If appropriate microorganisms are present in the oil reservoir, only nutrients are injected. In this approach, the injected or indigenous microorganisms can produce biosurfactants, biopolymers, biomass, acids, and gases directly into the oil reservoir (Figure 5.1). Furthermore, they can alter the properties of crude oil (e.g. by degrading the heavy oil fractions) to improve its mobility through the porous reservoir rock [2]. MEOR is a cost-effective technology as agro-industrial residues and byproducts can be used as inexpensive substrates for microbial growth. Additionally, its application does not require the construction of new facilities, as the existing ones can be used to inject the metabolites, nutrients, or microorganisms into the oil reservoir. It is also environmentally friendly because biosurfactants and biopolymers exhibit higher biodegradability and lower toxicity when compared with their chemical counterparts [4].

In situ MEOR is the simplest approach from an economic and technical point of view, as it just consists of the injection of nutrients (and eventually microorganisms) into the oil reservoir; after that, the oil production operations are stopped for several weeks to allow the production of metabolites; finally, the oil production operations are resumed. One of the advantages of *in situ* MEOR is that the metabolic activities can continue for long times without the need of further interventions. In addition, several beneficial activities can occur at the same time, which is advantageous to improve oil recovery [5, 13]. However, this approach exhibits several limitations. One of the main drawbacks is the difficulty of finding microorganisms able to grow and produce the desired metabolites or activities in the oil reservoir. Although oil reservoirs harbor different microbial communities, they represent harsh environments that usually exhibit extreme conditions [5]. The most relevant factor for the application of *in situ* MEOR is the oil reservoir temperature, due to its significant effect on microbial growth and the production of metabolites. This parameter limits the reservoirs where this technology can be applied to those with temperatures below 90 °C [14]. Accordingly, thermophilic microorganisms seem to be more appropriate for application in the *in situ* MEOR [15, 16]. Furthermore, oil reservoirs are characterized by the lack of oxygen, which limits the useful microorganisms to anaerobic

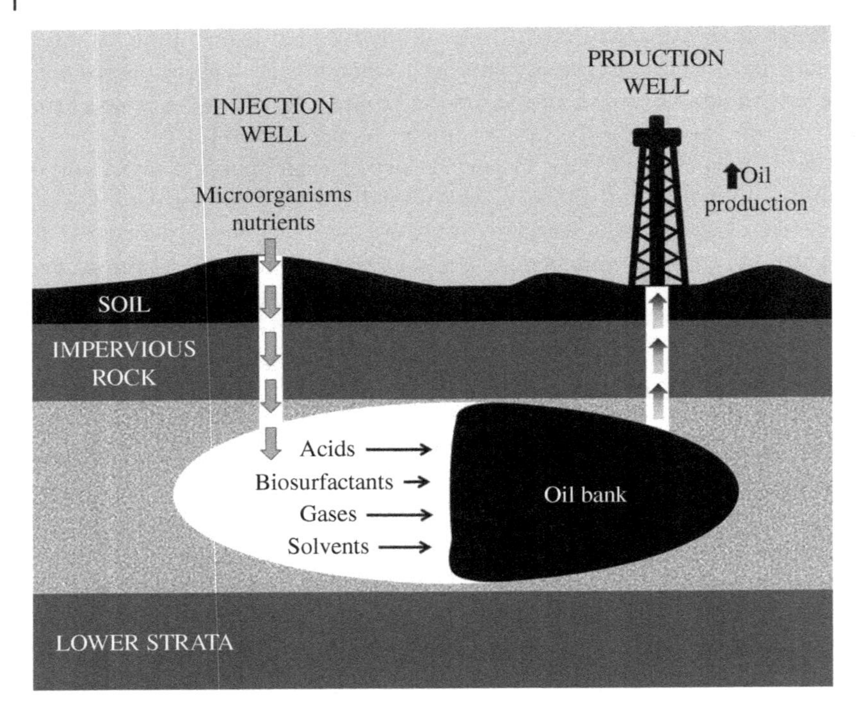

Figure 5.1 Schematic representation of *in situ* Microbial Enhanced Oil Recovery process (See insert for color representation of the figure). *Source:* Adapted from Sen [12].

or facultative anaerobic ones. Another parameter that must be taken into consideration is the effect of high pressure on microbial growth and metabolite production. Additionally, injected and formation water can contain different ions (Na^+, K^+, Ca^{2+}, Mg^{2+}, Cl^-, HCO_3^-, SO_4^{2-}), with salinities that can achieve 250 g/l [2]. Another drawback of *in situ* MEOR is that indigenous or injected microorganisms can be responsible for detrimental activities in the oil reservoir. Sulfate-reducing bacteria (SRB), manganese-oxidizing bacteria, or iron bacteria, which are common inhabitants of oil reservoirs, are responsible for the process known as microbiologically influenced corrosion (MIC), where the interaction of microorganisms with oil production and transportation infrastructures results in their corrosion and integrity reduction. This results in huge economic loses for the oil companies. Moreover, the production of H_2S by SRB results in the devaluation of crude oil [15]. These detrimental microorganisms can be activated by the injection of nutrients into the oil reservoir. Finally, *in situ* processes require more time as compared to *ex situ* ones, and the oil recovery operations must be stopped for several weeks. Besides, the development of accurate models to predict and follow the activity of microorganisms once injected into the oil reservoir is difficult. Although the *ex situ* approach is less favorable from an economic point of view, it allows a better control of the process and the operational times are shorter when compared with the *in situ* processes.

5.3 Mechanisms of Surfactant Flooding

Surfactant flooding is one of the most popular and effective EOR techniques. Two forces define the displacement efficiency of crude oil into the reservoir in oil recovery operations, namely viscous and capillary forces. Viscous forces have a positive effect in oil recovery, as they contribute to oil mobilization. In contrast, capillary forces have a negative effect, as they are responsible for trapping crude oil within rock pores. The capillary number (N_{ca}) is a dimensionless number that establishes the relationship between viscous and capillary forces, as expressed by Eq. (5.1):

$$Nca = \frac{Viscous\ forces}{Capillary\ forces} = \frac{V \times \mu}{\sigma \times \cos\theta} \tag{5.1}$$

where V and μ are the linear velocity (m/s) and the viscosity (mPa s) of the injected fluid, respectively; σ is the IFT between the injected and the displaced fluid (mN/m); and θ is the contact angle between the reservoir rock and the wetting phase [9, 17].

As N_{ca} increases, the ratio of viscous to capillary forces increases and the oil displacement efficiency also increases. For mature water-flooded reservoirs, N_{ca} values as low as 10^{-6}–10^{-7} have been reported, which results in high residual oil saturations. In order to substantially improve the mobilization of entrapped oil, N_{ca} needs to be increased by several orders of magnitude, up to 10^{-3}–10^{-1} [10].

N_{ca} can be increased by increasing the viscous forces, i.e. increasing the flow rate and/or increasing the viscosity of the injected fluid (through the addition of (bio)polymers to the injected water). However, these parameters can only be increased up to a certain limit, determined by the highest pressure that can be applied to the reservoir without causing damage to the formation. Taking into account these limitations, increasing the viscous forces usually allows increasing N_{ca} just by one order of magnitude [9]. The most feasible approach to increase N_{ca} is reducing the capillary forces through modification of the interactions between the injected and the displaced phases, and between the displaced phase and the reservoir rock. That can be achieved by reducing the high IFT values between the injected water and crude oil (10–40 mN/m), which is one of the main causes of low oil recoveries achieved during water flooding, to ultra-low values (10^{-2}–10^{-3} mN/m). This leads to a reduction of the capillary pressure and allows the deformation and mobilization of the oil droplets trapped into the rock pores. The other option is changing the wettability of the reservoir rock, from oil-wet ($105° < \theta < 180°$) to neutral-wet or water-wet ($0° < \theta < 75°$), which results in the spontaneous imbibition of water into the rock. This allows the desorption of crude oil from the rock and improves the sweep efficiency [17, 18]. Surfactants can be used to promote both effects: IFT reduction and rock wettability alteration. Examples of chemical surfactants commonly used in EOR include sulfate, sulfonate, carboxylate, ammonium bromide, and poly(ethylene/propylene) glycol derivatives [9].

5.4 Biosurfactants: An Alternative to Chemical Surfactants to Increase Oil Recovery

Microbial biosurfactants have attracted considerable attention as alternatives to the chemical surfactants used in EOR [5, 6]. They comprise a heterogeneous group of molecules that are usually classified into two main classes: low molecular weight and high molecular weight biosurfactants [19]. Low molecular weight biosurfactants are able to reduce the surface and IFT, and usually exhibit emulsifying activity. The most representative groups are glycolipids and lipopeptides. Among glycolipids, the most important are rhamnolipids (mainly produced by *Pseudomonas* sp. and *Burkholderia* sp.), sophorolipids (produced by yeasts such as *Candida bombicola*), trehalolipids (*Rhodococcus* sp.), and mannosylerythritol lipids (MELs, *Pseudozyma* sp.). In the case of glycolipids, the hydrophilic domain consists in a carbohydrate (one or two molecules of rhamnose in rhamnolipids; the disaccharides sophorose and trehalose in sophorolipids and trehalolipids, respectively; and 1/4-*O*-β-D-mannopyranosyl-erythritol in MELs), whereas the hydrophobic domain consists in one, two, or three (hydroxy) fatty acids, of variable length (8–22 carbons) and saturation degree, or mycolic acids in the case of trehalolipids. Lipopeptide biosurfactants comprise a cyclic peptide of 7 or 10 amino acids linked to a β-hydroxy or β-amino fatty acid of variable length (12–16 carbons). The most studied lipopeptide biosurfactant is surfactin, produced by *Bacillus subtilis* strains [20–24]. On the contrary, the high molecular weight biosurfactants (usually known as bioemulsifiers) exhibit remarkable emulsifying activity, but do not reduce the surface/IFT. They consist of macromolecules, usually polysaccharides, proteins, lipoproteins, lipopolysaccharides, or glycolipoproteins. Examples are emulsan and alasan, produced by *Acinetobacter* sp. [16, 25].

Lipopeptide biosurfactants (surfactin, lichenysin) and rhamnolipids are among the most studied biosurfactants for application in MEOR. Surfactin and rhamnolipids can reduce the IFT against crude oil up to 10^{-2}–10^{-3} mN/m. These biosurfactants exhibit critical micelle concentrations (CMC) as low as 10–20 mg/l, and are stable at high temperatures and salinities, making them suitable for application in oil reservoirs. Furthermore, they exhibit lower toxicity and higher biodegradability when compared with chemical surfactants [26–30]. Consequently, they have been widely studied as potential alternatives to the chemical surfactants to increase the productivity of oil reservoirs.

5.5 Biosurfactant MEOR: Laboratory Studies

In order to study the performance and the mechanisms of action of different compounds, metabolites, and microorganisms in oil recovery, different lab-scale models can be used to simulate the oil recovery operations, as recently reviewed by Rellegadla and co-workers [31]. One of the main advantages of using these models is that several assays can be performed simultaneously. The most common models are sand-pack columns and cores. Sand-pack columns are cylindrical devices usually made of glass, acrylic or stainless steel, of different volumes, which are filled with sand of different composition and size. Cores are cylindrical rock samples of variable size, usually collected from oil reservoirs or artificially constructed, which are placed into stainless steel containers (core holders). Both models are provided with inlet and outlet tubes to allow the injection and recovery of the different fluids. The utilization of these models usually occurs in five consecutive steps:

- The models are first saturated with water or formation brine, to determine parameters such as pore volume, porosity or permeability, and establish the flow paths.
- Crude oil is then injected into the models. The volume required to saturate the model with oil corresponds to the OOIP.
- Subsequently, the models are flooded with water or formation brine until no oil is observed in the effluent, simulating the secondary water flooding process. The oil that remains in the model is set as the residual oil after water flooding, which is the target of the tertiary oil recovery methods studied.
- The models are flooded with a solution of the selected EOR agent (e.g. (bio)surfactants, (bio) polymers, alkalis) (*ex situ* approach) or with selected microorganisms together with appropriate nutrients to allow their growth and the production of different metabolites (*in situ* approach). In

the last case, the system is closed after injection of the microorganisms and incubated at the reservoir conditions (temperature, pressure) for a variable shut-in time.
- Finally, another round of water flooding is conducted and the additional oil recovered (AOR) is quantified.

These studies are usually performed at controlled temperatures and different flow rates can be used. They can be performed under anaerobic conditions (e.g. through the injection of N_2), under oxygen-limiting conditions, or under aerobic conditions (e.g. through the injection of air). Moreover, they can be performed at atmospheric pressure or under pressurized conditions (to simulate the oil reservoir conditions).

Table 5.1 summarizes some of the laboratory-scale biosurfactant MEOR assays performed using the *ex situ* strategy. According to these studies, biosurfactants produced by different microorganisms resulted in AORs between 3 and 69%. However, the results obtained in the different assays are often not comparable due to the heterogeneity of the experimental conditions used (crude oil properties, biosurfactant concentration and purity, type of model and substrate chosen, temperature, pressure, and flow conditions). All of these variables result in a huge variety of results, thus limiting a clear perception of their value in real contexts.

Table 5.2 gathers some of the lab-scale biosurfactant MEOR assays performed using the *in situ* approach. In this case, AORs between 1 and 37% were reported. Again, comparison of the existing data is difficult due to the variability of the experimental conditions. Furthermore, in the *in situ* assays it becomes difficult to determine which factors lead to the mobilization of the residual oil, since several mechanisms can occur simultaneously to biosurfactant production [61]. For example, the bacterial consortium evaluated by Xia and co-workers [60] was found to produce methane besides producing biosurfactants, which can contribute to increase oil recovery by increasing the pressure inside the model. *Geobacillus toebii* R-32639 produced simultaneously biosurfactants and gases, was able to degrade heavy oil fractions, and increased oil recovery by selective plugging [51]. Other studies presented in Table 5.2 [13, 24, 56, 59] also reported the ability of the microorganisms used (*Luteimonas*, *Bacillus*, *Pseudomonas*, and *Chelatococcus* species) to degrade different crude oil fractions, which contributed to the reduction of the recovered oil's viscosity at the end of the assays.

Another important issue to take into consideration when applying *in situ* MEOR using exogenous strains is that the introduced microorganisms must be able to co-exist with the native bacterial communities present in the oil reservoir. Gao and co-workers [24] evaluated the interactions between *B. subtilis* M15-10-1 and the indigenous microbial population from the oil reservoir. The results obtained demonstrated that the exogenous strain was able to grow in the presence of the indigenous microbes and, at the same time, stimulated the growth of native hydrocarbon-degrading bacteria. On the other hand, the introduction of exogenous strains or their (bio)products into the oil wells may inhibit certain detrimental native microorganisms, such as SRB, which hinder the oil recovery process. El-Sheshtawy and co-workers [35] demonstrated that the lipopeptide biosurfactant produced by *Bacillus licheniformis* ATCC14580 at a concentration of 1% inhibited the growth of SRB.

The effectiveness of biosurfactants in EOR is highly dependent on the type of microorganism, the strategy (*in situ* or *ex situ*), and the experimental conditions used. Regarding the strategy used, some studies suggest that *in situ* oil recovery may yield better results [21, 44, 51], while others report the opposite [50]. In sand-pack column assays performed with the rhamnolipid-producing strain *Pseudomonas aeruginosa* WJ-1, a higher AOR was obtained in the *in situ* assays (5.3%) when compared with the *ex situ* ones (2.5%). According to the authors, this could be explained by

Table 5.1 Summary of different laboratory-scale MEOR *ex situ* assays performed using biosurfactants produced by different microorganisms. Interfacial tension (IFT) and emulsifying activity (E_{24}) measurements were performed using crude oil, unless stated otherwise. All the recovery assays were performed using crude oil, except those described in Liu et al. [27], Ashish [32], Câmara et al. [33], and de Araujo et al. [34], where a mixture of crude oil and kerosene (1 : 9 ratio), four stroke engine oil, diesel, or a mixture of crude oil and *n*-hexane (50%, w/w), respectively, were used.

Microorganism	Biosurfactant	CMC (mg/l)	ST (mN/m)	IFT (mN/m)	E_{24} (%)	API gravity (°)	Oil viscosity (cP)	Substrate	Porosity (%)	Temperature (°C)	Pressure (atm)	Flow rate (ml/min)	BS volume (PV)	Treatment - shut-in time	AOR (%)	Reference
Bacillus licheniformis ATCC14580	Lipopeptide	100.0	36.0	5.00^{1}	96.0^{6}	–	–	Sand (100 mesh)	21.0	35	1.0	2.50	0.6	Crude BS-1 day	11.0^{9}	[35]
B. licheniformis DS1	Lipopeptide	157.5	–	12.00	65.2	–	–	Berea sandstone	12.8	50	9.7	0.20	2.8	BS solution	5.1^{9}	[28]
Bacillus subtilis R1	Lipopeptide	20.0	29.0	–	–	-	–	Sand (100 mesh)	–	30	1.0	1.00	–	CFS	33.0	[36]
B. subtilis	Lipopeptide	–	–	≈1.10	≈95.0	–	598.0	Reservoir sandstone	15.6	65	1.0	0.05	0.5	BS-3 days	9.6	[37]
B. subtilis RI4914	Surfactin	–	–	0.20	–	20.3	–	Sand	27.3	–	1.0	–	5.0	Partially purified BS (2 g/l)	40.0	[38]
														CFS	69.0	
														CFS + biopolymer (4.2 g/l)	89.0	
B. subtilis 22.2	Surfactin	–	–	0.12	–	41.6	1.2	Berea sandstone	21.5	25	45.0	0.05	3.0	BS (1 g/l)	8.8	[39]
				0.06		27.3	2.4		21.0		24.0				3.1	
B. subtilis BR-15	Surfactin	–	20.2	–	69.4	–	5.1	Sand	18.1	–	1.0	–	0.6	CFS-2 days	66.3	[40]
Bacillus amyloliquefaciens SAS-1	Surfactin	–	22.9	–	68.6				19.0				0.6		56.9	
B. subtilis BS-37 (mutant strain)	Surfactin	20.0	27.7	–	≈60.0^{6}	–	–	Sand (20-30 mesh)	–	60	1.0	1.00	3.0	BS (30 mg/l)	10.7^{9}	[27]
Bacillus safensis J2	Surfactin	–	34.0	–	68.7^{6}	–	–	Sand (45 mesh)	–	–	1.0	–	–	BS	4.5^{9}	[26]
B. licheniformis ATCC10716	Surfactin	–	36.0	–	95.0^{6}	–	–	Sand	21.0	35	–	2.50	0.6	CFS-1 day	11.0^{9}	[23]

Candida tropicalis MTCC230 (adaptive strain)	Surfactin	32.5	32.0	–	62.0	–	–	Sand	15.1	–	1.0	2.50	0.6	CFS-1 day	39.8	[32]
B. licheniformis W16	Lichenysin A	–	24.3	2.47[3]	–	36.5	1.8	Berea sandstone	18.0-22.0	60	1.0	0.40	5.0	CFS	26.0	[41]
B. licheniformis Ali5	Lichenysin A	21.0	26.2	0.30[4]	66.4	35.0	–	Sand (100 mesh)	21.7	–	1.0	3.00	0.6	CFS-1 day	25.0[9]	[42]
Ochrobactrum pseudintermedium C1 and *Bacillus cereus* K1	Glycoprotein and Glycolipid	–	–	14.50	80.5	31.5	7.2 (70°C) 9.6 (40°C)	Sand (60–100 mesh)	20.5	70 40	–	–	1.0	Crude BS	46.9 40.9	[43]
Clostridium sp. N-4	Glycoprotein	100.0	28.3	–	100.0	18.2	1348.1	Sand	28.0	96	1.0	–	3.0	CFS	16.7[9]	[44]
Candida bombicola ATCC22214	Sophorolipid	–	28.6	2.13[3]	68.8	36.5	–	Berea sandstone	20.0	60	1.0	0.40	5.0	CFS	27.3	[22]
Candida albicans IMRU3669	Sophorolipid	–	45.0	–	65.0[6]	–	–	Sand	21.0	35	–	2.50	0.6	CFS-1 day	3.0[9]	[23]
Pseudomonas aeruginosa PBS	Rhamnolipid	–	23.8	–	70.0	–	–	Sand	16.9	–	1.0	–	0.7	CFS-1 day	56.2	[45]
P. aeruginosa HAK01	Rhamnolipid	120.0	28.1	2.50	≈67.0[7]	19.5	242	Glass micromodel	–	25	1.0	0.05	0.4	BS (120 mg/l)	27.0[9]	[46]
P. aeruginosa	Rhamnolipid	127.0	35.3	–	69.0	21.9	–	40% clay + 60% sand (65–100 mesh)	20.5	30	1.0	0.70	–	BS (254 mg/l)	11.9	[33]
P. aeruginosa Pa4	Rhamnolipid	–	–	32.40[5]	95.0	–	–	Sand (60–100 mesh)	–	34	1.0	–	–	BS (100 mg/l)	9.6	[47]
P. aeruginosa NCIM5514	Rhamnolipid	–	31.8	–	–	35.8	623.1	Berea sandstone	18.5	70	1.0	0.12	0.5	Partially purified BS	8.8	[30]
P. aeruginosa HATH	Rhamnolipid	120.0	26.0	2.00	85.0	34.0	–	Glass micromodel	41.5	80	1.0	0.50	2.0	BS (120 mg/l)	7.1	[20]
P. aeruginosa WJ-1	Rhamnolipid	–	–	5.50	–	32.7	67.0	Sand	27.0	35	1.0	–	0.5	CFS-10 days	2.5[9]	[21]
Ochrobactrum anthropi HM-1	Rhamnolipid	–	30.8	–	90.0	–	–	Sand	–	–	1.0	–	–	CFS	45.0[9]	[48]
Citrobacter freundii HM-2	Rhamnolipid	–	32.5	–	89.0										42.0[9]	

(Continued)

Table 5.1 (Continued)

Microorganism	Biosurfactant	CMC (mg/l)	ST (mN/m)	IFT (mN/m)	E_{24} (%)	API gravity (°)	Oil viscosity (cP)	Substrate	Porosity (%)	Temperature (°C)	Pressure (atm)	Flow rate (ml/min)	BS volume (PV)	Treatment - shut-in time	AOR (%)	Reference
Pseudoxanthomonas sp. G3	Rhamnolipid	730.0	–	11.30	72.9	22.0-28.0	58.9	Sand (50 mesh)	21.6	50	1.0	–	0.6	Crude BS-1 day	21.9^{9}	[49]
Acinetobacter junii BD	Rhamnolipid	–	30.3	–	57.0^{8}	–	–	Glass micromodel	–	35	1.0	0.48	–	CFS	13.4	[50]
B. safensis CCMA-560	Pumilacidin	96.2	56.8	11.30^{2}	–	–	11.3	Berea sandstone	20.7	–	68.0	–	2.7	Alternate injection BS solution (125 mg/l) and brine (30 g NaCl/l)	12.7	[34]
Escherichia coli W3110/pCA24N $OmpA^{+}$	Transmembrane protein (OmpA)	–	–	27.40^{5}	41.0–45.0	–	–	Sand (60–100 mesh)	–	34	1.0	–	–	OmpA (15.8 mg/l)	12.0	[47]
Geobacillus toebii R-32639	–	–	–	12.40	–	–	1.6	Cement and quartz sand (50 mesh) (1:4 ratio)	20.0-35.0	60	6.8	0.60	–	Bioproduct (followed by soaking over 7 days in brine)	5.4^{9}	[51]

-: Not reported. AOR: Additional oil recovery. BS: Biosurfactant. CFS: Cell-free supernatant. CMC: Critical micelle concentration. PV: Pore volume. ST: Surface tension.
[1,2,3,4,5] IFT measured against paraffin oil ([1]), a solution of crude oil with n-hexane (50%, w/w) ([2]), n-hexadecane ([3]), n-heptane ([4]), or a solution of crude oil with n-dodecane (25%, v/v) ([5]), respectively.
[6,7,8] Emulsifying activity measured using kerosene ([6]), n-hexane ([7]), or n-hexadecane ([8]), respectively.
[9] Additional oil recovery (AOR) values are the corrected values obtained after subtracting the AOR obtained in the control assays.

Table 5.2 Summary of different laboratory-scale MEOR *in situ* assays performed using different biosurfactant-producing microorganisms. Interfacial tension (IFT) and emulsifying activity (E_{24}) measurements were performed using crude oil, unless stated otherwise. All the recovery assays were performed using crude oil. Additional oil recovery (AOR) values are the corrected values obtained after subtracting the AOR obtained in the control assays, unless stated otherwise.

Microorganism	Biosurfactant	Yield (g/l)	CMC (mg/l)	ST (mN/m)	IFT (mN/m)	E_{24} (%)	API gravity (°)	Oil Viscosity (cP)	Substrate	Porosity (%)	Temperature (°C)	Pressure (atm)	Flow rate (ml/min)	Culture volume (PV)	Shut-in time (days)	AOR (%)	Reference
Luteimonas huabeiensis HB-2	Lipopeptide	0.20[a]	31.8	26.7	3.4	91.0	24.3	358.9	Artificial core	29.2	40	79.0	–	0.5	14.0	21.6[c]	[13]
Bacillus sp. W5	Lipopeptide	1.30[a]	–	35.3	–	–	–	45.4	Heterogeneous (three layers) core	19.8	60	1.0	–	0.4	7.0	9.8[c]	[52]
Bacillus subtilis M15-10-1	Lipopeptide	0.30–0.50	–	30.0	–	–	35.8	18.0	–	–	40	1.0	–	0.2	7.0	9.1	[24]
Fusant: *Bacillus mojavensis* JF-2 - *Pseudomonas stutzeri* DQ-1	Surfactin	0.40	60.0	31.2	–	58.6	–	–	Artificial core	25.0	39	1.0	1.000	1.0	10.0	4.2	[53]
Bacillus licheniformis	–	0.02[a]	–	23.8	0.9	–	37.0	21.6	Reservoir sandstone	16.0	50	1.0	0.200	0.3	7.0	22.2[c]	[54]
Bacillus amyloliquefaciens 702	–	1.60	–	25.9	26.1	84.7	28.2	5.6	Artificial core	20.1	39	1.0	0.500	0.5	10.0	1.1[d]	[55]
Clostridium sp. N-4	Glycoprotein	1.00	100.0	28.3	–	100.0	18.2	1348.1	Sand	31.4	96	1.0	–	1.3	14.0	36.7	[44]
Acinetobacter junii BD	Rhamnolipid	≈4.00	–	30.3	–	57.0[b]	–	–	Glass micromodel	–	35	1.0	0.008	–	1.5	9.6[c]	[50]
Pseudomonas aeruginosa WJ-1	Rhamnolipid	2.70[a]	–	–	1.7	–	32.7	67.0	Sand	27.0	35	1.0	–	0.5	10.0	5.3	[21]
Pseudomonas sp. SWP-4	Rhamnolipid	6.90[a]	–	22.7	–	58.3[b]	17.6	220000.0	Quartz sand (40–60 mesh)	30.2	30	1.0	5.000	–	1.0	24.3	[56]
P. aeruginosa DQ3	Rhamnolipid	0.06	–	33.8	–	58.0	28.2	10.0	Artificial core	15.3	42	1.0	0.200	0.3	15.0	4.4	[57]

(*Continued*)

Table 5.2 (Continued)

Microorganism	Biosurfactant	Yield (g/l)	CMC (mg/l)	ST (mN/m)	IFT (mN/m)	E_{24} (%)	API gravity (°)	Oil Viscosity (cP)	Substrate	Porosity (%)	Temperature (°C)	Pressure (atm)	Flow rate (ml/min)	Culture volume (PV)	Shut-in time (days)	AOR (%)	Reference
P. aeruginosa L6-1	–	2.20[a]	–	–	0.8	–	28.2	5.6	Oilfield sand	18.2	35	1.0	0.500	0.5	7.0	15.3	[58]
P. aeruginosa 709	–	8.20	–	26.7	6.0	93.2	28.2	5.6	Artificial core	19.1	39	1.0	0.500	0.5	10.0	7.1[d]	[55]
Chelatococcus daeguensis HB-4	–	–	–	66.2		89.0	–	500.0	Mixed silica sands (80–200 mesh)	–	38	79.0	–	0.5	14.0	35.5[c]	[59]
Methanogenic bacteria consortium	–	–	–	32.5	–	–	–	1823.9	–	26.8	50	1.0	–	1.0	100.0	15.0	[60]
Geobacillus toebii R-32639	–	–	–	–	12.4	–	–	1.6	Cement and quartz sand (50 mesh) (1:4 ratio)	20.0–35.0	60	6.8	0.600	–	7.0	11.2	[51]

-: Not reported. BS: Biosurfactant. CMC: Critical micelle concentration. PV: Pore volume. ST: Surface tension.

[a] Biosurfactant production using crude oil as the carbon source or in the presence of crude oil.

[b] Emulsifying activity measured using *n*-hexadecane.

[c] AOR value obtained without subtracting the AOR obtained in the control assays.

[d] AOR value calculated as the volume of displaced oil after MEOR divided by the volume of original oil in place.

the production of a higher amount of biosurfactant by the bacteria *in situ* (2.66 g rhamnolipid/l) when compared with the concentration injected in the *ex situ* assays (0.23 g rhamnolipid/l). Consequently, lower IFT values were achieved in the first case (1.7 mN/m *in situ* versus 5.5 mN/m *ex situ*). Furthermore, the diameter of the emulsified oil droplets was lower in the *in situ* assays (20–50 μm *versus* 40–80 μm *ex situ*), which facilitated the displacement of the entrapped oil [21]. In a similar way, a higher AOR was obtained with *G. toebii* R-32639 in the *in situ* assays (11.2%) when compared with the *ex situ* assays (5.4%), both of them performed at the same experimental conditions. This result was found to be the consequence of several oil recovery mechanisms occurring at the same time in the *in situ* assays [51]. On the other hand, lower AOR values were obtained in the *in situ* assays using *Acinetobacter junni* BD when compared with the *ex situ* assays, which can be due to a low biosurfactant production in the *in situ* assays performed under reservoir conditions [50].

The IFT values presented in Tables 5.1 and 5.2 are higher than the ultra-low IFT values necessary to increase N_{ca} to values that allow the mobilization of the entrapped oil. However, the efficiency of these biosurfactants in enhancing oil recovery has been widely demonstrated. In most cases, the IFT values determined in the lab are considerably higher when compared with the real IFT values achieved in the oil reservoir. One of the reasons is that the aging time necessary to allow the interaction of biosurfactants with the crude oil/aqueous phase/reservoir rock system is not always taken into consideration. Furthermore, the IFT values are usually determined as equilibrium IFT, which may differ from the dynamic IFT values achieved in the reservoir due to the fluid displacement. Moreover, besides reducing the IFT, biosurfactants also alter the wettability of the substrate from oil- or water-wet toward water-wet or more water-wet, which also contributes to mobilize the entrapped oil. This has been demonstrated for surfactin (θ reduction from 71° to 35°) [39], lichenysin A (56° to 19° and 47° to 17°) [41, 42], and rhamnolipids (168° to 56° and 106° to 7°) [21, 46].

The oil industry has taken advantage of the synergistic effect of combining polymers and surfactants in oil recovery. In a similar way, biosurfactants can be used in combination with (bio)polymers to improve the oil displacement efficiency. Qi and co-workers [52] studied the compatibility between the biosurfactant-producing strain *Bacillus* sp. W5 and a weak gel (a mixture of a polymer and a delayed cross-linker) for application in oil recovery. It was demonstrated that *Bacillus* sp. W5 did not change the rheological properties of the weak gel, and the weak gel did not affect the bacterial growth or the biosurfactant properties. Instead, it was found that oil recovery increased from 9.8 to 14.7% with the individual treatments, up to 23.6% when the weak gel and *Bacillus* sp. W5 were injected into the core simultaneously, demonstrating the potential of using biosurfactants and polymers together in MEOR. In a similar way, combining the biosurfactant-containing CFS from cultures of *B. subtilis* RI4914 with a partially purified biopolymer produced by the same strain increased oil recovery by 20% when compared with the effect of the CFS alone. AOR values obtained with the CFS were also higher than those obtained with solutions of the partially purified biosurfactant, probably due to the presence of compounds such as 2,3-butanediol, which acts synergistically with the biosurfactant in reducing the IFT [38].

Another alternative to improve the oil recovery rates is the use of engineered microorganisms better adapted to the oil reservoir conditions or with improved biosurfactant production yields. In that sense, the protoplasts fusion technology is an interesting tool that allows combining the desired properties from different microorganisms. The ability of *Bacillus mojavensis* JF-2 to produce a lipopeptide biosurfactant was combined with the ability of *Pseudomonas stutzeri* DQ1 to grow under anaerobic conditions. The resulting fusant produced a biosurfactant similar to that of the parental strain, under anaerobic conditions and temperatures up to 50 °C. *In situ* oil recovery assays confirmed the applicability of this engineered strain in MEOR, with AOR values up to

4.2% [53]. Álvarez Yela and co-workers [47] compared the performance of rhamnolipids (produced by *P. aeruginosa* Pa4) and the transmembrane protein OmpA (produced by an engineered *Escherichia coli* strain) in oil recovery. The highest AOR was obtained with OmpA (12%), which confirms the potential of using a protein with biosurfactant-like properties in MEOR. Mutant strains (*B. subtilis* BS-37 and *Candida tropicalis* MTCC230), with the ability of producing higher amounts of biosurfactant when compared with the respective parental strains, were also evaluated for application in *ex situ* MEOR [27, 32]. The results obtained demonstrated the feasibility of using the biosurfactants produced by these microorganisms in MEOR (Table 5.1).

Most of the studies presented in Tables 5.1 and 5.2 were conducted at temperatures up to 60 °C; however, most oil reservoirs exhibit higher temperatures. There is still a lack of information regarding microorganisms capable of growing and producing biosurfactants at temperatures usually found in oil reservoirs. Recently, a hyper-thermophilic *Clostridium* sp. strain, which can grow and produce biosurfactants at 96 °C, was evaluated for application in MEOR. The glycoprotein biosurfactant produced by this strain was stable at a wide range of temperatures (37–101 °C), pH values (5–10), and salinities (up to 13%), being a great candidate for MEOR applications. Furthermore, sand-pack column studies conducted at 96 °C yielded significant recovery rates in both *in situ* (37% AOR) and *ex situ* (17% AOR) assays [44].

Another limitation for the application of biosurfactants in MEOR is the high cost of the substrates used to grow the producing microorganisms, which results in high production costs. Whereas the price of synthetic surfactants can be around 1–3 US$/kg, the price of sophorolipid (the cheapest and most widely available biosurfactant) is around 35 US$/kg [62]. Consequently, several studies evaluated the use of alternative carbon and nitrogen sources for their production (e.g. sugarcane and date molasses, vegetable oils, waste cooking oils, starch, cornmeal, paraffin, corn steep powder), in order to reduce their production costs [13, 24, 26, 41, 48, 53, 56, 59]. In other cases, the composition of the culture medium or the operational conditions were optimized to increase their productivity [20, 46]. For instance, Gao et al. [24] used corn steep powder as the carbon source instead of glucose to grow *B. subtilis* M15-10-1 in the *in situ* MEOR assays, due to the high emulsifying activity obtained with this substrate. Sugarcane bagasse was used as the sole carbon source for the production of surfactin by *Bacillus safensis* J2 for application in the *ex situ* oil recovery assays. The produced biosurfactant was stable at different environmental conditions (30–90 °C, pH 7–12, salinities up to 8%), and resulted in 4.5% AOR. Furthermore, biodegradability and phyto-toxicity tests revealed that this biosurfactant is both biodegradable and non-toxic in the tested conditions [26]. This corroborates the potential application of biosurfactants in the petroleum industry as environmentally friendly alternatives to chemical surfactants.

5.6 Field Assays

Field trials often rely on *in situ* biosurfactant-mediated oil recovery, with several successful assays reported in different types of fields, the most recent ones occurring mainly in China. In Shengli oilfield, a number of recent trials were implemented where different nutrient solutions were injected to stimulate the growth of indigenous bacteria. It was found that, after treatment, *Pseudomonas* and *Bacillus* were among the dominant genera across the four blocks studied [63, 64]. Another trial reported in Xinjiang field also relied on the injection of nutrients to enhance oil recovery. Again, *Pseudomonas* was the dominant group during the trial, suggesting that biosurfactants were the primary oil recovery mechanism [65, 66] Additionally, bacteria from the genus

Acinetobacter, which have been reported as biosurfactant and bioemulsifier producers, were also enriched in the production wells after nutrient injection [67]. In Chunfeng oilfield, the recovery of ultra-heavy oil was enhanced through *in situ* MEOR. A selection of indigenous bacteria (*B. subtilis* XJZ2-1, *Pseudomonas* sp. XJZ3-1 and *Dietz coli* Z4M8-2) and one exogenous microbe (*B. subtilis* SLG5B10-17), together with nutrients and an activator solution, were injected into a production well over 7 days, followed by a shut-in period of 166 days. After 13 months, oil production increased from 0.4 to 4.7 tons/day, with a peak production of 17 tons/day and a water cut of 34%. Furthermore, the results obtained showed a longer effective oil recovery period in the wells where MEOR was applied when compared with the adjacent wells treated with steam [68].

More recently, a large-scale pilot test was performed in Baolige oilfield, which involved 78 injection wells and 169 production wells. Indigenous *B. licheniformis* LC and *Rhodococcus* sp. JH were selected in laboratory assays to be used in the field trial, due to the high surface activity of their biosurfactants. After four injection cycles, applied over 43 months, additional oil production reached a total of 2.1×10^5 tons. However, it was found that in 15% of the extraction wells, oil production did not increase significantly. This was attributed to the fact that some of the wells were located within formations with poor homogeneity or with poor connectivity in the underground network, which negatively affected the bacterial flow through the well [69]. In the same field, a smaller trial was initiated shortly after, this time involving only two injection wells and eight production wells. The biosurfactant-producing bacterium *Luteimonas huabeiensis* HB-2, which was found to reduce oil–water IFT and decrease oil viscosity in lab-scale assays, was used in this trial. The field test was carried out through two bacterial injection cycles, both followed by a period of water flooding. By the end of the two cycles 2300 tons of additional oil were recovered, enhancement that was attributed mainly to a reduction in oil viscosity [13].

These figures demonstrate the potential of *in situ* biosurfactant production to improve oil recovery. However, besides biosurfactants, both exogenous and indigenous microorganisms also produce other metabolites (gases, acids, and solvents) and biomass, which can contribute to the observed increase in oil production.

5.7 Current State of Knowledge, Technological Advances, and Future Perspectives

Biosurfactants have emerged as promising alternatives to the synthetic surfactants for application in several fields, including the oil industry. The remarkable surface-active properties exhibited by some of these biosurfactants, together with their stability at extreme environmental conditions and their environmentally friendly nature, make them potential candidates for application in EOR. Their ability to increase oil recovery has been demonstrated in numerous laboratory assays, using different models, oils with different properties, and a wide variety of experimental conditions. However, more filed assays are necessary to validate the results obtained in the lab.

Nowadays, the main drawback for the application of biosurfactants is their high production costs when compared with their chemical counterparts. Although in the last few years, important advances have been performed regarding the use of low-cost substrates for their biosynthesis, more studies are still necessary to optimize the operational conditions for their production and recovery, in order to make them competitive in the market. From this perspective, the application of *in situ* biosurfactant MEOR seems to be the most promising approach. However, the main limitation is the low availability of microorganisms with the ability of growing and producing biosurfactants at the oil reservoir conditions. Although some examples have been reported in the last few years, their efficiency has not been demonstrated yet in field assays. In that sense, genetic engineering could allow the construction of biosurfactant-producing microorganisms with the desired properties for this application.

Acknowledgements

This study was supported by PARTEX Oil and Gas and the Portuguese Foundation for Science and Technology (FCT) under the scope of the strategic funding of UIDB/04469/2020 unit and BioTecNorte operation (NORTE-01-0145-FEDER-000004) funded by the European Regional Development Fund under the scope of Norte 2020 – Programa Operacional Regional do Norte. The authors also acknowledge the Biomass and Bioenergy Research Infrastructure (BBRI)-LISBOA-01-0145-FEDER-022059, supported by the Operational Program for Competitiveness and Internationalization (PORTUGAL2020), by the Lisbon Portugal Regional Operational Program (Lisboa 2020), and by the North Portugal Regional Operational Program (Norte 2020) under the Portugal 2020 Partnership Agreement, through the European Regional Development Fund (ERDF).

References

1 IEA (2019). International Energy Agency: World Energy Outlook 2019. November 2019. https://www.iea.org/reports/world-energy-outlook-2019 (accessed 07/01/2020).
2 Nikolova, C. and Gutierrez, T. (2020). Use of microorganisms in the recovery of oil from recalcitrant oil reservoirs: current state of knowledge, technological advances and future perspectives. *Frontiers in Microbiology* 10: 2996. https://doi.org/10.3389/fmicb.2019.02996.
3 Zhang, J., Gao, H., and Xue, Q. (2020). Potential applications of microbial enhanced oil recovery to heavy oil. *Critical Reviews in Biotechnology* 40 (4): 459–474. https://doi.org/10.1080/07388551.2020.1739618.
4 Gudiña, E.J., Pereira, J.F.B., Rodrigues, L.R. et al. (2012). Isolation and study of microorganisms from oil samples for application in Microbial Enhanced Oil Recovery. *International Biodeterioration and Biodegradation* 68: 56–64. https://doi.org/10.1016/j.ibiod.2012.01.001.
5 Saravanan, A., Kumar, P.S., Vardhan, K.H. et al. (2020). A review on systematic approach for microbial enhanced oil recovery technologies: Opportunities and challenges. *Journal of Cleaner Production* 258: 120777. https://doi.org/10.1016/j.jclepro.2020.120777.
6 Niu, J., Liu, Q., Lv, J., and Peng, B. (2020). Review on microbial enhanced oil recovery: mechanisms, modelling and field trials. *Journal of Petroleum Science and Engineering* 192: 107350. https://doi.org/10.1016/j.petrol.2020.107350.
7 Couto, M.R., Gudiña, E.J., Ferreira, D. et al. (2019). The biopolymer produced by *Rhizobium viscosum* CECT908 is a promising agent for application in Microbial Enhanced Oil Recovery. *New Biotechnology* 49: 144–150. https://doi.org/10.1016/j.nbt.2018.11.002.
8 Pereira, J.F.B., Gudiña, E.J., Costa, R. et al. (2013). Optimization and characterization of biosurfactant production by *Bacillus subtilis* isolates towards microbial enhanced oil recovery applications. *Fuel* 111: 259–268. https://doi.org/10.1016/j.fuel.2013.04.040.
9 Negin, C., Ali, S., and Xie, Q. (2017). Most common surfactants employed in chemical enhanced oil recovery. *Petroleum* 3: 197–211. https://doi.org/10.1016/j.petlm.2016.11.007.
10 Tackie-Otoo, B.N., Ayoub-Mohammed, M.A., Yekeen, N., and Negash, B.M. (2020). Alternative chemical agents for alkalis, surfactants and polymers for enhanced oil recovery: Research trend and prospects. *Journal of Petroleum Science and Engineering* 187: 106828. https://doi.org/10.1016/j.petrol.2019.106828.

11 Beckman, J.W. (1926). The action of bacteria on mineral oil. *Industrial and Engineering Chemistry* 4: 23–26.

12 Sen, R. (2018). Biotechnology in petroleum recovery: The microbial EOR. *Progress in Energy and Combustion Science* 34 (6): 714–724. https://doi.org/10.1016/j.pecs.2008.05.001.

13 Ke, C.Y., Sun, W.J., Li, Y.B. et al. (2018). Microbial enhanced oil recovery in Baolige Oilfield using an indigenous facultative anaerobic strain *Luteimonas huabeiensis* sp. nov. *Journal of Petroleum Science and Engineering* 167: 160–167. https://doi.org/10.1016/j.petrol.2018.04.015.

14 Safdel, M., Anbaz, M.A., Daryasafar, A., and Jamialahmadi, M. (2017). Microbial enhanced oil recovery, a critical review on worldwide implemented field trials in different countries. *Renewable and Sustainable Energy Reviews* 74: 159–172. https://doi.org/10.1016/j.rser.2017.02.045.

15 Elumalai, P., Parthipan, P., Narenkumar, J. et al. (2019). Role of thermophilic bacteria (*Bacillus* and *Geobacillus*) on crude oil degradation and biocorrosion in oil reservoir environment. *3 Biotech* 9: 79. https://doi.org/10.1007/s13205-019-1604-0.

16 Tao, W., Lin, J., Wang, W. et al. (2020). Biodegradation of aliphatic and polycyclic aromatic hydrocarbons by the thermophilic bioemulsifier-producing *Aeribacillus pallidus* strain SL-1. *Ecotoxicology and Environmental Safety* 189: 109994. https://doi.org/10.1016/j.ecoenv.2019.109994.

17 Zulkifli, N.N., Mahmood, S.M., Akbari, S. et al. (2019). Evaluation of new surfactants for enhanced oil recovery applications in high-temperature reservoirs. *Journal of Petroleum Exploration and Production Technology* 10: 283–296. https://doi.org/10.1007/s13202-019-0713-y.

18 Ahmadi, M.A. and Shadizadeh, S.R. (2018). Spotlight on the new natural surfactant flooding in carbonate rock samples in low salinity condition. *Scientific Reports* 8: 10985. https://doi.org/10.1038/s41598-018-29321-w.

19 Gudiña, E.J., Pereira, J.F.B., Costa, R. et al. (2015). Novel bioemulsifier produced by a *Paenibacillus* strain isolated from crude oil. *Microbial Cell Factories* 14: 14. https://doi.org/10.1186/s12934-015-0197-5.

20 Amani, H. (2015). Study of enhanced oil recovery by rhamnolipids in a homogeneous 2D micromodel. *Journal of Petroleum Science and Engineering* 128: 212–219. https://doi.org/10.1016/j.petrol.2015.02.030.

21 Cui, Q.F., Sun, S.S., Luo, Y.J. et al. (2017). Comparison of *in-situ* and *ex-situ* microbial enhanced oil recovery by strain *Pseudomonas aeruginosa* WJ-1 in laboratory sand-pack columns. *Petroleum Science and Technology* 35 (21): 2044–2050. https://doi.org/10.1080/10916466.2017.1380042.

22 Elshafie, A.E., Joshi, S.J., Al-Wahaibi, Y.M. et al. (2015). Sophorolipids production by *Candida bombicola* ATCC22214 and its potential application in microbial enhanced oil recovery. *Frontiers in Microbiology* 6: 1324. https://doi.org/10.3389/fmicb.2015.01324.

23 El-Sheshtawy, H.S., Aiad, I., Osman, M.E. et al. (2016). Production of biosurfactants by *Bacillus licheniformis* and *Candida albicans* for application in microbial enhanced oil recovery. *Egyptian Journal of Petroleum* 25 (3): 293–298. https://doi.org/10.1016/j.ejpe.2015.07.018.

24 Gao, P., Li, G., Li, Y. et al. (2016). An exogenous surfactant-producing *Bacillus subtilis* facilitates indigenous microbial enhanced oil recovery. *Frontiers in Microbiology* 7: 186. https://doi.org/10.3389/fmicb.2016.00186.

25 Hongyan, H., Weiyao, Z., Zhiyong, S., and Ming, Y. (2017). Mechanisms of oil displacement by *Geobacillus stearothermophilus* producing bio-emulsifier for MEOR. *Petroleum Science and Technology* 35 (17): 1791–1798. https://doi.org/10.1080/10916466.2017.1320675.

26 Das, A.J. and Kumar, R. (2019). Production of biosurfactant from agro-industrial waste by *Bacillus safensis* J2 and exploring its oil recovery efficiency and role in restoration of diesel contaminated soil. *Environmental Technology and Innovation* 16: 100450. https://doi.org/10.1016/j.eti.2019.100450.

27 Liu, Q., Lin, J., Wang, W. et al. (2015). Production of surfactin isoforms by *Bacillus subtilis* BS-37 and its applicability to enhanced oil recovery under laboratory conditions. *Biochemical Engineering Journal* 93: 31–37. https://doi.org/10.1016/j.bej.2014.08.023.

28 Purwasena, I.A., Astuti, D.I., Syukron, M. et al. (2019). Stability test of biosurfactant produced by *Bacillus licheniformis* DS1 using experimental design and its application for MEOR. *Journal of Petroleum Science and Engineering* 183: 106383. https://doi.org/10.1016/j.petrol.2019.106383.

29 Shreve, G.S. and Makula, R. (2019). Characterization of a new rhamnolipid biosurfactant complex from *Pseudomonas* isolate DYNA270. *Biomolecules* 9: 885. https://doi.org/10.3390/biom9120885.

30 Varjani, S.J. and Upasani, V.N. (2016). Core flood study for enhanced oil recovery through *ex-situ* bioaugmentation with thermo- and halo-tolerant rhamnolipid produced by *Pseudomonas aeruginosa* NCIM5514. *Bioresource Technology* 220: 175–182. https://doi.org/10.1016/j.biortech.2016.08.060.

31 Rellegadla, S., Jain, S., and Agrawal, A. (2020). Oil reservoir simulating bioreactors: tools for understanding petroleum microbiology. *Applied Microbiology and Biotechnology* 104: 1035–1053. https://doi.org/10.1007/s00253-019-10311-5.

32 Ashish, D.M. (2018). Application of biosurfactant produced by an adaptive strain of *Candida tropicalis* MTCC230 in microbial enhanced oil recovery (MEOR) and removal of motor oil from contaminated sand and water. *Journal of Petroleum Science and Engineering* 170: 40–48. https://doi.org/10.1016/j.petrol.2018.06.034.

33 Câmara, J.M.D.A., Sousa, M.A.S.B., Barros-Neto, E.L., and Oliveira, M.C.A. (2019). Application of rhamnolipid biosurfactant produced by *Pseudomonas aeruginosa* in microbial-enhanced oil recovery (MEOR). *Journal of Petroleum Exploration and Production Technology* 9: 2333–2341. https://doi.org/10.1007/s13202-019-0633-x.

34 de Araujo, L.L.G.C., Sodré, L.G.P., Brasil, L.R. et al. (2019). Microbial enhanced oil recovery using a biosurfactant produced by *Bacillus safensis* isolated from mangrove microbiota – Part I biosurfactant characterization and oil displacement test. *Journal of Petroleum Science and Engineering* 180: 950–957. https://doi.org/10.1016/j.petrol.2019.06.031.

35 El-Sheshtawy, H.S., Aiad, I., Osman, M.E. et al. (2015). Production of biosurfactant from *Bacillus licheniformis* for microbial enhanced oil recovery and inhibition the growth of sulfate reducing bacteria. *Egyptian Journal of Petroleum* 24 (2): 155–162. https://doi.org/10.1016/j.ejpe.2015.05.005.

36 Jha, S.S., Joshi, S.J., and Geetha, S.J. (2016). Lipopeptide production by *Bacillus subtilis* R1 and its possible applications. *Brazilian Journal of Microbiology* 47 (4): 955–964. https://doi.org/10.1016/j.bjm.2016.07.006.

37 Song, Z., Zhu, W., Sun, G., and Blanckaert, K. (2015). Dynamic investigation of nutrient consumption and injection strategy in microbial enhanced oil recovery (MEOR) by means of large-scale experiments. *Applied Microbiology and Biotechnology* 99 (15): 6551–6561. https://doi.org/10.1007/s00253-015-6586-1.

38 Fernandes, P.L., Rodrigues, E.M., Paiva, F.R. et al. (2016). Biosurfactant, solvents and polymer production by *Bacillus subtilis* RI4914 and their application for enhanced oil recovery. *Fuel* 180: 551–557. https://doi.org/10.1016/j.fuel.2016.04.080.

39 Hadia, N.J., Ottenheim, C., Li, S. et al. (2019). Experimental investigation of biosurfactant mixtures of surfactin produced by *Bacillus subtilis* for EOR application. *Fuel* 251: 789–799. https://doi.org/10.1016/j.fuel.2019.03.111.

40 Sharma, R., Singh, J., and Verma, N. (2018). Production, characterization and environmental applications of biosurfactants from *Bacillus amyloliquefaciens* and *Bacillus subtilis*. *Biocatalysis and Agricultural Biotechnology* 16: 132–139. https://doi.org/10.1016/j.bcab.2018.07.028.

41 Joshi, S.J., Al-Wahaibi, Y.M., Al-Bahry, S.N. et al. (2016). Production, characterization, and application of *Bacillus licheniformis* W16 biosurfactant in enhancing oil recovery. *Frontiers in Microbiology* 7: 1853. https://doi.org/10.3389/fmicb.2016.0185.3.

42 Ali, N., Wang, F., Xu, B. et al. (2019). Production and application of biosurfactant produced by *Bacillus licheniformis* Ali5 in enhanced oil recovery and motor oil removal from contaminated sand. *Molecules* 24 (24): 4448. https://doi.org/10.3390/molecules24244448.

43 Bhattacharya, M., Guchhait, S., Biswas, D., and Singh, R. (2019). Evaluation of a microbial consortium for crude oil spill bioremediation and its potential uses in enhanced oil recovery. *Biocatalysis and Agricultural Biotechnology* 18: 101034. https://doi.org/10.1016/j.bcab.2019.101034.

44 Arora, P., Kshirsagar, P.R., Rana, D.P., and Dhakephalkar, P.K. (2019). Hyperthermophilic *Clostridium* sp. N-4 produced a glycoprotein biosurfactant that enhanced recovery of residual oil at 96°C in lab studies. *Colloids and Surfaces B: Biointerfaces* 182: 110372. https://doi.org/10.1016/j.colsurfb.2019.110372.

45 Sharma, R., Singh, J., and Verma, N. (2018). Optimization of rhamnolipid production from *Pseudomonas aeruginosa* PBS towards application for microbial enhanced oil recovery. *3 Biotech* 8 (1): 20. https://doi.org/10.1007/s13205-017-1022-0.

46 Khademolhosseini, R., Jafari, A., Mousavi, S.M. et al. (2019). Physicochemical characterization and optimization of glycolipid biosurfactant production by a native strain of *Pseudomonas aeruginosa* HAK01 and its performance evaluation for the MEOR process. *RSC Advances* 9 (14): 7932–7947. https://doi.org/10.1039/C8RA10087J.

47 Alvarez-Yela, A.C., Tibaquirá-Martínez, M.A., Rangel-Piñeros, G.A. et al. (2016). A comparison between conventional *Pseudomonas aeruginosa* rhamnolipids and *Escherichia coli* transmembrane proteins for oil recovery enhancing. *International Biodeterioration and Biodegradation* 112: 59–65. https://doi.org/10.1016/j.ibiod.2016.04.033.

48 Ibrahim, H.M.M. (2018). Characterization of biosurfactants produced by novel strains of *Ochrobactrum anthropi* HM-1 and *Citrobacter freundii* HM-2 from used engine oil-contaminated soil. *Egyptian Journal of Petroleum* 27 (1): 21–29. https://doi.org/10.1016/j.ejpe.2016.12.005.

49 Astuti, D.I., Purwasena, I.A., Putri, R.E. et al. (2019). Screening and characterization of biosurfactant produced by *Pseudoxanthomonas* sp. G3 and its applicability for enhanced oil recovery. *Journal of Petroleum Exploration and Production Technology* 9: 2279–2289. https://doi.org/10.1007/s13202-019-0619-8.

50 Dong, H., Xia, W., Dong, H. et al. (2016). Rhamnolipids produced by indigenous *Acinetobacter junii* from petroleum reservoir and its potential in enhanced oil recovery. *Frontiers in Microbiology* 7: 1710. https://doi.org/10.3389/fmicb.2016.01710.

51 Fulazzaky, M., Astuti, D.I., and Fulazzaky, M.A. (2015). Laboratory simulation of microbial enhanced oil recovery using *Geobacillus toebii* R-32639 isolated from the Handil reservoir. *RSC Advances* 5 (5): 3908–3916. https://doi.org/10.1039/C4RA14065F.

52 Qi, Y.B., Zheng, C.G., Lv, C.Y. et al. (2018). Compatibility between weak gel and microorganisms in weak gel-assisted microbial enhanced oil recovery. *Journal of Bioscience and Bioengineering* 126 (2): 235–240. https://doi.org/10.1016/j.jbiosc.2018.02.011.

53 Liang, X., Shi, R., Radosevich, M. et al. (2017). Anaerobic lipopeptide biosurfactant production by an engineered bacterial strain for *in situ* microbial enhanced oil recovery. *RSC Advances* 7 (33): 20667–20676. https://doi.org/10.1039/C7RA02453C.

54 Daryasafar, A., Jamialahmadi, M., Moghaddam, M.B., and Moslemi, B. (2016). Using biosurfactant producing bacteria isolated from an Iranian oil field for application in microbial enhanced oil recovery. *Petroleum Science and Technology* 34 (8): 739–746. https://doi.org/10.1080/10916466.2016.1154869.

55 Zhao, F., Shi, R., Cui, Q. et al. (2017). Biosurfactant production under diverse conditions by two kinds of biosurfactant-producing bacteria for microbial enhanced oil recovery. *Journal of Petroleum Science Engineering* 157: 124–130. https://doi.org/10.1016/j.petrol.2017.07.022.

56 Lan, G., Fan, Q., Liu, Y. et al. (2015). Effects of the addition of waste cooking oil on heavy crude oil biodegradation and microbial enhanced oil recovery using *Pseudomonas* sp. SWP-4. *Biochemical Engineering Journal* 103: 219–226. https://doi.org/10.1016/j.bej.2015.08.004.

57 Zhao, F., Li, P., Guo, C. et al. (2018). Bioaugmentation of oil reservoir indigenous *Pseudomonas aeruginosa* to enhance oil recovery through *in-situ* biosurfactant production without air injection. *Bioresource Technology* 251: 295–302. https://doi.org/10.1016/j.biortech.2017.12.057.

58 Cui, Q.F., Zheng, W.T., Yu, L. et al. (2017). Emulsifying action of *Pseudomonas aeruginosa* L6-1 and its metabolite with crude oil for oil recovery enhancement. *Petroleum Science and Technology* 35 (11): 1174–1179. https://doi.org/10.1080/10916466.2017.1315725.

59 Ke, C.Y., Lu, G.M., Wei, Y.L. et al. (2019). Biodegradation of crude oil by *Chelatococcus daeguensis* HB-4 and its potential for microbial enhanced oil recovery (MEOR) in heavy oil reservoirs. *Bioresource Technology* 287: 121442. https://doi.org/10.1016/j.biortech.2019.121442.

60 Xia, W., Shen, W., Yu, L. et al. (2016). Conversion of petroleum to methane by the indigenous methanogenic consortia for oil recovery in heavy oil reservoir. *Applied Energy* 171: 646–655. https://doi.org/10.1016/j.apenergy.2016.03.059.

61 Gudiña, E.J., Pereira, J.F.B., Costa, R. et al. (2013). Biosurfactant-producing and oil-degrading *Bacillus subtilis* strains enhance oil recovery in laboratory sand-pack columns. *Journal of Hazardous Materials* 261: 106–113. https://doi.org/10.1016/j.jhazmat.2013.06.071.

62 Roelants, S.L.K.W., Van Renterghem, L., Maes, K. et al. (2018). Microbial biosurfactants: From lab to market. In: *Microbial Biosurfactants and their Environmental and Industrial Applications* (eds. I.M. Banat and R. Thavasi), 341–363. Boca Raton, FL: CRC Press.

63 Du, C., Song, Y., Yao, Z. et al. (2019). Developments in *in-situ* microbial enhanced oil recovery in Shengli oilfield. *Energy Sources, Part A: Recovery, Utilization and Environmental Effects* 14: 1–11. https://doi.org/10.1080/15567036.2019.1648603.

64 Xingbiao, W., Yanfen, X., Sanqing, Y. et al. (2015). Influences of microbial community structures and diversity changes by nutrients injection in Shengli oilfield, China. *Journal of Petroleum Science and Engineering* 133: 421–430. https://doi.org/10.1016/j.petrol.2015.06.020.

65 Saikia, R.R., Deka, S., Deka, M., and Sarma, H. (2012). Optimization of environmental factors for improved production of rhamnolipid biosurfactant by *Pseudomonas aeruginosa* RS29 on glycerol. *Journal of Basic Microbiology* 52 (4): 446–457.

66 Sarma, H., Bustamante, K.L.T., and Prasad, M.N.V. (2018). Biosurfactants for oil recovery from refinery sludge: Magnetic nanoparticles assisted purification. In: *Industrial and Municipal Sludge* (eds. M.N.V. Prasad, P.J. de Campos, F. Meththika and V.S. Venkata Mohan). Butterworth-Heinemann, Oxford, UK Available at: https://doi.org/10.1016/B978-0-12-815907-1.00006-4.

67 Chai, L., Zhang, F., She, Y. et al. (2015). Impact of a microbial-enhanced oil recovery field trial on microbial communities in a low-temperature heavy oil reservoir. *Nature Environment and Pollution Technology* 14 (3): 455–462.

68 Xuezhong, W., Yuanliang, Y., and Weijun, X. (2016). Microbial enhanced oil recovery of oil-water transitional zone in thin-shallow extra heavy oil reservoirs: A case study of Chunfeng Oilfield in western margin of Junggar Basin, NW China. *Petroleum Exploration and Development* 43 (4): 689–694. https://doi.org/10.1016/S1876-3804(16)30080-5.

69 Ke, C.Y., Lu, G.M., Li, Y.B. et al. (2018). A pilot study on large-scale microbial enhanced oil recovery (MEOR) in Baolige Oilfield. *International Biodeterioration and Biodegradation* 127: 247–253. https://doi.org/10.1016/j.ibiod.2017.12.009.

6

Biosurfactant Enhanced Sustainable Remediation of Petroleum Contaminated Soil

Pooja Singh[1], Selvan Ravindran[1], and Yogesh Patil[2]

[1] *Symbiosis School of Biological Sciences, Symbiosis International (Deemed University), Pune, Maharashtra, India*
[2] *Symbiosis Centre for Research and Innovation, Symbiosis International (Deemed University), Pune, Maharashtra, India*

CHAPTER MENU

6.1 Introduction

Petroleum is one of the most widely used fuels and a critical global energy resource. Multiple components of petroleum, especially crude oil, are reported to be extremely carcinogenic and mutagenic and are reported to cause severe deleterious toxic effects to terrestrial and marine life [1, 2]. Petroleum finds its use not only as a fuel in various industries but also as a raw material for multiple industries including petrochemicals. These primarily include oil refineries and petroleum drilling, storage, handling, transportation, and refining processes, which not only result in oil leakages and spills but also generate large volumes of petroleum sludge. Such widespread use results in intentional and accidental introduction of petroleum hydrocarbons in the ecosystem, thereby threatening the survival and balance of biotic and abiotic components [3]. Unscientific release of waste from many industries like plastics, petrochemicals, and solvents adds to the oil-rich waste entering the environment. However, of all these releases, accidental oil spills in the ocean is one of the most detrimental occurrences in the petroleum industry, threatening large areas of oceans and

Biosurfactants for a Sustainable Future: Production and Applications in the Environment and Biomedicine, First Edition. Edited by Hemen Sarma and Majeti Narasimha Vara Prasad.
© 2021 John Wiley & Sons Ltd. Published 2021 by John Wiley & Sons Ltd.

shorelines [4, 5]. A number of cases of oil spills in marine, freshwater, and soil ecosystems in the last three decades has directed the attention of scientists and environmentalists toward petroleum pollution and has resulted in multiple research and attempts toward remediation of hydrocarbon polluted ecological niches [6, 7]. Considering the global climate crisis, use of biological means of remediation have emerged as the most preferred alternative. Sustainable remediation using biological agents like biosurfactants have been at the forefront of bio-based environment cleanup strategies. In this chapter we look at the current status of biosurfactants for the environmental cleanup of petroleum polluted niches. Biosurfactant assisted bioremediation technologies are a promising environmental tool for green applications. The recalcitrant nature and extreme toxicity of petroleum makes it imperative to critically assess the complex composition of petroleum and its interactions with biotic and abiotic soil systems for the enhanced effectiveness of various remediation technologies as well as future research directions.

6.1.1 Chemical Composition of Petroleum

Petroleum in Latin means "rock oil", which exists as a sticky, viscous, dark liquid formed in millions of years from natural soil activities. Petroleum is referred to as both, the unprocessed crude oil as well as various petroleum products, and contains a complex mixture of hydrocarbons, the complexity of which increases due to the action of various soil and topographical biotic and abiotic factors. The composition of crude oil varies substantially from source to source but in general crude oil is composed of various hydrocarbon compounds (aliphatics and aromatics), non-hydrocarbon compounds as well as organometallic and metallic compounds along with various metal impurities like vanadium, nickel, copper, iron, and sodium. The composition and distribution of hydrocarbons varies for different crude oils. Table 6.1 shows the major distribution of various major fractions of crude oil [8–10]. Petroleum or crude oil is subsequently processed into various petroleum products including gasoline, diesel, fuel oil, and kerosene, all with varying compositions. All of the above petroleum products are complex mixtures of olefinic, paraffinic, and aromatic hydrocarbons and are toxic to marine and terrestrial life. Multiple processes using petroleum as the raw material generate huge volumes of petroleum sludge, which is a complex mixture of oil, water, metals, and solids and their unique characteristics and varied composition makes them highly recalcitrant and very difficult to be readily treated and degraded. Treatment of oily sludge apart from crude oil is hence another major issue faced by the petroleum industry [11]. One of the most prominent petroleum-based fuels to be used extensively is motor oil (engine oil). Rapid urbanization and modernization have seen an unprecedented rise in the automobile sector across the globe [12], thereby increasing

Table 6.1 Components of crude oil [8, 9].

Sr.No.	Components	Examples	Fraction
1	Paraffins and iso-paraffins	n-Alkanes (methane, ethane, butane), branched alkanes (isooctane, isobutane)	15–60%
2	Aromatics	Single ring or multiple condensed ring compounds (benzene, toluene, xylene, naphthalene)	3–30%
3	Naphthenes	Cycloalkanes (methyl cyclopentane), 1,2-dimethyl-cyclohexane)	30–60%
4	Asphaltics (asphaltenes and resins)	O, S, N and metals containing compounds (naphthenic acid, phenol, mercaptans, quinoline pyridine, indole, carbazole)	6–10%

the use and subsequent disposal of used motor oil in the environment, without any appropriate treatment. Most of the motor oils used contain petroleum-based hydrocarbons, including toxic compounds like naphathalene, pyrene, floranthene, benzoanthracene, benzopyrene, etc., as well as various metals like vanadium, lead, nickel, chromium, lead, manganese, and cadmium [13]. Used motor oil also contains many polyalphaolefins as well as large amounts of other additives like metal phenoxides, anti-wear agents (e.g. zinc dialkyl dithiophosphate), and friction-reducing agents (like molybdenum disulfide), among others. Apart from oil spills, contamination of soil and water by used motor oil is emerging as a growing environmental concern in recent times [14].

Many physiochemical means have been used over the last few decades for the reduction in total petroleum hydrocarbons (TPHs) from sites contaminated with crude oils [15, 16]. However, a number of physical, chemical, and biological factors, in combination, affect the fate and rate of hydrocarbon removal from natural ecosystems and hence most of the physical and chemical processes used till date have proved to be inadequate for complete remediation of petroleum-contaminated ecological niches. Although the use of synthetic chemicals for removal of hydrocarbons and for enhancing the rate of degradation of petroleum hydrocarbons have been reported, their widespread use is questionable, keeping in mind the sensitivity of ecosystems to external chemical agents as well as environmental sustainability, especially in colder more pristine environments [16]. "Bioremediation" or use of biological agents for remediation, hence emerged as an alternative environmental cleanup technology that offered various advantages over their physical and chemical counterparts, including low operation cost, less use of chemicals, and, most importantly, complete degradation of petroleum pollutants, thereby leading to the complete absence of any residual product(s).

6.2 Microbial-Assisted Bioremediation of Petroleum Contaminated Soil

Crude oil is a mixture of hydrocarbon compounds, many of which can be utilized by various microorganisms as a carbon source and hence mineralize the hydrocarbon in the process [17]. Bioremediation using external hydrocarbon degrading microorganisms, referred to as "bioaugmentation," is an ecofriendly and attractive approach for reduction of TPH in contaminated soils, especially relevant in pristine environments not previously exposed to any hydrocarbons or exposed to a low hydrocarbon load, as in case of polar regions and areas with massive petroleum spills [18]. Bioaugmentation is not only effective and safe but also generates a low carbon footprint [19, 20]. Although a diverse range of microorganisms, more than 79 genera, have been reported for the above process, the presence, involvement, and success of organisms like *Oleispira, Oleiphilus, Alcanivorax, Pseudomonas, Acinetobacter*, and *Burkholderia* are unsolicited [2, 21, 22]. Since crude oil is a complex mixture of different hydrocarbons, use of consortia, consisting of microbes with varying metabolic potential, have proved to be most resilient, suitable, and successful for degradation of different fractions of petroleum hydrocarbons in soils [23–27]. Microorganisms have been reported to degrade TPHs either aerobically or anaerobically depending on the microbial population present and the soil physiological conditions. Additionally, since TPHs are diverse in nature, the degradation pathways at play are multitudinous. Figure 6.1 gives an overview of a few of the prominent pathways for major hydrocarbon fractions in crude oil. The various specific enzymes at play include monooxygenases, dioxygenases, peroxidases, reductase, hydroxylase, and dehydrogenases [18].

Apart from the presence and diversity of hydrocarbon degraders, other critical factors that need to be considered before applying bioremediation for a contaminated site cleanup are: (i) type,

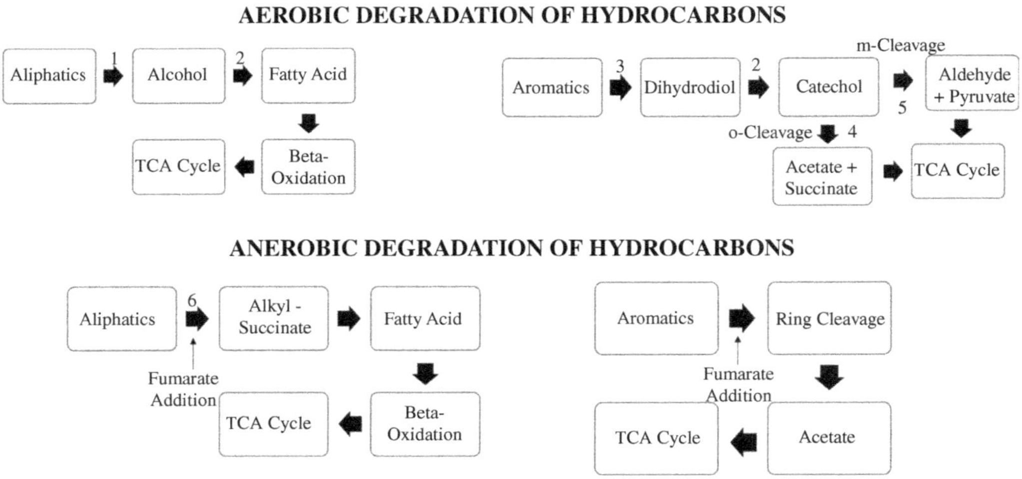

Figure 6.1 Overview of the prominent degradation pathways for aliphatic and aromatic petroleum hydrocarbons with major microbial enzymes used [18]. TCA: tricarboxylic acid; 1: monooxygenase enzyme; 2: dehydrogenase enzyme; 3: dioxygenase; 4: intradiol dioxygenase; 5: extradiol dioxygenase; 6: alkylsuccinate synthase.

concentration, and age of the oil; (ii) geotopography of contaminated shoreline/field; (iii) prevalent climatic conditions; (iv) nutrient content and pH of the soil/rocks; (v) soil organic matter (SOM); and (vi) soil texture and cation exchange capacity [5]. Use of external agents to enhance the survival and activity of hydrocarbon degraders *in situ* and to ensure an effective bioremediation process are of importance. Most relevant in such cases is amendment of soils with nutrients, especially nitrogen and phosphorus, as well as agents like surfactants that stimulate hydrocarbon mobility and availability, leading to enhanced solubility and microbial assisted degradation [28]. The United States Environment Protection Agency (US-EPA) has defined microbial cultures (degraders) and nutrient additives as bioremediation agents and has released a list of over nine commercial products as potential bioremediation stimulating agents, including Inipol EAP22, OSE II, and BIOREN [29]. One such stimulating agent to be used successfully are surface active agents, surfactants, that have shown most promising results for the removal of TPH from soils contaminated with different petroleum compounds.

6.3 Hydrocarbon Degradation and Biosurfactants

The presence of large amounts of particulate matter in soil increases the adsorption of hydrophobic hydrocarbons to humic substances, leading to the formation of persistent residues as well as partitioning in a non-aqueous phase liquid (NAPL) [30]. The primary challenge encountered in bioremediation is the removal of these sorbed fractions. Microbial attack on petroleum compounds requires a close contact of microorganisms with hydrocarbons since bacterial hydrocarbon degrading enzymes, like oxygenases, are membrane bound enzymes [2]. Biostimulation using biosurfactants has been reported to be used for the same purpose. Biosurfactants are surface-active multifunctional molecules from microorganisms with diverse applications [31, 32]. Their structural diversity, low toxicity, higher biodegradability and better environmental compatibility, emulsion forming and emulsion breaking ability, and activity at extreme conditions imparts them with prominent advantages over other synthetic surfactants [33]. Bioremediation with the help of efficient microbial populations and biological surface-active agents hence offer a sustainable wholesome approach to tackle the hydrocarbon menace [34]. Experimental evidences for biosurfactant production for the uptake of hydrophobic compounds by microbial populations have been obtained through numerous studies and most of the organisms isolated from oil-contaminated sites and found to be efficient hydrocarbon degraders are reported to be efficient biosurfactant producers as well [35, 36].

6.3.1 Mechanism of Biosurfactant Action

Among all of the known biosurfactants, rhamnolipids produced primarily by *Pseudomonas* sp. are the most effective in environmental remediation technology and offer a high potential for use in the area of total oil degradation and remediation of the polluted site [37]. They are therefore one of the most studied biosurfactants for enhanced hydrocarbon remediation worldwide [19, 38]. Biosurfactants have been known to interact with hydrocarbons and enhance remediation by various mechanisms [35]. One of the major mechanisms by which biosurfactants enhance degradation is by increasing the detachment of oil from rocks and sand particles, thus leading to their mobility and subsequent availability to the degrading microbial populations [39]. Reduction in surface tension by a biosurfactant strengthened this theory [40]. Enhanced microbial activity leads to complete degradation of petroleum hydrocarbon and hence complete removal of pollutants from the soil or water. Initially, emulsification and subsequent mobilization of hydrophobic substrate by the biosurfactant was the only mechanism proposed for enhanced hydrocarbon degradation since it increased the bioavailability of the substrate. Subsequently, formation of micelles by biosurfactants leading to enhanced solubility was proposed as an additional mechanism at play. Various reports present evidence for this mode of hydrocarbon uptake by microorganisms. These include remediation of diesel oil using indigenous microbial population [41], petroleum hydrocarbon degradation by *Burkholderia* [42], polycyclic aromatic hydrocarbon (PAH) utilization by *Pseudomonas* [43], crude oil hydrocarbon degradation by *Pseudomonas* [44] among others. A study by Noordman et al. [45] demonstrated that rhamnolipids enhanced two different processes that are relevant for the remediation of soil contaminated with NAPLs. These processes are the mass transfer of entrapped or residual substrate from matrices to the aqueous phase and subsequent biodegradation of the substrate present as a separate liquid phase. In another report, reduction in surface and interfacial tension by rhamnolipid solution washing resulted in mobilization and subsequent removal of crude oil from a contaminated soil by around 80% [46]. The type of biosurfactant influences the varying effects on different hydrocarbons. Low molecular weight biosurfactants like rhamnolipids and surfactins have been reported to enhance mobilization and solubilization while high molecular weight biosurfactants like alasan cause more emulsification [47]. Although solubilization and emulsification are still considered the primary means by which biosurfactants, especially rhamnolipids, enhance the degradation of hydrophobic compounds such as petroleum hydrocarbons, alterations in microbial cell surface properties by biosurfactants subsequently came to light and this led to another possible mode of hydrocarbon uptake: uptake of hydrocarbon by direct contact of a microbial cell with a hydrophobic carbon source [35]. Even though the bacteria/consortium used for hydrocarbon degradation is efficient, its effectiveness can be limited by bacterial adhesion to soil, sand, and rock particles and hence limits the spread of the bioaugmenting agent to the entire polluted site. Aging of the pollutant in soil aggravates this process and further decreases the interaction between hydrocarbons and the degrading microbial populations. Modifications of cell surface properties and lowering of the surface tension has been found to enhance mobilization and transport of oil degrading bacteria throughout the substrate zone (oil spill). This leads to a shortened path of travel for bacteria from adsorption to the substrate availability site, facilitating greater contact between degrading bacteria and available substrate for degradation, resulting in enhanced degradation [48]. Another aspect of altered cell surface properties is the increase in plasma membrane permeabilization that facilitates transmembrane transfer of hydrocarbon droplets into bacterial cells. In a series of work, altered cell hydrophobicity by rhamnolipids led to enhanced diesel oil degradation by *Pseudomonas alcaligenes* and *Pseudomonas stutzeri* by 92, 88, and 79% respectively [49–51]. Rhamnolipids were found to promote the interfacial uptake of hydrocarbons after the pseudo-solubilization process [52, 53]. In a recent study, improved oil bioavailability and altered cell surface hydrophobicity were ascertained to be responsible for enhanced oil degradation by a *Bacillus subtilis* strain [54]. In another interesting study, uptake of hydrocarbon by internalization of a surfactant covered substrate, in a process akin to "pinocytosis" was reported by our group for a hydrocarbon degrading biosurfactant producing *Pseudomonas* strain [52]. This process was after the rhamnolipid–hydrocarbon emulsion formation and subsequent dispersion of hydrocarbon to particles less than 0.22 μm in size. Bioremediation enhanced by biosurfactants can be summarized finally to be a combination of primarily four mechanisms: (i) mobilization and removal of petroleum components from rocks and soil (reduction in surface tension); (ii) an increase in solubility and hence increased bioavailability of organic compounds for microbial degradation; (iii) modifications in cell surface properties of microbial cell so as to increase hydrophobicity and subsequent decreased soil adhesion and increased hydrocarbon adherence; and (iv) internalization of biosurfactant-coated hydrocarbon droplets [55, 56]. Figure 6.2 summarizes the various complex mechanisms at play during hydrocarbon–biosurfactant interaction, leading to enhanced TPH degradation from contaminated soils.

Various studies have confirmed the active role of biosurfactants in enhancing the rate of degradation of various fractions of petroleum hydrocarbons in soil and water, and reduction in hydrocarbon by as much as 97% has been reported by many research groups [57–60]. A positive effect of biosurfactant application on motor oil hydrocarbons has also been successfully established [61]. When it comes to field applications, biosurfactant can be applied as pure suspension or even as cell free broth of biosurfactant-producing bacteria, which further makes the whole operation more cost effective. Use of biosurfactant-producing bacteria as one of the members of the oil degrading consortium offers another attractive cost-effective strategy. However, biosurfactant produced by one strain can be damaging to another member of the consortium and hence the compatibility of various consortium members needs to be worked out adequately [62]. Keeping in mind the specificity of interaction between biosurfactants and organisms, it might be beneficial to use biosurfactants produced by the indigenous population itself for enhancement of biodegradation in various polluted niches. In one study, rhamnolipid enhanced hexadecane degradation by the same producing organism (*Pseudomonas*), but the rate of hydrocarbon degradation by other organisms (*Rhodococcus* and *Acinetobacter*) was not significantly impacted [63]. Hence screening for an indigenous biosurfactant-producing microorganism and its subsequent incorporation as a consortium member is an attractive strategy for long-term *in situ* remediation process efficacy.

6.4 Soil Washing Using Biosurfactants

Soil washing as an ex-situ remediation technique has been successfully used to treat hydrocarbon contaminated soils. From among all the solutions used biosurfactant solutions have most potential to be used for soil washing of contaminated soil by increasing solubilization or mobilization of oil, thereby partitioning it in the aqueous phase for greater availability and recovery. Rhamnolipids

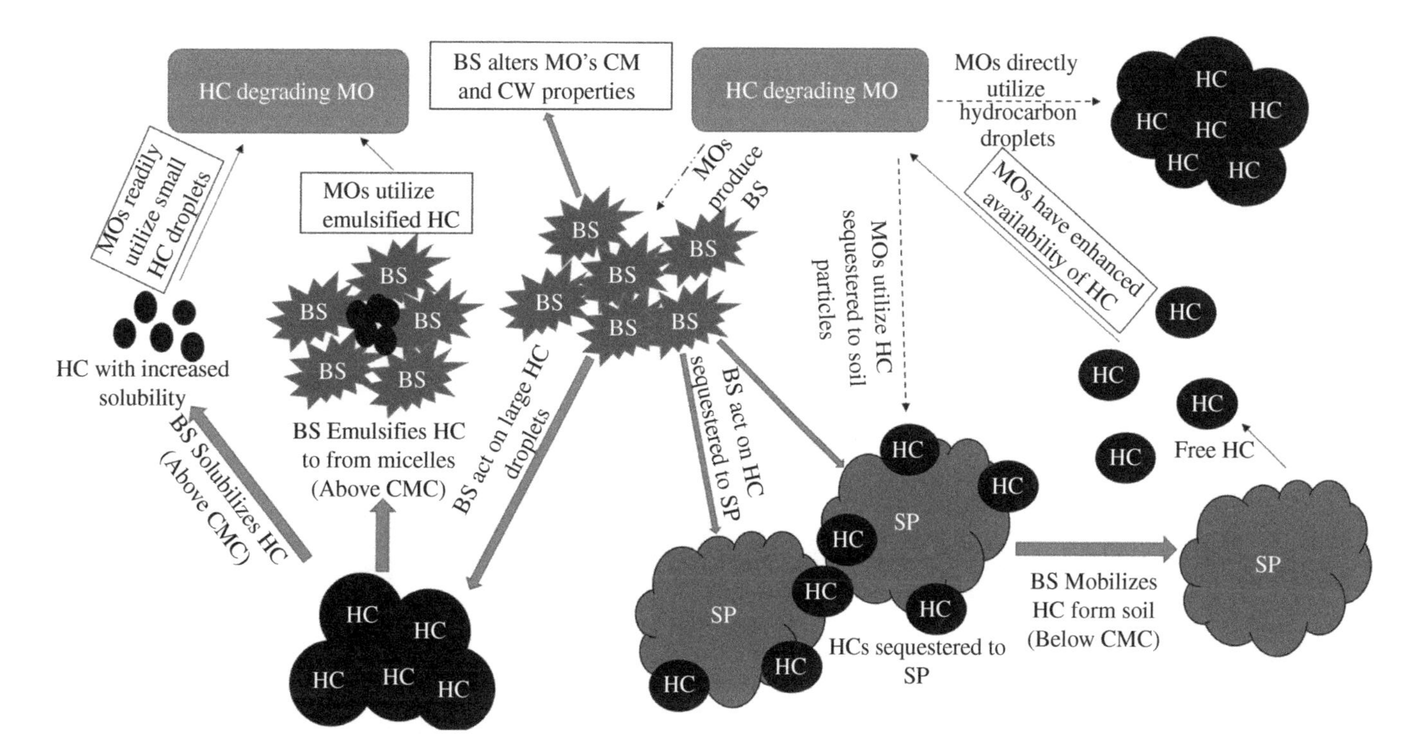

Figure 6.2 Mode of action of biosurfactants for enhanced hydrocarbon degradation. MO: microorganism, HC: hydrocarbon, SP: soil particle; BS: biosurfactant; CM: cell membrane; CW: cell wall (See insert for color representation of the figure).

have been reported to have most success in this process and can be used as foam, as purified solution, no-purified solutions, or as a cell-free culture broth [64, 65]. Biosurfactant by *Pseudomonas cepacia* demonstrated successful removal of around 81.3% of oil from motor oil coated porous rocks and 80% from clay soils [66], and 42.12% petroleum and 44.75% diesel oil removal was reported by rhamnolipid biosurfactant from a sandy soil contaminated with 5% of both hydrocarbons separately. Rhamnolipid was applied in the form of microfoam at a concentration of 0.1 g/l [67]. Solubilization of petroleum and mobilization of diesel oil was proposed as the possible mechanisms for oil removal. In another study, biosurfactant solution from various bacterial isolates recovered more than 40% of oil from oil sludge and exhibited potential to be used also for oil storage tank cleaning [68]. Use of biosurfactants decreased the sludge viscosity and led to the formation of emulsions that promoted better recovery of crude oil. Results like these establish the use of biosurfactant solution for clean up of oil contaminated soil and for sand washing, with subsequent reduction of TPH in soil and/subsequent oil recovery. Although biosurfactants have shown a better removal efficiency than other synthetic surfactants used, their effect is dependent on the type and concentration of pollutants, age of oil, type and concentration of the biosurfactant, and also on soil characteristics. Biosurfactants rhamnolipids and surfactin were tried for low (3000 mg/kg) and high (9000 mg/kg) petroleum contaminated soil washing. Rhamnolipids and surfactin (0.2%) successfully removed 23 and 14% of oil, respectively, from low contaminated soil and 63 and 62% of oil, respectively, from high contaminated soil [69]. Hallmann and Mędrzycka [70] studied the differential cleaning ability of rhamnolipids (30 g/kg) for different soils. Contaminated soil with a high clay and organic matter content was difficult to treat and desorb than oil from sandy soil (99% removal from sand as compared to 40–60% from clay loam). Increased adsorption of oil to clay and loam, rather than to sand particles, was the principle reason found. Although soil washing with biosurfactants like rhamnolipids and surfactins effectively remove oil from sand/soil, it still leaves the aspect of hydrocarbon removal from the biosurfactant solution a major issue. Selective adsorption has shown promising results but increases the treatment cost [71]. This area requires substantial research in finding innovative and cost-effective technologies for hydrocarbon removal and recovery from biosurfactants and subsequent biosurfactant reuse.

6.5 Combination Strategies for Efficient Bioremediation

Petroleum hydrocarbon polluted sites may or may not contain adequate amounts of hydrocarbon degraders. As mentioned earlier, one of the main limitations with hydrocarbon degradation is their hydrophobicity and substrate availability along with low nutrient availability for the degrading microorganisms. Hence the call of the day for enhanced TPH removal, keeping in mind extensive hydrocarbon pollution as well as a safe environment, is a combination approach, a bioremediation strategy devised with a combination of seed organisms (effective mix of hydrocarbon degraders), fertilizers, and nutrients and appropriate amounts and type of biosurfactants [72]. This approach has yielded the most successful results and is most promising for field-scale remediation of petroleum contaminated sites. Biosurfactants can be added externally as a crude extract (or even a cell-free broth), as foam, and also in the form of a biosurfactant-producing microorganism as a member of the oil degrading consortia used. Table 6.2 lists some of the successful studies using either only biosurfactants or biosurfactants along with other amendments for petroleum hydrocarbon degradation and site remediation. Adoption of a combination approach has yielded most successful results, as was proved in our earlier work on crude oil degradation, wherein from among all the combinations tried, application of an oil degrading

Table 6.2 Overview of some of the recent and successful biosurfactant assisted oil remediation studies fortified with other synergistic stimulators.

Organism(s) used	Type of oil	Activity and time of study	Concentration of biosurfactant used	Additional amendments (if any)	References
Sphingomonas sp.	Phenanthrene	99.5% degradation; 10 days	Rhamnolipid: 50 mg/l	–	Pei et al. [60]
Gordonia sp.	Oil tank bottom sludge	100% dispersion	High molecular weight polysaccharide with fatty acid esters: 1–10 g/l	–	Matsui et al. [73]
Candida sphaerica	Motor oil	95% removal (from sand)	Crude cell free extract	–	Luna et al. [74]
P. cepacia	Motor oil	81% recovery by cell free broth; 68% recovery from extracted BS (from rock)	0.5%	–	Silva et al. [66]
Bacillus licheniformis	Crude oil	85% removal (from sand and soil); 1 hour	Surfactin: 10%	–	Kavitha et al. [58]
Pseudomonas sp.	Motor oil	48% removal (stone) 63% (sand)	Rhamnolipid: 0.5%	–	Sarubbo et al. [75]
Micrococcus, Bacillus, Corynebacterium, Flavobacterium, Pseudomonas	Gasoline	78%; 60 days	Rhamnolipid: 1%	Poultry litter: 1% Coir pith: 1%	Rahman et al. [76]
Micrococcus sp., *Bacillus* sp. *Corynebacterium* sp., *Flavobacterium* sp., *Pseudomonas* sp.	Tank bottom sludge	100% C8-C11 and 73–98% C12 – C40 56 days	Rhamnolipid: 4 mg/100 g soil	NPK solution: 0.1 mg/100 g soil Inoculum: 1 ml/100 g soil	Rahman et al. [77]
Naturally occurring organisms	Crude oil	96% 18 days	Rhamnolipid: 10%	Uric acid, lecithin and molasses	Nikolopoulou and Kalogerakis [78]

(*Continued*)

Table 6.2 (Continued)

Organism(s) used	Type of oil	Activity and time of study	Concentration of biosurfactant used	Additional amendments (if any)	References
Pseudomonas aeruginosa, Rhodococcus equi	Crude oil	90%; 45 days	Cell free broth: 100 ml/kg soil	Nutrient solution: 50 ml/kg soil	Cameotra and Singh [79]
Pseudomonas	Crude oil	37.7%; 168 h	Lipopeptide: 0.1%	Fertilizer (urea+K_2HPO): 0.1% Inoculum: 1%	Thavasi et al. [80]
Ochrobactrum intermedium, Microbacterium oryzae, Pseudomonas sp., *Alcaligenes faecalis*	Crude oil	77.6%; 30 days (10 g/kg in 1 year)	Rhamnolipid: 0.1 g/kg	Consortia: 7.1 g/kg Nutrient solution (100 : 10 : 1): 2.75 g/kg soil N, 0.38 g/kg soil P, 0.4 g/kg soil K	Tahseen et al. [81]
Bacillus cereus UCP 1615	Motor oil	91% oil removal from marine rock; 70% oil dispersion in sea water	Crude biosurfactant: 0.9 g/l	0.2% potassium sorbate as preservative in biosurfactant formulation, increased dispersion capacity of extracted biosurfactant from 65 to 70%	Ostendorf et al. [82]

consortium along with nitrogen fertilizer and crude biosurfactant solution brought about a faster and more effective reduction in crude oil hydrocarbons under natural environmental conditions. The consortium was comprised of equal quantities of *Pseudomonas aeruginosa* and *Rhodococcus* sp. The biosurfactant used was a crude extract of rhamnolipid and the nutrient mixture was a blend of nitrate and phosphate salts. A 98% reduction in total hydrocarbons was observed in combined bioaugmentation and biostimulation as compared to only nutrient amendment (63%) or only a biosurfactant application (73%) [79].

Toxicity and effectiveness of a biosurfactant and its compatibility with an autochthonous microbial population needs to be assessed prior to field applications. A shorter lag phase and faster hydrocarbon degradation rates are critical in cases of oil spills at beaches and fields so as to avoid sequestration of oil to sand and rock particles. Further, the availability of the hydrocarbon for degradation becomes limiting at these sites and subsequently more than increased metabolism of degraders, a strategy to increase the bioavailability of hydrocarbons, assumes importance. Nevertheless, a combination approach using nutrients and surfactants has emerged as the most effective strategy to be worked out for field scale bioremediation.

6.6 Biosurfactant Mediated Field Trials

Excellent success stories of biosurfactant mediated enhanced remediation of TPHs have been well documented over the years, but not many established field scale successful trials have been reported. As discussed earlier, the varying eco-geological conditions and the nature, availability, and compatibility of bioaugmenting and biostimulating agents are at play in a dynamic relation at the polluted sites. Spread of hydrocarbons in soil is heterogeneous and extremely high concentrations at places that might prove to be toxic to the degrading microbes. This would affect the overall action of biosurfactants and affect the microbial hydrocarbon degradation rates. We document here some of the large-scale field applications that have been carried out using biosurfactants for cleaning oil-polluted soil and water.

i) The Exxon Valdez oil spill is one of the most prominently studied oil spills to date with respect to bioremediation and a wide ecological impact. A number of studies have been carried out over the years to test the efficacy of biosurfactant application on the removal of oil residues from contaminated soil and sand. In one study, Rhamnolipid surfactants from *Pseudomonas* were found to release three and a half times as much oil as water alone from the beaches in Alaska after the Exxon Valdez tanker spill [83]. This was tested at different biosurfactant concentrations and at different temperatures and positive results were obtained consistently.
ii) In a study on remediation of weathered crude oil from beach sands, a microbial derived, EPA approved biosurfactant formulation (PES-51®) was tested for its efficacy to clean subsurface of oil-contaminated rocky beaches on Sleepy Bay on LaTouche Island in Prince William Sound, Alaska. This site witnessed extensive oil pollution after the Exxon Valdez disaster and was treated initially by two synthetic fertilizers, Inipol and Customblen. However, extensive oil deposits were reported at subsurface levels. PES-51 (composed of biosurfactant, D-limonene, and biospersan) was applied for five days and the site was monitored for residual hydrocarbons and microbial activity. After three months, diesel range hydrocarbons were found to be reduced to below the detection limit of 0.5 mg/kg while semi-volatile petroleum hydrocarbons were found to be low by around 70% without any noticeable reduction in the microbial activity [84].

iii) Another field trial using BIOREN 1 (a fish meal derived EPA approved nutrient formulation containing biosurfactant) indicated faster and enhanced oil degradation rates on an oil-contaminated sandy beach in Brittany over four months. The experiment was carried out in $5m^2$ plots contaminated with 50 l of weathered light Arabian crude oil. BIOREN was applied twice at a concentration of 10% of the oil quantity. Permeability and grain size distribution in selected plots were monitored over the test period to assess their effect on degradation and it was found that oil decreased the permeability of sand and hence intermittent raking proved to be beneficial for prolonged bioremediation. The rate of supplement dilution in sea water and natural nutrient replenishment were concluded to be other critical factors that played a decisive role in a sea water covered oil contaminated zone [85].

The success of bioremediation of xenobiotics like petroleum hydrocarbons is dependent on, and determined by, a number of factors including site geochemistry, presence and availability of nutrients, biological interactions, and climatic effects. Hence extrapolating the data obtained in the laboratory to bioremediation in fields often leads to misleading predictions and failures [86]. Although most of the studies on the applications of biosurfactants for enhanced petroleum hydrocarbon remediation in soil under field conditions have yielded enhanced degradation, there are contradictory results as well. In a year-long study on remediation of diesel oil in the field, addition of biosurfactant was found to bring about no additional enhanced degradation. Bioaugmentation, using a specially developed consortium, reduced the TPH to 937 mg/kg from an initial concentration of 10 000 mg/kg. The developed inoculum consisted of *Pseudomonas putida* (33%), *Pseudomonas fluorescens* (21%), *Rhodococcus equi* (12%), *Alcaligenes xylosoxidans* (11%), *Stenotrophomonas maltophilia* (8%), *Aeromonas hydrophila* (7%), *Gordonia* sp. (6%), and *Xanthomonas* sp. (2%). Inoculum and biosurfactant solutions were added at a concentration of $1 l/m^2$ and 150 mg/kg, respectively. Augmented soil with and without rhamnolipids yielded a similar reduction in TPH, signifying no additional role of biosurfactant in the degradation of diesel hydrocarbons at the site [87]. Hence pertinent exhaustive laboratory studies assume importance before the technology is applied at a field scale.

6.7 Limitations, Strategies, and Considerations of Biosurfactant-Mediated Petroleum Hydrocarbon Degradation

A number of processes are at play when biosurfactants are applied for remediation and hence there have been varying reports over the last two decades with results varying from short-term to even no effect [88]. This could be due to multiple factors affecting the interaction between hydrocarbon, hydrocarbon degrading bacteria, and the biosurfactant [71]. In one study, only a short-term enhancement in bioremediation by biosurfactants was reported to be brought about, possibly due to an increase in hydrocarbon availability and its subsequent complete utilization by the degrading bacteria [87, 89]. Prolonged monitoring revealed no or insignificant enhancement as compared to bioaugmentation alone. This feature can also be attributed to the fact that biosurfactants are highly biodegradable and hence do not persist in soil for long durations. Additionally, biosurfactants may themselves act as an alternate carbon source, thus limiting hydrocarbon utilization as the carbon source by the degrading organisms [90]. Monitoring of soil for a longer duration during biosurfactant treatment hence would give a clearer picture of the effect of biosurfactant on bioremediation. Since biosurfactants are easily biodegraded, one strategy is to use biosurfactants not utilized by any member of the developed consortium so as to give a prolonged enhanced degradation effect.

Also critical for the process is the feasibility studies at lab scale in order to test the efficacy and compatibility of the consortium members and the externally added biosurfactant. Since petroleum is a mix of different hydrocarbons, one factor when using biosurfactants that needs to be kept in mind is the specificity of a biosurfactant to the specific class of petroleum hydrocarbons. In a study on *P. aeruginosa*, biosurfactants produced by different strains were found to differentially solubilize different portions of crude oils like phenanthrene and pyrene. Additionally, the biosurfactant produced was of lipoprotein and lipid-starch-protein in nature as opposed to glycolipids that are reported to be produced by most members of the *Pseudomonas* sp. [91]. In another study, varying effects of biosurfactants were observed on different hydrocarbon degraders. Three marine bacterial isolates were used, *Bacillus megaterium*, *Corynebacterium kutscheri*, and *P. aeruginosa*, and each had a differential effect on the degradation in the presence of equal amounts of biosurfactant [80]. However, an additional effect of nutrients on enhanced hydrocarbon degradation was ruled out and it was proposed that biosurfactants alone can enhance the biodegradation rate substantially (without any nutrients), thus reducing the overall treatment cost.

Another factor that limits and affects the applicability of biosurfactant is the concentration of the biosurfactant used. The critical micelle concentration (CMC) is the concentration above which a biosurfactant forms micelles, thereby solubilizing the oil. Below the CMC, mobilization of oil by reduction in surface tension is the primary feature at play [92]. An earlier study by Whang et al. [41] on the effect of surfactin and rhamnolipid solution on diesel oil degradation in soil, revealed the negative impact of high surfactin solutions while rhamnolipid biosurfactant had no such inhibitory effect. Surfactin when used below 40 mg/l was observed to enhance the rate of diesel biodegradation (94%) while inhibition of growth and biodegradation rates were observed above this concentration (complete inhibition occurred at 400 mg/l). Various theories have been proposed for the inhibitory effect of a biosurfactant. This limiting effect of biosurfactant concentration on biodegradation hinted at a possible toxicity of biosurfactants toward microorganisms at higher concentrations, possibly due to their antimicrobial activity [93, 94]. Further, high concentrations of biosurfactant might lead to sequestration of hydrocarbons, thus making them unavailable for microbial attack [95]. Increased mobilization of oil may increase the concentration of hydrocarbons in the vicinity of a microbial population, thereby increasing its toxicity to the cells [96]. Higher biosurfactant concentrations may be inhibiting to an autochthonous microbial population while lower concentrations might prove to be non-effective [41, 97]. In one study, concentration of up to 150 mg/kg of biosurfactant has been reported to be non-phytotoxic and can prove to be a suitable concentration for application in field studies [98]. Another theory, as also mentioned earlier, is that biosurfactants might themselves be preferentially used as a carbon source by the degrading bacteria subsequently excluding hydrocarbons from being utilized by the degrading microbes [99]. Thus, not only substrate or species-specific effects by biosurfactants are at play, but concentration of the biosurfactant used is also critical for effective bioremediation [30]. Soil characteristics are prominently decisive in the effect of the biosurfactant on hydrocarbon degradation as it primarily determines the extent of adsorption. Clay-rich soil tends to sorb hydrocarbons more, thus making it unavailable for bacterial degradation. Also, bacterial adhesion to particles in clay soil is enhanced, thereby sequestering the microbial population away from pollutants. Further, clay-rich soil can sorb biosurfactant itself on its surface, thereby reducing its availability for action and increasing remediation time, cost, and efforts [71]. An intermittent treatment regime using appropriate biosurfactant can take care of this limitation. SOM, age of the hydrocarbon, and age of the soil hence influence the fate and rate of biosurfactant-enhanced hydrocarbon degradation by affecting degradation activity, transport of hydrocarbon, and movement of bacteria in the soil. Soil with a high SOM was found to be most positively affected by addition of rhamnolipids as compared to other soils [100]. Desorption of hydrocarbon from soil is an integral part of the biosurfactant-mediated petroleum degradation in soil matrices. Thus 92% removal of motor oil from silty soil and 75% removal from clay soil by a biosurfactant solution from *Candida* confirms the effect of soil properties on biosurfactant effectiveness [101].

Use of biosurfactants are the most successful and sustainable approach, but its widespread use is limited by its high production cost and low yield. Hence one of the prime avenues for research is tailoring their production so as to increase yield and at the same time reduce the production cost [102]. This also integrates the aspect of sustainable waste management as a variety of feedstock including various agricultural wastes can be used for biosurfactant production [103]. Humic acid has also been reported to possess biosurfactant-like properties and hence its incorporation in the treatment regime can reduce the overall cost. Further, humic acids can also be obtained from waste agricultural biomass at a reduced cost [104]. This would make the whole bioremediation process using biosurfactants an economically viable and environmentally sustainable process [105]. The various considerations and strategies that need to be taken into consideration while using biosurfactants for a hydrocarbon polluted environmental cleanup are summarized here:

i) Biosurfactant application gives contradictory results: Study of site geochemistry, standardization of biosurfactant type, and concentration is a must before bioaugmentation or biostimulation. Use of a mixture of different biosurfactants can also give enhanced positive results.
ii) Biosurfactant action is species specific and short term: Use of a compatible and efficient biosurfactant producing strain as a consortium member is advantageous for long-term biosurfactant action and sustainability. Additional organic matter, humic and/organic fertilizers, along with bioaugmentation and biosurfactants, would prove to be beneficial at nutrient-depleted sites and reduce the process cost.
iii) Biosurfactant production is a cost-intensive process: Use of cell-free broth and/foam at low biosurfactant concentrations will reduce cost. Also use of nutrient solutions along with biosurfactants (the combination approach) would be equally effective albeit at a reduced treatment cost.

Lawniczak et al. [30] have summarized and described in detail the potential steps to be followed while designing a remediation process using biosurfactant. The most important aspects to be considered are identification of autochthonous bacterial isolates for degradation as well as biosurfactant production, biocompatibility of all the elements being used, the study of site geochemistry, as well as monitoring of the site for continued microbial activity and reduced toxicity. Nevertheless, enhancing substrate availability and compensating for unfavorable nutrient conditions form the basis of any hydrocarbon remediation strategy.

6.8 Conclusion

Petroleum hydrocarbons enter the environment by deliberate discharge or accidental spills. Remediation by microbial degradation of petroleum is effective but limited by the availability of the hydrophobic hydrocarbons to the degrading microorganisms. Biosurfactants tend to play a crucial role in taking care of this rate-limiting step in the whole remediation process. The success of an oil-degrading microorganism to colonize a polluted site that is often nutrient deficient is hence dependent on the ability of the microbes to produce biosurfactants. Choice and quantity of bacteria for consortium, concentration, and type of biosurfactant used, incorporation of additional nutrients, if at all (depending on soil characteristics), play a vital role in devising a combination strategy for successful petroleum contaminate soil remediation.

References

1 Sarma, H., Islam, N.F., Borgohain, P. et al. (2016). Localization of polycyclic aromatic hydrocarbons and heavy metals in surface soil of Asia's oldest oil and gas drilling site in Assam, Northeast India: Implications for the bio economy. *Emerging Contaminants* 2 (3): 119–127.

2 Xu, X., Liu, W., Tian, S. et al. (2018). Petroleum hydrocarbon-degrading bacteria for the remediation of oil pollution under aerobic conditions: A perspective analysis. *Frontiers in Microbiology* 9: 2885.

3 Varjani, S.J. (2017). Microbial degradation of petroleum hydrocarbons. *Bioresource Technology* 223: 277–286.

4 Sarma H, Prasad MNV (2015) Plant-microbe association-assisted removal of heavy metals and degradation of polycyclic aromatic hydrocarbons, In: S. Mukherjee (ed.), *Petroleum Geosciences. Indian Contexts*. Springer, Cham, pp. 219–236, doi: 10.1007/978-3-319-03119-4_10, ISBN: 978–3–319-03118-7.

5 Swannell, R.P.J., Lee, K., and Mcdonagh, M. (1996). Field evaluations of marine oil spill bioremediation. *Microbiological Reviews* 60 (2): 342–365.

6 Atlas, R.M. and Hazen, T.C. (2011). Oil biodegradation and bioremediation: A tale of the two worst spills in US history. *Environmental Science and Technology* 45: 6709–6715.

7 Sarma, H. and Prasad, M.N.V. (2016). Phytomanagement of polycyclic aromatic hydrocarbons and heavy metals-contaminated sites in Assam, North Eastern State of India, for boosting bioeconomy. In: *Bioremediation and Bioeconomy* (ed. M.N.V. Prasad), 609–626. Elsevier.

8 Hazen, T.C., Prince, R.C., and Mahmoudi, N. (2016). Marine oil biodegradation. *Environmental Science & Technology* 50 (5): 2121–2129.

9 Tissot, B.P. and Welte, D.H. (1978). Composition of crude oils. In: *Petroleum Formation and Occurrence*, 333–368. Berlin, Heidelberg.: Springer.

10 Mykhailova, L., Fischer, T., and Iurchenko, V. (2013). Distribution and fractional composition of petroleum hydrocarbons in roadside soils. *Applied and Environmental Soil Science* 2013: 1–6.

11 Hu, G., Li, J., and Zeng, G. (2013). Recent development in the treatment of oily sludge from petroleum industry: A review. *Journal of Hazardous Materials* 261 (4): 70–490.

12 Gupta, S., Huddar, N., Iyer, B. & Möller, T. (2018) The future of mobility in India's passenger vehicle market. https://http://www.mckinsey.com/industries/automotive-and-assembly/our-insights/the-future-of-mobility-in-indias-passenger-vehicle-market. Accessed May 25, 2020.

13 Ikhajiagbe, B., Anoliefo, G.O., Omoregbee, O., and Osigbemhe, P. (2014). Changes in the intrinsic qualities of a naturally attenuated waste engine oil polluted soil after exposure to different periods of heat shock. *Resources and Environment* 4 (1): 45–53.

14 Ritter, S.K. (2006). What's that stuff? Motor oil. *Chemical and Engineering News* 84 (11): 38.

15 Paria, S. (2008). Surfactant-enhanced remediation of organic contaminated soil and water. *Advances in Colloid and Interface Science* 138: 24–58.

16 Souza, E.C., Vessoni-Penna, T.C., and de Souza Oliveira, R.P. (2014). Biosurfactant-enhanced hydrocarbon bioremediation: An overview. *International Biodeterioration & Biodegradation* 89: 88–94.

17 Yakimov, M.M., Timmis, K.N., and Golyshin, P.N. (2007). Obligate oil-degrading marine bacteria. *Current Opinion in Biotechnology* 18 (3): 257–266.

18 Chaudhary, D. and Kim, J. (2019). New insights into bioremediation strategies for oil-contaminated soil in cold environments. *International Biodeterioration and Biodegradation* 142: 58–72.

19 Das, N. and Chandran, P. (2011). Microbial degradation of petroleum hydrocarbon contaminants: An overview. *Biotechnology Research International* 2011: 1–13.

20 Tremblay, J., Yergeau, E., Fortin, N. et al. (2017). Chemical dispersants enhance the activity of oil-and gas condensate-degrading marine bacteria. *ISME Journal* 11: 2793–2808.
21 Brooijmans, R.J.W., Pastink, M.I., and Siezen, R.J. (2009). Hydrocarbon-degrading bacteria: The oil-spill clean-up crew. *Microbial Biotechnology* 2 (6): 587–594.
22 Yang, Y., Wang, J., Liao, J. et al. (2015). Abundance and diversity of soil petroleum hydrocarbon-degrading microbial communities in oil exploring areas. *Applied Microbiology and Biotechnology* 99 (4): 1935–1946.
23 Kadali, K.K., Simons, K.L., Sheppard, P.J., and Ball, A.S. (2012). Mineralization of weathered crude oil by a hydrocarbonoclastic consortia in marine mesocosms. *Water, Air, & Soil Pollution* 223: 4283–4295.
24 Lee, Y., Lee, Y., and Jeon, C.O. (2019). Biodegradation of naphthalene, BTEX, and aliphatic hydrocarbons by *Paraburkholderia aromaticivorans* BN5 isolated from petroleum-contaminated soil. *Scientific Reports* 9 (1): 1–13.
25 Mukherjee, A.K. and Bordoloi, N.K. (2011). Bioremediation and reclamation of soil contaminated with petroleum oil hydrocarbons by exogenously seeded bacterial consortium: A pilot-scale study. *Environmental Science and Pollution Research* 18 (3): 471–478.
26 Qiao, J., Zhang, C., Luo, S., and Chen, W. (2014). Bioremediation of highly contaminated oilfield soil: Bioaugmentation for enhancing aromatic compounds removal. *Frontiers of Environmental Science & Engineering* 8 (2): 293–304.
27 Sarma, H., Sonowal, S., and Prasad, M.N.V. (2019). Plant-microbiome assisted and biochar-amended remediation of heavy metals and polyaromatic compounds – A microcosmic study. *Ecotoxicology and Environmental Safety* 176: 288–299.
28 Suja, F., Rahim, F., Taha, M.R. et al. (2014). Effects of local microbial bioaugmentation and biostimulation on the bioremediation of total petroleum hydrocarbons (TPH) in crude oil contaminated soil based on laboratory and field observations. *International Biodeterioration & Biodegradation* 90: 115–122.
29 Nichols, W.J. (2001) The U.S. Environmental Protect Agency: National Oil and Hazardous Substances Pollution Contingency Plan, Subpart J Product Schedule (40 CFR 300.900). *Proceedings of the International Oil Spill Conference* (1479–1483) American Petroleum Institute, Washington, DC, USA
30 Ławniczak, Ł., Marecik, R., and Chrzanowski, Ł. (2013). Contributions of biosurfactants to natural or induced bioremediation. *Applied Microbiology and Biotechnology* 97 (6): 2327–2339.
31 Marchant, R. and Banat, I.M. (2012a). Biosurfactants: A sustainable replacement for chemical surfactants? *Biotechnology Letters* 34 (9): 1597–1605.
32 Shekhar, S., Sundaramanickam, A., and Balasubramanian, T. (2015). Biosurfactant producing microbes and their potential applications: A review. *Critical Reviews in Environmental Science and Technology* 45 (14): 1522–1554.
33 Santos, D.K.F., Rufino, R.D., Luna, J.M. et al. (2016). Biosurfactants: multifunctional biomolecules of the 21st century. *International Journal of Molecular Sciences* 17 (3): 401.
34 Marchant, R. and Banat, I.M. (2012b). Microbial biosurfactants: challenges and opportunities for future exploitation. *Trends in Biotechnology* 30 (11): 558–565.
35 Chrzanowski, Ł., Ławniczak, Ł., and Czaczyk, K. (2012a). Why do microorganisms produce rhamnolipids? *World Journal of Microbial Biotechnology* 28: 401–419.
36 Ron, E.Z. and Rosenberg, E. (2002). Biosurfactants and oil bioremediation. *Current Opinion in Biotechnology* 13 (3): 249–252.
37 Saikia, R.R., Deka, S., Deka, M., and Sarma, H. (2012). Optimization of environmental factors for improved production of rhamnolipid biosurfactant by *Pseudomonas aeruginosa* RS29 on glycerol. *Journal of Basic Microbiology* 52 (4): 446–457.

38 Muller, M.M., Kügler, J.H., Henkel, M. et al. (2012). Rhamnolipids – Next generation surfactants? *Journal of Biotechnology* 162 (4): 366–380.

39 Mohanty, S., Jasmine, J., and Mukherji, S. (2013). Practical considerations and challenges involved in surfactant enhanced bioremediation of oil. *BioMed Research International* 2012: 328608.

40 Liu, Y., Zeng, G., Zhong, H. et al. (2017). Effect of rhamnolipid solubilization on hexadecane bioavailability: Enhancement or reduction? *Journal of Hazardous Materials* 322: 394–401.

41 Whang, L.M., Liu, P.W.G., Ma, C.C., and Cheng, S.S. (2008). Application of biosurfactants, rhamnolipid, and surfactin, for enhanced biodegradation of diesel-contaminated water and soil. *Journal of Hazardous Materials* 151 (1): 155–163.

42 Mohanty, S. and Mukherji, S. (2013). Surfactant aided biodegradation of NAPLs by *Burkholderia multivorans*: Comparison between Triton X-100 and rhamnolipid JBR-515. *Colloids and Surfaces, B: Biointerfaces* 102 (2): 644–652.

43 Wang, C.P., Li, J., Jiang, Y., and Zhang, Z.Y. (2014). Enhanced bioremediation of field agricultural soils contaminated with PAHs and OCPs. *International Journal of Environmental Research* 8 (4): 1271–1278.

44 Ma, K.Y., Sun, M.Y., Dong, W. et al. (2016). Effects of nutrition optimization strategy on rhamnolipid production in a *Pseudomonas aeruginosa* strain DN1 for bioremediation of crude oil. *Biocatalysis and Agricultural Biotechnology* 6: 144–151.

45 Noordman, W.H., Wachter, J.H., De Boer, G.J., and Janssen, D.B. (2002). The enhancement by surfactants of hexadecane degradation by *Pseudomonas aeruginosa* varies with substrate availability. *Journal of Biotechnology* 94 (2): 195–212.

46 Urum, K. and Pekdemir, T. (2004). Evaluation of biosurfactants for crude oil contaminated soil washing. *Chemosphere* 57 (9): 1139–1150.

47 Bustamante, M., Duran, N., and Diez, M.C. (2012). Biosurfactants are useful tools for the bioremediation of contaminated soil: a review. *Journal of Soil Science and Plant Nutrition* 12 (4): 667–687.

48 Zhong, H., Liu, G., Jiang, Y. et al. (2017). Transport of bacteria in porous media and its enhancement by surfactants for bioaugmentation: A review. *Biotechnology Advances* 35 (4): 490–505.

49 Kaczorek, E., Cieślak, K., Bielicka-Daszkiewicz, K., and Olszanowski, A. (2013). The influence of rhamnolipids on aliphatic fractions of diesel oil biodegradation by microorganism combinations. *Indian Journal of Microbiology* 53 (1): 84–91.

50 Kaczorek, E., Jesionowski, T., Giec, A., and Olszanowski, A. (2012). Cell surface properties of *Pseudomonas stutzeri* in the process of diesel oil biodegradation. *Biotechnology Letters* 34 (5): 857–862.

51 Kaczorek, E., Moszyńska, S., and Olszanowski, A. (2011). Modification of cell surface properties of *Pseudomonas alcaligenes* S22 during hydrocarbon biodegradation. *Biodegradation* 22 (2): 359–366.

52 Cameotra, S.S. and Singh, P. (2009). Synthesis of rhamnolipid biosurfactant and mode of hexadecane uptake by *Pseudomonas* species. *Microbial Cell Factories* 8 (1): 16.

53 Zhong, H., Liu, Y., Liu, Z. et al. (2014). Degradation of pseudo-solubilized and mass hexadecane by a *Pseudomonas aeruginosa* with treatment of rhamnolipid biosurfactant. *International Biodeterioration & Biodegradation* 94: 152–159.

54 Sharma, S. and Pandey, L.M. (2020). Production of biosurfactant by *Bacillus subtilis* RSL-2 isolated from sludge and biosurfactant mediated degradation of oil. *Bioresource Technology*: 123261.

55 Banat, I.M., Franzetti, A., Gandolfi, I. et al. (2010). Microbial biosurfactants production, applications and future potential. *Applied Microbiology and Biotechnology* 87 (2): 427–444.

56 Patricia, B. and Jean-Claude, B. (1999). Involvement of bioemulsifier in heptadecane uptake in *Pseudomonas nautica*. *Chemosphere* 38 (5): 1157–1164.

57 Aparna, A., Srinikethan, G., Hedge, S. (2011) Effect of addition of biosurfactant produced by *Pseudomonas* sps. on biodegradation of crude oil. *2nd International Proceedings of Chemical, Biological & Environmental Engineering*, vol. 6, IACSIT Press, Singapore, pp. 71e75
58 Kavitha, V., Mandal, A.B., and Gnanamani, A. (2014). Microbial biosurfactant mediated removal and/or solubilization of crude oil contamination from soil and aqueous phase: An approach with *Bacillus licheniformis* MTCC 5514. *International Biodeterioration & Biodegradation* 94: 24–30.
59 Pacwa-Płociniczak, M., Płaza, G.A., Poliwoda, A., and Piotrowska-Seget, Z. (2014). Characterization of hydrocarbon-degrading and biosurfactant-producing *Pseudomonas* sp. P-1 strain as a potential tool for bioremediation of petroleum-contaminated soil. *Environmental Science and Pollution Research* 21 (15): 9385–9395.
60 Pei, H., Xin-Hua, Z.H.A.N., Shi-Mei, W.A.N.G. et al. (2010). Effects of a biosurfactant and a synthetic surfactant on phenanthrene degradation by a *Sphingomonas* strain. *Pedosphere* 20 (6): 771–779.
61 Durval, I.J.B., Resende, A.H.M., Figueiredo, M.A. et al. (2019). Studies on biosurfactants produced using *Bacillus cereus* isolated from seawater with biotechnological potential for marine oil-spill bioremediation. *Journal of Surfactants and Detergents* 22 (2): 349–363.
62 Patel, S., Homaei, A., Patil, S., and Daverey, A. (2019). Microbial biosurfactants for oil spill remediation: Pitfalls and potentials. *Applied Microbiology and Biotechnology* 103 (1): 27–37.
63 Noordman, W.H. and Janssen, D.B. (2002). Rhamnolipid stimulates uptake of hydrophobic compounds by *Pseudomonas aeruginosa*. *Applied and Environmental Microbiology* 68 (9): 4502–4508.
64 Fenibo, E.O., Ijoma, G.N., Selvarajan, R., and Chikere, C.B. (2019). Microbial surfactants: The next generation multifunctional biomolecules for applications in the petroleum industry and its associated environmental remediation. *Microorganisms* 7: 581.
65 Torres, L.G., González, R., and Gracida, J. (2013). Production and application of no-purified Rhamnolipids in the soil-washing of TPHs contaminated soils. *Asian Soil Research Journal* 19: 1–2.
66 Silva, N.M.P.R., Rufino, R.D., Luna, J.M. et al. (2014a). Screening of *Pseudomonas* species for biosurfactant production using low-cost substrates. *Biocatalysis and Agricultural Biotechnology* 3 (2): 132–139.
67 da Rosa, C.F.C., Freire, D.M.G., and Ferraz, H.C. (2015). Biosurfactant microfoam: Application in the removal of pollutants from soil. *Journal of Environmental Chemical Engineering* 3 (1): 89–94.
68 Lima, T.M., Fonseca, A.F., Leão, B.A. et al. (2011). Oil recovery from fuel oil storage tank sludge using biosurfactants. *Journal of Bioremediation & Biodegradation* 2 (12): 10–4172.
69 Lai, C.C., Huang, Y.C., Wei, Y.H., and Chang, J.S. (2009). Biosurfactant enhanced removal of total petroleum hydrocarbons from contaminated soil. *Journal of Hazardous Materials* 167 (1–3): 609–614.
70 Hallmann, E. and Mędrzycka, K. (2015). Wetting properties of biosurfactant (rhamnolipid) with synthetic surfactants mixtures in the context of soil remediation. *Annales UMCS* 70 (1): 29–39.
71 Liu, G., Zhong, H., Yang, X. et al. (2018). Advances in applications of rhamnolipids biosurfactant in environmental remediation: A review. *Biotechnology and Bioengineering* 115 (4): 796–814.
72 Rahman, P.K.S.M. and Gakpe, E. (2008). Production, characterisation and applications of biosurfactants – A review. *Biotechnology* 7 (2): 360–370.
73 Matsui, T., Namihira, T., Mitsuta, T., and Saeki, H. (2012). Removal of oil tank bottom sludge by novel biosurfactant, JE 1058 BS. *Journal of the Japan Petroleum Institute* 55 (2): 138–141.
74 Luna, J.M., Rufino, R.D., Sarubbo, L.A., and Campos-Takaki, G.M. (2013). Characterization, surface properties and biological activity of a biosurfactant produced from industrial waste by *Candida sphaerica* UCP0995 for application in the petroleum industry. *Colloids and Surfaces B: Biointerfaces* 102: 202–209.

75 Sarubbo, L.A., Lunaa, J.M., and Rufinoa, R.D. (2015). Application of a biosurfactant produced in low-cost substrates in the removal of hydrophobic contaminants. *Chemical Engineering* 43: 295–300.

76 Rahman, K.S.M., Banat, I.M., Thahira, J. et al. (2002). Bioremediation of gasoline contaminated soil by a bacterial consortium amended with poultry litter, coir pith and rhamnolipid biosurfactant. *Bioresource Technology* 81 (1): 25–32.

77 Rahman, K.S.M., Rahman, T.J., Kourkoutas, Y. et al. (2003). Enhanced bioremediation of n-alkane in petroleum sludge using bacterial consortium amended with rhamnolipid and micronutrient. *Bioresource Technology* 90: 159–168.

78 Nikolopoulou, M. and Kalogerakis, N. (2008). Enhanced bioremediation of crude oil utilizing lipophilic fertilizers combined with biosurfactants and molasses. *Mar. Pollut. Bull.* 56 (11): 1855–1861.

79 Cameotra, S.S. and Singh, P. (2008). Bioremediation of oil sludge using crude biosurfactants. *International Biodeterioration & Biodegradation* 62 (3): 274–280.

80 Thavasi, R., Jayalakshmi, S., and Banat, I.M. (2011). Effect of biosurfactant and fertilizer on biodegradation of crude oil by marine isolates of *Bacillus megaterium*, *Corynebacterium kutscheri* and *Pseudomonas aeruginosa*. *Bioresource Technology* 102 (2): 772–778.

81 Tahseen, R., Afzal, M., Iqbal, S. et al. (2016). Rhamnolipids and nutrients boost remediation of crude oil-contaminated soil by enhancing bacterial colonization and metabolic activities. *International Biodeterioration & Biodegradation* 115: 192–198.

82 Ostendorf, T.A., Silva, I.A., Converti, A., and Sarubbo, L.A. (2019). Production and formulation of a new low-cost biosurfactant to remediate oil-contaminated seawater. *Journal of Biotechnology* 295: 71–79.

83 Harvey, S., Elashvili, I., Valdes, J.J. et al. (1990). Enhanced removal of *Exxon Valdez* spilled oil from Alaskan gravel by a microbial surfactant. *Bio/Technology* 8 (3): 228–230.

84 Tumeo, M., Braddock, J., Venator, T. et al. (1994). Effectiveness of a biosurfactant in removing weathered crude oil from subsurface beach material. *Spill Science & Technology Bulletin* 1 (1): 53–59.

85 Le Floch, S., Merlin, F.X., Guillerme, M. et al. (1999). A field experimentation on bioremediation: bioren. *Environmental Technology* 20 (8): 897–907.

86 Lee, K. G. H. Tremblay, J. Gauthier, S. E. Cobanli, and M. Griffin (1997) Bioaugmentation and biostimulation: A paradox between laboratory and field results. *Proceedings of the International Oil Spill Conference*, pp.697–705. American Petroleum Institute, Washington, DC, USA

87 Szulc, A., Ambrożewicz, D., Sydow, M. et al. (2014). The influence of bioaugmentation and biosurfactant addition on bioremediation efficiency of diesel-oil contaminated soil: Feasibility during field studies. *Journal of Environmental Management* 132: 121–128.

88 Franzetti, A., Gandolfi, I., Bestetti, G., and Banat, I.M. (2011). (Bio) surfactant and bioremediation, successes and failures. In: *Trends in Bioremediation and Phytoremediation* (ed. G. Plaza), 145–156. Kerala, India: Research Signpost.

89 Lin, T.C., Pan, P.T., Young, C.C. et al. (2011). Evaluation of the optimal strategy for ex situ bioremediation of diesel oil-contaminated soil. *Environmental Science and Pollution Research* 18 (9): 1487–1496.

90 Chrzanowski, Ł., Dziadas, M., Ławniczak, Ł. et al. (2012b). Biodegradation of rhamnolipids in liquid cultures: Effect of biosurfactant dissipation on diesel fuel/B20 blend biodegradation efficiency and bacterial community composition. *Bioresource Technology* 111: 328–335.

91 Bordoloi, N.K. and Konwar, B.K. (2009). Bacterial biosurfactant in enhancing solubility and metabolism of petroleum hydrocarbons. *Journal of Hazardous Materials* 170 (1): 495–505.

92 Usman, M.M., Dadrasnia, A., Lim, K.T. et al. (2016). Application of biosurfactants in environmental biotechnology; remediation of oil and heavy metal. *AIMS Bioengineering* 3 (3): 289–304.

93 Singh, P. and Cameotra, S.S. (2004). Potential applications of microbial surfactants in biomedical sciences. *Trends in Biotechnology* 22 (3): 142–146.

94 Vatsa, P., Sanchez, L., Clement, C. et al. (2010). Rhamnolipid biosurfactants as new players in animal and plant defense against microbes. *International Journal of Molecular Sciences* 11: 5095–5510.

95 Zeng, G., Liu, Z., Zhong, H. et al. (2011). Effect of monorhamnolipid on the degradation of n-hexadecane by *Candida tropicalis* and the association with cell surface properties. *Applied Microbiology and Biotechnology* 90 (3): 1155–1161.

96 Silva, R.D.C.F., Almeida, D.G., Rufino, R.D. et al. (2014b). Applications of biosurfactants in the petroleum industry and the remediation of oil spills. *International Journal of Molecular Sciences* 15 (7): 12523–12542.

97 Gottfried, A., Singhal, N., Elliot, R., and Swift, S. (2010). The role of salicylate and biosurfactant in inducing phenanthrene degradation in batch soil slurries. *Applied Microbiology and Biotechnology* 86 (5): 1563–1571.

98 Marecik, R., Wojtera-Kwiczor, J., Ławniczak, Ł. et al. (2012). Rhamnolipids increase the phytotoxicity of diesel oil towards four common plant species in a terrestrial environment. *Water, Air, & Soil Pollution* 223 (7): 4275–4282.

99 Cui, C.Z., Zeng, C., Wan, X. et al. (2008). Effect of rhamnolipids on degradation of anthracene by two newly isolated strains, *Sphingomonas* sp. 12A and *Pseudomonas* sp. 12B. *Journal of Microbiology and Biotechnology* 18 (1): 63–66.

100 Liu, P.W.G., Wang, S.Y., Huang, S.G., and Wang, M.Z. (2012). Effects of soil organic matter and ageing on remediation of diesel-contaminated soil. *Environmental Technology* 33 (23): 2661–2672.

101 Sobrinho, H.B.S., Rufino, R.D., Luna, J.M. et al. (2008). Utilization of two agroindustrial by-products for the production of a surfactant by *Candida sphaerica* UCP0995. *Process Biochemistry* 43: 912–917.

102 Singh, P., Patil, Y., and Rale, V. (2019). Biosurfactant production: emerging trends and promising strategies. *Journal of Applied Microbiology* 126 (1): 2–13.

103 Banat, I.M., Satpute, S.K., Cameotra, S.S. et al. (2014). Cost effective technologies and renewable substrates for biosurfactants' production. *Frontiers in Microbiology* 5: 697.

104. Salati, S., Papa, G., and Adani, F. (2011). Perspective on the use of humic acids from biomass as natural surfactants for industrial applications. *Biotechnology Advances* 29: 913–922.

105 Olasanmi, I.O. and Thring, R.W. (2018). The role of biosurfactants in the continued drive for environmental sustainability. *Sustainability* 10 (12): 4817.

7

Microbial Surfactants are Next-Generation Biomolecules for Sustainable Remediation of Polyaromatic Hydrocarbons

Punniyakotti Parthipan[1], Liang Cheng[2], Aruliah Rajasekar[3], and Subramania Angaiah[1]

[1] *Electro-Materials Research Lab, Centre for Nanoscience and Technology, Pondicherry University, Puducherry, India*
[2] *School of Environment and Safety Engineering, Jiangsu University, Zhenjiang, China*
[3] *Environmental Molecular Microbiology Research Laboratory, Department of Biotechnology, Thiruvalluvar University, Vellore, Tamilnadu, India*

CHAPTER MENU

7.1 Introduction

Biosurfactants are surface-active molecules produced by diverse microbial groups such as bacteria, fungi, and yeast and are also reported to be of plant and animal origin [1–3]. Biosurfactants are amphiphilic molecules that are usually composed of both hydrophilic and hydrophobic moieties [4–7]. This structural nature makes them more surface-active molecules with characteristics that reduce the surface and the interfacial tension of aqueous solutions and hydrocarbons [8]. Microbial biosurfactants are produced as secondary metabolites that may adhere to the surface of microbial cells or are found outside of microbes. Biosurfactants are compensated in a number of ways compared to chemical surfactants with less toxicity, are easily degradable, are stable with higher pH and temperature, and with higher foaming [9]. Also, they can be synthesized using cheap raw materials such as agro-wastes and industrial waste materials, and so are indirectly helpful in the reduction of other waste materials as well.

Biosurfactants for a Sustainable Future: Production and Applications in the Environment and Biomedicine, First Edition. Edited by Hemen Sarma and Majeti Narasimha Vara Prasad.
© 2021 John Wiley & Sons Ltd. Published 2021 by John Wiley & Sons Ltd.

Biosurfactants can be categorized into two major classes: low molecular weight biosurfactants, including glycolipids and lipopeptides, and high molecular weight biosurfactants including lipoproteins and polysaccharides, which are called bioemulsions or bioemulsifiers [10]. Overall, the most investigated biosurfactants are in the glycolipids group with rhamnolipids, sophorolipids, trehalolipids, and mannosylerythritol lipids [11]. Rhamnolipids from *Pseudomonas aeruginosa* are successfully manufactured by Jeneil Biosurfactant, USA, mainly as a fungicide agent for agricultural use and additionally as an improver for the bioremediation process [12].

Many of the indigenous microbial strains survive in a hydrocarbon-rich environment by producing surface-active biological molecules, which help them survive in extreme conditions. Competence and efficiency are important characteristics of a good biosurfactant. Competence is calculated by the critical micelle concentration (CMC), whereas efficiency is associated with surface and interfacial tensions. At low concentrations, biosurfactants can be found as distinct molecules. However, raising their concentration was accomplished where no further changes in their interfacial nature took place. The concentration needed to reach this level is named the CMC of that biosurfactant. The CMC of biosurfactants ranges from 1 to 2000 mg/l. Distilled water has a surface tension of about 72 mN/m. A potential biosurfactant can reduce surface tension below 30 mN/m, with interfacial tension of n-hexadecane reduced from 40 to 1 mN/m. The molecular weight of biosurfactants ranges from about 500 to 1500. Biosurfactants are used in many industries/applications including the detergent industry, the pharmaceutical industry, oil recovery, bioremediation, etc. Using biosurfactants in the bioremediation of hydrocarbon contaminants is gaining great attention due to their significant outcomes.

Polycyclic aromatic hydrocarbons (PAHs) are groups of hydrocarbons with two or even more fused benzene rings in their structure [13–15]. They are found naturally and formed by thermal decomposition of organic contents and their succeeding recombination process. Partial combustion of organic matter at high temperatures (i.e. 500–800 °C) or incapability of organic molecules at low temperatures (100–300 °C) for prolonged periods ends with the production of PAHs. They naturally occur as colorless contentments with white/pale yellow solids along with less solubility in water, higher melting and boiling points, and also low vapor pressure. If their molecular weight increases their solubility in water declines [16, 17]. PAH exposure is very common and is harmful to all living organisms in the environment if they are released as a result of anthropocentric activities [18–20].

PAHs are frequently encountered naturally or by man-made actions. Important natural causes are forest fires, oil spills, and volcanic eruptions. Anthropogenic resources for overaccumulation of PAHs are coal tar, burning of fossil fuel, garbage, wood, used lubricating oil and oil filters, waste incineration, and petroleum spill/discharge (Figure 7.1). PAHs from used motor oil makes a huge impact on the environment due to improper disposal into the soil. In a study by Paneque et al. [21] about 16 different types of PAHs were identified and all these PAHs are considered to be most toxic to nature since their disposal is closely relevant to human activities, which makes the chance of inhalation by human beings very common and may perhaps cause serious health problems. Partial burning of organic materials gives out about 100 different PAHs, which are the primary pollutants. Figure 7.1 illustrates different and common sources of PAHs abundant in the environment. PAHs are found in varying levels in soil from 1 μg to 300 g/kg, depending on the sources of the pollutants [22]. A few PAHs and their epoxides are extremely toxic and mutagenic even to microorganisms. About six particular PAHs are listed among the top 126 precedence pollutants by the US Environmental Protection Agency.

PAHs are very dangerous contaminants among other hydrocarbons since they are found anywhere with characteristics of mutagenic, toxic, and carcinogenic [23]. Table 7.1 describes the most

Figure 7.1 Major sources of PAHs in the environment (See insert for color representation of the figure).

dangerous PAHs and their basic properties. PAHs are considered as very important types of contaminants because once expressed their degradation or elimination from the environment is very difficult to achieve naturally. They present anywhere in the environment, have less biodegradation aptitude, have the capability to bioaccumulate, and importantly are of a carcinogenic nature. These things make them a primary source of a pollutant for soil, water, and atmosphere. The toxic nature of the PAHs can be sorted out by using conventional approaches like alteration, removal, or separation of the pollutant. These techniques have difficulty in separation from the original state to another one, are expensive, and are not eco-friendly. The alternative option for these conventional techniques is bioremediation, a greener approach to clean up the contaminant or hazardous material into nonhazardous material using less energy and also fewer chemicals [17]. Many microorganisms have the aptitude to utilize the PAHs as their carbon source for their growth and development and may belong to bacteria, algae, and fungi.

Bioremediation exploits the catabolic process of natural microorganisms to treat contaminated soil, water, and air [24, 25]. Figure 7.2 illustrates the biodegradation pathway of PAHs by different types of microorganisms and their enzymes. This is an eco-friendly approach and a cost-effective method. While bioremediation has been applied for decades, it is still in an emerging field that can be improved by innovative approaches with efficient microorganisms [26]. The bioremediation process has some limitations, such as less efficient microorganisms, hydrophobic nature of contaminants, and environmental conditions (pH and temperature). One of the key problems is the bioavailability of contaminants, because in many cases microbes cannot access the hydrophobic pollutant, which will lead to incomplete remediation. For substrate availability, the addition of emulsifiers is helpful if the contaminants have naturally low solubility that tends to partition into non-aqueous environments. Biosurfactants can be used as a stimulating material for the bioremediation process [27].

In this chapter, we will discuss the role and functions of biosurfactants in the biodegradation of toxic PAHs, particularly in the case of direct uptake of the hydrophobic compounds by microorganisms using the surface-active molecules with the help of the process called micellization.

Table 7.1 Important and most toxic PAHs with their key properties.

PAH	Structure	No. of rings	Chemical formula	Molecular weight	Solubility (mg/l)	Half-life in soil (days)
Naphthalene	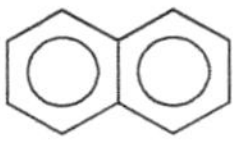	2	$C_{10}H_8$	128	31	80
Acenaphthene	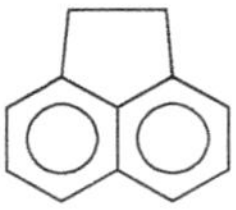	3	$C_{12}H_{10}$	154	3.8	102
Acenaphthylene		3	$C_{12}H_8$	152	16.1	60
Fluorene	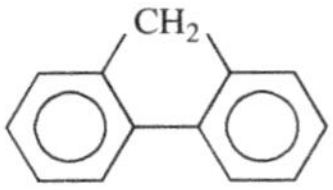 CH_2	3	$C_{13}H_{10}$	166	1.9	60
Phenanthrene	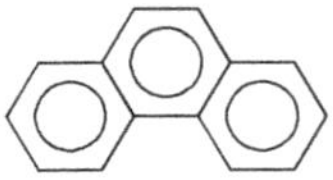	3	$C_{14}H_{10}$	178	1.1	16–200
Anthracene	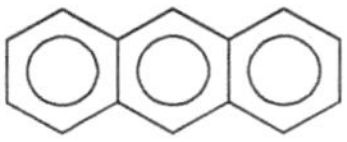	3	$C_{14}H_{10}$	178	0.045	50–460
Pyrene		4	$C_{16}H_{10}$	202	0.132	199–1870
Fluoranthene	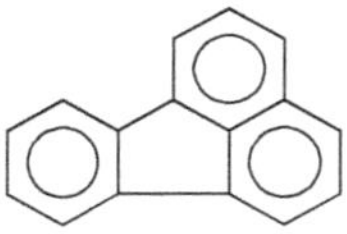	4	$C_{16}H_{10}$	203	0.26	44–440
Benzo(a) anthracene	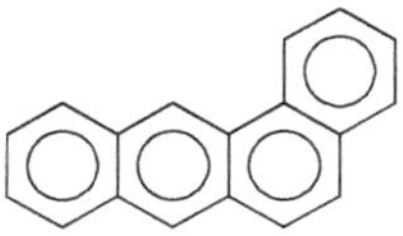	4	$C_{18}H_{12}$	228	0.011	162–670
Chrysene	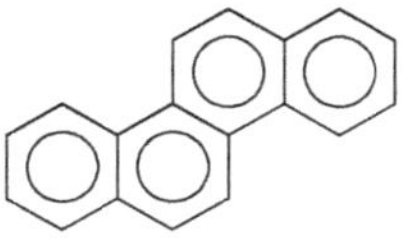	4	$C_{18}H_{12}$	228	0.0015	371–990
Benzo(a)pyrene	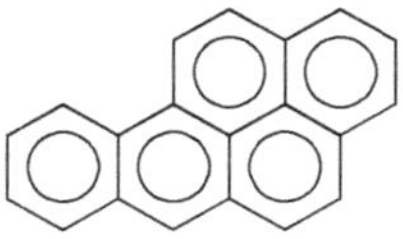	5	$C_{20}H_{12}$	252	0.0038	229–530
Benzo(b) fluoranthene	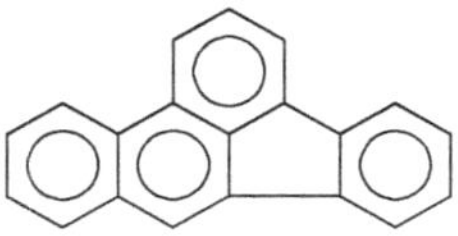	5	$C_{20}H_{12}$	252	0.0015	67–610

Table 7.1 (Continued)

PAH	Structure	No. of rings	Chemical formula	Molecular weight	Solubility (mg/l)	Half-life in soil (days)
Dibenzo(a,h) anthracene		5	$C_{22}H_{14}$	278	0.0005	361–940
Benzo(g,h,i) perylene		6	$C_{22}H_{12}$	276	0.00026	74–650
Indeno(1,2,3-cd) pyren		6	$C_{22}H_{12}$	276	0.062	288–730
Coronene		7	$C_{22}H_{12}$	300	0.00014	732

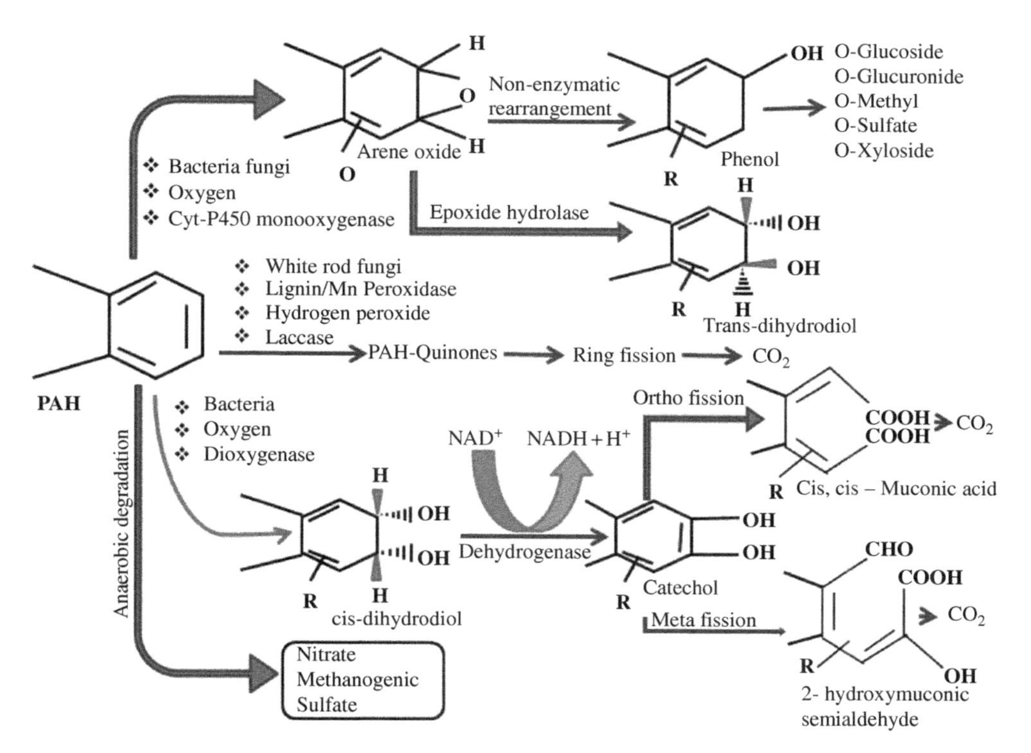

Figure 7.2 Biodegradation pathway of PAHs by microorganisms. *Source:* Adapted from Haritash and Kaushik [17].

We will discuss in more detail the importance of bioavailability and its impact on the biodegradation of the PAHs and also the surface adaptation/modification of the biosurfactant-producing microorganisms toward the utilization of the PAHs as the sole carbon source.

7.2 Biosurfactant-Enhanced Bioremediation of PAHs

Many physiochemical factors play a key role in bioremediation of PAHs, such as temperature, pH, electron donors/acceptors, nutrient availability, the concentration of the pollutant, the presence of degrading microorganisms, and, more importantly, the bioavailability of the contaminants [28]. Among these factors, bioavailability is crucial to enhance the degradation efficiency since other factors can be overcome very easily. Bioavailability can be defined as the number of pollutants available for the utilization of microorganisms [29]. The biodegradation rate is proportional to the bioavailability of the pollutants because microorganisms need to burn up energy to boost catabolic genes that were involved in the synthesis of degrading enzymes. If the deceptive pollutant level is low, encouragement will not occur, but perhaps if high, the process will be enhanced [30]. Indigenous soil microbial strains are typically slow-growing organisms usually exposed to oligotrophic conditions [31]. The presence of a contaminant can incorporate an imperative influence on the metabolic potential of such microorganisms. Three possible causes could happen if bioavailability is inconsistent. In the first situation, degradation will not occur if the level of accessible PAHs is insufficient to satisfy the energy expenses to promote biodegradation. In the second situation, if the bioavailability is much less, microorganisms can utilize pollutants that were insufficient for the development of the fresh cells. In the third situation, if a sufficient level of PAHs were found, the biodegradation rate could be boosted enormously. It is likely that the third situation could allow for the greatest complimentary rates of bioremediation. As mentioned earlier, a biosurfactant can be categorized into several classes such as low molecular weight and high molecular weight biosurfactants, as described in Table 7.2, where their role in PAHs degradation is given in detail.

7.2.1 Low Molecular Weight Biosurfactant and Their Role in PAH Degradation

7.2.1.1 Glycolipids and Their Role in PAH Degradation

Low molecular weight biosurfactants (LMWBs) can be mainly classified as rhamnolipids, trehalolipids, and sophorolipids. The diverse hydrophobic fatty acids of the glycolipids have the aptitude to lower the Kraft temperature to control their liquid crystalline arrangements [54]. One of the most important and widely studied and applied glycolipid biosurfactants are rhamnolipids, which are included with one or two rhamnoses along with up to three hydroxy fatty acids, each fatty acid having 8–22 carbons in its chain lengths [55]. Sophorolipid biosurfactants contain sophorose, a 1,2-disaccharide of glucose, which forms the hydrophilic head of the molecule and perhaps acetylated with one or two acetyl groups [56]. Trehalolipids are made of disaccharide trehalose, which link at C-6 and C-6′ of mycolic acids [8].

Li et al. [33] reported that rhamnolipid produced by a bacterial strain (*Bacillus* Lz-2) can increase the solubility of naphthalene, phenanthrene, and pyrene, and this solubility was increased with the augment of rhamnolipid dose above the bio-surfactant critical micelle concentration (CMC). They also found that the solubility of naphthalene, phenanthrene, and pyrene was increased up to 129, 131, and 372% from its original solubility in the aqueous phase. This much solubility was attained due to the fact that hydrophobic pollutants willingly partition into the hydrophobic core of a micelle, thus increasing that PAH aqueous concentration through micelle solubilization. This

Table 7.2 Low and high molecular weight biosurfactants with enhanced biodegradation of PAHs.

Biosurfactant	PAHs	Source microorganisms	Key outcomes/achievements	References
Low molecular weight biosurfactant-enhanced PAHs degradation				
Glycolipids enhanced PAHs degradation				
Rhamnolipid	Acenaphthene, fluorene and 13 more PAHs	Commercially purchased	Rhamnolipid was very useful to promote PAHs biodegradation	[32]
Rhamnolipids	Phenanthrene	*P. aeruginosa* ATCC9027	Rhamnolipids promoted phenanthrene degradation effectively	[76]
Rhamnolipid	Naphthalene, phenanthrene, and pyrene	*Bacillus* Lz-2	Maximum solubility obtained at pH 11, which shows biosurfactant was stable and active at high pH	[33]
Rhamnolipid	Phenanthrene, pyrene and benzo(a)pyrene	Commercially purchased	The addition of rhamnolipid increased the dissolution of PAH which helps to *Stenotrophomonas* sp. IITR87 to utilize them	[34]
Rhamnolipid	Phenanthrene and fluoranthene	*Pseudomonas aeruginosa*	Biosurfactant production ability of this strain enhanced the degradation efficiency of PAHs to 80–90%.	[35]
Biosurfactant	Phenanthrene and pyrene	*Pseudomonas stutzeri*	Higher emulsification index and reduced surface tension of the medium	[36]
Rhamnolipid	Indene, naphthalene, and 7 more PAHs	*Pseudomonas aeruginosa* SR17	60–80% of degradation efficiency was noticed in the presence of rhamnolipid	[37]
Rhamnolipid	Fluorene	*Pseudoxanthomonas* sp. PNK-04	96.28% of degradation efficiency obtained, which was almost equal to Tween 80	[38]
Rhamnolipid	Pyrene, anthracene, and phenanthrene	*Pseudomonas aeruginosa* PF2	Higher removal of PAH from soil was noticed while increasing the concentration of rhamnolipid	[39]
Rhamnolipid	Benzo(a)pyrene	Commercially provided by AGAE Technologies (Oregon, USA)	Biodegradation of slow-desorption PAHs during bioremediation	[40]
Mono and Di-Rhamnolipid	Anthracene, phenanthrene and pyrene	*P. sihuiensis*	This strain able to produce biosurfactant during PAHs degradation and enhance its removal	[41]
Glycolipid	PAH standard solution with 16 compounds	*Pseudomonas aeruginosa*	PAHs sludge removal was enhanced indigenous biosurfactant-producing strain	[42]

(Continued)

Table 7.2 (Continued)

Biosurfactant	PAHs	Source microorganisms	Key outcomes/achievements	References
Lipopeptide-enhanced PAH degradation				
Amphisin and viscosin	Naphthalene to fluorene	*P. fluorescens* DSS73 and *P. fluorescens* PfA7B	Actively enhance the solubility of low molecular weight PAHs	[43]
Surfactin, fengycin, and Lichenysin	Fluorene, naphthalene, phenanthrene, pyrene	*Pseudomonas* sp. WJ6	Enhanced degradation of PAHs, degradation and heavy oil sand washing	[44]
Lipopeptide	Phenanthrene and pyrene	*Paenibacillus dendritiformis*	Increased bioavailability also showed enhanced oil recovery activity	[45]
Lipopeptide	Naphthalene, phenanthren, pyrene, and 12 more PAHs	*P. aeruginosa CB1*	Degradation efficiency was enhanced significantly	[46]
Lipopeptide	Pyrene	*Paenibacillus dendritiformis* CN5	83.5% of degradation efficiency was improved from 16% in its absence	[47]
Lipopeptide	Naphthalene, acenaphthylene, and 11 more PAHs	*Bacillus cereus SPL-4*	Enhanced high molecular weight PAHs	[48]
Surfactin	Anthracene, fluorene, phenanthrene, and pyrene	*Aeribacillus pallidus,* *B. axarquiensis, B. siamensis*, and *B. subtilis* subsp.*inaquosorum*	This biosurfactant reduced surface tension of water up to 26.75 mN/m with enhanced biodegradation of PAHs	[49]
Surfactin and iturins	Anthracene, phenanthrene, and pyrene	*B. methylotrophicu*	This strain able to produced biosurfactant during PAHs degradation and enhance its removal	[41]
Surfactin	Alkyl phenanthrenes, methyl phenanthrene	*Bacillus amyloliquefaciens*	The enhanced degradation efficiency of PAHs along with bacterial *Achromobacter* sp., *Citrobacter* sp.	[50]
Biosurfactant	Anthracene	*Bacillus circulans*	Biosurfactant effectively entrap and solubilize PAH	[51]
Biosurfactant	Phenanthrene	*Brevibacillus* sp.	Biosurfactant produced by this strain enhanced degradation efficiency	[52]
Biosurfactant	Phenanthrene	*Acinetobacter calcoaceticus* BU03	Enhanced the efficiency of PAH degradation in bioslurry system	[77]
High molecular weight biosurfactant-enhanced PAHs degradation				
Glycoprotein	Naphthalene, phenanthren, pyrene	*Aeribacillus pallidus* SL-1	Enhanced the solubility of PAHs and increased the degradation efficiency	[53]

strain has been recognized as an indigenous potent biosurfactant producing a bacterial strain since it was isolated from the oil-polluted water from Dongying Shengli oilfield. Recently, Pourfadakari et al. [39] reported that a halotolerant bacterial strain, *Pseudomonas aeruginosa* PF2, produced a rhamnolipid type of biosurfactant with a CMC value of 60 mg/l with moderate emulsification activity. Mainly rhamnolipid was used for desorption of PAHs from the soil and that desorbed soil was then subjected to electrokinetic oxidation as an integrated approach to remediate PAH (pyrene, anthracene, and phenanthrene) contamination. From this integrated approach, they obtained 99% of PAH removal efficiency.

Similarly, very interesting findings were reported by Reddy et al. [38] with rhamnolipid biosurfactant. Briefly, they performed fluorene degradation efficiency by using the bacterial strain *Paenibacillus* sp. PRNK-6 along with synthetics (Tween-80, Tween-60, Tween-40, Tween-20, Triton X-100) and biosurfactants. During the degradation study, the growth of *Paenibacillus* sp. was enhanced along with enhanced fluorene utilization in the presence of biosurfactant and Tween-80, Tween-60, and Tween-40. More specifically, this bacterial strain consumed about 75% of fluorene in one day without any additional support, but at the same time inclusion of the surfactant and biosurfactant enhanced their degradation efficiency. The addition of Tween-80 led to complete removal within a day. Similarly, Tween-60 and Tween-40 showed 90.6 and 96.5% of degradation, respectively. Fascinatingly, biosurfactant inclusion showed 96.7% of fluorine degradation, which was higher than Tween-60 and Tween-40. These surfactant solutions have increased the cell surface hydrophobicity of the *Paenibacillus* sp. PRNK-6, which in turn reflects on their degradation efficiency. Almost equal degradation activity was achieved for both chemical and biosurfactant, which makes it clear that the biosurfactant is a significant alternative to the synthetic surfactant with potential equal degradation efficiency to use in environmental applications without any toxic effects.

Some researchers found even more biodegradation activity than synthetic surfactant, where strain *P. aeruginosa* SR17 was used for total petroleum hydrocarbon (TPH) degradation along with rhamnolipid biosurfactant and achieved 86 and 80.5% of 6800 ppm and 8500 ppm of TPH, respectively. At the same time, the addition of sodium dodecyl sulphate (SDS) achieved 70.8 and 68% degradation of TPH with the same concentration of the contaminant. Gas chromatography analysis of the same soil samples confirms the presence of PAHs such as naphthalene, anthracene, indene, chamazulene, phenanthrene, fluorene, fluoranthenefluoranthene, benz(d)anthracene, and benz(b)fluorene. Treatment of these PAHs with rhamnolipid eliminates fluoranthene, benz(d) anthracene, and benz(b)fluorine in six months of remediation time and remaining PAHs were eliminated in the range of 60–80% [37].

Zhao et al. [76] summarized the role of rhamnolipids on cell surface hydrophobicity (CSH) during the biodegradation of phenanthrene by rapid degrader (*Bacillus subtilis* BUM) and slow degrader (*P. aeruginosa* P-CG3). Interestingly, rhamnolipids produced by *P. aeruginosa* ATCC9027 significantly increased the CSH. Without rhamnolipids, mixed cultures of BUM and P-CG3 utilized about 82% of phenanthrene within 30 days, where strain BUM played a key role in that utilization and growth of strain P-CG3 was suppressed. At the same time, the inclusion of rhamnolipids enhanced degradation efficiency of P-CG3, but surprisingly inhibited the direct utilization by BUM. Due to the lower activity of strain BUM, strain P-CG3 dominated in biodegradation of phenanthrene at the initial stage and improved the biodegradation efficiency to 92.7%. From this research outcome, it was very clear that the biosurfactant played a key role in the co-degradation with bacterial strains, since many of the microorganisms are sensitive to the biosurfactants. Many researchers illustrated that biosurfactants are antimicrobial compounds so these factors also need to be considered for the selection of a biosurfactant for the bioremediation studies with other microorganisms. A recent study by Kumari et al. [57] discloses that the addition of rhamnolipid into the consortium-based degradation enhanced their original degradation efficiency by 10%. Also, they have disclosed that mixed consortium efficiently improved the degradation percentage rather than individual bacterial strains. In their study they summarized this as follows: strain *Microbacterium esteraromaticum* degraded 81.4% of naphthalene; *P. aeruginosa* degraded 67.1% of phenanthrene and 61% of benzo(b)fluoranthene; and *Stenotrophomonas maltophilia* degraded 47.9% of fluorene in 45 days. At the same time, degradation carried out with mixed consortiums enhanced the degradation efficiency to 89.1% for naphthalene, 63.8% for fluorene, 81% for phenanthrene, and 72.8% for benzo(b)fluoranthene.

Recent research by Posada-Baquero et al. [40] suggests that rhamnolipid biosurfactant can be applied in the slowly desorbed PAHs in previously remediated soil. The normal bioremediation process took several months to years to reduce the concentration of the pollutants; in such cases, the addition of a biosurfactant enhanced the degradation rate by increasing the bioavailability of the hydrophobic pollutants. Another work published by Sun et al. [42] describes the biosurfactant-producing bacterial strain *P. aeruginosa* promoting the *in situ* removal of high molecular weight PAHs from sludge samples. This bacterial strain has the aptitude to produce glycolipid biosurfactant with a CMC value of 96.5 mg/l; this biosurfactant highly reduced surface tension from 72.2 to 29.6 mN/m. This characteristic feature of the biosurfactant plays an important role in the removal of high molecular weight PAHs in the sludge phase by this bacterium.

Ortega Ramirez et al. [58] demonstrated dibenzothiophene biodegradation by strain *Burkholderia* sp. Dibenzothiophene is a sulfur-containing PAH frequently found in urbanized areas, and is also used as a model pollutant to measure the soil contamination level. Under an optimum condition, the dibenzothiophene degradation rate was increased to 18-fold and enhanced dibenzothiophene biodegradation by 25–30% on day 1. This much efficiency was achieved due to their bacterial growth and biosurfactant biosynthesis. Enzymes played a key role in biosurfactant biosynthesis. Similarly, Li et al. [59] also reported that dibenzothiophene degradation was enhanced by di-rhamnolipid biosurfactant produced by the bacterial strain *Pseudomonas* sp. LKY-5. Pereira et al. [41] summarized that anthracene, phenanthrene, and pyrene degradation was enhanced by biosurfactant produced by the bacterial strain *Pardanthus sihuiensis*. This strain produced mono- and di-rhamnolipids, which effectively reduced the surface tension and also showed good emulsifying activity of the cell-free supernatant. Biodegradation efficiency of this strain for anthracene, phenanthrene, and pyrene was about 46.8, 32.5, and 38.1, respectively. This efficiency percentage seemed to produce less activity but compared to the control it was considered as an effective one.

In 2018, fascinating research work was done by researchers Ma et al. [60]. In this work, they illustrated the role of biosurfactant in the bioavailability of hydrophobic PAH. The bacterial strain *Pseudomonas* sp. Ph6 was used to degrade phenanthrene with the addition of rhamnolipid. This glycolipid type of biosurfactant has the aptitude to change the cell surface hydrophobicity, cell-surface zeta potential, cell morphology, and functional groups. This modification mainly assisted in the biosorption of phenanthrene by strain *Pseudomonas* sp. Ph6. A higher level of phenanthrene was sorbed on to the cell envelopes, which led to more phenanthrene diffusing into cytochylema, which in turn favor *Pseudomonas* sp. Ph6 to utilize them. At the same time, it was suggested that optimum rhamnolipid needs to be used for efficient degradation, because 100 mg/l enhanced the degradation efficiency but efficiency was decreased while increasing the biosurfactant concentration to 200 mg/l. Hence this biosorption activity of the surfactant was specific with optimum concentration. Similarly, *P. aeruginosa* was used for degradation of different hydrocarbons including PAHs (pyrene and fluoranthene) and crude oil by Chebbi et al. [35]. This *Pseudomonas* strain showed about 90 and 99% of fluoranthene and crude oil degradation efficiency with the rhamnolipid type of biosurfactant production ability. A new bacterial strain *Brevibacillus* sp. PDM-3 was used for the degradation of phenanthrene. Different physicochemical factors such as pH, temperature, nutrient medium, rpm, and inoculum size were optimized for the optimal growth of this bacterium. After six days of degradation study, it was found that there was 93% of phenanthrene degradation. A key factor played in this degradation study is the capability of production of a glycolipid type of biosurfactant during the phenanthrene degradation [52]. A similar observation was also observed with the enhancement of PAHs degradation by other researchers [34]. Sponza and Gok [61] used three different biosurfactants, namely rhamnolipid, emulsan, and surfactin, for removal of PAHs in a laboratory-scale aerobic activated sludge reactor system and found that rhamnolipid was a more efficient PAHs remover than other studied biosurfactant [32].

7.2.1.2 Lipopeptides and Their Role in PAHs Degradation

Bezza and Chirwa [45] summarized that the *Paenibacillus dendritiformis* strain produced a lipopeptide type of biosurfactant that showed good emulsification activity toward hexane and cyclohexane. Interestingly, this biosurfactant showed higher stability at a diverse range of pH, temperature, and salinity. With these tremendous features the extracted biosurfactant remarkably desorbed the PAHs from spiked soils, particularly desorbing more than 96% of phenanthrene and 83% of pyrene from the contaminated soil within a short time (five days). The same researchers, Bezza and Chirwa [47], reported that biosurfactants increased the bioavailability of PAH. Microcosms amended with lipopeptides enhanced the effective removal of five- and six-ring PAHs up to 79%. The same researchers studied the impact of biosurfactant on pyrene biodegradation by *Pseudomonas* sp. In the presence of lipopeptides at concentrations of 300 and 600 mg/l, pyrene degradation was achieved by about 67 and 83.5%, respectively, which was much higher than that achieved in the absence of the biosurfactant (16% only). The noteworthy point from this study is that the degradation of pyrene was increased along with the increasing concentration of the lipopeptide level up to a threshold level [46].

PAH degradation by bio-surfactant is specific to every surfactant and PAHs used vice versa; interesting facts were described by a researcher Portet-Koltalo et al. [43]. In their research they have analyzed the influence of PAHs sorption behavior by a chemical surfactant SDS along with cyclolipopeptidic a biosurfactant synthesized from two *Pseudomonas fluorescens* strains. From that study, they have summarized that chemical surfactant SDS have influenced the release of almost all the used PAHs, and used cyclolipopeptidic biosurfactants are active against lower molecular weight PAHs such as naphthalene to fluorene. From this study, it was clear that bio-surfactant activity is specific to each hydrocarbon used. A research group from India used marine biosurfactant synthesizing bacterial strain *Bacillus circulans* to increase the bioavailability of anthracene. Based on the gas chromatography (GC) and high-performance thin-layer chromatography (HPTLC) analysis it was confirmed that biosurfactant efficiently solubilized that anthracene [42].

Besides, the biosurfactant nature is defined by the nature of the substrate being utilized by its producers. For instance, *Pseudomonas* sp., a common and well-known glycolipid producer, synthesized lipopeptide types of biosurfactant while grown with different hydrophobic carbon sources such as n-dodecane, fluorene, and pyrene. It implies that every microorganism has its metabolic pathways to produce secondary metabolites based on its primary carbon sources. Surfactin was confirmed in their three isoforms with n-dodecane as a carbon source but which was not formed while in the presence of PAHs (fluorine and pyrene). Nevertheless, the same bacteria produced fengycin and lichenysin with all three hydrophobic substrates. This observation shows the cyclic lipopeptide biosurfactant production capability of strain *Pseudomonas* sp. WJ6 was able to keep them active by utilizing n-alkanes and PAHs [44]. The addition of surfactin produced by bacterial strain *Bacillus amyloliquefaciens* plays a key role in the alkyl aromatic hydrocarbons by 1.5–87.2%. Biosurfactant addition increased the degradation efficiency of the bacterial strains (*Bacillus amyloliquefaciens*, *Achromobacter* sp., and *Brucella melitensis*) by enhancing the bioavailability of alkylphenanthrenes, alkyldibenzothiophenes, and methyltriaromatic steroids [50].

Pereira et al. [41] summarized the degradation by a biosurfactant producing a bacterial strain of *Bacillus methylotrophicus* which was used to degrade three different PAHs, such as anthracene, phenanthrene, and pyrene. This *Bacillus* strain produced lipopeptide types of biosurfactant, namely surfactins and iturins. The surfactant producing ability of this strain shows considerable degradation efficiency of anthracene, phenanthrene, and pyrene with 45.1, 36.0, and 31.8%, respectively.

PAH contaminations are common in the diverse environment, including the higher temperature regions, so it is necessary to make sure that microorganisms used for bioremediation purposes should be of a thermotolerant or thermophilic nature. Recently, Mehetre et al. [49] reported such bacterial strains for the bioremediation of phenanthrene, anthracene, fluorine, and pyrene. For that purpose, thermotolerant bacteria strains such as *Bacillus* axarquiensis (UCPD1), *Aeribacillus pallidus* (UCPS2), *Bacillus subtilis* subsp. *inaquosorum* (U277), and *Bacillus siamensis* (GHP76) were identified from Unkeshwar hot spring (India). A noteworthy point here is that the biodegradation experiment was conducted under two conditions for each strain, which was at the mesophilic condition (37 °C) and the thermophilic condition (50 °C). Higher degradation efficiency was obtained at thermophilic conditions only. The reason behind this potential activity of these thermophilic bacterial strains is that usually *Bacillus* species are spore formers, which helps them to exist at a higher temperature and also these strains produced a surfactin type of biosurfactant during the utilization of selected PAHs, so this biosurfactant producing capability was enhanced by their degradation efficiency at high temperatures.

In a recent research, Ibrar and Zhang [62] used multiple microbial strains, including *Lysinibacillus, Gordonia, Paenibacillus*, and *Cupriavidus* sp., which were isolated from the non-polluted area and confirmed that these strains are capable of utilizing naphthalene and anthracene as a primary sole carbon source with the production of lipopeptide biosurfactant. This biosurfactant was confirmed as heptapeptide and lipid moieties, which play a key role in decreasing the hydrophobicity of PAHs.

7.2.1.3 Emulsifier-Enhanced PAH Degradation

Bioemulsifiers are considered as high molecular weight surfactants with diverse properties and also play an important function in solubilizing hydrophobic contaminants [63]. The emulsan type of biosurfactant synthesized by *Acinetobacter calcoaceticus* RAG-1 and *A. calcoaceticus* BD4 consists of anionic lipopolysaccharide and protein [64, 65]. Similarly, biodispersan, an extracellular non-dialyzable dispersing agent produced by *A.* calcoaceticus A2, consists of anionic heteropolysaccharide with four reducing sugars [64]. Alasan produced by strain *A. radioresistens* KA-53 consists of anionic alanine-containing heteropolysaccharide protein [65].

Recently, Tao et al. [53] reported that thermophilic hydrocarbon utilizing strain *Aeribacillus pallidus* SL-1 has the potential to degrade aromatic hydrocarbons from crude oil at 60 °C. This strain can utilize 84% of naphthalene, 80% of phenanthrene, and 50% of pyrene. This much degradation efficiency was attained due to their bioemulsifier producing ability with the hydrophobic substrate at a higher temperature. The produced bioemulsifier was confirmed as a glycoprotein, with polysaccharides (65.6%) and proteins (13.1%). Among them, proteins playing a key role in the emulsifying properties and also this emulsifier enhanced the solubility of PAHs. The bioemulsifier produced by this thermophilic strain is highly active and stable at high temperatures, so can be used for high-temperature bioremediation purposes.

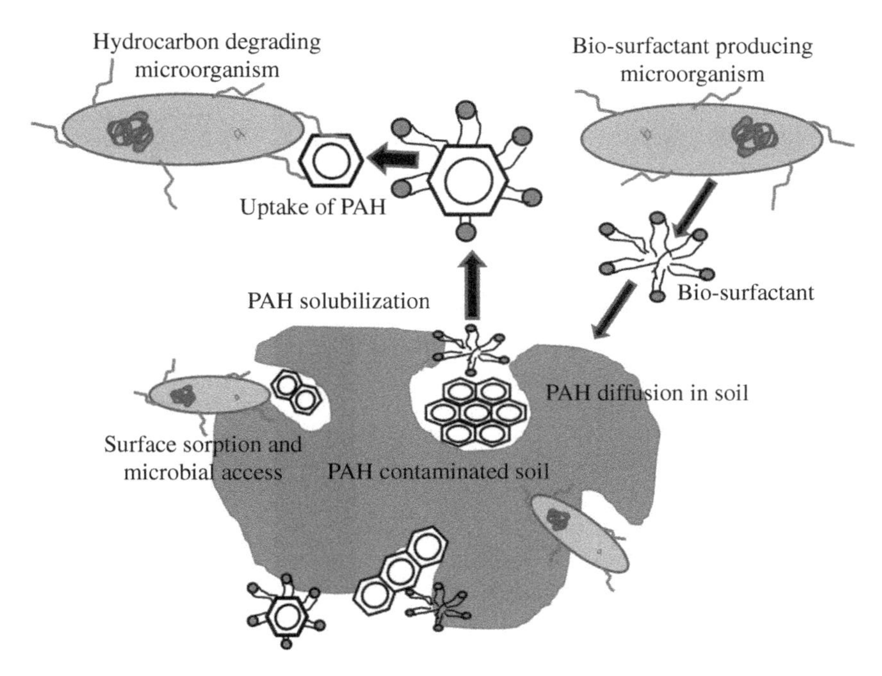

Figure 7.3 Biosurfactant-enhanced PAH degradation mechanism.

7.3 Microorganism's Adaptations to Enhance Bioavailability

Every microorganism has adopted a different approach to improving the bioavailability of organic pollutants. An important approach confirmed by many organisms is the enhancement of cell adhesion to hydrophobic surfaces. The hydrophobic nature of the PAHs has tended to make them immiscible compounds to the degrading organisms. To overcome this issue the microorganism needs to fix with the hydrophobic substrate and promptly entice it. Such attraction diminishes restrictions due to the rate-limited huge shift of a substrate into the aqueous phase. A second approach is by producing surface-active molecules in the presence of PAHs, as illustrated in Figure 7.3 [46].

Hydrocarbon degrading strains have an affinity with obligatory hydrophobic surfaces [66]. Indeed, several microbial strains attach over the surface and there are opportunities for specific and non-specific mechanisms for compliance to hydrophobic substances. As described above, many of the bacterial strains have the capability to synthesize glycolipid and lipopeptide types of biosurfactant in order to alter the surface properties of microorganisms and promote the biodegradation of PAHs. Biosurfactants can be involved in hydrocarbon degradation by increasing bioavailability of hydrocarbon through micellar solubilization, interaction with the cell surface, and modifying the cell surface hydrophobicity, which leads to increased association of microbial cells with the hydrocarbons. Usually, both mechanisms take place but the dominance of one or the other depends on the strain [46].

7.4 Influences of Micellization on Hydrocarbons Access

When biosurfactants are unconfined, their monomers organize spherically (micelles) in such a way that the hydrophobic segment is turned to the center, composing the nucleus, and the hydrophilic segment is turned to the sphere surface, making an interface with the aqueous medium.

Thus, the biosurfactant reduces the surface tension between water and oil and contributes to the micelle development, increasing hydrocarbon bioavailability to microorganisms as well as oxygen availability, thus favoring hydrocarbon biodegradation [28, 67].

Makkar and Rockne [68] have evaluated the functions of chemical surfactants and biosurfactants in the degradation of PAHs. In several instances, chemical surfactants have been found to inhibit PAH degradation, which could cause surfactant degradation, surfactant toxicity, or defensive moves on PAHs inside surfactant micelles. The stoppage of numerous biosurfactants to create true micelles has been described to encourage bioavailability of biosurfactant-coupled PAHs for the utilization by bacteria through direct transfer. Shin et al. [69] observed that pseudo-solubilization of phenanthrene by rhamnolipids does not cause it to form a utilizable or degrading strain, such as *Pseudomonas putida*.

7.5 Accession of PAHs in Soil Texture

To execute surfactant-mediated remediation successfully it is very important to consider a few things, including the adsorption behavior of the biosurfactants on to polluted soil, the solubilizing aptitude of the biosurfactants on the target pollutants, the biodegradability, and toxicity of the surfactant and biosurfactant. When biosurfactants are supplemented into the water–soil system, a confident level of surfactants will certainly be absorbed by soil particles. With more adsorbed surfactants, only lesser amounts can participate in the solubilization of PAH contaminants. Moreover, the hydrophobicity of the contaminated soil is enlarged as the surfactant is adsorbed on to the soil particles. As a result, desorbed soluble PAH will be readsorbed on to the soil surface [70]. Hence, the adsorption behavior of biosurfactants on to contaminated soil particles is a significant factor in choosing suitable surfactants. The molecular structure of biosurfactants, playing an important function in their characteristic behavior, is the most dominating factor for adsorption behavior.

Recently, Patowary et al. [66] reported that rhamnolipids possess high surface activity and demonstrated excellent emulsification performance against various hydrocarbon substrates. The phenanthrene degrading bacterial strain exhibits enhanced production of surface-active molecules in comparison with normal strains. Possibly, they liberate these amphoteric surfactant molecules from the surface of the bacterial cell, which affords a more hydrophobic nature to the bacterial surface.

7.6 The Negative Impact of Surfactant on PAH Degradations

A biosurfactant is the primary option needed to be used for bioremediation purposes rather than a chemical surfactant, since many of the chemical surfactants show adverse activity on the PAH degrading microorganisms and have suppressed their growth instead of increasing their bioavailability. Recent research by Ghosh and Mukherji [71] illustrates that the addition of triton ×100 surfactants for pyrene degradation turns into a toxic effect on the *P. aeruginosa* strain, which was used for biodegradation purposes. Another surfactant, Brij30, also showed an adverse effect on the PAH degrading bacteria [72]. CTAB and SDS also cause a negative impact on fluoranthene and anthracene biodegradation by *Pseudomonas* species [73]. Phenanthrene degradation by *Sphignomonas* species GF2B was found by about 33.5% only with the inclusion of Tween80, at the same time as it increased to 83.6% without Tween80 [74]. In some cases, a biosurfactant makes a negative impact on the biodegradation of hydrophobic contaminants [77]. Three factors may perhaps take place in this situation. The first factor deals with the bioavailability of pollutants inside a biosurfactant micelle, which could be insufficient to boost biodegradation [75]. The next deals with the direct impact of biosurfactants on cell surface characteristics, which might interfere with its attachment to hydrophobic substrates [46]. The third factor is the fact that biosurfactants are genus-specific and it has been believed that a given biosurfactant might enhance the activity of the synthesizing genus but hinder the activity of other microorganisms. From these observations, it is very important to choose the correct biosurfactant for the successful remediation of the PAH contaminants.

7.7 Conclusion and Future Directions

This chapter demonstrates the importance of biosurfactants on the PAH desorption. The magnitude of the different desorbing fractions present in the polluted soils is a useful benchmark for understanding the biosurfactant enhanced biodegradation of PAHs. Biosurfactants play a key role in decontaminating PAHs from polluted environments. PAH biodegradation efficiency is determined by a decrease in the hydrophobicity of PAHs. In this chapter, both glycolipid and lipopeptide types of biosurfactant roles in the increasing bioavailability were discussed in detail. Use of a biosurfactant increases the apparent solubility of PAHs, which assists in the mobility and biodegradability of PAHs. The micellar solubilizations of PAHs are more effective in the mixed system than in the single micellar system because of the reduction of the biosurfactant quantity and the level of surfactant pollution in the environment. Some factors such as the concentration of the biosurfactant, types of biosurfactant, temperature, type of bacterial strains used, and concentration of PAHs are playing a crucial role in the biosurfactant enhanced bioremediation of PAHs. Also, these are key factors that need to be considered with care for efficiency remediation.

Very few field studies have been done so far in contrast to many laboratory studies, but the reported field studies and many remediation projects show that biosurfactant-based remedial methods can be applied effectively to remediate different polluted sites. Although additional studies concerning the behavior of a biosurfactant in the destiny and movement of soil pollutants are still required, the biosurfactant emerges as a smart option for surfactant-based remediation methods. The biosurfactant requirement has enlarged enormously since their uses in the oil/gas industry, the food industry, and for bioremediation purposes. The major drawbacks in bio-surfactant usage are problems connected with their synthesizing, which includes a lack of efficient microorganisms, higher production costs, and low yield. Hence improvements in surfactant chemistry, biotechnology, and remedial technology are essential to urge the application of biosurfactants and other surfactant-based practices in the future.

References

1 Marchant, R. and Banat, I.M. (2012). Biosurfactants: a sustainable replacement for chemical surfactants. *Biotechnology Letters* 34 (9): 1597–1605.

2 Saikia, R.R., Deka, S., Deka, M., and Sarma, H. (2012). Optimization of environmental factors for improved production of rhamnolipid biosurfactant by *Pseudomonas aeruginosa* RS29 on glycerol. *Journal of Basic Microbiology* 52 (4): 446–457.

3 Sarma, H., Bustamante, K.L.T., and Prasad, M.N.V. (2018). Biosurfactants for oil recovery from refinery sludge: Magnetic nanoparticles assisted purification. In: *Industrial and Municipal Sludge*

(ed. M.N.V. Prasad). Elsevier Massachusetts, pp. 107–132. ISBN: 9780128159071, Editor Majeti Narasimha Vara Prasad, Paulo Jorge de Campos, Favas Meththika, Vithanage S. Venkata Mohan.

4 Parthipan, P., Elumalai, P., Karthikeyan, O.P. et al. (2017c). A review on biodegradation of hydrocarbon and their influence on corrosion of carbon steel with special reference to petroleum industry. *Journal of Environmental and Biotechnology Research* 6: 12–33.

5 Parthipan, P., Elumalai, P., Sathishkumar, K. et al. (2017). Biosurfactant and enzyme mediated crude oil degradation by *Pseudomonas stutzeri* NA3 and *Acinetobacter baumannii* MN3. *3 Biotech* 7: 278.

6 Parthipan, P., Preetham, E., Machuca, L.L. et al. (2017). Biosurfactant and degradative enzymes mediated crude oil degradation by bacterium *Bacillus subtilis* A1. *Frontiers in Microbiology* 8: 193.

7 Santos, D.K.F., Rufino, R.D., Luna, J.M. et al. (2016). Biosurfactants: Multifunctional biomolecules of the 21st century. *International Journal of Molecular Sciences* 17 (3): 401.

8 Jahan, R., Bodratti, A.M., Tsianou, M., and Alexandridis, P. (2020). Biosurfactants, natural alternatives to synthetic surfactants: Physicochemical properties and applications. *Advances in Colloid and Interface Science* 275: 102061.

9 Long, X., Sha, R., Meng, Q., and Zhang, G. (2016). Mechanism study on the severe foaming of rhamnolipid in fermentation. *Journal of Surfactants and Detergents* 19 (4): 833–840.

10 Parthipan, P., Sabarinathan, D., Angaiah, S., and Rajasekar, A. (2018). Glycolipid biosurfactant as an eco-friendly microbial inhibitor for the corrosion of carbon steel in vulnerable corrosive bacterial strains. *Journal of Molecular Liquids* 261: 473–479.

11 Singh, P. and Tiwary, B.N. (2016). Isolation and characterization of glycolipid biosurfactant produced by a *Pseudomonas otitidis* strain isolated from Chirimiri coal mines, India. *Bioresources and Bioprocessing* 3: 42.

12 Hazra, C., Kundu, D., and Chaudhari, A. (2012). Biosurfactant-assisted bioaugmentation in bioremediation. In: *Microorganisms in Environmental Management: Microbes and Environment*, 1e (eds. T. Satyanarayana, B.N. Johri and A. Prakash), 631–664. New York: Springer.

13 Sarma, H., Sonowa, S., and Prasad, M.N.V. (2019). Plant-microbiome assisted and biochar-amended remediation of heavy metals and polyaromatic compounds – A microcosmic study. *Ecotoxicology and Environmental Safety* 176: 288–299.

14 Sarma, H., Islam, N.F., Borgohain, P. et al. (2016). Localization of polycyclic aromatic hydrocarbons and heavy metals in surface soil of Asia's oldest oil and gas drilling site in Assam, Northeast India: implications for the bio economy. *Emerging Contaminants* 2 (3): 119–127.

15 Sarma, H., Nava, A.R., and Prasad, M.N.V. (2019). Mechanistic understanding and future prospect of microbe-enhanced phytoremediation of polycyclic aromatic hydrocarbons in soil. *Environmental Technology and Innovation* https://doi.org/10.1016/j.eti.2018.12.004.

16 Clar, E. (1964). *Polycyclic Hydrocarbons*. Academic Press, Heidelberg, Germany.

17 Haritash, A.K. and Kaushik, C.P. (2009). Biodegradation aspects of polycyclic aromatic hydrocarbons (PAHs): a review. *Journal of Hazardous Materials* 169l: 1–15.

18 Sarma H, Prasad MNV (2015) Plant-microbe association-assisted removal of heavy metals and degradation of polycyclic aromatic hydrocarbons, *Petroleum Geosciences: Indian Contexts*, Springer, Switzerland, pp. 219–236, doi:10.1007/978-3-319-03119-4_10, ISBN: 978–3–319-03118-7.

19 Sarma, H. and Prasad, M.N.V. (2016). Phytomanagement of polycyclic aromatic hydrocarbons and heavy metals-contaminated sites in Assam, North Eastern State of India, for boosting bioeconomy. In: *Bioremediation and Bioeconomy* (ed. M.N.V. Prasad), 609–626. Oxford, UK: Elsevier.

20 Sharma D, Sarma H, Hazarika S, Islam NF, Prasad MNV (2018) Agro-ecosystem diversity in petroleum and natural gas explored sites in Assam State, North-Eastern India: Socio-economic perspectives, *Sustainable Agriculture Reviews*, 27, 37–60, Series ISSN, 2210–4410.

21 Paneque, P., Caballero, P., Parrado, J. et al. (2020). Use of a biostimulant obtained from okara in the bioremediation of a soil polluted by used motor car oil. *Journal of Hazardous Materials* 389: 121820.

22 Bamforth, S.M. and Singleton, I. (2005). Bioremediation of polycyclic aromatic hydrocarbons: current knowledge and future directions. *Journal of Chemical Technology and Biotechnology* 80: 723–736.

23 Kaushik, C.P. and Haritash, A.K. (2006). Polycyclic aromatic hydrocarbons (PAHs) and environmental health. *Our Earth* 3: 1–7.

24 Sarma, H., Islam, N.F., and Prasad, M.N.V. (2017). Plant-microbial association in petroleum and gas exploration sites in the state of Assam, North-East India – Significance for bioremediation. *Environmental Science and Pollution Research* 24 (9): 8744–8758.

25 Speight, J.G. and Arjoon, K.K. (2012). *Bioremediation of Petroleum and Petroleum Products*, 1–38. Hoboken, Massachusetts, USA: Wiley.

26 Azubuike, C.C., Chikere, C.B., and Okpokwasili, G.C. (2016). Bioremediation techniques – classification based on site of application: principles, advantages, limitations and prospects. *World Journal of Microbiology and Biotechnology* 32: 180.

27 Laszlova, K., Dudasova, H., Olejnikova, P. et al. (2018). The application of biosurfactants in bioremediation of the aged sediment contaminated with polychlorinated biphenyls. *Water, Air, and Soil Pollution* 229: 219.

28 Souza, E.C., Vessoni-Penna, T.C., and de Souza Oliveira, R.P. (2014). Biosurfactant-enhanced hydrocarbon bioremediation: An overview. *International Biodeterioration and Biodegradation* 89: 88–94.

29 Megharaj, M., Ramakrishnan, B., Venkateswarlu, K. et al. (2011). Bioremediation approaches for organic pollutants: A critical perspective. *Environmental International* 37: 1362–1375.

30 Sarubbo, L., Luna, J., and Rufino, R. (2015). Application of a biosurfactant produced in low-cost substrates in the removal of hydrophobic contaminants. *Chemical Engineering Transactions* 43: 295–300.

31 Roszak, D.B. and Colwell, R. (1987). Survival strategies of bacteria in the natural environment. *Microbiological Reviews* 51: 365–379.

32 Sponza, D.T. and Gok, O. (2010). Effect of rhamnolipid on the aerobic removal of polyaromatic hydrocarbons (PAHs) and COD components from petrochemical wastewater. *Bioresource Technology* 101: 914–924.

33 Li, S., Pi, Y., Bao, M. et al. (2015). Effect of rhamnolipid biosurfactant on solubilization of polycyclic aromatic hydrocarbons. *Marine Pollution Bulletin* 101: 219–225.

34 Tiwari, B., Manickam, N., Kumari, S., and Tiwari, A. (2016). Biodegradation and dissolution of polyaromatic hydrocarbons by *Stenotrophomonas* sp. *Bioresource Technology* 216: 1102–1105.

35 Chebbi, A., Hentati, D., Zaghden, H. et al. (2017). Polycyclic aromatic hydrocarbon degradation and biosurfactant production by a newly isolated *Pseudomonas* sp. strain from used motor oil-contaminated soil. *International Biodeterioration & Biodegradation* 122: 128–140.

36 Singh, P. and Tiwary, B.N. (2017). Optimization of conditions for polycyclic aromatic hydrocarbons (PAHs) degradation by *Pseudomonas stutzeri* P2 isolated from Chirimiri coal mines. *Biocatalysis and Agricultural Biotechnology* 10: 20–29.

37 Patowary, R., Patowary, K., Kalita, M.C., and Deka, S. (2018). Application of biosurfactant for enhancement of bioremediation process of crude oil contaminated soil. *International Biodeterioration & Biodegradation* 129: 50–60.

38 Reddy, P.V., Karegoudar, T.B., and Nayak, A.S. (2018). Enhanced utilization of fluorene by *Paenibacillus* sp. PRNK-6: Effect of rhamnolipid biosurfactant and synthetic surfactants. *Ecotoxicology and Environmental Safety* 151: 206–211.

39 Pourfadakari, S., Ahmadi, M., Jaafarzadeh, N. et al. (2019). Remediation of PAHs contaminated soil using a sequence of soil washing with biosurfactant produced by *Pseudomonas aeruginosa* strain PF2 and electrokinetic oxidation of desorbed solution, effect of electrode modification with Fe_3O_4 nanoparticles. *Journal of Hazardous Materials* 379: 120839.

40 Posada-Baquero, R., Grifoll, M., and Ortega-Calvo, J. (2019). Rhamnolipid-enhanced solubilization and biodegradation of PAHs in soils after conventional bioremediation. *Science of the Total Environment* 668: 790–796.

41 Pereira, E., Napp, A.P., Allebrandt, S. et al. (2019). Biodegradation of aliphatic and polycyclic aromatic hydrocarbons in seawater by autochthonous microorganisms. *International Biodeterioration & Biodegradation* 145: 104789.

42 Sun, S., Wang, Y., Zang, T. et al. (2019). A biosurfactant-producing *Pseudomonas aeruginosa* S5 isolated from coking wastewater and its application for bioremediation of polycyclic aromatic hydrocarbons. *Bioresource Technology* 281: 421–428.

43 Portet-Koltalo, F., Ammami, M.T., Benamar, A. et al. (2013). Investigation of the release of PAHs from artificially contaminated sediments using cyclolipopeptidic biosurfactants. *Journal of Hazardous Materials* 261: 593–601.

44 Xia, W., Du, Z., Cui, Q. et al. (2014). Biosurfactant produced by novel *Pseudomonas* sp. WJ6 with biodegradation of n-alkanes and polycyclic aromatic hydrocarbons. *Journal of Hazardous Materials* 276: 489–498.

45 Bezza, F.A. and Chirwa, E.M.N. (2015). Biosurfactant from *Paenibacillus dendritiformis* and its application in assisting polycyclic aromatic hydrocarbon (PAH) and motor oil sludge removal from contaminated soil and sand media. *Process Safety and Environmental Protection* 98: 354–364.

46 Bezza, F.A. and Chirwa, E.M.N. (2016). Biosurfactant-enhanced bioremediation of aged polycyclic aromatic hydrocarbons (PAHs) in creosote contaminated soil. *Chemosphere* 144: 635–644.

47 Bezza, F.A. and Chirwa, E.M.N. (2017a). Pyrene biodegradation enhancement potential of lipopeptide biosurfactant produced by *Paenibacillus dendritiformis* CN5 strain. *Journal of Hazardous Materials* 321: 218–227.

48 Bezza, F.A. and Chirwa, E.M.N. (2017b). The role of lipopeptide biosurfactant on microbial remediation of aged polycyclic aromatic hydrocarbons (PAHs)-contaminated soil. *Chemical Engineering Journal* 309: 563–576.

49 Mehetre, G.T., Dastager, S.G., and Dharne, M.S. (2019). Biodegradation of mixed polycyclic aromatic hydrocarbons by pure and mixed cultures of biosurfactant producing thermophilic and thermo-tolerant bacteria. *Science of the Total Environment* 67: 52–60.

50 Wang, X., Cai, T., Wen, W. et al. (2020). Surfactin for enhanced removal of aromatic hydrocarbons during biodegradation of crude oil. *Fuel* 267: 117272.

51 Das, P., Mukherjee, S., and Sen, R. (2008). Improved bioavailability and biodegradation of a model polyaromatic hydrocarbon by a biosurfactant producing bacterium of marine origin. *Chemosphere* 72: 1229–1234.

52 Reddy, M.S., Naresh, B., Leela, T. et al. (2010). Biodegradation of phenanthrene with biosurfactant production by a new strain of *Brevibacillus* sp. *Bioresource Technology* 101: 7980–7983.

53 Tao, W., Lin, J., Wang, W. et al. (2020). Biodegradation of aliphatic and polycyclic aromatic hydrocarbons by the thermophilic bioemulsifier-producing *Aeribacillus pallidus* strain SL-1. *Ecotoxicology and Environmental Safety* 189: 109994.

54 Kitamoto, D., Morita, T., Fukuoka, T. et al. (2009). Self-assembling properties of glycolipid biosurfactants and their potential applications. *Current Opinion in Colloid & Interface Science* 14 (5): 315–328.

55 Abdel-Mawgoud, A.M., Lepine, F., and Deziel, E. (2010). Rhamnolipids: Diversity of structures, microbial origins and roles. *Applied Microbiology and Biotechnology* 86 (5): 1323–1336.

56 Mekala, S., Peters, K.C., Singer, K.D., and Gross, R.A. (2018). Biosurfactant-functionalized porphyrin chromophore that forms J-aggregates. *Organic & Biomolecular Chemistry* 16 (39): 7178–7190.

57 Kumari, S., Regar, R.K., and Manickam, N. (2018). Improved polycyclic aromatic hydrocarbon degradation in a crude oil by individual and a consortium of bacteria. *Bioresource Technology* 254: 174–179.

58 Ortega Ramirez, C.A., Kwan, A., and Li, Q.X. (2020). Rhamnolipids induced by glycerol enhance dibenzothiophene biodegradation in *Burkholderia* sp. C3. *Engineering* 2020 https://doi.org/10.1016/j.eng.2020.01.006.

59 Li, L., Shen, X., Zhao, C. et al. (2019). Biodegradation of dibenzothiophene by efficient *Pseudomonas* sp. LKY-5 with the production of a biosurfactant. *Ecotoxicology and Environmental Safety* 176: 50–57.

60 Ma, Z., Liu, J., Dick, R.P. et al. (2018). Rhamnolipid influences biosorption and biodegradation of phenanthrene by phenanthrene-degrading strain *Pseudomonas* sp. Ph6. *Environmental Pollution* 240: 359–367.

61 Sponza, D.T. and Gok, O. (2011). Effects of sludge retention time (SRT) and biosurfactant on the removal of polyaromatic compounds and toxicity. *Journal of Hazardous Materials* 197l: 404–416.

62 Ibrar, M. and Zhang, H. (2020). Construction of a hydrocarbon-degrading consortium and characterization of two new lipopeptides biosurfactants. *Science of the Total Environment* 714: 136400.

63 Camacho-Chab, J.C., Guezennec, J., Chan-Bacab, M.J. et al. (2013). Emulsifying activity and stability of a non-toxic bioemulsifier synthesized by microbacterium SP MC3B-10. *International Journal of Molecular Sciences* 14 (9): 18959–18972.

64 Desai, J.D. and Banat, I.M. (1997). Microbial production of surfactants and their commercial potential. *Microbiology and Molecular Biology Reviews* 61 (1): 47–64.

65 Ron, E.Z. and Rosenberg, E. (2001). Natural roles of biosurfactants. *Environmental Microbiology* 3 (4): 229–236.

66 Patowary, K., Patowary, R., Kalita, M.C., and Deka, S. (2017). Characterization of biosurfactant produced during degradation of hydrocarbons using crude oil as sole source of carbon. *Frontiers in Microbiology* 8: 279.

67 Soberon-Chavez, G. and Maier, R.M. (2010). Biosurfactants: a general overview. In: *Biosurfactants: From Genes to Applications* (ed. G. Soberon-Chavez), 1–11. Berlin, Heidelberg, Germany: Springer.

68 Makkar, R.S. and Rockne, K.J. (2003). Comparison of synthetic surfactants and biosurfactants in enhancing biodegradation of polycyclic aromatic hydrocarbons. *Environmental Toxicology and Chemistry* 22: 2280–2292.

69 Shin, K.H., Ahn, Y., and Kim, K.W. (2005). Toxic effect of biosurfactant addition on the biodegradation of phenanthrene. *Environmental Toxicology and Chemistry* 24: 2768–2774.

70 Paria, S. (2008). Surfactant-enhanced remediation of organic contaminated soil and water. *Advances in Colloid and Interface Science* 138: 24–58.

71 Ghosh, I. and Mukherji, S. (2016). Diverse effect of surfactants on pyrene biodegradation by a *Pseudomonas* strain utilizing pyrene by cell surface hydrophobicity induction. *International Biodeterioration & Biodegradation* 108: 67–75.

72 Llado, S., Covino, S., Solanas, A.M. et al. (2013). Comparative assessment of bioremediation approaches to highly recalcitrant PAH degradation in a real industrial polluted soil. *Journal of Hazardous Materials* 248-249: 407–414.

73 Rodrigues, A., Nogueira, R., Melo, L.F., and Brito, A.G. (2013). Effect of low concentrations of synthetic surfactants on polycyclic aromatic hydrocarbons (PAH) biodegradation. *International Biodeterioration & Biodegradation* 83: 48–55.

74 Pei, X., Zhan, X., Wang, S. et al. (2010). Effects of a biosurfactant and a synthetic surfactant on phenanthrene degradation by a *Sphingomonas* strain. *Pedosphere* 20: 771–779.

75 Guha, S. and Jaffe, P.R. (1996). Bioavailability of hydrophobic compounds partitioned into the micellar phase of nonionic surfactants. *Environmental Science and Technology* 30: 1382–1391.

76 Zhao, Z., Selvam, A., and Wong, J.W. (2011). Effects of rhamnolipids on cell surface hydrophobicity of PAH degrading bacteria and the biodegradation of phenanthrene. *Bioresource Technology* 102: 3999–4007.

77 Zhao, Z., Selvam, A., and Wong, J.W. (2011). Synergistic effect of thermophilic temperature and biosurfactant produced by *Acinetobacter calcoaceticus* BU03 on the biodegradation of phenanthrene in bioslurry system. *Journal of Hazardous Materials* 190: 345–350.

8

Biosurfactants for Enhanced Bioavailability of Micronutrients in Soil

A Sustainable Approach

Siddhartha Narayan Borah[1], Suparna Sen[2], and Kannan Pakshirajan[3]

[1] *Royal School of Biosciences, Royal Global University, Guwahati, Assam, India*
[2] *Environmental Biotechnology Laboratory, Resource Management and Environment Section, Life Sciences Division, Institute of Advanced Study in Science and Technology, Guwahati, Assam, India*
[3] *Department of Biosciences and Bioengineering, Indian Institute of Technology Guwahati, Guwahati, Assam, India*

CHAPTER MENU

8.1 Introduction

Micronutrients are essential elements that are vital for the functioning and maintenance of physiological processes in all organisms [1, 2]. Depending on the type of organism, the requirement of micronutrients varies; for the animals, they are comprised of minerals and vitamins [3]. Plants, on the other hand, require a different set of minerals obtained exclusively from the soil. These micronutrients are copper (Cu), zinc (Zn), manganese (Mn), iron (Fe), nickel (Ni), molybdenum (Mo), boron (B), and chlorine (Cl) [1].

Biosurfactants for a Sustainable Future: Production and Applications in the Environment and Biomedicine, First Edition. Edited by Hemen Sarma and Majeti Narasimha Vara Prasad.
© 2021 John Wiley & Sons Ltd. Published 2021 by John Wiley & Sons Ltd.

Since plants constitute the primary component of the diet of animals and humans, any micronutrient deficiency in the soil that affects the plants ultimately affects the animals. Although required in minimal amounts, micronutrients are indispensable for the sustenance of optimal growth and crop production in plants [4]. Micronutrients play critical roles in biological processes like protein synthesis, gene expression, auxin metabolism, maintenance of biological membranes, protection against photooxidative damage, heat stress, and disease [5]. Hence, agricultural soil deficient in micronutrients negatively affect the yield and quality of the crop, and, thus, human health [6, 7]. Planet Earth currently sustains a global human population of more than 7.63 billion (http://faostat.fao.org/beta/en/#data/OA), which is putting severe pressure on the food resources of our planet. The global population has increased exponentially after the industrial revolution, fueling the rapid modernization of cultivation practices around the world to meet the ever-increasing pressure of an adequate food supply. In this regard, an optimum balance of water, nutrients, and environmental factors play a pivotal role [8]. Extensive irrigation plans have been worked out to ensure an adequate water supply to crop fields, whereas to maintain and improve the nutrient levels and plant health, the widespread use of chemical fertilizers and pesticides has been adopted. Because of these modernized agricultural practices, the production of food grains has steadily increased over the years [9]. However, the indiscriminate use of these chemical fertilizers and pesticides has resulted in detrimental effects on humans, as these chemicals can enter the food chain [10, 11]. Moreover, chemical fertilizers cause severe degradation of the physical, chemical, and biological status of topsoils, indicating counterproductive effects in the long run [12, 13]. In this context, recent reports highlighting the increasing micronutrient deficiency in agricultural soil throughout the world are noteworthy [14–16]. The over-reliance on chemical fertilizers has made the traditional practice of the application of organic matter, such as bioslurry and farmyard manure, redundant in recent times, thereby compounding the problem further. The insufficient quantities of the essential nutrients in agrarian soil will lead to malnutrition among the masses [17].

The micronutrient levels in a particular plant can indicate the status of micronutrients in the soil where it was grown, notwithstanding the variations in the uptake efficiency of individual plants [18]. The micronutrient level in a plant is also influenced by the fact that the micronutrients are usually bound to soil particles, which reduces their availability for the plants [18]. To mitigate this nutrient scarcity, chemical fertilizers contain a balanced proportion of nutrients and are highly water-soluble, making them readily available for biosorption [19, 20]. Nevertheless, the performance of synthetic fertilizers is not uniform across different agrarian soil types [21]. Reported estimates put the overall efficiency of applied fertilizers at around 50% or lower for N, lower than 10% for P, and approximately 40% for K [22]. Therefore, enhancing the bioavailability of soil micronutrients by detachment from soil particles becomes essential. The critical factor in determining the bioavailability of micronutrients is their mobility, which, in turn, is dependent on their solubility [23]. The inherent solubility and the bioavailability of micronutrients are affected by several properties of the soil, e.g. type and texture, moisture content, temperature, pH, organic content, extent of calcification, cation exchange capacity (CEC), and structure [5, 24].

The use of surfactants in the remediation of contaminated soils has been a well-studied and established approach [25]. The surfactants efficiently reduce the surface tension (ST) of insoluble contaminants in soil, thereby enhancing their solubility and bioavailability for bioremediation. Synthetic surfactants are used as adjuvants in the formulation of pesticides and as such are an integral part of modern agricultural practice [26]. However, synthetic surfactants have been described to persist for a long time in the soil and, despite their proven benefits, bear implications for the environment and biota [27, 28]. In this regard, biological surface-active agents or biosurfactants appear to be suitable alternatives to their chemical counterparts, owing to their superior surface activity, selectivity, and usability over a wide range of temperature, pH, and salinity [29, 30]. Although reports are abundant regarding the use of biosurfactants in the remediation of metal contaminated soil, targeted studies for their application to mitigate micronutrient deficiency in agricultural soil are scanty. This chapter discusses the possibility of their use in this regard based on the available literature. The factors that play critical roles in limiting the micronutrient availability in the soil, as well as properties of biosurfactants that make them suitable candidates to mitigate them, are also discussed.

8.2 Micronutrient Deficiency in Soil

When it comes to the relationship between micronutrient concentration in soil and their bioavailability, the correlation is not direct and often influenced by many factors. Plants absorb micronutrients only from the soil solution, which effectively restricts the availability to the soluble portion of the total micronutrients in soil [31]. Hence, the soil is considered deficient in a particular micronutrient when its external supplementation causes increased growth and yield, irrespective of the actual levels of that nutrient in that soil. As previously reviewed by others [32, 33], the use of micronutrients as fertilizers in numerous studies has led to increased growth, substantiating the unavailability of sufficient micronutrients to plants. Micronutrient deficiency in plants leads to suboptimal growth, as well as lower crop yield with poor nutritional value. On the other hand, research has shown that the optimum ranges of all the micronutrients are very narrow and excessive concentrations cause toxicity, which also results in lower crop production [34]. Hence, it is important to monitor the agrarian soil and crops to ensure the maintenance of optimum levels of micronutrients in the soil. Centuries of farming have led to differential levels of scarcity of the micronutrients in the agricultural soils throughout the world [14–16]. The percentage of agricultural soil at the global as well as national (India) level deficient in the soil micronutrients are listed in Table 8.1.

8.3 Factors Affecting the Bioavailability of Micronutrients

8.3.1 Effect of Soil pH, Moisture, and Temperature

Micronutrients are absorbed by plants in the form of ions [1]. Soil pH has been described to play a decisive role in the solubility, mobility, and bioavailability of micronutrients. Except for Mo, the bioavailability of all other micronutrients is inversely proportional to pH [37]. Notwithstanding the effect of soil type, the solubility of micronutrients is usually high at low pH due to high desorption. However, with increasing pH, the rate of desorption starts to decline as adsorption

Table 8.1 Status of the essential micronutrients in the soil throughout the world and India.

	Micronutrient deficiency (%)						
Region	**Boron (B)**	**Copper (Cu)**	**Iron (Fe)**	**Manganese (Mn)**	**Molybdenum (Mo)**	**Zinc (Zn)**	**References**
India	23.4	4.2	12.8	7.1	11	36.5	[35, 36]
World	31	14	3	10	15	49	[14]

increases and completely predominates within a narrow range of pH, the pH-adsorption edge. Beyond this point, micronutrients are completely absorbed [38, 39]. As micronutrients are mostly absorbed as cations, except Mo and Cl [1], their availability in acidic soil is higher than that in alkaline soils. Alkaline soils with high pH are rich in hydroxyl ions (OH^-), which react with cationic micronutrients causing precipitation, thereby making them unavailable for absorption by plants. For example, Fe and Mn are soluble in the reduced state but are readily oxidized in soils with alkaline pH, which effectively nullifies their availability in such soils [14]. The soil moisture and temperature have also been reported to significantly affect the distribution and uptake of micronutrients [40]. Najafi-Ghiri and co-workers evaluated the effect of different soil moisture and temperature regimes on the availability of trace elements in calcareous soils. They reported the highest micronutrient concentrations in soils with aquic moisture and mesic temperature, and lowest in soils with aridic and ustic moisture and hyperthermic temperature regimes. Nevertheless, too high moisture content may negatively affect soil micronutrients, as reported in the case of B, which, due to its high solubility, often leaches out in regions receiving higher than average rainfall [14, 41].

8.3.2 Effect of Soil Organic Matter

Organic matter in the soil is another important factor that affects the availability of micronutrients. The decomposition of organic matter leads to the formation and accumulation of humic acid (HA) in the soil, which can chelate unavailable trace elements and also buffer pH levels to promote plant growth [42]. An optimum organic matter content results in high cation exchange capacity (CEC) of the soil, which increases the bioavailability of micronutrients. Incessant cultivation of crops, coupled with the unsystematic use of synthetic fertilizers and lack of organic supplements, has caused significant damage to the organic matter content in agricultural soils [12]. Consequently, the micronutrient availability has been affected as well. Widespread deficiencies of micronutrients in Indian soil have been correlated to diminishing levels of organic carbon [36].

8.3.3 Interactions with Other Nutrients and Environmental Factors

Micronutrients, in addition to other factors, also interact with each other, which makes their diagnosis and supplementation in a particular soil quite complex. Due to the mutual interactions, straightforward supplementation of a deficient micronutrient may not yield the expected benefits. Theoretically, the supplementation of agricultural soil with a particular deficient nutrient will promote healthier growth and enhance yields. However, in practice, indiscriminate supplementation of a nutrient often disturbs the optimal balance critical for maximum yield, thereby proving counterproductive [32]. Hence, agronomists advocate the application of combinations of multiple nutrients over the single most deficient one for sustained higher yields [14]. Environmental factors that interactively affect the nutrient levels are drought [43], soil salinity [44], dry topsoil, nutrient-deficient subsoil [45], and seasonal variations [46]. Plants are highly resilient and may not exhibit significant yield losses under the effect of a single environmental factor. In an experiment involving oats, it was shown that prolonged drying (42 days) of topsoil did not have any significant effect on the micronutrient concentration in the shoots of the experimental plants [47]. However, when combined with the added stress of nutrient-deficient subsoil, drying of topsoil rapidly induced micronutrient deficiency of Mn, in particular, resulting in significant yield depression [45].

8.3.4 Uptake Efficiency of Plants

Micronutrient levels in the soil are usually much higher than that required for optimum plant growth. Thus micronutrient deficiency in crops is the result of the combined effects of insufficiency of soil micronutrients in bioabsorbable form and inefficiency of plants in micronutrient assimilation [31]. Indeed, both intra- and interspecies variation in nutrient assimilation efficiency has been widely reported in the literature [48, 49]. The majority of the crop cultivars used by farmers have been developed via experimental breeding programs in agricultural soils under carefully optimized and controlled nutrient levels. Lack of exposure to any nutrient stress during the life cycle minimizes the chances of expression of micronutrient efficiency traits, thereby limiting their selection. The targeted breeding and selection of highly efficient cultivars for micronutrient assimilation under low or severe nutrient-deficient conditions will resolve this issue to a large extent and ensure sustained high yields. The possible reasons to explain the variations in the efficiency of micronutrient absorption between cultivars have been concisely summarized by Alloway [34]. The author attributed these variations to differences in:

- Root length and volume
- Presence or absence of proteoid roots
- Variations in rhizospheric pH induced by roots
- Presence or absence of symbiotic associations
- Secondary metabolites that facilitate uptake (e.g. phytosiderophores, under Fe or Zn limitation)
- Utilization efficiency of micronutrients after absorption by individual plants
- Tissue-specific utilization of nutrients during plant growth
- Tolerance toward a particular inhibitory factor

8.4 Effect of Micronutrient Deficiency on the Biota

Micronutrients, as previously described, are essential elements for all organisms [2]. Arnon and Stout proposed three basic criteria for any element to be considered an essential one for organisms: [1] its deficiency disrupts normal growth and reproduction, [2] its action is specific that cannot be substituted by any other element, and [3] it must have a direct action [50]. So, it becomes fundamental that the lack or deficiency of any micronutrient will have observable repercussions on any organism.

8.4.1 Effect on Plants

The specific functions, common deficiency symptoms, and the critical limits in the soil of all the micronutrients for plants have been extensively reviewed [34, 51–58]. A concise summary of each factor has been presented in Table 8.2.

8.4.2 Effect on Animals

The Food and Agricultural Organization, in their annual report, revealed that a total of about 2 billion people in the world experience moderate to severe levels of food insecurity [59]. The moderate risk level includes people who, although not hungry, do not consume sufficiently nutritious food. This leads to "hidden hunger" (deficiency of micronutrients) and puts them at greater risk of

Table 8.2 A summary of the essential functions, deficiency symptoms of micronutrients in plants, and their critical levels in the soil.

Element	Functions	Critical level in the soil (mg/kg)	Deficiency symptoms and effects
B	DNA synthesis; the formation of pollen and pollination; synthesis and maintenance of cell wall and membranes; carbohydrate metabolism and transport; elongation of roots	0.5	Chlorosis of terminal leaves; the death of terminal buds and shoot tips; deformed young leaves and fruits; root lesions
Cl	Photosynthesis, osmotic regulation in all plant cells, viz. opening and closure of stomata, regulates enzyme activity	8.0	Chlorosis and burning of leaf tips; shriveling and over-wilting of leaves
Cu	Photosynthesis, respiration, protein and carbohydrate metabolism; a constituent of many enzymes; ethylene formation; pollen formation and lignification	0.2	Rosetting, chlorosis, and wilting of leaves; pollen sterility; whitening and twisting of leaves; reduced density of ear production in cereals
Fe	Chlorophyll synthesis; a constituent of cytochromes and metalloenzymes; enzyme and RNA metabolism; regulates photosynthesis, fixation and metabolism of N, redox reactions	4.5	Interveinal chlorosis (lime induced iron chlorosis) of young leaves, in severe cases, the entire leaf may become bleached
Mn	Photolysis of water in chloroplasts; regulation of enzyme activities; protection against oxidative damage of membranes	2.0	Interveinal chlorosis with necrotic spots of young leaves; stunted root development
Mo	Nitrogen fixation; a constituent of enzymes like nitrate reductase and sulfite oxidase	0.1	Necrotic spots over the leaf, affected areas may extrude a resinous substance from the undersurface of the leaf; restricted flowering; leaf deformation due to NO_3^- excess and destruction of embryonic tissues
Ni	A constituent of urease enzyme; role in N assimilation and Fe absorption, prohibits the inactivation of nitrate reductase; role in seed germination and growth	0.1	Urea toxicity that reduces germination, growth, vigor and flowering, internode length, and kernel filling; dwarf foliage production; reddish pigmentation in young leaves
Zn	A constituent of several enzymes with roles in carbohydrate and protein synthesis; maintenance of membrane integrity; regulates auxin synthesis, pollen formation, N and P assimilation	0.6	Interveinal chlorotic leaves, reduced leaf size, stunted growth, random necrotic spots on the leaves

various forms of malnutrition and poor health. As per the report, sizeable portions of the population of advanced countries also suffer from nutrient deficiency and food insufficiency; an estimated 8% of North Americans and Europeans are moderately food insecure. In contrast to plants, human beings need several added trace elements to satiate their nutritional requirements [3]. These are summarized in Table 8.3.

Micronutrient deficiencies of iron, zinc, iodine, vitamin A, vitamin D, and folate are the most widespread, each of which is a contributor to poor growth, intellectual impairments, perinatal

Table 8.3 The essential micronutrients in humans [3].

Group	Components
Trace elements	Iron (Fe), Zinc (Zn), Copper (Cu), Manganese (Mn), Iodine (I), Fluorine (F), Selenium (Se), Molybdenum (Mo), Cobalt (Co), Boron (B), Nickel (Ni), Chromium (Cr), Vanadium (V), Silicon (Si), Arsenic (As), Lithium (Li), Tin (Sn)
Vitamins	**A** (Retinoid); **D** (D_3 – Cholecalciferol, D_2 – Ergocalciferol); **E**; **K** (K_1 – Phylloquinone, K_2 – Menaquinone)
	C (Ascorbic acid)
	B complex (B_1 – Thiamin, B_2 – Riboflavin, B_3 – Niacin, B_5 – Pantothenic acid, B_6 – Pyroxidine, B_7 – Biotin, B_9 – Folic acid or Folate, B_{12} – Cobalamin)

complications, and increased risk of morbidity and mortality [60]. The main functions and deficiency symptoms are described below:

- Iron – Iron is the most deficient micronutrient worldwide. It constitutes a major component of hemoglobin. Its deficiency leads to microcytic anemia, impaired cognitive and motor development, reduced working capacity, decreased immune and endocrine function [60].
- Zinc – Zinc is involved in immune regulations, glucose homeostasis (storage and processing of insulin), infant growth and development [61, 62]. Its deficiency leads to respiratory ailments, gastrointestinal disturbances, and increased risk of Type 2 diabetes.
- Iodine – Iodine is essential for the production of thyroid hormones. Optimum iodine levels are critical during pregnancy as maternal thyroxine (T4) is indispensable for the survival of the fetus [63]. Major deficiency disorders include goiter, mental retardation, or reduced cognitive function [64].
- Vitamin A – Vitamin A promotes immune function, growth, and cell differentiation and is also involved in the regulation of gene transcription. Its deficiency leads to malnutrition and symptoms include night blindness (total blindness in severe deficiency), loss of appetite, and skin flaking [65].
- Vitamin D – Vitamin D plays an essential role in the maintenance of healthy bones and teeth. It also has potential therapeutic applications in the prevention and treatment of cardiovascular diseases [66]. The most common deficiency symptoms are bone diseases, including rickets in children and osteoporosis in adults [67].
- Folic acid – Folate is crucial for DNA synthesis and repair. It is the second most important cause of anemia after nutritional iron deficiency. Other deficiency symptoms include chronic diseases and congenital disabilities; deficiency during pregnancy is implicated in neural tube defects [68].

Over the years, dietary supplements have become the favored option to address the nutrient deficiency in masses. The inter-relationships among agricultural practice, nutritional deficiency, and human health have not been at the forefront of targeted areas to alleviate malnutrition and dietary diseases [3]. Nevertheless, the increasing nutrient deficiency in the global population underlines the inadequacies and flaws in the current agricultural practices, as the majority of the food that caters to our nutritional needs comprise agricultural produce [69]. Biofortification, in recent years, has come across as a promising way of increasing the nutrient levels in food crops, especially cereals, legumes, vegetables, and fruits, to mitigate the micronutrient deficiency in masses [70, 71]. However, market penetration, availability, and affordability of fortified crops, as well as dietary supplements, remain very limited [72]. Hence, the primary focus needs to be given to the root of the problem and improve the agricultural practices, including crop rotation, irrigation, and the use of fertilizers to supplement macro- and micronutrients to achieve a sustainable solution to nutrient deficiency.

8.5 The Role of Surfactants in the Facilitation of Micronutrient Biosorption

The bioavailability of any micronutrient in the soil is dependent upon the capacity of the roots to absorb it. The successful absorption, on the other hand, depends on the solubility of nutrients, which facilitates their mobility from the soil to plants [73]. The majority of the agricultural land throughout the world is deficient in one or more micronutrients. Consequently, farmers supplement their fields with micronutrient-rich fertilizers to ensure satisfactory yields. However, any supplemented micronutrient is subject to interactive factors that may negatively affect their bioavailability [74]. The addition of P fertilizer during wheat cultivation has been shown to reduce the Zn content in the grains, despite no Zn deficiency in the cultivation area [75, 76]. Such antagonistic interactions lead to economic losses, the reasons for which, due to the underlying complexity, may not be apparent to the farmers. Therefore, alternative approaches must be considered to enhance the performance of supplements. Surfactants, in this regard, have been recognized as viable candidates to increase soil wetting as well as improve nutrient uptake efficiency and quality of crops [77]. Surfactants can release soil-bound micronutrients for sequestration by chelating agents and transport them across plant membranes for increased biosorption [78, 79]. The positive effect of surfactant in nutrient utilization and crop yield has been demonstrated across a variety of crops and geographic locations. The application of non-ionic surfactant to the irrigation water during soybean cultivation has been reported to increase the yield [80]. The leaching of nitrate [81] and nitrogen [82] from potato fields can be reportedly reduced by applying surfactants at the planting stage. Chaichi and co-workers in their study involving corn under a Mediterranean climate observed increased water use efficiency and higher yield and dry matter [83]. Higher efficiency and yields guarantee additional economic benefits for the farmers, attributable to reduced water and fertilizer inputs. The authors estimated a 19.7% increase in profits, which more than supplemented for the 4.7% increase in irrigation costs incurred for the application of the surfactant.

8.6 Surfactants

Surface-active agents or surfactants are amphiphilic compounds comprising both a hydrophilic and a hydrophobic moiety. Their structure confers the ability to accumulate at the interface between two immiscible fluids, reducing the repulsive forces between two different phases and allowing these phases to mix and interact more efficiently, thereby decreasing surface (liquid–air) and interfacial (liquid–liquid) tension [84]. Due to their dual structural properties, the surfactant monomers can develop micelles in solution, giving them the ability to solubilize metals and compounds in the soil that are otherwise insoluble. Depending on the net charge of the hydrophilic head group, surfactants can be non-ionic, cationic, anionic, or amphoteric [29]. Likewise, based on their ionic character and the polarity of the solution they are in, surfactants can form four basic types of micelles, i.e. normal, reverse, mixed, and liposomal (Figure 8.1). Based on the source of origin, surfactants are of two types – synthetic or chemical surfactants and natural or biosurfactants.

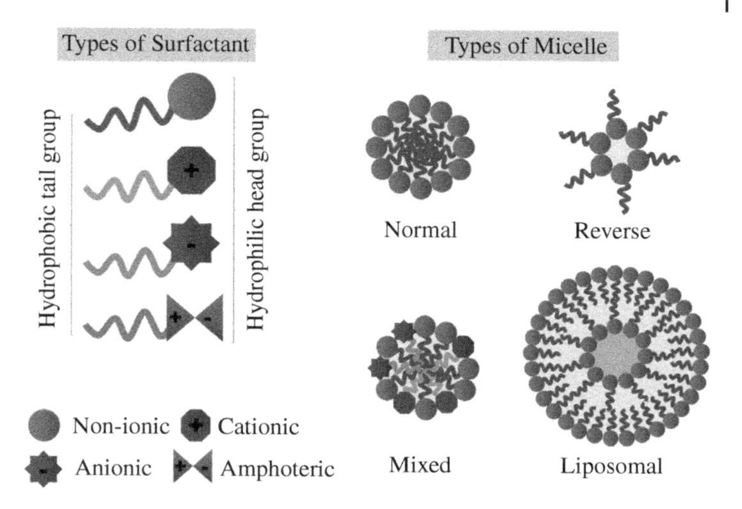

Figure 8.1 Schematic representing the types of biosurfactant based on ionic character and the subsequent different micelles formed in solutions (See insert for color representation of the figure).

8.6.1 Synthetic Surfactants

Despite a similar basic (amphiphilic) structure, synthetic surfactants are quite different in the individual compositions of the hydrophilic and hydrophobic moieties than those of the biosurfactants. The commonly observed hydrophobic groups in synthetic surfactants comprise alcohols, alkylphenols, olefins, and paraffins [85]. The hydrophilic head groups exhibit a higher degree of variation [25]:

- In the case of non-ionic surfactants, the hydrophilic head groups comprise polyoxyethylene or polypeptides that contribute to the solubility by the formation of H-bonds. Examples include ethoxylates, propylene oxide copolymers, and sorbitan esters.
- Anionic surfactants, on the other hand, contain sulfates, sulfonates, phosphates, or carboxylates as the head group. Common examples are alkyl sulfates, alkyl ether phosphates, and carboxylate salts.
- Cationic surfactants contain quaternary ammonium salts or amines as head groups. Examples include cetrimonium bromide, cetyl pyridinium chloride, etc.
- In contrast to the above three types, amphoteric surfactants comprise head groups containing both cationic (quaternary ammonium salts) and anionic (sulfonates) groups. Examples are hydroxysultaines, imidazoline carboxylates, and N-alkyl betaines. The ionic nature of some amphoteric surfactants is sensitive to pH changes, being cationic at low pH and anionic at high pH.

The diverse structural compositions of surfactants have found applications in a wide variety of industries involving the use of emulsifiers, foaming agents, detergents, wetting agents, and dispersents or solubilizers [86]. Despite their widespread use, the majority of the synthetic surfactants have been deemed hazardous for the environment and organisms [87, 88]. The synthetic surfactants most commonly used in the solubilization of metals and trace elements in the soil are – Triton X-100, Tween 80, sodium dodecyl sulfate (SDS), and cetyl trimethyl ammonium bromide (CTAB) [89–92]. All these surfactants have been reported to have a negative effect on the biota (Table 8.4). The eco-toxicity of synthetic surfactants is mainly associated with their propensity to accumulate in the environment. Due to their recalcitrance toward degradation [102], they

Table 8.4 The properties and hazardous effects of the synthetic surfactants most commonly used in the solubilization of soil micronutrients.

Surfactant			Molecular				
Tradename	**Chemical name**	**Ionic nature**	**Formula**	**Weight**	**Structure**	**Hazardous effects**	**References**
TX-100	*P*-tertiary-octylphenoxy polyethyl alcohol	Non-ionic	$C_{14}H_{22}O(C_2H_4O)_{10}$	646.85		Endocrine disrupters in aquatic organisms, wildlife, and humans as well due to its capacity to mimic natural hormones; highly recalcitrant in the environment; resistant to biodegradation because of its toxicity to microorganisms	[93–95]
Tween 80	Polyoxyethylene sorbitan monooleate	Non-ionic	$C_{64}H_{124}O_{26}$	1309.65	$w + x + y + z = 20$	Activates the complement system leading to acute hypersensitivity and systemic immunostimulation; alters the human gut microbiota leading to intestinal inflammation; acute toxicity in green algae	[96–98]
SDS	Sodium dodecyl sulfate	Anionic	$CH_3(CH_2)_{11}OSO_3Na$	288.38		Structural changes of DNA or polypeptide chains; alteration of the surface charge of biomacromolecules like peptides, enzymes, and DNA that may hinder normal molecular functions	[99, 100]
CTAB	Cetyl trimethyl Ammonium Bromide	Cationic	$C_{16}H_{33}(CH_3)_3NBr$	364.45		Toxic to aquatic microalgae, diatoms, and crustaceans	[101]

often lead to serious ramifications for the aquatic and terrestrial biota [103]. This has led to increased interest at the global scale to find eco-friendly and sustainable biological alternatives to the synthetic surfactants.

8.6.2 Biosurfactants

Biosurfactants are surface-active metabolites belonging to a structurally diverse group of biochemical compounds produced predominantly by microbes (bacteria, fungi, yeasts) on cell surfaces or secreted extracellularly [104, 105]. Just like their synthetic counterparts, all biosurfactants are

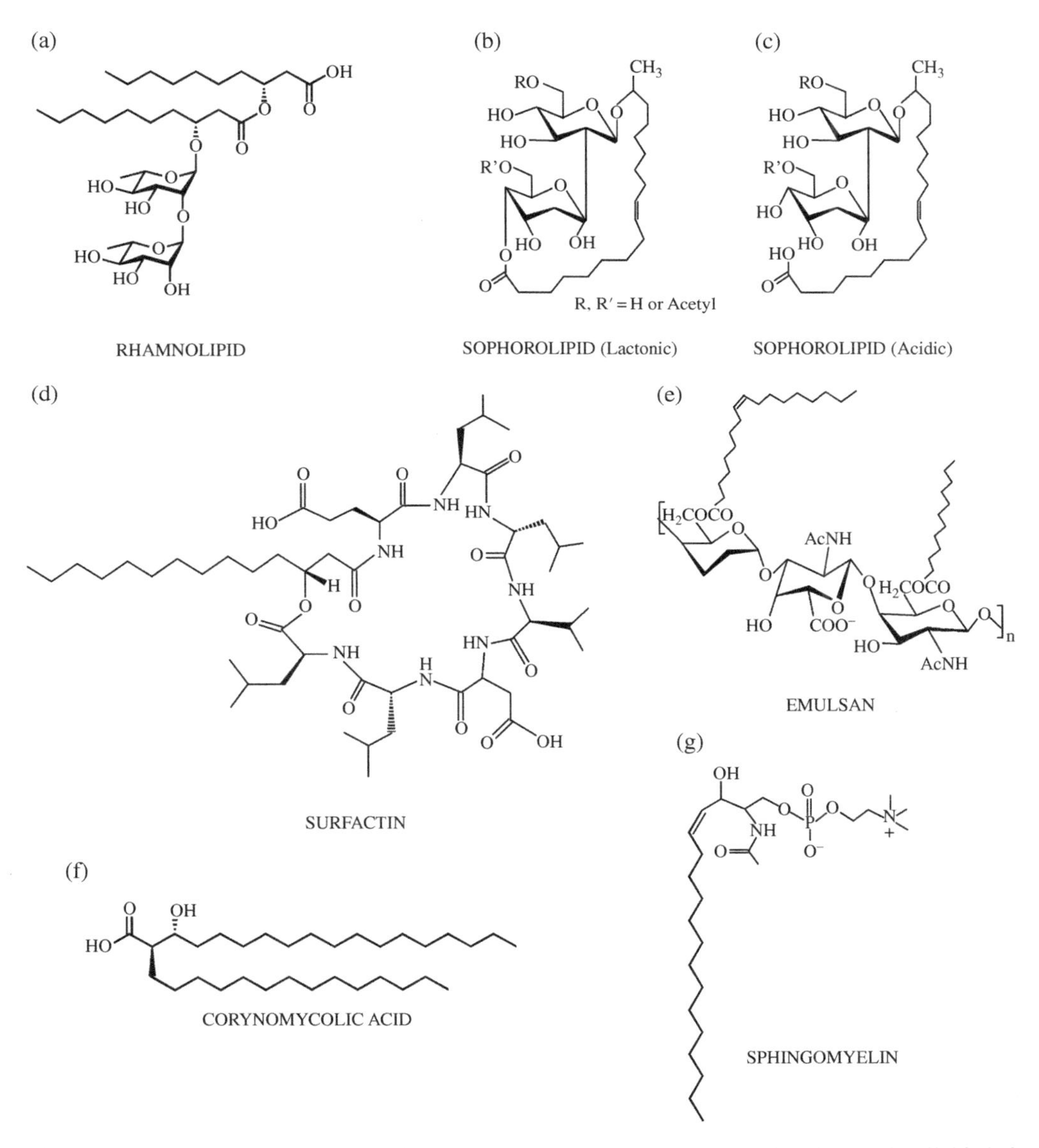

Figure 8.2 The structures of some of the most well-studied biosurfactants belonging to glycolipids (a, b, c), lipopeptide (d), polymeric biosurfactant (e), fatty acid (f), and phospholipid (g) types.

Table 8.5 The most well-studied biosurfactants, their typical producer organisms, and the commonly reported applications.

Biosurfactant				
Class	**Name**	**Producers**	**Applications**	**References**
Glycolipid	Rhamnolipid	*Pseudomonas aeruginosa, P. putida, P. fluorescens, Burkholderia* sp.	Bioremediation, enhanced oil recovery (EOR), antimicrobials, pharmaceuticals, cosmetics, biocontrol	[109–115]
	Sophorolipid	*Candida bombicola, Torulopsis bombicola, T. apicola, C. bogoriensis*	Antimicrobial, anti-HIV, household cleaning, cosmetics	[114, 116–119]
	Trehaloselipid	*Rhodococcus erythropolis, Arthrobacter* sp.	Bioremediation, MEOR, cancer therapeutics, antiviral	[120–124]
	Mannosyerythritol lipid	*Pseudozyma antarctica, Pseudozyma* sp.	Bioremediation, cancer therapeutics, cosmetic, detergent	[125–127]
Lipopeptide	Surfactin	*Bacillus subtilis, B. pumilus*	Bioremediation, EOR, biocontrol, antimicrobial, antiviral	[110, 128–131]
Polymeric	Emulsan	*Acinetobacter calcoaceticus, A. venetianus, Alcaligenes faecalis*	Emulsifier, EOR	[132, 133]
	Alasan	*A. radioresistens*	Emulsifier	[134]
Fatty acid	Corynomycolic acid	*Corynebacterium lepus*	Surfactant	[135]
Phospholipid	Sphingolipid	*Saccharomyces cerevisiae, Wickerhamomyces ciferrii*	Antibacterial, antifungals, cosmetics, cancer therapeutics	[136–139]

amphiphiles containing both hydrophilic and hydrophobic groups. The hydrophilic part of a biosurfactant molecule consists of mono-, oligo-, or polysaccharides, peptides or proteins, a phosphate group, carboxylic acid, alcohol, or some other compounds. The hydrophobic group comprises an unsaturated or saturated hydrocarbon chain or long chain of fatty acids, hydroxy fatty acids, or α-alkyl-β-hydroxy fatty acids [105, 106]. Additionally, the hydrophobic moiety is usually a C8 to C22 alkyl chain or an alkyl aryl derivative that may be linear or branched [107]. These structures confer a wide range of properties, in addition to their ability to lower surface and interfacial tension of liquids and to form micelles and microemulsions between two different phases [88].

Biosurfactants are categorized by their chemical composition, molecular weight, physicochemical properties, mode of action, and microbial origin, unlike that of the polar grouping based classification followed for their synthetic counterparts [104, 108]. Biosurfactants are classified broadly into glycolipids, lipopeptides, polymeric biosurfactants, fatty acids, and phospholipids based on their molecular structure [86] (Figure 8.2). Some of the most well-studied biosurfactants are detailed in Table 8.5.

In recent years, interest in microbial surfactants has increasingly amplified, primarily because of their natural origin and as a sustainable alternative to synthetic surfactants. Synthetic surfactants are an indispensable ingredient in the formulation of pesticides, where they are used as adjuvants [26] and applied randomly, not to mention their severe implications to the environment. Biosurfactants represent natural alternatives with distinct advantages over their synthetic counterparts, such as biodegradability [103], possible production from renewable and cheap resources [140], lower or non-toxicity [141], high specificity [142], and stability over a wide range of temperature, pH, and salinity [143]. Biosurfactants, being readily biodegradable, do not pose much threat to the environment and thus are referred to as "green chemicals" [144]. Importantly, these properties have compelling applications among others in the food [30], pharmaceutical [145], environmental [146], and cosmetic industries [147]. Reportedly, biosurfactants also increase the bioavailability of nutrients for beneficial plant-associated microbes [26].

8.6.2.1 Properties of Biosurfactants Critical for Enhancement of Nutrient Bioavailability

The mobility and bioavailability of micronutrients in the soil are mainly governed by the interfacial interactions between the soil particles and the trace elements present in the ground. The following properties of the biosurfactants play a critical role in the enhancement of solubilization and bioavailability of micronutrients in the soil.

Surface and Interfacial Activities The ST reducing activity is the primary characteristic of a biosurfactant. In this context, a surface designates the phase boundary between a condensed and gaseous phase (e.g. air–water), whereas interphase denotes that between two condensed phases (e.g. solid–liquid). This distinction is based on the fact that interfacial tension (IFT) involves significant adhesive forces between two different substances as well, in addition to the cohesive forces involved in ST, wherein adhesive forces are not significant [148]. An efficient biosurfactant can reduce the ST between water and air from 72 to 35 mN/m and the IFT between water and n-hexadecane from 40 to 1 mN/m [149]. A reduction in the surface and interfacial tensions facilitate the desorption of the bound nutrients from the hydrophobic soil particles, thereby enhancing their dispersal in the water in the rhizospheric soil.

Critical Micelle Concentration (CMC) Critical micelle concentration (CMC) is defined as the concentration of a surfactant in a solution at which the surfactant monomers start to aggregate and form micelles. At the CMC, the solution surface gets saturated with surfactant monomers. Consequently, above the CMC, no significant reduction in ST or IFT is observed. The CMC depends on the structure of the surfactant and the pH, ionic strength, and the temperature of the solution. The polarity of the solvent dictates the type of micelle formed by the surfactant dissolved in it. In an aqueous solution, surfactant monomers form normal (water in oil) micelle with the polar head groups oriented outward and the nonpolar tails toward the core. In contrast, in oil, reverse (oil in water) micelles occur with the lipophilic tails pointing outward and the hydrophilic heads toward the core. In addition to simple micelles, surfactant monomers also form vesicles and lamellae at concentrations higher than the CMC. These aggregates form as a result of numerous weak chemical interactions like hydrophobic, van der Waals, and hydrogen bonding between the polar head groups and the non-polar tail groups [150]. From the standpoint of trace element solubilization, the CMC is the most critical parameter. Beyond the CMC, surfactants can sequester metal ions dispersed from the soil by encapsulation in the core of the micelle and mobilize them for higher bioavailability [108].

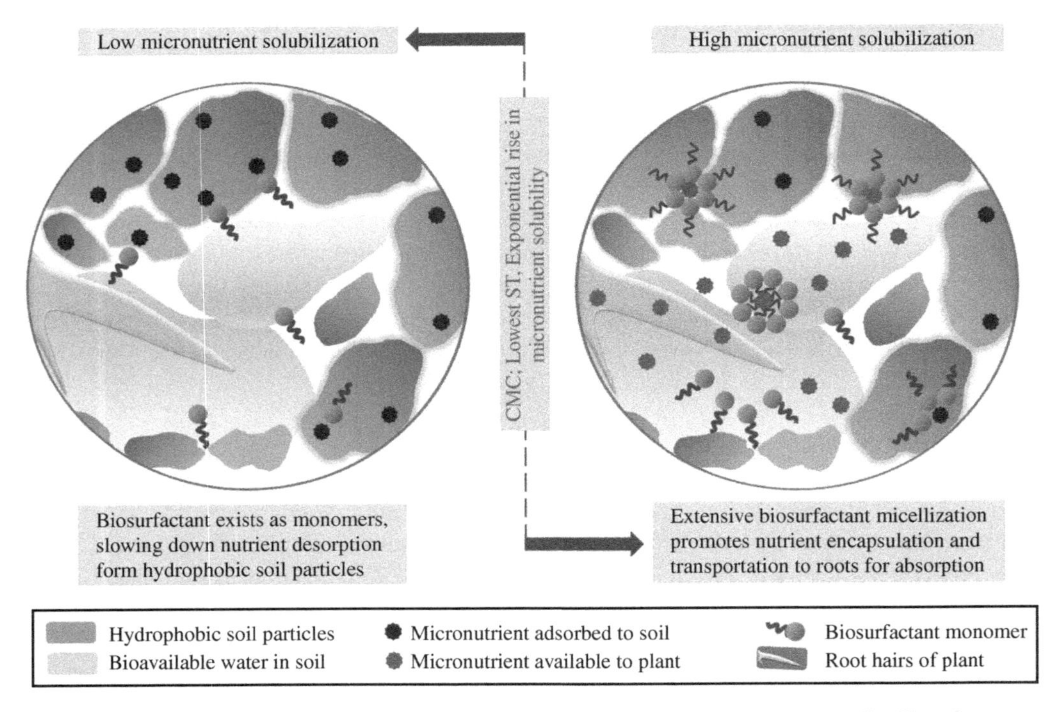

Figure 8.3 Mechanism of action of biosurfactants to enhance the solubility and bioavailability of micronutrients in deficient soils. CMC = critical micelle concentration, ST = surface tension.

Tolerance to Changes in pH, Temperature, and Ionic Strength Biosurfactants retain their surface activity over a wide range of pH, temperature, and ionic strength, as has been demonstrated with several biosurfactants. For example, sophorolipid produced by *Rhodotrorula babjevae* YS3 has been reported to retain its surface activity over a pH range of 2–10, NaCl concentrations (2–10%), and after two hours of heating at 120 °C [143]. Similarly, a lipopeptide from *Bacillus subtilis* LB5a retained activity over variations in pH (5–11), NaCl concentrations (up to 20%), and temperature (121 °C for 20 minutes, and −18 °C for six months) [151].

Biodegradability and Toxicity Biosurfactants are favored over their chemical counterparts because of their biodegradability and low-toxicity [102, 152]. Lower toxicity minimizes adverse effects on the soil microbiota, whereas biodegradability ensures the removal of the compounds without any damage to the environment. For example, rhamnolipid has been successfully commercialized as the active component of the biofungicide Zonix™, approved by the US Environmental Protection Agency (US-EPA) [153]. The US-EPA has also approved rhamnolipid for use in confectionery, pharmaceuticals, and cosmetics owing to its low toxicity [30]. Research indicates that rhamnolipids cause no mutagenic effect and comparable or lower levels of acute toxicity than chemical surfactants in daphnids and zebrafish embryos [154].

8.6.2.2 Mechanism of Action of Biosurfactants

An intricate interplay of all the properties described above ensures that biosurfactants can effectively minimize the interactions between the soil particles and the trace elements to enhance the bioavailability of the micronutrients in the soil. Due to their surface activity, biosurfactants can lower the hydrophobicity of soil particles and establish ionic interactions with the metal ions bound to the soil particles, consequently facilitating the desorption of the bound ions into the soil solution, leading to their sequestration and increased mobility [155]. In the absence of any other negative factor (like interactions with other trace elements), roots can directly absorb the solubilized nutrients. However, under field conditions, this is unlikely, and hence the encapsulation and transport of the ions become important. At concentrations above their CMC, biosurfactants can form micelles that can chelate the metal ions to keep them in a soluble form and prevent further adsorption or fixation (Figure 8.3) [156]. Soil pH plays an important role in the bioavailability of micronutrients and alkaline soils are the most deficient in micronutrients [14]. Except for Cl and Mo, all other micronutrients are absorbed by plants as cations [1]. These metals have the propensity to form phosphates or carbonates at higher pH, reducing their bioavailability [157]. Nevertheless, the biosurfactants retain their activity over a wide pH range that gives them the flexibility to sequester metals even from alkaline soils to promote uptake by roots [158]. The case of rhamnolipids is noteworthy in this regard. At $pH \leq 5$, the carboxyl group of the rhamnolipid molecule is protonated, thus having a nonionic character in aqueous solution. This reduces the electrostatic repulsion among the rhamnolipid monomers, facilitating their aggregation and a lower CMC value. On the other hand, at $pH > 5$, the head group becomes negatively charged, leading to an increase in the CMC as well as the size of the micelle (vesicles predominate over normal micelle) [159]. However, the anionic character enables rhamnolipids to preferentially bind to metal cations in alkaline soil ($pH > 8$) and form lipophilic metal–rhamnolipid aggregations. At the same time, the encapsulation of these metal ions in bi-layered vesicles avoids counter-interactions. In their experiment involving canola (*Brassica napus*), bread wheat (*Triticum aestivum*), and durum wheat (*Triticum turgidum*), Stacey and co-workers demonstrated that both mono- and di-rhamnolipid could form lipophilic metal–rhamnolipid complexes with Zn, Cu, and Mn. Due to their lipophilicity, these complexes could easily be transported across root membranes, enhancing the content of these micronutrients in the plants [158]. In addition to the application of biosurfactants alone, agricultural soils can be supplemented with biosurfactant-producing bacteria as well, for sustained bioavailability of the micronutrients in deficient soil. For example, rhamnolipid-producing *Pseudomonas* strains have been successfully used to enhance the growth and antioxidant activity of *Withania somnifera* in hydrocarbon-contaminated soil [160]. Although many microbe-producing biosurfactants have been known [161], their production needs further optimization of various factors [162].

8.7 Conclusion

Micronutrient deficiency in soil is a major concern throughout the world. Soil deficient in micronutrients affects the nutritional value of crops and consequently has ramifications across the food chain. The use of micronutrient supplements to address the problem has not been very effective. Biosurfactants can sequester metal ions bound to hydrophobic soil particles, increasing their solubility and bioavailability. Traditionally, synthetic surfactants are used as supplements alongside fertilizers to enhance micronutrient solubilization. However, their efficiency, specificity, and ecotoxicity have been areas of concern. In contrast, biosurfactants are structurally diverse with high specificity and are functionally efficient across a wide range of pH, temperature, and ionic stress, all the while being completely biodegradable. This chapter provides a detailed case for the use of biosurfactants as eco-friendly supplements to enhance the bioavailability of micronutrients in the soil. Several factors like soil pH, temperature, interactions with counterions, as well as uptake efficiency of plants, influence the bioavailability of micronutrients. These parameters vary widely from region to region and, therefore, no strategy can be generalized for guaranteed success. A target-specific and careful evaluation of all the parameters will be essential to mitigate the deficiency of micronutrients sustainably across different geographic regions.

References

1 Hänsch, R. and Mendel, R.R. (2009). Physiological functions of mineral micronutrients (Cu, Zn, Mn, Fe, Ni, Mo, B, Cl). *Curr. Opin. Plant Biol.* 12 (3): 259–266.
2 Gernand, A.D., Schulze, K.J., Stewart, C.P. et al. (2016). Micronutrient deficiencies in pregnancy worldwide: Health effects and prevention. *Nat. Rev. Endocrinol.* 12 (5): 274–289.
3 Welch, R.M. (2008). Linkages between trace elements in food crops and human health. In: *Micronutrient Deficiencies in Global Crop Production* (ed. B.J. Alloway), 287–309. Dordrecht: Springer.
4 Shukla, R., Sharma, Y., and Shukla, A. (2014). Molecular mechanism of nutrient uptake in plants. *Int. J. Curr. Res. Acad. Res.* 2 (12): 142–154.
5 Singh, R., Glick, B.R., and Rathore, D. (2018). Biosurfactants as a biological tool to increase micronutrient availability in soil: A review. *Pedosphere* 28 (2): 170–189.
6 Welch, R.M. (2002). The impact of mineral nutrients in food crops on global human health. *Plant Soil* 247 (1): 83–90.
7 White, P. and Brown, P. (2010). Plant nutrition for sustainable development and global health. *Ann. Bot.* 105 (7): 1073–1080.
8 Baligar, V. and Fageria, N. (2015). Nutrient use efficiency in plants: An overview. In: *Nutrient Use Efficiency: From Basics to Advances* (eds. R.A. SHB and A. Sen), 1–14. New Delhi: Springer.
9 Aktar, W., Sengupta, D., and Chowdhury, A. (2009). Impact of pesticides use in agriculture: Their benefits and hazards. *Interdiscip. Toxicol.* 2 (1): 1–12.
10 Uriu-Adams, J.Y. and Keen, C.L. (2005). Copper, oxidative stress, and human health. *Mol. Asp. Med.* 26 (4–5): 268–298.
11 Boxall, A.B., Hardy, A., Beulke, S. et al. (2009). Impacts of climate change on indirect human exposure to pathogens and chemicals from agriculture. *Environ. Health Perspect.* 117 (4): 508–514.
12 Fageria, N., Slaton, N., and Baligar, V. (2003). Nutrient management for improving lowland rice productivity and sustainability. *Adv. Agron.* 80 (1): 63–152.
13 Fageria, N., Dos Santos, A., and Moreira, A. (2010). Yield, nutrient uptake, and changes in soil chemical properties as influenced by liming and iron application in common bean in a no-tillage system. *Commun. Soil Sci. Plant Anal.* 41 (14): 1740–1749.
14 Graham, R.D. (2008). Micronutrient deficiencies in crops and their global significance. In: *Micronutrient Deficiencies in Global Crop Production* (ed. B.J. Alloway), 41–61. Dordrecht: Springer.
15 Jones, D.L., Cross, P., Withers, P.J. et al. (2013). Nutrient stripping: The global disparity between food security and soil nutrient stocks. *J. Appl. Ecol.* 50 (4): 851–862.
16 Dhaliwal, S., Naresh, R., Mandal, A. et al. (2019). Dynamics and transformations of micronutrients in soil environment as influenced by organic matter build-up: A review. *Environ. Sustain. Ind.* 1-2: 100007.
17 Miller, D.D. and Welch, R.M. (2013). Food system strategies for preventing micronutrient malnutrition. *Food Policy* 42: 115–128.
18 Knez, M. and Graham, R.D. (2013). The impact of micronutrient deficiencies in agricultural soils and crops on the nutritional health of humans. In: *Essentials of Medical Geology* (ed. O. Selinas), 517–533. Dordrecht: Springer.

19 Uprety, D., Hejcman, M., Száková, J. et al. (2009). Concentration of trace elements in arable soil after long-term application of organic and inorganic fertilizers. *Nutr. Cycl. Agroecosyst.* 85 (3): 241–252.

20 Nkebiwe, P.M., Weinmann, M., Bar-Tal, A., and Müller, T. (2016). Fertilizer placement to improve crop nutrient acquisition and yield: A review and meta-analysis. *Field Crop Res.* 196: 389–401.

21 Osman, K.T. (2013). *Plant nutrients and soil fertility management. Soils: Principles, Properties and Management*, 129–159. Dordrecht: Springer.

22 Baligar, V., Fageria, N., and He, Z. (2001). Nutrient use efficiency in plants. *Commun. Soil Sci. Plant Anal.* 32 (7–8): 921–950.

23 Chatzistathis, T. (2014). *Micronutrient Deficiency in Soils and Plants*, 204. Sharjah: Bentham Science Publishers.

24 Schoonover, J.E. and Crim, J.F. (2015). An introduction to soil concepts and the role of soils in watershed management. *J. Contemp. Water Res. Educ.* 154 (1): 21–47.

25 Mao, X., Jiang, R., Xiao, W., and Yu, J. (2015). Use of surfactants for the remediation of contaminated soils: A review. *J. Hazard. Mater.* 285: 419–435.

26 Sachdev, D.P. and Cameotra, S.S. (2013). Biosurfactants in agriculture. *Appl. Microbiol. Biotechnol.* 97 (3): 1005–1016.

27 Jessop, P., Ahmadpour, F., Buczynski, M. et al. (2015). Opportunities for greener alternatives in chemical formulations. *Green Chem.* 17 (5): 2664–2678.

28 Kang, S. and Jeong, H.Y. (2015). Sorption of a nonionic surfactant Tween 80 by minerals and soils. *J. Hazard. Mater.* 284: 143–150.

29 Desai, J.D. and Banat, I.M. (1997). Microbial production of surfactants and their commercial potential. *Microbiol. Mol. Biol. Rev.* 61 (1): 47–64.

30 Nitschke, M. and Costa, S. (2007). Biosurfactants in food industry. *Trends Food Sci. Technol.* 18 (5): 252–259.

31 White, J.G. and Zasoski, R.J. (1999). Mapping soil micronutrients. *Field Crop Res.* 60 (1–2): 11–26.

32 Holloway, R., Graham, R., and Stacey, S. (2008). Micronutrient deficiencies in Australian field crops. In: *Micronutrient deficiencies in global crop production* (ed. B.J. Alloway), 63–86. Dordrecht: Springer.

33 Montalvo, D., Degryse, F., Da Silva, R. et al. (2016). Agronomic effectiveness of zinc sources as micronutrient fertilizer. In: *Advances in Agronomy*, vol. 139 (ed. D.L. Sparks), 215–267. Elsevier.

34 Alloway, B.J. (2008). Micronutrients and crop production: An introduction. In: *Micronutrient Deficiencies in Global Crop Production* (ed. B.J. Alloway), 1–39. Dordrecht: Springer.

35 Shukla, A.K., Behera, S.K., Pakhre, A., and Chaudhari, S. (2018). Micronutrients in soils, plants, animals and humans. *Indian J. Fertil.* 14 (3): 30–54.

36 Singh, M.V. (2008). Micronutrient deficiencies in crops and soils in India. In: *Micronutrient Deficiencies in Global Crop Production* (ed. B.J. Alloway), 93–125. Dordrecht: Springer.

37 Gupta, U.C., Kening, W., and Liang, S. (2008). Micronutrients in soils, crops, and livestock. *Earth Sci. Front.* 15 (5): 110–125.

38 Bradl, H.B. (2004). Adsorption of heavy metal ions on soils and soils constituents. *J. Colloid Interface Sci.* 277 (1): 1–18.

39 Neina, D. (2019). The role of soil pH in plant nutrition and soil remediation. *Appl. Environ. Soil Sci.* 2019: 1–9.

40 Najafi-Ghiri, M., Ghasemi-Fasaei, R., and Farrokhnejad, E. (2013). Factors affecting micronutrient availability in calcareous soils of Southern Iran. *Arid Land Res. Manag.* 27 (3): 203–215.

41 Kumar, M., Jha, A., Hazarika, S. et al. (2016). Micronutrients (B, Zn, Mo) for improving crop production on acidic soils of Northeast India. *Natl. Acad. Sci. Lett.* 39 (2): 85–89.

42 Mackowiak, C., Grossl, P., and Bugbee, B. (2001). Beneficial effects of humic acid on micronutrient availability to wheat. *Soil Sci. Soc. Am. J.* 65 (6): 1744–1750.

43 Bista, D.R., Heckathorn, S.A., Jayawardena, D.M. et al. (2018). Effects of drought on nutrient uptake and the levels of nutrient-uptake proteins in roots of drought-sensitive and-tolerant grasses. *Plants* 7 (2): 28.

44 Fageria, N., Gheyi, H., and Moreira, A. (2011). Nutrient bioavailability in salt affected soils. *J. Plant Nutr.* 34 (7): 945–962.

45 Nambiar, E. (1977). The effects of drying of the topsoil and of micronutrients in the subsoil on micronutrient uptake by an intermittently defoliated ryegrass. *Plant Soil* 46 (1): 185–193.

46 Provin, T.L., Wright, A.L., Hons, F.M. et al. (2008). Seasonal dynamics of soil micronutrients in compost-amended bermudagrass turf. *Bioresour. Technol.* 99 (7): 2672–2679.

47 Nambiar, E. (1977). The effects of water content of the topsoil on micronutrient availability and uptake in a siliceous sandy soil. *Plant Soil* 46 (1): 175–183.

48 Jhanji, S. and Sadana, U.S. (2014). Genotypic variation in partitioning of dry matter and manganese between source and sink organs of rice under manganese stress. *Plant Cell Rep.* 33 (8): 1227–1238.

49 Fageria, N.K., de Brito Ferreira, E.P., and Knupp, A.M. (2015). Micronutrients use efficiency in tropical cover crops as influenced by phosphorus fertilization. *Revista Caatinga.* 28 (1): 130–137.

50 Arnon, D.I. and Stout, P. (1939). The essentiality of certain elements in minute quantity for plants with special reference to copper. *Plant Physiol.* 14 (2): 371–375.

51 Barker, A.V. and Eaton, T.E. (2015). Zinc. In: *Handbook of Plant Nutrition*, 2e (eds. A.V. Barker and D.J. Pilbeam), 537–564. Boca Raton: CRC Press.

52 Barker, A.V. and Stratton, M.L. (2015). Iron. In: *Handbook of Plant Nutrition*, 2e (eds. A.V. Barker and D.J. Pilbeam), 399–426. Boca Raton: CRC Press.

53 Eaton, T.E. (2015). Manganese. In: *Handbook of Plant Nutrition*, 2e (eds. A.V. Barker and D.J. Pilbeam), 427–486. Boca Raton: CRC Press.

54 Kopsell, D.A., Kopsell, D.E., and Hamlin, R.L. (2015). Molybdenum. In: *Handbook of Plant Nutrition*, 2e (eds. A.V. Barker and D.J. Pilbeam), 487–510. Boca Raton: CRC Press.

55 Kopsell, D.E. and Kopsell, D.A. (2015). Chlorine. In: *Handbook of Plant Nutrition*, 2e (eds. A.V. Barker and D.J. Pilbeam), 347–366. Boca Raton: CRC Press.

56 Wimmer, M.A., Goldberg, S., and Gupta, U.C. (2015). Boron. In: *Handbook of Plant Nutrition*, 2e (eds. A.V. Barker and D.J. Pilbeam), 305–346. Boca Raton: CRC Press.

57 Wood, B.W. (2015). Nickel. In: *Handbook of Plant Nutrition*, 2e (eds. A.V. Barker and D.J. Pilbeam), 511–536. Boca Raton: CRC Press.

58 Yruela, I. (2015). Copper. In: *Handbook of Plant Nutrition*, 2e (eds. A.V. Barker and D.J. Pilbeam), 367–398. Boca Raton: CRC Press.

59 FAO, IFAD, WFP, WHO, UNICEF (2019). *The State of Food Security and Nutrition in the World 2019: Safeguarding against Economic Slowdowns and Downturns*. Rome: FAO.

60 Bailey, R.L., West, K.P. Jr., and Black, R.E. (2015). The epidemiology of global micronutrient deficiencies. *Ann. Nutr. Metab.* 66 (Suppl. 2): 22–33.

61 Ackland, M.L. and Michalczyk, A.A. (2016). Zinc and infant nutrition. *Arch. Biochem. Biophys.* 611: 51–57.

62 Chabosseau, P. and Rutter, G.A. (2016). Zinc and diabetes. *Arch. Biochem. Biophys.* 611: 79–85.

63 Lazarus, J.H. (2015). The importance of iodine in public health. *Environ. Geochem. Health* 37 (4): 605–618.

64 Andersson, M., Karumbunathan, V., and Zimmermann, M.B. (2012). Global iodine status in 2011 and trends over the past decade. *J. Nutr.* 142 (4): 744–750.

65 Combs, G.F. Jr. and McClung, J.P. (2017). *Vitamin A. The Vitamins: Fundamental Aspects in Nutrition and Health*, 5e, 109–159. Westborough, MA, USA: Academic Press.

66 Gardner, D.G., Chen, S., and Glenn, D.J. (2013). Vitamin D and the heart. *Am. J. Phys. Regul. Integr. Comp. Phys.* 305 (9): R969–R977.

67 Combs, G.F. Jr. and McClung, J.P. (2017). *Vitamin D. The Vitamins: Fundamental Aspects in Nutrition and Health*, 5e, 161–206. Westborough, MA, USA: Academic Press.

68 Combs, G.F. Jr. and McClung, J.P. (2017). Folate. In: *The Vitamins: Fundamental Aspects in Nutrition and Health*, 5e, 399–429. Westborough, MA, USA: Academic Press.

69 Graham, R.D., Welch, R.M., and Bouis, H.E. (2001). Addressing micronutrient malnutrition through enhancing the nutritional quality of staple foods: Principles, perspectives and knowledge gaps. *Adv. Agron.* 70: 77–142.

70 Gilani, G.S. and Nasim, A. (2007). Impact of foods nutritionally enhanced through biotechnology in alleviating malnutrition in developing countries. *J. AOAC Int.* 90 (5): 1440–1444.

71 Garg, M., Sharma, N., Sharma, S. et al. (2018). Biofortified crops generated by breeding, agronomy, and transgenic approaches are improving lives of millions of people around the world. *Front. Nutr.* 5: 12.

72 Pérez-Massot, E., Banakar, R., Gómez-Galera, S. et al. (2013). The contribution of transgenic plants to better health through improved nutrition: opportunities and constraints. *Genes Nutr.* 8 (1): 29–41.

73 Fageria, N., Baligar, V., and Clark, R. (2002). Micronutrients in crop production. In: *Advances in Agronomy*, vol. 77 (ed. D. Sparks), 185–268. San Diego, CA, USA: Academic Press.

74 Rietra, R.P., Heinen, M., Dimkpa, C.O., and Bindraban, P.S. (2017). Effects of nutrient antagonism and synergism on yield and fertilizer use efficiency. *Commun. Soil Sci. Plant Anal.* 48 (16): 1895–1920.

75 Zhang, Y.-Q., Deng, Y., Chen, R.-Y. et al. (2012). The reduction in zinc concentration of wheat grain upon increased phosphorus-fertilization and its mitigation by foliar zinc application. *Plant Soil* 361 (1–2): 143–152.

76 Sacristán, D., González-Guzmán, A., Barrón, V. et al. (2019). Phosphorus-induced zinc deficiency in wheat pot-grown on noncalcareous and calcareous soils of different properties. *Arch. Agron. Soil Sci.* 65 (2): 208–223.

77 Trinchera, A. and Baratella, V. (2018). Use of a non-ionic water surfactant in lettuce fertigation for optimizing water use, improving nutrient use efficiency, and increasing crop quality. *Water* 10 (5): 613.

78 Moore RA, Kostka SJ, Mane S, Miller CM, inventors; Aquatrols Holding Co. Inc. Texas Instruments Inc., assignee. Fully compatible surfactant-impregnated water-soluble fertilizer; concentrate; and use. United States of America Grant No. 6460290. 2002.

79 Mclaughlin M, Stacey S, Lombi E, inventors; Adelaide Research and Innovation Pty. Ltd., assignee. Sequestering agent for micronutrient fertilisers. United States of America Patent US 8,217,004 B2. 2012.

80 McCauley, G. (1993). Nonionic surfactant and supplemental irrigation of soybean on crusting soils. *Agron. J.* 85 (1): 17–21.

81 Kelling, K., Speth, P., Arriaga, F., and Lowery, B. (2003). Use of a nonionic surfactant to improve nitrogen use efficiency of potato. *Acta Hortic.* 619: 225–232.

82 Arriaga, F.J., Lowery, B., and Kelling, K.A. (2009). Surfactant impact on nitrogen utilization and leaching in potatoes. *Am. J. Potato Res.* 86 (5): 383–390.

83 Chaichi, M.R., Nurre, P., Slaven, J., and Rostamza, M. (2015). Surfactant application on yield and irrigation water use efficiency in corn under limited irrigation. *Crop Sci.* 55 (1): 386–393.

84 Karanth, N., Deo, P., and Veenanadig, N. (1999). Microbial production of biosurfactants and their importance. *Curr. Sci.* 77 (1): 116–126.

85 Volkering, F., Breure, A., and Rulkens, W. (1997). Microbiological aspects of surfactant use for biological soil remediation. *Biodegradation* 8 (6): 401–417.

86 Cameotra, S.S., Makkar, R.S., Kaur, J., and Mehta, S. (2010). Synthesis of biosurfactants and their advantages to microorganisms and mankind. In: *Biosurfactants. Advances in Experimental Medicine and Biology*, vol. 672 (ed. R. Sen), 261–280. New York: Springer.

87 Mungray, A.K. and Kumar, P. (2008). Anionic surfactants in treated sewage and sludges: risk assessment to aquatic and terrestrial environments. *Bioresour. Technol.* 99 (8): 2919–2929.

88 Banat, I.M., Franzetti, A., Gandolfi, I. et al. (2010). Microbial biosurfactants production, applications and future potential. *Appl. Microbiol. Biotechnol.* 87 (2): 427–444.

89 Shin, M. (2004). *Surfactant/Ligand Systems for the Simultaneous Remediation of Soils Contaminated with Heavy Metals and Polychlorinated Biphenyls*. Canada: McGill University.

90 Chang, S.-H., Wang, K.-S., Kuo, C.-Y. et al. (2005). Remediation of metal-contaminated soil by an integrated soil washing-electrolysis process. *Soil Sediment Contam. Int. J.* 14 (6): 559–569.

91 Sun H, Wang H, Qi J, Shen L, Lian X. Study on surfactants remediation in heavy metals contaminated soils. In: *2011 International Symposium on Water Resource and Environmental Protection*; 2011: Xi'an, China: IEEE.

92 Dong, Z.-Y., Huang, W.-H., Xing, D.-F., and Zhang, H.-F. (2013). Remediation of soil co-contaminated with petroleum and heavy metals by the integration of electrokinetics and biostimulation. *J. Hazard. Mater.* 260: 399–408.

93 Saien, J., Ojaghloo, Z., Soleymani, A., and Rasoulifard, M. (2011). Homogeneous and heterogeneous AOPs for rapid degradation of Triton X-100 in aqueous media via UV light, nano titania hydrogen peroxide and potassium persulfate. *Chem. Eng. J.* 167 (1): 172–182.

94 Chen, H.-J., Tseng, D.-H., and Huang, S.-L. (2005). Biodegradation of octylphenol polyethoxylate surfactant Triton X-100 by selected microorganisms. *Bioresour. Technol.* 96 (13): 1483–1491.

95 Perkowski, J., Bulska, A., and Józwiak, W. (2005). Titania-assisted photocatalytic decomposition of Triton X-100 detergent in aqueous solution. *Environ. Prot. Eng.* 31 (2): 61–75.

96 Weiszhár, Z., Czúcz, J., Révész, C. et al. (2012). Complement activation by polyethoxylated pharmaceutical surfactants: Cremophor-EL, Tween-80 and Tween-20. *Eur. J. Pharm. Sci.* 45 (4): 492–498.

97 Chassaing, B., Van de Wiele, T., De Bodt, J. et al. (2017). Dietary emulsifiers directly alter human microbiota composition and gene expression ex vivo potentiating intestinal inflammation. *Gut* 66 (8): 1414–1427.

98 Ma, J., Lin, F., Zhang, R. et al. (2004). Differential sensitivity of two green algae, *Scenedesmus quadricauda* and *Chlorella vulgaris*, to 14 pesticide adjuvants. *Ecotoxicol. Environ. Saf.* 58 (1): 61–67.

99 Cserháti, T. (1995). Alkyl ethoxylated and alkylphenol ethoxylated nonionic surfactants: interaction with bioactive compounds and biological effects. *Environ. Health Perspect.* 103 (4): 358–364.

100 Ivanković, T. and Hrenović, J. (2010). Surfactants in the environment. *Arch. Ind. Hyg. Toxicol.* 61 (1): 95–110.

101 Kaczerewska, O., Martins, R., Figueiredo, J. et al. (2020). Environmental behaviour and ecotoxicity of cationic surfactants towards marine organisms. *J. Hazard. Mater.* 392: 122299.

102 Mohan, P.K., Nakhla, G., and Yanful, E.K. (2006). Biokinetics of biodegradation of surfactants under aerobic, anoxic and anaerobic conditions. *Water Res.* 40 (3): 533–540.

103 Poremba, K., Gunkel, W., Lang, S., and Wagner, F. (1991). Toxicity testing of synthetic and biogenic surfactants on marine microorganisms. *Environ. Toxicol. Water Qual.* 6 (2): 157–163.

104 Muthusamy, K., Gopalakrishnan, S., Ravi, T.K., and Sivachidambaram, P. (2008). Biosurfactants: properties, commercial production and application. *Curr. Sci.* 94 (6): 736–747.

105 Rahman, P.K. and Gakpe, E. (2008). Production, characterisation and applications of biosurfactants-review. *Biotechnology* 7: 360–370.

106 Lang, S. (2002). Biological amphiphiles (microbial biosurfactants). *Curr. Opin. Colloid Interface Sci.* 7 (1): 12–20.

107 Van Ginkel, C. (1996). Complete degradation of xenobiotic surfactants by consortia of aerobic microorganisms. *Biodegradation* 7 (2): 151–164.

108 Pacwa-Płociniczak, M., Płaza, G.A., Piotrowska-Seget, Z., and Cameotra, S.S. (2011). Environmental applications of biosurfactants: recent advances. *Int. J. Mol. Sci.* 12 (1): 633–654.

109 Rahman, K.S., Rahman, T.J., Kourkoutas, Y. et al. (2003). Enhanced bioremediation of n-alkane in petroleum sludge using bacterial consortium amended with rhamnolipid and micronutrients. *Bioresour. Technol.* 90 (2): 159–168.

110 Amani, H., Sarrafzadeh, M.H., Haghighi, M., and Mehrnia, M.R. (2010). Comparative study of biosurfactant producing bacteria in MEOR applications. *J. Pet. Sci. Eng.* 75 (1–2): 209–214.

111 Borah, S.N., Goswami, D., Lahkar, J. et al. (2015). Rhamnolipid produced by *Pseudomonas aeruginosa* SS14 causes complete suppression of wilt by *Fusarium oxysporum* f. sp. pisi in *Pisum sativum*. *BioControl* 60 (3): 375–385.

112 Borah, S.N., Goswami, D., Sarma, H.K. et al. (2016). Rhamnolipid biosurfactant against *Fusarium verticillioides* to control stalk and ear rot disease of maize. *Front. Microbiol.* 7: 1505.

113 Sen, S., Borah, S.N., Kandimalla, R. et al. (2019). Efficacy of a rhamnolipid biosurfactant to inhibit *Trichophyton rubrum* in vitro and in a mice model of dermatophytosis. *Exp. Dermatol.* 28 (5): 601–608.

114 Lourith, N. and Kanlayavattanakul, M. (2009). Natural surfactants used in cosmetics: Glycolipids. *Int. J. Cosmet. Sci.* 31 (4): 255–261.

115 Vatsa, P., Sanchez, L., Clement, C. et al. (2010). Rhamnolipid biosurfactants as new players in animal and plant defense against microbes. *Int. J. Mol. Sci.* 11 (12): 5095–5108.

116 Van Bogaert, I.N., Zhang, J., and Soetaert, W. (2011). Microbial synthesis of sophorolipids. *Process Biochem.* 46 (4): 821–833.

117 Sen, S., Borah, S.N., Kandimalla, R. et al. (2020). Sophorolipid biosurfactant can control cutaneous dermatophytosis caused by *Trichophyton mentagrophytes*. *Front. Microbiol.* 11: 329.

118 Shah, V., Doncel, G.F., Seyoum, T. et al. (2005). Sophorolipids, microbial glycolipids with anti-human immunodeficiency virus and sperm-immobilizing activities. *Antimicrob. Agents Chemother.* 49 (10): 4093–4100.

119 Develter, D.W. and Lauryssen, L.M. (2010). Properties and industrial applications of sophorolipids. *Eur. J. Lipid Sci. Technol.* 112 (6): 628–638.

120 Kretschmer, A., Bock, H., and Wagner, F. (1982). Chemical and physical characterization of interfacial-active lipids from *Rhodococcus erythropolis* grown on n-alkanes. *Appl. Environ. Microbiol.* 44 (4): 864–870.

121 Peng, F., Liu, Z., Wang, L., and Shao, Z. (2007). An oil-degrading bacterium: *Rhodococcus erythropolis* strain 3C-9 and its biosurfactants. *J. Appl. Microbiol.* 102 (6): 1603–1611.

122 Nazina, T., Sokolova, D.S., Grigor'yan, A. et al. (2003). Production of oil-releasing compounds by microorganisms from the Daqing oil field. *China. Microbiology.* 72 (2): 173–178.

123 Sudo, T., Zhao, X., Wakamatsu, Y. et al. (2000). Induction of the differentiation of human HL-60 promyelocytic leukemia cell line by succinoyl trehalose lipids. *Cytotechnology* 33 (1–3): 259–264.

124 Azuma, M., Suzutani, T., Sazaki, K. et al. (1987). Role of interferon in the augmented resistance of trehalose-6, 6′-dimycolate-treated mice to influenza virus infection. *J. Gen. Virol.* 68 (3): 835–843.

125 Morita, T., Fukuoka, T., Imura, T., and Kitamoto, D. (2013). Production of mannosylerythritol lipids and their application in cosmetics. *Appl. Microbiol. Biotechnol.* 97 (11): 4691–4700.

126 Sajna, K.V., Sukumaran, R.K., Jayamurthy, H. et al. (2013). Studies on biosurfactants from *Pseudozyma* sp. NII 08165 and their potential application as laundry detergent additives. *Biochem. Eng. J.* 78: 85–92.

127 Zhao, X., Geltinger, C., Kishikawa, S. et al. (2000). Treatment of mouse melanoma cells with phorbol 12-myristate 13-acetate counteracts mannosylerythritol lipid-induced growth arrest and apoptosis. *Cytotechnology* 33 (1–3): 123–130.

128 Whang, L.M., Liu, P.-W.G., Ma, C.-C., and Cheng, S.-S. (2008). Application of biosurfactants, rhamnolipid, and surfactin, for enhanced biodegradation of diesel-contaminated water and soil. *J. Hazard. Mater.* 151 (1): 155–163.

129 Cawoy, H., Mariutto, M., Henry, G. et al. (2014). Plant defense stimulation by natural isolates of *Bacillus* depends on efficient surfactin production. *Mol. Plant-Microbe Interact.* 27 (2): 87–100.

130 Heerklotz, H. and Seelig, J. (2001). Detergent-like action of the antibiotic peptide surfactin on lipid membranes. *Biophys. J.* 81 (3): 1547–1554.

131 Kracht, M., Rokos, H., Özel, M. et al. (1999). Antiviral and hemolytic activities of surfactin isoforms and their methyl ester derivatives. *J. Antibiot.* 52 (7): 613–619.

132 Su, W.-T., Chen, W.-J., and Lin, Y.-F. (2009). Optimizing emulsan production of A. venetianus RAG-1 using response surface methodology. *Appl. Microbiol. Biotechnol.* 84 (2): 271–279.

133 Salehizadeh, H. and Mohammadizad, S. (2009). Microbial enhanced oil recovery using biosurfactant produced by *Alcaligenes faecalis*. *Iran. J. Biotechnol.* 7: 216–223.

134 Toren, A., Navon-Venezia, S., Ron, E.Z., and Rosenberg, E. (2001). Emulsifying activities of purified alasan proteins from *Acinetobacter radioresistens* KA53. *Appl. Environ. Microbiol.* 67 (3): 1102–1106.

135 Duvnjak, Z. and Kosaric, N. (1985). Production and release of surfactant by *Corynebacterium lepus* in hydrocarbon and glucose media. *Biotechnol. Lett.* 7 (11): 793–796.

136 Dickson, R.C. and Lester, R.L. (1999). Yeast sphingolipids. *Biochim. Biophys. Acta Gen. Subj.* 1426 (2): 347–357.

137 Schorsch, C., Boles, E., and Schaffer, S. (2013). Biotechnological production of sphingoid bases and their applications. *Appl. Microbiol. Biotechnol.* 97 (10): 4301–4308.

138 Ekiz, H.A. and Baran, Y. (2010). Therapeutic applications of bioactive sphingolipids in hematological malignancies. *Int. J. Cancer* 127 (7): 1497–1506.

139 Bibel, D., Aly, R., and Shinefield, H. (1995). Topical sphingolipids in antisepsis and antifungal therapy. *Clin. Exp. Dermatol.* 20 (5): 395–400.

140 Borah, S.N., Sen, S., Goswami, L. et al. (2019). Rice based distillers dried grains with solubles as a low cost substrate for the production of a novel rhamnolipid biosurfactant having anti-biofilm activity against *Candida tropicalis*. *Colloids Surf. B: Biointerfaces* 182: 110358.

141 Edwards, K.R., Lepo, J.E., and Lewis, M.A. (2003). Toxicity comparison of biosurfactants and synthetic surfactants used in oil spill remediation to two estuarine species. *Mar. Pollut. Bull.* 46 (10): 1309–1316.

142 Liu, Y., Ding, S., Dietrich, R. et al. (2017). A biosurfactant-inspired heptapeptide with improved specificity to kill MRSA. *Angew. Chem. Int. Ed.* 56 (6): 1486–1490.

143 Sen, S., Borah, S.N., Bora, A., and Deka, S. (2017). Production, characterization, and antifungal activity of a biosurfactant produced by *Rhodotorula babjevae* YS3. *Microb. Cell Factories* 16 (1): 95.

144 Abdel-Mawgoud, A.M., Lépine, F., and Déziel, E. (2010). Rhamnolipids: diversity of structures, microbial origins and roles. *Appl. Microbiol. Biotechnol.* 86 (5): 1323–1336.

145 Katiyar, S.S., Kushwah, V., Dora, C.P., and Jain, S. (2019). Novel biosurfactant and lipid core-shell type nanocapsular sustained release system for intravenous application of methotrexate. *Int. J. Pharm.* 557: 86–96.

146 Hasani Zadeh, P., Moghimi, H., and Hamedi, J. (2018). Biosurfactant production by *Mucor circinelloides*: Environmental applications and surface-active properties. *Eng. Life Sci.* 18 (5): 317–325.

147 Vecino, X., Cruz, J., Moldes, A., and Rodrigues, L. (2017). Biosurfactants in cosmetic formulations: Trends and challenges. *Crit. Rev. Biotechnol.* 37 (7): 911–923.

148 Bustamante, M., Duran, N., and Diez, M. (2012). Biosurfactants are useful tools for the bioremediation of contaminated soil: A review. *J. Soil Sci. Plant Nutr.* 12 (4): 667–687.

149 Banat, I.M. (1995). Biosurfactants production and possible uses in microbial enhanced oil recovery and oil pollution remediation: A review. *Bioresour. Technol.* 51 (1): 1–12.

150 Soberón-Chávez, G. and Maier, R.M. (2011). Biosurfactants: a general overview. In: *Biosurfactants: From Genes to Applications. Microbiology Monographs* (ed. G. Soberón-Chávez), 1–11. Berlin, Germany: Springer.

151 Nitschke, M. and Pastore, G.M. (2006). Production and properties of a surfactant obtained from *Bacillus subtilis* grown on cassava wastewater. *Bioresour. Technol.* 97 (2): 336–341.

152 Diniz Rufino, R., Moura de Luna, J., de Campos Takaki, G.M., and Asfora Sarubbo, L. (2014). Characterization and properties of the biosurfactant produced by *Candida lipolytica* UCP 0988. *Electron. J. Biotechnol.* 17 (1): 34–38.

153 Müller, M.M., Kügler, J.H., Henkel, M. et al. (2012). Rhamnolipids – Next generation surfactants? *J. Biotechnol.* 162 (4): 366–380.

154 Johann, S., Seiler, T.-B., Tiso, T. et al. (2016). Mechanism-specific and whole-organism ecotoxicity of mono-rhamnolipids. *Sci. Total Environ.* 548: 155–163.

155 Wang, S. and Mulligan, C.N. (2004). Rhamnolipid foam enhanced remediation of cadmium and nickel contaminated soil. *Water Air Soil Pollut.* 157 (1–4): 315–330.

156 Mnif, I. and Ghribi, D. (2015). Lipopeptides biosurfactants: Mean classes and new insights for industrial, biomedical, and environmental applications. *Pept. Sci.* 104 (3): 129–147.

157 Olaniran, A.O., Balgobind, A., and Pillay, B. (2013). Bioavailability of heavy metals in soil: Impact on microbial biodegradation of organic compounds and possible improvement strategies. *Int. J. Mol. Sci.* 14 (5): 10197–10228.

158 Stacey, S.P., McLaughlin, M.J., Çakmak, I. et al. (2008). Root uptake of lipophilic zinc–rhamnolipid complexes. *J. Agric. Food Chem.* 56 (6): 2112–2117.

159 Raza, Z.A., Khalid, Z.M., Khan, M.S. et al. (2010). Surface properties and sub-surface aggregate assimilation of rhamnolipid surfactants in different aqueous systems. *Biotechnol. Lett.* 32 (6): 811–816.

160 Kumar, R., Das, A.J., and Juwarkar, A.A. (2015). Reclamation of petrol oil contaminated soil by rhamnolipids producing PGPR strains for growing *Withania somnifera*, a medicinal shrub. *World J. Microbiol. Biotechnol.* 31 (2): 307–313.

161 Sarma, H., Bustamante, K.L.T., and Prasad, M.N.V. (2018). Biosurfactants for oil recovery from refinery sludge: Magnetic nanoparticles assisted purification. In: *Industrial and Municipal Sludge* (ed. M.N.V. Prasad), 107–132. Cambridge, MA, USA: Elsevier ISBN: 9780128159071, Editor Majeti Narasimha Vara Prasad, Paulo Jorge de Campos, Favas Meththika, Vithanage S. Venkata Mohan.

162 Saikia, R.R., Deka, S., Deka, M., and Sarma, H. (2012). Optimization of environmental factors for improved production of rhamnolipid biosurfactant by *Pseudomonas* aeruginosa RS29 on glycerol. *J. Basic Microbiol.* 52 (4): 446–457.

9

Biosurfactants

Production and Role in Synthesis of Nanoparticles for Environmental Applications

Ashwini N. Rane[1], *S.J. Geetha*[2], *and Sanket J. Joshi*[3]

[1] *Department of Environmental Science, Savitribai Phule Pune University, Pune, Maharashtra, India*
[2] *Department of Biology, College of Science, Sultan Qaboos University, Muscat, Oman*
[3] *Oil & Gas Research Center, Central Analytical and Applied Research Unit, Sultan Qaboos University, Muscat, Oman*

CHAPTER MENU

9.1 Nanoparticles

Nanoscience has been an evolving branch of biotechnology over the last few decades. Potential applications of nanoparticles have attracted scientists from nearly every branch of science. The synthesis of nanoparticles using physiochemical methods and its long-term harmful impacts on the ecosystem have been realized over the last few years, which has led scientists across the globe to find out "green" ways to synthesize nanoparticles. Particles having at least one dimension of <100 nm are called nanoparticles. Their small size imparts unique chemical and chemical properties to the nanoparticles, which are different from their parent synthesis materials. In ancient Rome, artisans

Biosurfactants for a Sustainable Future: Production and Applications in the Environment and Biomedicine, First Edition. Edited by Hemen Sarma and Majeti Narasimha Vara Prasad.
© 2021 John Wiley & Sons Ltd. Published 2021 by John Wiley & Sons Ltd.

were believed to use films of nanoparticles of copper and silver by heating their oxides with clay, ochre, and vinegar to give the lustrous texture to the art work. The Lycurgus cup is one such example. Modern nanotechnology is said to have originated with Michael Faraday's work [1], wherein he observed a change in the optical properties of gold in its nanoform. The chemical synthesis of nanoparticles dated back to the thirteenth and fourteenth centuries when Egyptians and Mesopotamians used metals in the glass-making process. Chaudhary et al. [2] reported that the ancient bhasmas in Indian Ayurveda were also nanomedicine formulations of metals like silver, gold, and copper, free from toxicity of these metals. Nanoparticles are classified into two main classes: organic (liposomes, dendrimers, micelles, ferritin, and compact polymeric) nanoparticles and inorganic (quantum dots, fellerenes, polystyrene, magnetic, ceramic, and metallic) nanoparticles.

9.1.1 Organic Nanoparticles

Micelles: These are amphiphilic molecules comprised mainly of lipids or polymers.

Dendrimers: Monometallic, bimetallic, and semiconductor nanoparticles are synthesized in dendrimer form. These are highly branched, nearly monodisperse polymeric systems with three main parts, i.e. core, branch, and surface.

Liposomes: These are phospholipid vesicles that offer biocompatibility and good entrapment efficiency, mainly in drug delivery applications.

Compact polymeric nanoparticles: These are nanostructures made up completely of polymers, natural or synthetic. Their compact nature offers continued localized drug delivery for weeks with minimal leakage of drugs.

Ferritin nanoparticles: Owing to a unique architecture and surface properties, ferritin nanoparticles may impart functionalities to their surfaces owing to their genetic or chemical modifications.

9.1.2 Inorganic Nanoparticles

Nanoparticles having uniform dimensions and those of less than 100 nm are called zero-dimensional (0D) nanoparticles. Thin films or monolayers of the nanoparticles are called one-dimensional nanoparticles. Carbon nanotubes are classified as two-dimensional nanoparticles and dendrimers, quantum dots, and fullerenes are classified as three-dimensional nanoparticles. Furthermore, another way of classifying nanoparticles depends on the material used for their synthesis. This approach classifies nanoparticles as gold, silver, magnetic, and alloy nanoparticles [3].

Gold nanoparticles: Gold nanoparticles are used mostly in immunochemical and DNA fingerprinting investigations. Their application in the detection of cancer cells is proving a boon in the medical field of diagnostics [4, 5].

Silver nanoparticles: Silver nanoparticles are proven to have good antimicrobial properties against the vast array of microorganisms and applications in textile and cosmetic industries are also well known [6, 7].

Alloy nanoparticles: Properties of bimetallic nanoparticles are influenced by the components of the alloy and showed more advantages over their components when used individually [8].

9.2 Synthesis of Nanoparticles

Generally, nanoparticles are synthesized by physical or chemical methods, where chemical synthesis involves chemicals and solvents for synthesis, capping, and stabilization of the nanoparticles. This particular method is easy to perform and fast, but the harmful effects of the chemical reagents used in the process are long lasting. The reagents used for the nanoparticle synthesis by this method are usually hazardous, toxic, and have long-term effects on the ecosystem. The method is tedious because of the high pressure and energy requirements most of the time; also, the separation of the nanoparticles from the parent solutions is a time and money consuming process.

9.2.1 Biogenesis of Nanoparticles

Realizing the harmful effects of the toxic and highly reactive reducing agents used for nanoparticles synthesis, there has been considerable interest in employing eco-friendly approaches for nanoparticles synthesis [9, 10]. This chapter focuses mainly on the biogenic synthesis of the nanoparticles. Nanoparticles are synthesized by the mineralization of the metals, also known as "Biomineralization" due to the biological entities involved [11]. Nanoparticles synthesized by the "green approach" possess the advantage of mostly having monodispersity of the nanoparticles; moreover, they are reported to be more stable as compared to their chemical counterparts. Green synthesis involves the use of biological agents such as plants or microbes, or their byproducts. These "Bio-nano factories" are environmentally safe, affordable, and uniquely structured [12–15]. Biogenesis of different metal nanoparticles like gold, silver, selenium, copper oxide, tellurium, zinc oxide, iron oxide, palladium, uraninite, platinum, nickel oxide, and magnesium oxide has been reported [16–18].

Biogenesis of nanoparticles is becoming a sensational area of research, as it involves the lesser/negligible or no use of hazardous substances during the synthesis process. In the huge laboratory of nature, many biomolecules, generated by different biotic systems, such as prokaryotes and eukaryotes, are being explored for the biogenesis of nanoparticles. Production of nanoparticles with the biological systems in the presence of chemical reducing agents was the first step toward the green synthesis of nanoparticles. Consequently, biosynthesis of the nanoparticles in the complete absence of any chemical reagents has been achieved by a few scientists (Table 9.1). Moreover, the use of biological systems for nanoparticle synthesis could also impart better stability and control over the size, shape, and structure [37]. Even though the vast array of biological systems has been explored for nanoparticles synthesis, the intracellular synthesis of the nanoparticles becomes a tedious and costly process due to additional steps for the extraction. Thus, microbes synthesizing nanoparticles with the help of their extracellular metabolites is being extensively studied [38].

9.2.2 Nanoparticle Synthesis by Plant Extracts

Use of plant extracts has attracted tremendous attention as it offers a single-step biosynthesis of nanoparticles [39]. Different plant metabolites like alkanoids, phenolics, terpenoids, and enzymes act as reducing agents in the process of plant extract mediated nanoparticle synthesis [21]. The diverse range of plant species have been studied for their potential to act as a reducing agent for nanoparticle synthesis (Table 9.1). Plant extract mediated nanoparticle synthesis is environmentally benign, free from toxic contaminants as needed in pharmaceuticals, can easily be scaled up, and provides nanoparticles of controlled size and morphology [21, 40].

9.2.3 Nanoparticle Synthesis by Fungi

Ahmad et al. [41] investigated suitability of fungi for nanoparticle synthesis and found that fungi are suitable candidates for producing nanoparticles of different shapes and sizes from different metals and oxides. Various genera of fungi were exploited to test their candidacy for nanoparticle synthesis (Table 9.1), and it was found that fungi are "extremely good" candidates for gold and silver nanoparticle synthesis [41–43].

Table 9.1 Biogenic synthesis of nanoparticles using different plants, microbes, or their metabolites.

Biogenic material	Type of nanoparticle	References
Plant origin		
Sargassum muticum	Magnetic iron oxide nanoparticles	Mahdavi et al. [19]
Lingonberry and cranberry juices	Silver nanoparticles	Puiso et al. [20]
Pomegranate peel extract	Silver nanoparticles	Joshi et al. [21]
Rosa brunonii Lindl	Silver nanoparticles	Bhagat et al. [22]
Berberis vulgaris	Silver nanoparticles	Behravan et al. [23]
Eucalyptus citriodora	Silver nanoparticles	Paosen [24]
Rosemary leaves	Silver sulphide nanoparticles	Awwad et al. [25]
Pistachio leaf extract	Magnetic nanoparticles with Ni complex	Tamoradi and Mousavi [26]
Microbial origin		
Schwanniomyces occidentalis NCIM 3459	Silver and gold nanoparticles	Mohite et al. [27]
Bacillus subtilis ANR 88	Silver and gold nanoparticles	Rane et al. [28]
Bacillus cereus	Silver nanoparticles	Gurunathan [29]
Monascus purpureus	Cobalt ferrite	El-Sayed et al. [30]
Fusarium oxysporum	Gold nanoparticles	Naimi-Shamel et al. [31]
Yarrowia lipolytica MTCC 9520	Silver nanoparticles	Radha et al. [32]
Portieria hornemannii	Silver nanoparticles	Fstima et al. [33]
Gelidium corneum	Silver nanoparticles	Öztürk et al. [34]
Nocardiopsis dassonvillei NCIM 5124	Gold nanoparticles	Bennur et al. [35]
Marine actinomycetes	Silver nanoparticles	Hamed et al. [36]

9.2.4 Nanoparticle Synthesis by Algae

The field of nanobiotechnology has employed yet another group of the plant kingdom, i.e. algae for nanoparticles synthesis. The ability of the algal biosystems to remediate toxic metals and convert these harmful metals to value-added nanoparticles is now being used to produce model systems (Table 9.1). These photoautotrophic oxygenic microbes have the ability to accumulate heavy metals and scientists are hence exploiting all the possibilities to use these bionano factories as an abundant raw material source [10].

9.2.5 Nanoparticle Synthesis by Yeasts

Yeasts act as a source of reducing agents for nanoparticle synthesis (Table 9.1). *Hansenula anomola* is one of the yeasts that have the ability to donate electrons and act as a biocatalyst for biofuel production [10].

9.2.6 Nanoparticle Synthesis by Actinomycetes and Bacteria

These prokaryotic cells can be easily genetically engineered to achieve a desired size of polydispersed nanoparticles [9]. Bacteria, when exposed to metal stress, exhibit the mechanism to convert toxic heavy metals into non-toxic value-added nanoparticles, either intracellular or extracellular, employing various products, such as biosurfactants, biopolymers, emulsifiers, enzymes, etc. (Table 9.1).

9.3 Biosurfactants

Biosurfactants are amphiphilic biomolecules produced by plants, animals, and microorganisms, with diverse functional groups having both hydrophilic and hydrophobic parts (Figure 9.1) and imparting those surface-active properties [44–46]. Van Hamme et al. [47] postulated that biosurfactants may be produced intracellularly and/or by being excreted by microorganisms for the purpose of motility and quorum sensing activities.

Several prokaryotic and eukaryotic microorganisms are reported to produce biosurfactants, of which some highly researched and reported biosurfactants, their yield, carbon and nitrogen sources used, and methods of extractions are mentioned in Table 9.2. *Bacillus* spp. are mainly reported for producing a lipopeptide type of biosurfactant, while *Pseudomonas* spp., *Rhodococcus* spp., and *Candida* spp. are reported to produce a glycolipid (few researchers also reported a lipopeptide type of biosurfactants produced by *Pseudomonas* spp.) type of biosurfactant [38, 46 68, 72, 73, 131, 140, 142].

Biosurfactants are classified based on their size or its chemical structure (Figure 9.2). Widely known biosurfactant types are: lipopeptides (small molecules), glycolipids (small to medium size molecules), lipoproteins (small to medium size molecules), phospholipids (medium size

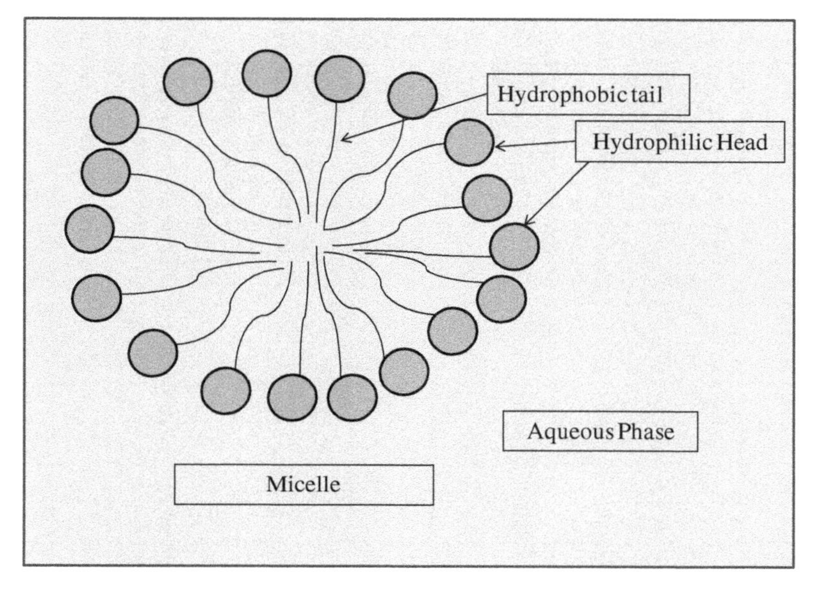

Figure 9.1 Graphic representation of typical (bio)surfactant molecules, such as a micelle.

Table 9.2 Highly researched microbial biosurfactants [45, 46].

Microorganism/ type of biosurfactant	Biosurfactant yield (g/l)	Different carbon sources used	Different nitrogen sources used	Extraction methods	References
Pseudomonas spp./ Rhamnolipids	0.20–24	Soybean oil, soybean waste frying oil, vegetable oil, Karanja oil, olive oil, molasses, waste motor lubricant oil, peanut oil, crude oil, glycerol, glucose, Soapstock, naphthalene, peanut oil cake	No separate N_2 sources, $NaNO_3$ and yeast extract, corn steep liquor, NH_4Cl and peptone, $NaNO_3$, NH_4NO_3, beef extract	Ethyl acetate, chloroform: methanol 2:1, silica gel/adsorption chromatography with chloroform: methanol/ methanol as gradient/ solvent, cold acetone precipitation, dichloromethane, diethyl ether	Deziel et al. [48]; Patel and Desai [49]; Sim et al. [50]; Bennincasa et al. [51]; Abouseoud et al. [52]; Nie et al. [53]; Thavasi et al. [54]; Abbasi et al. [55]; Aparna et al. [56]; Darvishi et al. [57]; Onwosi and Odibo [58]; Saikia et al. [59]; Silva et al. [60]; Deepika et al. [61]; dos Santos et al. [62]; Varjani and Upasani [63]
Bacillus spp./ surfactin, lichenysin, iturin, fengycin	0.031–6.9	Molasses supernatant (16%); date/cane molasses (8%); Glucose/date molasses; glucose + pyrene; raw glycerol (2–5%); orange peels (1.55% and soybean meal (1%); corn steep liquor (10%); tuna fish cooking residue (4%), sesame peel flour (3.3%); potato process effluent; glucose (1–4%); cassava waste water; sucrose (1–2%); palm oil (2%); clarified cashew apple juice; petroleum oil (2%); brain heart infusion media + pharmamedia; yeast extract; oily sludge	No separate N_2 source; $NaNO_3$; NH_4NO_3; urea; yeast extract and tryptone; urea and ammonium sulphate; tryptophan; beef extract, yeast extract and peptone; KNO_3; nutrient broth; meat extract, yeast extract; $(NH_4)_2SO_4$ and $NaNO_3$; NH_4NO_3 and tryptone; $(NH_4)_2SO_4$	Acid precipitation; ultra filtration-HPLC/ microfiltration; chloroform; methanol: 65:15 or 2:1; HPLC/ RP-HPLC; dichloromethane/ methanol	Cooper et al. [64]; Lin et al. [65]; Makkar and Cameotra, [66–68]; Davis et al. [69]; Mulligan et al. [70]; Vater et al. [71–73]; Mukherjee and Das [74]; Noah et al. [75]; Nitschke and Pastore [76]; Yeh et al. [77]; Al Ajlani et al. [78]; Das and Mukherjee [79]; Fernandese et al. [80]; Abdel-Mawgoud et al. [81]; Barros et al. [82]; Pornsunthorn-tawee et al. [83]; de Faria et al. [84]; Ghribi et al. [85]; Pemmaraju et al. [86]; Al Bahry et al. [87]; Coutte et al. [88]; de Oliveira et al. [89]; Mnif et al. [90]; Pereira et al. [91]; Al Wahaibi et al. [92]; Sousa et al. [93]; Gudina et al. [94]; Liu et al. [95]; Fernandes et al. [96]; Jha et al. [97]; Joshi et al. [98]
Candida spp./ sophorolipids	2.42–>400	Glucose and corn oil; soybean oil refinery residue; glucose/ safflower oil; glycerol; sugar cane molasses (5%), residual soybean oil (5%); restaurant food waste	No separate N_2 source; yeast extract and urea; NH_4NO_3; peptone; urea; NH_4Cl; $NaNO_3$; 3% corn steep liquor	Ethyl acetate; chloroform	Cooper and Paddock [99]; Pekin et al. [100]; Accorsini et al. [101]; Rufino et al. [102]; Elshafie et al. [103]; Kaur et al. [104]; Silva et al. [105]

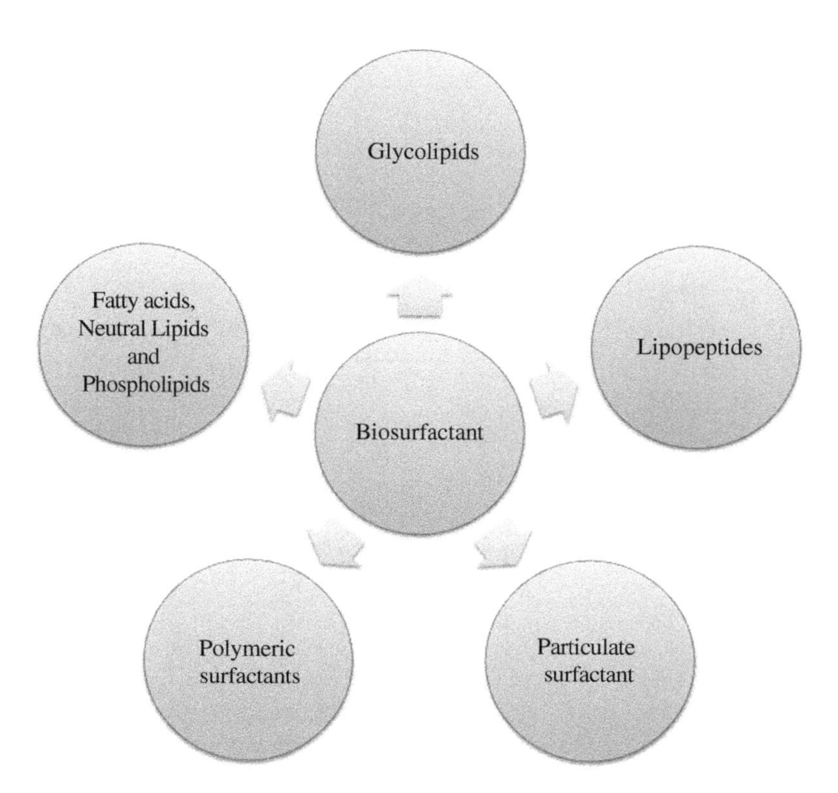

Figure 9.2 Different types of biosurfactants.

molecules), fatty acids (medium size molecules), polymeric surfactants (big molecules), and particulate surfactants (big molecules) [45, 46].

Even though biosurfactants showed superior traits over chemical surfactants, such as lower toxicity, an eco-friendly nature, better stability and specificity, cost and economics is the biggest hurdle for widespread applications. Continuous research is on-going in several parts of the world to reduce the production costs and to increase the yield for industrial biosurfactant production. Some of those strategies are: isolation and selection of hyperproducing strains; use of cheaper substrates; statistical methods for optimization of the nutrient concentrations, process parameters, environmental conditions, and the downstream process [45, 46].

9.3.1 Isolation and Selection of Biosurfactant-Producing Microbes

Isolation and selection of hyperproducing microbial strains are two of the major criteria for any biotechnological industrial production process, as the entire process depends on the production capability of wild-type isolates. The biosurfactant production process is also no exception and several types of hyperproducing wild-type and mutated strains of bacteria, fungi, and yeast are reported. The source of isolation and the method used for screening (such as blood agar plate, oil spreading method, surface and interfacial tension reduction analysis, etc.) are two important criteria for isolating biosurfactant-producing microbes [45, 106–108]. Therefore, continuous search for isolating and selecting biosurfactant hyperproducing microbes is on-going [108].

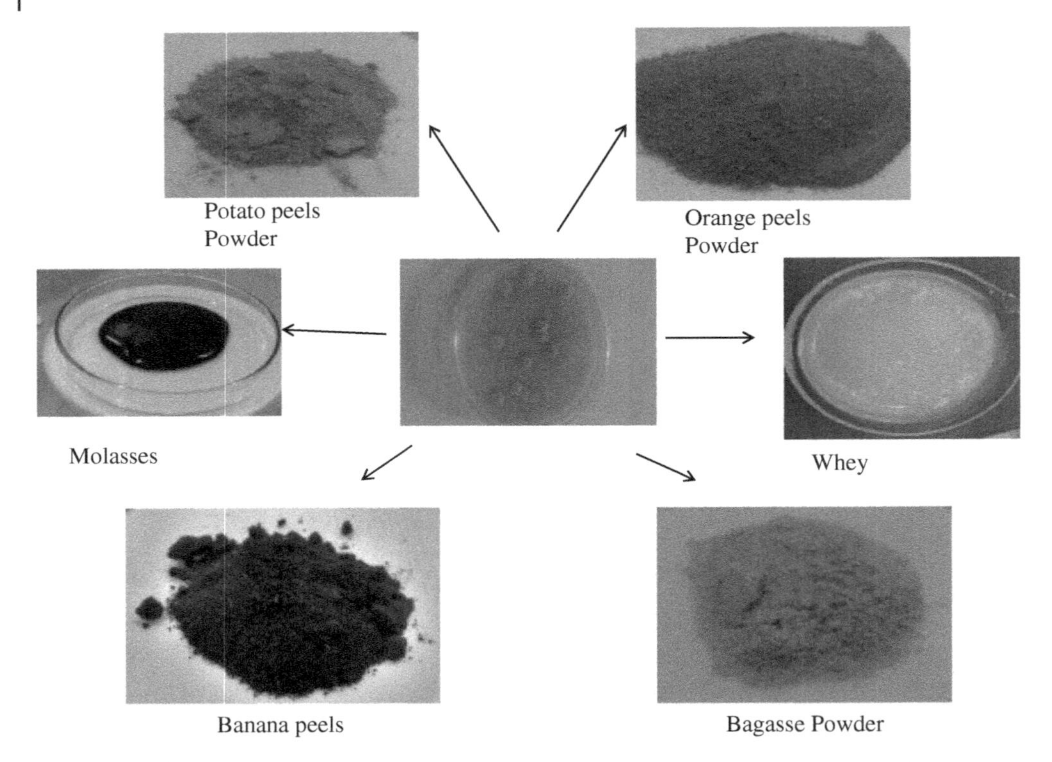

Figure 9.3 Different types of agro-industrial wastes used as carbon/nitrogen sources for biosurfactant production [46] (See insert for color representation of the figure).

9.3.2 Use of Cheaper Substrates

Like any other type of bioproducts, carbon and nitrogen sources are two main substrates needed for biosurfactant production [37, 137]. Several types of substrates are reported to be used for biosurfactant production, such as carbohydrates, lipids, fats, oils, and hydrocarbons [66, 109]. In general, such pure substrates are quite expensive, adding to the production costs, so various types of agro-industrial waste products (Figure 9.3), such as molasses, cheese whey, corn steep liquor, waste water, citrus fruit peels, potato peels, banana peels, waste frying oil, animal fats, etc., are reported to be useful for economical biosurfactant production [46, 66, 75, 110–124]. The use of such waste substrates serves both purposes: remediation and utilization of such waste and reducing the production cost of such valuable biological products.

9.3.3 Statistical Methods for Optimization of the Media Components, Process Parameters, Environmental Conditions, and Downstream Process

An optimized and efficient production process is the founding-stone for any profit-making industries. Biotech industries are no exception and optimized media components, bioprocess, environmental parameters, and upstream and downstream processes are some of the key parameters for successful commercialization of products such as biosurfactants [119]. For better yields, optimal concentrations of substrates, microbial load (inoculum), and environmental parameters (pH, temperature, agitations, aeration, dissolved oxygen concentration, k_La, tip speed of the bioreactor, etc.)

are of the utmost importance [37, 137]. Different types of statistical methods are employed for such optimizations, both on lab-scale and industrial-scale applications: one-variable-at a-time (although it is a time-consuming method), Plackett-Burman design, three factorial methods, five factorial methods, response surface methodology, central composite design, artificial neuron network, etc. [116, 125, 126]. For the majority of biotechnological industrial-scale processes, >60% of the total production cost is due to downstream processing. Several methods are reported for biosurfactant recovery, depending on solubility and stability in the presence of different solvents, intracellular or extracellular product, mass and charge, etc. [45, 127].

9.4 Biosurfactant Mediated Nanoparticles Synthesis

The use of biosurfactants in the nanotechnology field has introduced an environmentally benign alternative to harmful chemicals, which tend to accumulate in the environment and are responsible for the disturbance in an ecosystem. Moreover, the nanoparticles obtained with the help of biosurfactants are unique in their structural, physical, and magnetic properties along with their stability [128, 129]. The probable modes of action are as a capping agent, reducing agent, and/or through adsorbing on to nanoparticles faces, thus leading to surface stabilization, preventing clumps formation, and aggregation [18, 130, 131]. The photo-elicited (sunlight irradiation induced) technique and the microemulsion-based technique are two of the recent nanoparticle synthesis techniques having some clear advantages: they are cost-effective; an excessive reducing agent is not needed for metal reduction; and/or the synthesis through a reverse-micellar process leads to mixing of two reverse-micellar solutions (reducing agent and metal salt solution) [132, 133]. A probable process of biosurfactant-mediated nanoparticle synthesis is depicted in Figure 9.4.

Xie et al. [134] reported rhamnolipid reverse micelles as a stabilizing agent for the silver nanoparticles synthesized using sodium tetrahydridoborate as the reducing agent. Thus synthesized nanoparticles are spherical and quite stable for up to two months. Kasture et al. [135] reported the use of sophorolipid obtained from *Candida bombicola* ATCC 22214 as a capping agent for cobalt nanoparticles. Palanisamy [136] reported the use of rhamnolipid in nickel nanoparticle synthesis and made use of pH-dependent morphology alterations in rhamnolipid for the alterations in morphology of the nanoparticles. Physiological properties of metal nanoparticles strongly depend on their size and shape, which in turn are responsible for their applications in various fields. Kasture et al. [128] for the first time reported the production of silver nanoparticles using sophorolipids derived from oleic acid and linoleic acid, both as reducing as well as capping agents. This was the milestone in the studies of biosurfactant nanoparticles. They observed the temperature-dependent size variations in silver nanoparticles when sophorolipid was used as a reducing and capping agent. Reddy et al. [137] in two separate studies reported the use of surfactin as the stabilizing agent in the synthesis of silver and gold nanoparticles.

Kiran et al. [138] reported the use of glycolipid biosurfactant for the stabilization of the silver nanoparticles. Kumar et al. [129] synthesized silver nanoparticles using sodium borohydrate as the reducing agent and rhamnolipid biosurfactant as the stabilizing agent. Ravi Kumar [139] synthesized silver nanoparticles with the sophorolipid biosurfactants derived from oleic acid and stearic acid and the use of KOH as a hydrolysing agent. Potassium hydroxide also maintained the pH alkaline and the high temperature accelerated synthesis of the silver nanoparticles. Monodispersed and smaller sized nanoparticles (5.5 nm) were obtained at a higher temperature and at a lower temperature the size of the nanoparticles was 20 nm. Further, these experiments indicated that the

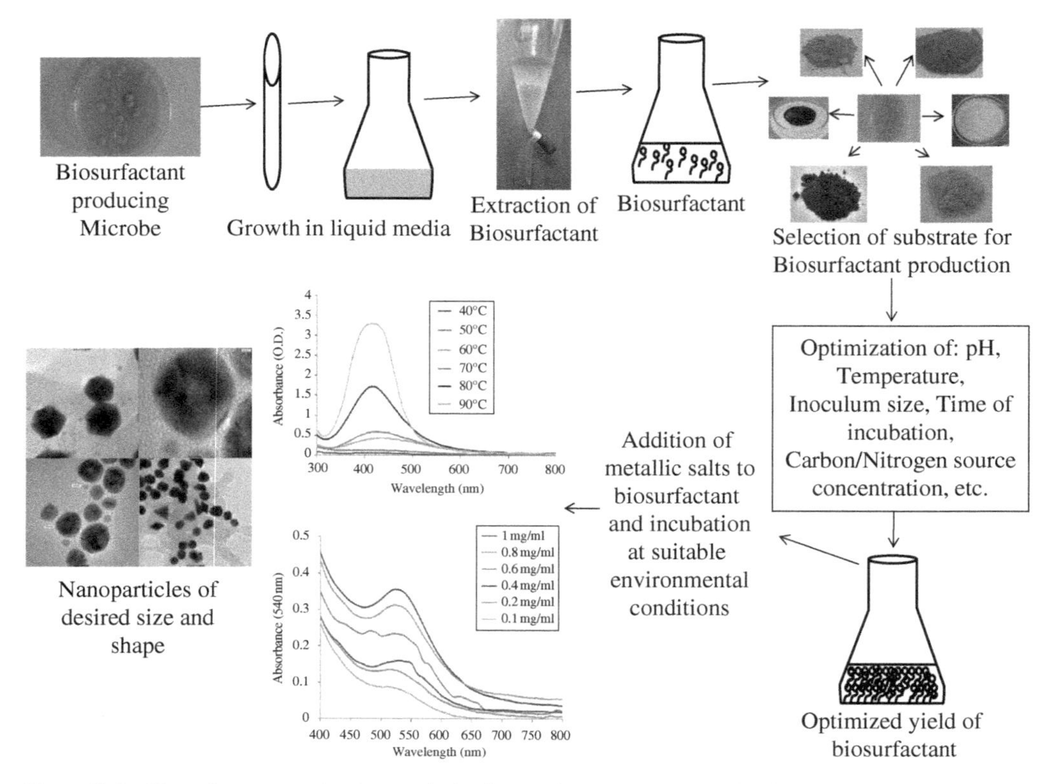

Figure 9.4 Biosurfacatnt production, optimization, and nanoparticle synthesis using a biosurfactant [46].

use of stearic acid derived sophorolipid biosurfactant yielded fast synthesis and the silver nanoparticles that formed were more uniform and relatively smaller. Narayanan et al. [140] used rhamnolipid biosurfactant as the capping agent to control the size uniformity of the ZnS nanoparticles. They found that the capped nanoparticles were stable and water soluble. Water solubility of the nanoparticles was highlighted in the study of their increased potential utility. Reddy et al. [141] used a 72-hour old culture of glucose grown *Bacillus subtilis* for silver and gold nanoparticles synthesis. Silver nanoparticles were formed extracellularly, whereas the gold nanoparticles were found to be formed both intra- and extracellularly. They also reported that, using a 72-hour old culture of *B. subtilis*, the synthesis of gold nanoparticles was faster than that of silver nanoparticles. Kumar and Mamidyala [142] reported the silver nanoparticle synthesis using a culture supernatant of *Pseudomonas aeruginosa* and concluded that the reduction of the silver ions was due to nitrate reductase and the nanoparticles were stabilized by the rhamnolipid present in the culture supernatant. Worakitsiri et al. [143] demonstrated the use of rhamnolipid as a template for the synthesis of polyaniline nanofibers and nano-tubes. They further showed that, as compared to a conventional polyaniline synthesis process, rhamnolipid mediated synthesis yielded a uniform size and morphology with improved electrical conductivity and crystallinity. They concluded that the rhamnolipids were better candidates as compared to synthetic surfactants for the template synthesis of conductive polymeric nanoparticles.

Hazra et al. [144] reported the synthesis of ZnS nanoparticles using rhamnolipid and elucidated the dual role of the biosurfactant as capping as well as stabilizing agent in the reaction. They further suggested the use of such nanoparticles as nanophotocatalysts in textile industries for azo dye degradation. Basnet et al. [145] used rhamnolipid biosurfactant as a surface modifier to achieve better stability of the Pd-doped zerovalent nanoparticles. Farias et al. [146] synthesized silver nanoparticles using $NaBH_4$ and a biosurfactant obtained from *P. aeruginosa* grown on a low-cost medium was used as the reducing agent. Das et al. [147] exposed glycolipid biosurfactant and silver nitrate mixture to sunlight (42–45 °C) and observed nanoparticle synthesis within 15–20 minutes of exposure, yielding nanoparticles of size 70–90 nm. The silver nanoparticle synthesis was obtained within 25–30 minutes and the added biosurfactant acted both as a reducing as well as a stabilizing agent. Rane et al. [28] reported the synthesis of silver and gold nanoparticles in the complete absence of any other chemical reducing or stabilizing agent. *B. subtilis* was grown on a low-cost renewable substrate and extracted biosurfactant was used for nanoparticle synthesis. Production of silver and gold nanoparticles was achieved at elevated temperatures (90 °C). Different types of nanoparticles synthesized by biosurfactants are listed in Table 9.3.

9.4.1 Environmental Applications of Nanoparticles

Nanoparticles are reported to be useful in several applications (Figure 9.5), out of which this chapter will focus on possible environmental applications. Environmental pollution is one of the main problems humankind is facing and several innovative technologies are relentlessly being studied for the environmental remediation. Some of those persistent contaminates are released due to industrial practices (waste heavy metals, crude oil spills, polychlorinated biphenyls, polycyclic aromatic hydrocarbons), agricultural practices (chemical pesticides, herbicides, fertilizers), and municipal and industrial effluents and sewage. Nanoparticles are one of those different types of materials having a potential role in environmental remediation. However effective those nanoparticles are, they are still plagued by issues such as cost effectiveness, toxicity, non-biodegradability, recyclability, how to prevent agglomeration, enhance monodispersity, and recovery costs, which limits its application in environmental remediation. Different types of metal nanomaterials and nanocomposites reported for the environmental remediation are effective by different mechanisms: absorption or adsorption, photocatalytic and/or chemical reactions, and filtration [156]. Metal-based and metal oxide-based nanoparticles are reported to be efficient for removal of heavy metals and chlorinated organic pollutants, for water treatment and a disinfectant (as a photocatalyst), air purification, dye degradation, etc. [156]. It has been reported that nanoparticles could be properly dispersed by ultrasonication, but such nanoparticles tend to re-aggregate, something that could be avoided by the addition of surfactants. Such an improved stability and monodispersity would increase the rate/efficiency, with an accelerated degradation capacity. Especially, biosurfactant mediated nanoparticles are potentially more reactive and hence would offer enhanced reactivity and effectiveness as compared to their metallic counterparts alone, thus making them more promising candidates in environmental remediation processes. Another advantage of using biosurfactant-based nanoparticles is the amphiphilic nature of biosurfactant coating – with the hydrophilic surface promoting the mobility in water/soil and the hydrophobic interior providing affinity for the hydrophobic organic contaminants.

Bioremediation of radioactive wastes resulting from nuclear power plants involve quite hazardous material, with a very long half-life, which needs to be either dumped in some secure location or site and preferably remediate to less hazardous intermediates. Among several reported bioremediation methods for dealing with such radioactive wastes, nanoparticles are one of the possible solutions. Choi et al. [157] reported efficient removal (>99%) of radioactive material (iodine) in aqueous solutions within 30 minutes, using biogenic gold nanomaterial containing *Deinococcus radiodurans* R1 bacteria and a possible application to treat radioactive isotopes/wastes. Biologically

Table 9.3 Different types of nanoparticles synthesized by biosurfactants.

Biosurfactant	Type of nanoparticles	References
Rhamnolipid	Silver nanoparticles	Xie et al. [134]
Sophorolipid	Cobalt nanoparticles	Kasture et al. [135]
Rhamnolipid	Nickel nanoparticles	Palanisamy [136]
Sophorolipids	Silver nanoparticles	Kasture et al. [128]
Surfactin	Silver and gold nanoparticles	Reddy et al. [137]
Glycolipid biosurfactant	Silver nanoparticles	Kiran et al. [138]
Rhamnolipid	Silver nanoparticles	Kumar et al. [129]
Sophorolipid	Silver nanoparticles	Ravi Kumar et al. [139]
Rhamnolipid	ZnS nanoparticles	Narayanan et al. [140]
Culture supernatant of *P. aeruginosa*	Silver nanoparticles	Kumar and Mamidyala [142]
Rhamnolipid	Polyaniline nano fibers and nano-tubes	Worakitsiri et al. [143]
Rhamnolipid	Zinc sulfide nanoparticles	Hazra et al. [144]
Rhamnolipid	Pd-doped zerovalent nanoparticles	Basnet et al. [145]
P. aeruginosa UCP0992 biosurfactant	Silver nanoparticles	Farias et al. [146]
Glycolipids obtained from a *Pseudomonas* sp.	Silver nanoparticles	Das et al. [147]
B. subtilis biosurfactant	Silver and gold nanoparticles	Rane et al. [28]
Surfactin	Silver nanoparticles	Reddy et al. [137]
Rhamnolipids	Spherical nickel oxide nanoparticles	Palanisamy and Raichur [148]
Rhamnolipids	Silver and iron oxide nanoparticles	Khalid et al. [149]
Yarrowia lipolytica MTCC 9520 lipoprotein biosurfactant	Silver nanoparticles	Radha et al. [32]
B. subtilis strain I′-1a biosurfactant	Silver nanoparticles	Joanna et al. [150]
Surfactin	Brushite nanoparticles	Maity et al. [151]
Biosurfactants	Inorganic hybrid ($Cu_3[PO_4]_2.3H_2O$) nanoflowers	Jiao et al. [152]
Bacillus subtilis CN2 lipopeptide biosurfactant	Silver nanoparticles	Bezza et al. [153]
Mannosylerythritol lipids	Silver nanoparticles	Ga'al et al. [154]
Three different types of biosurfactants	Ti_2CT_X nanosheets	Wang et al. [155]
Biosurfactant extracted from corn steep liquor	Gold and silver nanoparticles	Gómez-Graña et al. [130]

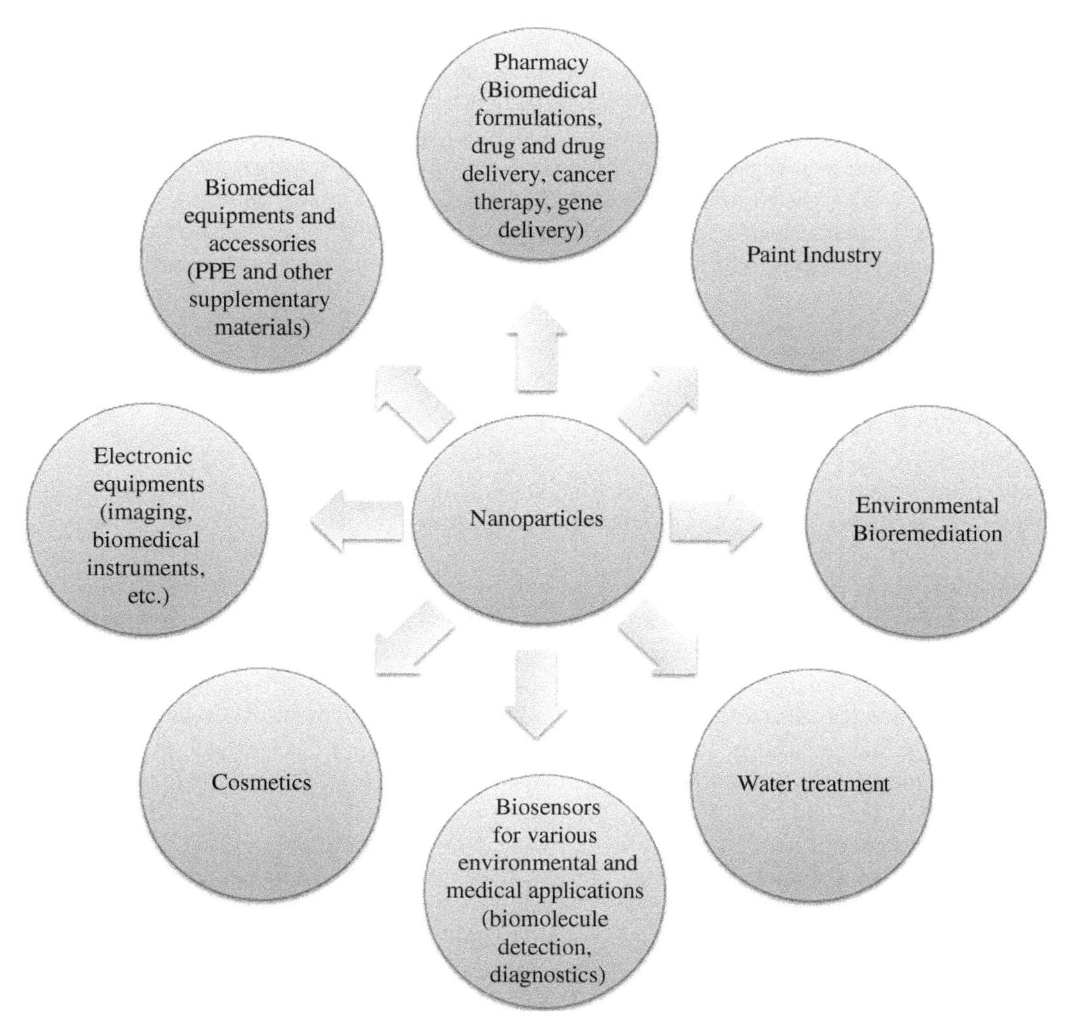

Figure 9.5 Applications of nanoparticles in various potential fields.

synthesized silver nanoparticles were reported for wastewater treatment and as a catalyst for reduction of 4-nitrophenol [158, 159]. Green synthesized nanomaterials (such as an iron-based and TiO_2 nanoparticle) are reported to be quite useful in environmental bioremediation, especially for organic dye degradation, polychlorinated biphenyls, removal of heavy metals, and treatment of wastewater and contaminated groundwater [160].Carbon nanotube membranes and nanofilters are touted as a possible solution to reduce desalination costs, and could also be used for remediation of contaminated ground or surface water. Nanosensors have also been developed to detecting water-borne contaminants. Silver nanoparticles doped with TiO_2 are reported for the photodegradation of two different types of direct azo dyes [161]. Rhamnolipid capped zinc sulfide nanoparticles were reported to be quite stable and biocompatible, with possible application in the textile industry wastewater and effluent treatment, where 94–96% degradation of "direct brown MR dye" at 25 °C has been reported [144]. Mandal et al. [162] reported benzo[α]pyrene degradation using a yeast consortium (consisting of *Rhodotorula* sp. NS01, *Hanseniaspora opuntiae* NS02, and *Debaryomyces hansenii* NS03) in the presence of ZnO nanoparticles and anionic biosurfactant. Jiao et al. [152] reported production of "biosurfactant–$Cu_3(PO_4)_2.3H_2O$ nanoflowers" having high stability and the degradation of cationic dyes, with potential applications in industrial biocatalysis, biosensors, and environmental chemistry. Although a well-reported neurotoxic compound, lead is still extensively used in the construction sector, paints, gasoline, plumbing, batteries, weights, and many other sectors. Recently Wang et al. [155] reported (non-ionic) biosurfactants-functionalized Ti_2CT_X removal of lead ions, with the potential as an environmental adsorption material for such toxic metals.

9.5 Challenges in Environmental Applications of Nanoparticles and Future Perspectives

Nanotechnology is a rapidly advancing field that has potential diverse applications in engineering science and technology (electronics, mechanical, construction, aeronautics, and petroleum), health and medicinal sciences, chemistry and surface sciences, life sciences, molecular biology, and environmental sciences, etc. that is leading to the need for its increased production. However economical, those traditionally followed chemical synthesis techniques are raising widespread concerns about toxicity and environmental impact, especially for those products meant to be used for human consumption or applications. Green synthesis approaches, such as biological synthesis by plant and microbial sources, could lead the path in the development of a reliable and eco-friendly nanoparticle synthesis. Both plant-based and microbe-based processes have their own merits and demerits, such as: exact mechanistic aspects of the biologically synthesized nanoparticles are still not clear and need further research; downstream processing and the purification of synthesized nanoparticles are rather difficult, especially when microbes are used; scale-up and industrial feasibility is still a big challenge; and cost effectiveness with chemical processes is still a bottleneck. Stability and monodispersity are some of the big concerns for nanoparticles, which are somewhat sorted by using biosurfactants, thus providing better stability, in an environmentally friendly manner, without the need for toxic-hazardous chemicals. Albeit a bit expensive, biosurfactants could be produced from renewable substrates (including wastes) and nanoparticles could be synthesized at lower temperatures, giving less waste products, in a sustainable manner. Further reducing the manufacturing costs, increasing biosurfactant yields and having a better understanding of the process will eventually lead to an efficient and economic biosurfactant mediated nanoparticle synthesis on an industrial scale.

Despite the tremendous potential of nanoparticle applications in the environmental sector, concerns regarding a risk it may pose when applied for environmental remediation purposes still lingers, especially on ground or surface water and soil, where it can be easily transferred from one source to another. Another missing link is how it may affect the environment as a whole in a real case scenario (on a much bigger scale), as most of the published research/information is for laboratory scale studies, and the fate of these materials after application is underexplored. Therefore, extensive research is still necessary to elucidate the fate of these materials in order to avoid "the possibility of these nanomaterials becoming themselves a source of environmental contamination." The presence of nanoparticles in the environment may induce different types of stress/toxicological impacts on water/soil microorganisms and plant species. However, few studies reported that biogenic nanoparticles are less toxic, so a comprehensive risk assessment of such green fabricated nanoparticles is needed, highlighting its fate, transport, aggregation, dissolution, and end products. In conclusion, biosurfactant mediated nanoparticle synthesis could serve as a building block in the development of a new variety of functionalized nanoparticles, which can be successfully applied in environmental remediation and restoration sectors.

Acknowledgements

AR would like to kindly acknowledge the support and research facility provided by Savitribai Phule Pune University, Pune, India, during her PhD study; GSJ and SJJ would like to kindly acknowledge the support and facility provided by Sultan Qaboos University, Oman.

References

1 Faraday, M. (1857). X. The Bakerian lecture – Experimental relations of gold (and other metals) to light. *Philosophical Transactions of the Royal Society of London* 147: 145–181.
2 Choudhary, N. and Sekhon, B.S. (2011). An overview of advances in the standardization of herbal drugs. *Journal of Pharmaceutical Education and Research* 2 (2): 55.
3 Hasan, S. (2015). A review on nanoparticles: their synthesis and types. *Research Journal of Recent Sciences, ISSN* 2277: 2502.
4 Baban, D.F. and Seymour, L.W. (1998). Control of tumour vascular permeability. *Advanced Drug Delivery Reviews* 34 (1): 109–119.
5 Tomar, A. and Garg, G. (2013). Short review on application of gold nanoparticles. *Global Journal of Pharmacology* 7 (1): 34–38.
6 Gong, P., Li, H., He, X. et al. (2007). Preparation and antibacterial activity of Fe_3O_4@ Ag nanoparticles. *Nanotechnology* 18 (28): 285604.
7 Rai, M., Yadav, A., and Gade, A. (2009). Silver nanoparticles as a new generation of antimicrobials. *Biotechnology Advances* 27 (1): 76–83.
8 Mohl, M., Dobo, D., Kukovecz, A. et al. (2011). Formation of CuPd and CuPt bimetallic nanotubes by galvanic replacement reaction. *The Journal of Physical Chemistry C* 115 (19): 9403–9409.
9 Edison, L.K. and Pradeep, N.S. (2020). Actinobacterial nanoparticles: green synthesis, evaluation and applications. In: *Green Nanoparticles* (eds. J. Patra, L. Fraceto, G. Das and E. Campos). Cham, pp.371–384: Nanotechnology in the Life Sciences. Springer.
10 Menon, S., Rajeshkumar, S., and Kumar, V. (2017). A review on biogenic synthesis of gold nanoparticles, characterization, and its applications. *Resource-Efficient Technologies* 3 (4): 516–527.
11 Maheshwari, V., Kane, J., and Saraf, R.F. (2008). Self-assembly of a micrometers-long one-dimensional network of cemented au nanoparticles. *Advanced Materials* 20 (2): 284–287.
12 Du, L., Jiang, H., Liu, X., and Wang, E. (2007). Biosynthesis of gold nanoparticles assisted by *Escherichia coli* DH5α and its application on direct electrochemistry of hemoglobin. *Electrochemistry Communications* 9 (5): 1165–1170.
13 Dumur, F., Guerlin, A., Dumas, E. et al. (2011). Controlled spontaneous generation of gold nanoparticles assisted by dual reducing and capping agents. *Gold Bulletin* 44 (2): 119–137.
14 Sarkar, D., Mandal, M., & Mandal, K. (2012). Domain controlled magnetic and electric properties of variable sized magnetite nano-hollow spheres. *Journal of Applied Physics*, 112(6), 064318, 1–12.
15 Cai, C., Zhu, X.B., Zheng, G.Q. et al. (2011). Electrodeposition and characterization of nano-structured Ni–SiC composite films. *Surface and Coatings Technology* 205 (11): 3448–3454.

16 Anandaradje, A., Meyappan, V., Kumar, I., and Sakthivel, N. (2020). Microbial synthesis of silver nanoparticles and their biological potential. In: *Nanoparticles in Medicine* (ed. A. Shukla), 99–133. Singapore: Springer.

17 Das, R.K., Pachapur, V.L., Lonappan, L. et al. (2017). Biological synthesis of metallic nanoparticles: plants, animals and microbial aspects. *Nanotechnology for Environmental Engineering* 2 (1): 18.

18 Płaza, G.A., Chojniak, J., and Banat, I.M. (2014). Biosurfactant mediated biosynthesis of selected metallic nanoparticles. *International Journal of Molecular Sciences* 15 (8): 13720–13737.

19 Mahdavi, M., Namvar, F., Ahmad, M.B., and Mohamad, R. (2013). Green biosynthesis and characterization of magnetic iron oxide (Fe_3O_4) nanoparticles using seaweed (Sargassum muticum) aqueous extract. *Molecules* 18 (5): 5954–5964.

20 Puišo, J., Jonkuvienė, D., Mačionienė, I. et al. (2014). Biosynthesis of silver nanoparticles using lingonberry and cranberry juices and their antimicrobial activity. *Colloids and Surfaces B: Biointerfaces* 121: 214–221.

21 Joshi, S.J., Geetha, S.J., Al-Mamari, S., and Al-Azkawi, A. (2018). Green synthesis of silver nanoparticles using pomegranate peel extracts and its application in photocatalytic degradation of methylene blue. *Jundishapur Journal of Natural Pharmaceutical Products* 13 (3): e67846.

22 Bhagat, M., Anand, R., Datt, R. et al. (2019). Green synthesis of silver nanoparticles using aqueous extract of *Rosa brunonii* Lindl and their morphological, biological and photocatalytic characterizations. *Journal of Inorganic and Organometallic Polymers and Materials* 29 (3): 1039–1047.

23 Behravan, M., Panahi, A.H., Naghizadeh, A. et al. (2019). Facile green synthesis of silver nanoparticles using Berberis vulgaris leaf and root aqueous extract and its antibacterial activity. *International Journal of Biological Macromolecules* 124: 148–154.

24 Paosen, S., Jindapol, S., Soontarach, R., and Voravuthikunchai, S.P. (2019). *Eucalyptus citriodora* leaf extract-mediated biosynthesis of silver nanoparticles: Broad antimicrobial spectrum and mechanisms of action against hospital-acquired pathogens. *APMIS* 127 (12): 764–778.

25 Awwad, A.M., Salem, N.M., Aqarbeh, M.M., and Abdulaziz, F.M. (2020). Green synthesis, characterization of silver sulfide nanoparticles and antibacterial activity evaluation. *Chemistry International* 6 (1): 42–48.

26 Tamoradi, T. and Mousavi, S.M. (2020). In situ biogenic synthesis of functionalized magnetic nanoparticles with Ni complex by using a plant extract (pistachio leaf) and its catalytic evaluation towards polyhydroquinoline derivatives in green conditions. *Polyhedron* 175: 114211.

27 Mohite, P., Apte, M., Kumar, A.R., and Zinjarde, S. (2016). Biogenic nanoparticles from Schwanniomyces occidentalis NCIM 3459: Mechanistic aspects and catalytic applications. *Applied Biochemistry and Biotechnology* 179 (4): 583–596.

28 Rane, A.N., Baikar, V.V., Ravi Kumar, V., and Deopurkar, R.L. (2017). Agro-industrial wastes for production of biosurfactant by *Bacillus subtilis* ANR 88 and its application in synthesis of silver and gold nanoparticles. *Frontiers in Microbiology* 8: 492.

29 Gurunathan, S. (2019). Rapid biological synthesis of silver nanoparticles and their enhanced antibacterial effects against *Escherichia fergusonii* and *Streptococcus mutans*. *Arabian Journal of Chemistry* 12 (2): 168–180.

30 El-Sayed, E.S.R., Abdelhakim, H.K., and Zakaria, Z. (2020). Extracellular biosynthesis of cobalt ferrite nanoparticles by *Monascus purpureus* and their antioxidant, anticancer and antimicrobial activities: yield enhancement by gamma irradiation. *Materials Science and Engineering: C* 107: 110318.

31 Naimi-Shamel, N., Pourali, P., and Dolatabadi, S. (2019). Green synthesis of gold nanoparticles using Fusarium oxysporum and antibacterial activity of its tetracycline conjugant. *Journal de Mycologie Medicale* 29 (1): 7–13.

32 Radha, P., Suhazsini, P., Prabhu, K. et al. (2020). Chicken tallow, a renewable source for the production of biosurfactant by *Yarrowia lipolytica* MTCC9520, and its application in silver nanoparticle synthesis. *Journal of Surfactants and Detergents* 23 (1): 119–135.

33 Fatima, R., Priya, M., Indurthi, L. et al. (2020). Biosynthesis of silver nanoparticles using red algae *Portieria hornemannii* and its antibacterial activity against fish pathogens. *Microbial Pathogenesis* 138: 103780.

34 Öztürk, B.Y., Gürsu, B.Y., and Dağ, İ. (2020). Antibiofilm and antimicrobial activities of green synthesized silver nanoparticles using marine red algae *Gelidium corneum*. *Process Biochemistry* 89: 208–219.

35 Bennur, T., Javdekar, V., Tomar, G.B., and Zinjarde, S. (2020). Gold nanoparticles biosynthesized by Nocardiopsis dassonvillei NCIM 5124 enhance osteogenesis in gingival mesenchymal stem cells. *Applied Microbiology and Biotechnology* 104: 4081–4092.

36 Hamed, A.A., Kabary, H., Khedr, M., and Emam, A.N. (2020). Antibiofilm, antimicrobial and cytotoxic activity of extracellular green-synthesized silver nanoparticles by two marine-derived actinomycete. *RSC Advances* 10 (17): 10361–10367.

37 Sriram, M.I., Kalishwaralal, K., and Gurunathan, S. (2012). Biosynthesis of silver and gold nanoparticles using *Bacillus licheniformis*. In: *Nanoparticles in Biology and Medicine* (ed. M. Soloviev), 33–43. Totowa, NJ: Humana Press.

38 Ovais, M., Khalil, A.T., Ayaz, M. et al. (2018). Biosynthesis of metal nanoparticles via microbial enzymes: A mechanistic approach. *International Journal of Molecular Sciences* 19 (12): 4100.

39 Mittal, A.K., Chisti, Y., and Banerjee, U.C. (2013). Synthesis of metallic nanoparticles using plant extracts. *Biotechnology Advances* 31 (2): 346–356.

40 Kulkarni, N. and Muddapur, U. (2014). Biosynthesis of metal nanoparticles: A review. *Journal of Nanotechnology* 2014, 1–8.

41 Ahmad, A., Mukherjee, P., Senapati, S. et al. (2003). Extracellular biosynthesis of silver nanoparticles using the fungus *Fusarium oxysporum*. *Colloids and Surfaces B: Biointerfaces* 28 (4): 313–318.

42 Ahmad, A., Mukherjee, P., Mandal, D. et al. (2002). Enzyme mediated extracellular synthesis of CdS nanoparticles by the fungus, *Fusarium oxysporum*. *Journal of the American Chemical Society* 124 (41): 12108–12109.

43 Mukherjee, P., Ahmad, A., Mandal, D. et al. (2001). Fungus-mediated synthesis of silver nanoparticles and their immobilization in the mycelial matrix: a novel biological approach to nanoparticle synthesis. *Nano Letters* 1 (10): 515–519.

44 Geetha, S.J., Banat, I.M., and Joshi, S.J. (2018). Biosurfactants: production and potential applications in microbial enhanced oil recovery (MEOR). *Biocatalysis and Agricultural Biotechnology* 14: 23–32.

45 Joshi, S. J. (2008). Isolation and characterization of biosurfactant producing microorganisms and their possible role in microbial enhanced oil recovery (MEOR). PhD Thesis, Maharaja Sayajirao University of Baroda, Vadodara, India.

46 Rane, A. N. (2017). Production of biosurfactants using unconventional substrates. PhD Thesis, Savitribai Phule Pune University, Pune, India.

47 Van Hamme, J.D., Singh, A., and Ward, O.P. (2003). Recent advances in petroleum microbiology. *Microbiology and Molecular Biology Reviews* 67 (4): 503–549.

48 Déziel, É., Paquette, G., Villemur, R. et al. (1996). Biosurfactant production by a soil *Pseudomonas* strain growing on polycyclic aromatic hydrocarbons. *Applied and Environmental Microbiology* 62 (6): 1908–1912.

49 Patel, R.M. and Desai, A.J. (1997). Biosurfactant production by *Pseudomonas aeruginosa* GS3 from molasses. *Letters in Applied Microbiology* 25 (2): 91–94.

50 Sim, L., Ward, O.P., and Li, Z.Y. (1997). Production and characterisation of a biosurfactant isolated from *Pseudomonas aeruginosa* UW-1. *Journal of Industrial Microbiology and Biotechnology* 19 (4): 232–238.

51 Benincasa, M., Abalos, A., Oliveira, I., and Manresa, A. (2004). Chemical structure, surface properties and biological activities of the biosurfactant produced by *Pseudomonas aeruginosa* LBI from soapstock. *Antonie Van Leeuwenhoek* 85 (1): 1–8.

52 Abouseoud, M., Yataghene, A., Amrane, A., and Maachi, R. (2008). Biosurfactant production by free and alginate entrapped cells of *Pseudomonas fluorescens*. *Journal of Industrial Microbiology & Biotechnology* 35 (11): 1303–1308.

53 Nie, M., Yin, X., Ren, C. et al. (2010). Novel rhamnolipid biosurfactants produced by a polycyclic aromatic hydrocarbon-degrading bacterium *Pseudomonas aeruginosa* strain NY3. *Biotechnology Advances* 28 (5): 635–643.

54 Thavasi, R., Nambaru, V.S., Jayalakshmi, S. et al. (2011). Biosurfactant production by *Pseudomonas aeruginosa* from renewable resources. *Indian Journal of Microbiology* 51 (1): 30–36.

55 Abbasi, H., Hamedi, M.M., Lotfabad, T.B. et al. (2012). Biosurfactant-producing bacterium, *Pseudomonas aeruginosa* MA01 isolated from spoiled apples: Physicochemical and structural characteristics of isolated biosurfactant. *Journal of Bioscience and Bioengineering* 113 (2): 211–219.

56 Aparna, A., Srinikethan, G., and Smitha, H. (2012). Production and characterization of biosurfactant produced by a novel *Pseudomonas* sp. 2B. *Colloids and Surfaces B: Biointerfaces* 95: 23–29.

57 Darvishi, P., Ayatollahi, S., Mowla, D., and Niazi, A. (2011). Biosurfactant production under extreme environmental conditions by an efficient microbial consortium, ERCPPI-2. *Colloids and Surfaces B: Biointerfaces* 84 (2): 292–300.

58 Onwosi, C.O. and Odibo, F.J.C. (2012). Effects of carbon and nitrogen sources on rhamnolipid biosurfactant production by *Pseudomonas nitroreducens* isolated from soil. *World Journal of Microbiology and Biotechnology* 28 (3): 937–942.

59 Saikia, R.R., Deka, S., Deka, M., and Sarma, H. (2012). Optimization of environmental factors for improved production of rhamnolipid biosurfactant by *Pseudomonas aeruginosa* RS29 on glycerol. *Journal of Basic Microbiology* 52 (4): 446–457.

60 e Silva, N.M.P.R., Rufino, R.D., Luna, J.M. et al. (2014). Screening of *Pseudomonas* species for biosurfactant production using low-cost substrates. *Biocatalysis and Agricultural Biotechnology* 3 (2): 132–139.

61 Deepika, K.V., Kalam, S., Sridhar, P.R. et al. (2016). Optimization of rhamnolipid biosurfactant production by mangrove sediment bacterium *Pseudomonas aeruginosa* KVD-HR42 using response surface methodology. *Biocatalysis and Agricultural Biotechnology* 5: 38–47.

62 dos Santos, A.S., Pereira, N. Jr., and Freire, D.M. (2016). Strategies for improved rhamnolipid production by *Pseudomonas aeruginosa* PA1. *PeerJ* 4: e2078.

63 Varjani, S.J. and Upasani, V.N. (2016). Carbon spectrum utilization by an indigenous strain of *Pseudomonas aeruginosa* NCIM 5514: Production, characterization and surface active properties of biosurfactant. *Bioresource Technology* 221: 510–516.

64 Cooper, D.G., Macdonald, C.R., Duff, S.J.B., and Kosaric, N. (1981). Enhanced production of surfactin from *Bacillus subtilis* by continuous product removal and metal cation additions. *Applied and Environmental Microbiology* 42 (3): 408–412.

65 Lin, S.C. and Jiang, H.J. (1997). Recovery and purification of the lipopeptide biosurfactant of Bacillus subtilis by ultrafiltration. *Biotechnology Techniques* 11 (6): 413–416.

66 Makkar, R.S. and Cameotra, S.S. (1997). Biosurfactant production by a thermophilic *Bacillus subtilis* strain. *Journal of Industrial Microbiology and Biotechnology* 18 (1): 37–42.

67 Makkar, R.S. and Cameotra, S.S. (1998). Production of biosurfactant at mesophilic and thermophilic conditions by a strain of *Bacillus subtilis*. *Journal of Industrial Microbiology and Biotechnology* 20 (1): 48–52.

68 Makkar, R.S. and Cameotra, S.S. (2002). Effects of various nutritional supplements on biosurfactant production by a strain of *Bacillus subtilis* at 45°C. *Journal of Surfactants and Detergents* 5 (1): 11–17.

69 Davis, D.A., Lynch, H.C., and Varley, J. (1999). The production of surfactin in batch culture by *Bacillus subtilis* ATCC 21332 is strongly influenced by the conditions of nitrogen metabolism. *Enzyme and Microbial Technology* 25 (3–5): 322–329.

70 Mulligan, C.N., Yong, R.N., Gibbs, B.F. et al. (1999). Metal removal from contaminated soil and sediments by the biosurfactant surfactin. *Environmental Science & Technology* 33 (21): 3812–3820.

71 Vater, J., Kablitz, B., Wilde, C. et al. (2002). Matrix-assisted laser desorption ionization-time of flight mass spectrometry of lipopeptide biosurfactants in whole cells and culture filtrates of *Bacillus subtilis* C-1 isolated from petroleum sludge. *Applied and Environmental Microbiology* 68 (12): 6210–6219.

72 Wei, Y.H. and Chu, I.M. (2002). Mn^{2+} improves surfactin production by *Bacillus subtilis*. *Biotechnology Letters* 24 (6): 479–482.

73 Wei, Y.H., Lai, C.C., and Chang, J.S. (2007). Using Taguchi experimental design methods to optimize trace element composition for enhanced surfactin production by *Bacillus subtilis* ATCC 21332. *Process Biochemistry* 42 (1): 40–45.

74 Mukherjee, A.K. and Das, K. (2005). Correlation between diverse cyclic lipopeptides production and regulation of growth and substrate utilization by *Bacillus subtilis* strains in a particular habitat. *FEMS Microbiology Ecology* 54 (3): 479–489.

75 Noah, K.S., Bruhn, D.F., and Bala, G.A. (2005). Surfactin production from potato process effluent by *Bacillus subtilis* in a chemostat. In: *Twenty-Sixth Symposium on Biotechnology for Fuels and Chemicals* (eds. B.H. Davison, B.R. Evans, M. Finkelstein and J.D. McMillan), 465–473. Humana Press.

76 Nitschke, M. and Pastore, G.M. (2006). Production and properties of a surfactant obtained from *Bacillus subtilis* grown on cassava wastewater. *Bioresource Technology* 97 (2): 336–341.

77 Yeh, M.S., Wei, Y.H., and Chang, J.S. (2006). Bioreactor design for enhanced carrier-assisted surfactin production with *Bacillus subtilis*. *Process Biochemistry* 41 (8): 1799–1805.

78 Al-Ajlani, M.M., Sheikh, M.A., Ahmad, Z., and Hasnain, S. (2007). Production of surfactin from *Bacillus subtilis* MZ-7 grown on pharmamedia commercial medium. *Microbial Cell Factories* 6 (1): 17.

79 Das, K. and Mukherjee, A.K. (2007). Comparison of lipopeptide biosurfactants production by *Bacillus subtilis* strains in submerged and solid state fermentation systems using a cheap carbon source: some industrial applications of biosurfactants. *Process Biochemistry* 42 (8): 1191–1199.

80 Fernandes, P.A.V., Arruda, I.R.D., Santos, A.F.A.B.D. et al. (2007). Antimicrobial activity of surfactants produced by *Bacillus subtilis* R14 against multidrug-resistant bacteria. *Brazilian Journal of Microbiology* 38 (4): 704–709.

81 Abdel-Mawgoud, A.M., Aboulwafa, M.M., and Hassouna, N.A.H. (2008). Characterization of surfactin produced by *Bacillus subtilis* isolate BS5. *Applied Biochemistry and Biotechnology* 150 (3): 289–303.

82 Barros, F.F.C., Ponezi, A.N., and Pastore, G.M. (2008). Production of biosurfactant by *Bacillus subtilis* LB5a on a pilot scale using cassava wastewater as substrate. *Journal of Industrial Microbiology & Biotechnology* 35 (9): 1071–1078.

83 Pornsunthorntawee, O., Arttaweeporn, N., Paisanjit, S. et al. (2008). Isolation and comparison of biosurfactants produced by *Bacillus subtilis* PT2 and *Pseudomonas aeruginosa* SP4 for microbial surfactant-enhanced oil recovery. *Biochemical Engineering Journal* 42 (2): 172–179.

84 de Faria, A.F., Teodoro-Martinez, D.S., de Oliveira Barbosa, G.N. et al. (2011). Production and structural characterization of surfactin (C14/Leu7) produced by *Bacillus subtilis* isolate LSFM-05 grown on raw glycerol from the biodiesel industry. *Process Biochemistry* 46 (10): 1951–1957.

85 Ghribi, D., Mnif, I., Boukedi, H. et al. (2011). Statistical optimization of low-cost medium for economical production of *Bacillus subtilis* biosurfactant, a biocontrol agent for the olive moth Prays oleae. *African Journal of Microbiology Research* 5 (27): 4927–4936.

86 Pemmaraju, S.C., Sharma, D., Singh, N. et al. (2012). Production of microbial surfactants from oily sludge-contaminated soil by *Bacillus subtilis* DSVP23. *Applied Biochemistry and Biotechnology* 167 (5): 1119–1131.

87 Al-Bahry, S.N., Al-Wahaibi, Y.M., Elshafie, A.E. et al. (2013). Biosurfactant production by *Bacillus subtilis* B20 using date molasses and its possible application in enhanced oil recovery. *International Biodeterioration & Biodegradation* 81: 141–146.

88 Coutte, F., Lecouturier, D., Leclère, V. et al. (2013). New integrated bioprocess for the continuous production, extraction and purification of lipopeptides produced by *Bacillus subtilis* in membrane bioreactor. *Process Biochemistry* 48 (1): 25–32.

89 de Oliveira, D.W.F., França, Í.W.L., Félix, A.K.N. et al. (2013). Kinetic study of biosurfactant production by *Bacillus subtilis* LAMI005 grown in clarified cashew apple juice. *Colloids and Surfaces B: Biointerfaces* 101: 34–43.

90 Mnif, I., Besbes, S., Ellouze-Ghorbel, R. et al. (2013). Improvement of bread dough quality by *Bacillus subtilis* SPB1 biosurfactant addition: optimized extraction using response surface methodology. *Journal of the Science of Food and Agriculture* 93 (12): 3055–3064.

91 Pereira, J.F., Gudiña, E.J., Costa, R. et al. (2013). Optimization and characterization of biosurfactant production by *Bacillus subtilis* isolates towards microbial enhanced oil recovery applications. *Fuel* 111: 259–268.

92 Al-Wahaibi, Y., Joshi, S., Al-Bahry, S. et al. (2014). Biosurfactant production by *Bacillus subtilis* B30 and its application in enhancing oil recovery. *Colloids and Surfaces B: Biointerfaces* 114: 324–333.

93 Sousa, M.D., Dantas, I.T., Felix, A.K.N. et al. (2014). Crude glycerol from biodiesel industry as substrate for biosurfactant production by *Bacillus subtilis* ATCC 6633. *Brazilian Archives of Biology and Technology* 57 (2): 295–301.

94 Gudiña, E.J., Fernandes, E.C., Rodrigues, A.I. et al. (2015). Biosurfactant production by *Bacillus subtilis* using corn steep liquor as culture medium. *Frontiers in Microbiology* 6: 59.

95 Liu, J.F., Mbadinga, S.M., Yang, S.Z. et al. (2015). Chemical structure, property and potential applications of biosurfactants produced by *Bacillus subtilis* in petroleum recovery and spill mitigation. *International Journal of Molecular Sciences* 16 (3): 4814–4837.

96 Fernandes, P.L., Rodrigues, E.M., Paiva, F.R. et al. (2016). Biosurfactant, solvents and polymer production by *Bacillus subtilis* RI4914 and their application for enhanced oil recovery. *Fuel* 180: 551–557.

97 Jha, S.S., Joshi, S.J., and Geetha, S.J. (2016). Lipopeptide production by *Bacillus subtilis* R1 and its possible applications. *Brazilian Journal of Microbiology* 47 (4): 955–964.

98 Joshi, S.J., Al-Wahaibi, Y.M., Al-Bahry, S.N. et al. (2016). Production, characterization, and application of *Bacillus licheniformis* W16 biosurfactant in enhancing oil recovery. *Frontiers in Microbiology* 7: 1853.

99 Cooper, D.G. and Paddock, D.A. (1984). Production of a biosurfactant from *Torulopsis bombicola*. *Applied and Environmental Microbiology* 47 (1): 173–176.

100 Pekin, G., Vardar-Sukan, F., and Kosaric, N. (2005). Production of sophorolipids from *Candida bombicola* ATCC 22214 using Turkish corn oil and honey. *Engineering in Life Sciences* 5 (4): 357–362.

101 Accorsini, F.R., Mutton, M.J.R., Lemos, E.G.M., and Benincasa, M. (2012). Biosurfactants production by yeasts using soybean oil and glycerol as low cost substrate. *Brazilian Journal of Microbiology* 43 (1): 116–125.

102 Diniz Rufino, R., Moura de Luna, J., de Campos Takaki, G.M., and Asfora Sarubbo, L. (2014). Characterization and properties of the biosurfactant produced by *Candida lipolytica* UCP 0988. *Electronic Journal of Biotechnology* 17 (1): 6–6.

103 Elshafie, A.E., Joshi, S.J., Al-Wahaibi, Y.M. et al. (2015). Sophorolipids production by *Candida bombicola* ATCC 22214 and its potential application in microbial enhanced oil recovery. *Frontiers in Microbiology* 6: 1324.

104 Kaur, G., Wang, H., To, M.H. et al. (2019). Efficient sophorolipids production using food waste. *Journal of Cleaner Production* 232: 1–11.

105 Silva, I.A., Veras, B.O., Ribeiro, B.G. et al. (2020). Production of cupcake-like dessert containing microbial biosurfactant as an emulsifier. *PeerJ* 8: e9064.

106 Karanth, N.G.K., Deo, P.G., and Veenanadig, N.K. (1999). Microbial production of biosurfactants and their importance. *Current Science*: 116–126.

107 Muriel, J.M., Bruque, J.M., Olías, J.M., and Jiménez-Sánchez, A. (1996). Production of biosurfactants by *Cladosporium resinae*. *Biotechnology Letters* 18 (3): 235–240.

108 Walter, V., Syldatk, C., and Hausmann, R. (2010). Screening concepts for the isolation of biosurfactant producing microorganisms. In: *Biosurfactants* (ed. R. Sen), 1–13. New York, NY: Springer.

109 Hommel, R. and Kleber, H.P. (1990). Properties of the quinoprotein aldehyde dehydrogenase from "*Acetobacter rancens*". *Microbiology* 136 (9): 1705–1711.

110 Banat, I.M., Satpute, S.K., Cameotra, S.S. et al. (2014). Cost effective technologies and renewable substrates for biosurfactants' production. *Frontiers in Microbiology* 5: 697.

111 Belligno, A., Di Leo, M. G., Marchese, M., & Tuttobene, R. (2005). Effects of industrial orange wastes on soil characteristics and on growth and production of durum wheat.

112 Chooklin, C.S., Maneerat, S., and Saimmai, A. (2014). Utilization of banana peel as a novel substrate for biosurfactant production by *Halobacteriaceae archaeon* AS65. *Applied Biochemistry and Biotechnology* 173 (2): 624–645.

113 Fox, S.L. and Bala, G.A. (2000). Production of surfactant from *Bacillus subtilis* ATCC 21332 using potato substrates. *Bioresource Technology* 75 (3): 235–240.

114 George, S. and Jayachandran, K. (2013). Production and characterization of rhamnolipid biosurfactant from waste frying coconut oil using a novel *Pseudomonas aeruginosa* D. *Journal of Applied Microbiology* 114 (2): 373–383.

115 Ghurye, G.L., Vipulanandan, C., and Willson, R.C. (1994). A practical approach to biosurfactant production using nonaseptic fermentation of mixed cultures. *Biotechnology and Bioengineering* 44 (5): 661–666.

116 Joshi, S., Yadav, S., and Desai, A.J. (2008). Application of response-surface methodology to evaluate the optimum medium components for the enhanced production of lichenysin by *Bacillus licheniformis* R2. *Biochemical Engineering Journal* 41 (2): 122–127.

117 Kosaric, N. (1992). Biosurfactants in industry. *Pure and Applied Chemistry* 64 (11): 1731–1737.

118 Makkar, R.S., Cameotra, S.S., and Banat, I.M. (2011). Advances in utilization of renewable substrates for biosurfactant production. *AMB Express* 1 (1): 5.

119 Muthusamy, K., Gopalakrishnan, S., Ravi, T.K., and Sivachidambaram, P. (2008). Biosurfactants: properties, commercial production and application. *Current Science*: 94, 736–94, 747.

120 Thompson, D.N., Fox, S.L., and Bala, G.A. (2000). Biosurfactants from potato process effluents. In: *Twenty-First Symposium on Biotechnology for Fuels and Chemicals* (eds. M. Finkelstein and B.H. Davison), 917–930. Totowa, NJ: Humana Press.

121 Camilios-Neto, D., Bugay, C., de Santana-Filho, A.P. et al. (2011). Production of rhamnolipids in solid-state cultivation using a mixture of sugarcane bagasse and corn bran supplemented with glycerol and soybean oil. *Applied Microbiology and Biotechnology* 89 (5): 1395–1403.

122 Jain, R.M., Mody, K., Joshi, N. et al. (2013). Production and structural characterization of biosurfactant produced by an alkaliphilic bacterium, *Klebsiella* sp.: Evaluation of different carbon sources. *Colloids and Surfaces B: Biointerfaces* 108: 199–204.

123 Daniel, H.J., Reuss, M., and Syldatk, C. (1998). Production of sophorolipids in high concentration from deproteinized whey and rapeseed oil in a two stage fed batch process using *Candida bombicola* ATCC 22214 and *Cryptococcus curvatus* ATCC 20509. *Biotechnology Letters* 20 (12): 1153–1156.

124 Dubey, K. and Juwarkar, A. (2001). Distillery and curd whey wastes as viable alternative sources for biosurfactant production. *World Journal of Microbiology and Biotechnology* 17 (1): 61–69.

125 Joshi, S., Yadav, S., Nerurkar, A., and Desai, A.J. (2007). Statistical optimization of medium components for the production of biosurfactant by *Bacillus licheniformis* K51. *Journal of Microbiology and Biotechnology* 17 (2): 313.

126 Mukherjee, S., Das, P., Sivapathasekaran, C., and Sen, R. (2008). Enhanced production of biosurfactant by a marine bacterium on statistical screening of nutritional parameters. *Biochemical Engineering Journal* 42: 254–260.

127 Desai, J.D. and Banat, I.M. (1997). Microbial production of surfactants and their commercial potential. *Microbiology and Molecular Biology Reviews* 61 (1): 47–64.

128 Kasture, M.B., Patel, P., Prabhune, A.A. et al. (2008). Synthesis of silver nanoparticles by sophorolipids: Effect of temperature and sophorolipid structure on the size of particles. *Journal of Chemical Sciences* 120 (6): 515–520.

129 Kumar, C.G., Mamidyala, S.K., Das, B. et al. (2010a). Synthesis of biosurfactant-based silver nanoparticles with purified rhamnolipids isolated from *Pseudomonas aeruginosa* BS-161R. *Journal of Microbiology and Biotechnology* 20 (7): 1061–1068.

130 Gómez-Graña, S., Perez-Ameneiro, M., Vecino, X. et al. (2017). Biogenic synthesis of metal nanoparticles using a biosurfactant extracted from corn and their antimicrobial properties. *Nanomaterials* 7 (6): 139.

131 Priyadarshini, E., Pradhan, N., Pradhan, A.K., and Pradhan, P. (2016). Label free and high specific detection of mercury ions based on silver nano-liposome. *Spectrochimica Acta Part A: Molecular and Biomolecular Spectroscopy* 163: 127–133.

132 Kumar, R. and Das, A.J. (2018). Rhamnolipid-assisted synthesis of stable nanoparticles: A green approach. In: *Rhamnolipid Biosurfactant: Recent Trends in Production and Application*, 111–124. Singapore: Springer.

133 Rangarajan, V., Majumder, S., and Sen, R. (2014). Biosurfactant-mediated nanoparticle synthesis: a green and sustainable approach. In: *Biosurfactants* (eds. C.N. Mulligan, S.K. Sharma and A. Mudhoo), 232–245. CRC Press, Boca Raton, FL.

134 Xie, Y., Ye, R., and Liu, H. (2006). Synthesis of silver nanoparticles in reverse micelles stabilized by natural biosurfactant. *Colloids and Surfaces A: Physicochemical and Engineering Aspects* 279 (1–3): 175–178.

135 Kasture, M., Singh, S., Patel, P. et al. (2007). Multiutility sophorolipids as nanoparticle capping agents: Synthesis of stable and water dispersible co nanoparticles. *Langmuir* 23 (23): 11409–11412.

136 Palanisamy, P. (2008). Biosurfactant mediated synthesis of NiO nanorods. *Materials Letters* 62 (4–5): 743–746.

137 Reddy, A.S., Chen, C.Y., Baker, S.C. et al. (2009). Synthesis of silver nanoparticles using surfactin: A biosurfactant as stabilizing agent. *Materials Letters* 63 (15): 1227–1230.

138 Kiran, G.S., Sabu, A., and Selvin, J. (2010). Synthesis of silver nanoparticles by glycolipid biosurfactant produced from marine *Brevibacterium casei* MSA19. *Journal of Biotechnology* 148 (4): 221–225.

139 Kumar, D.R., Kasture, M., Prabhune, A.A. et al. (2010b). Continuous flow synthesis of functionalized silver nanoparticles using bifunctional biosurfactants. *Green Chemistry* 12 (4): 609–615.

140 Narayanan, J., Ramji, R., Sahu, H., and Gautam, P. (2010). Synthesis, stabilisation and characterisation of rhamnolipid-capped ZnS nanoparticles in aqueous medium. *IET Nanobiotechnology* 4 (2): 29–34.

141 Sarma, H., Bustamante, K.L.T., and Prasad, M.N.V. (2019). Biosurfactants for oil recovery from refinery sludge: Magnetic nanoparticles assisted purification. In: *Industrial and Municipal Sludge* (ed. M.N.V. Prasad). Elsevier, Massachusetts, USA, pp. 107-132. ISBN: 9780128159071, Editors: Majeti Narasimha Vara Prasad, Paulo Jorge de Campos Favas, Meththika, Vithanage, S. Venkata Mohan.

142 Kumar, C.G. and Mamidyala, S.K. (2011). Extracellular synthesis of silver nanoparticles using culture supernatant of *Pseudomonas aeruginosa*. *Colloids and Surfaces B: Biointerfaces* 84 (2): 462–466.

143 Worakitsiri, P., Pornsunthorntawee, O., Thanpitcha, T. et al. (2011). Synthesis of polyaniline nanofibers and nanotubes via rhamnolipid biosurfactant templating. *Synthetic Metals* 161 (3–4): 298–306.

144 Hazra, C., Kundu, D., Chaudhari, A., and Jana, T. (2013). Biogenic synthesis, characterization, toxicity and photocatalysis of zinc sulfide nanoparticles using rhamnolipids from *Pseudomonas aeruginosa* BS01 as capping and stabilizing agent. *Journal of Chemical Technology & Biotechnology* 88 (6): 1039–1048.

145 Basnet, M., Ghoshal, S., and Tufenkji, N. (2013). Rhamnolipid biosurfactant and soy protein act as effective stabilizers in the aggregation and transport of palladium-doped zerovalent iron nanoparticles in saturated porous media. *Environmental Science & Technology* 47 (23): 13355–13364.

146 Farias, C.B., Ferreira Silva, A., Diniz Rufino, R. et al. (2014). Synthesis of silver nanoparticles using a biosurfactant produced in low-cost medium as stabilizing agent. *Electronic Journal of Biotechnology* 17 (3): 122–125.

147 Das, A.J., Kumar, R., Goutam, S.P., and Sagar, S.S. (2016). Sunlight irradiation induced synthesis of silver nanoparticles using glycolipid bio-surfactant and exploring the antibacterial activity. *Journal Bioengineering and Biomedical Science* 6 (05): 1–5.

148 Palanisamy, P. and Raichur, A.M. (2009). Synthesis of spherical NiO nanoparticles through a novel biosurfactant mediated emulsion technique. *Materials Science and Engineering: C* 29 (1): 199–204.

149 Khalid, H.F., Tehseen, B., Sarwar, Y. et al. (2019). Biosurfactant coated silver and iron oxide nanoparticles with enhanced anti-biofilm and anti-adhesive properties. *Journal of Hazardous Materials* 364: 441–448.

150 Joanna, C., Marcin, L., Ewa, K., and Grażyna, P. (2018). A nonspecific synergistic effect of biogenic silver nanoparticles and biosurfactant towards environmental bacteria and fungi. *Ecotoxicology* 27 (3): 352–359.

151 Maity, J.P., Lin, T.J., Cheng, H.P.H. et al. (2011). Synthesis of brushite particles in reverse microemulsions of the biosurfactant surfactin. *International Journal of Molecular Sciences* 12 (6): 3821–3830.

152 Jiao, J., Xin, X., Wang, X. et al. (2017). Self-assembly of biosurfactant–inorganic hybrid nanoflowers as efficient catalysts for degradation of cationic dyes. *RSC Advances* 7 (69): 43474–43482.

153 Bezza, F.A., Tichapondwa, S.M., and Chirwa, E.M. (2020). Synthesis of biosurfactant stabilized silver nanoparticles, characterization and their potential application for bactericidal purposes. *Journal of Hazardous Materials* 393: 122319.

154 Ga'al, H., Yang, G., Fouad, H. et al. (2020). Mannosylerythritol lipids mediated biosynthesis of silver nanoparticles: An eco-friendly and operative approach against Chikungunya vector *Aedes albopictus*. *Journal of Cluster Science*: 1–9.

155 Wang, S., Liu, Y., Lü, Q.F., and Zhuang, H. (2020). Facile preparation of biosurfactant-functionalized Ti2CTX MXene nanosheets with an enhanced adsorption performance for Pb (II) ions. *Journal of Molecular Liquids* 297: 111810.

156 Guerra, F.D., Attia, M.F., Whitehead, D.C., and Alexis, F. (2018). Nanotechnology for environmental remediation: materials and applications. *Molecules* 23 (7): 1760.

157 Choi, M.H., Jeong, S.W., Shim, H.E. et al. (2017). Efficient bioremediation of radioactive iodine using biogenic gold nanomaterial-containing radiation-resistant bacterium, *Deinococcus radiodurans* R1. *Chemical Communications* 53 (28): 3937–3940.

158 Otari, S.V., Patil, R.M., Nadaf, N.H. et al. (2014). Green synthesis of silver nanoparticles by microorganism using organic pollutant: its antimicrobial and catalytic application. *Environmetal Science and Pollution Research* 21: 1503–1513.

159 Sathiyanarayanan, G., Kiran, G.S., and Selvin, J. (2013). Synthesis of silver nanoparticles by polysaccharide bioflocculant produced from marine *Bacillus subtilis* MSBN17. *Colloids and Surfaces. B, Biointerfaces* 102: 13–20.

160 Saif, S., Tahir, A., and Chen, Y. (2016). Green synthesis of iron nanoparticles and their environmental applications and implications. *Nanomaterials* 6 (11): 209.

161 Sobana, N., Muruganadham, M., and Swaminathan, M. (2006). Nano-Ag particles doped TiO_2 for efficient photodegradation of direct azo dyes. *Journal of Molecular Catalysis A: Chemical* 258 (1–2): 124–132.

162 Mandal, S.K., Ojha, N., and Das, N. (2018). Optimization of process parameters for the yeast mediated degradation of benzo [a] pyrene in presence of ZnO nanoparticles and produced biosurfactant using 3-level Box-Behnken design. *Ecological Engineering* 120: 497–503.

163 Reddy, A.S., Chen, C.Y., Chen, C.C. et al. (2010). Biological synthesis of gold and silver nanoparticles mediated by the bacteria *Bacillus subtilis*. *Journal of Nanoscience and Nanotechnology* 10 (10): 6567–6574.

164 Saharan, B.S., Sahu, R.K., and Sharma, D. (2011). A review on biosurfactants: fermentation, current developments and perspectives. *Genetic Engineering and Biotechnology Journal* 2011 (1): 1–14.

165 Shekhar, S., Sundaramanickam, A., and Balasubramanian, T. (2015). Biosurfactant producing microbes and their potential applications: A review. *Critical Reviews in Environmental Science and Technology* 45 (14): 1522–1554.

10

Green Surfactants

Production, Properties, and Application in Advanced Medical Technologies

Ana María Marqués[1], Lourdes Pérez[2], Maribel Farfán[1], and Aurora Pinazo[2]

[1]*Department of Biology, Healthcare and the Environment, Section of Microbiology, University of Barcelona, Barcelona, Spain*
[2]*Department of Surfactant and Nanobiotechnology, IQAC, CSIC, Barcelona, Spain*

CHAPTER MENU

10.1 Environmental Pollution and World Health

Contamination of air, water, and soil, a consequence of accelerated industrialization and urbanization worldwide, produces human diseases and death. The world's poorest countries suffer particularly from indoor air pollution and contaminated drinking water. The World Health

Biosurfactants for a Sustainable Future: Production and Applications in the Environment and Biomedicine, First Edition. Edited by Hemen Sarma and Majeti Narasimha Vara Prasad.
© 2021 John Wiley & Sons Ltd. Published 2021 by John Wiley & Sons Ltd.

Organization (WHO) estimates that each year 7.9 million people in middle- and low-income countries die of pollution-related diseases, typically pneumonia and diarrhea [1, 2].

In richer countries, pollution hazards are predominantly toxic chemicals, pesticides, ambient air pollution, and hazardous wastes, exposure to which leads to birth defects, heart disease, and cancer. A huge number of new chemicals have been produced in the last half-century and are widely disseminated, both in the environment and in human bodies. The relocation of polluting industries in poorer countries with lax safety controls is endangering an increasing number of workers [1]. The desire for economic growth and consumption clashes with a growing awareness in society of the need to recover, recycle, and reuse to care for the environment.

As a consequence, to prevent environmental pollution, there is a strong demand for biodegradable products developed by the green chemistry (the elimination of the use or generation of hazardous substances in the design, manufacture, and application of chemical products) [3]. Microorganisms, which are responsible for large-scale transformations in the biosphere, have the ability to degrade toxic chemicals with a complete aerobic or anaerobic mineralization.

Surfactants are amphipathic compounds in which the apolar moiety is often a hydrocarbon chain, whereas the polar group may be ionic, nonionic, or amphoteric [4]. The versatility of surfactants makes them useful in many industrial processes and in our day-to-day life. Synthetic surfactants are used in abundance by a wide range of industries such as pharmaceutical, food and feed, chemical, paint, and petroleum as well as in agriculture and environmental remediation [5]. It is now accepted that the widespread use of non-degradable synthetic surfactants is negatively affecting the environment and is associated with the growing incidence of cancer worldwide. The increasing legal and societal pressure for these substances to be biodegradable and sustainably produced has stimulated research into new degradable surfactants of synthetic or biological origin. As a consequence, manufacturers are increasingly interested in eco-friendly and cost-effective production processes [6].

10.2 Amino Acid-Derived Surfactants

10.2.1 Surfactants, Definition and Applications

"Surfactant" is an abbreviation for surface active agent, which are active at the interface of a solid and liquid, air and liquid, or between two liquids, one immiscible. Colloids are heterogeneous mixtures made of tiny particles or droplets with a size range of 10^{-9}– 10^{-6} m and the appearance of a homogeneous solution. At a low concentration, surfactants exist as monomers of subcolloidal size and at a high concentration they form micelles or aggregates of colloidal size. Colloidal systems have a wide variety of potential applications (Figure 10.1), and considerable attention has been focused on their design and fabrication in recent years. Many innovations related to biological activity such as drug delivery systems and medical imaging are based on colloid and interface science. Colloids are used as biofunctional products or serve as man-made cell-like compartments in which biological activity takes place.

Colloidal vesicular systems such as niosomes are uni- or multilamellar structures made up of surfactant molecules surrounded by an aqueous bulk solution. They are useful for the delivery of both hydrophilic and hydrophobic drugs, which are encapsulated in the interior hydrophilic compartment or outer bilayer, respectively. Many of these systems currently under development, particularly those for bio-applications, draw their inspiration from naturally occurring structures. In this context, the chemical structure of amino acid-based surfactants, which mimic that of natural

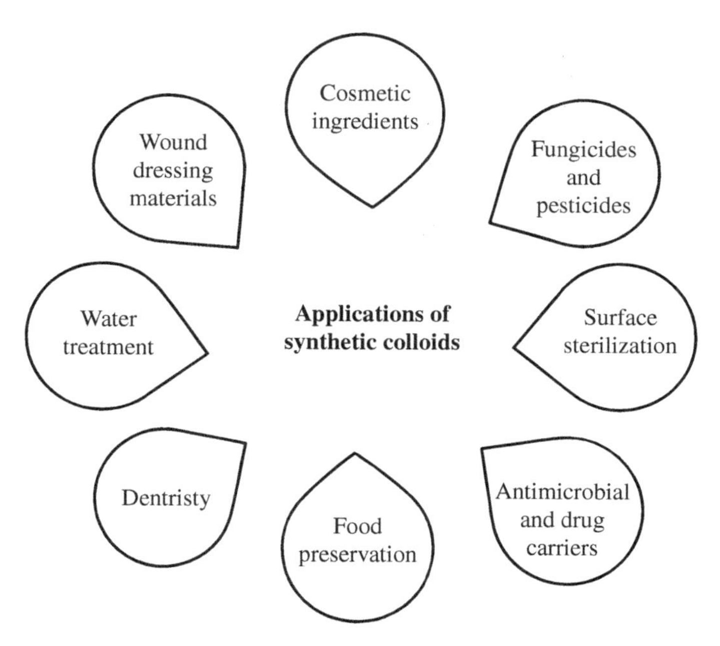

Figure 10.1 Applications of synthetic colloids. *Source:* Aurora Pinazo.

lipoamino acids, makes them an attractive alternative to conventional surfactants. Highly biodegradable, milder, and less irritant, they meet three fundamental requirements of green chemistry: multifunctionality, low toxicity, and production from renewable raw materials.

The hydrophilic part of amino acid surfactants is obtained by enzymatic synthesis [7] and the hydrophobic part from natural oils [8]. These natural surface-active molecules are of great interest to organic and physical chemists and biologists, and have an unpredictable number of basic and industrial applications [9–14].

Although amino acid surfactants are structurally a very heterogeneous group of compounds, they share a common advantage in that they are relatively easy to design and synthesize. The 20 types of α-amino acids commonly present in proteins have a general structure of NH_2–CHR–COOH, where R is the side chain. Depending on the ionic nature of the side chain, amino acids can be classified as neutral, with a nonionic side chain that can be weakly hydrophilic, hydrophobic, or aromatic (i.e. proline, leucine, and phenylalanine); acidic, with a negatively charged side chain (i.e. glutamic acid, aspartic acid); and basic, with a positively charged side chain (i.e. lysine, arginine). This wide structural range forms the basis of the diversity of amino acid surfactants and their physicochemical and biological properties [15, 16]. Depending on the position of the hydrophobic chain and the side chain charge, amino acid surfactants can be anionic, cationic, or zwitterionic. Depending on the number of hydrophobic chains (Figure 10.2) allows linear and double chain structures.

10.2.2 Linear Amino Acid-Based Surfactants

Linear amino acid–based surfactants consist of an amino acid as a head group linked to a hydrophobic tail (Figure 10.2a). In some surfactants, the head group carries a charge, whereas the tail

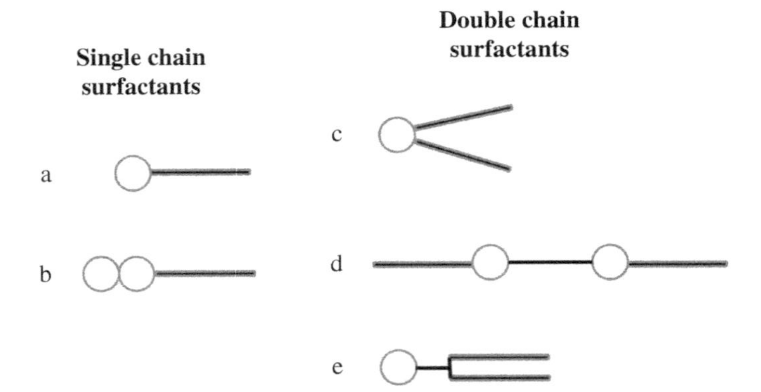

Figure 10.2 Structures of amino acid-based surfactants. (a) Linear amino acid surfactants; (b) Linear amino acid-based surfactants with two amino acids on the polar head; (c) Double chain amino acid-based surfactants; (d) Amino acid Gemini surfactants; (e) Glycero lipid amino acid-based surfactants. *Source:* Aurora Pinazo.

varies in structure and length and can come from fatty acids, fatty alcohols, fatty amines, or alkyl halides. Underlying the diverse structures and properties of amino acid surfactants is the nature and multifunctionality of the amino acid residue. Different types of linkage (acyl, ester, amide, or alkyl) can connect the hydrophobic chain [17, 18], resulting in *O*-alkyl ester, *N*-alkyl, *N*-acyl, and amide amino acid surfactants. Their structural simplicity means these surfactants can be readily synthesized in only a few reaction steps, using chemical or enzymatic methodologies. Therefore, their production is economically feasible and meets environmental requirements [19–22].

The amino acid arginine is an excellent raw material to prepare linear surfactants with antimicrobial properties, the cationic. ***N*-acyl arginine methyl ester** derivatives being particularly effective [23]. Essential structural factors for antimicrobial activity include the length of the fatty chain and the protonated guanidine group of arginine.

The essential amino acid lysine belongs to the aspartate biosynthetic pathway and can be produced by bacterial fermentation [7]. The presence of one carboxyl and two amino groups at positions α and ε in the lysine molecule makes it a versatile starting material in surfactant synthesis, leading to many ionic surfactant structures [24, 25]. **Single-chain lysine-based cationic** surfactants with a fatty acyl chain attached to the α- or ε-amino group of lysine through an amide linkage have been synthesized and studied [26, 27]. Lysine-based surfactants have the cationic charge on one of the protonated amino groups, and they present an acid–base equilibrium in aqueous solutions. As their cationic character depends on the pH of the medium, these surfactants behave as pH-sensitive compounds. N^{α}-acyl-lysine surfactants exhibit a wide spectrum of antimicrobial activity against Gram-positive and Gram-negative bacteria, comparable to arginine analogs. Moving the alkyl chain from the α-amino to the ε-amino group of lysine resulted in a significant decrease in antimicrobial properties, and none of the homologs showed activity against Gram-negative bacteria [23].

The synthesis of monocatenary cationic histidine-based surfactants of different alkyl chain length was recently reported. The polar head of these surfactants consists of the histidine amino acid with methylene groups at the $N(\pi)$ and $N(\tau)$ positions on the imidazole group [28]. These compounds showed remarkable growth inhibition against several Gram-positive and Gram-negative bacteria as well as fungus. Antimicrobial activity, which is influenced by the alkyl chain length, was at its maximum in compounds bearing 14 carbon atoms in the fatty chain. Hemolytic activity was found to increase with the lengthening of the alkyl chain. Promisingly, the activity against Gram-positive bacteria occurred at lower concentrations than erythrocyte toxicity. Structurally simple, these new surfactants are easy and cheap to prepare.

The numerous different types of amino acids allow the design of a wide range of cationic structures with diverse specifications. In general, these kinds of surfactants can be considered as readily biodegradable compounds. An advantage of antimicrobial amino acid surfactants is their double functionality, they have the typical properties of surface-active molecules, reduction surface tension and forming molecular aggregates such as micelles or vesicles, and they can also act as antimicrobial agents. For example, cationic surfactants based on L-tryptophan and L-tyrosine have proved to be excellent gelling agents and exhibit remarkable bactericidal properties. The presence of aromatic groups in these surfactants favors the formation of vesicles [29–31]. Linear amino acid surfactants are biocompatible, which makes them highly suitable for use as emulsifiers, detergents, wetting and foaming agents, dispersing compounds, or antimicrobials.

10.2.3 Linear Amino Acid-Based Surfactants with Two Amino Acids on the Polar Head

Cationic monocatenary surfactants have been synthesized, combining a fatty chain of 12 carbon atoms and a polar head of two basic amino acids with a double positive charge (Figure 10.2b). The two positive charges are provided by two basic amino acids joined by peptide linkages: **α-arginine-ε-lysine** (N^{α}-lauroyl-arginine-N^{ε}-lysine methyl ester), with one positive charge on the guanidine group of arginine and the other positive charge on the ε-amino group of lysine; **α-arginine-α-lysine** (N^{α}-lauroyl-arginine-N^{α}-lysine methyl ester), with one positive charge on the guanidine group of arginine and the other on the α-amino group of lysine; and **lysine-lysine** (N^{ε}-lauroyl-lysine-N^{ε}-lysine methyl ester) in which the two positive charges lie on the ε-amino group of the lysines.

The amino acid surfactants with two positive charges in the polar head were designed with the aim of improving antimicrobial activity, which in cationic surfactants is generally attributed to the hydrophilic–lipophilic balance and the net positive charge. Also playing an important role are the position and density of the cationic charge. The presence of lysine as a second amino acid on the polar head reduces the antimicrobial activity, because the two cationic charges increase the hydrophilic character and solubility of the molecule, which in turn reduces interfacial activity.

Palmitoyl-lipopeptides consisting of one alkyl chain of 16 carbon atoms and two, three, or four amino acids have been described as potent antimicrobial and antifungal agents [24]. These compounds assemble in solution into nanostructures with different morphologies, which could partially explain the differences in their antimicrobial activity [32], these being highest with the **lysine–lysine–lysine** sequence as the polar head. The lysine–alanine–lysine sequence resulted in inactivity, **lysine–leucine–lysine** was active only toward Gram-positive bacteria and **lysine–glycine–lysine** showed activity against all the tested microorganisms.

10.2.4 Double-Chain Amino Acid-Based Surfactants

Numerous structural modifications have been implemented to increase surfactant hydrophobicity in an effort to enhance their properties. This section describes three types of amino acid-based surfactants designed to be environmentally acceptable and with high performance: cationic double chain (Figure 10.2c), gemini (Figure 10.2d), and cationic glycerolipid (Figure 10.2e).

1) *Cationic double chain amino acid surfactants*
   Cationic double chain surfactants bearing two fatty chains and arginine as the polar head have attracted a great deal of interest because they can produce cationic vesicles for biomedical applications [33]. They are less effective as antimicrobial agents, as the formation of large vesicles implies a low monomer concentration in the media, which is insufficient to inhibit bacterial growth. However, these new arginine derivatives show very low hemolytic activity and weaker cytotoxic effects than conventional cationic surfactants. Studies show that a rational design applied to cationic double chain surfactants is the way forward to the development of safe cationic vesicular systems.
2) *Cationic gemini amino acid surfactants*
   Two single surfactants can be chemically joined at or near the head group by a rigid or flexible spacer chain to form a bifunctional gemini surfactant with unusual physicochemical properties. Having more hydrophobic groups per molecule, they can aggregate at concentrations at least one order of magnitude lower than the corresponding monomeric surfactant. As lower amounts are required, they have less of an impact on the environment. Amino acid gemini surfactants have generated a considerable amount of literature [23], but here we restrict our focus to those derived from arginine, cystine, and lysine.

- *Arginine gemini surfactants.* The amino acid arginine has a side chain of three carbon atoms with a guanidinium group attached to the end. This group has a high pKa value, which confers a positive charge in neutral, acidic and even in basic media. Structurally, **arginine gemini surfactants** have two fatty chains and two arginines linked by amide covalent bonds to an alkenediamine spacer chain. They can be prepared by chemical and enzymatic technologies [34, 35]. The antimicrobial properties of arginine gemini surfactants depend on the length of the spacer chain and the length of the alkyl chain [23]. Cationic colloidal systems consisting of arginine gemini surfactants and membrane additive compounds such as dilauroyl phosphatidylcholine or cholesterol have been explored [36]. All formulations assayed had positive zeta-potential and colloidal stability, essential properties for surfactants with cosmetic and pharmaceutical application.
- *Cationic gemini surfactants from cystine.* Cystine consists of two amino acid residues linked by a disulphide bond, which makes it a useful starting material in the synthesis of cationic gemini surfactants. Cationic gemini surfactants derived from cystine are prepared by condensing *N*-lauroyl glycine betaine to cystine dimethyl ester hydrochloride [37]. They contain a reactive disulphide bond, which plays a key role in modifying the surface and biological properties of thiol-containing substrates via a disulphide–thiol interchange reaction [38].
- *Cationic gemini surfactants from lysine.* The interest of lysine as a surfactant building block lies in its three reactive groups, carboxylic, α-amino, and ε-amino, which allow the synthesis of a variety of gemini structures [26, 39]. Considerable differences in antimicrobial activity have been found between the structures. As in their single chain counterparts, higher activity was obtained for compounds with a cationic charge on the trimethylated ε-amino group.
- *Gemini surfactants from histidine.* Histidine-based gemini cationic surfactants were prepared with imidazolium groups as polar heads and alkyl chains, whose variable lengths influence the surfactant microbial and hemolytic activities [40]. As they are active against a wide range of bacteria, these gemini surfactants have potential application in the cosmetic, food, and pharmaceutical industries.

3) *Glycero lipid amino acid-based surfactants*
   Arginine glycero conjugates form a class of lipoamino acids, which are analogs of partial glycerides and phospholipids. Structurally, they consist of two aliphatic chains and one amino acid

comprising the polar head linked through a glycerol moiety (Figure 10.2e). The hydrophilic part is formed by either arginine or acetyl-arginine [41–43]. By combining the physicochemical properties of glycerol derivatives and arginine-based surfactants, these glycerol conjugates provide certain advantages over conventional phospholipids. The presence of arginine increases surfactant solubility in aqueous solutions and provides antimicrobial activity, whereas the two fatty chains endow the molecule with the ability to spontaneously form vesicles in water.

10.3 Biosurfactants

Biosurfactants (BS) are a group of microbial structurally diverse amphiphilic molecules with the ability to reduce surface and interfacial tension of liquids and form micelles and microemulsions [6, 44, 45]. BS production by bacteria, yeast, and fungi, organisms characterized by a large surface-to-volume ratio, is a survival mechanism that allows adaptation in otherwise inhospitable habitats [46]. By interacting at interphases, BS play an important role in nutrient bioavailability through emulsification, cell adhesion to biotic and abiotic surfaces, and defense mechanisms. They are involved in microorganism interrelationships and quorum sensing responses, and act as biocides in competition for habitats [47, 48]. There is considerable research interest in developing BS with high added value properties.

10.3.1 Biosurfactant Types and Classification

Their diverse structures and functional properties make BS an attractive group of compounds for numerous applications. They can be classified according to the chemical structure of the hydrophilic moiety as (i) glycolipids, (ii) lipopeptides or lipoproteins, (iii) phospholipids, fatty acids or natural lipids, and (iv) polymeric and (v) particulate surfactants [45].

Classification of BS by their molecular weight, which is generally between 500 and 15000 Da [4], gives rise to two groups. Those of low molecular weight, which include chemical groups 1, 2, and 3, efficiently reduce surface tension, and those of high molecular weight, including chemical groups 4 and 5, are known for their emulsifying properties [6, 47, 49].

1) *Glycolipids*, which are among the most studied BS, consist of a carbohydrate moiety linked to an aliphatic acid or hydroxyl aliphatic fatty acid. The four most widely described groups are the following:
   - *Rhamnolipids*, produced as mono- or di-rhamnolipids by *Pseudomonas aeruginosa* and *Pseudomonas* sp., reduce surface tension to 29 mN/m and have been extensively studied and reported for various industrial applications [50, 51].
   - *Trehalose lipids* (THL), which consist of disaccharide trehalose linked to mycolic acids (β-hydroxy-branched fatty acids), reduce surface tension to 25 mN/m and are mainly produced by Gram-positive bacteria with a high GC content, such as *Nocardia* sp., *Mycobacterium* sp., and *Corynebacterium* sp. [52].
   - *Sophorolipids*, composed of a dimeric sophorose linked to a hydroxy fatty acid, reduce surface tension to 33 mN/m and are mainly produced by non-pathogenic yeast strains such as *Stamerella bombicola* or *Candida apicola*.
   - *Mannosylerythritol lipids* (MEL) consist of the mannose molecule linked to an erythritol residue. The hydrophilic head can be acylated with a short-chain (C_2–C_8) or a medium chain (C_{10}–C_{18}) fatty acid. These glycolipids are produced by yeast such as *Pseudozyma aphidis*,

Pseudozyma rugulosa, or *Pseudozyma antarctica* and are of high interest for their properties and abundant production [53].

2) *Lipopeptides* are composed of a linear or cyclic amino acid chain linked to fatty acids. The producers are mainly species of the genus *Bacillus*, but also include *Pseudomonas and Actinomycetes* and fungus. *Bacillus* LPs, which are synthesized non-ribosomally via large multienzymes, have remarkable heterogeneity. The same strain can produce several isoforms or different homologous compounds of all the lipopeptide families, which are classified by their amino acid sequence as follows:
   - *Surfactin*, the most studied lipopeptide, is formed by a cyclic heptapeptide and a β-hydroxy fatty acid and is produced by several strains of *Bacillus subtilis* during the stationary growth phase. Several surfactin isoforms usually coexist in the cell as a mixture of peptidic variants. One of the most powerful BS, surfactin, lowers water surface tension to 27 mN/m at very low concentrations (10 μM). Included in the same family are *lichenysin* (produced universally by *Bacillus licheniformis*), *esperin*, and *pumacilin* [54].
   - *Iturins* are a group of lipopeptides that share a sequence of three amino acids (Asx-Tyr-Asx) and contain a variable moiety of four amino acids, linked to a β-amino fatty acid by amide bonds. This group also includes *bacillopeptin* and *mycosubtilin*.
   - *Fengycins*, also known as pliplastatins, are lipodecapeptides with an internal lactone ring in the peptidic moiety linked with a β-hydroxy fatty acid chain. Fengycins are poorly hemolytic but have strong anti-fungal properties due to their capacity to alter membrane permeability in a dose-dependent way [55].
   - *Kurstatins* are low-molecular-weight heptapeptides featuring a constant amino acid sequence with lactone linkage and different fatty acids linked to the *N*-terminal amino acid residue via an amide bond. They are mainly produced by *Bacillus thuringensis kurstaki* HD-1.
   - Lipopeptide biosurfactants can also be produced by a wide variety of *Pseudomonas*, which are composed of a short oligopeptide with a linked fatty acid tail with antimicrobial properties. *Friulimicin*, *laspartomycins*, and *amphomyucis* are cyclic LPs produced by *Actimomycetes* and are undecapeptides that form lactose rings [44].
3) *Phospholipids and fatty acids* are produced in high amounts by bacteria such as *Sphingobacterium detergens* growing on n-alkanes. Among other phospholipids, the microbial membrane component phosphatidylethanolamine can reduce water surface tension to 33 mN/m [56]. *Acinetobacter* sp. is able to produce phosphatidylethanolamine vesicles that form microemulsions of alkanes in water [4].
4) *Polymeric biosurfactants* are a group of polysaccharide and protein complexes that include *emulsan*, *lipomanan*, *alasan*, and *liposan*. Emulsan is an extracellular polymeric bioemulsifier (anionic heteropolysaccharide with protein) produced in relatively large amounts by *Acinetobacter calcoaceticus* RAG-1 during the early exponential growth phase, and is able to emulsify hydrocarbons in water at very low concentrations (0.0001–0.01%) [49, 57].
5) *Particulate biosurfactants* are cellular components, such as extracellular membrane vesicles, and are composed of proteins, phospholipids, and lipo-polysaccharides. They form stable microemulsions required for alkane uptake in microbial cells [52].

10.3.2 Biosurfactant Production Using Low-Cost Raw Materials

The main factor that impedes the widespread use of BS, especially in high-volume applications, is the economics of their production [58]. The high cost of BS production can be reduced by two basic strategies: (i) the use of cheap or waste substrates to lower the initial outlay on raw materials and

(ii) the development of efficient bioprocesses, including optimization of culture conditions and strategies [59]. Thus, large-scale BS production needs cheap raw material, non-pathogenic microorganisms, a simple and quick fermentation process, an easy and cheap purification step, and the study of physicochemical and biological properties for potential applications [4]. High production costs are a limiting factor, especially in high-volume applications, but in compensation, BS are efficient in biological systems at low concentrations (μM), which could facilitate their usage in the biomedical, cosmetic, therapeutic, and healthcare sectors [46, 58].

Around 10–30% of the BS production cost is accounted for by the raw materials, which should be renewable. In producing food for a growing world population, the food industry is generating increasingly large amounts of untreated waste. Elimination of these byproducts can be complex (for example, olive oil mill effluent) and the sheer volume of production can lead to an immense accumulation of waste in the environment. Agro-industrial byproducts and wastes usually contain nutritional elements (carbohydrates, fats, oils, or vitamins) that could be recycled as low-cost substrates for BS producer microorganisms [60]. Valorization of industrial residues is an interesting concept, as the transformation of waste into a new added-value product simultaneously eliminates or reduces environmental pollution.

The use of molasses, a byproduct of sugarcane processing with a 48–56% sugar content, as a substrate for BS industrial production is feasible after optimization. Another candidate is olive oil mill effluent, a concentrated black liquor produced during the pressing of olives for oil extraction, which was successfully employed by Mercadé et al. [61] to produce rhamnolipids. This application would have environmental benefits, as the effluent is an abundant residue in Mediterranean countries and its high polyphenol concentration has a polluting impact. A waste substance produced in large quantities by the dairy industry is whey, which contains a high amount of lactose (75%) together with proteins, vitamins, and organic acids, and can also be used as a substrate for bacterial growth and BS production. Other nutritionally suitable substrates include corn steep liquor, a byproduct of corn wet-milling, which is rich in proteins and lactic acid, and starchy wastes from the potato processing industry [4]. With a mixture of corn steep liquor (9%) and ground-nut oil refinery residue (9%), *Candida sphaerica* was able to produce 9 g/l of a glycolipid BS with high surface tension-reducing activity (25 mN/m) [62]. *Oleomonas sagaranensis* produced an active glycolipid after growth in recycled lubricating oil, glucose, and soybean oil, but the most effective substrate for BS production was found to be molasses [63]. Environmental pollution can be resolved by microbial BS production with a pollutant serving as a carbon source, as occurs with *Halomonas pacifica* Cnaph3 [64].

Whereas some studies report that BS are primary metabolites, whose production is associated with the log [63] or stationary phases of growth [49], others suggest they are secondary metabolites, produced in the stationary phase. It is well established that BS structure and production is dependent on the composition of the microbial growth medium [45, 65] and environmental factors such as pH, temperature, agitation, and oxygen tension.

In BS production, 50–80% of the total cost is accounted for by downstream processing, especially the frequent requirement for BS purification from the fermentation broth [4]. Thus, although increasing environmental awareness is focusing new attention on green surfactants, their commercialization is still fairly limited due to high production costs and relatively low yields [44, 53].

10.3.3 Biosurfactant Properties and Applications

The highly variable chemical structures of BS endow them with different properties with potential application in many industries, including biotechnology and environmental clean-up. To determine suitable applications for BS, it is first necessary to study their physicochemical and biological properties.

A characteristic BS feature is the hydrophilic–lipophilic balance (HLB), which specifies the proportion of hydrophilic and hydrophobic constituents in surface-active substances. HLB values indicate whether the BS functions as a water-in-oil or an oil-in-water emulsifier. A low or lipophilic HLB (3–6) stabilizes water-in-oil emulsions, whereas a high HLB (8–18) confers better water solubility. The function of an emulsifier (an amphiphilic molecule) is to facilitate and stabilize the formed emulsion without micelle clustering, which ultimately leads to the mixing of two immiscible liquids [45, 47].

The emulsion capacity of BS has been widely studied. A small proportion of THLs is able to produce a stable emulsion of isopropyl myristate and paraffin [66]. The ability of a low concentration of rhamnolipids to emulsify isopropyl myristate, which is used in cosmetics, olive and soybean oil used in food products, and Casablanca oil as a model of environmental pollutant, has been described by Torrego-Solana et al. [67]. In the pharmaceutical field, the antimicrobial activity of essential oils of *Melaleuca alternifolia*, *Cinnamomum verum*, *Origanum compactum*, and *Lavandula angustifolia* against pathogens *Candida albicans* and *Staphylococcus aureus* increased after emulsion with rhamnolipids due to an enhanced dispersion [50]. BS also characteristically maintain their activity in environmental conditions of extreme temperature, pH (values of 2–12), and salt concentration (up to 10%), whereas 2% NaCl is enough to inactivate synthetic surfactants [4].

BS biodegradability and low toxicity open up a range of applications in pharmacy (dosage reducers), the food industry (stabilizers or consistency modifiers), cosmetics (cream stabilizers), and environmental technology (bioavailability of pollutant oils) [44, 60, 68]. As BS are easily degraded by microorganisms in water and soil, they are suitable for bioremediation and waste treatment. Although rhamnolipids are considered biodegradable, a few reports have demonstrated their degradation through acting as a carbon and energy source for microorganisms. In a recent study, the biodegradation of a mono-rhamnolipid (98–99%) and four rhamnolipids functionalized with arginine and lysine (85–61%) was compared, and the functionalization was found to have a delaying effect [69]. Low toxicity allows BS to be used not only in environmental applications but also in more demanding fields such as food, cosmetics, and pharmaceuticals, in which contact with living beings requires biocompatibility and digestibility [4].

The level of toxicity varies among surfactants and can be measured using cells as fibroblasts and mouse keratinocytes. Sodium dodecyl sulphate (SDS) is reported to have the greatest irritant capacity, followed by rhamnolipids and THLs, all with low IC_{50} values. Surfactants produced by *Sphingobacterium detergens* are the least toxic, despite the excellent surface activity of the phospholipid mixture [66, 67, 70].

The amphiphilic nature of surfactants favors their interaction with cell membranes, whose physicochemical properties and function are altered in the process. To ascertain BS properties for biological applications, BS–membrane interaction can be studied by preparing phosphatidylcholine (POPC) vesicles, the main constituent of lipid bilayers, and using carboxyfluorescein as a marker of integrity loss. Depending on the BS, cytotoxicity occurs at different rates and can be associated with permeation (THLs and lipopeptides) or solubilization (dirhamnolipids). The effect of membrane lipid composition has also been studied. When cholesterol and POPE (phosphoethanolamine) were incorporated into the membrane, the initial rate of leakage was considerably reduced. In particular, cholesterol (30% or 50%) had a strong protective effect, reducing or practically eliminating carboxyfluorescein leakage, in contrast with the effect of THL, rhamnolipid, and lichenysin [71–73]. This fact is of great importance since bacterial cell membranes, unlike eukaryotic membranes, do not possess cholesterol. For drug administration applications, molecules need the capacity to traverse the lipid membrane [71].

Erythrocytes represent a very simple model for the study of the interaction of bioactive substances such as surfactants with cell membranes. Proposed mechanisms of lysis include: (i) solubilization of membrane proteins by the formation of mixed micelles (at high surfactant concentrations, close to the critical micellar concentration [cmc]), and (ii) permeabilization to small solutes by surfactant incorporation into the membrane, with subsequent colloid-osmotic lysis. Studies of hemolysis$_{50}$ (H_{50}) show marked differences among the surfactant families, indicating that different processes are at work. The lipopeptides lichenysin and iturin A are the most hemolytic surfactants, with the lowest H_{50} values (5.6 and 18 μM, respectively), followed by THLs (28 μM). The least hemolytic surfactants, causing only partial hemolysis in blood agar, are the flavolipids produced by *S. detergens* [70–72, 74, 75]. Studies with different sized osmotic protectors such as polyethylene glycol indicate that THLs, lichenysin, and iturin A produce hemolysis by a 32–34 Å pore formation. This effect may be a consequence of surfactant molecule grouping and the formation of enhanced permeability domains surrounded by phospholipid molecules, which allows the influx of small solutes [65, 75]. Hemolysis induced by these surfactants occurs gradually, suggesting the development of small membrane lesions leading to a colloid-osmotic process [74]. The hemolysis mechanism of rhamnolipids is different, being characterized by a rapid membrane solubilisation not affected by osmotic protectors (>72 Å).

The hemolytic activity of most reported BS and the scarcity of clinical data on the use and validation of such molecules in animal models and human volunteers constitutes a major challenge for their application in preparing safe drugs or drug delivery formulations. Nevertheless, some BS have proven efficacy in cosmetic and pharmaceutical formulations and fulfill the requirements of drug regulatory bodies worldwide for biocompatible and nontoxic excipients [47].

In conclusion, BS have many remarkable advantages, including high biodegradability and biocompatibility, and low ecotoxicity and cmc values. They can be produced from renewable energy sources, including industrial waste and byproducts, in abidance with green chemistry principles. Moreover, they remain active under extreme conditions (pH, temperature, and salinity concentration) and are highly structurally diverse, which furnishes BS with different properties and a wide range of applications [6].

10.3.4 Importance of Biofilms and the Effect of Biosurfactants on their Development

Biofilm formation is an important stage in the establishment of a microorganism in the environment or in living tissues. Sogdagari et al. [76] divided biofilm formation in three steps. First, the bacteria adhere to a surface, a process that depends on the type of microorganism, the surface hydrophobicity and electrical charge, and the production and interaction of extracellular polymers (proteins, polysaccharides, and lipids) on the surface. The second stage is microfouling, which involves the colonization and spread of the bacteria and other small microorganisms on the surface, and the third is macrofouling, or the attachment and growth of larger microorganisms such as microalgae.

Surfactants participate in the development and maintenance of the organized biofilm structure. BS such as rhamnolipids promote the initial microcolony establishment [77] and participate in the formation of channels by an active intercellular communication mechanism during the latter stages of biofilm development [78]. The function of rhamnolipids determined in pure-culture biofilm is to preserve its shape and act as a defensive mechanism, preventing other bacteria from colonizing open spaces [79]. It is also thought that BS play an important role in maintaining channels for gas and nutrient diffusion in biofilm communities. They achieve this by partially disrupting the biofilm, a function they perform effectively at the appropriate concentration.

Biofilms pose a serious challenge for industry, as microbial growth in canalizations can impede fluid mobility or cause contamination during manufacture, severely impacting on the final product quality. Bacteria growing as a biofilm are also an important concern in the biomedical field, as adherence on abiotic material such as catheters and prostheses creates a bacterial reservoir. The formation of biofilm, a hydrated polymeric matrix, plays a key role in the survival of pathogenic and non-pathogenic microorganisms in hostile environments, allowing survival under stressful conditions such as nutrient limitation, anaerobiosis, and heat shock. Biofilm communities, which are usually composed of multiple species, show strong resistance to antimicrobial agents and avoid the host immune response, thus favoring the development of persistent and chronic infections [80]. Many nosocomial infections are biofilm-associated diseases, including infective endocarditis and especially cystic fibrosis pneumonia [6, 81].

The highly compact polysaccharide structure of biofilms makes it difficult for antibiotics to access and eliminate the bacteria present, which may already have resistance. Pathogen implantation in equipment or products is generally controlled by cleaning and disinfection procedures, but microorganisms may also have a certain degree of resistance to the chemical products used. There is therefore an urgent need for novel strategies that specifically target the biofilm mode of growth [65].

To prevent biofilm formation, it is of great interest to design an eco-friendly strategy. In this context, BS are promising candidates, due to their low toxicity and biodegradability as well as antiadherent activity. BS are able to modify bacterial surface hydrophobicity and, consequently, microbial adhesion to solid surfaces. On the other hand, BS adsorption on solid surfaces can inhibit or reduce bacterial adhesion by altering the surface properties. These traits have potential application as a controlling method in industrial installations or medical devices to reduce nosocomial infection development and antibiotic use [4, 47].

As an example of BS effectiveness in this field, when the Calgary Biofilm Device was precoated with fengycin solution (5120–80 μg/ml), a reduction of 97% of *Escherichia coli* and 90% of *S. aureus* biofilms was observed. The postulated mechanism of action is that by binding to the microbial cell surface or to its components, the BS affected the membrane hydrophobicity [82]. Studies carried out with hydrophilic glass and hydrophobic octadecyltrichlorosilane-modified (OTS-modified) glass found that 10–200 mg/l of rhamnolipids significantly reduced the initial attachments of *E. coli, Pseudomonas putida*, and *P. aeruginosa*. In the case of *Staphylococcus epidermidis*, rhamnolipids inhibited the attachment only on OTS-modified glass. Moreover, rinsing with a 200 mg/l rhamnolipid solution was ineffective in detaching bacterial cells previously attached. No conclusion was drawn about the responsible mechanisms [76].

Immunocompromised, transplanted patients and those with medical implants are highly susceptible to fungal infections. Coronel et al. [71] showed that a pre-treatment of a polystyrene surface with lichenysin efficiently and rapidly reduced growth of attached bacteria, especially Methicillin-Resistant *Staphylococcus aureus* or MRSA (68%) and *C. albicans* (74%). The detergent action on pre-formed biofilm was also studied, showing a medium cellular growth elimination of 55–37%, with the best results obtained with *S. aureus* (55%) and *Yersinia enterocolitica* (46%). Lichenysin was effective above its cmc (15 mg/l), suggesting that the surface became covered with micelles, resulting in an electrostatic repulsion between the negative charges of the bacterial surface and the pre-treated polystyrene surface. Sophorolipid was found to reduce *C. albicans* biofilm formation as well as the viability of pre-formed biofilms, acting synergistically when applied together with amphotericin B or fluconazole. Additionally, the morphology of *C. albicans* biofilm cells was altered by sophorolipids [83].

10.4 Antimicrobial Resistance

The development of antibiotics, vaccines, and food hygiene has dramatically improved the quality of life in developed countries and lengthened life expectancy. Since the discovery of sulfonamides in 1935, antibiotics have helped save millions of lives, but the rapid emergence of multiresistant bacteria, especially in the last decade, is now a source of worry for health workers [84]. In this scenario, infectious diseases continue to be one of the major causes of morbidity and mortality in many areas of the world [85, 86].

Antibiotic resistance is a natural phenomenon of bacterial evolution and adaptation, but after the golden age of antibiotic discovery it has grown to alarming levels [87]. The abuse or misuse of antibiotics in medicine, veterinary medicine, and agriculture has favored bacterial adaptation, resistance acquisition, and the consequent selection of resistant bacteria [84]. It is clear that resistances are appearing and spreading faster than our ability to contain them. With the development of modern medicine, the population's life expectancy has improved, but in turn, there are more patients at risk of acquiring infections. One of the recommendations of the Organisation for Economic Co-operation and Development (OECD) to reduce antibiotic usage is to prevent 30–40% of infections by improvements in public health. This implies prevention measures, knowledge of the microbiome, and the development and proper use of vaccines, together with more investment in research into new antimicrobials in order to have tools to treat infections due to resistant bacteria.

Despite the urgent need worldwide for new antibiotics or strategies to control resistant pathogenic bacteria, investment by the pharmaceutical industry in this area is in decline. The development of a new drug has become more time-consuming and expensive, due to stricter and more complex regulations, especially if the probabilities of achieving a successful product are low. Another dissuading factor for investment is that a new antibiotic may be restricted for hospital consumption only. Consequently, the distance between the clinical demand for new antibiotics and the speed of discovery and development of new antimicrobials is widening [86].

10.4.1 New Strategies to Fight Antimicrobial Resistance

The rise of antibiotic resistance is driving investigations of novel strategies against microbial infections [88]. Antibiotic resistance is inheritable among microorganisms and its acquisition allows survival even when high drug concentrations are used. Widely extended in hospitals, this trait is also spreading in the wider environment, resulting in a reservoir of resistance that can be extended horizontally by gene transfer [84]. Resistance mechanisms involve: (i) reduction of antibiotic uptake due to changes in the outer membrane permeability, (ii) antibiotic secretion by activation of efflux pumps, (iii) loss of affinity or bypass development, (iv) production of inactivating enzymes, and (v) biofilm formation [87, 89, 90].

Although both Gram-positive and Gram-negative bacteria with multiple antibiotic resistance have been described, two thirds of deaths from infectious diseases are the result of Gram-negative bacteria. The treatment of infections caused by antibiotic-resistant pathogens is a challenge, but especially so in the case of Gram-negative bacteria, whose outer membrane is an effective permeability barrier against many antibiotics that might otherwise be effective [85]. The WHO, based on various criteria, has created a global priority list of antibiotic-resistant pathogens. The situation is highly critical for infections associated with hospitalization, which are mainly caused by the Gram-negative ESKAPE (*Enterococcus faecium*, *Klebsiella pneumoniae*, *Acinetobacter baumannii*,

P. aeruginosa, *S. aureus*, *and Enterobacter* sp.). These bacteria are resistant to carbapenems, which are often considered the last option in reserve therapy [87, 91].

While antibiotic resistance is difficult and perhaps impossible to overcome, different strategies can be used to minimize its emergence: (i) more effective antimicrobials, based on those currently in use, (ii) new chemical compounds with novel mechanisms of action (a highly challenging option), and (iii) the development of antibiotic adjuvants (or enhancers) that can block the mechanism of antibiotic resistance or increase antimicrobial action [87, 90].

The strategy of using adjuvants in combination with an antibiotic can prolong the useful life of the available antibiotic arsenal, but it has certain drawbacks, including a risk of adverse effects due to possible pharmacological interactions and the complexity of establishing regimens of effective co-doses [87]. Antibiotic adjuvants or enhancers are compounds with little or no antimicrobial activity, but they act by improving antibiotic action or blocking resistance when co-administered [87, 90]. They may also suppress the emergence of resistance and play a role in the discovery of new antibiotics [90].

There are two main classes of antibiotic adjuvants:

- *Class I*: Agents acting on the pathogen that work with antibiotics against bacterial targets by inactivating enzymes, flow pumping systems or targeting bypass mechanisms. Currently in clinical use, three main types of adjuvants of this class have been developed [87, 90].
  a) *Beta-lactamase inhibitors*. The most successful adjuvants are β-lactamase inhibitors such as *clavulanic acid*, *sulbactam*, and *tazobactam*. They have little antibiotic activity but are potent inactivators of ser-β-lactamases and have been in clinical use for 70 years [90].
  b) *Efflux pump inhibitors*. Considerable efforts have been dedicated to the discovery of active efflux pump inhibitors as a strategy to restore the activity of existing antibiotics. However, studies have shown that this resistance mechanism is not specific, which makes the development and identification of efficient efflux pump inhibitors very difficult [87].
  c) *Membrane permeabilizers*. The use of permeabilizers has proven to be an effective method to improve the absorption of antibiotics. The outer membrane of Gram-negative bacteria, now extremely resistant to most antibiotics, is characterized by a highly charged bilayer, which restricts the entry of many compounds into the bacterial cell, the location of many antimicrobial targets. Thus, compounds that alter the negative charge of the outer membrane, such as polymyxins or chelating agents, can increase permeability. Usually cationic and amphiphilic compounds destabilize the outer layer of the membrane by interacting with polyanionic lipopolysaccharides or by capturing cations. Permeabilizing agents also alter the integrity of the cytoplasmic membrane and have intracellular targets that affect essential processes in the bacterial cell [85]. Polymyxins are used to treat Gram-negative bacterial infections and today represent the last line of defence against multidrug-resistant Gram-negative bacteria, but their toxicity, especially for the kidneys, is a drawback [87, 89, 90]. Attempts are currently being made to reduce the side effects by producing nanoparticulate polymyxins and developing new analogues with reduced toxicity [92].
- *Class II*: These agents, which are still in a preclinical phase, enhance antibiotic action by acting on host properties. Host defense mechanisms can be improved by immunomodulatory peptides, which boost the innate immune system [90].

To sum up, combining antibiotics with enhancing compounds constitutes a promising strategy to address the widespread emergence of antibiotic-resistant strains and to extend the life of antimicrobials in use [93].

10.4.2 Biosurfactants as Antimicrobial Agents

In this scenario of increasing drug resistance in pathogenic bacteria and the need for new lines of therapy, attention is turning to BS as alternative antimicrobial agents or as adjuvants for classical antibiotics due to their membrane-destabilizing properties.

Natural surfactants can play an important role in defense mechanisms. For example, pulmonary surfactant, a lipopeptide complex synthesized and secreted by lung epithelial cells, lowers the surface tension at the air/liquid interface of the lung and is fundamental in host defense [46]. Bacterial capacity to produce BS with antimicrobial properties is part of a survival strategy in a competitive environment [94]. Permeabilized cells can be a source of insoluble enzymes or make bacteria more sensitive to antibiotics. By inducing pore and ion channel formation in lipid bilayer membranes, BS can cause death or arrest growth of bacteria and fungi [44].

The antimicrobial activity of rhamnolipids has been extensively studied and some results are shown in Table 10.1. *Pseudomonas aeruginosa* AT10, isolated by Ábalos et al. [95] from oil-contaminated soil, produced a mixture of at least seven rhamnolipid homologues, which reduced surface tension to 26.8 mN/m and has a cmc of 120 μg/ml. The mixture showed excellent antifungal properties against *Aspergillus niger* and *Gliocadium virens* as well as antibacterial activity. It was more effective against Gram-positive (*S. epidermidis* and *Mycobacterium phlei*) than Gram-negative bacteria, *Serratia marcescens* being the most sensitive. RL_{47T2}, a new BS formed by 11 rhamnolipid homologs produced by *P. aeruginosa* 47T2, which was obtained using waste cooking oils as the sole carbon source, decreased surface tension to 32.8 mN/m and has a cmc of 108.8 μg/ml. This BS displayed remarkable antimicrobial properties against *K. pneumoniae*, *Enterobacter aerogenes*,and *S. marcescens* (Table 10.1), as well as some anti-phytopathogenic fungi activity. Biocompatibility studies demonstrated a non-irritant behavior (category IV of EPA classification) [50].

Pseudomonas aeruginosa LBI cultivated on soapstock as the sole carbon source produced rhamnolipids (six homologs) that reduced surface tension to 24 mN/m with a cmc of 120 μg/ml.

Table 10.1 The minimum inhibitory concentration (MIC) of rhamnolipids against different microorganisms.

Publications	Ábalos [95]	Haba [50]	Benincasa [96]	Lotfabad [97]	Bharali [94]	Ramos [69]
			MIC (μg/ml)			
Gram-negative bacteria						
Escherichia coli	32	64	250	>512		>250
Pseudomonas aeruginosa	>256	256	32	>512		>250
Klebsiella pneumoniae				>512	250	>250
Gram-positive bacteria						
Bacillus subtilis	64	16	8	128		>250
Staphylococcus epidermidis	8	32	250	128		>250
Staphylococcus aureus	128			128	250	>250
Fungi						
Aspergillus niger	16					
Candida albicans	>256					

Source: Data obtained from [50, 69, 94–97].

This product showed good antibacterial activity with a minimum inhibitory concentration (MIC) of 4 μg/ml for *Streptococcus faecalis* (reclassified as *Enterococcus faecalis*), 8 μg/ml for *B. subtilis*, *S. aureus* and *Proteus vulgaris* and 32 μg/ml for *P. aeruginosa* and it was also active against phytophatogenic fungal species (32 μg/ml) (Table 10.1) [96]. Analysis of a rhamnolipid produced by *P. aeruginosa* MP01 revealed up to 17 different congeners, principally di-Rha (77.2%), and with Rha-Rha-C_{10}-C_{10} as the major component; a MIC of 128 μg/ml was reported against clinical and ATCC strains of *Streptococcus pneumoniae*, *S. aureus*, and *B. subtillis* and > 512 μg/ml against Gram-negative strains of *E. coli*, *P. aeruginosa*, and others [97].

Rhamnolipids at 500 and 250 μg/ml were found to inhibit the growth of *S. aureus* and *K. pneumoniae*. Below the cmc, antibacterial activity against the Gram-negative *K. pneumoniae* was insignificant, but distinct activity was observed against the Gram-positive *S. aureus* (Table 10.1). Concentrations beyond the cmc were strongly bactericidal against both species. Such inhibitory activity might be due to an increase in cell permeability, which causes the efflux of intracellular protein, ions, and other intracellular components, leading to cell death [94]. After rhamnolipid treatment, *S. aureus* and *K. pneumoniae* showed a significant increase in the uptake of crystal violet compared to the control cells, demonstrating that BS directly enhanced cell envelope permeability, which was also manifested by a higher extracellular protein level [94]. Using scanning-electron microscopy, Sotirova et al. [98] observed that the permeabilizing effect of rhamnolipids on *B. subtilis* and *P. aeruginosa* resulted in extracellular protein but without cell disruption. However, Ramos da Silva et al. [69] describe a high level of bacterial resistance to a mixture of rhamnolipids and monorhamnolipids (>250 μg/ml), both in Gram-positive and Gram-negative bacteria (Table 10.1).

A crude glycolipid produced by *O. sagaranensis* AT18 showed antimicrobial properties, qualitatively measured, against *C. albicans*, Gram-positive bacteria such as *B. subtilis* and *S. aureus* and the resistant Gram-negative *P. aeruginosa*. The antimicrobial effect of BS was more marked in Gram-positive bacteria, which have a simpler cell wall structure. The presence of an outer membrane confers more resistance to Gram-negative bacteria, especially after changes in porin structures. The result obtained with *P. aeruginosa* is of particular interest, as it suggests a remarkable permeabilizing property [59, 63]. Although rhamnolipids inhibit the growth of many bacteria, yeast, and fungi, a high concentration is usually necessary. The exact mechanism of rhamnolipid antimicrobial action is still not clearly understood, but it is attributed to their permeabilizing effect and disruption of the bacterial cell plasma membrane. Reported MIC values vary considerably among studies, which can be attributed to the variable homolog composition of rhamnolipid mixtures, a consequence of using different strains, substrates, and production strategies. It may also be affected by differences in the methodology employed or assay pH, especially in the case of Gram-positive bacteria [69, 94, 99].

As a consequence of its amphiphilic nature, surfactin securely attaches itself to lipid layers and disrupts membrane integrity. At a low concentration, surfactin inserts in the outer leaflet of the membrane, chelating divalent cations and causing only limited perturbation. At a higher concentration, it induces pore formation, leading to complete solubilization of the lipid bilayer and mixed micelle formation [46, 55]. In contrast, the membrane permeabilization properties of iturins are based on the formation of ion-conducting pores and subsequent osmotic perturbation. Iturins have a strong *in vitro;* effect against a wide variety of yeast and fungi but only limited antibacterial and no antiviral activity [74].

The extensive antibacterial activities of BS open new prospects for their use in biomedical and pharmaceutical sciences against microorganisms responsible for infectious diseases [44]. BS are also reported to enhance the effect of other antimicrobial agents when co-administered. In studies with *P. aeruginosa, B. subtilis, Alcaligenes faecalis, and Rhizopus nigricans*, Sotirova et al. reported a synergistic effect between rhamnolipids and methyl and ethyl esters of thiosulfonic acid. The co-administration of rhamnolipids at concentrations below the cmc enhanced microbial cell permeability for propidium iodide and reduced the effective bactericidal and fungicidal dosage [88]. A mixture of tetracycline and sophorolipids was more active against *S. aureus* and *E. coli* than the antibiotic alone and the treated bacteria showed cell membrane damage and pore formation. This self-assembling BS penetrates the structurally similar cell membrane and thus facilitates the entry of drug molecules. The treatment reduced the likelihood of bacterial survival and probably also the development of resistance, as the bacteria are engaged in combat against two agents [100]. Despite the myriad of interesting antimicrobial properties of BS revealed by studies, some of which are outlined above, their application in the pharmaceutical, biomedical, and health industries remains quite limited.

10.5 Catanionic Vesicles

Vesicular nanosystems have attracted considerable attention due to their promising therapeutic applications. The most widely studied are liposomes, discovered by Bangham et al. in the 1960s [101]. Liposomes are spherical aggregates consisting of one or more lipid bilayers that entrap a fraction of the dispersing solvent, generally water. They can encapsulate drugs of variable hydrophobicity with a high loading capacity, thereby improving the drug therapeutic index and the circulation time *in vivo;* [102]. Furthermore, the liposomal carrier can be directed to a specific tissue by adding targeting agents [103, 104]. Also, cationic liposomes interact with the negatively charged DNA to form lipoplexes that can be used as vectors in gene therapy [105]. Nevertheless, liposomes also have some important drawbacks: (i) their raw materials, usually phospholipids, are costly; (ii) their production can require a high energy input and lengthy preparation methods with multiple steps (sonication, extrusion, homogenization) [106]; and (iii) they have poor colloidal stability and usually aggregate into large multilamellar vesicles [107].

A strategy to overcome these impediments is the formulation of catanionic vesicles [108], which spontaneously self-assemble without any energy requirements after the mixing of inexpensive cationic and anionic surfactant solutions [109]. The development of these catanionic aggregates is due to both a strong attractive electrostatic interaction between the polar heads of the monocatenary surfactants and repulsive hydrophobic interaction between the alkyl chains.

Notably, catanionic vesicles are thermodynamically stable for long periods of time. Compared to the corresponding single chain surfactants, they have enhanced interfacial and aggregation properties, cmc values of one or two orders of magnitude lower, and greater efficiency in lowering surface tension [110, 111].

The spontaneous formation of stable vesicles can be explained by using the critical packing parameter (CCP), that is, the ratio of the volume of the fully extended length of the hydrophobic tail (v) to the product of the area of hydrophilic headgroup (a) and the hydrophobic tail length (l) [112]. CCP values greater than 0.5 and less than 1 indicate the formation of vesicles:

$$\mathrm{CCP} = \mathrm{v/al}$$

The addition of a monomeric cationic surfactant to a solution of an anionic single-chain surfactant results in the formation of tight ion pairs than can be regarded as a pseudo double-tailed zwitterionic surfactant (Figure 10.3). In these zwitterionic surfactants the effective head group area is lower compared to the individual surfactants, while a relative increase in the area of the hydrophobic tail ensues. As a result, the P value of the pseudo double-chain surfactant increases, which favors structures with low curvature-like vesicles.

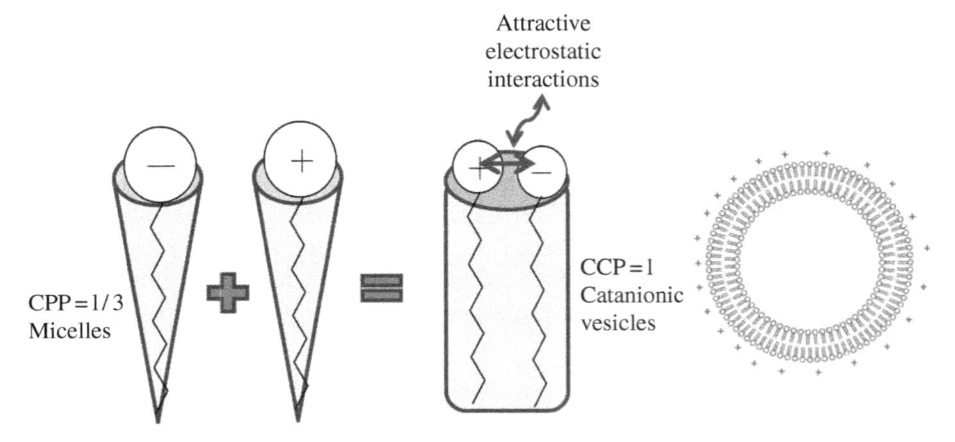

Figure 10.3 Formation of pseudo double-chain surfactants from monocatenary amphiphiles. *Source:* Lourdes Pérez.

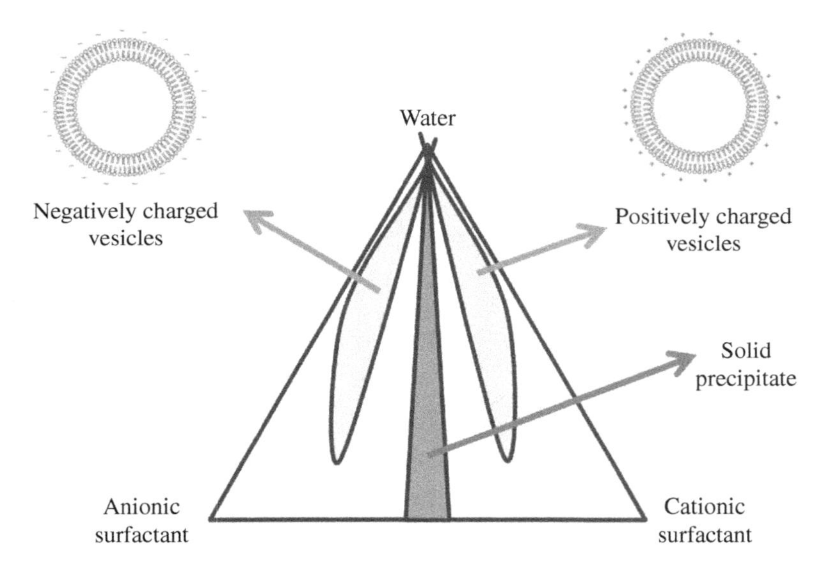

Figure 10.4 Typical phase diagrams of catanionic mixtures. *Source:* Lourdes Pérez.

The mixture of oppositely charged surfactants can give rise to a range of molecular aggregates, including micelles, worm-like micelles, vesicles, bilayers, and discs. Figure 10.4 shows a typical phase diagram for catanionic mixtures [113]. These formulations usually precipitate when the two oppositely charged surfactants are mixed in an equimolecular ratio. However, by changing the molar mixing ratio of the two amphiphiles or their total concentration, it is possible to obtain two regions corresponding to different catanionic vesicles. Interestingly, the size, surface density, flexibility, and permeability of these catanionic vesicles can be tailored by adjusting the temperature, concentration, and chain length of the components and the addition of salts or co-solvents [108].

Due to their physicochemical and biological properties, catanionic vesicles are of special interest for applications in nanotechnology and biomedicine. Numerous references describe the potential use of these formulations for the controlled and targeted delivery of drugs [114], in gene therapy [115], in the synthesis of novel materials [116], for *in vitro;* technology as diagnosis agents [117], and for the development of novel analytical methods [118].

10.5.1 Biocompatible Catanionic Mixtures

Until now, the cationic surfactants most frequently used to prepare catanionic mixtures have been alkyltrimethylammonium halides (CnTAX), dialkyldimethylammonium halides, benzalkonium halides, alkylpyridinium chlorides, and alkylimidazolinium surfactants [119]. Recently, due to their enhanced properties, catanionic mixtures containing gemini surfactants have also been investigated. Among the most studied systems are also those containing bis-quaternary ammonium salts with alkyl spacer chains. All these systems present several drawbacks: (i) cationic surfactants based on quaternary ammonium are prepared using non-renewable raw materials; (ii) these cationic surfactants are not biodegradable; and (iii) they usually show high toxicity against aquatic microorganisms and mammalian cells.

Current EU regulations and public demand for non-toxic and ecological products has generated considerable interest in biocompatible surfactants, and biosurfactants, amino acid–based surfactants, sugar-based surfactants, and peptides have been formulated to prepare eco-friendly catanionic systems.

10.5.2 Catanionic Mixtures from Amino Acid-Based Surfactants

Amino acid-based surfactants consisting of one or more amino acids as the hydrophilic moiety attached to one or two alkyl chains are attractive, as they are usually more biocompatible and environmentally friendly than conventional surfactants. Additionally, they can be prepared via green methodologies using renewable starting materials [120]. Among the amino acid-based surfactants, arginine derivatives stand out, because the presence of a guanidine group confers potent antimicrobial activity without reducing biocompatibility and biodegradability [121].

Monocatenary arginine derivatives LAM and ALA (Figure 10.5) were used to prepare biocompatible catanionic vesicles. These cationic surfactants were combined with two types of anionic

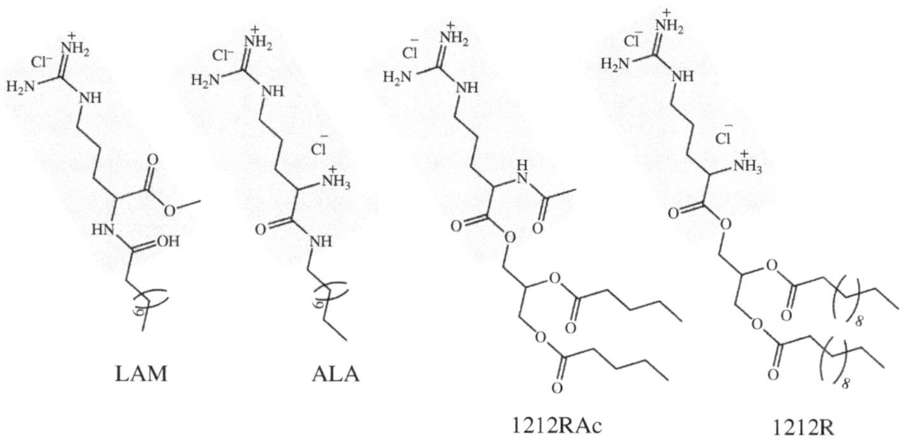

Figure 10.5 Single chain and double chain arginine-based surfactants used to prepare catanionic mixtures: lauroyl arginine methyl ester hydrochloride (LAM); arginine lauroyl amide dihydrochloride (ALA); 1,2-dilauroyl-glycero-3-*O*-L-arginine-hydrochloride (1212RAc) and 1,2-dilauroyl-glycero-3-*O*-(Nα-acetyl-L-arginine)-dihydrochloride (1212R). *Source:* Lourdes Pérez.

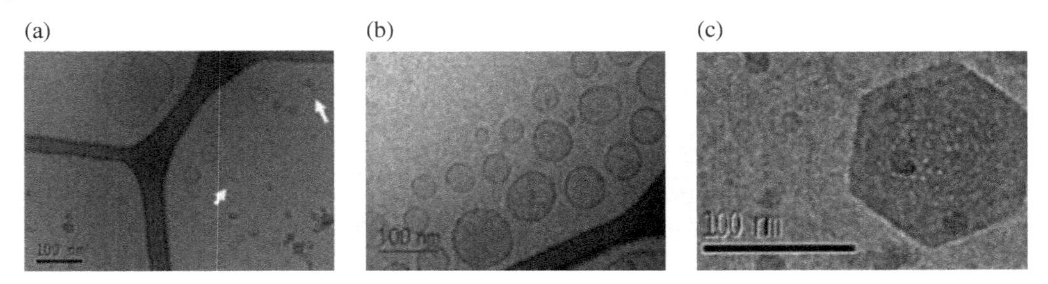

Figure 10.6 Cryo-TEM micrographs of samples prepared with ALA. (a) Coexistence of vesicles with rod-like micelles; (b) Polydisperse vesicles; (c) Cubosomes. *Source:* Adapted from [122] (Figures 6 and 10). Reproduced with permission of ACS publications.

surfactants, the conventional sodium cetyl sulphate and an amino acid–based surfactant prepared with glutamic acid. Both cationic surfactants spontaneously formed polydisperse catanionic vesicles when mixed with the tested anionic surfactants. All the formulations prepared with LAM contained vesicles coexisting with other kinds of structures, such as rod-like micelles or disks. Some of the samples prepared with ALA showed a one-phase region corresponding to polydisperse vesicles (Figure 10.6) in which the size distribution became narrower with time. Remarkably, in some samples prepared with ALA and the glutamate derivative, stable densely packed structures corresponding to hexasomes and cubosomes were observed by cryo-TEM microscopy (Figure 10.6). It is also notable that no stabilizer was required to form these particles [122].

Subsequently, DNA was added to the stable catanionic vesicles prepared with ALA at different charge ratios (R from 0.2 to 2.5) to study its compaction. As expected, the positively charged catanionic vesicles showed a strong association with the DNA. Phase separation was observed in all the samples, the precipitate phase increasing with the R until the isoelectric point, after which a partial redissolution of the precipitate occurred. The aggregation depended on the quantity of DNA added. Small quantities (R = 0.2) resulted in the visualization of mainly small globular DNA complexes with a lamellar structure. When the R was greater, the size of the aggregates increased considerably, with concomitant changes in morphology. After adding yet more DNA (R = 2.5), smaller complexes with a hexagonal internal structure predominated. These results show that it is possible to control the internal structure, morphology, and size of the aggregates by changing the ratio between the DNA and the catanionic vesicles [122–124].

Diacyl glycerol arginine-based surfactants (1212R, 1212RAc, 1414R, and 1414RAc, Figure 10.5) were used by Lozano et al. to formulate catanionic vesicles using phosphatidylglycerol as the anionic amphiphile. Pure surfactants showed antimicrobial activity against *Acinobacter baumannii* and *S. aureus*, although these cationic compounds were also hemolytic. Penetration studies using DPPC-spread monolayers as model mammalian membranes provided useful information on the degree of deformation undergone by these membranes when they interact with catanionic mixtures, and made it possible to tune the hemolytic/antimicrobial activity ratio of these systems. It was found that by formulating negatively charged catanionic vesicles, the hemolytic activity was significantly reduced, while maintaining or even increasing antimicrobial activity [125].

In a study on positively charged formulations prepared with the 1010R and 1414R derivatives, Tavano et al. found that these cationic surfactants form by themselves stable cationic vesicles with very good antimicrobial activity against representative Gram-positive and Gram-negative bacteria. The addition of a zwitterionic lipid such as DPPC diminished the mean diameter of the vesicles and increased their stability. Ciprofloxacin and 5-FU (Fluoracilo) were loaded on to the cationic

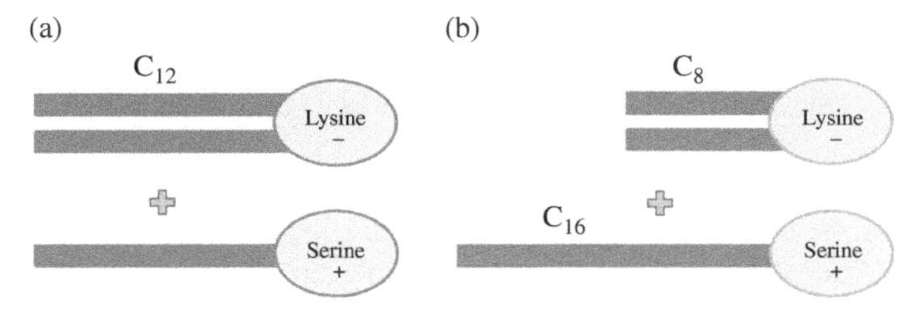

Figure 10.7 Schematic representation of the molecular structure of lysine- and serine-based surfactants used to prepare catanionic mixtures. *Source:* Lourdes Pérez.

systems and their *in vitro;* release from all formulations was effectively delayed with respect to the corresponding free drug solutions. Notably, the antimicrobial activity of the encapsulated ciprofloxacin was similar to or higher than that of the free drug solution [126].

Two catanionic mixtures of anionic dicatenary lysine-based surfactants and single chain cationic serine-based surfactants with different alkyl chains were prepared and studied (Figure 10.7).

In these mixtures the catanionic ion pair contained three alkyl chains, one of these systems involves molecules with the same hydrophobic moiety (C_{12} alkyl chains) while in the other there were pronounced differences in chain length (C_8/C_{16}). Catanionic vesicles were formed spontaneously in both mixtures. In the symmetric system, phase separation occurred between the micellar and vesicle solutions, whereas the asymmetric mixtures presented a continuous transition between the micellar and vesicular phase, as well as a greater area of vesicular stability. This behavior was ascribed to the greater flexibility associated with the mismatch in alkyl chains of amphiphiles [127].

Catanionic vesicles based on $C_{12}C_{12}$Lys/DTAB have also been prepared. The formation of fairly polydisperse vesicles, with an average size of 30–40 nm, was observed in the DTAB-rich region, co-existing with micelles [128]. The biocompatibility and environmental behavior of the catanionic systems derived from amino acid-based surfactants was evaluated by studying immobilization in the fish *Daphnia magna* and haemolytic analysis. The ecotoxicity of the catanionic mixtures against the *D. magna* is significatively lower than that corresponding to the pure cationic surfactants. Moreover, the catanionic system is 25 times less hemolytic than the corresponding pure surfactants [129]. Using a single-chain cationic surfactant and double-chain anionic surfactants based on the same amino acid serine, both negatively and positively charged catanionic vesicles with long-term stability were obtained. The positive vesicles had a mean size of 10 nm, whereas the negative ones were larger and more polydisperse. Vesicle size could be tuned by changing the composition, pH, and the preparation method [130]. Their biocompatibility was tested using the A549 cell line at a concentration range of 2–64 μM, and at least 80% of the cells were viable at 32 μM. The anticancer drug DOX was successfully encapsulated in the serine-based catanionic vesicles after being mixed with the two surfactants before vesicle formation. The DOX nanocarrier's enhanced cell uptake compared to that of the free drug, and resulted in high intracellular accumulation around the nucleus [131].

Catanionic systems sensitive to pH conditions can be useful for therapeutic applications. The pH-sensitive catanionic vesicles were prepared using amino acid-based surfactants with different alkyl chains (sodium *N*-alkanoyl-L-sarcosinate) and commercial cationic surfactants such as cetyltrimethyl ammonium hydroxide and decyltrimethyl ammonium hydroxide [132]. Compared to the individual surfactants, the surface tension and cmc were much lower in the mixed systems, which displayed long-term stability at room temperature. At pH values lower than 5, a vesicle-to-micelle transition occurred. Interestingly, these catanionic systems were neither hemolytic nor cytotoxic against three T3 cells.

The self-assembly of catanionic vesicles by six different amino acid-based cationic surfactants and two commercial ones (sodium dodecylbenzene sulfonate [SDBS] and sodium dodecylsulfate [SDS]) were investigated by Shome et al. Amino acids with aliphatic and aromatic side chains were used to establish the role of the aromatic moiety in the vesicle formation. The presence of an aromatic amino acid on the polar head favored the spontaneous formation of catanionic vesicles. The polydispersity index of the systems was below 0.3, indicating a narrow size distribution and the vesicles were stable at temperatures as high as 65 °C. Using the MTT-based cell viability assay in NH3T3 cells, the catanionic vesicles were found to have low cellular toxicity. Moreover, these colloidal systems can be applied for the entrapment of gold nanoparticles in vesicles [31].

The formation of catanionic mixtures in water using a cationic alanine-based surfactant and conventional anionic surfactants was investigated by Olutas using conductivity, density, and dynamic light scattering (DLS). As expected, the DLS measurements showed that these catanionic mixtures formed aggregates larger than those of pure surfactants, suggesting a transformation from small micelles to vesicles. However, a synergistic effect on the micellization was not observed and the cmc values lay within those of the pure surfactants [133]. A synergistic effect on foaming was apparent in catanionic mixtures prepared with an anionic amino acid-based surfactant, lauroyl glutamate (LGS). It was found that the distance between the LGS polar head groups at the air/water interface was greater compared to the LGS/DTAB mixtures, where a reduction in the electric repulsion between the LGS molecules resulted in a more compact arrangement. All these changes favor foam stability. It was also found that the LGS/DTAB combination could overcome the propensity of catanionic systems to precipitate, which was mainly attributed to the two negative charges in the chemical structures of LGS [134].

A synergistic formulation of a cationic surfactant from lysine, N^{ϵ}-myristoyl lysine methyl ester, and an anionic polysaccharide (hyaluronic acid) was developed by Bračič et al. Viscose fabric treated with this formulation exhibited excellent antimicrobial activity toward Gram-positive and Gram-negative bacteria and fungi, pointing to its application in the preparation of functionalized fibers for textile medicine [135].

10.5.3 Catanionic Mixtures from Gemini Surfactants

Given their chemical structure as well as their interesting properties [136], gemini surfactants provide more options than monocatenary surfactants for forming mixed surfactant systems. A catanionic mixture composed of an anionic gemini surfactant containing a triazine ring and cetyltrimethyl-ammonium bromide studied by Hu et al. had synergistic effects in surface-tension reduction and mixed micelle formation [137]. Catanionic mixtures prepared with gemini surfactants of the bisquat type and *N*-dodecanoyl glutamic acid were characterized by Ji et al. The surfactants were strongly bound by electrostatic interactions as each carried a double charge. The aggregates formed by mixing these compounds could be controlled by the temperature. For example, at higher temperatures, spherical micelles were transformed into large vesicles and large vesicles into solid spherical aggregates. All these transitions were thermally reversible. These results suggest that surfactants with multiple head groups facilitate the preparation of temperature-sensitive surfactant mixtures [138].

Catanionic mixtures were formulated by Barai et al. using amino acid-based anionic surfactants (AAS) with two negative charges in the polar head (disodium *N*-dodecylaminomalonate, disodium-*N*-dodecyl aspartate, and disodium *N*-dodecyl glutamate). Due to the presence of two

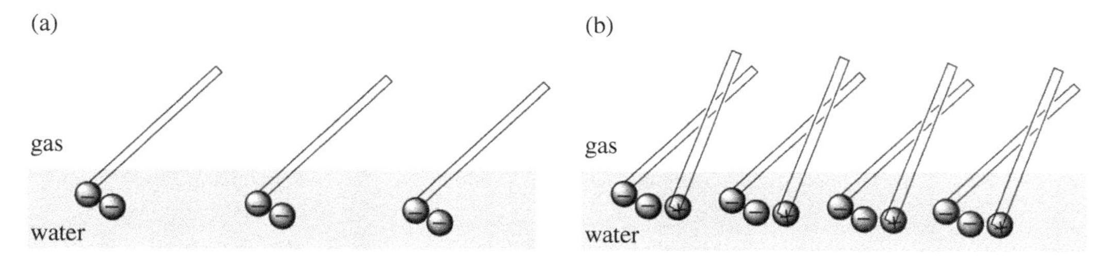

Figure 10.8 Schematic representation of the head groups' distance in an anionic surfactant with two cationic charges (a) and in catanionic mixtures (b). *Source:* Lourdes Pérez.

anionic charges in the AAS, the mixtures did not precipitate at stoichiometric ratios (Figure 10.8). AAS/HTAB systems were prepared, systematically varying the α_{AAS} as well as the AAS type, and different physicochemical parameters (cmc, surface pressure at the cmc, minimum area of the surfactant molecules at the air–water interface, fraction of counterion binding, hydrodynamic diameter, polydispersion index, and viscosity) were determined. Compared to the corresponding pure components, the mixtures had far lower cmc values and formed large aggregates that although less compact were rigid. Interestingly, the mixtures containing 20–40 mol% of AAS resulted in the formation of temperature-sensitive viscous or gel-like materials [139].

10.5.3.1 Antimicrobial Properties of Catanionic Mixtures from Gemini Amino Acid-Based Surfactants and Biosurfactants

In the search for a new generation of antimicrobials that are biodegradable, obtained by green chemistry, and slow down the appearance of resistant strains, compounds that act by destabilizing membranes are of particular interest. Polymyxins, classic antibiotics discovered in 1940, are active against *P. aeruginosa* and other Gram-negative bacteria. The changes in the membrane lipid composition induced by polymyxins results in an osmotic imbalance and subsequent bacterial stasis and cell death. Due to their nephro- and neurotoxicity, polymyxins had limited use, but there have been several attempts to improve their pharmacological properties through synthetic strategies. Sp-85 is a synthetic cyclic lipopeptide based on the structure of natural polymyxin. This analog, which disrupts the cytoplasmatic membrane by altering its permeability, has shown remarkable antimicrobial action against *S. aureus* (4 μg/ml), *Enterococcus faecalis* (8 μg/ml), *E. coli* (8 μg/ml), and *P. aeruginosa* (16 μg/ml) [140].

A new line in antimicrobial research is the development of catanionic mixtures or the functionalization of surfactants such as rhamnolipids or sophorolipids, with membrane destabilization as the mechanism of action. In comparison with the individual components, catanionic mixtures often show improved properties due to synergistic effects [125]. Four catanionic mixtures were prepared, combining four cationic amino acid-derived surfactants ($C_3(CA)_2$, CAM, $C_3(LA)_2$ and $C_6(LL)_2$) with the anionic biosurfactant lichenysin in a ratio of 8 : 2 (Table 10.2). The antimicrobial properties of the mixtures and individual surfactants were studied against Gram-positive and Gram-negative bacteria and a yeast, *C. albicans*.

$C_3(LA)_2$, a geminal arginine surfactant, showed efficient antimicrobial activity, especially against Gram-positive bacteria, and when combined with lichenysin, a synergistic antimicrobial effect was observed against the Gram-negative bacteria studied, as well as *Listeria monocytogenes* and *C. albicans*. Based on these results, *L. monocytogenes* and *E. coli* O157: H7 were selected for further studies [141]. $C_3(CA)_2$, another geminal arginine surfactant, was highly effective against all the microorganisms studied, especially Gram-positive bacteria, and synergistic effects with lichenysin

Table 10.2 Surfactants used in catanionic mixtures.

Acronym	Surfactant	Characteristics
$C_3(CA)_2$	Arginine geminal surfactant (2+) Nα,Nω-bis(Nαcaproilarginina)α,ω-propildiamina	MW: 766.5 g/mol Surface tension: 32 cmc: 4.3 mM
CAM	Arginine simple strain chain surfactant (1+) Nα-caproil-arginina metil ester	MW: 378 g/mol Surface tension: 40 mN/m cmc: 16 mM
$C_3(LA)_2$	Arginine geminal surfactant, (2+) Nα,Nω-bis(Nα-lauroil-arginina)α,ω dipropilamina	MW: 822.5 g/mol Surface tension: 35 mN/m cmc: 5×10^{-3} mM
$C_6(LL)_2$	Lisine geminal surfactant, (2+) Nα,Nω-bis(Nα-lauroil-lisina) α,ω-hexilendiamida	PM: 808.6 g/mol T. sup: 30 mN/m cmc: $5\text{x}10^{-3}$ mM
Lich	Cyclic peptide with 7 amino acids as the polar head (Gln-Leu-Leu-Val-Asp(-)-Leu-Ile) and β-hydroxy fatty acid with 14–16 carbons as the apolar part, (1-)	PM: 1021.6 g/mol T. sup: 28.5 mN/m cmc: 14.4 μM

Source: Data obtained from [141, 142].

were apparent against four microorganisms, *E. coli* O157: H7, *Y. enterocolitica*, *B. subtilis*, and *C. albicans*. When the mechanism of action was studied [142], not all the geminal surfactants were found to act as antimicrobials. $C_6(LL)_2$, derived from lysine, and CAM, derived from arginine, were poorly antimicrobial when tested alone or in a mixture, as a consequence of the resulting charge or the non-formation of a catanionic surfactant.

The antimicrobial effect of catanionic mixtures on the cytoplasmic membranes was studied by flow cytometry and staining with BOX (stains cells with a depolarized membrane, a reversible process) and IP (stains cells with a permeabilized membrane due to pore formation indicating cell death), and comparing viable cells obtained by plate count.

After *E. coli* O157:H7 and *L. monocytogenes* were exposed to $C_3(LA)_2$, 39 and 41% of the initial cells were recovered by culture, respectively. Flow cytometry results indicated a growth inhibition of 52 and 40% of cells (population between the indicator bars), but their cytoplasmic membrane was not altered, suggesting the surfactant antimicrobial activity may have had an alternative target. In fact, Sierra et al. [89] suggest that in addition to membrane permeabilization, antimicrobial peptides act intracellularly by altering the synthesis of different macromolecules such as DNA, RNA, and proteins (Figure 10.9). The catanionic mixture also showed a membrane-associated antimicrobial action against both microorganisms [141].

The antimicrobial mechanism of action of the geminal surfactant $C_3(CA)_2$, alone and in a mixture with lichenysin (8 : 2), was also studied (Figure 10.10). When *Y. enterocolitica* and *B. subtilis* were treated, the populations could not be recovered by culturing but did not show any membrane alterations (population between the indicator bars), once again suggesting an alternative target to the membrane. With *E. coli* O157: H7, the contact time with antimicrobials was insufficient to inhibit population growth and the membranes were scarcely affected. On the contrary, *C. albicans* quickly showed membrane permeabilization as a result of the surfactant antimicrobial activity [142].

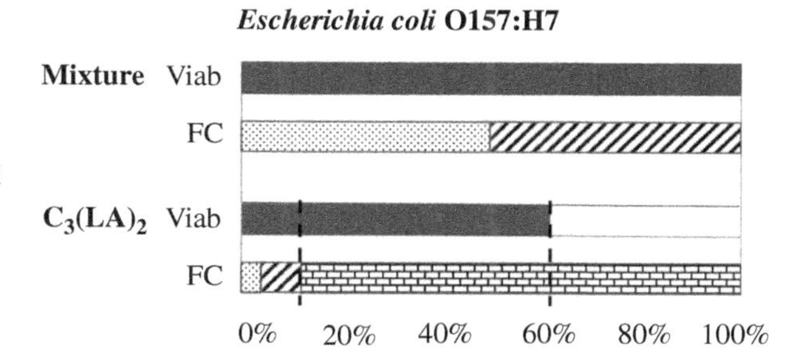

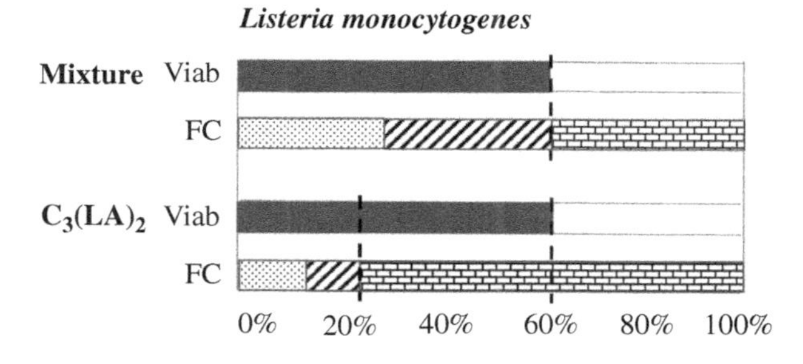

Figure 10.9 Reduction viability (Viab) and flow cytometry (FC) results of treatment with mixture $C_3(LA)_2$: lichenysin, 8 : 2 (mol:mol) and $C_3(LA)_2$ at the corresponding MIC. Viability assay: non-cultivable (black) and cultivable (white). Flow cytometry: BOX and PI stained (small points), BOX stained (lines), and non-stained (squares). *Source:* Ana M. Marqués. Data obtained from [141].

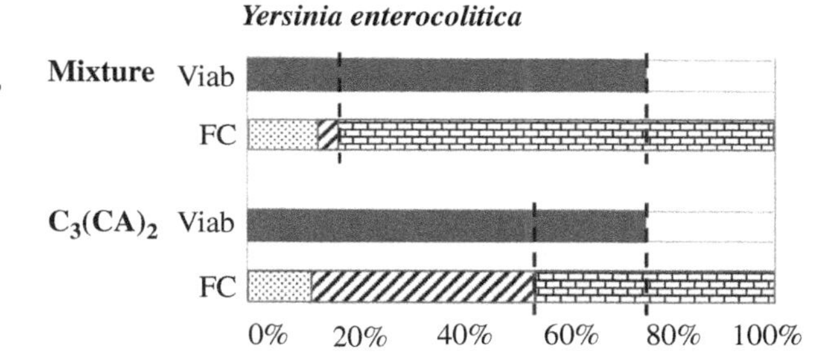

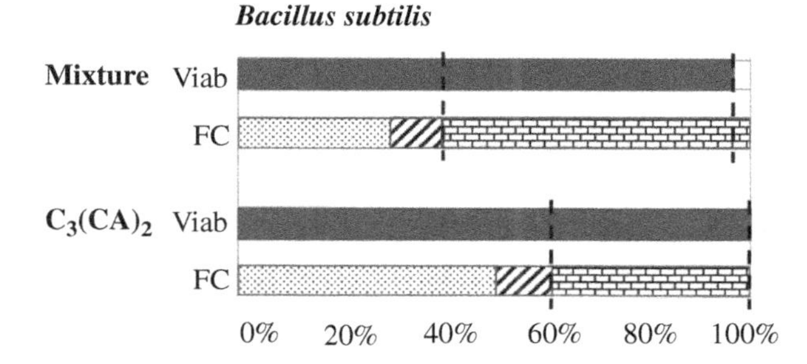

Figure 10.10 Reduction viability (Viab) and flow cytometry (FC) results of treatment with mixture $C_3(CA)_2$: lichenysin, 8 : 2 (mol:mol) and $C_3(CA)_2$ at the corresponding MIC. Viability assay: non-cultivable (black) and cultivable (white). Flow cytometry: BOX and PI stained (small points), BOX stained (lines), and non-stained (squares). *Source:* Adapted from [142] (Figure 4). Reproduced with permission of ACS Publications.

Cationic antimicrobial agents can damage cell membranes by promoting the formation of membrane domains [143]. After applying $C_3(LA)_2$ and the catanionic mixture, the effects on the microorganisms were studied by transmission electron microscopy (TEM). In *C. albicans* and the studied Gram-positive bacteria, no alteration in cell envelopes was visible, perhaps because the membrane permeability was affected without changes in the peptidoglycan structural function. In *E. coli* O157:H7 and $C_3(LA)_2$, the organization of the cytoplasm was clearly disturbed by $C_3(LA)_2$ and breaks in the envelopes were observed after treatment with the catanionic mixture. A disturbed

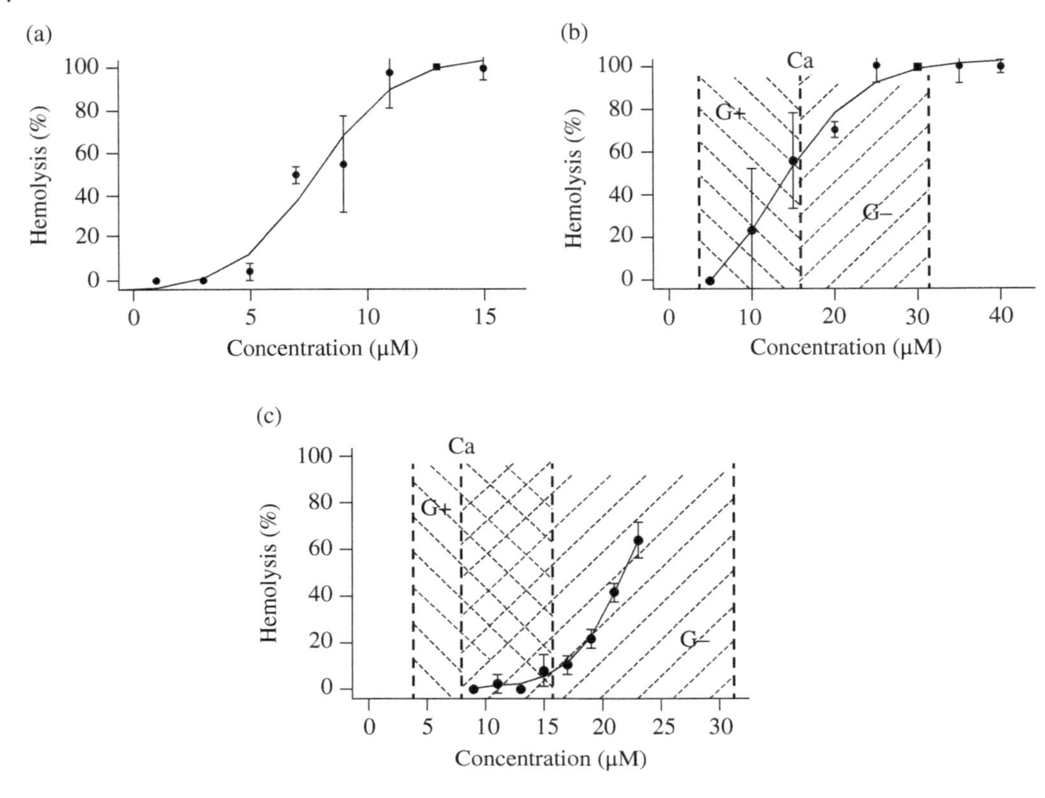

Figure 10.11 Relationship between surfactant hemolysis obtained with Lichenysin (a), $C_3(CA)_2$ (b), and the catanionic mixture (c) and MIC. Hemolysis is shown in curves in black. MIC ranges are shown in the gray shaded area. (G+) Gram-positive; (Ca) *C. albicans*; (G-) Gram-negative. *Source:* [142] (Figure 6). Reproduced with permission of ACS Publications.

cytoplasm was also apparent in *Y. enterocolitica* after the two treatments, together with abundant cell debris [141, 142].

The limitations of membrane-active antimicrobials include their toxicity and lack of biocompatibility. In the few studies on the biocompatibility or catanionic systems, it is usually reported to decrease with the cationic character. Using erythrocytes as a plasma membrane model, hemolysis was measured after exposure to the geminal surfactant $C_3(CA)_2$ and the catanionic mixture, and the results were compared with the MIC values (Figure 10.11). Among the compounds tested, lichenysin was the most hemolytic (H_{50}: 8 μM), but shows no antimicrobial activity (Figure 10.11a). Regarding $C_3(CA)_2$, the H_{50} value was higher (13.3 μM) and the MIC values were lower than the H_{50} for Gram-positive bacteria and *C. albicans* (Figure 10.11b). The least hemolytic compound was the catanionic mixture, with an H_{50} of 22.1 μM, and the MIC values for the Gram-positive bacteria, *C. albicans*, and many of the Gram-negative bacteria were lower H_{50}, indicating a low hemolytic capacity at the concentration with antimicrobial activity (Figure 10.11c).

Additionally, the catanionic mixture had a better therapeutic index, that is, the ratio between the doses producing a toxic effect (H_{50}) and a therapeutic effect (MIC). To overcome biocompatibility issues, Rosa et al. [122] recommended using low toxicity surfactants, such as the positively charged amino acid-based compound ALA (arginine-*N*-lauroyl amide hydrochloride) or LAM

($N\alpha$-lauroyl-arginine-methyl ester hydrochloride), combined with negatively charged ones. Individually, these surfactants are well tolerated in cell viability tests.

As reflected by the studies described here, the focus of research interest is moving away from the potential antimicrobial effects of BS to how they act in mixtures with currently available antibiotics with the aim of maintaining or improving antibiotic efficacy.

10.5.4 Catanionic Mixtures from Sugar-Based Surfactants

Renewed interest has been focused on the development of environmentally friendly, salt-free catanionic systems to avoid an excessive production of small counterions. In this context, different sugar-derived catanionic surfactants have been reported, produced by simple salt-free acid-base reactions using cheap bio-based starting materials. Additionally, a disaccharidic-derived polar head group gives the structure both biocompatibility and a certain hydrosolubility, which allows stable vesicles to be prepared at equimolecular ratios while remaining biocompatible.

Two-chain and gemini catanionic analogs of galactopiranose catanionic surfactants were prepared via an acid–base reaction in water by the addition of a carboxylic or dicarboxylic acid to the sugar derivative (Figure 10.12). Compounds with short alkyl chains were not toxic but without anti-HIV activity, whereas surfactants with large hydrophobic groups were far more active, but had to be discarded because of their high toxicity. However, the most hydrophobic gemini compound exhibited both low toxicity and strong anti-HIV activity, with a selective index of more than 200. Dynamic light scattering studies showed that all of these catanionic surfactants resulted in the

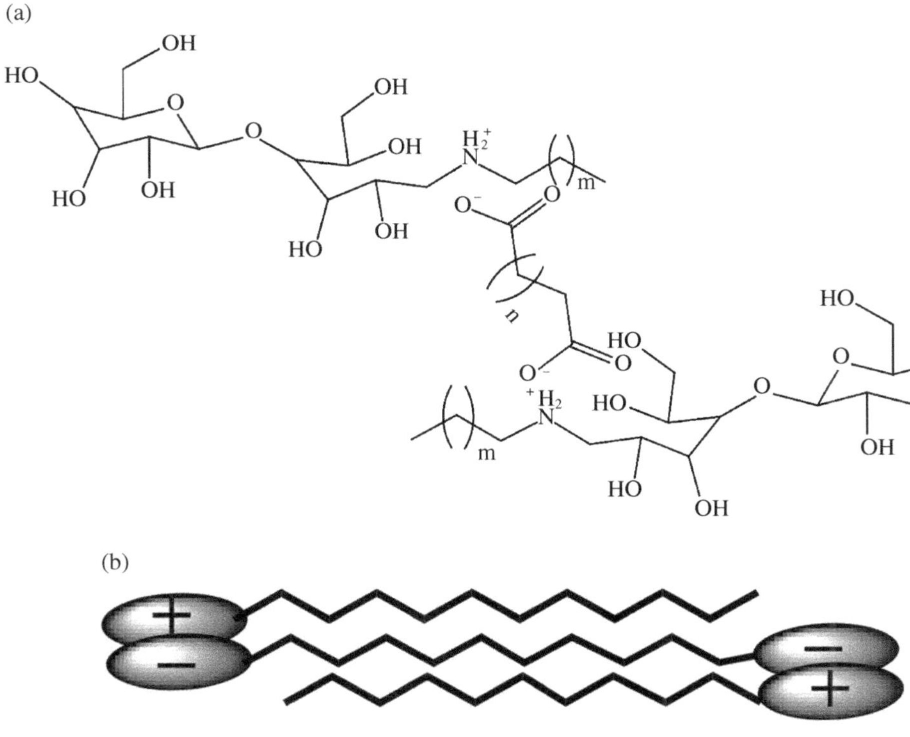

(a)

(b)

Figure 10.12 Chemical structure and schematic representation of catanionic gemini surfactants analogs to galactopiranose. *Source:* Lourdes Pérez.

spontaneous formation of vesicles of a broad size at concentrations of 1–4 mM and lamellar phases at high concentrations [144, 145].

The vesicles formed by tricatenary sugar-based surfactants developed by Soussan et al. efficiently encapsulated arbutin, a hydrophilic fluorescent dye, and displayed a high capacity to retain drugs in their aqueous cavity [146]. Richard et al. used these trimeric sugar surfactants to prepare new potential carriers for dermal drug delivery. They found that catanionic vesicles were also formed in the presence of ethanol and glycerol, compounds usually employed to enhance the dermal penetration of active substances. Remarkably, glycerol stabilized the vesicles, probably by developing a protective layer around them. Finally, a water-soluble hydroxyethyl cellulose polymer was added to the vesicular dispersion to give the formulations a texture that facilitates their application [147].

10.6 Biosurfactant Functionalization: A Strategy to Develop Active Antimicrobial Compounds

Inefficient production of pure BS limits their exploration and commercial application. Although chemical synthetic pathways have been developed, they are complex and costly. In an alternative approach, different authors have proposed the functionalization of BS to design new structures that retain the characteristic surfactant properties. Such modifications could include the insertion or exchange of functional chemical groups to shift the ratio between hydrophilic and hydrophobic molecule domains, which might create unexpected properties [148].

In the context of a growing market for bio-based products, sophorolipids produced by S. *bombicola* are being explored as possible building blocks for chemical derivatization. Azim et al. [149] prepared functionalized sophorolipids by linking the α-NH_2 group of amino acids to the carboxyl terminus of the fatty acid. The highest antimicrobial activity against Gram-positive and Gram-negative bacteria was found in the leucine-conjugated sophorolipid (MIC of 5 mg/ml or less), although it needs improvement before being considered a candidate for antimicrobial therapy. Four quaternary ammonium salt sophorolipid derivatives displayed strong antimicrobial activity (MIC 10–5 μg/ml) against Gram-positive bacteria such as *S. aureus*, *E. faecalis*, *B. subtilis*, and *S. pneumoniae*, but not against the more recalcitrant Gram-negative bacteria. Microscopic analysis of bacterial cells revealed cellular lysis after surfactant contact. These quaternary ammonium salts represent an interesting new class of surfactants, whose potential should be confirmed by toxicity and action spectrum studies [150].

The functionalization of rhamnolipids with arginine or lysine produces cationic surfactants with new antimicrobial properties. In the polymyxin model, the destabilization and permeabilization of the bacterial envelope is favoured by the interaction between the polyanionic cell surface and an oppositely charged antimicrobial. The rhamnolipid derivatives obtained with arginine showed improved antimicrobial properties against Gram-positive bacteria such as *B. subtilis* (16 μg/ml), *S. epidermidis* (4 μg/ml), *S. aureus* (8 μg/ml), MRSA (32–16 μg/ml), and *L. monocytogenes* (16 μg/ml) compared with the original surfactant (>250 μg/ml). In contrast, lysine-derived rhamnolipids did not exhibit antimicrobial activity, which was attributed to the lower pK_a of lysine (7.5) compared to arginine (9.5) molecules, the latter resulting in 100% protonation. The arginine-based surfactants showed an enhanced tendency to disturb bacterial membranes [69].

Jovanovic et al. [151] replaced the sugar component of rhamnolipids with amino acids and peptides because of their reported antimicrobial properties. Mono-amino acid derivative 15 (MIC = 50 mg/ml), with a C-8 carboxylic side chain, was the most promising candidate in reducing *C. albicans* biofilm formation, achieving a 96.5% reduction. The most active antimicrobial compounds were benzyl esters, with a noticeable predominance of non-polar amino acid residues such as leucine. As the replacement of rhamnose in the ramnolipids causes a loss of surface activity, the antimicrobial action must be associated with a different mechanism, most likely related to an interaction with specialized proteins such as adhesins.

10.7 Conclusions

Surfactants with eco-friendly and sustainable characteristics are in demand due to a heightened public awareness of the need to reduce environmental pollution. Amino acid-based surfactants and biosurfactants, with their wide variety of structures and physicochemical properties, have an enormous potential in industrial applications.

Bacterial infections, especially those caused by Gram-negative bacteria, are a growing health challenge. The outer membrane of Gram-negative bacteria is an efficient permeability barrier due to the acquisition of multidrug efflux pumps and/or porin loss. The outer leaflet of the outer membrane, formed by lipopolysaccharides, is a highly polar barrier stabilized by divalent cations. Cationic compounds, such as polymyxins and cationic peptides, can act as antibiotic adjuvants by disrupting this barrier and forming channels. An adjuvant may or may not have antimicrobial activity, but its properties can recover the use of classical antibiotics by increasing permeability, providing access to the action site [152].

By altering surfaces, surfactants can also prevent bacterial adhesion. The formation of biofilms is an essential stage in the colonization of a surface by microorganisms, providing protection from environmental and antimicrobial factors, and biosurfactants participate in their development and maintenance.

Geminal surfactants derived from arginine interact with lichenysin to form a positively charged pseudo-surfactant or catanionic system with improved antimicrobial activity. The positive charge of the catanionic mixture favours interaction with the bacterial envelopes, altering the permeability of the cytoplasmic membrane and possibly acting on other intracellular targets. Notably, the surface activity and vesicle formation capacity of co-administered biosurfactants can facilitate antimicrobial penetration through biofilms.

The development of biocompatible and biodegradable catanionic mixtures or tailor-made surfactants with new or improved properties is of interest because their enhanced activity at a lower concentration favors a more controlled usage. Catanionic surfactants are membrane-active agents and represent a promising alternative to current antimicrobials, partly due to their own antimicrobial activity as membrane permeabilizers but even more so as adjuvants to classical antibiotics.

Before catanionic mixtures can be applied in biomedical and health-related areas, more research is required on their antimicrobial activity, mode of action, and toxicity for human cells and natural microbiota. However, it is only a matter of time before the true potential of catanionic mixtures is fully exploited and used in medical science.

References

1 Landrigan, P.J. and Fuller, R. (2015). Global health and environmental pollution. *International Journal of Public Health* 60: 761–762.

2 Xu, X., Nie, S., Ding, H. et al. (2018). Environmental pollution and kidney diseases. *Nature Reviews Nephrology* 14: 313–324.

3 Dvorak, P., Nikel, P.I., Damborsky, J. et al. (2017). Bioremediation 3.0: Engineering pollutant-removing bacteria in the times of systemic biology. *Biotechnology Advances* 35: 845–866.

4 Santos, D.K., Rufino, R.D., Luna, J.M. et al. (2016). Biosurfactants: Multifunctional biomolecules of the 21st century. *International Journal of Molecular Sciences* 17: 1–31.
5 Sarma, H., Bustamante, K.L.T., and Prasad, M.N.V. (2018). Biosurfactants for oil recovery from refinery sludge: Magnetic nanoparticles assisted purification. In: Industrial and Municipal Sludge (ed. M.N.V. Prasad), 107–132. Massachusetts: Elsevier. ISBN: 9780128159071, Editor Majeti Narasimha Vara Prasad, Paulo Jorge de Campos, Favas Meththika, Vithanage S. Venkata Mohan.
6 Naughton, P.J., Marchant, R., Naughton, V. et al. (2019). Microbial biosurfactants: current trends and applications in agricultural and biomedical industries. *Journal of Applied Microbiology* 127: 12–28.
7 Kumagai, H. (2013). Amino acid production. In: The Prokaryotes (eds. E. Rosenberg, E.F. DeLong, S. Lory, et al.), 169–177. Berlin, Heidelberg: Springer-Verlag.
8 Williams, M.A. (1997). Extraction of lipids from natural sources. In: Lipid Technologies and Applications (eds. F.D. Gunstone and F.B. Padley), 113–135. New York;: Marcel Dekker.
9 Ananthapadmanabhan, K.P. (2019). Amino acid surfactants in personal cleansing (review). *Tenside, Surfactants, Detergents* 56 (5): 378–386.
10 Bernal, C., Guzman, F., Illanes, A. et al. (2018). Selective and eco-friendly synthesis of lipoaminoacid-based surfactants for food, using immobilized lipase and protease biocatalysts. *Food Chemistry* 239: 189–195.
11 Burnett, C.L., Heldreth, B., Bergfeld, W.F. et al. (2017). Safety assessment of amino acid alkyl amides as used in cosmetics. *International Journal of Toxicology* 36: 17S–56S.
12 Calejo, M.T., KjØniksen, A.L., Pinazo, A. et al. (2012). Thermoresponsive hydrogels with low toxicity from mixtures of ethyl(hydroxyethyl) cellulose and arginine-based surfactants. *International Journal of Pharmaceutics* 436: 454–462.
13 Muriel-Galet, V., López-Carballo, G., Hernández-Muñoz, P. et al. (2014). Characterization of ethylene-vinyl alcohol copolymer containing lauril arginate (LAE) as material for active antimicrobial food packaging. *Food Packaging and Shelf Life* 1: 10–18.
14 Xia, J. and Nnanna, I.A. (2001). Protein Based-Surfactants: Chemistry, Synthesis and Properties, Surfactant Science Series Vol. 101. New York;: Marcel Dekker.
15 Ohta, A., Toda, K., Morimoto, K.Y. et al. (2008). Effect of the side chain of N-acyl amino acid surfactants on micelle formation: an isothermal titration calorimetry study. *Colloids and Surfaces A: Physicochemical and Engineering Aspects* 317: 316–322.
16 Pinheiro, L. and Faustino, C. (2017). Amino Acid-Based Surfactants for Biomedical Applications, *Application and Characterization of Surfactants* (ed. R. Najjar). IntechOpen https://doi.org/10.5772/67977.
17 Pinazo, A., Pons, R., Pérez, L. et al. (2011). Amino acids as raw material for biocompatible surfactants. *Industrial and Engineering Chemistry Research* 50: 4805–4817.
18 Pinazo, A., Pérez, L., Morán, M.C. et al. (2019a). Arginine-based surfactants: Synthesis, aggregation properties, and applications. In: Biobased Surfactants, 2e (eds. D.G. Hayes, D.H.Y. Solaiman and R.D. Ashby), 413–445. AOCS Press https://doi.org/10.1016/C2016-0-03179-0.
19 Infante, M.R., Pinazo, A., and Seguer, J. (1997). Non-conventional surfactants from amino acids and glycolipids: Structure, preparation and properties. *Colloids and Surfaces A: Physicochemical and Engineering Aspects* 123-124: 49–70.
20 Infante, M.R., Pérez, L., Morán, M.C. et al. (2010). Biocompatible surfactants from renewable hydrophiles. *European Journal of Lipid Science and Technology* 112: 110–121.
21 Joondan, N., Laulloo, S.J., and Caumul, P. (2018). Amino acids: Building blocks for the synthesis of greener amphiphiles. *Journal of Dispersion Science and Technology* 39 (11): 1550–1564.

22 Pinazo, A., Infante, M.R., Izquierdo, P. et al. (2000). Synthesis of arginine-based surfactants in highly concentrated water-in-oil emulsions. *Journal of the Chemical Society, Perkin Transactions* 2: 1535–1539.

23 Pinazo, A., Manresa, M.A., Marqués, A.M. et al. (2016a). Amino acid-based surfactants: New antimicrobial agents. *Advances in Colloid and Interface Science* 228: 17–39.

24 Gryc, W., Dabrowska, B., Tomicka, B. et al. (1979). Antibacterial peptide derivatives. Part V. Hydrochlorides of 2-aminoethyl esters of N^{α}–palmitoil-L-lysine and its peptides. *Polish Journal of Chemistry* 53: 1085–1093.

25 Nakamiya, T., Mizumo, H., Meguro, T. et al. (1976). Antibacterial activity of lauryl ester of DL-lysine. *Journal of Fermentation Technology* 54: 369–373.

26 Colomer, A., Pinazo, A., Manresa, M.A. et al. (2011). Cationic surfactants derived from lysine: Effects of their structure and charge type on antimicrobial and hemolytic activities. *Journal of Medicinal Chemistry* 54: 989–1002.

27 Mezei, A., Pérez, L., Pinazo, A. et al. (2012). Self assembly of pH-sensitive cationic lysine based surfactants. *Langmuir* 28: 16761–16771.

28 Bustelo, M., Pinazo, A., Manresa, M.A. et al. (2017). Monocatenary histidine-based surfactants: Role of the alkyl chain length in antimicrobial activity and their selectivity over red blood cells. *Colloids and Surfaces A: Physicochemical and Engineering Aspects* 532: 501–509.

29 Dutta, S., Shome, A., Kar, T. et al. (2011). Counterion-induced modulation in the antimicrobial activity and biocompatibility of amphiphilic hydrogelators: Influence of in-situ-synthesized Ag-nanoparticle on the bactericidal property. *Langmuir* 27 (8): 5000–5008.

30 Roy, S. and Das, P.K. (2008). Antibacterial hydrogels of amino acid-based cationic amphiphiles. *Biotechnology and Bioengineering* 100: 756–764.

31 Shome, A., Kar, T., and Das, P.K. (2011). Spontaneous formation of biocompatible vesicles in aqueous mixtures of amino acid-based cationic surfactants and SDS/SDB. *ChemPhysChem* 12 (2): 369–378.

32 Makovitzki, A., Baram, J., and Shai, Y. (2008). Antimicrobial lipopolypeptides composed of palmitoyl di- and tricationic peptides: *in vitro;* and *in vivo;* activities, self-assembly to nanostructures, and a plausible mode of action. *Biochemistry* 47: 10630–10636.

33 Pinazo, A., Petrizelli, V., Bustelo, M. et al. (2016b). New cationic vesicles prepared with double chain surfactants from arginine: Role of the hydrophobic group on the antimicrobial activity and cytotoxicity. *Colloids and Surfaces B: Biointerfaces* 41: 19–27.

34 Pérez, L., Torres, J.L., Manresa, M.A. et al. (1996). Synthesis, aggregation and biological properties of a new class of gemini cationic amphiphilic compounds from arginine: bis (Args). *Langmuir* 12: 5296–5301.

35 Piera, E., Infante, M.R., and Clapés, P. (2000). Chemo-enzymatic synthesis of arginine-based gemini surfactants. *Biotechnology and Bioengineering* 70: 323–331.

36 Tavano, L., Infante, M.R., Abo Riya, M. et al. (2013). Role of aggregate size in the hemolytic and antimicrobial activity of colloidal solutions based on a single and gemini surfactants from arginine. *Soft Matter* 9: 306–319.

37 Pinazo, A., Diz, M., Solans, C. et al. (1993). Synthesis and properties of cationic surfactants containing a disulfide bond. *Journal of the American Oil Chemists' Society* 70: 37–42.

38 Infante, M.R., Diz, M., Manresa, M.A. et al. (1996). Microbial resistance of wool fabric treated with bis-Quats compounds. *Journal of Applied Bacteriology* 81: 212–216.

39 Pérez, L., Pinazo, A., García, M.T. et al. (2009). Cationic surfactants from lysine: Synthesis, micellization and biological evaluation. *European Journal of Medicinal Chemistry* 44: 1884–1892.

40 Pinazo, A., Pons, R., Bustelo, M. et al. (2019b). Gemini histidine based surfactants: characterization; surface properties and biological activity. *Journal of Molecular Liquids* 289: 111156.

41 Morán, M.C., Pinazo, A., Pérez, L. et al. (2004b). Enzymatic synthesis and physicochemical characterization of glycero arginine-based surfactants. *Comptes Rendus Chimie* 7: 169–176.

42 Pérez, L., Pinazo, A., Vinardell, M.P. et al. (2002). Synthesis and biological properties of dicationic arginine-diglycerides. *New Journal of Chemistry* 26: 1221–1227.

43 Pérez, L., Infante, M.R., Pons, R. et al. (2004). A synthetic alternative to natural lecithins with antimicrobial properties. *Colloids and Surfaces B: Biointerfaces* 35: 235–242.

44 Mnif, I. and Ghribi, D. (2015). Review lipopeptides biosurfactants: Mean classes and new insights for industrial, biomedical and environmental applications. *Biopolimers* 104: 129–147.

45 Varjani, S.J. and Upasani, V.N. (2017). Critical review on biosurfactant analysis, purification and characterization using rhamnolipid as a model biosurfactant. *Bioresource Technology* 232: 389–397.

46 Seydlová, G. and Svobodová, J. (2008). Review of surfactin chemical properties and the potential biomedical applications. *Central European Journal of Medicine* 3: 123–133.

47 Gudiña, E.J., Rangarajan, V., Sen, R. et al. (2013). Potential therapeutic applications of biosurfactants. *Trends in Pharmacological Sciences* 34: 667–675.

48 Jirku, V., Cejkova, A., Schreiberova, O. et al. (2015). Multicomponent biosurfactants – A "Green Toolbox" extension. *Biotechnology Advances* 33: 1272–1276.

49 Ron, E.Z. and Rosenberg, E. (2001). Natural roles of biosurfactants. *Environmental Microbiology* 3: 229–236.

50 Haba, E., Pinazo, A., Jauregui, O. et al. (2003). Physicochemical characterization and antimicrobial properties of rhamnolipids produced by *Pseudomonas aeruginosa* 47T2 NCBIM 40044. *Biotechnology and Bioengineering* 81: 316–322.

51 Saikia, R.R., Deka, S., Deka, M., and Sarma, H. (2012). Optimization of environmental factors for improved production of rhamnolipid biosurfactant by *Pseudomonas aeruginosa* RS29 on glycerol. *Journal of Basic Microbiology* 52 (4): 446–457.

52 Sivapathasekaran, C. and Sen, R. (2017). Origin, properties, production and purification of microbial surfactants as molecules with immense commercial potential. *Tenside, Surfactants, Detergents* 54: 92–107.

53 Jezierska, S., Claus, S., and Van Bogaert, I. (2018). Yeast glycolipid biosurfactants. *FEBS Letters* 592: 1312–1329.

54 Madslien, E.H., Ronning, H.T., Lindback, T. et al. (2013). Lichenysin is produced by most *Bacillus licheniformis* strains. *Journal of Applied Microbiology* 115: 1068–1080.

55 Ongena, M. and Jacques, P. (2008). *Bacillus* lipopeptides: Versatile weapon for plant disease biocontrol. *Trends in Microbiology* 16: 115–125.

56 Burgos-Díaz, C., Pons, R., Espuny, M.J. et al. (2011). Isolation and partial characterization of a biosurfactant mixture produced by *Sphingobacterium* sp. isolated from soil. *Journal of Colloid and Interface Science* 361: 195–204.

57 Goldman, S., Shabati, Y., Rubinovitz, C. et al. (1982). Emulsan in *Acinetobacter calcoaceticus* RAG1: Distribution of cell-free and cell-associated cross-reacting material. *Applied and Environmental Microbiology* 44: 165–170.

58 Cameotra, S.S. and Makkar, R.S. (2004). Recent applications of biosurfactants as biological and immunological molecules. *Current Opinion in Microbiology* 7: 262–266.

59 Ghribi, D., Abdelkefi-Mesrati, L., Mnif, I. et al. (2012). Investigation of antimicrobial activity and statistical optimization of *Bacillus subtilis* SPB1 biosurfactant production in solid-state fermentation. *Journal of Biomedicine and Biotechnology* 2012: 1–12.

60 Mnif, I. and Ghribi, D. (2016). Glycolipid biosurfactants: Main properties and potential applications in agriculture and food industry. *Journal of the Science of Food and Agriculture* 96: 4310–4320.

61 Mercadé, M.E., Manresa, M.A., Robert, M. et al. (1993). Olive oil mill effluent (OOME). New substrate for biosurfactant production. *Bioresource Technology* 43: 1–6.

62 Luna, J.M., Rufino, R.D., Campos-Takaki, G. et al. (2012). Properties of the biosurfactant produced by *Candida sphaerica* cultivated in low-cost substrates. *Chemical Engineering Transactions* 27: 67–72.

63 Saimmai, A., Rukadee, O., Onlamool, T. et al. (2012). Isolation and functional characterization of a biosurfactant produced by a new and promising strain of *Oleomonas sagaranensis* AT18. *World Journal of Microbiology and Biotechnology* 28: 2973–2986.

64 Cheffi, M., Hentati, D., Chebbi, A. et al. (2020). Isolation and characterization of a newly naphthalene-degrading *Halomonas pacifica*, strain Cnaph3: Biodegradation and biosurfactant production studies. *3 Biotech* 10: 89–103.

65 Coronel-León, J., Marqués, A.M., Bastida, J. et al. (2016). Optimizing the production of the biosurfactant lichenysin and its application in biofilm control. *Journal of Applied Microbiology* 120: 99–111.

66 Marqués, A.M., Pinazo, A., Farfan, M. et al. (2009). The physicochemical properties and chemical composition of trehalose lipids produced by *Rhodococcus erythropolis* 51T7. *Chemistry and Physics of Lipids* 158: 110–117.

67 Torrego-Solana, N., García-Celma, M.J., Garreta, A. et al. (2014). Rhamnolipids obtained from a PHA-negative mutant of *Pseudomonas aeruginosa* 47T2 ΔAD: Composition and emulsifying behaviour. *Journal of the American Oil Chemists' Society* 91: 503–511.

68 Kanlayavattanakul, M. and Lourith, N. (2010). Lipopeptides in cosmetics. *International Journal of Cosmetic Science* 32: 1–8.

69 da Silva Ramos, A., Manresa, M.A., Pinazo, A. et al. (2019). Rhamnolipids functionalized with basic amino acids: Synthesis, aggregation behaviour, antibacterial activity and biodegradation studies. *Colloids and Surfaces B: Biointerfaces* 181: 234–243.

70 Burgos-Díaz, C., Martín-Venegas, R., Martínez, V. et al. (2013). *in vitro;* study of the cytotoxicity and antiproliferative effects of surfactants produced by *Sphingobacterium detergens*. *International Journal of Pharmaceutics* 453: 433–440.

71 Coronel, J.R., Aranda, F.J., Teruel, J.A. et al. (2016). Kinetic and structural aspects of the permeabilization of biological and model membranes by lichenysin. *Langmuir* 32: 78–87.

72 Sánchez, M., Aranda, F.J., Teruel, J.A. et al. (2010). Permeabilization of biological and artificial membranes by a bacterial dirhamnolipid produced by *Pseudomonas aeruginosa*. *Journal of Colloid and Interface Science* 341: 240–247.

73 Zaragoza, A., Aranda, F.J., Espuny, M.J. et al. (2009). Mechanism of membrane permeabilization by a bacterial trehalose lipid biosurfactant produced by *Rhodococcus* sp. *Langmuir* 25: 7892–7898.

74 Aranda, F.J., Teruel, J.A., and Ortiz, A. (2005). Further aspects on the hemolytic activity of the antibiotic lipopeptide iturin A. *Biochimica et Biophysica Acta* 1713: 51–56.

75 Zaragoza, A., Aranda, F.J., Espuny, M.J. et al. (2010). Hemolytic activity of a bacterial trehalose lipid biosurfactant produced by *Rhodococcus* sp.: Evidence for a colloid-osmotic mechanism. *Langmuir* 26: 8567–8572.

76 Sodagari, M., Wang, H., Newby, B.M. et al. (2013). Effect of rhamnolipids on initial attachment of bacteria on glass and octadecyltrichlorozilane-modified glass. *Colloids and Surfaces, B: Biointerfaces* 103: 121–128.

77 Pamp, S.J. and Tolker-Nielsen, T. (2007). Multiple roles of biosurfactants in structural biofilm development by *Pseudomonas aeruginosa*. *Journal of Bacteriology* 189: 2531–2539.

78 Davey, M.E., Caiazza, N.C., and O'Toole, G.A. (2003). Rhamnolipid surfactant production affects biofilm architecture in *Pseudomonas aeruginosa* PAO1. *Journal of Bacteriology* 185: 1027–1036.

79 Espinosa-Urgel, M. (2003). Resident parking only: Rhamnolipids maintain fluid channels in biofilms. *Journal of Bacteriology* 185: 699–700.

80 Costerton, J.W., Stewart, P.S., and Greenberg, E.P. (1999). Bacterial biofilms: A common cause of persistent infections. *Science* 284: 1318–1322.

81 de la Fuente-Núñez, C., Reffuveille, F., Fernández, L. et al. (2013). Bacterial biofilm development as a multicellular adaptation: antibiotic resistance and new therapeutic strategies. *Current Opinion in Microbiology* 16 (5): 580–509.

82 Rivardo, F., Turner, R.J., Allegrone, G. et al. (2009). Anti-adhesion activity of two biosurfactants produced by *Bacillus* spp. prevents biofilm formation of human bacterial pathogens. *Applied Microbiology and Biotechnology* 83: 541–553.

83 Haque, F., Alfatah, M., Ganesan, K. et al. (2016). Inhibitory effect of sophorolipid on *Candida albicans* biofilm formation and hyphal growth. *Scientific Reports* 6: 23575.

84 Durao, P., Balbontín, R., and Gordo, I. (2018). Evolutionary mechanisms shaping the maintenance of antibiotic resistance. *Trends in Microbiology* 26: 677–691.

85 Rabanal, F. and Cajal, Y. (2016). Therapeutic potential of antimicrobial peptides. In: New Weapons to Control Bacterial Growth (eds. T., T. Villa and M. Viñas), 433–451. Cham: Springer.

86 Roca, I., Akova, M., Baquero, F. et al. (2015). The global threat of antimicrobial resistance: science for intervention. *New Microbes and New Infections* 6: 22–29.

87 González-Bello, C. (2017). Antibiotics adjuvants – a strategy to unlock bacterial resistance to antibiotics. *Bioorganic and Medicinal Chemistry Letters* 27 (18): 4221–4228.

88 Sotirova, A., Avramova, T., Stoitsova, S. et al. (2012). The importance of rhamnolipid-biosurfactant-induced changes in bacterial membrane lipids of *Bacillus subtilis* for the antimicrobial activity of thiosulfonates. *Current Microbiology* 65: 534–541.

89 Sierra, J.M., Fusté, E., Rabanal, F. et al. (2017). An overview of antimicrobial peptides and the latest advances in their development. *Expert Opinion on Biological Therapy* 17: 663–676.

90 Wright, G.D. (2016). Antibiotic adjuvants: Rescuing antibiotics from resistance. *Trends in Microbiology* 24: 862–871.

91 Taconelli, E., Carmeli, Y., Harbart, S. et al. (2017). Global Priority List of Antibiotic-Resistant Bacteria to Guide Research, Discovery, and Development of New Antibiotics. WHO.

92 Rabanal, F., Grau-Campistany, A., Vila-Farrés, X. et al. (2015). A bioinspired peptide scaffold with high antibiotic activity and low in vivo; toxicity. *Scientific Reports* 5: 10558–10568.

93 Tyers, M. and Wright, G.D. (2019). Drug combinations: A strategy to extend the life of antibiotics in the 21st century. *Nature Reviews Microbiology* 17: 141–155.

94 Bharali, P., Saikia, J.P., Ray, A. et al. (2013). Rhamnolipid (RL) from *Pseudomonas aeruginosa* OBP1: A novel chemotaxis and antibacterial agent. *Colloids and Surfaces B: Biointerfaces* 103: 502–509.

95 Ábalos, A., Pinazo, M., Infante, M. et al. (2001). Physicochemical and antimicrobial properties of new rhamnolipids produced by *Pseudomonas aeruginosa* AT10 from soybean oil refinery wastes. *Langmuir* 17: 1367–1371.

96 Benincasa, M., Ábalos, A., Oliveira, I. et al. (2004). Chemical structure, surface properties and biological activities of the biosurfactant produced *by Pseudomonas aeruginosa* LBI from soapstock. *Antonie van Leeuwenhoek* 85: 1–8.

97 Lotfabad, T.B., Abassi, H., Ahmadkhaniha, R. et al. (2010). Structural characterization of a rhamnolipid-type biosurfactant produced by *Pseudomonas aeruginosa* MR01: Enhancement of di-rhamnolipid proportion using gamma irradiation. *Colloids and Surfaces B: Biointerfaces* 81: 397–405.

98 Sotirova, A.V., Spasova, D.I., Galabova, D.N. et al. (2008). Rhamnolipid-biosurfactant permeabilizing effects on gram-positive and gram-negative bacteria strains. *Current Microbiology* 56: 639–644.

99 de Freitas Ferreira, J., Alan Vieira, E., and Nitschke, M. (2019). The antibacterial activity of rhamnolipid biosurfactant is pH dependent. *Food Research International* 116: 737–744.

100 Joshi-Navare, K. and Prabhune, A. (2013). A biosurfactant-sophorolipid acts in synergy with antibiotics to enhance their efficiency. *BioMed Research International* 2013: 1–8.

101 Bangham, A.D. and Horne, R.W. (1964). Negative staining of phospholipids and their structural modification by surface-active agents as observed in the electron microscope. *Journal of Molecular Biology* 8 (5): 660–668.

102 Barratt, G. (2003). Colloidal drug carriers: achievements and perspectives. *Cellular and Molecular Life Sciences* 60 (1): 21–37.

103 Cheng, Z., Al Zaki, A., Hui, J.Z. et al. (2012). Multifunctional nanoparticles: Cost versus benefit of adding targeting and imaging capabilities. *Science* 338 (6109): 903–910.

104 Zhu, J., Xue, J., Guo, Z. et al. (2007). Biomimetic glycoliposomes as nanocarriers for targeting P-selectin on activated platelets. *Bioconjugate Chemistry* 18 (5): 1366–1369.

105 Kocer, A. (2010). Functional liposomal membranes for triggered release. *Methods in Molecular Biology* 605: 243–255.

106 Samad, A., Sultana, Y., and Aqil, M. (2007). Liposomal drug delivery systems: An update review. *Current Drug Delivery* 4 (4): 297–305.

107 Winterhalter, M. and Lasic, D.D. (1993). Liposome stability and formation: Experimental parameters and theories on the size distribution. *Chemistry and Physics of Lipids* 64 (1–3): 35–43.

108 Dhawan, V.V. and Nagarsenker, M.S. (2017). Catanionic systems in nanotherapeutics – Biophysical aspects and novel trends in drug delivery applications. *Journal of Controlled Release* 266: 331–345.

109 Marques, E.F. (2000). Size and stability of catanionic vesicles: Effects of formation path, sonication, and aging. *Langmuir* 16 (11): 4798–4807.

110 Kaler, E.W., Herrington, K.L., Lampietro, D.J. et al. (2004). Phase behaviour and microstructure in aqueous mixtures of cationic and anionic surfactants. In: Mixed Surfactants Systems, 2e (eds. M. Abe and J.F. Scamehorn). Boca Ratón: CRC Press and Taylor & Francis Group, pp. 289–338.

111 Khan, A. and Marques, E.F. (2000). Synergism and polymorphism in mixed surfactants systems. *Current Opinion in Colloid and Interface Science* 4: 402–410.

112 Israelachvili, J.N. (2011). Intermolecular and Surface Forces, 3e. San Diego: Academic Press 978-0-12-375182-9. https://doi.org/10.1016/C2009-0-21560-1.

113 Jokela, P., Joensson, B., and Khan, A. (1987). Phase equilibria of catanionic surfactant-water systems. *The Journal of Physical Chemistry* 91 (12): 3291–3298.

114 Jiang, Y., Li, F., Luan, Y. et al. (2012). Formation of drug/surfactant catanionic vesicles and their application in sustained drug release. *International Journal of Pharmaceutics* 436: 806–814.

115 Dias, R.S., Lindman, B., and Miguel, M.G. (2002). DNA interaction with catanionic vesicles. *The Journal of Physical Chemistry B* 106 (48): 12600–12607.

116 Zhao, M., Yuan, J., and Zheng, L. (2012). The formation of vesicles by *N*-dodecyl-*N*-methylpyrrolidinium bromide ionic liquid/copper dodecyl sulfate and application in the synthesis of leaflike CuO nanosheets. *Colloid and Polymer Science* 290: 1361–1369.

117 Dowling, M.B., Javvaji, V., Payne, G.F. et al. (2011). Vesicle capture on patterned surfaces coated with amphiphilic biopolymers. *Soft Matter* 7 (3): 1219–1226.

118 Kahe, H., Chamsaz, M., and Zavar, M.H.A. (2016). A novel supramolecular aggregated liquid-solid microextraction method for the preconcentration and determination of trace amounts of lead in saline solutions and food samples using electrothermal atomic absorption spectroscopy. *RSC Advances* 6: 49076–49082.

119 Jurašin, D.D., Šegota, S., Čadež, V. et al. (2017). Recent advances in catanionic mixtures. In: Application and Characterization of Surfactants (ed. R. Najjar), 33–73. IntechOpen https://doi.org/10.5772/67998.

120 Morán, M.C., Pinazo, A., Pérez, L. et al. (2004a). "Green" amino acid-based surfactants. *Green Chemistry* 6: 233–240.

121 Morán, C., Clapés, P., Comelles, F. et al. (2001). Chemical structure/property relationship in single-chain arginine surfactants. *Langmuir* 17 (16): 5071–5075.

122 Rosa, M., Infante, M.R., Miguel, M.G. et al. (2006). Spontaneous formation of vesicles and dispersed cubic and hexagonal particles in amino acid-based catanionic surfactant systems. *Langmuir* 22: 5588–5596.

123 Rosa, M., Miguel, M.G., and Lindman, B. (2007a). DNA encapsulation by biocompatible catanionic vesicles. *Journal of Colloid and Interface Science* 312 (1): 87–97.

124 Rosa, M., Morán, M.C., Miguel, M.G. et al. (2007b). The association of DNA and stable catanionic amino acid-based vesicles. *Colloids and Surfaces A: Physicochemical and Engineering Aspects* 301 (1–3): 361–375.

125 Lozano, N., Pérez, L., Pons, R. et al. (2011). Diacyl glycerol arginine-based surfactants: Biological and physicochemical properties of catanionic formulations. *Amino Acids* 40: 721–729.

126 Tavano, L., Pinazo, A., A., and Abo-Riya, M. (2014). Cationic vesicles based on biocompatible diacyl glycerol-arginine surfactants: Physicochemical properties, antimicrobial activity, encapsulation efficiency and drug release. *Colloids and Surfaces B: Biointerfaces* 120: 160–167.

127 Marques, E.F., Brito, R.O., Silva, S.G. et al. (2008). Spontaneous vesicle formation in catanionic mixtures of amino acid-based surfactants: Chain length symmetry effects. *Langmuir* 24: 11009–11017.

128 Brito, R.O., Marqués, E.F., Gomes, P. et al. (2006). Self-assembly in a catanionic mixture with an aminoacid-derived surfactant: From mixed micelles to spontaneous vesicles. *The Journal of Physical Chemistry B* 110: 18158–18165.

129 Brito, R.O., Marqués, E.F., Silva, S.G. et al. (2009). Physicochemical and toxicological properties of novel amino acid-based amphiphiles and their spontaneously formed catanionic vesicles. *Colloids and Surfaces B: Biointerfaces* 72: 80–87.

130 Silva, S.G., do Vale, M.L.C., and Marques, E.F. (2015). Size, charge, and stability of fully serine-based catanionic vesicles: Towards versatile biocompatible nanocarriers. *Chemistry – A European Journal* 21 (10): 4092–4101.

131 Gonçalves Lopes, R.C.F., Silvestre, O.F., Faria, A.R. et al. (2019). Surface charge tunable catanionic vesicles based on serine-derived surfactants as efficient nanocarriers for the delivery of the anticancer drug doxorubicin. *Nanoscale* 11: 5932–5941.

132 Ghosh, S., Ray, A., Pramanik, N. et al. (2016). Can a catanionic surfactant mixture act as a drug delivery vehicle? *Comptes Rendus Chimie* 19: 951–954.

133 Olutaş, E.B. (2019). Interactions in mixed micellar systems comprising chiral cationic amino acid based and conventional anionic surfactants. *Journal of Molecular Liquids* 275: 126–135.

134 Xue, C., Zhao, H., Wang, Q. et al. (2018). Interfacial molecular array behaviours of mixed surfactant systems based on sodium laurylglutamate and the effect on the foam properties. *Journal of Dispersion Science and Technology* 39: 1427–1434.

135 Bračič, M., Pérez, L., Infante, R. et al. (2014). A novel synergistic formulation between a cationic surfactant from lysine and hyaluronic acid as an antimicrobial coating for advanced cellulose materials. *Cellulose* 21: 2647–2663.

136 Zana, R. and Xia, J. (2003). Gemini Surfactants: Synthesis, Interfacial and Solution-Phase Behaviour, and Applications, Surfactant Science Series, vol. 117. New York;: CRC Press.

137 Hu, Z., Wang, L., Guo, J. et al. (2015). Interaction of a novel anionic gemini surfactant containing a triazine ring with cetyltrimethylammonium bromide in aqueous solution. *Journal of Surfactants and Detergents* 18: 17–24.

138 Ji, X., Tian, M., and Wang, Y. (2016). Temperature-induced aggregate transitions in mixtures of cationic ammonium gemini surfactant with anionic glutamic acid surfactant in aqueous solution. *Langmuir* 32: 972–981.

139 Barai, M., Mandal, M.K., Karak, A. et al. (2019). Interfacial and aggregation behaviour of dicarboxylic amino acid-based surfactants in combination with a cationic surfactant. *Langmuir* 35: 15306–15314.

140 Grau-Campistany, A., Pujol, M., Marqués, A.M. et al. (2015). Membrane interaction of a new synthetic antimicrobial lipopeptide sp-85 with broad spectrum activity. *Colloids and Surfaces A: Physicochemical and Engineering Aspects* 480: 307–317.

141 Coronel-León, J., Pinazo, A., Pérez, L. et al. (2017). Lichenysin-geminal amino acid-based surfactants: Synergistic action of an unconventional antimicrobial mixture. *Colloids and Surfaces B: Biointerfaces* 149: 38–47.

142 Ruiz, A., Pinazo, A., Pérez, L. et al. (2017). Green catanionic gemini surfactant-lichenysin mixture: Improved surface, antimicrobial, and physiological properties. *ACS Applied Materials and Interfaces* 9: 22121–22131.

143 Epand, R.M. and Epand, R.F. (2009). Lipid domains in bacterial membranes and the action of antimicrobial agents. *Biochimica et Biophysica Acta* 1788: 289–294.

144 Blanzat, M., Pérez, E., Rico-Lattes, I. et al. (1999a). Synthesis and anti-HIV activity of catanionic analogs of galactosylceramide. *New Journal of Chemistry* 23: 1063–1065.

145 Blanzat, M., Pérez, E., Rico-Lattes, I. et al. (1999b). New Catanionic glycolipids. 1. Synthesis, characterization, and biological activity of double-chain and gemini catanionic analogues of galactosylceramide. *Langmuir* 15: 6163–6169.

146 Soussan, E., Mille, C., Blanzat, M. et al. (2008). Sugar-derived tricatenar catanionic surfactant: Synthesis, self-assembly properties, and hydrophilic probe encapsulation by vesicles. *Langmuir* 24: 2326–2330.

147 Richard, C., Souloumiac, E., Jestin, J. et al. (2018). Influence of dermal formulation additives on the physicochemical characteristics of catanionic vesicles. *Colloids and Surfaces, A: Physicochemical and Engineering Aspects* 558: 373–383.

148 Wittgens, A. and Rosenau, F. (2018). On the road towards tailor-made rhamnolipids: Current state and perspectives. *Applied Microbiology and Biotechnology* 102: 8175–8185.

149 Azim, A., Shah, V., Doncel, G.F. et al. (2006). Amino acid conjugated sophorolipids: A new family of biologically active functionalized glycolipids. *Bioconjugate Chemistry* 17: 1523–1529.

150 Delbeke, E.I.P., Roman, B.I., Marin, G.B. et al. (2015). A new class of antimicrobial biosurfactants: Quaternary ammonium sophorolipids. *Green Chemistry* 17: 3373–3377.

151 Jovanovic, M., Radivojevic, J., O'Connor, K. et al. (2019). Rhamnolipid inspired lipopeptides effective in preventing adhesion and biofilm formation of *Candida albicans*. *Bioorganic Chemistry* 87: 209–217.

152 Zabawa, T.P., Pucci, M.J., Parr, T.R. et al. (2016). Treatment of Gram-negative bacterial infections by potentiation of antibiotics. *Current Opinion in Microbiology* 33: 7–12.

11

Antiviral, Antimicrobial, and Antibiofilm Properties of Biosurfactants

Sustainable Use in Food and Pharmaceuticals

Kenia Barrantes[1], Juan José Araya[2], Luz Chacón[1], Rolando Procupez-Schtirbu[3], Fernanda Lugo[4], Gabriel Ibarra[4], and Víctor H. Soto[2]

[1] *Nutrition and Infection Section, Health Research Institute, University of Costa Rica, San Jose, Costa Rica*
[2] *Escuela de Química, Centro de Investigaciónen Electroquímica y, Energía Química (CELEQ), Universidad de Costa Rica, San José, Costa Rica*
[3] *General Chemistry, Department of Chemistry, University of Costa Rica, San Jose, Costa Rica*
[4] *Department of Public Health Sciences, College of Health Sciences, University of Texas at El Paso, El Paso, TX, USA*

CHAPTER MENU

11.1 Introduction

Antimicrobial resistance (AMR) is an important public health problem of broad concern and is considered as a crisis for humanity [1]. Annually, AMR diseases already cause at least 700 000 deaths worldwide [2]. A study assessed in India confirmed that infections with multidrug resistant

Biosurfactants for a Sustainable Future: Production and Applications in the Environment and Biomedicine, First Edition. Edited by Hemen Sarma and Majeti Narasimha Vara Prasad.
© 2021 John Wiley & Sons Ltd. Published 2021 by John Wiley & Sons Ltd.

pathogens (MDR) and extensively drug-resistant pathogens such as *Escherichia coli*, *Klebsiella pneumonie*, and *Acinetobacter baumannii* are associated with two to three times higher mortality compared to infections with non-resistant bacterial strains [3].

The threat posed by AMR to public health has turned attention to the role of the environment as a source and dissemination route for antibiotic resistance [4, 5]. Studies worldwide have reported an increased level of antibiotic residues in the aquatic environment, such as rivers, lakes, and drinking water [6], which may lead to adverse effects in terms of selection for resistant bacteria communities and the emergence and spread of antibiotic resistance genes [4, 5].

Under this scenario, there is a need for the discovery of new compounds with antimicrobial activities whose mechanisms of action should be distinct from common antimicrobial agents with known resistance profiles.

Biosurfactants are a diverse group of compounds whose characteristics, like biodegradability, low toxicity, biocompatibility, and even production from cheap renewable raw materials, make them an attractive alternative for various industrial processes in terms of releasing less pollution and contaminants into the environment [7].

Although most biosurfactants are considered to be secondary metabolites, some of them play important roles in the survival of biosurfactant-producing microorganisms [8]. Biosurfactants show specific biological activities such are bacterial quorum sensing, biofilm formation and maintenance, and cellular differentiation [9]. For the above, biosurfactants could be considered ideal candidates to substitute the traditional antimicrobial agents for medical, veterinary, and agriculture purposes [8, 10]. Biosurfactants also have shown therapeutic potential, mainly due to their antimicrobial and antibiofilm properties.

In this chapter, we discuss the various antimicrobial applications of biosurfactants in the pharmaceutical, therapeutic, and food industries. It is important to mention that despite their advantages, the high cost of producing biosurfactants still limits their extensive use as an alternative in the medical, pharmaceutical, and food industries.

11.2 Antimicrobial Properties

11.2.1 Biosurfactants Affect Microbial Adhesion and Motility

The ability of biosurfactants to reduce surface tension and osmotic pressure relief enables microbial mobility in a multitude of environments, allowing for better growth, reproduction, and colonization [11]. These compounds play an essential role in swarming motility, such as the case of surfactin in *Bacillus subtilis* [11, 12]. Other studies had described similar results for swarm cells of *E. coli, Pseudomonas aeruginosa, Burkholderia thailandensis, Serratia marcescens*, and *B. subtilis*, showing higher resistance to more than 10 antibiotics tested, including ampicillin, piperacillin, meropenem, nalidixic acid, ciprofloxacin polymyxin B, colistin, trimethoprim, gentamicin, erythromycin, streptomycin, kanamycin, and tobramycin [13].

The list of some biosurfactants with known antimicrobial activity includes Polymyxin A produced by *Bacillus polymyxa* [14, 15], surfactin, iturin, Bacillomycin D and Bacillomycin Lc produced by *B. subtilis* strains [16], mannosyl erythritol lipids and Rufisan from *Candida* [17, 18], rhamnolipids from *P. aeruginosa* [19], and biosurfactants isolated from *Bacillus amyloliquefaciens, Lactococcus lactis* 53, *Streptomyces roseosporus*, and *Streptococcus thermophilus* A [18, 20], to mention a few.

Table 11.1 describes some examples of biosurfactants used as antimicrobial agents produced by other microorganisms and their mechanisms of action.

Table 11.1 Examples of biosurfactants used as antimicrobial agents.

Biosurfactans (BS)	Microorganism	Action mechanisms	References
Polymyxin A	*Bacillus polymyxa*	Binds to the lipid A component of lipopolysaccharide (LPS) and disrupting the outer membrane, followed by permeabilizing and disrupting the inner membrane	[14, 21]
Surfactin iturin	*Bacillus subtilis*	Disturbs the integrity and permeability of membranes; generates physical, structural changes or disrupts protein conformations; can change some central membrane functions such as the generation of energy and transport; inserts in the viral membrane and to reduce the probability of viral membrane fusion with the host	[22, 23]
WH1 fungin	*Bacillus amyloliquefaciens*	To elicit pores on cell membrane at high concentration and low concentration induces apoptosis in fungal cell; binds to ATPase on the mitochondrial membrane and decreases ATPase activity in fungal cell	[24]
Iturin A Bacillomycin D Bacillomycin Lc	*Bacillus subtilis*	Disruption of the plasma membrane by the formation of small vesicles and the aggregation of intramembranous particles in fungi	[8]
Daptomycin	*Streptomyces roseosporus*	Disrupting multiple aspects of cell membrane function and inhibiting protein, DNA, and RNA synthesis	[120]
Pumilacidin	*Bacillus subtilis*	Antiviral activity against herpes simplex virus type 1	[25]

Recent studies have evaluated the antibiofilm and antiadhesive activities of the biosurfactants of *Lactobacillus* against *Streptococcus mutans* and multidrug-resistant strains of *A. baumannii*, *E. coli*, and methicillin-resistant *Staphylococcus aureus* (MRSA) [26, 27].

Walencka et al. [28] described the influence of biosurfactants produced by *Lactobacillus* on bacterial-surface interactions related to the changes in surface tension and bacterial cell wall in *S. aureus* and *Staphylococcus epidermidis* [28]. These compounds may affect cell-to-cell and also cell-to-surface interactions. For this reason, the surface is made less supportive for bacterial deposition; some proteins known as hydrophobins decrease, and their adhesion and/or co-aggregation ability is affected [28, 29].

11.2.2 Biosurfactants Affect Microbial Membranes and Proteins

Figure 11.1 describes the main (known) mechanisms of action of biosurfactants in microbial and fungi cells. In bacterial cells, disruption of microbial membranes is caused by biosurfactants, with significant alterations on cell surface structure by interactions with phospholipid membranes and proteins. Also, biosurfactants can alter the microbial adhesion and the motility. These compounds affect the adhesion and detachment of bacteria and biofilm formation. In fungi cells, it is described as apoptosis. WH1 fungin induce apoptosis at low concentrations by binding with ATPase on the mitochondrial membrane.

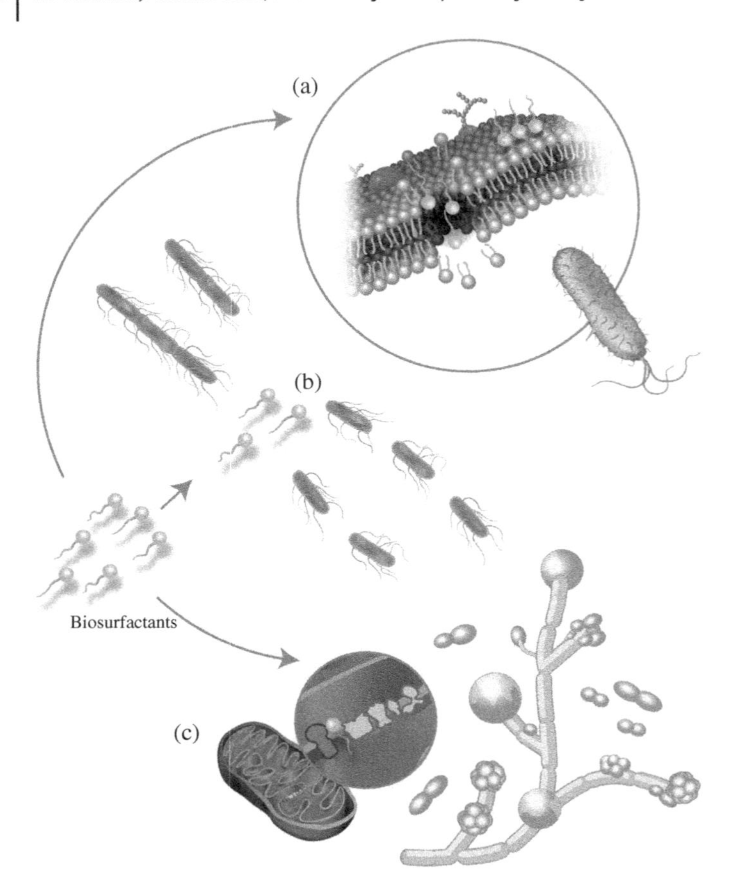

Figure 11.1 Antimicrobial effects of biosurfactants: (a) **Disruption of microbial membranes.** Biosurfactants cause significant alterations on cell surface structures by interactions with phospholipid membranes and proteins. (b) **Alteration of microbial adhesion and motility.** Biosurfactants affect the adhesion and detachment of bacteria and biofilm formation. (c) **Apoptosis in fungi.** WH1 fungin induce apoptosis at low concentrations by binding with ATPase on the mitochondrial membrane. See the text for more details (See insert for color representation of the figure).

11.2.2.1 Lipopeptides

The primary mechanism for the action of antimicrobial biosurfactants is their ability to come in close contact with the microbial cell wall structures, inducing them to release extracellular substances [30]. Biosurfactants cause alterations in cell morphology, surface hydrophobicity, surface functional groups, and electrokinetic potential [30, 31]. This mechanism varies according to the biosurfactant structure, as not all surfactants interact with the phospholipid bilayer [32]. Surfactin, for instance, interacts with artificial and biomembrane systems such as enveloped virus and bacterial protoplasts [8]. This biosurfactant is cyclic lipopeptide biomolecule and was originally isolated from *B. subtilis* and known as an efficient inhibitor of fibrin clotting [15].

This lipopeptide is used commercially for curing of cell cultures and cleansing of biotechnological product contamination [17], even though the elevated cost of surfactin production in low yields limits its range of therapeutic and environmental applications [33].

Antifungal agents in the form of other lipopeptides such as bacillomycin F and fengycin, isolated from *B. subtilis* spp. *inaquosorum*, have shown potent inhibitory effects on mycelial growth *in vitro* in a wide spectrum of fungi [34]. The ability of biosurfactants to modify membrane surfaces appears to be the primary mechanism of action through which bacteria developed antifungal properties of lipopeptides, mainly as a biological control of plant diseases [8].

Sambanthamoorthy et al. [35] described biosurfactants isolated from probiotic lactobacilli (*Lactobacillus jensenii* and *Lactobacillus rhamnosus*) with antimicrobial properties that were able to inhibit the formation of biofilms by several multidrug-resistant pathogens, including *E. coli, Acinetobacter baumanii, S. aureus,* and methicillin-resistant *S. aureus* (MRSA). The authors observed through transmission electron microscopy that the microbicidal action of these biosurfactants is to damage the bacterial cell membrane, resulting in leakage and lysis [35].

Daptomycin is one of the few biosurfactants that has reached the level of commercial development and was approved for the treatment of Gram-positive infections, including right-sided infective endocarditis, skin and surgical structure infections, meningitis, bacteraemia, sepsis, and urinary tract infections [36]. This lipopeptide is produced by *S. roseosporus* and is highly active against MDR bacteria such as vancomycin-resistant *Enterococcus faecium* (VRE) and methicillin-resistant *S. aureus* (MRSA) [37].

This commercial expansion and widespread use of daptomycin has led to yet more bacterial resistance. There are reports of resistance in *S. aureus, E. faecium,* and *Enterococcus faecalis* [37]. It is reported that the antimicrobial activity of daptomycin increases with the presence of a lipid tail length of 10–12 carbons atoms. Conversely, an enhanced antibacterial activity is presented in lipopeptides with a lipid tail length of 14 or 16 carbon atoms [38].

The polymyxins, another type of antibiotic biosurfactant, were discovered in 1947, isolated from *B. polymyxa* (reclassified as *Paenibacillus* in 1993) [14, 15]. Polymyxins are effective against Gram-negative bacteria by binding to a component of lipopolysaccharide (LPS), the lipid A. After this binding, the outer membrane is disrupted, followed by permeabilizing and disrupting the inner membrane [14, 21]. Although polymyxins are very potent antibiotics, their high toxicity had limited their clinical application [39].

Polymyxin B and E are the two forms of polymyxins that have been used clinically. Colistin is a mixture of polymyxin E1 and E2 and is an effective antibiotic for the treatment of MDR Gram-negative bacteria. They are typically administered to prevent infections and are used in low concentrations in combination with other antibiotics [39, 40].

Polymyxins show good antimicrobial activity; however, resistance to these antibiotics had also been reported worldwide [13]. The emergence of colistin-resistant bacterial pathogens is considered a significant clinical and public health concern [1, 14].

11.2.2.2 Glycolipids

Another class of biosurfactants are glycolipids. These biosurfactants are one of the best-studied and commercially most important groups. Studies worldwide had described their antiviral, antibacterial, antiadhesive, and antifungal activities [8, 41].

Glycolipids have great structural diversity. There are subclasses including trehalolipids, succinoyl-trehalolipids, sophorolipids, rhamnolipids, lipids of cellobiose or ustilagic acid, ustilipids, xylolipids, and lipids of oligosaccharides [42]. Among the glycolipids group, sophorolipids, mannosylerythritol lipids, trehalolipids, and rhamnolipids are the best known [43], while rhamnolipids have the greatest number of patents and scientific publications [44].

Rhamnolipids are glycolipids, produced mainly by *Burkholderia* sp. and *P. aeruginosa*. More than 60 types of rhamnolipids produced by different strains of bacteria have been identified and studied, chiefly those synthesized by *P. aeruginosa* [45, 46]. Rhamnolipids are composed of one or two rhamnose sugars linked to one or two β-hydroxyl fatty acids through a glycosidic bond [44].

These biosurfactants show useful characteristics and versatility, and they have been considered to be used in the food industry [47, 48].

Besides the notable rhamnolipids, other glycolipid biosurfactant-producing bacterial groups are *Acinetobacter calcoaceticus*, *Pseudoxanthomonas* sp., *Enterobacter* sp., *Pantoea* sp., *Serratia rubidaea*, and *Lysinibacillus sphaericus* [48]. The development of new analytical techniques has led to discovering more and more of these biosurfactants.

Magalhães and Nitschke [49] described a bacteriostatic effect of rhamnolipids over 90% of 32 isolates of *Listeria monocytogenes*, which is considered an important food-borne pathogen. The study demonstrated that these compounds showed a synergistic effect when combined with nisin, in a combination of 156.2 mg/ml of rhamnolipids and 320 IU/ml of nisin, with a complete reduction of the *L. monocytogenes* L17 population after two hours of incubation [48, 49].

In another study, de Freitas Ferrera et al. [47] also demonstrated antimicrobial activity of rhamnolipids against *L. monocytogenes* and other Gram-positive pathogens such are *Bacillus cereus* and *S. aureus*. They described the antimicrobial activity to be pH-dependent, where Gram-positive pathogens were favored at more acidic conditions while Gram-negative *Salmonella enterica* and enterohemorrhagic *E. coli* (EHEC) showed resistance at all pH levels studied [47].

Sophorolipids are an interesting class of biosurfactants and have promising antimicrobial activity against Gram-positive and Gram-negative bacteria, acting mechanistically through the rupture in the cellular membrane [50]. Because of their antibiofilm properties, this group of biosurfactants is finding increasing use in personal care and pharmaceutical products as an alternative to conventional drugs [51].

Lydon et al.[51] described a sophorolipid produced by the yeast *Starmerella bombicola* with antimicrobial activities against the pathogens *P. aeruginosa* and *E. faecalis*, with significant reductions in microbial populations at concentrations of as low as 5 mg/ml [51]. In a similar study, Zhang et al. [52] reported that lactonic stearic and oleic sophorolipids were more effective in reducing five strains of pathogenic *E. coli* O157:H7 than the free-acid counterparts. The *E. coli* O157:H7 strains showed different susceptibilities to this biosurfactant, although all inhibitions were due to membrane damage [52].

Trehalolipids are disaccharide trehalose-mycolic acid linked glycolipid surfactants that have a hydrophilic backbone, specifically consisting of two α,α-1,1 glycosidic linked glucose units forming the trehalose moiety [53, 54]. They are mainly produced by Actinomycetales, a Gram-positive group of bacteria that includes *Rhodococcus*, *Nocardia*, *Mycobacterium*, and *Corynebacterium*. Less frequently described are the genera *Tsukamurella*, *Brevibacterium*, and *Micrococcus* [54]. Besides the environmental and industrial uses of this group of biosurfactant, potential applications as therapeutic agents are emerging [8]. Janek et al. [55] assessed the antimicrobial and antiadhesive activity of trehalose lipids produced by *Rhodococcusfascians* BD8 against several pathogenic microorganisms, primarily against *Proteus vulgaris* and *Vibrio harveyi*. The highest concentration tested (0.5 mg m/l) caused a partial (11–34%) inhibition of other Gram-positive and Gram-negative bacteria and 30% inhibition of *Candida albicans* growth [55].

Biosurfactants, in association with antibiotics, lead to a synergistic increase in the efficacy of antibiotics against pathogenic microorganisms. This relationship had been extensively described and included lipopeptides, glycolipids, and combinations of biosurfactants [50, 51, 56].

Rivardo et al. [56] reported that V9T14 lipopeptide biosurfactant from *Bacillus licheniformis* was effective combined with various antibiotics, ampicillin, cefazolin, ciprofloxacin, ceftriaxone, tobramycin, piperacillin, and trimethoprim/sulfamethoxazole (SXT), in eradicating a uropathogenic *E. coli* strain. The results demonstrated a significant synergistic effect in the reduction of bacterial growth, both planktonically and as a biofilm [56].

Samadi et al. [57] described the effect of rhamnolipids produced by *P. aeruginosa* MN1 against methicillin-resistant *Staphylococcus aureus* (MRSA). Biosurfactants combined with oxacillin were able to inhibit the MRSA population. Interestingly, authors described an enhancing activity of oxacillin combined with *Pseudomonas* biosurfactants against the MRSA strain. This biosurfactant and antibiotic combination lowered the minimum inhibitory concentrations of oxacillin, but it was not effective against Gram-negative bacteria (*E. coli*, *Klebsiella pneumoniae*, *Salmonella Typhi*, and *Shigella flexneri*) [57].

Similar results are described for sophorolipids and tetracycline combinations against *S. aureus* with a total inhibition before four hours of exposure, while tetracycline alone could not achieve complete inhibition till the end of six hours [50].

Combinations of different groups of biosurfactants have proven to be more effective than the use of only one group of biosurfactant against Gram-negative and Gram-positive bacteria. Lipopeptide-type biosurfactants, BS15 produced by *Bacillus stratosphericus* A15 and rhamnolipids produced by *P. aeruginosa* C2 used in combination, disrupted cell membranes of the target bacteria, as observed by Sana et al. [58]. This biosurfactant combination was examined against *S. aureus* ATCC 25923 and *E. coli* K8813, showing synergistic activity. The survival curve of both bacteria showed bactericidal activity after treating with biosurfactants as it killed more than 90% of the initial population. Cell membrane disruption by enhancing the bacterial cell membrane permeability was also confirmed by scanning electron microscopy (SEM) [58].

In past years, novel synthetic pathways allowed the production of a broad range of useful biomolecules, starting from microbially produced biosurfactants, such as the case of sophorolipid amines and sophorolipid quaternary ammonium salts [59, 60]. The results regarding antimicrobial properties of this new generation of biosurfactants are significant in terms of medical therapy of multidrug-resistant infections. Delbeke et al. [59] evaluated sophorolipid quaternary ammonium salts against *K. pneumoniae* LMG 2095, *E. coli* LMG 8063, *S. aureus* LMG 8064, and *B. subtilis* LMG 13579. Derivatives with an octadecyl group on the nitrogen atom proved to be more active than the antibiotic gentamicin sulfate against all tested Gram-positive strains, but they did not inhibit the Gram-negative strains [59]. In another study, the synthesis of new long-chained quaternary ammonium sophorolipids was evaluated for their antimicrobial, transfection, and self-assembly properties, showing modest to high activities against the Gram-positive strains of *S. aureus* ATCC 6538 and methicillin-resistant *S. aureus* (MRSA), which also is a vancomycin-resistant strain (*S.aureus* Mu50). Several of the derivatives showed modest activity against the Gram-negative strains of *E. coli* LMG 8063, *K. pneumoniae* LMG 2095, and *P. aeruginosa* PAO1.

Authors have described a good correlation between the antimicrobial activity and transfection efficiency of these novel derivatives and the property of self-assembly [60].

11.2.2.3 Nucleolipids

Nucleolipids are biosurfactants with their lipid chain connected directly to the nucleoside or with the linker of glycerol. They are found in bacterial and marine sponges [9, 61]. These biosurfactants show potential as gene vectors and drug carriers due to the properties of excellent biocompatibility and specific base interaction [9].

11.2.3 Biosurfactants Induce Apoptosis in Fungi

Biosurfactant lipopeptides produced by some species of *Bacillus* (*B. licheniformis*, *B. cereus*, *Bacillus Mycoides*, and *B. amyloliquefaciens*) had shown antifungal properties by inducing apoptosis or programmed cell death (PCD) [8]. *B. amyloliquefaciens* produces a new surfactin called WH1 fungin, which can inhibit the growth of fungi. At high concentrations, this biosurfactant is able to elicit pores on cell membranes and at low concentration to induce apoptosis [24]. In fungi, the mitochondrial component of mammalian apoptosis is more conserved [62]. In filamentous fungal species, apoptosis appears to occur in response to WH1 fungin by binding with ATPase on the mitochondrial membrane. As a result, there is decreased ATPase activity in fungal cells [24]. WH1 fungin also inhibits glucan synthesis in *C. albicans*, resulting in reduced callose levels in the fungal cell wall [24].

11.3 Biofilms

A biofilm is defined as an assemblage or surface-associated microbial cell that is enclosed in an extracellular polymeric substance matrix. This definition includes microbial aggregates and floccules as well as adherent populations within spaces of a porous media. Noncellular materials such as minerals, clay or silt particles, and even blood components (depending on where the biofilm has developed) may also be found in a biofilm matrix. At a higher level of organization, bacteria within a biofilm benefit from physiological cooperativity in which such bacteria constitute a coordinated functional community [63, 64].

Microorganisms in general form biofilms as an action to preserve themselves from environmental challenges [65]. A biofilm formation follows five consecutive stages known as:

- initial reversible attachment,
- irreversible attachment,
- maturation stage I,
- maturation stage II, and
- dispersion.

Initial reversible attachment. This is the initial event in biofilm development and occurs when there is contact between the surface of the cell and an interface. The presence of flagella during the planktonic phase of the biofilm growth cycle appears to enhance reversible attachment. These adherent cells are not at this time in the differentiation process, and many may quit the surface and turn back to a planktonic lifestyle. During the reversible adhesion stage, an essentially instantaneous attraction of bacteria to a surface develops even though Brownian motion is still detected [66], along with other certain species-specific behavior, such as creeping, rolling, "windrow" formation, and aggregate formation. Small amounts of exopolymeric material surrounded the single adherent cells that start biofilm formation on a surface and then begin to adhere to it irreversibly [66, 67].

As biofilms mature, they develop a basic microcolony/water channel architecture commonly found in natural and *in vitro* biofilms.

Irreversible attachment: In a continuous flow, irreversible attachment is indicated by the activation of a group of genes and commenced cell cluster. Motility ceased during this developmental stage, possibly due to the loss of flagella. The bacterial cell must maintain contact with the substratum and grow to develop a mature biofilm. The main difference between bacterial biofilms and bacteria that are simply attached to a substratum lies in the extracellular polymeric substances (EPSs) that surround the resident bacteria in the biofilm. In the early events of the biofilm formation, this structure is rigidly attached to the surface [66].

Maturation stage I: During this stage, the complex architecture of the biofilm is generated. It has been postulated that during this stage, the biofilm becomes oxygen-limited, at least in selected zones (presumably at or near the substratum). It is also at this stage that matrix polymer production has been detected [67].

Maturation stage II: The biofilm reaches its maximum thickness during this stage of development and also the biofilm bacteria are profoundly different from planktonic bacteria regarding the number of differentially expressed proteins [67].

Dispersion: During this, the final, stage, bacteria within cell clusters can be seen to actively detach from the biofilm (either individually or in groups), leaving behind structures that appear shell-like with a hollow center and walls of non-motile bacteria. Starvation may lead to the detachment that permits bacteria to find nutrient-rich habitats, while at the same time allowing better access to nutrients for the cells remaining in the biofilm [66].

Although currently there are several new published perspectives in biofilm eradication, including the use of metallic cations, special antibiotics, electrochemical methods, and diverse antimicrobial compounds [68], here the focus will be on the effect of biosurfactants in biofilms.

In specific conditions, biosurfactants can be more effective than other conventional strategies to inhibit and/or disrupt biofilms; for example, these substances may affect the development of flagella, suggesting changes in the way bacteria attach to surfaces [69]. Since a lot of health-related diseases are due to the formation of biofilms by pathogens (including biomedical devices, body parts, food, medical facilities, houses, etc.), there is an interest in using biosurfactants to stop the emergence of such biofilms. Researchers have found that surfaces preconditioned with biosurfactants can prevent the formation of biofilms, among other ways by triggering dispersal of the biofilm [68].

Critical micellar concentration (CMC) and different surfaces from animate to inanimate, rough to smooth, hydrophobic to hydrophilic, as well as different phases, and so forth, are important when determining the quality and effect of a biosurfactant for dispersing, eliminating, or inhibiting the formation of a biofilm – the latter being the focus of most modern research.

Electrostatic (double-layer) interactions, van der Walls forces, and steric interactions are the physical forces responsible for bacterial adhesion to surfaces. Such an antiadhesive property of a biosurfactant can be requisite to wetting properties (small contact angle), and so the hydrophobicity is altered and the initial attachment of biofilms to the surface is prevented [70].

Disrupting the multicellular structure of bacterial biofilms has been proposed as the most promising strategy for increasing the sensitivity of pathogens in a biofilm to bactericidal agents and host immune systems. Efficient biofilms as eradicating agents must diffuse through EPS and penetrate between the sessile cells and adhering surface, and modify cell surface properties to induce disruption or dislodging of the biofilm [71].

The inherent resistance of biofilms and their pervasive involvement in implant-related infections has prompted research toward the development of antibiofilm, antiadhesive agents. Biosurfactants can disrupt biofilm on medical implants due to their antimicrobial, antibiofilm, and antiadhesive potential [72]. Hydrophobicity is modified by the adsorption of biosurfactants to a substratum surface. This modification affects the processes of microbial adhesion and desorption [18].

Cell associated biosurfactants (CABSs) are used for their antiadhesive properties rather than antimicrobial activity in concentrations between 4 and 50 mg/ml [33]. The activity of cell-free biosurfactants (CFBSs) on growth and biofilm formation, as well as adhering inhibition, is significant for the derived glycolipid biosurfactant of *Lactobacillus acidophilus*, and testing on polydimethylsiloxane (PDMS) surfaces strongly inhibits the biofilm formation. This is important since this polymeric material is used to manufacture medical devices, such as catheters. The wetting properties of the biosurfactant are essential as it leads to adhesion of bacterial cells, including pathogens, to biomaterial implants in the human body [70].

Most of the microorganisms are negatively charged; therefore, the antiadhesion property shown by biosurfactants to inhibit pathogenic bacterial biofilms in PDMS could be due to a negative charge or anionic nature of the compound, which could have driven away the bacterial adherence to surfaces [69].

Rufisan, a biosurfactant produced by *Candida lipolytica* UCP 0988, showed antiadhesive activity to several microorganisms tested, but the effect depends on the concentration and the microorganism tested itself [18].

DispersinB® is a biofilm-releasing glycoside hydrolase isolated from periodontal bacteria. This enzyme inhibits biofilm formation and disrupts/disperses preformed biofilm by depolymerizing poly-β-1,6-*N*-acetyl-D-glucosamine (PNAG/PGA), which is essential for the biofilm formation of *E. coli*, *A. baumannii*, and *Staphylococcus* sp. Although DispersinB has not shown antimicrobial activity, it could inhibit the biofilm formation and increase the KSL W peptide activity by making them more sensitive to this antimicrobial agent [73].

Vibrio natriegens MK3 produces a biosurfactant that is negatively charged, which contributes to its efficient biofilm disruption potential, as explained previously. This biosurfactant has also been found to be active against the biofilm of *V. harveyi*. The study revealed disintegrated biofilm architecture due to the biosurfactant, by disturbance of the interbacterial adhesion facilitated by the negative charge of the MK3 biosurfactant. Moreover, the study also demonstrated that the biosurfactant could damage the cell wall and was able to destroy the entire cell structure, a much more acceptable strategy against *V. harveyi* [74].

Pediococcus acidilactici and *Lactobacillus plantarum* produce a biosurfactant effective against the biofilm of *S. aureus* GMCC26003. Nevertheless, comparing the biosurfactant of *L. plantarum* CFR 2194 to the biosurfactant produced by *L. plantarum* 27172 shows that the latter exhibits a weaker antiadhesive activity compared to the former. Based on this, the antimicrobial and antiadhesion activities of the biosurfactant may be specific to species and strains [75].

A lipopeptide biosurfactant produced by *B. cereus* NK1 strongly inhibits the biofilm formation by pathogenic microorganisms *P. aeruginosa* and *S. epidermidis* [36], while a biosurfactant isolated from *B. subtilis* VSG4 and *B. licheniformis* VS16 exhibited considerable antibiofilm (antiadhesion) activity against *S. aureus* ATCC 29523, *Salmonella Typhimurium* ATCC 14028 and *B. cereus* ATCC 11788, although to varying degrees [76].

Another lipopeptide biosurfactant produced *by Bacillus tequilensis* SDS21 has shown antiadherence activity on glass, stainless steel, and polystyrene surfaces. The most efficient biofilm removals were observed for *S. aureus* MTCC 3160, where 0.5 mg/ml of biosurfactant removed more than 90% of the biofilm. Removal of biofilm by the biosurfactant was observed to be by concentration, time, and microorganism dependency. The study also showed that bactericidal and biofilm eradication activity by the biosurfactant was not pH sensitive in the range of 5–12, nor was affected by the use of hard water or conformational changes induced by divalent metals [77].

Recently a lipopeptide biosurfactant produced by *Acinetobacter junii* B6 showed the potential to decrease the biofilm formation of several pathogens (*Proteus mirabilis*, *S. aureus*, and *P. aeruginosa*) [78]. Previous studies showed that biosurfactants were more effective than traditional biofilm inhibitors and affected expressions of biofilm-related genes (*cidA*, *icaA*, *dltB*, *agrA*, *sortaseA*, and *sarA*) and interfered with the release of signaling molecules in quorum-sensing systems [78].

Nowadays, a considerable amount of information has been generated by various researchers worldwide toward finding suitable natural developed agents that can act against biofilm formation and/or eradication. Biosurfactants have been proved to be among such agents, but in many cases the substance was a crude mixture of microorganisms. In other cases, the activity was limited to certain species and/or strains. Nevertheless, biosurfactants are a promising field of research and development in the battle toward stopping biofilm formation and/or elimination. Table 11.2 summarizes the main surfactants used as antibiofilm agents.

Table 11.2 Main surfactants used as antibiofilm agents.

Biosurfactants[a]	Microorganism	Action mechanisms	References
BSM	*Lactobacillus acidophilus* NCIM 2903	Antiadhesive on PDMS	[72, 66]
Rufisan	*Candida lipolytica* UCP 0988	Antiadhesive	[63]
Dispersin B	*Aggregatibacter actinomyacetemcomitans*	Inhibits biofilm formation and disrupts/disperses preformed BF by depolymerising poly-β-1,6-*N*-acetyl-D-glucosamine (PNAG/PGA)	[1, 21, 73]
BSM	*Vibrio natriegens* MK3	Anionic BS active against the BF of *Vibrio harveyi*	[74]
BSM	*Lactobacillus plantarum* CFR 2194	Antiadhesive against BF of *Staphylococcus aureus* GMCC26003	[75]
BSM	*Bacillus subtilis* VSG4	Antiadhesion activity against *Staphylococcus aureus* ATCC 29523, *Salmonella typhimurium* ATCC 14028, and *Bacillus cereus* ATCC 11788 at various degrees	[76]
BSM	*Bacillus tequilensis* SDS21	Antiadherence activity on glass, stainless steel, and polystyrene surfaces	[77]
BSM	*Acinetobacter junii* B6	Potential to decrease the BF formation of several pathogens (*Proteus mirabilis*, *Staphylococcus aureus*, and *Pseudomonas aeruginosa*)	[78]

[a] = BSM = crude mixture of biosurfactant molecules.

11.4 Antiviral Properties

In 1990 Naruse and coworkers [25] analyzed the antiviral potential of pumilacidin, a biosurfactant produced by *B. subtilis* on herpes simplex virus type 1 (HSV-1). They reported a strong antiviral activity of this compound, which is a complex of biosurfactant molecules related to surfactin, also produced by *B. subtilis*.

Vollenbroichet al.[79] described *in vitro* assays of surfactin by antiviral activity against some enveloped viruses: Semliki Forest virus (SFV), HSV-1 and HSV-2, suid herpes virus (SHV-1), vesicular stomatitis virus (VSV), simian immunodeficiency virus (SIV), feline calicivirus, murine encephalomyocarditis virus (EMCV), and human immunodeficiency virus (HIV).

For all virus tested, surfactin was effective, and the proposed action mechanism was related to a partial disruption of the lipidic viral membrane [79].

In 1999, some experiments analyzed the efficiency of surfactin with fatty acid carbon atoms of different lengths (13–15) on the inactivation of VSV, SHV-1, and SFV. The results showed a better activity of the surfactin with 15 carbon atoms compared to the molecule with 13 carbon atoms, so the conclusion was that the antiviral efficiency of the biosurfactant molecules was related to the number of carbon atoms of its fatty acid (hydrophobicity) [80].

The tests in animal pathogenic viruses are remarkable. In 2006, the antiviral activity of surfactin was evaluated against Pseudorabies virus (PRV), Porcine Parvo virus (PPV), Newcastle Disease virus (NDV), and Infectious Bursal Disease virus (IBDV), showing a potent effect on NDV and IBDV (more than 95%), and to a lesser degree of inactivation for PRV and PPV (around 70%). The antiviral mechanisms at this moment could not be described [81]. Similar results were found relating to antiviral activity against NDV using biosurfactants produced by *B. cereus* [82].

Most recently, studies with an animal pathogenic virus, theporcine epidemic diarrhoea virus (PEDV), and the swine transmissible gastroenteritis virus (TGEV), described the capability of surfactin to insert in the viral membrane and to reduce the probability of viral membrane fusion with the host. This biosurfactant is the first naturally wedged lipid membrane fusion inhibitor identified and can be effective against the structurally similar virus [22, 23]. Nevertheless, surfactin cannot affect the dengue virus infection capability [83].

Pseudomonas is another important bacterial microorganism related to biosurfactant production. Some of the most potent related biosurfactants are the rhamnolipids; these molecules have shown *in vitro* antiviral activity against HSV-1 and HSV-2 [84]. *Candida* sp. can also produce biosurfactants as sophorolipids. These natural molecules and some modified series have been tested for antiviral efficiency and demonstrated virucidal activity against HIV[85], Epstein Barr virus (EBV), Coxsackie viruses (CVB1-6, CVA7, and CVA9), Murine Herpes virus (MHV-68), and Influenza Virus A (H3N2 subtype) [86].

11.5 Therapeutic and Pharmaceutical Applications of Biosurfactants

Biosurfactants have shown therapeutic potential mainly due to their antimicrobial and antibiofilm properties. With a growing need for additional chemotherapies to treat multiresistant infectious diseases, more considerable attention has been given to this group of compounds [87, 88]. Additionally, other properties of biosurfactants such as its tense activity and biodegradability, also make them attractive as components in pharmaceutical formulations [89, 90].

11.5.1 Therapeutic Applications

Despite the growing number of publications describing the biological properties of biosurfactants, actual applications as therapeutics remain very limited [87, 88]. In Table 11.3, examples of biosurfactants with potential clinical use are listed. Biological properties of biosurfactants have been previously discussed; in the following sections, we focus on current and potential applications.

11.5.1.1 Antibiotics

Several biosurfactants exhibit bactericidal, bacteriostatic, antibiofilm, and antiadhesive activity; consequently, a direct application as antibiotics have been pursued since its discovery [8, 93]. The first biosurfactant, the lipopeptide Polymyxin A, was discovered and isolated in 1949 from a soil bacterium *B. polymyxa* [14, 15]. Although polymyxins are very potent antibiotics, their clinical applications are limited by their high toxicity [39]. Colistin is a mixture of polymyxin E1 and E2 and is an effective antibiotic for the treatment of most MDR Gram-negative bacteria [40]. Daptomycin is another biosurfactant that reached the level of commercial development in 2003 and was approved for the treatment of skin infections caused by Gram-positive microorganisms, particularly for the methicillin-resistant endocarditis and bacteraemia caused by *S. aureus* [91].

Although antibacterial properties of biosurfactants have been growing in the scientific literature, few examples such as mentioned above have reached preclinical and clinical studies [94, 95].

Table 11.3 Biosurfactants used in clinic or with clinic use potential.

Biosurfactants	Microorganism	Type	Application	References
Daptomycin	*Streptomyces roseosporus*	Lipopeptides	Treat serious skin and blood infections caused by certain Gram-positive bacteria	[91, 122]
Surfactin	*Bacillus subtilis*	Lipopeptide	Inhibit the growth of viruses and bacteria.	[22, 92]
Polymixin	*Bacillus polymyxa*	Lipopeptide	Wide spectrum of antibacterial activity	[123]
Laspartomycin	*Streptomyces viridochromogenes*	Lipopeptides	Calcium-dependent antibiotic for treatment of *Staphylococcus aureus*,vancomycin intermediate *S. aureus* (VISA), vancomycin-resistant *S.aureus* (VRSA), vancomycin-resistant *Enterococci* (VRE), and methicillin-resistant strains of *S. aureus* (MESA)	[124]
Fengycin	*Bacillus subtilis*	Lipopeptide	Activity against filamentous fungi (but cannot inhibit bacteria and yeast)	[125, 34]
Iturin	*Bacillus subtilis*	Lipopeptide	Antifungal activity	[16, 126]
Tridecaptin A	*Bacillus polymyxa*	Lipopeptide	Selective and strong antimicrobial activity against some Gram-negative bacteria which includes multidrug-resistant strains of *Escherichia coli*, *Acinetobacter baumannii*, and *Klebsiella pneumoniae*	[127]

In addition, antibiofilm and antiadhesive properties of biosurfactants are also attractive for the pre-treatment of solid surfaces, such as implants or catheters, to prevent colonization by pathogenic microorganisms [94].

11.5.1.2 Antifungal

Antifungal investigation of biosurfactants has grown over the past few years. Screening of biosurfactants against several pathogenic yeasts associated with human mycoses or plant diseases is showing the potential application of biosurfactants against these types of pathogens [96, 97].

Other studies described the use of biosurfactant to biocontrol plant fungal pathogens and as a potential alternative to replace harmful agrochemicals [96, 97].

Another study described a purified rhamnolipid biosurfactant from *P. aeruginosa* isolated from oil-contaminated mangrove sediments with significant antifungal activity against *Fusarium oxysporum* wilt disease in tomato plants [96].

11.5.2 Pharmaceutical Applications

The amphiphilic properties of biosurfactants, in addition to their biodegradability and biocompatibility, have attracted the attention for pharmaceutical applications. Biosurfactants have been explored as emulsifiers, foaming, wetting and solubilizing agents, and stabilizers in different pharmaceutical applications [98]. Here, we briefly described some promising applications for this group of molecules.

11.5.2.1 Drug Delivery

The discovery of new drug delivery systems is crucial to improve current therapies and to facilitate new ones. In order to optimize drug loading capacity and to control drug release, several carriers such as polymers, particles, macromolecules, and cells are currently being developed and tested. Biosurfactants are suitable molecules for microemulsion formulation, a type of particulate carrier that has multiple advantages. These have the property for use for transdermal, topical, oral, nasal, ocular, intravenous, and parenteral drug delivery routes [88].

Recent publications with potential drug delivery application of biosurfactants include drug delivery in tumors, sustained-release for intravenous applications [99], dermal applications [100], wound healing [101], and application in photodynamic therapy [102], to cite a few examples.

11.5.2.2 Gene Delivery

Gene delivery is a safe method for introducing exogenous nucleotides into mammalian cells for clinical purposes. The use of cationic liposomes for gene transfection is considered a potential approach to deliver a foreign gene into the target cells efficiently, especially for cancer treatment [93, 103].

11.5.2.3 Immunological Adjuvants

In order to modify the immune response, immunological agents may be added to vaccines to boost the antibody productions and ensure longer-lasting protection. In the last few years, several potent non-toxic, non-pyrogenic bacterial lipopeptides have been investigated as immunological adjuvants [92].

11.5.2.4 Cosmetics

The broad potential cosmetic applications of biosurfactants include anti-aging skincare products, shower gels, hair shampoo, conditioning hair mask, skin-nourishing cosmetics, moisturizing skin cleansers, and toothpaste. In this field, biosurfactants offer a unique combination of surface tension and antimicrobial properties [100, 104].

11.6 Biosurfactants in the Food Industry: Quality of the Food

Among the large number of compounds used in the food industry, biosurfactants, due to the characteristics mentioned previously in this chapter, have established themselves as important agents in food processing, having various functions. For example, they have been used as emulsifiers, stabilizers, defoamers, and also to improve the texture of some formulations and their rheological properties [105, 106].

Other uses of the biosurfactants are related to ensuring the quality of food so that they are safe for manipulation and marketing. Due to their antimicrobial, antibiofilm/antiadhesive properties, biosurfactants prevent the development of microorganisms and biofilms in different food products [107, 108].

In this regard, Kubicki et al. [109] and Kiran et al. [110] recently mentioned that the biosurfactant from the actinobacterial strain *Nesterenkonia* sp. MSA31, isolated from a marine sponge

Fasciospongia cavernosa, could be used to sanitize production equipment and prevent food spoilage because the lipopeptide surfactant was non-toxic and showed antibiofilm activity against the food pathogen *S. aureus*.

Sharma and Satpute [111] and Nitsche and Silva [48] cite the use of biosurfactant in the materials used for food processing and packaging, to prevent, inhibit, and remove biofilms of dangerous organisms. It is known that inanimate surfaces used in food processing could be pretreated with biosurfactants like rhamnolipid and surfactin to prevent the adhesion of food spoiling and pathogenic bacteria. Both types of biosurfactant effectively disrupt preformed biofilms. On polystyrene surfaces, it has been proved that rhamnolipid and surfactin inhibit the adhesion of *L. monocytogenes*. Surfactin also reduce the adhesion of *L. monocytogenes* on stainless steel and polypropylene. Surfactin and rhamnolipid were also useful against attachment of *S. aureus*, *Micrococcus luteus*, and *L. monocytogenes* on polystyrene surfaces.

The adhesion of spores of *B. cereus* to Teflon and stainless steel surfaces was reduced using surfactin and iturin. In another study, polypropylene films were pretreated with air plasma and oxygen to adsorb a rhamnolipid and various authors demonstrated that the bacterial adhesion was reduced using this surface attachment of biosurfactants. The development of coatings or films combining with biosurfactants is a promising area of research [111].

Lactic acid bacteria are found in the human body and prevent the proliferation of pathogens. These microorganisms act in various ways in the fight against microbes. One of these is the generation of biosurfactants. The nature of these biosurfactants is complex, and although some have been characterized, others are known to be mixtures of proteins and polysaccharides. Lactic acid bacteria biosurfactants interfere with the adhesion mechanisms of pathogens to epithelial cells of intestinal and urogenital tracts. Biosurfactants isolated from *Lactobacillus* sp. can function as antibiofilm agents [112].

Abdalsadiq et al. [113] cited that biosurfactants from Lactobacilli decrease the adhesion of bacterial pathogens to silicone rubber, glass, voice prostheses, and surgical implants. Other microorganisms against which the Lactobacilli biosurfactants have shown activity are *P. aeruginosa* ATCC 278, *P. fluorescens*, *E. coli*, *E. faecium*, *Yersinia enterocolitica*, *S. Typhi*, and *S. Typhimurium* [113].

Ranasalva, Sunil, and Poovarasan [114] mentioned antimicrobial activity in two biosurfactants from probiotic bacteria (*L. lactis* and *S. thermophilus*) in biofilm development by *S. epidermidis*, *Staphylococcus salivarius*, *S. aureus*, *Rothiadentocariosa*, *C. albicans*, and *C. tropicalis* in voice prostheses [114].

The addition of biosurfactants to packaging materials is another option to study their antiadhesive properties. It has been reported that mixtures of biosurfactants isolated from *L. rhamnosus* with polyvinyl alcohol (PVA) is able to inhibit the formation of a biofilm produced by *P. aeruginosa* and *S. aureus* [48].

Due to the hemolytic activity of surfactin against red blood cells, it cannot be added directly to food. This fact causes the industry to avoid adding some of the biosurfactants to food products. Despite that, some attempts have been made concerning applications of biosurfactants to control pathogens in food. It was reported that milk treated with μg/ml of surfactin/polylysine combination (1 : 1) significantly reduced the *Salmonella enteritidis* population [115].

In the food industry other alternatives are reported, such as biosurfactants derived from *Lactobacillus* and yeasts. The growth of *E. coli* and methicillin-resistant *S. aureus* (MRSA) was inhibited with biosurfactants produced by *L. jensenii* and *L. rhamnosus* [35].

Candida apicola and *Candidabombicola* produce asophorolipid that showed lower toxicity compared to other biosurfactants such are surfactin. This biosurfactant demonstrates antimicrobial activity and can be used for vegetable and fruit sanitization [48].

Wickerhamiella domercqiae is another sophorolipid producing yeast. It was reported that this compound was useful to preserve fruit at room temperature. The biosurfactant demonstrated bacteriostatic activity and also a synergistic effect when combining with the nisin [48].

The application of biosurfactants from *Saccharomyces cerevisiae* and *P. aeruginosa* to salad dressings showed sufficient emulsion stability against control, which showed rapid separation of oil. The results of the study concluded that the new biosurfactants from microorganisms can contribute to the detection of different molecules in terms of structure and properties. Still, the toxicological aspects of these unique and current molecules should be emphasized to guarantee the safety of these compounds for food utilization [116].

Potential applications were suggested some time ago to decrease microbial fouling in dairy processing caused by attachment to a heat exchanger. Such attachments are troublesome because they can cause contamination of pasteurized milk [117].

A recent study has described a potential application of biosurfactants from *Candida utilis* UFPEDA1009 in cookie formulation. This compound showed antioxidant capacity and absence of cytotoxicity. Therefore, this biomolecule is a promissory ingredient in flour-based sweet food formulations [118].

Biosurfactants are not used as additives in the food industry despite their potential applications. Many of their properties, a wide range of types and nature of the producing microorganisms, in addition to the availability of use of food-approved downstream processing are all parameters that are constraining applications [117, 119].

Finally, a lot of research regarding food and consumer safety, as well as the required changes in the law related to the use of novel food additives, also need to be addressed before biosurfactants can openly be used in the food industry [117, 119].

11.7 Conclusions

Biosurfactants are a very diverse group of amphiphilic compounds. Due to their antiviral, antibacterial, antiadhesive, and antifungal effects, they have become the focus of a highly specific area of research.

A primary effect of biosurfactants in bacterial cells is the disruption of microbial membranes, inducing microbial cells to release extracellular substances. Additionally, biosurfactants alter microbial adhesion and motility, affecting the adhesion and detachment of bacteria and biofilm formation.

At low concentrations, biosurfactants cause apoptosis in fungi cells after binding with ATPase on the mitochondrial membrane. Also, they showed potent inhibitory effects on mycelial growth *in vitro*.

Because of its antimicrobial and antibiofilm properties, biosurfactants have a high potential as therapeutic agents and as components in pharmaceutical formulations. However, although there is a growing body of evidence and support in the literature for its antibacterial properties, only a few studies have reached preclinical and clinical phases.

Food processing is another field in which biosurfactants have established themselves as important agents. Here, they are used as emulsifiers, stabilizers, defoamers, and to improve the texture of some formulations. Other uses of these biosurfactants are related to ensuring the quality of food, making them safe for manipulation and marketing.

Because of its low toxicity, high biodegradability, and specific bioactivity properties, these biomolecules have the enormous potential to substitute the traditional antimicrobial agents for medical, veterinary, and agriculture purposes. Moreover, although we must consider the elevated cost of production and purification, high-purity biosurfactants can still be obtained at an acceptable cost. Future research in this field must be amplified and improved to obtain biosurfactants as a sustainable alternative in specific applications.

Acknowledgements

The authors thank the Health Research Institute (INISA) –UCR, School of Chemistry – UCR, and the University of Texas at El Paso (UTEP) for their support, giving us the time and access to library resources to write this chapter.

References

1 World Health Organization (2018) *Global Antimicrobial Resistance Surveillance System (GLASS). The Detection and Reporting of Colistin Resistance*, WHO Report. Available at: https://apps.who.int/iris/bitstream/handle/10665/279656/9789241515061-eng.pdf.

2 Neill, J. O. (2014) 'Antimicrobial resistance: Tackling a crisis for the health and wealth of nations. *Review on Antimicrobial Resistance. Tackling a Crisis for the Health and Wealth of Nations*' (December), Wellcome Trust Report.

3 Gandra, S. et al. (2019). The mortality burden of multidrug-resistant pathogens in India: Aretrospective, observational study. *Clinical Infectious Diseases* 69 (4): 563–570. https://doi.org/10.1093/cid/ciy955.

4 An, X.L. et al. (2018). Impact of wastewater treatment on the prevalence of integrons and the genetic diversity of integron gene cassettes. *Applied and Environmental Microbiology* 84 (9): 1–15. https://doi.org/10.1128/AEM.02766-17.

5 Gillings, M.R. and Stokes, H.W. (2012). Are humans increasing bacterial evolvability? *Trends in Ecology & Evolution* 27 (6): 346–352. https://doi.org/10.1016/j.tree.2012.02.006.

6 Ahmad, K.F., Bo, S., and Jana, J. (2019). Prevalence and diversity of antibiotic resistance genes in Swedish aquatic environments impacted by household and hospital wastewater. *Frontiers in Microbiology* 10 (April): 1–12. https://doi.org/10.3389/fmicb.2019.00688.

7 Upmanyu, N. et al. (2011). Synthesis and anti-microbial evaluation of some novel 1,2,4-triazole derivatives. *Acta Poloniae Pharmaceutica* 68 (2): 213–221.

8 Rodrigues, L. et al. (2006). Biosurfactants: Potential applications in medicine. *Journal of Antimicrobial Chemotherapy* 57 (4): 609–618. https://doi.org/10.1093/jac/dkl024.

9 Liu, K. et al. (2020). Rational design, properties, and applications of biosurfactants: A short review of recent advances. *Current Opinion in Colloid and Interface Science* 45: 57–67. https://doi.org/10.1016/j.cocis.2019.12.005.

10 Rodrigues, L.R. (2015). Microbial surfactants: Fundamentals and applicability in the formulation of nano-sized drug delivery vectors. *Journal of Colloid and Interface Science* 449: 304–316. https://doi.org/10.1016/j.jcis.2015.01.022.

11 Van Hamme, J.D., Singh, A., and Ward, O.P. (2006). Physiological aspects. Part 1 in a series of papers devoted to surfactants in microbiology and biotechnology. *Biotechnology Advances* 24 (6): 604–620. https://doi.org/10.1016/j.biotechadv.2006.08.001.

12 Sharma, D. and Saharan, B.S. (2016). Functional characterization of biomedical potential of biosurfactant produced by *Lactobacillus helveticus*. *Biotechnology Reports* 11: 27–35. https://doi.org/10.1016/j.btre.2016.05.001.

13 Lai, S., Tremblay, J., and Déziel, E. (2009). Swarming motility: A multicellular behaviour conferring antimicrobial resistance. *Environmental Microbiology* 11 (1): 126–136. https://doi.org/10.1111/j.1462-2920.2008.01747.x.

14 Cochrane, S.A. and Vederas, J.C. (2016). Lipopeptides from *Bacillus* and *Paenibacillus* spp.: A gold mine of antibiotic candidates. *Medicinal Research Reviews* 36 (1): 4–31. https://doi.org/10.1002/med.21321.

15 Meena, K.R., Sharma, A., and Kanwar, S.S. (2017). Microbial Lipopeptides and their medical applications. *Annals of Pharmacology and Pharmaceutics* 2 (21): 1–5.

16 Ahimou, F., Jacques, P., and Deleu, M. (2000). Surfactin and iturin A effects on *Bacillus subtilis* surface hydrophobicity. *Enzyme and Microbial Technology* 27 (10): 749–754. https://doi.org/10.1016/S0141-0229(00)00295-7.

17 Arutchelvi, J.I. et al. (2008). Mannosylerythritol lipids: A review. *Journal of Industrial Microbiology and Biotechnology* 35 (12): 1559–1570. https://doi.org/10.1007/s10295-008-0460-4.

18 Rufino, R.D. et al. (2011). Antimicrobial and anti-adhesive potential of a biosurfactant Rufisan produced by *Candida lipolytica* UCP 0988. *Colloids and Surfaces B: Biointerfaces* 84 (1): 1–5. https://doi.org/10.1016/j.colsurfb.2010.10.045.

19 Benincasa, M. et al. (2004). Chemical structure, surface properties and biological activities of the biosurfactant produced by *Pseudomonas aeruginosa* LBI from soapstock. *Antonie van Leeuwenhoek* 85 (1): 1–8. https://doi.org/10.1023/B:ANTO.0000020148.45523.41.

20 Rodrigues, L., Van Der Mei, H.C. et al. (2004). Influence of biosurfactants from probiotic bacteria on formation of biofilms on voice prostheses. *Applied and Environmental Microbiology* 70 (7): 4408–4410. https://doi.org/10.1128/AEM.70.7.4408-4410.2004.

21 Velkov, T. et al. (2010). Structure – Activity relationships of Polymyxin antibiotics. *Journal of Medicinal Chemistry* 53 (5): 1898–1916. https://doi.org/10.1021/jm900999h.

22 Wang, X. et al. (2017). *Bacillus subtilis* and surfactin inhibit the transmissible gastroenteritis virus from entering the intestinal epithelial cells. *Bioscience Reports* 37 (2): 1–10. https://doi.org/10.1042/BSR20170082.

23 Yuan, L. et al. (2018). Surfactin inhibits membrane fusion during invasion of epithelial cells by enveloped viruses. *Journal of Virology* 92 (21): 1–19. https://doi.org/10.1128/jvi.00809-18.

24 Qi, G. et al. (2010). Lipopeptide induces apoptosis in fungal cells by a mitochondria-dependent pathway. *Peptides* 31 (11): 1978–1986. https://doi.org/10.1016/j.peptides.2010.08.003.

25 Naruse, N., Tenmyo, O., Kobaru, S. et al. (1990). Pumilacidin, a complex of new antiviral antibiotics: Production, isolation, chemical properties, structure and biological activity. *The Journal of Antibiotics* 43 (3): 267–280. https://doi.org/10.7164/antibiotics.43.267.

26 Saravanakumari, P. and Mani, K. (2010). Structural characterization of a novel xylolipid biosurfactant from *Lactococcus lactis* and analysis of antibacterial activity against multi-drug resistant pathogens. *Bioresource Technology* 101 (22): 8851–8854. https://doi.org/10.1016/j.biortech.2010.06.104.

27 Tahmourespour, A., Kasra-Kermanshahi, R., and Salehi, R. (2019). *Lactobacillus rhamnosus* biosurfactant inhibits biofilm formation and gene expression of caries-inducing *Streptococcus mutans*. *Dental Research Journal* 16 (2): 87–94.

28 Walencka, E., Różalska, S., Sadowska, B., and Różalska, B. (2008). The influence of *Lactobacillus acidophilus*-derived surfactants on staphylococcal adhesion and biofilm formation. *Folia Microbiologica* 53 (1): 61–66. https://doi.org/10.1007/s12223-008-0009-y.

29 Rodrigues, L., Van Der Mei, H. et al. (2004). Biosurfactant from *Lactococcus lactis* 53 inhibits microbial adhesion on silicone rubber. *Applied Microbiology and Biotechnology* 66 (3): 306–311. https://doi.org/10.1007/s00253-004-1674-7.

30 Kaczorek, E. et al. (2018). The impact of biosurfactants on microbial cell properties leading to hydrocarbon bioavailability increase. *Colloids and Interfaces* 2 (3): 35. https://doi.org/10.3390/colloids2030035.

31 Liu, G. et al. (2018). Advances in applications of rhamnolipids biosurfactant in environmental remediation: A review. *Biotechnology and Bioengineering* 115 (4): 796–814. https://doi.org/10.1002/bit.26517.

32 Otzen, D.E. (2017). Biosurfactants and surfactants interacting with membranes and proteins: Same but different? *Biochimica et Biophysica Acta – Biomembranes* 1859 (4): 639–649. https://doi.org/10.1016/j.bbamem.2016.09.024.

33 Chen, W.-C., Juang, R.-S., and Wei, Y.-H. (2015). Applications of a lipopeptide biosurfactant, surfactin, produced by microorganisms. *Biochemical Engineering Journal* 103 (November): 158–169. https://doi.org/10.1016/j.bej.2015.07.009.

34 Knight, C.A. et al. (2018). The first report of antifungal lipopeptide production by a *Bacillus subtilis* subsp. inaquosorum strain. *Microbiological Research* 216 (April): 40–46. https://doi.org/10.1016/j.micres.2018.08.001.

35 Sambanthamoorthy, K., Feng, X., Patel, R. et al. (2014). Antimicrobial and antibiofilm potential of biosurfactants isolated from *Lactobacilli* against multi-drug-resistant pathogens. *BMC Microbiology* 14 (1) https://doi.org/10.1186/1471-2180-14-197.

36 Giuliani, A., Pirri, G., and Nicoletto, S.F. (2007). *Antimicrobial peptides: An overview of a promising class of therapeutics. Central European Journal of Biology* https://doi.org/10.2478/s11535-007-0010-5.

37 Miller, W.R., Bayer, A.S., and Arias, C.A. (2016). Mechanism of action and resistance to daptomycin in *Staphylococcus aureus* and enterococci. *Cold Spring Harbor Perspectives in Medicine* 6 (11) https://doi.org/10.1101/cshperspect.a026997.

38 Mandal, S.M., Barbosa, A.E.A.D., and Franco, O.L. (2013). Lipopeptides in microbial infection control: scope and reality for industry. *Biotechnology Advances* 31 (2): 338–345. https://doi.org/10.1016/j.biotechadv.2013.01.004.

39 Li, J. et al. (2006). Colistin: the re-emerging antibiotic for multidrug-resistant gram-negative bacterial infections. *Lancet Infectious Diseases* 6 (9): 589–601. https://doi.org/10.1016/S1473-3099(06)70580-1.

40 Falagas, M.E. and Kasiakou, S.K. (2006). Toxicity of polymyxins: A systematic review of the evidence from old and recent studies. *Critical Care* 10 (1) https://doi.org/10.1186/cc3995.

41 Rodrigues, L.R. and Teixeira, J.A. (2010). Biomedical and therapeutic applications of biosurfactants. *Advances in Experimental Medicine and Biology* 672: 75–87. https://doi.org/10.1007/978-1-4419-5979-9_6.

42 Kitamoto, D., Isoda, H., and Nakahara, T. (2002). Functions and potential applications of glycolipid biosurfactants –From energy-saving materials to gene delivery carriers. *Journal of Bioscience and Bioengineering* 94 (3): 187–201. https://doi.org/10.1263/jbb.94.187.

43 Desai, J.D. and Banat, I.M. (1997). Microbial production of surfactants and their commercial potential. *Fuel and Energy Abstracts* 38 (4): 221. https://doi.org/10.1016/S0140-6701(97)84559-6.

44 Lovaglio, R.B. et al. (2015). Rhamnolipids know-how: Looking for strategies for its industrial dissemination. *Biotechnology Advances* 33 (8): 1715–1726. https://doi.org/10.1016/j.biotechadv.2015.09.002.

45 Abdel-Mawgoud, A.M., Lépine, F., and Déziel, E. (2010). Rhamnolipids: Diversity of structures, microbial origins and roles. *Applied Microbiology and Biotechnology* 86 (5): 1323–1336. https://doi.org/10.1007/s00253-010-2498-2.

46 Drakontis, C.E. and Amin, S. (2020). Biosurfactants: Formulations, properties, and applications. *Current Opinion in Colloid & Interface Science* https://doi.org/10.1016/j.cocis.2020.03.013.

47 de Freitas Ferreira, J., Vieira, E.A., and Nitschke, M. (2019). The antibacterial activity of rhamnolipid biosurfactant is pH dependent. *Food Research International* 116: 737–744. https://doi.org/10.1016/j.foodres.2018.09.005.

48 Nitschke, M. and Silva, S.S.E. (2018). Recent food applications of microbial surfactants. *Critical Reviews in Food Science and Nutrition* 58 (4): 631–638. https://doi.org/10.1080/10408398.2016.1208635.

49 Magalhães, L. and Nitschke, M. (2013). Antimicrobial activity of rhamnolipids against *Listeria* monocytogenes and their synergistic interaction with nisin. *Food Control* 29 (1): 138–142. https://doi.org/10.1016/j.foodcont.2012.06.009.

50 Joshi-Navare, K. and Prabhune, A. (2013). A biosurfactant-sophorolipid acts in synergy with antibiotics to enhance their efficiency. *BioMed Research International* 2013 https://doi.org/10.1155/2013/512495.

51 Lydon, H.L., Baccile, N., Callagaghan, B. et al. (2017). Adjuvant antibiotic activity of acidic Sophorolipids with potential for facilitating wound healing. *Antimicrobial Agents and Chemotherapy* 61 (5) https://doi.org/10.1128/AAC.02547-16.

52 Zhang, X., Ashby, R.D., Solaiman, D.K.Y. et al. (2017). Antimicrobial activity and inactivation mechanism of lactonic and free acid sophorolipids against *Escherichia coli* O157:H7. *Biocatalysis and Agricultural Biotechnology* 11: 176–182. https://doi.org/10.1016/j.bcab.2017.07.002.

53 Franzetti, A. et al. (2010). Production and applications of trehalose lipid biosurfactants. *European Journal of Lipid Science and Technology* 112 (6): 617–627. https://doi.org/10.1002/ejlt.200900162.

54 Kügler, J.H. et al. (2015). Surfactants tailored by the class Actinobacteria. *Frontiers in Microbiology* 6 (March) https://doi.org/10.3389/fmicb.2015.00212.

55 Janek, T., Krasowski, A., Czyznikowska, Z., and Łukaszewicz, M. (2018). Trehalose lipid biosurfactant reduces adhesion of microbial pathogens to polystyrene and silicone surfaces: An experimental and computational approach. *Frontiers in Microbiology* 9 (OCT): 1–14. https://doi.org/10.3389/fmicb.2018.02441.

56 Rivardo, F., Turner, R.J., Allegrone, G. et al. (2009). Anti-adhesion activity of two biosurfactants produced by *Bacillus* spp. prevents biofilm formation of human bacterial pathogens. *Applied Microbiology and Biotechnology* 83 (3): 541–553. https://doi.org/10.1007/s00253-009-1987-7.

57 Samadi, N., Abadian, N., Ahmadkhanika, R. et al. (2012). Structural characterization and surface activities of biogenic rhamnolipid surfactants from *Pseudomonas aeruginosa* isolate MN1 and synergistic effects against methicillin-resistant *Staphylococcus aureus*. *Folia Microbiologica* 57 (6): 501–508. https://doi.org/10.1007/s12223-012-0164-z.

58 Sana, S., Datta, S., Biswas, D., and Sengupta, D. (2018). Assessment of synergistic antibacterial activity of combined biosurfactants revealed by bacterial cell envelop damage. *Biochimica et Biophysica Acta – Biomembranes* 1860 (2): 579–585. https://doi.org/10.1016/j.bbamem.2017.09.027.

59 Delbeke, E.I.P., Roman, B.I., Marin, G.B. et al. (2015). A new class of antimicrobial biosurfactants: Quaternary ammonium sophorolipids. *Green Chemistry* 17 (6): 3373–3377. https://doi.org/10.1039/c5gc00120j.

60 Delbeke, E.I.P. et al. (2019). Lipid-based quaternary ammonium sophorolipid amphiphiles with antimicrobial and transfection activities. *ChemSusChem* 12 (15): 3642–3653. https://doi.org/10.1002/cssc.201900721.

61 Capon, R.J. and Faulkner, D.J. (1984). Antimicrobial metabolites from a Pacific sponge, *Agelas* sp. *Journal of the American Chemical Society* 106 (6): 1819–1822. https://doi.org/10.1021/ja00318a045.

62 Gonçalves, A.P. et al. (2017). Regulated forms of cell death in fungi. *Frontiers in Microbiology* 8 (SEP) https://doi.org/10.3389/fmicb.2017.01837.

63 Costerton, J.W. et al. (1995). Microbial biofilms. *Annual Review of Microbiology* 49: 711–745.

64 Donlan, R.M. (2002). Biofilms: Microbial life on surfaces. *Emerging Infectious Diseases* 8 (9): 881–890. https://doi.org/10.3201/eid0809.020063.

65 Banat, I.M., De Rienzo, M.A.D., and Quinn, G.A. (2014). Microbial biofilms: Biosurfactants as antibiofilm agents. *Applied Microbiology and Biotechnology* 98 (24): 9915–9929. https://doi.org/10.1007/s00253-014-6169-6.

66 Marshall, K.C., Stout, R., and Mitchell, R. (1971). Mechanism of the initial events in the sorption of marine bacteria to surfaces. *Journal of General Microbiology* 68 (3): 337–348. https://doi.org/10.1099/00221287-68-3-337.

67 Sauer, K. et al. (2002). Displays multiple phenotypes during development as a biofilm. *Society* 184 (4): 1140–1154. https://doi.org/10.1128/JB.184.4.1140.

68 Wolfmeier, H. et al. (2018). New perspectives in biofilm eradication. *ACS Infectious Diseases* 4 (2): 93–106. https://doi.org/10.1021/acsinfecdis.7b00170.

69 Splendiani, A., Livingston, A.G., and Nicolella, C. (2006). Control of membrane-attached biofilms using surfactants. *Biotechnology and Bioengineering* 94 (1): 15–23. https://doi.org/10.1002/bit.20752.

70 Wicken, A.J. and Knox, K.W. (1980). Bacterial cell surface amphiphiles. *Biochimica et Biophysica Acta (BBA) – Reviews on Biomembranes* 604: 1–26. https://doi.org/10.1016/0304-4157(80)90002-7.

71 Oppenheimer-Shaanan, Y., Steinberg, N., and Kolodkin-Gal, I. (2013). Small molecules are natural triggers for the disassembly of biofilms. *Trends in Microbiology* 21 (11): 594–601. https://doi.org/10.1016/j.tim.2013.08.005.

72 Satpute, S.K. et al. (2019). Inhibition of pathogenic bacterial biofilms on PDMS based implants by *L. acidophilus* derived biosurfactant. *BMC Microbiology* 19 (1): 1–15. https://doi.org/10.1186/s12866-019-1412-z.

73 Gawande, P.V., Leung, K.P., and Madhyastha, S. (2014). Antibiofilm and antimicrobial efficacy of Dispersinb®-KSL-w peptide-based wound gel against chronic wound infection associated bacteria. *Current Microbiology* 68 (5): 635–641. https://doi.org/10.1007/s00284-014-0519-6.

74 Kannan, S. et al. (2019). Effect of biosurfactant derived from *Vibrio natriegens* MK3 against *Vibrio harveyi* biofilm and virulence. *Journal of Basic Microbiology* 59 (9): 936–949. https://doi.org/10.1002/jobm.201800706.

75 Yan, X. et al. (2019). Antimicrobial, anti-adhesive and anti-biofilm potential of biosurfactants isolated from *Pediococcus acidilactici* and *Lactobacillus plantarum* against *Staphylococcus aureus* CMCC26003. *Microbial Pathogenesis* 127: 12–20. https://doi.org/10.1016/j.micpath.2018.11.039.

76 Giri, S.S. et al. (2019). Antioxidant, antibacterial, and anti-adhesive activities of biosurfactants isolated from *Bacillus* strains. *Microbial Pathogenesis* 132: 66–72. https://doi.org/10.1016/j.micpath.2019.04.035.

77 Singh, A.K. and Sharma, P. (2020). Disinfectant-like activity of lipopeptide biosurfactant produced by *Bacillus tequilensis* strain SDS21. *Colloids and Surfaces B: Biointerfaces* 185 (September 2019): 110514. https://doi.org/10.1016/j.colsurfb.2019.110514.

78 Ohadi, M. et al. (2020). Antimicrobial, anti-biofilm, and anti-proliferative activities of lipopeptide biosurfactant produced by *Acinetobacter junii* B6. *Microbial Pathogenesis* 138 (February 2019): 103806. https://doi.org/10.1016/j.micpath.2019.103806.

79 Vollenbroich, D., Özel, M., Vater, J. et al. (1997). Mechanism of inactivation of enveloped viruses by the biosurfactant surfactin from *Bacillus subtilis*. *Biologicals* 25 (3): 289–297. https://doi.org/10.1006/biol.1997.0099.

80 Kracht, M. et al. (1999). Antiviral and hemolytic activities of surfactin isoforms and their methyl ester derivatives and a hydrophobic/J-hydroxy fatty acid varying in length. *Journal of Antibiotics* 52 (7): 613–619.

81 Huang, X. et al. (2006). Antiviral activity of antimicrobial lipopeptide from *Bacillus subtilis* fmbj against pseudorabies virus, porcine parvovirus, Newcastle disease virus and infectious bursal disease virus in vitro;. *International Journal of Peptide Research and Therapeutics* 12 (4): 373–377. https://doi.org/10.1007/s10989-006-9041-4.

82 Basit, M. et al. (2018). Biosurfactants production potential of native strains of *Bacilluscereus* and their antimicrobial, cytotoxic and antioxidant activities. *Pakistan Journal of Pharmaceutical Sciences* 31 (1): 251–256.

83 Lima, W.G. et al. (2018). Absence of antibacterial, anti-candida, and anti-dengue activities of surfactin isolated from *Bacillus subtilis*. *Journal of Pharmaceutical Negative Results* 9 (43): 39. https://doi.org/10.4103/jpnr.JPNR.

84 Remichkova, M. et al. (2008). Anti-herpesvirus activities of *Pseudomonas* sp. S-17 rhamnolipid and its complex with alginate. *Zeitschrift fur Naturforschung – Section C Journal of Biosciences* 63 (1–2): 75–81. https://doi.org/10.1515/znc-2008-1-214.

85 Shah, V. et al. (2005). Sophorolipids, microbial glycolipids with anti-human immunodeficiency virus and sperm-immobilizing activities. *Antimicrobial Agents and Chemotherapy* 49 (10): 4093–4100. https://doi.org/10.1128/AAC.49.10.4093-4100.2005.

86 Borsanyiova, M. et al. (2016). Biological activity of sophorolipids and their possible use as antiviral agents. *Folia Microbiologica* 61: 85–89. https://doi.org/10.1007/s12223-015-0413-z.

87 Fracchia, L. et al. (2015). Potential therapeutic applications of microbial surface-active compounds. *AIMS Bioengineering* 2 (3): 144–162.

88 Gudiña, E.J., Rangarajan, V., Sen, R., and Rodrigues, L.R. (2013). Potential therapeutic applications of biosurfactants. *Trends in Pharmacological Sciences* 34 (12): 667–675. Available at: https://doi.org/10.1016/j.tips.2013.10.002.

89 Bhattacharya, B., Ghosh, T.K., and Das, N. (2017). Application of bio-surfactants in cosmetics and pharmaceutical industry. *Scholars Academic Journal of Phamracy* 6 (7): 320–329.

90 Robbel, L. and Marahiel, M.A. (2010). Daptomycin, a bacterial lipopeptide synthesized by a nonribosomal machinery. *Journal of Biological Chemistry* 285 (36): 27501–27508. https://doi.org/10.1074/jbc.R110.128181.

91 Tally, F.P. et al. (1999). Daptomycin: A novel agent for Gram-positive infections. *Expert Opinion on Investigational Drugs* 8 (8): 1223–1238. https://doi.org/10.1517/13543784.8.8.1223.

92 Park, S.Y. and Kim, Y.H. (2009). urfactin inhibits immunostimulatory function of macrophages through blocking NK-κB, MAPK and Akt pathway. *International Immunopharmacology* 9 (7–8): 886–893. https://doi.org/10.1016/j.intimp.2009.03.013.

93 Gharaei-Fathabad, E. (2011). Biosurfactants in pharmaceutical industry. *American Journal of Drug Disocvery and Development* 1 (1): 58–69.

94 Ceresa, C. et al. (2017). Synergistic activity of antifungal drugs and lipopeptide AC7 against *Candida albicans* biofilm on silicone. *AIMS Bioengineering* 4 (2): 318–334.

95 Itapary dos Santos, C. et al. (2019). Antifungal and antivirulence activity of vaginal *Lactobacillus* spp. products against *Candida* vaginal isolates. *Pathogens* 8 (3): 150. https://doi.org/10.3390/pathogens8030150.

96 Deepika, K.V., Ramu Sridhar, P., and Bramhachari, P.V. (2015). Characterization and antifungal properties of rhamnolipids produced by mangrove sediment bacterium *Pseudomonas aeruginosa* strain KVD-HM52. *Biocatalysis and Agricultural Biotechnology* 4 (4): 608–615. https://doi.org/10.1016/j.bcab.2015.09.009.

97 Wu, S. et al. (2019). Characterization of antifungal lipopeptide biosurfactants produced by marine bacterium *Bacillus* sp. CS30. *Marine Drugs* 17 (4) https://doi.org/10.3390/md17040199.

98 Santos, D.K.F. et al. (2016). Biosurfactants: Multifunctional biomolecules of the 21st century. *International Journal of Molecular Sciences* https://doi.org/10.3390/ijms17030401.

99 Katiyar, S.S. et al. (2019). Novel biosurfactant and lipid core-shell type nanocapsular sustained release system for intravenous application of methotrexate. *International Journal of Pharmaceutics* 557 (December 2018): 86–96. https://doi.org/10.1016/j.ijpharm.2018.12.043.

100 Knoth, D. et al. (2019). Evaluation of a biosurfactant extract obtained from corn for dermal application. *International Journal of Pharmaceutics* 564 (January): 225–236. https://doi.org/10.1016/j.ijpharm.2019.04.048.

101 Gupta, S. et al. (2017). Accelerated *in vivo;* wound healing evaluation of microbial glycolipid containing ointment as a transdermal substitute. *Biomedicine & Pharmacotherapy* 94: 1186–1196.

102 Gliszczyńska, A. et al. (2016). Synthesis and biological evaluation of novel phosphatidylcholine analogues containing monoterpene acids as potent antiproliferative agents. *PLoS One* 11 (6): 1–18. https://doi.org/10.1371/journal.pone.0157278.

103 Ribeiro, M.H.L. et al. (2019). Lipoaminoacids enzyme-based production and application as gene delivery vectors. *Catalysts* 9 (12): 1–18. https://doi.org/10.3390/catal9120977.

104 Sandeep, L. (2017). Biosurfactant: Pharmaceutical perspective. *Journal of Analytical & Pharmaceutical Research* 4 (3): 19–21. https://doi.org/10.15406/japlr.2017.04.00105.

105 De, S. et al. (2015). A review on natural surfactants. *RSC Advances* 5 (81): 65757–65767. https://doi.org/10.1039/c5ra11101c.

106 Md, F. (2012). Biosurfactant: Production and application. *Journal of Petroleum & Environmental Biotechnology* 03 (04) https://doi.org/10.4172/2157-7463.1000124.

107 Diniz, R., Soares, A., Pereira, A., and Naves, E. (2013). Antimicrobial and anti-adhesive potential of a biosurfactants produced by *Candida* species. In: Practical Applications in Biomedical Engineering (ed. A. Andrade), 246–256. London, UK: InTech https://doi.org/10.5772/52578.

108 Karthik, R. et al. (2013). Efficacy of Bacteriocin from *Lactobacillus* sp. (AMET 1506) as a biopreservative for seafood' s under different storage temperature conditions. *Journal of Modern Biotechnology* 2 (3): 59–65.

109 Kubicki, S., Bollinger, A., Katzke, N. et al. (2019). Marine biosurfactants: Biosynthesis, structural diversity and biotechnological applications. *Marine Drugs* 17 (7): 1–30. https://doi.org/10.3390/md17070408.

110 Kiran, G.S., Privadharsini, S., Sajayan, A. et al. (2017). Production of lipopeptide biosurfactant by a marine *Nesterenkonia* sp. and its application in food industry. *Frontiers in Microbiology* 8 (JUN): 1–11. https://doi.org/10.3389/fmicb.2017.01138.

111 Sharma, D. and Satpute, S.K. (2018). Recent updates on biosurfactants in the food industry. *Microbial Cell Factories*: 1–20. https://doi.org/10.1201/b22219-1.

112 Meena, K.R., Sharma, A., and Kanwar, S.S. (2020). Antitumoral and antimicrobial activity of surfactin extracted from *Bacillus subtilis* KLP2015. *International Journal of Peptide Research and Therapeutics* 26: 423–433. Available at: https://doi.org/10.1007/s10989-019-09848-w.

113 Abdalsadiq, N., Hassan, Z., Nilai, B.B. et al. (2018). Characterization of the physicochemical properties of the biosurfactant produced by *L. acidophilus* and *L. pentosus*. *EPH – International Journal of Science And Engineering* 4 (7): 1–17. https://doi.org/10.13140/RG.2.2.23831.42409.

114 Ranasalva, N., Sunil, R., and Poovarasan, G. (2014). Importance of Biosurfactantinfood industry. *IOSR Journal of Agriculture and Veterinary Science* 7 (5): 06–09. https://doi.org/10.9790/2380-07540609.

115 Huang, X., Suo, J., and Cui, Y. (2011). Optimization of antimicrobial activity of surfactin and polylysine against *Salmonella enteritidis* in milk evaluated by a response surface methodology. *Foodborne Pathogens and Disease* 8 (3): 439–443. https://doi.org/10.1089/fpd.2010.0738.

116 Sridhar, B., Karthik, R., Pushpam, A.C., and Vanitha, M.C. (2015). Biosurfactants from production and purification. *International Journal of Advanced Research in Engineering and Technology* 6 (10): 97–104.

117 Campos, J.M. et al. (2013). Microbial biosurfactants as additives for food industries. *Biotechnology Progress* 29 (5): 1097–1108. https://doi.org/10.1002/btpr.1796.

118 Ribeiro, B.G. et al. (2020). Biosurfactant produced by *Candida utilis* UFPEDA1009 with potential application in cookie formulation. *Electronic Journal of Biotechnology*: 116544. https://doi.org/10.1016/j.ejbt.2020.05.001.

119 Luxminarayan, L. et al. (2017). A review on chromatography techniques. *Asian Journal of Pharmaceutical Research and Development* 5 (2): 1–8.

120 Heidary, M., Khosravi, A.D., Khoshnood, S. et al. (2018). Daptomycin. *J. Antimicrob. Chemother.* 73 (1): 1–11. https://doi.org/10.1093/jac/dkx349.

121 Itoh, Y., Wang, X., Hinnebusch, B.J. et al. (2005). Depolymerization of b-1,6-*N*-acetyl-D-glucosamine disrupts the integrity of diverse bacterial biofilms. *J. Bacteriol.* 187: 382–387.

122 Nakhate, P.H., Yadav, V.K., and Pathak, A.N. (2013). A Review on Daptomycin: The first US-FDA approved Lipopeptide antibiotics. *J. Sci. Innov. Res.* 2 (5): 970–980.

123 Shaheen, M., Li, J., Ross, A.C. et al. (2011). *Paenibacilluspolymyxa* PKB1 produces variants of polymyxin B-type antibiotics. *Chem. Biol.* 18 (12): 1640–1648. https://doi.org/10.1016/j.chembiol.2011.09.017.

124 Kleijn, L.H.J., Oppedijk, S.F., Hart, P. et al. (2016). *Journal of Medicinal Chemistry* 59 (7): 3569–3574. https://doi.org/10.1021/acs.jmedchem.6b00219.

125 Meena, K.R. and Kanwar, S.S. (2015). Lipopeptides as the antifungal and antibacterial agents: Applications in food safety and therapeutics. *Biomed. Res. Int.* 2015 https://doi.org/10.1155/2015/473050.

126 Yuan, J., Raza, W., Huang, Q., and Shen, Q. (2011). Quantification of the antifungal lipopeptide iturin A by high performance liquid chromatography coupled with aqueous two-phase extraction. *J. Chromatogr. B Anal. Technol. Biomed. Life Sci.* 879 (26): 2746–2750. https://doi.org/10.1016/j.jchromb.2011.07.04.

127 Satchwell, K.L. (2016). Application of Microcin N and Tridecaptin A 1 to control bacterial. *pathogens.* Available at: https://era.library.ualberta.ca/items/2ac688d4-d0a1-403a-935a-59ee6c2fb4e5/view/4dab6bec-b2b4-4c3b-afe2-652e63d31315/Satchwell_Katherine_L_201605_MSc.pdf.

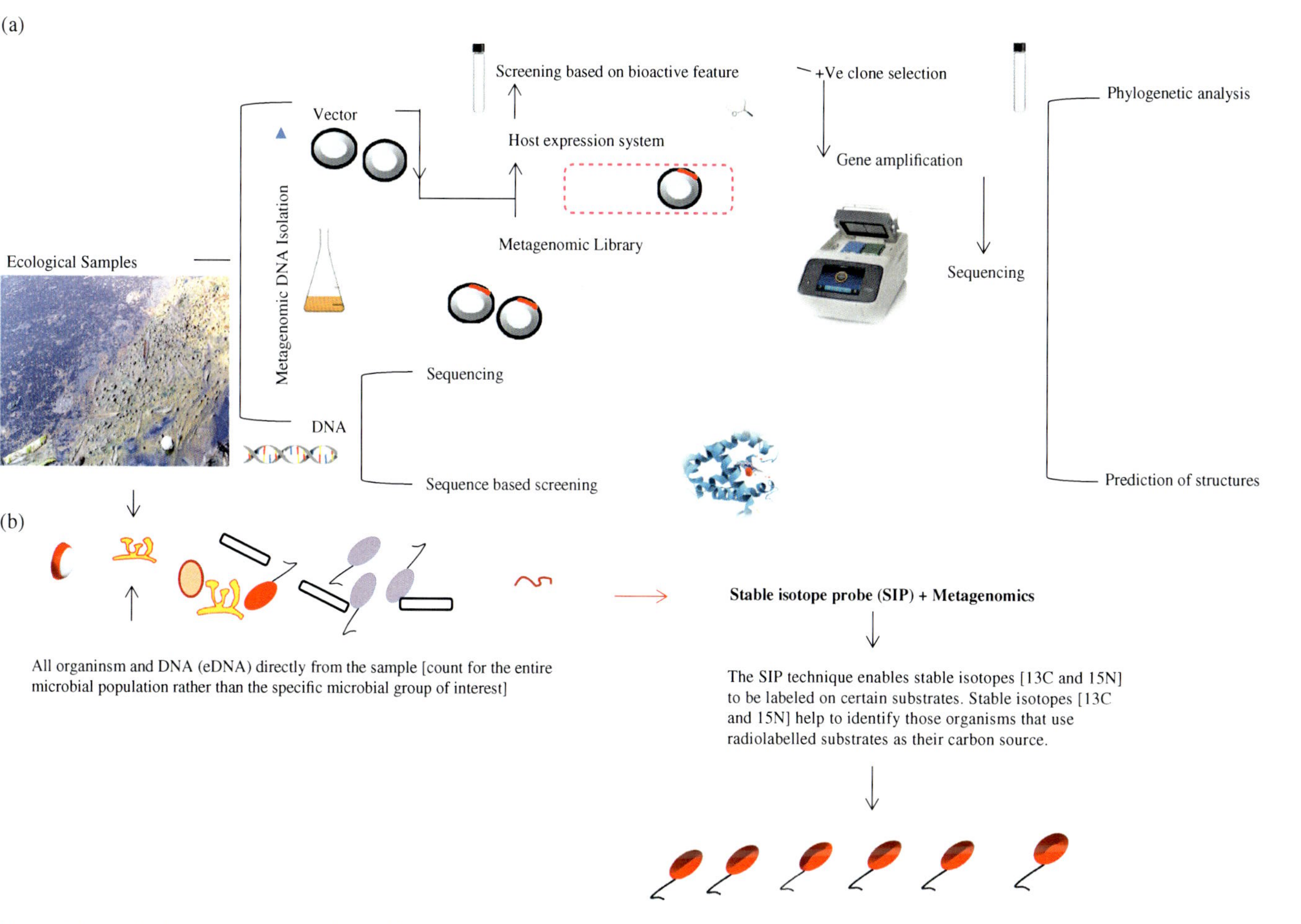

Figure 2.2 Collation of metagenomics and DNA stable isotope probe (DNA-SIP): (a) Isolation of metagenomic DNA from environmental samples. Metagenomic DNA is subject to either sequence-based or functional-based screening. Sequence-based screening involves NGS and PCR amplification by means of designed probes and primers based on known gene sequences. Functional-based screening involves the cloning of environmental DNA into an appropriate vector for the construction of the metagenomic clone library. Single or multiple host systems are used to express genes in a library. (b) Environmental samples are exposed to stable isotope-labeled substrates, with incorporation of heavier isotopes in the DNA of microorganism-consuming substrates during incubation and separation of radiolabeled microbial DNA from non-labeled gradient density centrifugation. Isolated DNA is subject to PCR amplification for microbial phylogeny or to the construction of a metagenomics library for functional screening.

Biosurfactants for a Sustainable Future: Production and Applications in the Environment and Biomedicine, First Edition. Edited by Hemen Sarma and Majeti Narasimha Vara Prasad.
© 2021 John Wiley & Sons Ltd. Published 2021 by John Wiley & Sons Ltd.

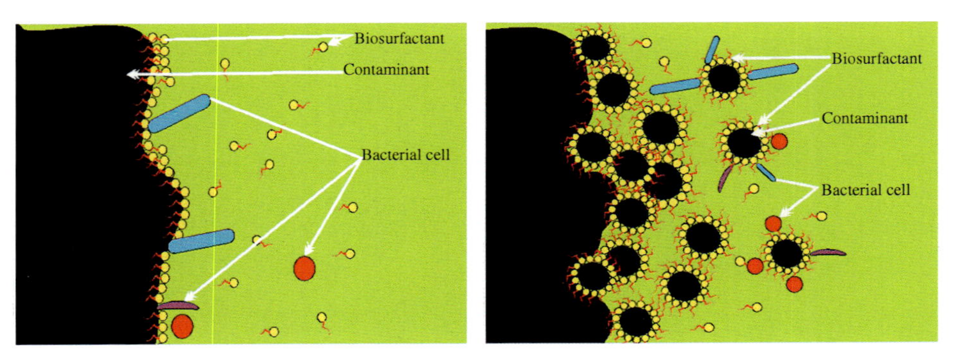

Figure 3.1 Schematic representation of the adhesion of bio surfactant molecules to the containment in which bacterial cell is associated (Guerra-Santos et al., 1984).

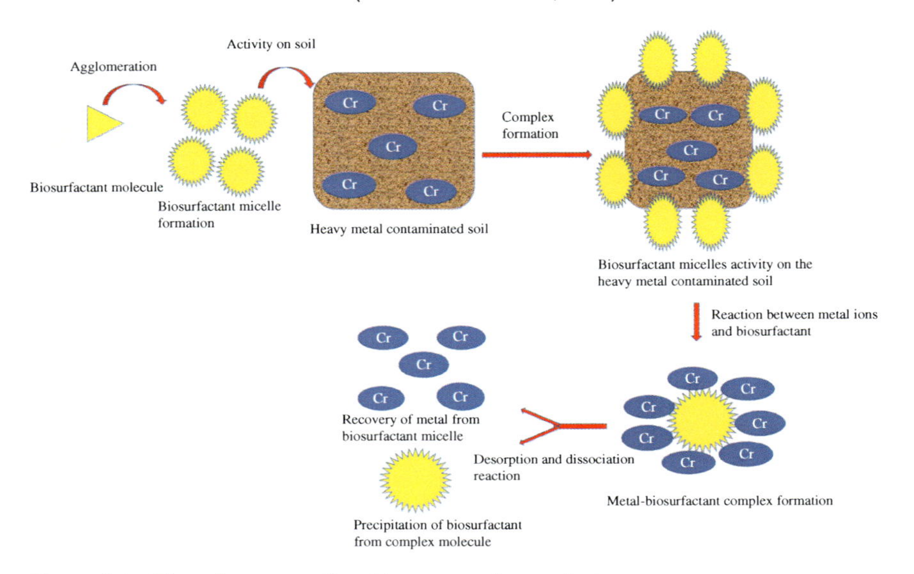

Figure 4.1 Biosurfactant mediated heavy metal remediation.

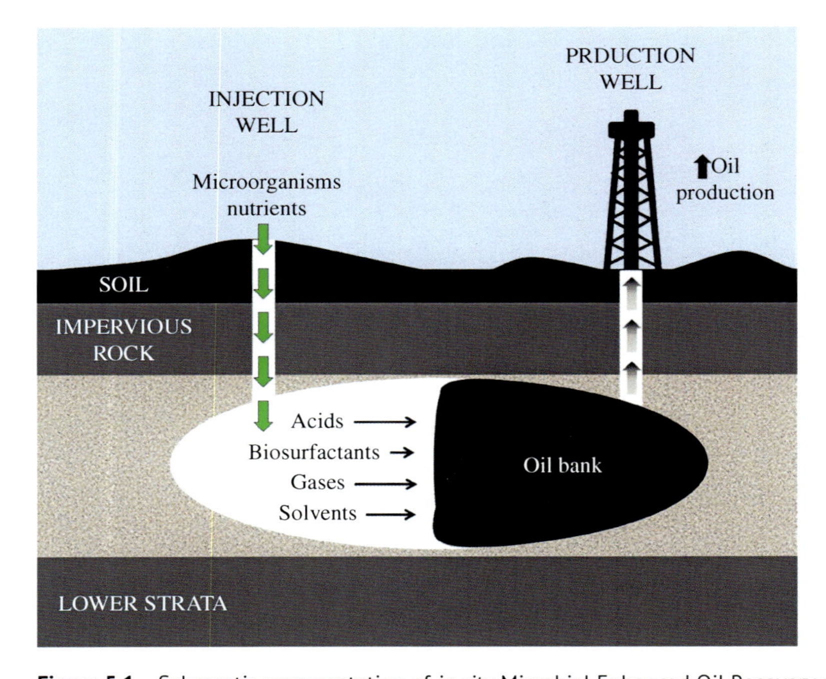

Figure 5.1 Schematic representation of *in situ* Microbial Enhanced Oil Recovery process. *Source:* Adapted from Sen [12].

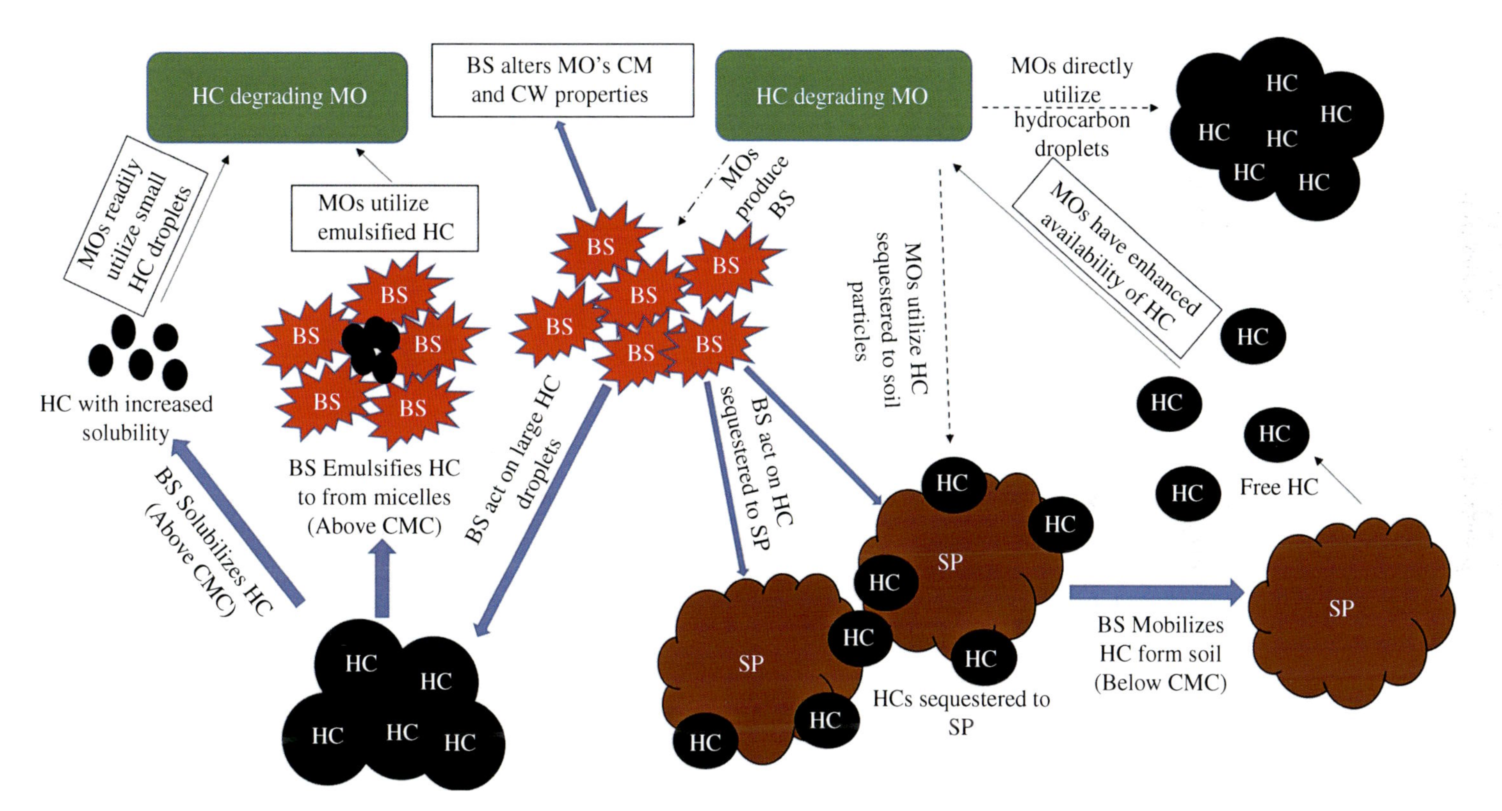

Figure 6.2 Mode of action of biosurfactants for enhanced hydrocarbon degradation. MO: microorganism, HC: hydrocarbon, SP: soil particle; BS: biosurfactant; CM: cell membrane; CW: cell wall.

Figure 7.1 Major sources of PAHs in the environment.

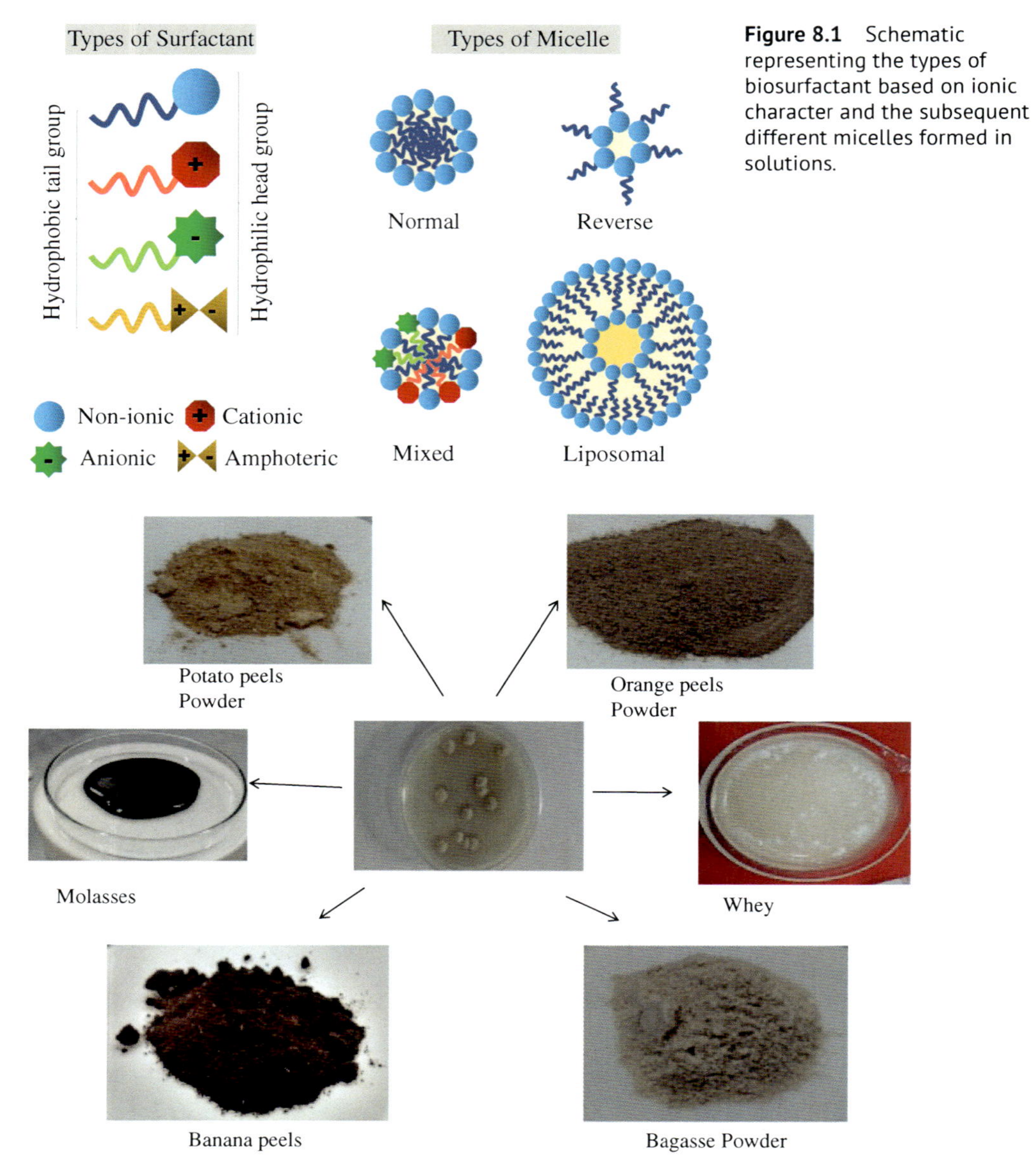

Figure 8.1 Schematic representing the types of biosurfactant based on ionic character and the subsequent different micelles formed in solutions.

Figure 9.3 Different types of agro-industrial wastes used as carbon/nitrogen sources for biosurfactant production [46].

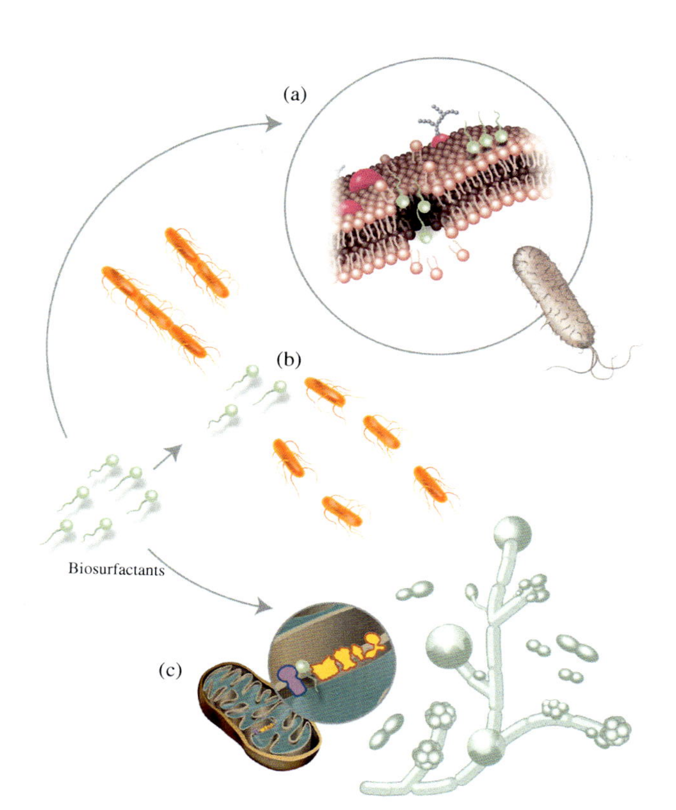

Figure 11.1 Antimicrobial effects of biosurfactants: (a) **Disruption of microbial membranes.** Biosurfactants cause significant alterations on cell surface structures by interactions with phospholipid membranes and proteins. (b) **Alteration of microbial adhesion and motility.** Biosurfactants affect the adhesion and detachment of bacteria and biofilm formation. (c) **Apoptosis in fungi.** WH1 fungin induce apoptosis at low concentrations by binding with ATPase on the mitochondrial membrane. See the text for more details.

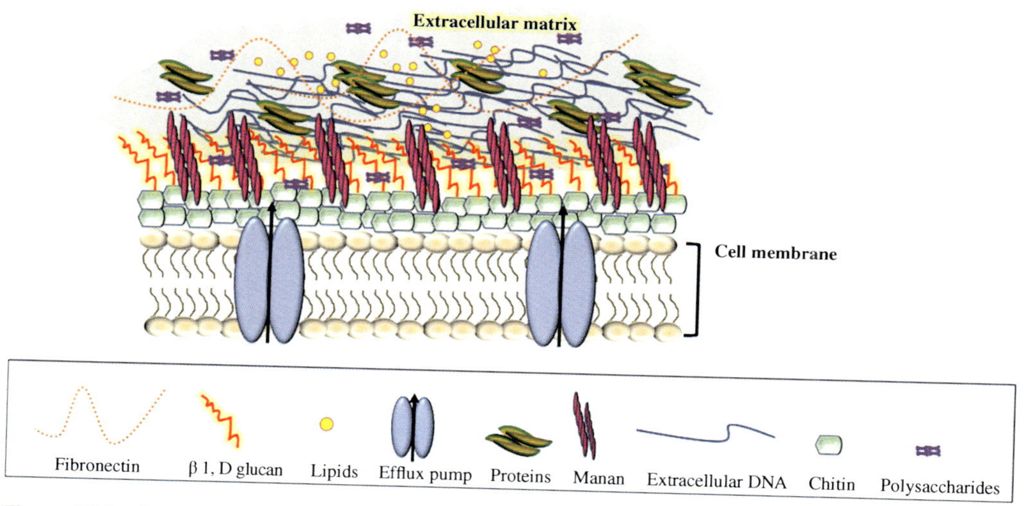

Figure 12.1 Representation of *Candida* biofilm virulence structure.

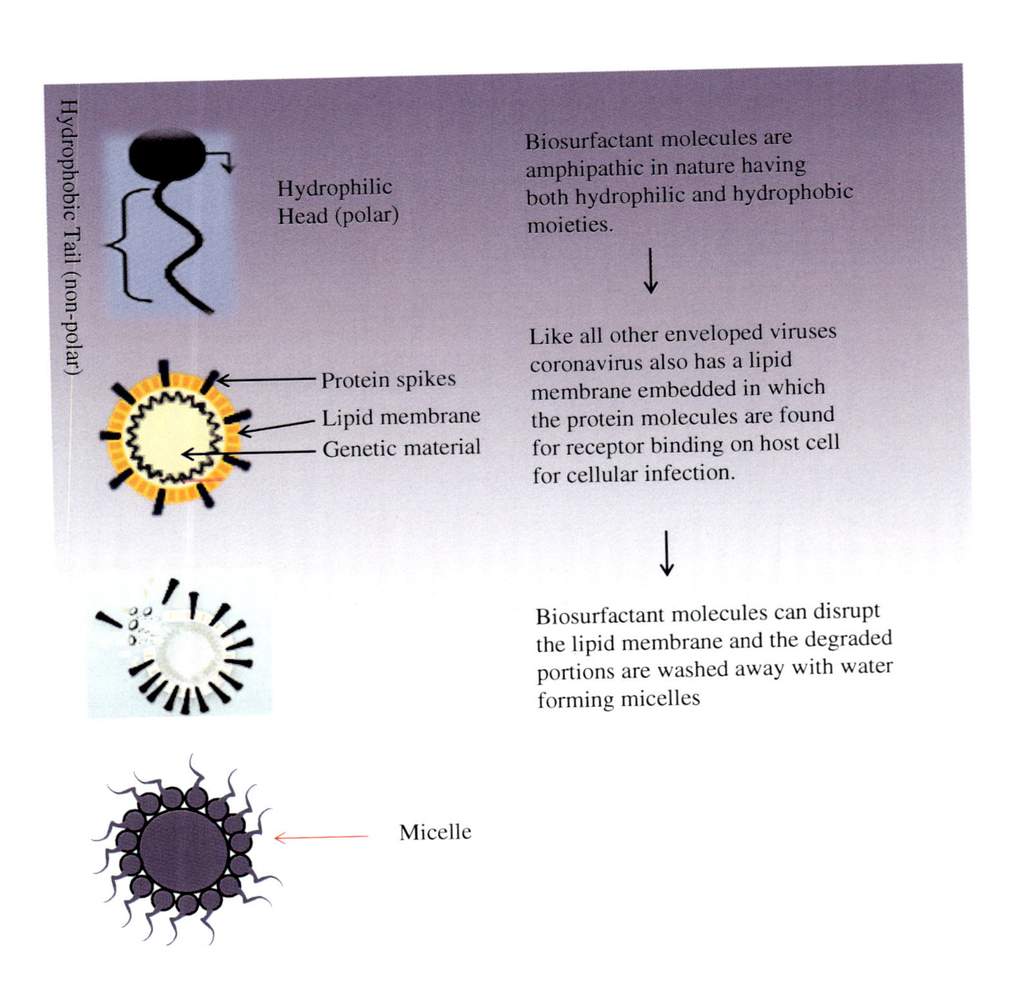

Figure 13.3 Biosurfactant molecule working mechanism for viruses like coronavirus.

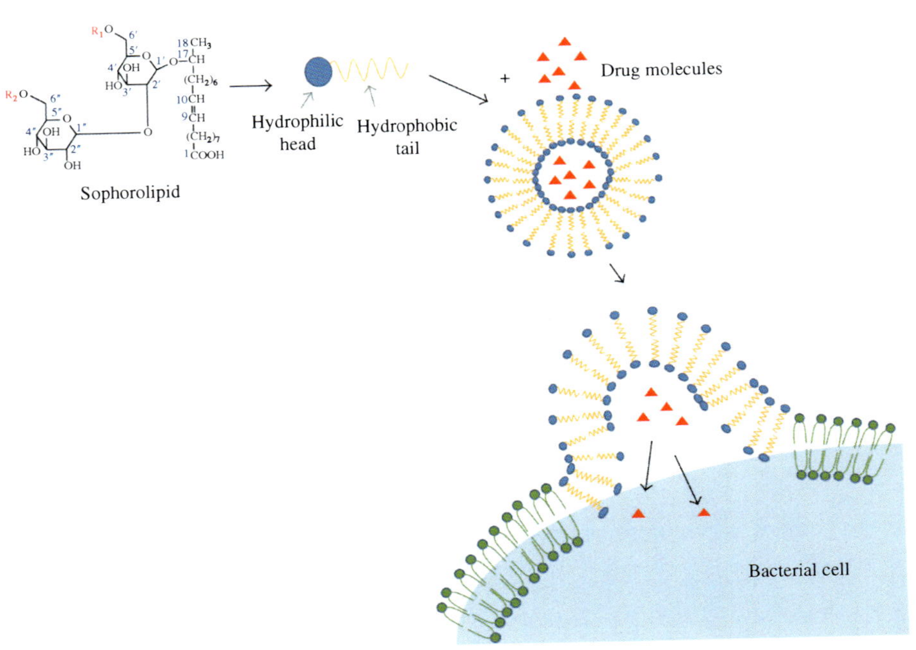

Figure 14.1 A schematic diagram from Joshi-Navarre and Prabhune [60] revealing the proposed mechanism of action. The amphipathic biosurfactant liposomes loaded with tetracycline allow for bypass of the cell membrane and access to the bacterial cytosol.

Figure 15.2 Steps involved in wound healing progression, which are inflammation, migration, proliferation and remodeling.

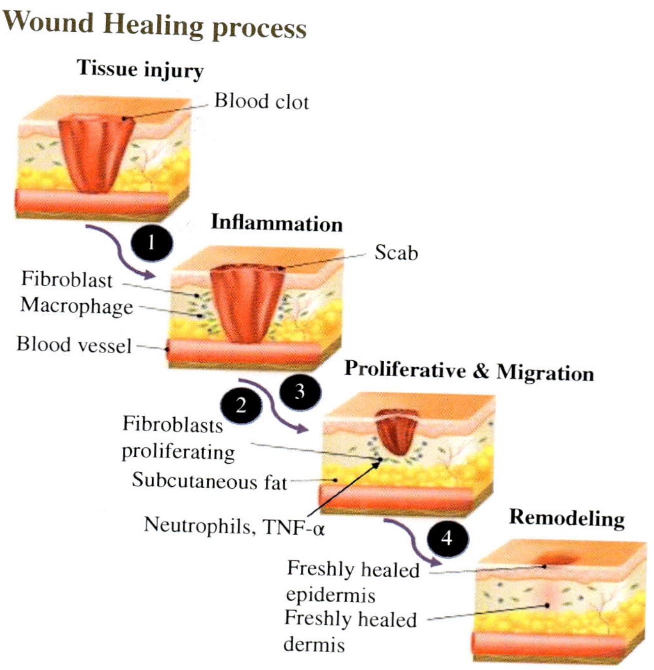

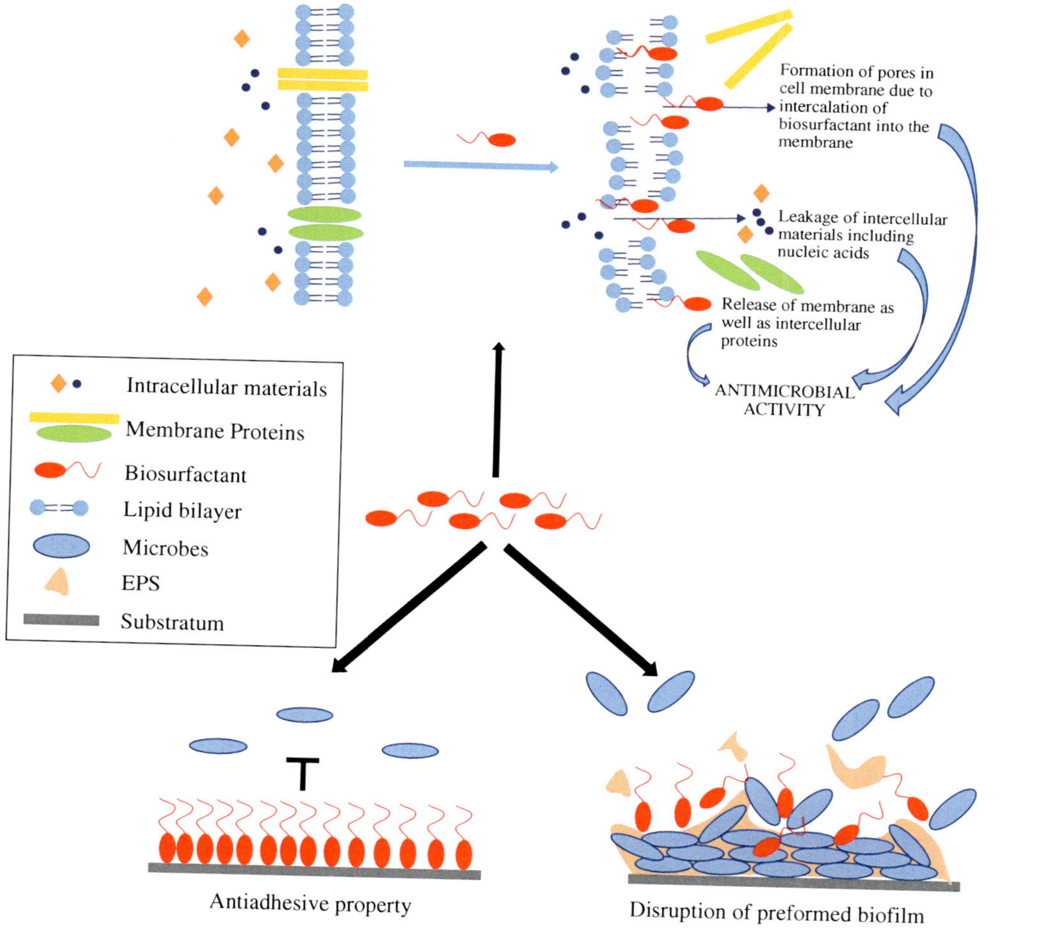

Figure 16.2 Schematic representation of antimicrobial and antibiofilm activity prompted by biosurfactant.

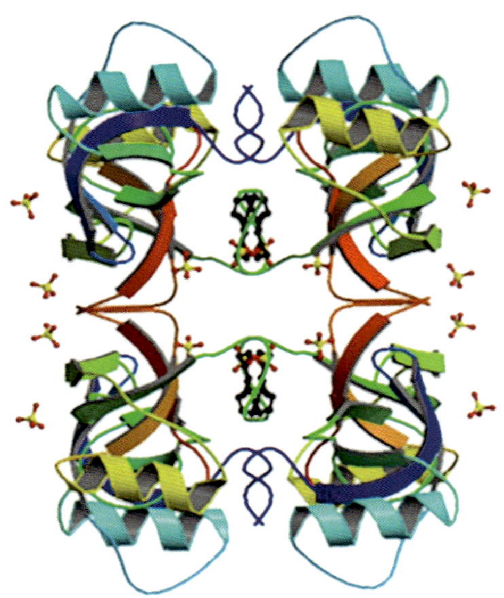

Figure 17.4 Hydrophobins. Amphiphilic nanotubes in the crystal structure of a biosurfactant protein hydrophobin HFBII (3QQT). From: https://www.rcsb.org/structure/3QQT.

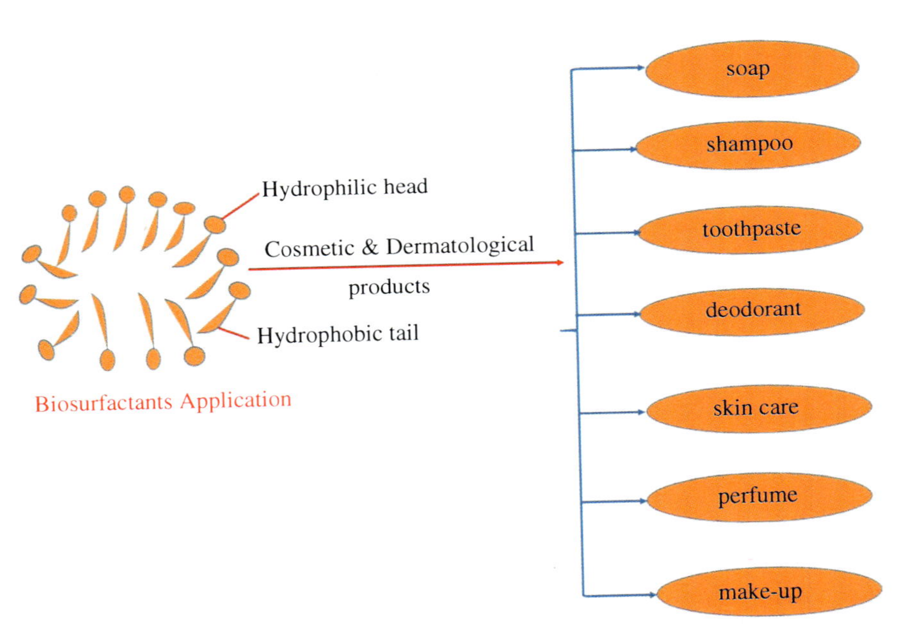

Figure 18.5 Different applications of biosurfactants in cosmetic formulation.

12

Biosurfactant-Based Antibiofilm Nano Materials

Sonam Gupta

Department of Biotechnology, National Institute of Technology, Raipur, Chhattisgarh, India

CHAPTER MENU

12.1 Introduction

Despite several available conventional antibiofilm drug therapies, most of the currently available drugs have failed or they are either less effective in the treatment of biofilm and associated infections due to the virulence and inherent drug-resistant property. The characteristic multidrug-resistant property of the biofilm-forming pathogens has led to the emergence of new more virulent biofilm-forming pathogens, most of which are hospital-acquired infections. Thus, they have become apparently a noteworthy threat to human society and have been associated with a high mortality rate worldwide [1–4]. These deadly diseases need to be tackled using new effective

Biosurfactants for a Sustainable Future: Production and Applications in the Environment and Biomedicine, First Edition. Edited by Hemen Sarma and Majeti Narasimha Vara Prasad.
© 2021 John Wiley & Sons Ltd. Published 2021 by John Wiley & Sons Ltd.

approaches and nanomedicines have been emerged as an effective therapy in this regard. The nano-medicinal approach have emerged as a recent advancement in the drug development to combat the recalcitrant biofilm infections due to the presence of its novel, non-invasive, and efficient strategy. Various critical constituents of biofilm such as the extracellular matrix (ECM), virulence genes, and quorum sensing (QS), a phenomenon involving a protective covering of the biofilm, play a critical role in the development of multidrug resistant biofilms [5, 6]. In this context, various studies have been done on the nano-formulations synthesis from chemical and bioactive compounds, but bioactive compounds such as biosurfactants have been gaining significant attention in recent times. These are microbial surface-active agents containing versatile biomedical properties including antibiofilm activity, antimicrobial activity, and therapeutic properties [7–10]. The major obstacle in the clinical translation of inhalable nanomedicine is being hampered by our lack of understanding about their deposition and clearance from the body parts and applicability. However, amphiphatic biosurfactants are wonder molecules, which have deciphered outstanding responses against biofilm infections. Interestingly, when they have been synthesized on the nano-scale in the form of nanoparticles or nano-drug delivery carriers apparently their activity has been increased several-fold [11, 12]. Thus, they are suitable bioactive compounds to be used as antibiofilm nanofactories to combat recalcitrant biofilm infections. This chapter addresses the recent progress and scientific investigations in the treatment of recalcitrant bacterial and fungal biofilm infections specifically by biosurfactants and their nanoformulations.

12.2 Emerging Biofilm Infections

Microbial cells exhibit two types of growth mode, one is a planktonic cell and another is called a microbial sessile clustered structure, which is best known as the biofilm. In other words, biofilm can be defined as a multifactorial aetiological complex structure of unicellular microorganisms formed on a living or non-living surface and encased within a thick ECM. The first microbial biofilm discovery was led by Antoine van Leeuwenhoek who observed the microbial growth on the surface of a tooth by using a simple microscope and called it “animalcule” [109]. Later, William J. Costerton, in 1978, coined the term “biofilm” and made the world aware about the significance of the biofilm. Biofilm is a kind of microbial community that is present everywhere in the surroundings, including hospital settings, natural environments, and labs, and remains either submerged on the hard surfaces or exposed to an aqueous solution [13].

Pathogenicity of a microbial biofilm in terms of stress tolerance and virulence behavior is specifically dependent on the species diversity and environmental abundance of the pathogen. Biofilm development is considered as the result of three consequent events, including adherence to a biotic/abiotic surface, attachment, proliferation and development of a mature clustered structure named as “biofilm,” followed by dispersion in the environment under favorable conditions to spread their virulence [14]. However, biofilm infections may also develop independently of medical implants, such as intravascular catheters, vascular grafts, prosthetic joints, valve endocarditis, cardiac devices, open wounds, breakage body parts, or dental plaque and shunts. Coagulase-negative *Staphylococci, Staphylococci epidermidis, Candida albicans*, and non-*albicans Candida* (NAC) are also involved in the biofilm development on the indwelling devices [15, 16].

Among fungal pathogens, *C. albicans* is the most notorious fungus, causing infections by both planktonic growth and in the form of a multicellular biofilm architecture. *C. albicans* belongs to class of Ascomycetous yeast, which consists of 150 species. Among all of these, about 20 species are found to have more virulent properties for causing biofilm infections and associated diseases [17]. Besides *C. albicans, Candida krusei, Candida glabrata, Candida parapsilosis*, and *Candida tropicalis* are other well-known NACs for causing clinical biofilm infections and emerging multidrug resistance behavior. Among them, *C. albicans* is the largely identified (50–80%) opportunistic pathogen and as per reports about 30–50% of humans are asymptomatic carriers of *C. albicans* 2009 [18]. Studies demonstrated that immunosuppressive patients, such as those with cancer, AIDS, or organ transplant patients, are at a very high risk of nosocomial or blood stream infections [19]. In healthy individuals, it colonizes asymptomatically in various body parts like the reproductive tract, oral cavity, gastrointestinal (GI) tract, skin, as well as vaginal surfaces by keeping equilibrium with other members of the local microbiota as long as it finds a host immune system is functioning well. Moreover, a weak hostile immune response due to chemotherapy, usage of immunosuppressive drugs, and emergence of multidrug resistance offer favorable conditions for growth of these harmful pathogens throughout the body. Studies revealed that *C. albicans* is a predominant pathogen of most of the diseases and in fact the second highest in colonization to human hostile niches. Additionally, it is found to be ranked fourth for causing nosocomial infections, third in catheter-associated infections and highest overall in crude mortality with neonatal bloodstream infections [20]. On the other hand, studies showed that these invasive candidiasis of *C. albicans* have been epidemiologically sifted to NAC species at an alarming rate and have contributed to 35–65% of total systemic and invasive candidiasis [19].

The modern scientific approaches such as use of indwelling medical devices, prophylaxis, and organ transplantation have significantly contributed to increased biofilm infections and their futile multidrug-resistant properties worldwide [17]. In fact, the morbidity rate of biofilm infections is maximum in organ transplant and cancer patients. Reports on *Candida dubliniensis* showed that this species is rarely found in the normal oral cavity of individuals having a strong immune system, but is exclusively present in the oral cavities of immunocompromised patients. The major shortcoming of these diseases is that in both cases, the host immune system becomes very weak, which prompts microbial adherence and their biofilm development at a very high rate. Therefore, rapid emergence of these most prevalent infections has raised concern of the scientific community throughout the world. Moreover, NAC species also evolved with multidrug-resistant properties by altering their inherent resistance nature, which led to making them resistant against a wide class of antifungals. These NAC species are mainly associated with nosocomial and bloodstream infections, and most commonly are called Fungaemia. These characteristic virulence properties of NAC species make them clinically important pathogens as they offer a major challenge to scientific communities in terms of their high mortality rate and emergence of multidrug resistance, which tends to limit therapeutic options and also creates economic issues due to extended hospital stays and expensive medical treatment [21].

Besides *Candida* sp., *Aspergillus* sp. also contribute to systemic and invasive fungal biofilm infections. It specifically colonizes in the respiratory tract of asthma and cystic fibrosis patients and facilitates biofilm development. It possesses high concentrations of conidia and thus exhibits a very high sporulating activity [22, 23]. Virulence attributes of *Aspergillus* sp. involve production of toxic fungal proteins and extensive mycelial growth into the lung parenchyma. Conidia of this fungus is apparently a most important phenotypic characteristic, which contributes to its antifungal resistance against a different class of antifungal drugs and a host immune system as well. Therefore, during the course of infection, initially it produces toxin proteins and proteases, which encounter the epithelial tissues and invade the lung parenchyma tissue later in order to develop infection. In the chronic conditions of infection, this fungal pathogen develops biofilm and causes several diseases such as invasive pulmonary aspergillosis and hypersensitive asthma. In particular, immunocompromised and AIDS patients are more susceptible to these fungal pathogens as spores of conidia remain present in the air after dispersion.

Use of indwelling medical devices associated with biofilm infections have been increasing day by day and causative agents specifically belong to different genera of bacteria or yeast. Most commonly known pathogens are *Staphylococci*, *Streptococci*, *Pseudomonas*, and *Candida* sp., which predominantly reside in the environment in variable numbers. *Staphylococcus* is specifically found in the skin of immunocompetent individuals but remains inactive/suppressed due to a strong and quick immune response of the host cells. Apparently, some pathogens are also found in contaminated tap water or are due to cross-contamination from other disease-prone people. Depending upon the surrounding hostile environmental conditions, biofilm development involves unicellular cells of single species or multispecies.

12.3 Challenges and Recent Advancement in Antibiofilm Agent Development

Biofilms play an enormously vital role in human health, as they protect microorganisms from host-defensive systems and antibiofilm drugs during the infection [24]. Infection occurrence is variable depending upon various factors such as pathogen types, host niche, and environmental conditions. Evidence from different agencies documented different estimates on the biofilm infections, such as that from the Centers for Disease Control, which was demonstrated to be approximately 65%. On the other hand, the National Institutes of Health have estimated about an 80% occurrence of biofilm infections to date, which were caused mainly by *Staphylococci* and *Pseudomonas aeruginosa* [25]. These pathogens are commensal microbes that usually live on human body surfaces, such as the outer skin layer and tracheal tract, and become activated during a weak host immune system or during environmental changes [25]. A complex multifactorial architecture of the biofilm restricts the penetration and transport of a drug due to a reaction–diffusion barrier. This leads to the interaction of a lethal drug dose with the biofilm surface layer only and limits drug activity [26]. Although some antibiotics managed to enter the matrix, due to the viscosity and complex composition of the biofilm, there was poor diffusion of the antibiotics [26]. In this way, sessile cells of the biofilm notoriously decreased their susceptibility and started to develop a multidrug resistance property against several antibiotics, which could be increased up to 1000-fold [5]. On the other hand, secretion of an antibiotic degrading enzyme like β-lactamase or poor diffusion of antibiotics within the biofilm also limited the antibiotic effectiveness [27–29].

12.3.1 Inherent Resistance

Inherent resistance indicates an inherent property of drug resistance, which limits the antibiotic susceptibility of biofilm-forming pathogens. For instance, in Gram-negative bacteria, the outer lipid membrane is associated with this activity, which is made of a thick layer as compared to the Gram-positive bacteria and therefore exhibits increased antibiotic resistance. An extensive study of *Pseudomonas* sp. pathogenesis revealed that it has an inherent resistant property against different classes of antibiotics such as β-lactam antibiotics, carbapenems, and monobactum [30].

12.3.2 Adaptive Resistance

Recent studies showed that, due to an extensive exposure of antibiotics, the multidrug resistant nature of pathogens have evolved by genetic manipulation containing more virulent properties as compared to their wild strains [30]. Physical and mechanical stress induced by different antibiotic classes predominantly play a crucial role in the emergence of multidrug-resistant pathogens, which is commonly known as adaptive resistance. Studies also elucidated that adaptive resistance is a result of a direct gene transfer between strains and species [31].

12.4 Impact of Extracellular Matrix and Their Virulence Attributes

ECM is a characteristic feature of a microbial biofilm, which plays a pivotal role in holding all microbial cells together within the biofilm under dynamic conditions and provides them with shelter as well as nutrients. It makes a protective shelter around the microbial cells of the biofilm and hence provides resistance to them from external insults, including antibiotics and hostile immunological factors, and thus promotes biofilm-invasive infections. The thickness of the EPS matrix is approximately 0.2–1.0 mm, whereas the size of the biofilm does not exceed more than 10–30 nm [32]. Primarily, 5–35% of the biofilm volume is constituted by the sessile cells and the rest of the volume is ECM. It mainly consists of glycoproteins (55%), polysaccharides (25%), ions (bound and free), nucleic acids (5%) and lipids (15%), RNA molecules (<1%), DNA molecules (<1%), and 97% of water and formation of ECM occurs in the attachment stage of the biofilm to the surface [17, 33]. The major function of a matrix protein is known to be involved in the form of enzymes that can break down complex macromolecules and provide nutrients that can be readily utilized [34]. Extracellular DNA (eDNA) plays an important role in biofilm establishment and structural integrity [5, 35]. The composition of matrix polymers varies significantly according to the biofilm growth phase [36].

Studies on ECM revealed that its structure is as diverse as biofilm-forming pathogens themselves and contributes in cell-to-cell communication, nutrient recycling, and a significantly high level of drug resistance [37]. Further, it also helps in the development of biofilm of a heterogenous system having multispecies microbes. The schematic illustration of a typical *Candida* biofilm virulence structure is represented in Figure 12.1.

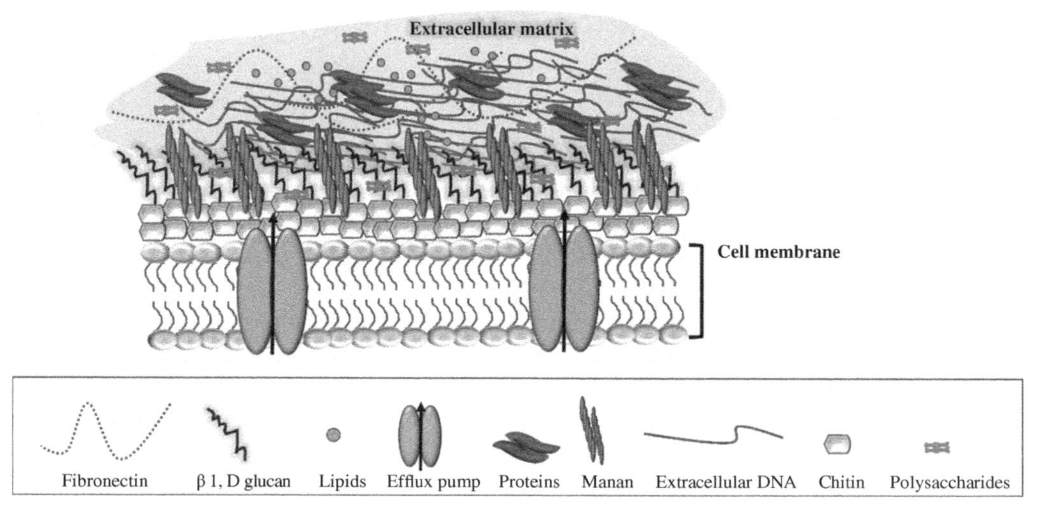

Figure 12.1 Representation of *Candida* biofilm virulence structure (See insert for color representation of the figure).

According to recent studies, many biofilm-producing microorganisms can produce amyloid-like fibers (ALFs) such as in *Pseudomonas* sp. [38]. On the other hand, Gram-positive bacteria such as *Staphylococci* sp. produce teichoic acids that act a crucial role in biofilm formation and are major components of the bacterial cell surfaces [39]. Major properties of ECM include the formation of scaffolds for protein attachment and its interaction with other surface polymers [39]. The ECM of *P. aeruginosa* consists of three different types of exopolysaccharides including Pel, Psl, and alginate polysaccharides, which are rich in glucose, mannose, and alginate, respectively [40, 41]. Unlike the other two, alginate is an acylated polysaccharide comprised of guluronic acid (GulUA) and mannuronic acid (ManUA) monomers [41].

12.5 Role of Indwelling Devices in Emerging Drug Resistance

Indwelling devices are specifically prone to biofilm infections if not replaced or maintained during clinal usage, leading to chronic infections development. This could be due to poor immune responses and a pathogenic intrinsic drug resistance property [42]. The host immune system is very effective in clearing toxins and foreign materials from the body, but shows an extensively reduced activity in close proximity to inserted indwelling devices, which leads to the adherence of microbial cells on them. In short, indwelling devices promote microbial growth leading to biofilm development. Biofilm growth in the catheters, like the urinary or central vascular systems, are fatal, which initially may be composed of single species, but prolonged time exposure can promote heterogenous biofilm formation [43]. In addition, biofilm formation on indwelling devices hinders the activity of the particular implanted devices, causing their failure, which in turn increases the life risk of patients many times higher. For instance, biliary stents should be removed regularly within two months of insertion due to bacterial biofilm formation on them [44]. Hence, it can be conferred that microbial biofilms may cause public health threats for those persons requiring medical implants for their life support. Therefore, versatility of medical devices in terms of variable designs and use characteristics cannot prevent these biomaterials from microbial contaminations and biofilm formation.

12.6 Role of Physiological Factors (Growth Rate, Biofilm Age, Starvation)

Biofilm is comprised of metabolically active and inactive microbial cell subpopulations. In particular, those cells that reside very deep inside the biofilm matrix receive very low amounts of nutrients so tend to show growth of biofilm. [45, 46]. Therefore, metabolically active and slow growing cells show different susceptibility behavior against a wide range of drugs. Moreover, most commercial drugs require metabolically active cells to promote their activity. In this way, slow growing or metabolically inactive cells of biofilm escape and develop resistance against drugs. It has been reported that metabolically active cells that are present on the surface are susceptible to ciprofloxacin, tetracycline, and tobramycin, while metabolically inactive cells present very deep inside the biofilm are resistant to these antibiotics [45].

12.6.1 Quorum Sensing

In order to develop biofilm structure cells from intra- and interspecies interact with each other through a cell to cell signaling system, which is known as a QS signaling pathway [2]. This signaling pathway is governed by a different class of signaling molecules, which are produced by cells and accepted by neighbor cells to trigger a controlled expression of specific genes that are very essential for biofilm development [2, 42, 47, 48]. For instance, acylated homoserine lactones (AHLs) for Gram-negative bacteria [49] and specific peptides for Gram-positive bacteria are known QS signal molecules [50].

Evidence showed that during the adhesion phase of biofilm development, transcription of specific genes are switched on, leading to the secretion of signal molecules such as algD, algC, and algU::lacZ reporter constructs in the case of *P. aeruginosa*, which are important for ECM synthesis, specifically alginate [2, 51]. The *Escherichia coli* QS system depends on the production of receptor-like protein (SdiA) while *Staphylococcus aureus* consists of two component QS systems containing GraS (HK)/GraR (RR) [52, 53].

12.7 Impact of Efflux Pump in Antibiotic Resistance Development

Efflux pumps comprise a single component or multicomponent system and contribute to sessile recalcitrance to different classes of drugs [6]. A gene encoding efflux pump is either found on plasmids or on chromosomes [42]. Most virulent pathogens possess multidrug resistant efflux pumps and the very important intriguing feature of these pumps is their binding ability with a wide range of structurally unrelated drugs and tremendous efficiency of drug extrusion. Studies showed that the inherent drug-resistant property of Gram-negative bacteria have been associated with the expression of efflux pump genes [54]. Bacterial efflux pumps are composed of either a single or multicomponent system [6, 55]. Gram-negative bacteria exclusively comprise a single component efflux pump relative to the Gram-positive bacteria. The bacterial efflux pump is classified into five families, namely a resistance–nodulation–division (RND) family, the major facilitator superfamily (MFS), the ATP (adenosine triphosphate)-binding cassette (ABC) superfamily, the small multidrug resistance (SMR) family (a member of the much larger drug/metabolite transporter (DMT) superfamily), and the multidrug and toxic compound extrusion (MATE) family. Among them, the RND family is specific to Gram-negative bacteria ([6, 42, 55]. In addition to this, their classification is based on a different component, the number of transmembrane regions and energy source used by the pump [163].

Fungal pathogens such as the virulent *Candida* sp. possess two different efflux pumps consisting of ABC and MFS transporters, which specifically confer resistance against the azole class of antifungals [56]. ABC transporters are a multigene family, which consist of CDR genes and have been associated with drug resistance against different classes of antifungals. Most importantly, after a drug-efflux pump interaction CDR genes overexpressed, which led to the emergence of cross-resistance among different drugs [6]. On the other hand, among MFS transporters, CaMDR1 genes are very important, which exclusively account for fluconazole resistance [6, 56, 57]. Hence, due to the cross-resistance characteristic of virulent bacterial and fungal efflux pumps, it is mandatory to develop effective efflux-pump inhibitor (EPI) drugs. Besides, efflux pumps also play an imperative role in the traverse movement of quorum-sensing biomolecules across the cell membrane in a controlled manner, leading to biofilm formation by the recalcitrant microbes [48, 58]. Interruption in the functioning of the efflux pumps either by a host defensive system or an effective drug treatment can trigger the inhibition of the transit movement of QS biomolecules across the cell membrane [48]. For instance, in *Burkholderia pseudomallei*, a BpeAB-OprB efflux pump controls the QS signaling pathway [59]. Similarly, the RND efflux system containing an lecA-lux expression system is involved in *P. aeruginosa* biofilm infections [58]. During biofilm formation, cells secrete signal molecules such as 3OC12-HSL, which also require an efflux pump to be actively traverse out of the cell membrane. Therefore, an increase in efflux pump activity is required for biofilm formation in order to increase transportation of QS signal molecules [58]. In this context, current investigations have been centered on the development of EPIs using combination strategies with the aim of hinderance in the functioning of efflux pumps and which impair the biofilm development on biotic or abiotic surfaces and t resolves their associated challenges. Therefore, nanoformulation synthesis, especially using bioactive compounds, such as biosurfactants, represents an alternative candidate to target efflux pumps of microbial pathogen cells as they are biocompatible, nontoxic, and highly effective over their chemical counterparts [9, 60].

12.8 Nanotechnology-Based Approaches to Combat Biofilm

Owing to grave health concerns associated with the biofilm infections caused by both bacteria and/fungus, nanotechnology, a modern scientific approach, has emerged recently to combat these drug-resistant and device-centered chronic infections. Recent applications of emerging nanotechnology-based approaches have been shown to remove the limitations of conventional antibiofilm drugs. It is a broad-spectrum area consisting of different strategies of nanoformulation preparations such as nanotubes, nanofibers, nanoemulsion, nanoparticles, nanocomposite, microemulsion and nanoflowers, etc. [61]. Nanomaterials can be formulated using various approaches, such as laser ablation, microemulsion, sol–gel, electrospinning, solvothermal, co-precipitation, and pyrolysis [62]. Conventional techniques like the sol–gel method, microemulsion, and co-precipitation can produce a wide range of nanomaterials due to uncontrolled growth of the nuclei.

12.8.1 Parameters Affecting Nanomaterial Fabrication

Nanomaterial fabrication is affected by several parameters, including temperature, pH of the solution, pressure, time, particle size, shape, environment, proximity, pore size, and preparation cost [63–67].

Temperature is the key factor for the proposed nature of the nanoparticle, which influences the kinetics of nanoparticle synthesis to a very great extent [60, 68, 69]. For instance, the highest temperature (>350 °C) is required in the case of the physical method of nanomaterial synthesis while less than <350 °C is necessary for chemical methods. On the other hand, bioactive compound-based nanomaterial synthesis requires ambient temperatures approximately less than 100 °C. The temperature of the reaction mixture determines the nature of the nanoparticle formed [69]. Besides temperature, nanomaterial synthesis is also affected by the pressure used for the reaction mixture, causing variable shapes and sizes of nanomaterials [63]. Ambient pressure conditions are the optimum parameter for synthesis of metallic nanoparticles using biological agents [63, 70]. Studies showed that when biosurfactant is used as a reducing agent for the synthesis of silver and gold nanoparticles the rate of metal ion reduction becomes very fast [12].

The type and quality of a synthesized nanoparticle are significantly affected by time duration of the reaction medium proceeded [71]. Additionally, the properties of the fabricated nanoparticles were also transformed with a change in time length and are found as a function of the synthesis process, for conditions and chemical/light or dark exposure [65, 72]. This exposure leads to the variation in the nanoparticle shape and size, such as aggregation due to a long time for incubation and shrinkage in the size due to long storage, which in turn also affect the stability and functional behavior of the fabrication nanomaterial [72]. Therefore, it can be conferred that shape and particle size are the two critical factors for effectiveness and proper functional behavior of the synthesized nanomaterials in terms of bioavailability, transport, fate, cellular uptake across the spectrum of *in vitro* and *in vivo* nano–bio interfaces [73]. The chemical properties of nanomaterials are markedly affected by the dynamic nature, surface area, and shape of the synthesized nanomaterials. Hence, nano-sized particles are biologically very effective as compared to materials with a larger surface area of the same chemistry. However, cells possess highly tuned and specific functions and follow the scale rule to some extent to regulate the uptake and transportation of biological nanomaterials. For example, a bilayer cell membrane is 4–10 nm in size whereas a nuclear pore complex of vertebrate is only 80–120 nm in size and thus function as a regulatory barrier for the entrance and exit of the biological nanomaterials [74]. Therefore, synthesis of biological nanomaterial should be well designed with the main emphasis on size and shape so that nanomaterials can easily penetrate within the biological systems. Besides size, shape, and aspect ratio, particle charge and porosity of the fabricated biological nanomaterials should also be considered [66]. These parameters also make an additional independent impact on the uptake and transportation of biological nanomaterials across the cellular system [73]. The effectiveness of a synthesized nanomaterial also markedly influenced by the vicinity of its surroundings often result in an alteration of its morphological and functional characteristics [72]. Particle charge, their physical and magnetic properties, and substrate binding affinity are the other proximity factors involved in the synthesis of very fine nanoformulations to be effectively used for the nanodrug delivery systems and in therapeutics [72].

Apart from these important physicochemical properties, the cost-effective synthesis of nanomaterials is a very critical parameter for their large-scale usage in therapeutics and clinical tests. Unlike a chemical process, which has a short duration procedure, a biological synthetic procedure is very less costly and also offers several eco-friendly advantages.

12.9 Biosurfactants: A Promising Candidate to Synthesize Nanomedicines

Biosurfactants are known as amphiphatic molecules having two moieties, hydrophobic as well as hydrophilic [9, 75, 76]. They possess versatile biological properties such as anticancerous activity, antimicrobial activity, tailor-made multifaceted diversity antibiofilm activity, effectiveness at extreme pH and temperature, suitability with an eco-friendly nature, low toxicity, and higher biodegradable ability [7, 9, 77]. These biomolecules are found in different structural forms such as lipopeptides, glycolipids, acyl-glycerols, and biopolymers. The polar, water-soluble part of a biosurfactant is comprised of carboxylate or hydroxyl or a complex mixture of phosphate, mono-, di-, or polysaccharides, anions or cations, amino acids or peptides. The hydrophobic part of a biosurfactant is made up of long-chain, saturated or unsaturated, hydroxyl, or α-alkyl-β-hydroxy fatty acids, a hydrocarbon tail, and may also contain cyclic structures [7]. Glycosidic ester or amide bond is involved in the linking of a fatty acid chain of a hydrophobic part to the hydrophilic moiety [78]. Biosurfactants can be found in either negatively charged or neutral forms. The negatively charged biosurfactants are anionic in nature due to the presence of a carboxylate, phosphate, or sulfate group. The least number of cationic biosurfactants contains an amine group [9, 79]. Different microorganisms producing biosurfactants are shown in Table 12.1.

Depending on the presence of the carbon source, biosurfactant biosynthesis involves three different strategies; both hydrophobic and hydrophilic parts can be synthesized de novo; lipid moiety can be synthesized from the carbon source if hydrocarbon is used as the carbon source; and sugar moiety is synthesized de novo [107]. On the other hand, sugar moiety can be synthesized from a used carbon source and lipid moiety is synthesized de novo in the case where a hydrophilic carbon source is used [108].

Table 12.1 Biosurfactant and their microbial source.

Biosurfactant types	Microorganisms	References
Glycolipids		
Rhamnolipids	*P. aeruginosa*	[80]
	Pseudomonas sp.	[81]
Cellobiolipids	*U. maydis*	[82]
	U. zeae	[83]
Sophorolipids	*T. apicola*	[84]
	T. bombicola	[85]
	T. petrophilum	[86]
Trehalolipids	*N. erythropolis*	[87]
	R. erythropolis	[88]
	Mycobacterium sp.	[89]
Lipopeptides and lipoproteins		
Viscosin	*P. fluorescens*	[90]
Peptide-lipid	*B. licheniformis*	[91]
Polymyxins	*B. polymyxa*	[92]
Gramicidins	*B. brevis*	[93]
Serrawettin	*S. marcescens*	[94]
Subtilisin	*B. subtilis*	[95]
Surfactin	*B. subtilis*	[95]
Fatty acids, neutral lipids, and phospholipids		
Fatty acids	*C. lepus*	[96]
Neutral lipids	*N. erythropolis*	[97]
Phospholipids	*T. thiooxidans*	[98]
Polymeric surfactants		
Protein PA	*P. aeruginosa*	[99]
Biodispersan	*A. calcoaceticus*	[100]
Emulsan	*A. calcoaceticus*	[101]
Liposan	*C. lipolytica*	[102]
Carbohydrate-protein-lipid	*P. fluorescens*	[103]
	D. polymorphis	[104]
Mannan-lipid-protein	*C. tropicalis*	[105]
Particulate biosurfactants		
Vesicles and fimbriae	*A. calcoaceticus*	[106]

12.10 Synthesis of Nanomaterials

Nanomaterials are usually synthesized by employing two different approaches, which include "top-down and "bottom-up" methods that can be further used for the functionalization of nanomaterials. Top-down approaches are known for the bulk material "cut away material" and its

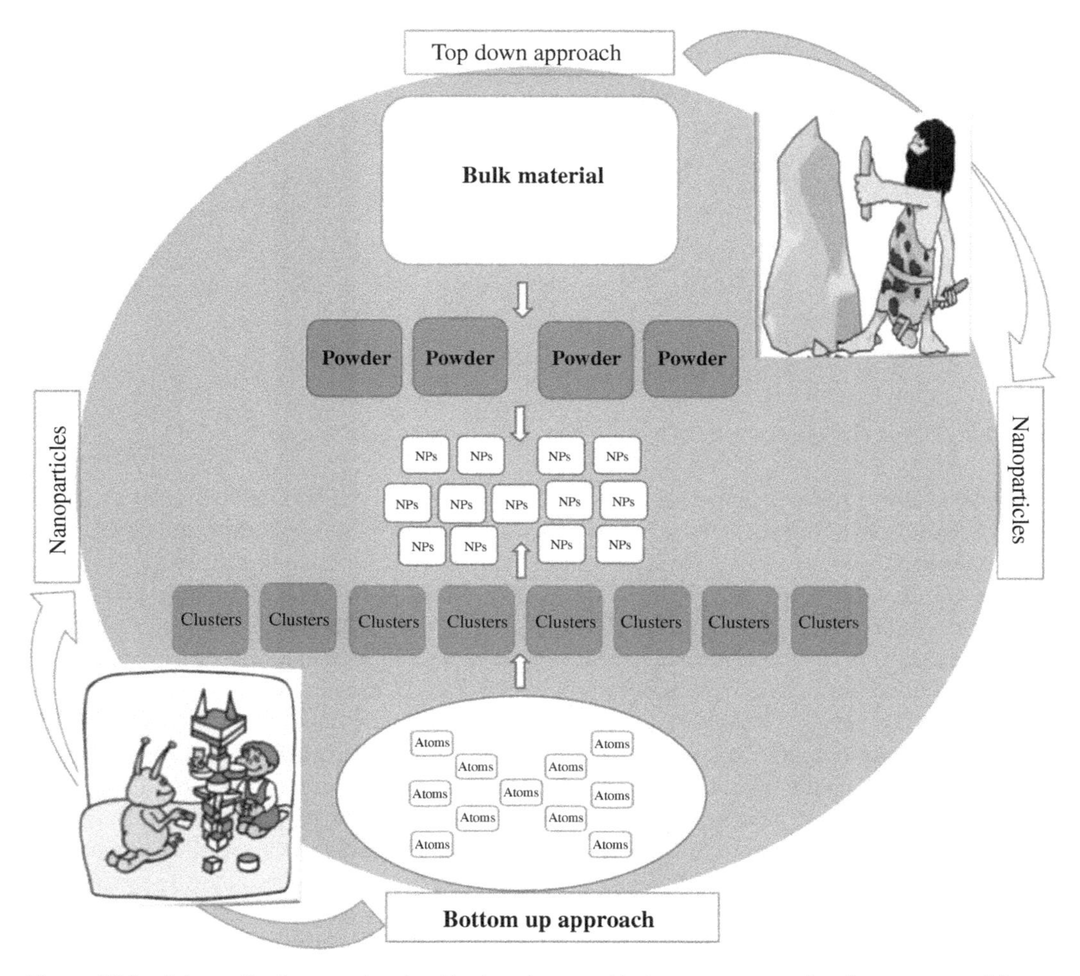

Figure 12.2 Schematic diagram showing the top-down and bottom-up approaches for a nanoparticle.

conversion to make what you want although retaining the original integrity. Top-down approaches involve the attrition process, ball mill technique, or planetary, vibratory, rod, tumbler ball, or lithographic pattern technique and are usually used for the synthesis of structural, catalytic, and magnetic nanoparticles [109, 110]). The "bottom-up approach" involves the assembling of the atom and small molecules to make nanostructures (atom by atom, molecule by molecule, or cluster by cluster), as shown in Figure 12.2. The major drawbacks of the top-down approaches include the low surface area, high polydisperse size distributions, and the partially amorphous state of the prepared powders [111]. On the other hand, the bottom-up approach is found to be more beneficial over the top-down approach as it has a better chance of producing nano-structures with fewer defects, equal composition of chemicals, and better short- and long-range ordering.

12.10.1 Microemulsion Technique

Based on the "top-down and "bottom-up" approaches several nanomaterial synthetic techniques have been derived and categorized into two different categories, including a physical process (attrition, sputter deposition, thermal decomposition, spray pyrolysis, laser ablation, vapor deposition,

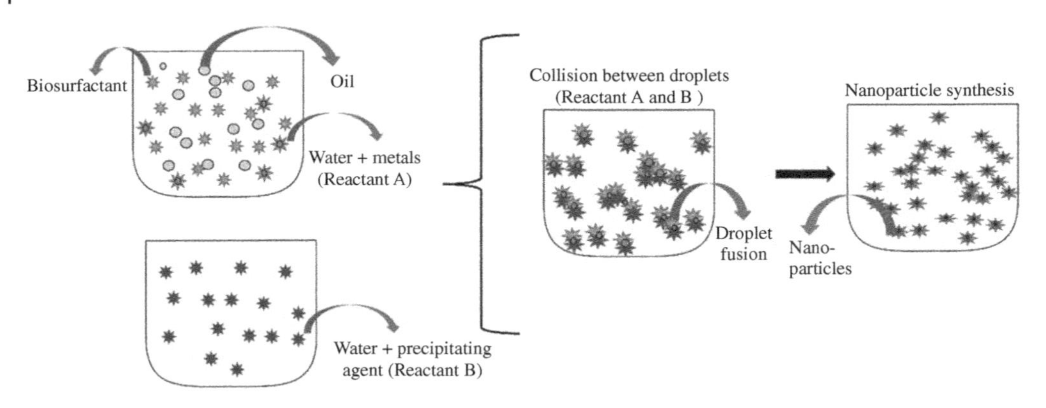

Figure 12.3 Microemulsion technique of biosurfactant-based nanoparticle synthesis.

and ultrasonication) and a chemical process (metal salt reduction, sol–gel chemistry, co-precipitation, microemulsion, photo-reduction, thermolysis, and spray pyrolysis). However, among all techniques, microemulsion is the most commonly used technique due to its wide scale of applications in frontiers of nanomedicine, starting from biosensors, microfluidics, and microarray and tissue engineering. A microemulsion generally consists of a surfactant, a lipophilic phase, and a hydrophilic phase [23]. Different biogenic nanoparticles of natural products including both plant-derived and those obtained from microbial sources in the range of 1–100 nm have been synthesized using this technique [12, 60, 112]. Therefore, chemical surfactants can be replaced by biosurfactants to make biocompatible and non-toxic nanomaterial designs [113]. Schematic representation of nanoparticle synthesis using the microemulsion technique has been represented in Figure 12.3.

There are several toxic effects of synthetic nanodrugs that have been reported for clinical applications. Therefore, to counter this problem biogenic nanomaterials have emerged as a potent alternative [11, 60, 114]. Advantages of microemulsions have also been employed to synthesize biosurfactant-based nano-formulations, including biosurfactant conjugated nanoparticles, lipid–polymer–hydrid nanoparticles (LPHN), and nano- and self-nano-emulsifying drug delivery systems (SNEDDs). Further, produced biosurfactant-based nanomaterials using a microemulsion approach revealed that they effectively induce a triggered drug release response for various antimicrobial activities. Therefore, biosurfactant-based nanomaterial synthesis has been emphasized by several researchers and modern scientists in this era, which offers many advantages to technological and environmental challenges of biomedical sciences over their chemical counterparts. However, despite the broad range of applications, not all bioactive compounds are applicable for nanomaterial synthesis due to poor solubility in aqueous solutions and poor stability, which limit their clinical efficiency and bioavailability [11, 61, 114]. To overcome this problem, utility of biosurfactant containing nanoformulation synthesis have been deciphered by various researchers [12]. Among different approaches of nanomaterial synthesis, nanoparticle, nano-rod, lipid–polymer hybrid nanoparticles, microemulsion, nano and self-nano-emulsifying drug delivery systems, and nanoflower synthesis are some of the recently developed biosurfactant-based nanoformulations [115].

12.10.2 Biosurfactant-Based Nanoparticles

Synthesis of nanomaterials using biosurfactants has emerged as a safer and non-toxic approach to overcome the environmental and technical issues associated with their chemical counterparts. The most widely used biosurfactants are glycolipids and lipopeptides for the synthesis of metallic nanoparticles of different metals such as Au, Ag, Cu, Ni, Zn, Fe, Co, etc., where they act as capping or reducing agents [11, 61]. These biosurfactants offer effectually a cost-effective and easily scalable-up synthetic process unlike conventional capping agents, including fatty acids like oleic acid, linoleic acid, and their derivatives [12, 116]. The use of a glycolipids type of biosurfactant, such as rhamnolipids and sophorolipids, resolves the poor solubility issues often associated with the fatty acid capping agents and promotes uniform dispersion of synthesized nanoparticles in the liquid medium [76, 116–119]. Moreover, they also avoid nanoparticle aggregation, which is most often a tendency associated with nanoparticle synthesis [76, 119, 120]. There are several techniques available that are used for the synthesis of metallic nanoparticles, but biosurfactant conjugated nanoparticles are synthesized using the microemulsion technique with a bottom-up approach that has shown tremendous activity. Nguyen et al. [89] illustrated the significance of glycolipids such as rhamnolipids and sophorolipids for nanomaterial preparations using a microemulsion technique to be used in therapeutic applications. Spherical NiO nanoparticles were synthesized using rhamnolipids in a microemulsion technique in the range of 47 ± 7 nm [115, 121]. The mechanism of a biosurfactant-dependent nanoformulation synthesis involving a self-aggregation phenomenon leading to micelle or vesicles formation is the result of noncovalent interactions between the hydrophobic moiety of the biosurfactant. Moreover, several different morphological structures are produced due to variations in the self-aggregation pattern of these surface-active biomolecules, such as the reverse micelle, lamellar structure, rod-shaped, hexagonal, ellipsoidal, cylindrical, spherical, etc., as shown in Figure 12.4. On the other hand, vesicles are hollow spheres covered by the bilayer of biosurfactants [122, 123].

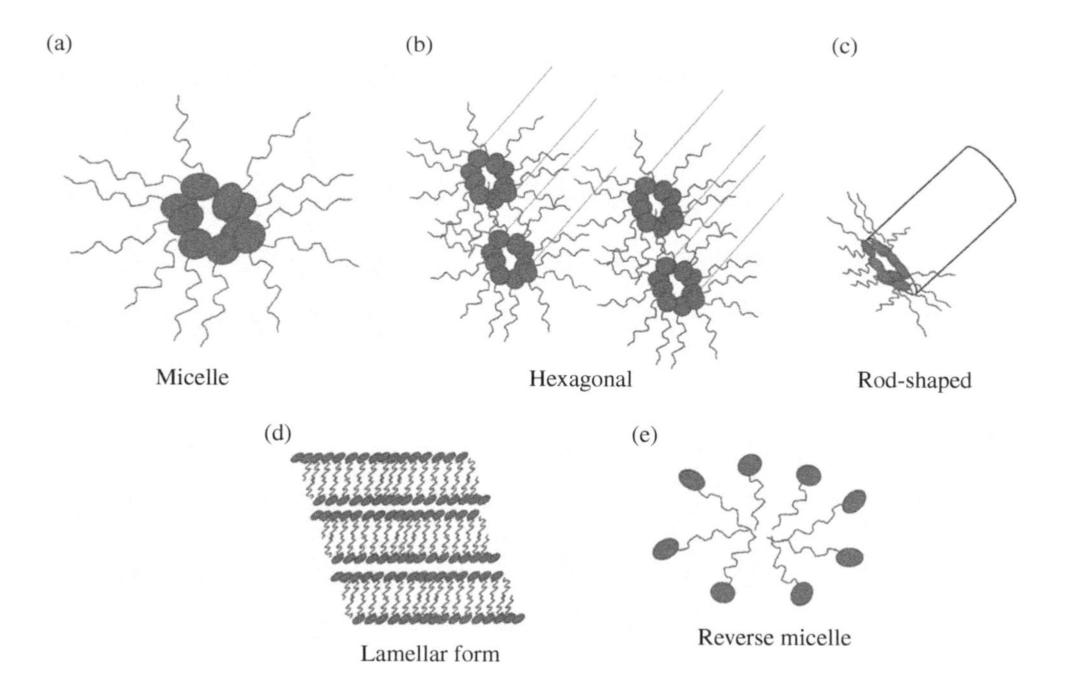

Figure 12.4 Different types of self-aggregation pattern of a biosurfactant during nanomaterial synthesis.

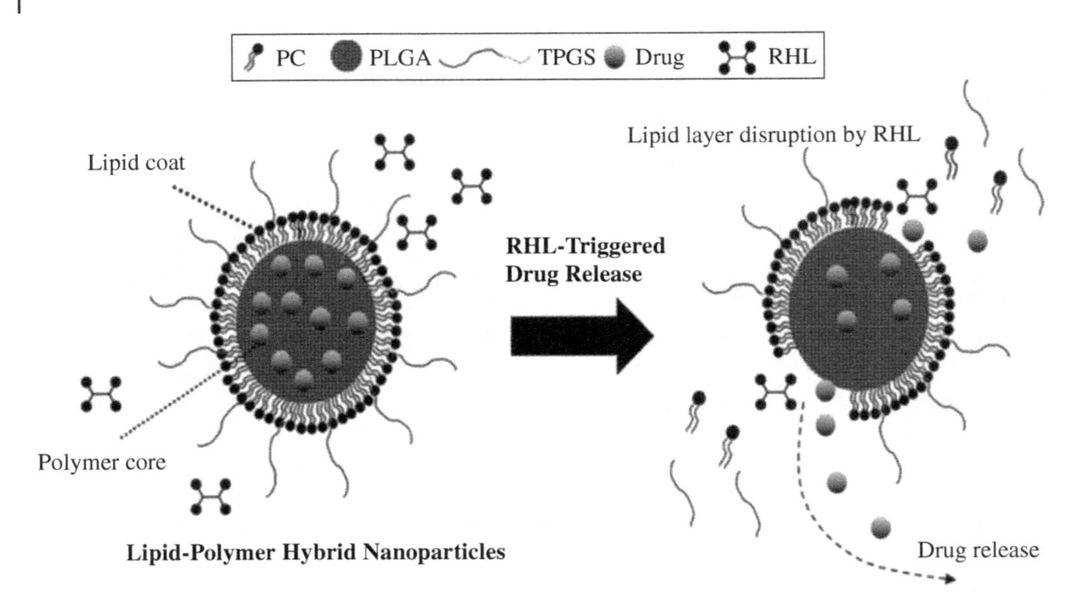

Figure 12.5 Schematic representation of a rhamnolipid-based lipid hybrid polymer nanoparticle [114].

12.10.3 Lipid–Polymer Hybrid Nanoparticles (LPHN)

Despite having effective penetration efficiency, liposome synthesis possesses several drawbacks, which limit their clinical usage. The major problems associated with a liposome-mediated drug delivery system is its poor stability at variable pH and temperature and lack of physical robustness nature, which cause drug leakage problems and hinder drug activity [124]. To counter these issues, a modified approach involving LPHN synthesis has been employed for various clinical translation processes such as an antibiofilm drug coating in medical implants and a nanocarrier-based inhaler for lung-biofilm infections. LPHN is designed in such a way that it can enhance the physical robustness of liposome nanocarriers to overcome liposome-associated issues. It consists of a polymer–nanoparticle core and a lipid coat. As a result, they become more effective in penetrating the thick ECM of biofilm as compared to the liposomes. An emerging green technology approach that prompts the usage of biosurfactants such as rhamnolipids has also shown a significance response for the synthesis of LPHN in terms of excellent drug encapsulation, better stability, morphology, and a specific target [114]. Use of rhamnolipids have shown a tremendous triggered and controlled encapsulated drug release response within the vicinity of biofilm colonies formed on the solid surface [114, 125], as shown in Figure 12.5.

12.11 Self-Nanoemulsifying Drug Delivery Systems (SNEDDs)

Poor gastrointestinal adsorption remains a challenge for a number of commercial drugs that limit their clinical and therapeutic usage. In this context, biosurfactants, which are well known amphiphatics and self-emulsifying biomolecules, have emerged as an alternative candidate [113, 114, 126]. Due to their varied functional properties, such as cleansing, emulsification, foaming, wetting, and tensioactive natures, biosurfactants offer several micro- and nanoemulsion-based applications in the range of 10–1000 nm, they can be used as nanodrug delivery systems [113, 114]. In a heterogenous system, biosurfactants act at the interfacial tension of two immiscible solutions such as water and oil and therefore affect the wettability and surface tension of each solution, causing mixing there by reducing their surface energy. Besides interfacial tension, they also affect the mass transfer by lowering the interfacial boundary existing between the immiscible phases. SNEDDs are another recent approach used in nanotechnology for drug delivery using biodegradable and biocompatible polymers in which biosurfactants have emerged as an alternative to their chemical counterparts to be used in SNEDDs synthesis, which produced nanomaterials in the range of 20–200 nm [113, 127, 128]. Besides facilitating a self-assembling property, they have good adsorption potential and also act as good carriers to be used in nano drug delivery systems. Therefore, their great potential to elicit improved therapeutic and pharmacological responses is envisaged [113, 128].

12.12 Biosurfactant-Based Antibiofilm Nanomaterials

Nanomaterials target pathogens by such methods as alteration in the chemical and physical properties, which account for cell membrane disruption, cell wall lysis, permeability, osmoregulation, electron transport, and respiration [129]. Their implementation in aseptic surgical and procedural techniques could prevent infections that require medical apparatus removal, i.e. implants, and long-term systemic antibiotic therapy, thus reducing healthcare costs [130]. The fundamental mechanism of their antimicrobial activity involves induction of reactive oxygen species (ROS) generation underlying interactions with specific functional groups of proteins (e.g. thiol and amino groups) as documented earlier [131]. Studies also revealed that the antimicrobial potential of nanoparticles is a shape- and size-dependent phenomenon and if size is smaller (20–200 nm) and of a specific shape (e.g. truncated-triangular nanoparticles) it shows better antimicrobial activity [132, 133]. However, most widely synthesized and clinically used nanoparticles are spherical in shape, having a colloidal or immobilized state. They are best-suited nanomaterials due to their ease of preparation and effectiveness [134, 135]. However, the morphology of these nanomaterials can be extensively varied with a change in pH, biosurfactant concentration, and the solution's ionic strength. Different biosurfactant-conjugated nanoparticle fabrication, such as lipopeptide-conjugated nanoparticles (surfactin-silver nanoparticles and surfactin-gold nanoparticles) and glycolipid-conjugated nanoparticles (rhamnolipid-silver nanoparticles) were found to have potential antimicrobial activity against *Staphylococci*, *E. coli*, *P. aeruginosa*, *S. aureus* having sizes in the range from 5 to 60 nm [11, 136]. Different nanoformulations of biosurfactants showing antimicrobial and antibiofilm activity are shown in Table 12.2. Based on the above-mentioned studies, schematic representation of the antibiofilm activity of biosurfactant-based nanoformulations are depicted in Figure 12.6.

12.13 Conclusions and Future Prospects

Despite the advances made in the development of novel antibiofilm agents, devised biofilm treatment strategies are limited by their high costs and complexities, which means urgent development is required to identify cost-efficient substitutes. Nanoformulations and specifically those fabricated using bioactive compounds have been found to be effective and also economically valuable. In this context, biosurfactant-based nanoformulation synthesis with potential bactericidal and fungicidal

Table 12.2 Different biological potential of biosurfactant-based nano materials.

S·No.	Biogenetic NPs of biosurfactants	Biosurfactant as reducing/ capping agent	Biosurfactant producer	Applications	Susceptibility of microorganisms	References
1	Microemulsion synthesis of silver nanoparticles using biosurfactant	Capping agent	*Pseudomonas aeruginosa* MKVIT3	antimicrobial and cytotoxic activities	—	[137]
2	Silver nanoparticles synthesis with biosurfactant	Capping agent	*Bacillus subtilis*	Toxic effect on environmental bacteria and fungi	—	[138]
3	Surfactin mediated gold nanoparticles and silver nanoparticles	Reducing agent	Lactic acid bacteria	Antibacterial activity and ROS generation	*E. coli, Pseudomonas aeruginosa, S. aureus*	[136]
4	Rhamnolipid stabilized zinc nanoparticle	Capping agent	*Pseudomonas aeruginosa*	Stabilizing agent for nanoparticle synthesis	—	[12]
5	Rhamnolipid based silver nanoparticles	Capping agent	*Pseudomonas aeruginosa* BS-161R	Antibiotic and antimicrobial activity	Broad-spectrum antibiofilm activity	[11]

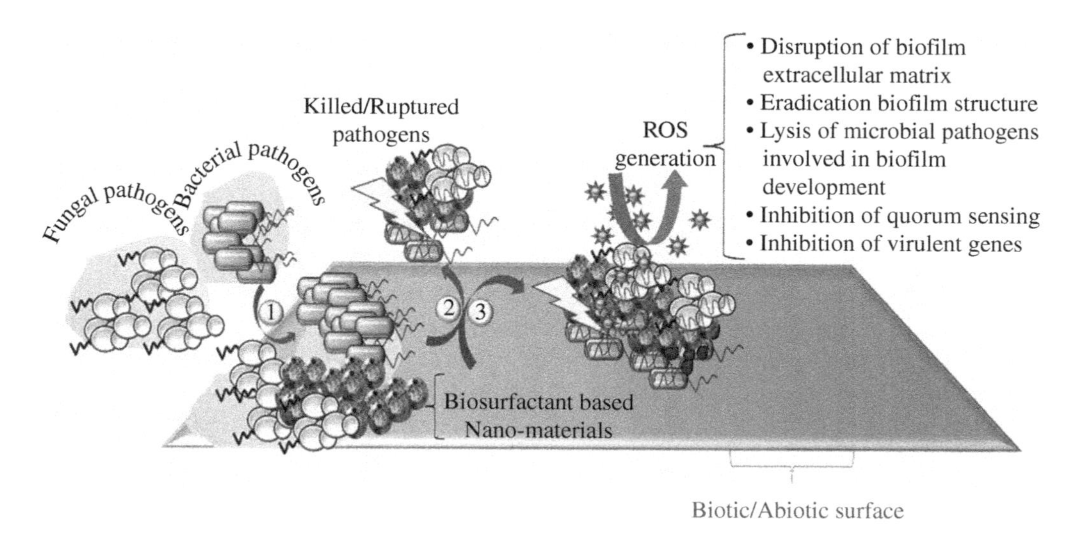

Figure 12.6 Schematic representation of the antibiofilm activity of biosurfactant-based nanoformulations.

properties have been shown to be efficient alternatives to their chemical counterparts in terms of wound care and related technical biomedical issues. In particular, these bio-based nanomaterials are worthy of serious consideration, especially in the case of biomedical devices. Nanomaterials are used as constituents of coatings, biomedical agents, and drug-delivery vehicles and implant materials and research remain active in these areas. However, key issues like nanoparticle resistance and surface interactions between nanoparticles, biofilms, and hosts need to be resolved to ensure successful clinical applications. It is hoped that this review of the literature persuades the reader that nanomaterials and nanomaterial-based biomedical devices with broad-spectrum antibiofilm activities will be produced such that they are potent, nontoxic, biocompatible, and cost-effective and that these novel materials will establish new standards for the treatment and prevention of pathogenic biofilms.

Acknowledgement

The author is very grateful to the Department of Science and Technology, Government of India (Registration no. IF130678) for their financial assistance.

References

1 Abbasi, H., Hamedi, M.M., Lotfabad, T.B. et al. (2012). Biosurfactant-producing bacterium, *Pseudomonas aeruginosa* MA01 isolated from spoiled apples: Physicochemical and structural characteristics of isolated biosurfactant. *J. Biosci. Bioeng.* 113: 211–219.
2 Gupta, P., Gupta, S., Sharma, M. et al. (2018). Effectiveness of phytoactive molecules on transcriptional expression, biofilm matrix, and cell wall components of *Candida glabrata* and its clinical isolates. *ACS Omega.* 3: 12201–12214.
3 Pasternak, G., Askitosari, T.D., and Rosenbaum, M.A. (2020). Biosurfactants and synthetic surfactants in bioelectrochemical systems: A mini-review. *Front. Microbiol.*

4 Arciola, C.R., Campoccia, D., and Montanaro, L. (2018 Jul). Implant infections: adhesion, biofilm formation and immune evasion. *Nat. Rev. Microbiol.* 16 (7): 397–409. https://doi.org/10.1038/s41579-018-0019-y. PMID: 29720707.
5 Pemmaraju, S.C., Kumar, P., Pruthi, P.A. et al. (2016). Impact of oxidative and osmotic stresses on *Candida albicans* biofilm formation. *Biofouling* 32: 897–909.
6 Ramage, G., Bachmann, S., Patterson, T.F. et al. (2002 June). Investigation of multidrug efflux pumps in relation to fluconazole resistance in *Candida albicans* biofilms. *J. Antimicrob. Chemother.* 49 (6): 973–980. https://doi.org/10.1093/jac/dkf049. PMID: 12039889.
7 Banat, I.M., Franzetti, A., Gandolfi, I. et al. (2010). Microbial biosurfactants production, applications and future potential. *Appl. Microbiol. Biotechnol.* 87: 427–444.
8 Gudina, K.E.J., Rangarajan, V., Sen, R., and Rodrigues, L.R. (2013). Potential therapeutic applications of biosurfactants. *Trends Pharmacol. Sci.* 34: 667–675.
9 Gupta, S., Raghuwanshi, N., Varshney, R. et al. (2017). Accelerated *in vivo* wound healing evaluation of microbial glycolipid containing ointment as a transdermal substitute. *Biomed. Pharmacother.* 94: 1186–1196.
10 Płaza, G. and Achal, V. (2020). Biosurfactants: Eco-friendly and innovative biocides against biocorrosion. *Int. J. Mol. Sci.* 2020 (21): 2152.
11 Kumar, C.G., Mamidyala, S.K., Das, B. et al. (2010). Synthesis of biosurfactant-based silver nanoparticles with purified rhamnolipids isolated from *Pseudomonas aeruginosa* BS-161R. *J. Microbiol. Biotechnol.* 20: 1061–1068.
12 Narayanan, J., Ramji, R., Sahu, H., and Gautam, P. (2010). Synthesis, stabilisation and characterization of rhamnolipid-capped ZnS nanoparticles in aqueous medium. *IET Nanobiotech.* 4: 29–34.
13 Romanova, I.M. and Gintsburg, A.L. (2011 May-June). Bacterial biofilms as a natural form of existence of bacteria in the environment and host organism. *Zh. Mikrobiol. Epidemiol. Immunobiol.* 3: 99–109. Russian. PMID: 21809653.
14 O'Toole, G., Kaplan, H.B., and Kolter, R. (2000). Biofilm formation as microbial development. *Annu. Rev. M.* 54: 49–79.
15 Otto, M. (2009). *Staphylococcus epidermidis* – The "accidental" pathogen. *Nat. Rev. Microbiol.* 7 (8): 555–567. https://doi.org/10.1038/nrmicro2182.
16 Rogers, C.L., Gibson, C., Mitchell, S.L. et al. (2009). Disseminated candidiasis secondary to fungal and bacterial peritonitis in a young dog. *J. Vet. Emerg. Crit. Care. (San Antonio)* 19: 193–198.
17 Nobile, C.J. and Johnson, A.D. (2015). *Candida albicans* biofilms and human disease. *Annu. Rev. Microbiol.* 69: 71–92.
18 Pappas, G., Kiriaze, I.J., Giannakis, P., and Falagas, M.E. (2009). Psychosocial consequences of infectious diseases. *Clinical Microbiology and Infection* 15 (8): 743–747.
19 Diekema, D., Arbefeville, S., Boyken, L. et al. (2012). The changing epidemiology of healthcare-associated candidemia over three decades. *Diagn. Microbiol. Infect. Dis.* 73: 45–48.
20 Fu, J., Ding, Y., Wei, B. et al. (2017). Epidemiology of *Candida albicans* and non-*C. albicans* of neonatal candidemia at a tertiary care hospital in western China. *BMC Infect. Dis.* 17: 329.
21 Diekema, D. and Pfaller, M. (2015). Prevention of Health Care-Associated Infections. In: *Manual of Clinical Microbiology*, 11the (eds. J. Jorgensen, M. Pfaller, K. Carroll, et al.), 106–119. Washington, DC: ASM Press https://doi.org/10.1128/9781555817381.ch8.
22 Amitani, R., Taylor, G., Elezis, E.N. et al. (1995). Purification and characterization of factors produced by *Aspergillus fumigatus* which affect human ciliated respiratory epithelium. *Infect. Immun.* 63: 3266–3271.
23 Latge, J.P. (2001). The pathobiology of *Aspergillus fumigatus*. *Trends Microbiol.* 9: 382–389.

24 Costerton, J.W., Stewart, P.S., and Greenberg, E.P. (1999 May 21). Bacterial biofilms: A common cause of persistent infections. *Science.* 284 (5418): 1318–1322. https://doi.org/10.1126/science.284.5418.1318. PMID: 10334980.

25 Joo, H.-S. and Otto, M. (2013). Molecular basis of in-vivo biofilm formation by bacterial pathogens. *Chem. Biol.* 19: 1503–1513.

26 Anderl, J.N., Franklin, M.J., and Stewart, P.S. (2000). Role of antibiotic penetration limitation in *Klebsiella pneumoniae* biofilm resistance to ampicillin and ciprofloxacin. *Antimicrob. Agents Chemother.* 44: 1818–1824.

27 Giwercman, B., Jensen, E.T., Hoiby, N. et al. (1991). Induction of b-lactamase production in *Pseudomonas aeruginosa* biofilm. *Antimicrob. Agents Chemother.* 35: 1008–1010.

28 Kumon, H., Tomochika, K., Matunaga, T. et al. (1994). A sandwich cup method for the penetration assay of antimicrobial agents through *Pseudomonas* exopolysaccharides. *Microbiol. Immunol.* 38: 615–619.

29 Stewart, P.S. (1996). Theoretical aspects of antibiotic diffusion into microbial biofilms. *Antimicrob. Agents Chemother.* 40: 2517–2522.

30 Taylor, P., Roy, S., Leal, S. et al. (2014). Available at). Activation of neutrophils by autocrine IL-17A–IL-17RC interactions during fungal infection is regulated by IL-6, IL-23, RORγt and dectin-2. *Nat. Immunol.* 15: 143–151. https://doi.org/10.1038/ni.2797.

31 Molin, S. and Tolker-Nielsen, T. (2003). Gene transfer occurs with enhanced efficiency in biofilms and induces enhanced stabilisation of the biofilm structure. *Curr. Opin. Biotech.* 14: 255–261.

32 Jamal, M., Ahmad, W., Andleeb, S. et al. (2017). Bacterial biofilm and associated infections. *J. Chin. Med. Assoc.* 81: 7–11.

33 Zarnowski, R., Westler, W.M., Lacmbouh, G.A. et al. (2014). Novel entries in a fungal biofilm matrix encyclopedia. *M. Bio.* 5: e01333–e01314.

34 Dignac, M.F., Urbain, V., Rybacki, D. et al. (1998). Chemical description of extracellular polymers: Implication on activated sludge floc structure. *Water Sci. Technol.* 38: 45–53.

35 Whitchurch, C.B., Tolker-Nielsen, T., Ragas, P.C., and Mattick, J.S. (2002). Extracellular DNA required for bacterial biofilm formation. *Science* 295: 1487.

36 Mukherjee, P.K. and Chandra, J. (2004). *Candida* biofilm resistance. *Drug Resist. Updat.* 7: 301–309.

37 Al-Fattani, M.A. and Douglas, L.J. (2006 Aug). Biofilm matrix of *Candida albicans* and *Candida tropicalis*: Chemical composition and role in drug resistance. *J. Med. Microbiol.* 55 (Pt 8): 999–1008. https://doi.org/10.1099/jmm.0.46569-0. PMID: 16849719.

38 Dueholm, M.S., Petersen, S.V., Sønderkær, M. et al. (2010). Functional amyloid in *Pseudomonas*. *Mol. Microbiol.* 77: 1009–1020.

39 Gross, M., Cramton, S.E., Götz, F., and Peschel, A. (2001). Key role of teichoic acid net charge in *Staphylococcus aureus* colonization of artificial surfaces. *Infect. Immun.* 69: 3423–3426.

40 Friedman, L. and Kolter, R. (2004). Genes involved in matrix formation in *Pseudomonas aeruginosa* PA14 biofilms. *Mol. Microbiol.* 51: 675–690.

41 Periasamy, S., Nair, H.A., Lee, K.W. et al. (2015). *Pseudomonas aeruginosa* PAO1 exopolysaccharides are important for mixed species biofilm community development and stress tolerance. *Front. Microbiol.* 6: 851.

42 Sun, J., Deng, Z., and Yan, A. (2014). Bacterial multidrug efflux pumps: mechanisms, physiology and pharmacological exploitations. *Biochemical and Biophysical Research Communications* 453 (2): 254–267.

43 Pittet, D., Monod, M., Suter, P.M. et al. (1994). *Candida* colonization and subsequent infections in critically ill surgical patients. *Ann. Surg.* 220: 751–758.

44 Leung, J.W. et al. (1998). Is there a synergistic effect between mixed bacterial infection in biofilm formation on biliary stents? *Gastrointestinal Endoscopy* 48 (3): 250–257.

45 Duguid, I.G., Evans, E., Brown, M.R., and Gilbert, P. (1992). Effect of biofilm culture upon the susceptibility of *Staphylococcus epidermidis* to tobramycin. *J. Antimicrob. Chemother.* 1992 (30): 803–810.

46 Evans, D.J., Allison, D.G., Brown, M.R.W., and Gilbert, P. (1991). Susceptibility of *Pseudomonas aeruginosa* and *Escherichia coli* biofilms towards ciprofloxacin: effect of specific growth rate. *J. Antimicrob. Chemother.* 27: 177–184.

47 Xu, G. (2016). Relationships between the regulatory systems of quorum sensing and multidrug resistance. *Front. Microbiol.* 7: 958.

48 Soto, S.M. (2013). Role of efflux pumps in the antibiotic resistance of bacteria embedded in a biofilm. *Virulence* 4 (3): 223–229.

49 Davies, D.G., Parsek, M.R., Pearson, J.P. et al. (1998). The involvement of cell-to-cell signals in the development of a bacterial biofilm. *Science* 280: 295–298.

50 Yarwood, J.M., Bartels, D.J., Volper, E.M., and Greenberg, E.P. (2004). Generation of virulence factor variants in Staphylococcus aureus biofilms. *J. Bacteriol.* 186: 1838–1850.

51 Davies, D.G. and Geesey, G.G. (1995). Regulation of the alginate biosynthesis gene algC in *Pseudomonas aeruginosa* during biofilm development in continuous culture. *Appl. Environ. Microbiol.* 61: 8607.

52 Ahmer, B.M. (2004). Cell-to-cell signalling in *Escherichia coli* and *Salmonella enterica*. *Mol. Microbiol.* 52: 933–945.

53 Boles, B.R. et al. (2010). Identification of genes involved in polysaccharide-independent Staphylococcus aureus biofilm formation. *PloS One* 5 (4): e10146.

54 Webber, M.A. and Piddock, L.J.V. (2003). The importance of efflux pumps in bacterial antibiotic resistance. *J. Antimicrob. Chemother.* 51: 9–11.

55 Du, D., Wang-Kan, X., Neuberger, A. et al. (2018). Multidrug efflux pumps: structure, function and regulation. *Nat. Rev. Microbiol.* 16: 523–539.

56 Cannon, R.D. et al. (2009). Efflux-mediated antifungal drug resistance. *Clinical Microbiology Reviews* 22 (2): 291–321.

57 Kohli, A., Gupta, V., Krishnamurthy, S. et al. (2001 Sept). Specificity of drug transport mediated by CaMDR1: A major facilitator of *Candida albicans*. *J. Biosci.* 26 (3): 333–339. https://doi.org/10.1007/BF02703742. PMID: 11568478.

58 Diggle, S.P., Winzer, K., Lazdunski, A. et al. (2002). Advancing the quorum in *Pseudomonas aeruginosa*: MvaT and the regulation of *N*-acylhomoserine lactone production and virulence gene expression. *J. Bacteriol.* 184: 2576–2586.

59 Chan, Y.Y. and Chua, K.L. (2005). The *Burkholderia pseudomallei* BpeAB-OprB efflux pump: expression and impact on quorum sensing and virulence. *J. Bacteriol.* 187: 4707–4719.

60 Raghuwanshi, N., Kumari, P., Srivastava, A.K. et al. (2017). Synergistic effects of *Woodfordia fruticosa* gold nanoparticles in preventing microbial adhesion and accelerating wound healing in Wistar albino rats in vivo. *Mater. Sci. Eng. C Mater. Biol. Appl.* 80: 252–262.

61 Kiran, G.S., Sabarathnam, B., and Selvin, J. (2010). Biofilm disruption potential of a glycolipid biosurfactant from marine *Brevibacterium casei*. *FEMS Immunol. Med. Microbiol.* 59: 432–438. https://doi.org/10.1111/j.1574-695X.2010.00698.x.

62 Zhao, J., Zhang, X.Y., Yonzon, C.R. et al. (2006). Localized surface plasmon resonance biosensors. *Nanomedicine* 1: 219–228.

63 Abhilash, B.D.P. (2012). Synthesis of zinc-based nanomaterials: A biological perspective. *IET Nanobiotech.* 6: 144–148.

64 Baer, D.R., Engelhard, M.H., Johnson, G.E. et al. (2013). Surface characterization of nanomaterials and nanoparticles: important needs and challenging opportunities. *J. Vac. Sci. Technol.* 31: 050820.

65 Kuchibhatla, S.V.N.T., Karakoti, A.S., Baer, D.R. et al. (2012). Influence of aging and environment on nanoparticle chemistry: implication to confinement effects in nanoceria. *J. Phys. Chem.* 116: 14108–14114.S.

66 Ruckenstein, E. and Kong, X.Z. (1999). Control of pore generation and pore size in nanoparticles of poly(styrene-methyl methacrylateacrylic acid). *J. Appl. Polym. Sci.* 72: 419–426.

67 Torresdey, J.L., Tiemann, K.J., Gamez, G. et al. (1999). Recovery of gold (III) by alfalfa biomass and binding characterization using X-ray microfluorescence. *Adv. Environ. Res.* 3: 83–93.

68 Kasture, M.B., Patel, P., Prabhune, A.A. et al. (2008). Synthesis of silver nanoparticles by sophorolipids: Effect of temperature and sophorolipid structure on the size of particles. *J. Chem. Sci.* 120: 515–520.

69 Rai, A., Singh, A., Ahmad, A., and Sastry, M. (2006). Role of halide ions and temperature on the morphology of biologically synthesized gold nanotriangles. *Langmuir* 22: 736–741.

70 Tran, Q.H., Nguyen, V.Q., and Le, A.T. (2013). Silver nanoparticles: Synthesis, properties, toxicology, applications and perspectives. *Adv. Nat. Sci. Nanosci. Nanotechnol.* 4: 43001–45018.

71 Darroudi, M., Ahmad, M.B., Zamiri, R. et al. (2011). Time-dependent effect in green synthesis of silver nanoparticles. *Int. J. Nanomedicine* 6: 677–681.

72 Baer, M., Sawa, T., Flynn, P. et al. (2009). An engineered human antibody fab fragment specific for *Pseudomonas aeruginosa* PcrV antigen has potent antibacterial activity. *Infect. Immun.* 77: 1083–1090. https://doi.org/10.1128/IAI.00815-08.

73 Zhu, X., Yates, M.D., and Logan, B.E. (2012). Set potential regulation reveals additional oxidation peaks of *Geobacter sulfurreducens* anodic biofilms. *Electrochemistry Communications* 22: 116–119.

74 Alber, F., Dokudovskaya, S., Veenhoff, L.M. et al. (2007). Determining the architectures of macromolecular assemblies. *Nature* 2007 (450): 695–701.

75 Saikia, R.R., Deka, S., Deka, M., and Sarma, H. (2012). Optimization of environmental factors for improved production of rhamnolipid biosurfactant by Pseudomonas aeruginosa RS29 on glycerol. *J. Basic Microbiol.* 52 (4): 446–457.

76 Sarma, H., Bustamante, K.L.T., and Prasad, M.N.V. (2018). Biosurfactants for oil recovery from refinery sludge: magnetic nanoparticles assisted purification. In: *Industrial and Municipal Sludge* (ed. M.N.V. Prasad). Elsevier Massachusetts, USA, pp.107–132. ISBN: 9780128159071, Editor Majeti Narasimha Vara Prasad, Paulo Jorge de Campos, Favas Meththika, Vithanage S. Venkata Mohan.

77 Rodrigues, L., Banat, I.M., Teixeira, J., and Oliveira, R. (2006). Biosurfactants: potential applications in medicine. *J. Antimicrob. Chemother.* 57: 609–618.

78 Rosenberg, E. and Ron, E.Z. (1999). High- and low-molecular mass microbial surfactants. *Appl. Microbiol. Biotechnol.* 52: 154–162.

79 Cooper, D.G. (1986). Biosurfactants. *Microbiological Sciences* 3.5: 145.

80 Robert, M., Mercade, M.E., Bosch, M.P. et al. (1989). Effect of the carbon source on biosurfactant production by *Pseudomonas aeruginosa* 44T. *Biotechnol. Lett.* 11: 871874.

81 Lang, S., Katsiwela, E., and Wagner, F. (1989). Antimicrobial effects of biosurfactants. *Fat. Sci. Technol.* 91: 363–366.

82 Syldatk, C., Lang, S., and Wagner, F. (1985). Chemical and physical characterization of four interfacial-active rhamnolipids from *Pseudomonas* sp. DSM 2874 grown on n-alkanes. *Z. Naturforsch.* 40C: 51–60.

83 Boothroyd, B., Thorn, J.A., and Haskins, R.H. (1956). Biochemistry of the ustilaginales: XII. *Characterization of extracellular glycolipids produced by Ustilago sp. Can. J. Biochem. Physiol.* 34: 10–14.

84 Gobbert, U., Lang, S., and Wagner, F. (1984). Sophorose lipid formation by resting cells of *Torulopsis bombicola*. *Biotechnol. Lett.* 6: 225–230.

85 Cooper, D.G., Liss, S.N., Longay, R., and Zajic, J.E. (1989). Surface activities of *Mycobacterium* and *Pseudomonas*. *J. Ferment. Tech.* 59: 97–101.

86 Hommel, R., Stiiwer, O., Stuber, W. et al. (1987). Production of water-soluble surface-active exolipids by *Torulopsis apicola*. *Appl. Microbiol. Biotechnol.* 26: 199–205.

87 Margaritis, A., Kennedy, K., and Zajic, J.E. (1980). Applications of an air lift fermenter in the production of biosurfactants. *Dev. Ind. Microbiol.* 21: 285–294.

88 Rapp, P., Bock, H., Wray, V., and Wagner, F. (1979). Formation, isolation and characterization of trehalose dimycolates from *Rhodococcus erythropolis* grown on n-alkanes. *J. Gen. Microbiol.* 115: 491–503.

89 Nguyen, T.T.L., Edelen, A., Neighbors, B., and Sabatini, D.A. (2010). Biocompatible lecithin-based microemulsions with rhamnolipid and sophorolipid biosurfactants: Formulation and potential applications. *J. Colloid Interface Sci.* 348: 498–504.

90 Neu, T.R. and Poralla, K. (1990). Emulsifying agent from bacteria isolated during screening for cells with hydrophobic surfaces. *Appl. Microbiol. Biotechnol.* 32: 521–525.

91 Yakimov, M.M., Timmis, K.N., Wray, V., and Fredrickson, H.L. (1995). Characterization of a new lipopeptide surfactant produced by thermotolerant and halotolerant subsurface *Bacillus licheniformis* BAS50. *Appl. Environ. Microbiol.* 61: 1706–1713.

92 Suzuki, T., Hayashi, K., Fujikawa, K., and Tsukamoto, K. (1965). The chemical structure of polymyxin E. The identities of polymyxin E1 with colistin A and polymyxin E2 with colistin B. *J. Biol. Chem.* 57: 226–227.

93 Marahiel, M., Denders, W., Krause, M., and Kleinkauf, H. (1977). Biological role of gramicidin S in spore functions. Studies on gramicidin-S negative mutants of *Bacillus brevis* 9999. *Eur. J. Biochem.* 99: 49–52.

94 Matsuyama, T., Sogawa, M., and Yano, I. (1991). Direct colony thin-layer chromatography and rapid characterization of *Serratia marcescens* mutants defective in production of wetting agents. *Appl. Environ. Microbiol.* 53: 1186–1188.

95 Bernheimer, A.W. and Avigad, L.S. (1970). Nature and properties of a cytolytic agent produced by *Bacillus subtilis*. *J. Gen. Microbiol.* 61: 361–369.

96 Cooper, D.G. and Paddock, D.A. (1983). *Torulopsis petrophilum* and surface activity. *Appl. Environ. Microbiol.* 46: 1426–1429.

97 Lin, S.C., Minto, M.A., Sharma, M.M., and Georgiou, G. (1994). Structural and immunological characterization of a biosurfactant produced by *Bacillus licheniformis* JF-2. *Appl. Environ. Microbiol.* 60: 31–38.

98 Koch, A.K., Reiser, J., Kappeli, O., and Fiechter, A. (1988). Genetic construction of lactose-utilizing strains of *P. aeruginosa* and their application in biosurfactant production. *Nat. Biotechnol.* 6: 1335–1339.

99 Hisatsuka, K., Nakahara, T., Minoda, Y., and Yamada, K. (1977). Formation of protein-like activator for *n*-alkane oxidation and its properties. *Agric. Biol. Chem.* 41: 445–450.

100 Rosenberg, E., Rubinovitz, C., Gottlieb, A. et al. (1988). Production of biodispersan by *Acinetobacter calcoaceticus* A2. *Appl. Environ. Microbiol.* 54: 317–322.

101 Zosim, Z., Gutnick, D.L., and Rosenberg, E. (1982). Properties of hydrocarbon-in-water emulsions stabilized by *Acinetobacter* RAG-1 emulsan. *Biotechnol. Bioeng.* 24: 281–292.

102 Cirigliano, M.C. and Carman, G.M. (1985). Purification and characterization of liposan, a bioemulsifier from *Candida lipolytica*. *Appl. Environ. Microbiol.* 50: 846–850.

103 Persson, A., Oesterberg, E., and Dostalek, M. (1988). Biosurfactant production by *Pseudomonas fluorescens* 378: Growth and product characteristics. *Appl. Microbiol. Biotechnol.* 29: 1–4.

104 Singh, M. and Desai, J.D. (1989). Hydrocarbon emulsification by *Candida tropicalis* and *Debaryomyces polymorphus*. *Indian J. Exp. Biol.* 27: 224–226.

105 Kappeli, O., Walther, P., Mueller, M., and Fiechter, A. (1984). Structure of cell surface of the yeast *Candida tropicalis* and its relation to hydrocarbon transport. *Arch. Microbiol.* 138: 279–282.

106 Gutnick D L, Shabtai Y. 1987. Exopolysaccharide bioemulsifiers. In: *Biosurfactants and Biotechnology*, Kosaric N, Cairns W L, Gray N C C (eds), Marcel Dekker, New York;, pp 211–246.

107 Soberón-Chávez, G. and Maier, R.M. (2011). *Biosurfactants: A General Overview. Biosurfactants*, 1–11. Springer.

108 Varjani, S.J. and Upasani, V.N. (2017). A new look on factors affecting microbial degradation of petroleum hydrocarbon pollutants. *Int. Biodeterior. Biodegrad.* 120: 71–83.

109 Devadasu, V., Bhardwaj, R., Kumar, V., and M. N. V. R. (2013). Can controversial nanotechnology promise drug delivery. *Chem. Rev.* 113: 1686–1735.

110 Loh, X.J., Lee, T.C., Dou, Q., and Deen, G.R. (2016). *Biomater. Sci.* https://doi.org/10.1039/c5bm00277j.

111 Sarikaya, M., Tamerler, C., Jen, A.-Y. et al. (2003). Molecular biomimetics: nanotechnology through biology. *Nat. Mater.* 2: 577–585.

112 Iravani, S., Korbekandi, H., Mirmohammadi, S.V., and Zolfaghari, B. (2014). Synthesis of silver nanoparticles: Chemical, physical and biological methods. *Res. Pharma. Sci.* 9: 385–406.

113 Kiran, G.S., Sabu, A., and Selvin, J. (2010). Synthesis of silver nanoparticles by glicololid biosurfactant produced from marine *Brevibacterium casei* MSA 19. *J. Biotechnol.* 148: 221–225.

114 Cheow, W.S. and Hadinoto, K. (2012). Lipid-polymer hybrid nanoparticles with rhamnolipid-triggered release capabilities as antibiofilm drug delivery vehicles. *Particuology* 10: 327–333.

115 Palanisamy, P. and Raichur, A.M. (2009). Synthesis of spherical NiO nanoparticles through a novel biosurfactant mediated emulsion technique. *Mater. Sci. Eng. C.* 29: 199–204.

116 Kasture, M., Singh, S., Patel, P. et al. (2007). Multiutility sophorolipids as nanoparticle capping agents: Synthesis of stable and water dispersible Co nanoparticles. *Langmuir* 23: 11409–11412.

117 Biswas, M. and Raichur, A.M. (2008). Electrokinetic and rheological properties of nano zirco-nia in the presence of rhamnolipid biosurfactant. *J. Am. Ceram. Soc.* 91: 197–3201.

118 Raichur, A.M. (2007). Dispersion of colloidal alumina using a rhamnolipid -biosurfactant. *J. Dispers. Sci. Technol.* 28: 1272–1277.

119 Raveendran, P., Fu, J., and Wallen, S.L. (2003). Completely "green" synthesis and stabilization of metal nanoparticles. *J. Am. Chem. Soc.* 125: 13940–13941.

120 Cushing, B.L., Kolesnichenko, V.L., and O'Connor, C.J. (2004). Recent advances in the liquid-phase syntheses of inorganic nanoparticles. *Chem. Rev.* 104: 3893–3946.

121 Palanisamy, P. (2008). Biosurfactant mediated synthesis of NiO nanorods. *Mater. Lett.* 62: 743–746.

122 Engberts, J.B.F.N. and Kevelam, J. (1996). Formation and stability of micelles and vesicles. *Curr. Opin. Colloid Interface Sci.* 1: 779–789.

123 Davies, G., Wells, A.U., Doffman, S. et al. (2006 Nov). The effect of *Pseudomonas aeruginosa* on pulmonary function in patients with bronchiectasis. *Eur. Respir. J.* 28 (5): 974–979. https://doi.org/10.1183/09031936.06.00074605. Epub. 2006 Aug 9. PMID: 16899482.

124 Drulis-Kawa, Z. and Dorotkiewicz-Jach, A. (2010). Liposomes as delivery systems for antibiotics. *Int. J. Pharm.* 387: 187–198.

125 Meers, P., Neville, M., Malinin, V. et al. (2008). Biofilm penetration, triggered release and in vivo activity of inhaled liposomal amikacin in chronic *Pseudomonas aeruginosa* lung infections. *Journal of Antimicrobial Chemotherapy* 61 (4): 859–868.

126 Morita, T., Kawamura, D., Morita, N. et al. (2013). Characterization of mannosylerythritol lipids containing hexadecatetraenoic acid produced from cuttlefish oil by *Pseudozyma churashimaensis* OK96. *J. Oleo. Sci.* 62: 319–327.

127 Balakumara, K., Raghavana, C.V., Tamilselvana, N. et al. (2013). Self-nanoemulsifying drug delivery system (SNEDDS) of Rosuvastatin calcium: Design, formulation, bioavailability and pharmacokinetic evaluation. *Colloids Surf. B: Biointerfaces* 112: 337–343.

128 Kohli, K., Chopra, S., Dhar, D. et al. (2010). Self-emulsifying drug delivery systems: An approach to enhance oral bioavailability. *Drug Discov. Today* 15: 965–968.

129 Ibrahim, H.M.M. (2015). Green synthesis and characterization of silver nanoparticles using banana peel extract and their antimicrobial activity against representative microorganisms. *J. Radiat. Res.* 8: 265–275.

130 Vincent, J.L. (2003). Nosocomial infections in adult intensive-care units. *Lancet* 361: 2068–2077.

131 Banerjee, M., Mallick, S., Paul, A. et al. (2010). Heightened reactive oxygen species generation in the antimicrobial activity of a three-component iodinated chitosan–silver nanoparticle composite. *Langmuir* 26: 5901–5908.

132 Morones, J.R., Elechiguerra, J.L., Camacho, A. et al. (2005). The bactericidal effect of silver nanoparticles. *Nanotechnology* 16: 2346–2353.

133 Pal, S., Tak, Y.K., and Song, M. (2007). Does the antibacterial activity of silver nanoparticles depend on the shape of the nanoparticle? A study of the gram-negative bacterium *Escherichia coli*. *Appl. Environ. Microbiol.* 73: 1712–1720.

134 Mukherji, S., Ruparelia, J.P., and Agnihotri, S. (2012). Progress and prospects. In: *NanoAntimicrobials* (eds. N. Cioffi and M. Rai), 225–251. Berlin, Heidelberg: Springer Verlag.

135 Shekhar, A., Mukherji, S., and Mukherji, S. (2012). Antimicrobial chitosan–PVA hydrogel as a nanoreactor and immobilizing matrix for silver nanoparticles. *Applied Nanoscience* 2 (3): 179–188.

136 Gómez-Graña, S., Perez-Ameneiro, M., Vecino, X. et al. (2017). Biogenic synthesis of metal nanoparticles using a biosurfactant extracted from corn and their antimicrobial properties. *Nanomaterials (Basel)* 7: 139.

137 Das, M. et al. (2016). Microemulsion synthesis of silver nanoparticles using biosurfactant extracted from *Pseudomonas aeruginosa* MKVIT3 strain and comparison of their antimicrobial and cytotoxic activities. *IET Nanobiotechnology* 10 (6): 411–418.

138 Chojniak, J. et al. (2018). A nonspecific synergistic effect of biogenic silver nanoparticles and biosurfactant towards environmental bacteria and fungi. *Ecotoxicology* 27 (3): 352–359.

139 Malik, M.A., Wani, M.Y., and Hashim, M.A. (2012). Microemulsion method: a novel route to synthesize organic and inorganic nanomaterials: 1st nano update. *Arab. J. Chem.* 5: 397–417.

140 Płaza, G.A., Chojniak, J., and Banat, I.M. (2014). Biosurfactant mediated biosynthesis of selected metallic nanoparticles. *Int. J. Mol. Sci.* 2014 (15): 13720–13737.

141 Ramíreza, I.M., Tsaousib, K., Ruddenb, M. et al. (2015). Rhamnolipid and surfactin production from olive oil mill waste as sole carbon source. *Bioresour. Technol.* 198: 231–236.

142 Reddy, A.S., Hao, Y.K., Atla, S.B. et al. (2011). Low-temperature synthesis of rose-like ZnO nanostructures using surfactin and their photocatalytic activity. *J. Nanosci. Nanotechnol.* 11: 5034–5041.

143 Rodrigues, L.R., Banat, I.M., van der Mei, H.C. et al. (2004). Interference in adhesion of bacteria and yeasts isolated from explanted voice prostheses to silicone rubber by rhamnolipid biosurfactants. *Appl. Microbiol. Biotechnol.* 66: 306–311.

144 Rodrigues, R.C., Pocheron, A.-L., Hernould, M. et al. (2015). Description of *Campylobacter jejuni* Bf, an atypical aero-tolerant strain. *Gut Pathog.* 7: 1. https://doi.org/10.1186/s13099-015-0077-x.

145 Deziel, E., Comeau, Y., and Villemur, R. (2001). Initiation of biofilm formation by *Pseudomonas aeruginosa* 57RP correlates with emergence of hyperpiliated and highly adherent phenotypic variants deficient in swimming, swarming, and twitching motilities. *J. Bacteriol.* 183: 1195–1204.

13

Biosurfactants from Bacteria and Fungi

Perspectives on Advanced Biomedical Applications

Rashmi Rekha Saikia[1], *Suresh Deka*[2], *and Hemen Sarma*[3]

[1] *Department of Zoology, Jagannath Barooah College (Autonomous), Jorhat, Assam, India*
[2] *Environmental Biotechnology Laboratory, Life Sciences Division, Institute of Advanced Study in Science and Technology (IASST), Guwahati, Assam, India*
[3] *Department of Botany, N N Saikia College, Titabar, Assam, India*

CHAPTER MENU

13.1 Introduction

Health and medical research is at an enormous pace today. There have been different biomolecules, tools, and techniques that have been used to deal with various human diseases. Biosurfactants are one of the best biomolecules to be used in medical sciences. Biosurfactants are surface-active molecules synthesized by microorganisms. They may be synthesized either on cell surfaces or excreted extracellularly. The biosurfactants possess both hydrophilic and hydrophobic domains, making it possible to decrease interfacial and surface tensions by them [1]. Some biosurfactants are anionic and some are neutral, but microorganisms also produce cationic biosurfactants. The hydrophilic component may be an amino acid, a carbohydrate, a phosphate group, etc., and the hydrophobic component is primarily a fatty acid with a long carbon chain [2].

Biosurfactants for a Sustainable Future: Production and Applications in the Environment and Biomedicine, First Edition. Edited by Hemen Sarma and Majeti Narasimha Vara Prasad.
© 2021 John Wiley & Sons Ltd. Published 2021 by John Wiley & Sons Ltd.

There are a large number of microorganisms that produce biosurfactants, which are of many types, like glycolipids, phospholipids, neutral lipids, lipopeptides, etc. [3]. These compounds are produced while growing in natural environments [4].

Biosurfactants are in great demand since they have a wide range of properties and became a very important biotechnological candidate for industrial, biomedical, and pharmaceutical applications. They may be the alternatives to chemically produced surfactants and have tremendous commercial prospects [5].

Antimicrobial activities of biosurfactants have made them capable of combating many diseases [6]. Research into biosurfactants has discovered many interesting biological and chemical properties that have possible use in biomedical, prophylactic, and therapeutic applications [7].

The importance of biosurfactants has therefore increased steadily over the last few years. Recent research carried out in ScienceDirect showed a continual increase in publications featuring biosurfactant–antifungal, biosurfactant–antibacterial, and biosurfactant–biomedical subjects, where a total of 1954 articles were indexed in scopus starting from only 64 in 2010 (Figure 13.1).

In 2001, Ryll et al. [8] reported that mycolic acid-containing glycolipid biosurfactants have properties to stimulate humoral and cellular immunity, formation of granuloma and anti-tumor activity. Similarly, Rodrigues et al. [9] highlighted some very interesting applications of biosurfactants in biomedical and clinical sciences. Banat et al. [10] discussed the production, applications, and potential for the sustainable future for microbial biosurfactants. This means that this chapter cannot but be a rundown on the main points of such published works. However, we have been discussing many of the fascinating areas of application of biosurfactant in recent years.

Although there are reports describing biosurfactants as therapeutic agents, this chapter seeks to combine the attractive combinations of biosurfactants with medical and pharmaceutical applications and focuses most of the applications of biosurfactant in these fields.

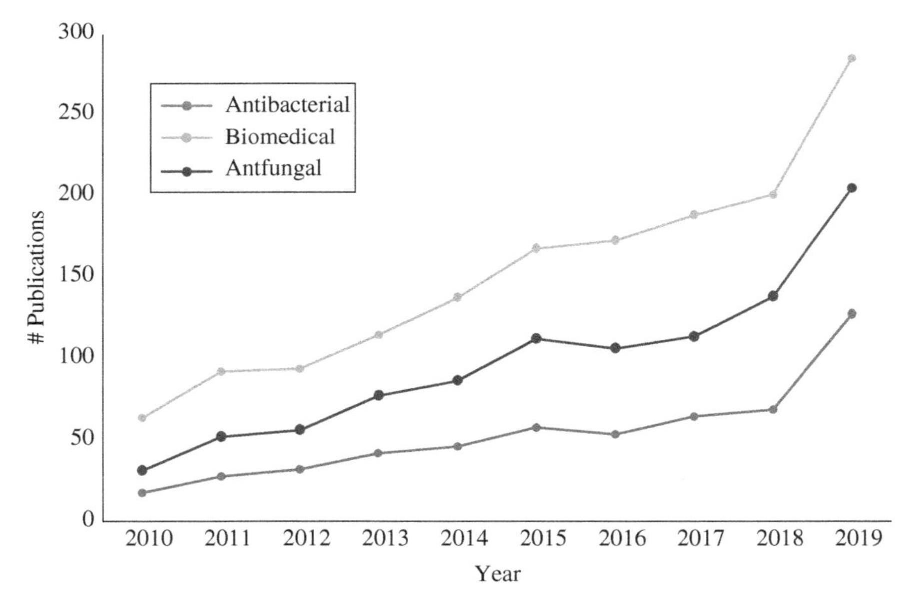

Figure 13.1 Publications with continuous increases featuring biosurfactant–antifungal, biosurfactant–antibacterial, and biosurfactant–biomedical subjects. *Source:* www.sciencedirect.com; search parameter used = biosurfactants, antifungal, antibacterial, and biomedical.

13.2 Biomedical Applications of Biosurfactants: Recent Developments

The extensive research that has been conducted over the last few decades on biosurfactants to be used as high-value chemicals for pharmaceutical and biomedical industries has shown very good prospects. Some of the most important and promising biomedical applications are discussed in the following nine subsections (Figure 13.2).

13.2.1 Biosurfactants Used to Control Bacteria, Fungi and Viruses

The antimicrobial activity is one of the major functions of many biosurfactants [11]. The mechanisms of action of biosurfactants against pathogenic bacteria are the affinity of some of the components such as bacterial lipopolysaccharides. Because of the affinity of the components, cell walls get disrupted killing the microbe. A lot of biosurfactant-producing bacteria possessing antibacterial activity are reported in the literature (see Table 13.1).

Even in the human body, certain bacteria have been sustained that also produce biosurfactants. For example, in premenopausal women, *Lactobacilli* is predominant in the vagina and properties like adhesion and production of bacteriocins and/or biosurfactants are important for the protection of the host organism [40].

Rani et al. [41] investigated antimicrobial activity of a biosurfactant producing *Bacillus methylotrophicus* OB9 strain and found that the strain could inhibit a diverse of pathogens and suppress the growth of *Salmonella* and *Xanthomonas* while co-cultured with *B. methylotrophicus* OB9.

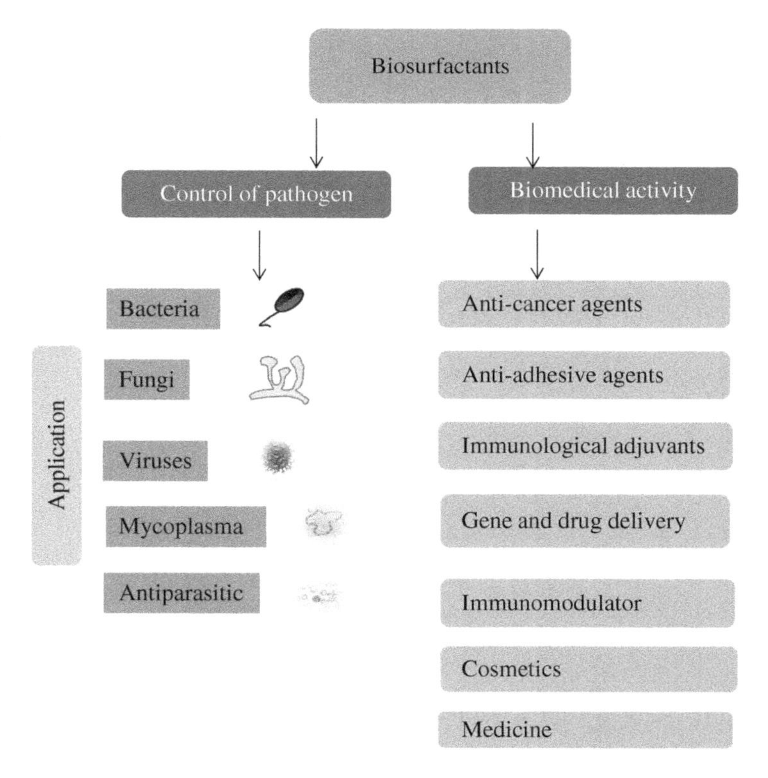

Figure 13.2 Biosurfactants help in the control of pathogenic microbes and demonstrate extensive activities that are beneficial for biomedical applications.

Table 13.1 Glycolipids biosurfactants and their economic importance in the biomedical field.

Biosurfactant				
Group	**Class**	**Microorganism**	**Economic importance**	**References**
Glycolipids	Rhamnolipids	*Pseudomonas aeruginosa, Pseudomonas* sp., *Burkholderia glumae, Burkholderia plantarii, Burkholderia thailandensis,Lactobacillus pentosus*	Antimicrobial activity against *Mycobacterium tuberculosis,* antimicrobial action against bacteria, fungi, algae, and viruses, induce cell differentiation, drug delivery system, stimulation of innate immunity in animal cells, antiadhesive activity against several bacterial and yeast strains	Isoda et al. [12], Kitamoto et al. [13], Howe et al., [14], Sifour et al. [15], Nguyen et al. [16], Yoo et al. (2015), Vecino et al. [17]
	Trehalose lipids	*Rhodococcus erythropolis, Rhodococcus ruber, Nocardia erythropolis, Mycobacterium* sp., *Arthobacter* sp.	Antiviral activity against HSV and influenza virus, permeabilization of biological and artificial membranes, immunomodulator	Franzetti et al. [18], Sánchez et al. [19]
	Sophorolipids	*Torulopsis bombicola, Torulopsis apicola, Torulopsis petrophilum Starmerella bombicola Rhodotorula babjevae YS3, Candida bombicola*	Spermicidal and virucidal agent, modulators of the immune response, anti-inflammatory agent, in cosmetic applications, stimulating skin dermal fibroblast cell, exhibiting moisturizing, antibacterial, and antioxidant properties, antifungal activity	Borzeix and Frederique [20], Shah et al. [21], Hagler et al. [22], Muthusamy et al. [23], Williams [24], Yoshihiko et al. [25], Gross et al. [26], Mayri et al. [27], Yoo et al. [28], Borsanyiova et al. [29], Lydon et al. [30], Sen et al. [31], Sen et al. [32]
	Mannosylerythritol lipid	*Candida antartica*	Antimicrobial, immunological, and neurological properties, induce cell differentiation, act as antiaging and skin smoothing agent, gene transfection	Kitamoto et al. [13], Isoda et al. [12], Wakamatsu et al. [33], Kitamoto et al. [34], Lourith and Kanlayavattanakul [35]
	Cellobiolipids	*Ustilago zeae, Ustilago maydis, Pseudozyma fusiformata, Sympodiomycopsis paphiopedili*	Fungicidal activity	Desai and Banat [2], Golubev et al. [36], Golubev et al. [37]
	Cybersan	*Cyberlindnera saturnus*	Antibacterial activity	Balan et al. [38]
	Ustiligic acid	*Ustilago maydis*	Antifungal activity	Teichmann et al. [39]

The lipopeptide biosurfactant of *Acinetobacter junii* (AjL) has shown non-selective activity against both Gram-negative and Gram-positive bacterial strains. It has been shown that AjL has an effective antibacterial activity at concentrations almost below the critical micelle concentration [42] (see Table 13.2). *Planococcus maritimus* SAMP MCC 3013 produces glycolipid biosurfactant (Table 13.1a). Biological potential of the biosurfactant was evaluated against *Mycobacterium tuberculosis* and *Planococcus falciparum*. Glycolipid biosurfactant derived from *Planococcus* may inhibit *M. tuberculosis*. Biosurfactant also demonstrated a growth inhibition of *P. falciparum* [59]. Pumilacidin, lichenysin, and polymyxin B [51, 56] are other antimicrobial lipopeptide biosurfactants synthesized by *Bacillus pumilus*, *Bacillus licheniformis*, and *Bacillus polymyxa*, respectively (Table 13.2).

Biosurfactants can also help control some of the antibiotic resistance clinical bacteria that have received widespread attention (Liu et al. [55]). The probiotic, *Lactobacillus fermentum*, and its secreted biosurfactant significantly inhibit infections on surgical implants caused by *Staphylococcus aureus*, which is a common cause of hospital and community acquired infections [60]. The surfactin biosurfactant demonstrates antibacterial activity against *Salmonella enteritidis* in milk [61]. Biosurfactant from *Acinetobacter calcoaceticus* in a form of supernatant show antibacterial effect against a number of microorganisms, namely *Bacillus subtilis*, *Candida tropicalis*, *Candida albicans*, *Candida utilis*, *Escherichia coli* and *Saccharomyces cerevisiae* [62]. The detailed application of biosurfactant to control methicillin-resistant *Staphylococcus aureus* (MRSA) is discussed in Chapter 14.

Balan et al. [38] examined the marine yeast, *Cyberlindnera saturnus* SBPN-27, for the bactericidal effect of its biosurfactant. The yeast produces a glycolipid biosurfactant, Cybersan (see Table 13.1), which showed significant inhibition of growth toward clinical bacterial pathogens without significant cytotoxicity against 3T3 fibroblast cells in mammals. The yeast *Starmerella bombicola* produces sophorolipids that have antimicrobial activity against *Enterococcus faecalis* and *P. aeruginosa*. The biosurfactant could significantly reduce CFU at concentrations as low as 5 mg/ml. The possibility of a biosurfactant being a potential constituent of topically applied creams for the cure of wound infections has also been identified [30]. Sophorolipid and rhamnolipid have been tested as antifungal agents against plant pathogenic fungi (*Phytophthora* sp. and *Pythium* sp.). Concentration of 200 mg/l of rhamnolipid or 500 mg/l of sophorolipid could inhibit 8% of mycelial growth of plant pathogens. The motility of the Zoospore of *Phytophthora* sp. was also reduced by 90% at 50 mg/l of rhamnolipid and 80% at 100 mg/l of sophorolipid [28] (see Table 13.1). *Rhodotorula babjevae*, a relatively less studied yeast, produces sophorolipid biosurfactants, which exhibited promising antifungal activity against a very large group of pathogenic fungi viz. *Colletotrichum gloeosporioides*, *Fusarium verticiliodes*, *Fusarium oxysporum*, *Corynespora cassiicola*, and *Trichophyton rubrum* [31]. Ohadi et al. [42] found that AjL had almost 100% inhibition of *C. utilis*, and the minimum inhibitory concentration (MIC) of *C. utilis* was lower than that of standard antifungal agents. The phytopathogenic fungus *Ustilago maydis* produces and secretes a large amount of the biosurfactant ustilagic acid (UA), which is a glycolipid (see Table 13.1). The plant pathogenic fungus, *Botrytis cinerea* infects tomato leaves and this infection can be prevented by inoculating wild-type *U. maydis* sporidia. Investigation with *U. maydis* mutants defective in UA biosynthesis has confirmed the importance of UA for inhibition of *B. cinerea* [39]. Biosurfactant from *Streptomyces* sp. SNJASM6 may inhibit many pathogenic fungi [63]. For application of biosurfactant to control mycotoxicogenous fungi, see Chapter 21.

Biosurfactants have a strong antiviral activity. Inhibition of HIV growth in leukocytes by biosurfactants has been cited in the literature [21]. Antiviral action appears to be due to the physicochemical interaction of the biosurfactant with the lipid membrane of the virus. The incidence of

Table 13.2 Lipopeptides and lipoproteins classes of biosurfactants and their economic importance in the biomedical field.

Class	Biosurfactant	Microorganism	Economic importance	References
Lipopeptides and lipoproteins	Surfactin/iturin/fengycin	*Bacillus subtilis, Bacillus licheniformis*	Immunosuppressive agent, cosmetic ingredient, antimicrobial and antifungal activities, inhibition of fibrin clot formation, haemolysis and formation of ion channels in lipid membranes, antitumour activity against Ehrlich's ascite carcinoma cells, antiviral activity against HIV-1, antibiotic activity, antimicrobial activity and antifungal activity against profound mycosis effect on the morphology and membrane structure of yeast cells, increase in the electrical conductance of biomolecular lipid membranes, non-toxic and non-pyrogenic immunological adjuvant, anti-inflammatory agents, anti-mycoplasma activity	Awashti et al. [43], Ahimou et al. [44], Carrillo et al. [45], Hwang et al. [46], Park and Kim [47], Lourith and Kanlayavattanakul [35], Tang et al. [48], Zhao et al. [49]
	Viscosin	*Pseudomonas fluorescens*	Antimicrobial activity	Saini et al. [50]
	Lichenysin	*Bacillus licheniformis*	Antibacterial activity, membrane disrupting effect	Grangemard et al. [51]
	Serrawettin	*Serratia marcescens*	Chemorepellent	Pradel et al. [52]
	Subtilism	*Bacillus subtilis*	Antimicrobial activity	Chen et al. [53]
	Gramicidin	*Brevibacterium brevis*	Antibiotic activity, disease control	Elad and Stewart [54]
	Polymixin	*Bacillus polymyxa*	Bactericidal and fungicidal activity, antibacterial activity	http://www.cyberlipid.org/simple/simp0005.htm, Liu et al. [55], Grangemard et al. [51], Landman et al. [56]
	Antibiotic TA	*Myxococcus xanthus*	Bactericidal activity, chemotherapeutic applications	Karanth et al. [57], Xiao et al. [58]
		Acinetobacter junii	Antimicrobial, anti-biofilm activity	Ohadi et al. [42]

HIV in women has increased on a daily basis, increasing the demand for an effective and safe female vaginal microbicide. Sophorolipid surfactants from *Candida bombicola* and its analogs, such as sophorolipid diacetate ethyl ester, can act as the most powerful spermicidal and virucidal agent. This biosurfactant carries virucidal activity against human semen [23]. The virucidal activity of the diacetate-ethyl ester form of the sophorolipid biosurfactant was observed at a concentration of 0.3 mg/ml in all sophorolipid derivatives tested. The biosurfactant showed a reduction in human immunodeficiency virus type 1 titer after only two minutes of incubation with the biosurfactant [21]. Furthermore, biosurfactant trehalose possesses activity against herpes simplex and influenza viruses [18] (see Table 13.1). The lipopeptides produced by *B. subtilis* fmbj effectively inhibit the replication as well as the infectivity of the new castle disease and bursal disease viruses [64]. Rhamnolipid biosurfactant PS-17 shows antiviral activity against HSV types 1 and 2. These compounds may therefore be considered as promising substances for the development of anti-herpetic agents (Remichkova et al. 2008) [65]. *Tolypocladium inflatum* produces a bio-peptide cyclosporin A, which inhibits the proliferation of the influenza virus by interfering with its life cycle [66, 67]. It is very clear from the above discussion that the biosurfactant may be a powerful antiviral agent for many infectious viral diseases. It may inhibit viral spread by affecting either their viral membrane or the process of viral protein synthesis, or it may interfere with the assembly of protein molecules. The molecule may play a significant role in coronavirus control during the Covid-19 pandemic (Figure 13.3). A biosurfactant molecule can directly disrupt the lipid membrane that destroys the virus. They may be added as handwashing components, in antiviral face masks, and in cleaning products [68]. However, huge research in this area is mandatory for that purpose. Borsanyiova et al. [29] extensively reviewed the biological activity and possible use of sophorolipids as antiviral agents and found that although sophorolipids may be used as disinfectants, immunomodulators, antivirals, and antimicrobial agents, there are few reports of the use of sophorolipids as antiviral drugs. See Chapters 11, 16, and 22 for further details on antiviral, antimicrobial, and biofilm properties and their sustainable use in pharmaceutical products.

13.2.2 Application Against Mycoplasma

There are reports describing biosurfactants for their antimycoplasmal activity. The most important biosurfactant that shows antimycoplasmal activity is surfactin (see Table 13.2). Contamination of cell culture with mycoplasma is a common problem leading to a serious challenge for biomedical research. Contamination of irreplaceable cell lines with mycoplasma causes cell destruction. Studies have shown that mammalian cells contaminated with mycoplasma can be treated with surfactin specifically for mycoplasma inactivation without destroying cell metabolism significantly in culture [69]. Surfactin can eliminate mycoplasma from a highly infected, irreplaceable cell line of hybridoma. There is a report describing the use of surfactin for mycoplasma decontamination [70]. However, the toxicity of surfactin on a number of cell lines has been observed. Pre-testing of the biosurfactant for its toxicity is therefore a necessary precautionary measure. Surfactin has been effective in eliminating mycoplasma without depending on the target cell [71]. Surfactin exhibits a synergistic effect to enhance mycoplasma killing activity of about two times of enrofloxacin when used in combination with the molecule. Hwang et al. [46] investigated surfactins isolated from *B. subtilis* complex BC1212 for an antimycoplasmal effect, against *Mycoplasma hyopneumoniae*. The researchers studied minimal inhibiting concentrations of different antibacterials and surfactins (surfactins A, B, C, and D) prior to study the combination effect of surfactins, against *M. hyopneumoniae*. It was found that surfactin C was the most potent of all surfactants and it may be useful as a preventive or therapeutic adjuvant in the treatment of mycoplasmal infection. Rhamnolipids of different strains of *P. aeruginosa* exhibit mycoplasmacidal activity [72].

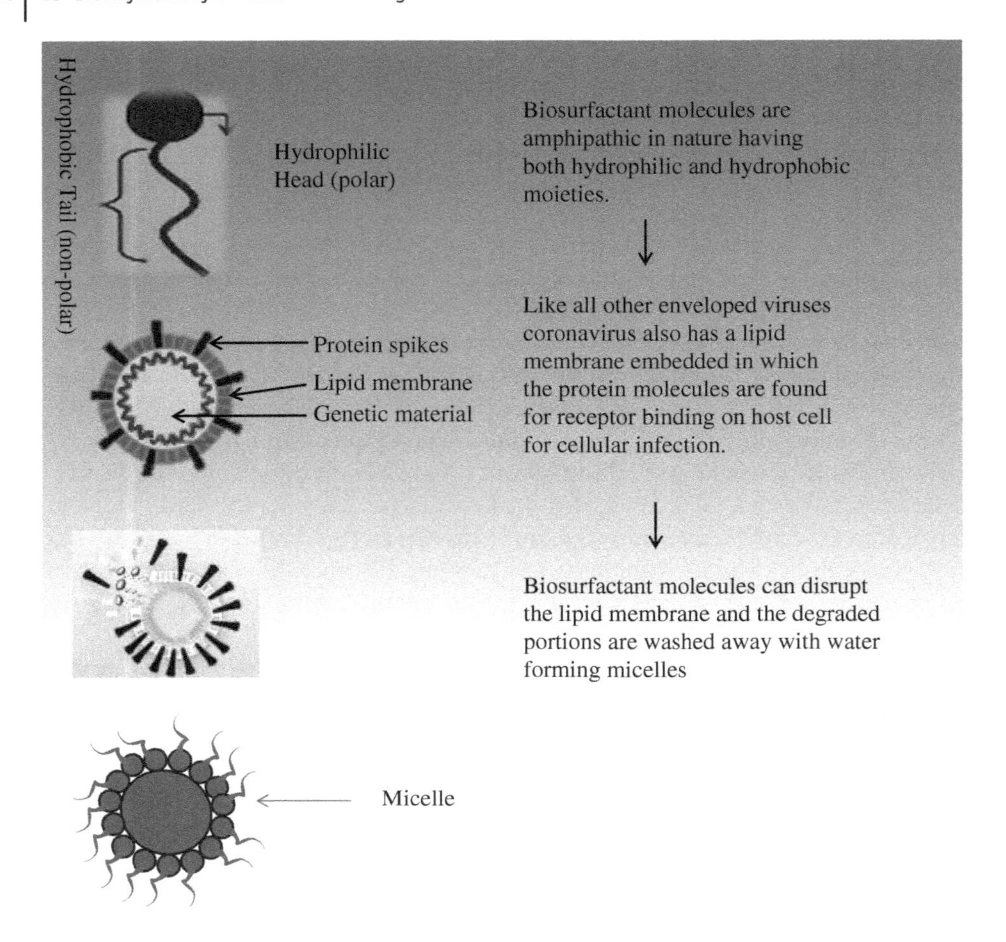

Figure 13.3 Working mechanism of biosurfactant molecule on viruses like coronavirus (See insert for color representation of the figure).

13.2.3 Biosurfactants as Anti-Cancer Agents

Extensive research has been conducted around the world to explore the role of biosurfactant as an anticancer agent and a number of publications have reported on the anticancer activity of biosurfactants. Some microbes secrete extracellular glycolipids, which can induce cell differentiation rather than cell proliferation in the human promyelocytic leukemia cell line. The exposure of PC12 pheochromocytoma cells to mannosylerythritol (MEL) increases the activity of acetylcholine esterase and interrupts the cell cycle during the G1 phase and results inhibition of cell multiplication. This, in turn, increases the partial differentiation of the cells indicating the role of MEL in anticancer activity [23] (see Table 13.1). In addition to the PC12 cells, MEL-A, MEL-B, and sophorose lipids cause significant neurite growth. MEL-A can increase the activity of acetylcholine esterase and acts as a growth factor for nerves [12]. It is also reported that MEL increases galactoceramide and neurite growth from PC12 cells [73]. The biosurfactants MEL-A, MEL-B, rhamnolipid, polyol lipid, succinoyl trehalose lipid (STL)-1, lipid sophorose, and succinoyl trehalose lipid-3 induce cell differentiation rather than cell proliferation in human promyelocytic leukemia cell line HL60 [12]. Both STL and MEL significantly increase the differentiation process of monocytes and granulocytes. This differentiating property of STL and MEL is due to a specific action on the cell plasma membrane. In an investigation, exposure of B16 cells to MEL results in chromatin condensation, DNA fragmentation, and arrest of sub-G1 [74]. The MEL exercises growth inhibition and differentiation-inducing activity against human leukemia cell lines. For this purpose, it directly affects the intracellular signal transduction pathway through the phosphate cascade system [75, 76]. Kameda et al. [77] reported that surfactin has antitumor activity against Ehrlich's ascite carcinoma cells and induce apoptosis in human breast cancer cells (MCF-7) through the mitochondrial/caspase pathway [78]. Chen et al. [79] reported that sophorolipid produced by *Wickerhamiella domercqiae* also has anticancer activity. STL-3 analogs inhibit growth and induce differentiation of human promyelocytic leukemia cells (HL60) and the activity is influenced by the biosurfactant trehalose lipid [80]. Trehalose dicorinomycolate from *Corynebacterium glutamicum* exhibits *in vivo* macrophage priming effects [81]. Meticulous research in this field is needed to achieve complete success in the use of molecules for cancer treatment. Recently, Zhao et al. [49] showed that lipopeptides of *B. subtilis*, mainly iturin, had a promising potential to inhibit chronic myelogenous leukemia *in vitro* (see Table 13.2). Apoptosis, paraptosis, and inhibition of autophagy have also been shown. In the investigation, it was found that the glycolipid biosurfactant extracted from *P. maritimus* SAMP MCC 3013 showed cytotoxic behavior against HeLa (IC50 41.41 ± 4.21 μg/ml), MCF-7 (IC50 42.79 ± 6.07 μg/ml) and HCT (IC50 31.233 ± 5.08 μg/ml) cell lines [59].

13.2.4 Biosurfactants as Antiadhesive Agents

Biosurfactants have been found to inhibit the adherence of pathogenic organisms to solid surfaces or to sites of infection [82]. This, in turn, makes them useful candidates for the treatment of many diseases and their use as therapeutic and probiotic agents. Biofilms are microcolonies of microbes that adhere to biotic and abiotic surfaces. These biofilms are often responsible for chronic infection and contamination of medical devices. Since adhesion is the first stage of biofilm formation, the best time for the action of antiadhesive and antibiofilm compounds is before adhesion of the cells. Surface treatment with proper biosurfactant with antiadhesive and antibiofilm properties is one of the most important techniques [83]. The study of this technique has shown very good prospects. Pre-coating vinyl urethral catheter with surfactin has been shown to result in a decrease in the biofilm formation of *Salmonella typhimurium, Salmonella enterica, E. coli*, and *Proteus mirabilis* [84–86]. Pre-treatment of silicone rubber with a surfactant produced by *Streptococcus thermophilus* can inhibit 85% adhesion of *C. albicans* and surfactants from *L. fermentum* and *L. acidophilus* exhibits a 77% reduction of adhesion of *E. faecalis* [23]. If somehow the adherence of microorganisms to food-contact surfaces can be controlled along with surface detachment of microorganisms, the possibility of providing consumers with safe and quality products increases. Sophorolipids have been reported to be capable of disrupting biofilms at concentrations above 5% (v/v) of single and mixed cultures of *B. subtilis* and *S. aureus* [27]. Bioconditioning of surfaces through the use of microbial surfactants has been recommended as a new strategy to reduce microbial adhesion. Surfactin-conditioned polystyrene surfaces are capable of inhibiting bacterial adhesion at different temperatures, resulting in a 63–66% reduction in the adhesion of bacterial strains *S. aureus, Listeria monocytogenes*, and *Micrococcus luteus* at 4 °C [87]. Rhamnolipid promotes a slight decrease in the attachment of *S. aureus*. Biosurfactant on surface conditioning becomes effective in removing microbial adhesion at a concentration as low as 0.1 g/l. The percentage of inhibition of bacterial adhesion ranges from 15 to 89% using 0.1–10 g/l of purified biosurfactant [82]. The yeast-derived mannoprotein accelerates the detachment of mature *Staphylococcal* biofilms [88]. Pseudofactin II is a newly identified biosurfactant excreted by *Pseudomonas fluorescens*. Adsorption of the biosurfactant to the surface, e.g. glass, polystyrene, and silicone, interferes with microbial adhesion and disorption processes [89]. Released biosurfactant by *S. thermophilus* inhibits the adherence and growth of several bacterial and yeast strains isolated from explanted voice prosthesis [90]. The biosurfactant produced by the strain of *Lactobacillus* causes a significant dose-related inhibition of the initial deposition rate of the pathogenic enteric bacteria *E. faecalis* [91]. Biosurfactant made by *B. licheniformis* inhibits the biofilm formation of *E. coli*, one of the most common pathogens involved in uropathogenic infections and catheter contamination. Biosurfactant studies in association with various antibiotics (ampicillin, cefazolin, ceftriaxone, ciprofloxacin, piperacillin, tobramycin, and trimethoprim/sulfamethoxazole at 19 : 1) against 24 hours grown *E. coli* biofilm shows that the use of antibiotics in combination with a biosurfactant has a significant increase in the removal of adherent cells from a pre-formed biofilm. The produced biosurfactant alone was unable to remove the adhesive cells from the biofilm. Antibiotics also have a minimal effect on biofilm removal while being tested on their own. This shows the synergistic effect of the biosurfactant to increase the efficacy of antibiotics in the killing of biofilms and, in some combinations, leads to the total eradication of *E. coli* CFT073 biofilm [92]. Rodrigues et al. [86] demonstrated the ability of rhamnolipid biosurfactants to inhibit the adherence of micro-organisms including *Streptococcus salivarius*, *C. tropicalis, Staphylococcus epidermidis*, and *S. aureus* to silicone rubber. Zeraik and Nitschke [87] and de Araujo et al. [93] argued that rhamnolipid biosurfactants could inhibit the adhesion of *S. aureus* and *L. monocytogenes*. Balan et al. [38] reported promising biofilm inhibition and dislodging activities of the glycolipid biosurfactant of the marine *Staphylococcus saprophyticus* against a wide range of biofilms forming bacteria at both single and multi-species levels, such as *P. aeruginosa*, *Serratia liquefaciens*, *Acinetobacter beijerincki*, *Marinobacter lipolyticus*, *Marinobacter liquefaciens*, *Acinetobacter beijerinckii*, *and M. lipolyticus*. Vecino et al. [17] investigated the antimicrobial and anti-adhesive properties of the biosurfactant synthesized by *Lactobacillus pentosus* (PEB) against several microorganisms found in human skin flora. The cell-bound biosurfactant was characterized as a glycolipide molecule. PEB demonstrated antimicrobial activity against *Streptococcus agalactiae*, *P. aeruginosa*, *S. aureus*, *Streptococcus pyogens*, *E. coli*, *and C. albicans*, which was comparable to the results of *Lactobacillus paracasei* (PAB). Both PEB and PAB had antiadhesive properties against all the tested microorganisms except for *C. albicans and E. coli*. Surfactin and rhamnolipid biosurfactants have been tested for their biofilm activity. Investigation was conducted to determine their effect on the reduction of the adhesion of *Streptococcus mutans* to the polystyrene surfaces and the disruption of its pre-formed biofilms. The results showed that rhamnolipid-conditioned surfaces had higher antiadhesive and antibiofilm activity compared to surfactin-conditioned surfaces [94]. Lipopeptide biosurfactants produced by *A. junii* may interfere with the biofilm of *P. aeruginosa*, *P. mirabilis*, and *S. aureus* at 1250 μg/ml and 2500 μg/ml concentrations [42] (Table 13.2). Biosurfactants have been shown to be one of the best eco-friendly anticorrosion substances to inhibit the biocorrosion process and protect microbial-induced corrosion-proof materials, mostly known as microbiologically induced corrosion [95].

13.2.5 Immunological Adjuvants

Research on biosurfactants to be used as immunological adjuvants to increase the immune response to microbe-causing diseases has shown promising results. Adjuvants induce immune reactions by forming a depot in the tissues from which the antigen is slowly released. This process stimulates immune mediating cells (e.g. B-lymphocytes and/or T-lymphocytes). Bacterial lipopeptides may act as potent non-toxic, non-pyrogenic immunological adjuvants when combined with conventional antigens. Significant enhancement of humoral immune response is achieved in rabbits and chickens with low-molecular – mass antigens iturin AL, herbicolin A, and microcystin when combined with poly-L-lysine [96] (see Table 13.2). A suitable and effective adjuvant is needed to produce conventional or monoclonal antibodies to drugs, antibodies, toxins, and other low molecular mass substances. Lipopeptide-Th-cell epitope conjugates are effective adjuvants for either human mononuclear cells or mouse B cells with microcystin-poly-L-lysine *in vitro* immunization and result in significantly higher yields of antibody secreting hybridoma cells [97]. An immunization formulation can be prepared by an antigen and a biosourfactant, emulsan, or emulsan analog and can be injected to a host. The biosurfactant emulsan or emulsan analog is an adjuvant in the immunization formulation secreted by the *A. calcoaceticus* (see Table 13.3).

The biosurfactant stimulates the release of cytokine to the host without inducing toxicity to the host. They may also generate immune responses in the host by producing antigen antibodies or by providing antigen prophylaxis while administered with the antigen to the host. They thus play the role of an effective adjuvant [100]. The acidic form of the sophorolipid compound and the deacetylated sophorolipidic acid ester may be therapeutically active substances in the therapeutic treatment of the human or animal body. They can act as macrophages activators, fibrinolytic agents, and healers [107].

13.2.6 Use of Biosurfactant for Gene and Drug Delivery

A biosurfactant can be used as a delivery agent in the techniques of gene and drug delivery. Establishing an effective and safe method of introducing exogenous nucleotides in mammalian cells is essential for medical and clinical applications, e.g. gene therapy [96]. There are various methods of transfection of genes and the lipofection methods using cationic liposomes are considered as a potential way to supply foreign genes without side effects to the target cells [108]. While several kinds of cationic liposomes have been developed for lipofection, viral vectors are more efficient in gene transfection. Further studies are therefore needed to develop a non-viral vector that is comparable to viral vector performance [109]. In comparison to commercially available cationic liposomes, biosurfactants-based liposomes show an increased gene transfection efficiency [34]. New liposome-based transfection techniques and methodologies with biosurfactants have been reported [110]. MEL-A is the most common biosurfactant used for gene transfection, while MEL's cationic liposome bearing increases gene transfection effectiveness in mammalian cells [111]. In the case of liposome–biosurfactant mediated gene transfection, injectable complexes known as lipoplexes are prepared with a high transfection efficiency of genes. Liposomes made up of 3 ([*N*-(*N*′, *N*′–dimethylaminoethane)-carbamoyl] cholesterol (DC-Chol), L-dioleoylphosphatidylethanolamine were used with a biosurfactant. Biosurfactants used in this study for Sit-G-liposomes were β-sitosterol β-D-glucoside (Sit-G) and mannosylerythrytol lipid A (MEL), for MEL-liposomes with approximately 300 nm lipoplexes. The transfection efficacy of the luciferase marker gene was increased in sit-G- and MEL-liposomes. Sit-G-liposome is a potential vector of gene therapy for HSV-tk [111]. The cationic MEL-liposome (MEL-L) reveals its transfection efficacy in Hela cells. MEL-L induces a significantly higher level of gene expression than Tfx20 and a liposome with no MEL-A available in the market. The cell-associated DNA also increased rapidly with MEL-A addition to the liposome [112]. Glycolipid biosurfactants may be used in gels, niosomes, hexosomes, and cubosomes, in addition to their use as surfactants or absorption enhancements in basic formulations. These systems permit the solubilization and trapping of medicines. The benefit of glycolipids is that their sugar moisture can be specifically recognized through carbs exposed to cell surface proteins [113]. Fengycin and surfactin have a twofold

Table 13.3 Major classes of biosurfactants and their economic importance in the biomedical field.

Fatty acids/neutral lipids/phospholipids	Spiculisporic acid	*Penicillium spiculisporum*	Superfine microcapsules (vesicles or liposomes)	Ishigami et al. [98]
	Phosphati-dylethanolamine	*Acinetobacter* sp., *Rhodococcus, Erythropolis, Mycococcus* sp.	Increase the tolerance of bacteria to heavy metals, membrane interaction with tumor cells	Appanna et al. [99]
Polymeric surfactants	Emulsan	*Acinetobacter calcoaceticus*	Adjuvant	Kaplan et al. [100]
	Alasan	*Acinetobacter radioresistens*	–	Toren et al. [101]
	Biodispersan	*Acinetobacter calcoaceticus* A2	Effectively disperses calcium carbonate and titanium dioxide used in cosmetic and pharmaceutical industries	Rosenberg and Ron [102], Kaplan [103], Ron and Rosenberg [104]
	Polysaccharide protein complex	*Acinebacter calcoaceticus*	–	Karanth et al. [57]
	Liposan	*Candida lipolytica*	High emulsification activity	Nerurkar et al. [105]
	Mannoprotein	*Saccharomyces cerevisiae*	Inhibit biofilm development of *Staphylococcus aureus* and *S. epidermidis* while used with 3-(4,5-dimethylthiazol-2-yl)-2,5-diphenyltetrazolium bromide	Walencka et al. [88]
	Protein PA	*Pseudomonas aeruginosa*	–	–
Particulate biosurfactants	Vesicles	*Acinetobacter calcoaceticus, Pseudomonas marginalis*	Antimicrobial activity	Pirog et al. [62]
	Whole microbial cells	*Cyanobacteria*	Antioxidant activity	Aydaş et al. [106]

increased aciclovir concentration in the epidermis to improve the transdermal penetration and skin accumulation of the aciclovir [114] (see Table 13.2). Microemulsions composed of lecithin/sophorolipid/rhamnolipid biosurfactant mixtures for the supply of medicinals show that they are practically temperature-insensitive and are not significantly affected by changes in electrolyte concentration. Rhamnolipid liposomes have been patented as useful tools for delivery of medicines, proteins, nucleic acids, dyes, etc. They were treated as biomimetic models and sensors for the detection of pH changes for biological membranes. These newer liposomes are safe and biodegradable and have a stable and long-life affinity to biological organisms. Thermodynamically stable and simple preparations with high solubility make microemulsions produced with biosurfactants very promising and very efficient for delivery of drugs ([115], cited in Faivre and Rosilio [113]). Calabrese et al. [116] have experimented on preparation and characterization of new surfactant-modified clays. The applicability of modified clays as drug delivery systems for the oral administration of cinnamic acid (CA) was also investigated. Their investigation revealed that release of the drug from both the clay as well as the organoclay was prolonged as compared to the free drug delivery kinetics. In addition, the surfactant intercalation into the nano-carrier confirmed the full release of cinnamic acid following oral drug administration. See Chapter 17 for details.

13.2.7 Immuno Modulatory Action of Biosurfactants

Biosurfactants can make a significant contribution as an immune modulator. They induce the production of different cytokines leading to the activation of the immune response. Sophorolipids are reported as promising immune modulators. They can reduce mortality due to sepsis at 36 hours *in vivo* in a rat septic peritonitis model by cytokine production and by decreasing IgE production *in vitro* in U266 cells. Production of IgE in U266 cells was reduced by regulating important genes involved in IgE production in a synergistic manner by the biosurfactant. These data continue to support the use of sophorolipids as an anti-inflammatory agent and novel potential therapy for IgE-modified diseases [22]. Glycolipids extracted from *Rhodococcus ruber* can induce the production of interleukin IL-12, IL-18, and reactive oxygen species, by cells of innate immunity [117]. Glycolipids from *R. ruber* were studied for their immunomodulatory action and it was reported that the produced biosurfactant stimulated production of tumor necrosis factor-III, IL-1β, and IL-6 when applied to the cell culture of a human peripheral blood monocyte [118]. Rhamnolipids have long been known as exotoxins produced by the human pathogen *P. aeruginosa* [14] and several recent papers have highlighted their role in stimulating innate immunity in animal cells. The bacterium *Burkholderia plantarii* produces the heat stable Rha-Rha-C14-C14 biosurfactant, which have been specifically investigated along with some synthetic derivatives [14]. This rhamnolipid has strong stimulating power for mononuclear cells for the production of TNF5-007, pleiotropic inflammatory cytokine [119]. Lipopeptides of *Streptosporangium amethystogenes* can induce activity of the immune system [120]. Significant stimulation of bone marrow cell proliferation from BALB/c female mice has been achieved by *S. amethystogens* lipopeptides (Takizawa et al. [121]).

The biosurfactant surfactin is a very efficient immunosuppressive agent. The immunosuppressive agents are very important for the transplantation process and to control diseases such as allergies, arthritis and diabetes since it can inhibit the immunostimulatory function of macrophages of the host body [47]. Kuyukina et al. [122] investigated the immunological activities of *Rhodococcus*-produced trehalolipids and compared their biomedical efficiency with the immunological functions of the trehalose extracted from *M. tuberculosis*. High immunomodulatory and antitumor activity and lower cytotoxicity of *Rhodococcus* trehalolipids were reported compared to mycobacterial derivatives.

13.2.8 Biosurfactants for Cosmetics and Dermatological Repair

Biosurfactants have recently been extensively studied for use in clinical cosmetology or cosmetic dermatology, and positive instances are reported from various parts of the world. These compounds are very much effective and have excellent skin compatibility not shown by chemical surfactants [24]. This biosurfactant possesses antiradical and hygroscopic properties, can stimulate dermal fibroblast metabolism, particularly collagen neosynthesis, in order to support healthy skin physiology and therefore remains an enormous prospect for sophorolipid-based products in the future. The list includes different types of cosmetics, body lotions, beauty washes, hair products, etc. [123]. See Chapters 18 to 20 for details.

13.2.9 Other Applications in Pharmacology

In addition to the applications discussed above, biosurfactants have a broad range of other biomedical applications. Some of the applications that do not fall within the categories mentioned above are discussed here. The biosurfactant MEL-A from *Candida antartica* has binding affinity to human immunoglobulin G. Binding of a natural human antibody to a yeast glycolipid makes it possible to use MEL-A as an alternative ligand for immunoglobulins [124]. Investigation of MELs has shown that they play a central role in allergic reactions. In allergic conditions, inflammatory mediators are secreted from mast cells that cause serious illness and MELs inhibit the secretion of these inflammatory mediators from mast cells [125]. Surfactin may be used as a good anti-inflammatory agent. Nuclear factor-kappa B activation may be inhibited to reduce the production of nitric oxide induced by lipopolysaccharide in RAW264.7 cells and primary macrophages [126]. As a natural probiotic, *B. subtilis* PB6 has an inhibitory action against phospholipase A2 inflammatory bowel disease [127]. Tang et al. [48] demonstrated that surfactin isomers extracted from *Bacillus* sp. (No.061341) may be used as anti-inflammatory agents to treat allergic conditions. Surfactin and its synthetic analogs can alter the organization of supported bilayers that is important for biomedical applications [128–130]. Fengycin, a lipopeptide biosurfactant, is capable of causing membrane disruption [131]. Rhamnolipids can inhibit the growth of some harmful algal species [132]. The biosurfactant trehalose lipid may get incorporated into the lipid bilayer causing disturbances in the membrane structure that can affect the function of the membranes. It is also reported that dirhamnolipid from *P. aeruginosa* can make the biological and artificial membranes permeable [19]. Biosurfactants obtained from *Bacillus* sp. elicit dose-dependent cardiac depressant activity (Aminabee et al. [133]). Antimicrobial activity and toxicity of oral washes formulated with a biosurfactant, microbial chitosan, and peppermint (*Mentha piperita*) essential oil (POE) were investigated. Three biosurfactants extracted from *P. aeruginosa* (PB), *Bacillus cereus* (BB), and *C. bombicola* (CB) have also been investigated. All the substances investigated showed an MIC for carcinogenic microorganisms and confirmed that mouthwashes having biosurfactants and other natural products are safe and effective. They are a natural alternative to commercially available mouthwashes to control microorganisms in the mouth cavity [134]. Ohadi et al. [42] experimented on the antioxidant property of surfactin and rhamnolipid derived from *Bacillus amyloliquefaciens* NS6 and *P. aeruginosa* MN1, respectively. Their antioxidant activity was tested by ferric antioxidant reduction and by 1, 1-diphenyl-2-picrylhydrazyl methods. The results showed that both biosurfactants had antioxidant properties, but surfactin showed a higher antioxidant activity than rhamnolipids. Da Silva et al. [135] studied the antioxidant activities of the biosurfactant produced by *C. bombicola* URM 3718 on low-cost substances and found that the biosurfactant showed very good antioxidant capacity and an increase in the concentration of the biosurfactant contributed to an increase in the percentage of the total antioxidant capacity (TAC) of the biomolecule. The TAC of the biomolecule was evaluated using the method of phosphomolybdenum.

13.3 Conclusion

Interestingly, biosurfactants exhibit extensive antimicrobial, anticancer, antiparasitary, and immunomodulatory activity. Biosurfactants currently hold a highly competent position in biomedicine and pharmaceuticals and are the strongest contenders of the existing therapeutic molecules in the near future, but the economic viability of their production is a very challenging issue in relation to large-scale use of the product. Bioindustrial manufacturing processes depend generally on the use of microbial strains of hyper producing nature, even though very cheap raw materials are used to reduce production costs and efficient recovery processes. Therefore, selection/creation of hypersecreting organisms is significant. Microbial surfactants are high-value, highly potential lipid derivatives that hardly access the market due to their high production costs. In addition to their interesting antimicrobial activity, many are excellent ecological detergents.

Acknowledgements

The authors would like to express their sincere thanks to Bimal Barah, Principal of J. B. College (Autonomous), Jorhat, Assam, for his continued encouragement and inspiration to contribute to the fraternity of science and research. They would also like to acknowledge that the Institute of Advanced Study in Science and Technology (IASST), Guwahati, Assam, provided them with a platform to carry out their research work on biosurfactants.

References

1 Saikia, R.R., Deka, S., Deka, M., and Sarma, H. (2012). Optimization of environmental factors for improved production of rhamnolipid biosurfactant by *Pseudomonas aeruginosa* RS29 on glycerol. *J. Basic Microbiol.* 52: 446–457. John Wiley.

2 Desai, J.D. and Banat, I.M. (1997). Microbial production of surfactants and their commercial potential. *Microbiol. Mol. Biol. Rev.* 61: 47–64.

3 Sarma H, Bustamante KLT, Prasad MNV (2018) Biosurfactants for oil recovery from refinery sludge: Magnetic nanoparticles assisted purification. In: M.N.V. Prasad (ed,), *Industrial and Municipal Sludge*, Elsevier, Cambridge, MA, USA, Butterworth-Heinemann, Osford, UK, pp. 107–132. ISBN: 9780128159071, Editor Majeti Narasimha Vara Prasad, Paulo Jorge de Campos, Favas Meththika, and Vithanage S. Venkata Mohan.

4 Ron, E.Z. and Rosenberg, E. (2001). Natural roles of biosurfactants. *Environ. Microbiol.* 3: 229–236.

5 Cohen, R. and Exerowa, D. (2007). Surface forces and properties of foam films from rhamnolipid biosurfactants. *Adv. Colloid Interf.* 134: 24–34.

6 Kalyani, R., Bishwambhar, M., and Suneetha, V. (2011). Recent potential usage of surfactant from microbial origin in pharmaceutical and biomedical arena: A perspective. *IRJP* 2: 11–15.

7 Fracchia, L., Cavallo, M., Martinotti, M.G., and Banat, I.M. (2012). Biosurfactants and bioemulsifiers biomedical and related applications – Present status and future potentials. In: *Biomedical Science, Engineering and Technology* (ed. D.N. Ghista). UK: InTech https://doi.org/10.5772/23821.

8 Ryll, R., Kumazawa, Y., and Yano, I. (2001). Immunological properties of trehalose dimycolate (cord factor) and other mycolic acid-containing glycolipids – A review. *Microbiol. Immunol.* 45 (12): 801–811.

9 Rodrigues, L., Banat, I.M., Teixeira, J., and Oliveira, R. (2006b). Biosurfactants: Potential applications in medicine. *J. Antimicrob. Chemother.* 57: 609–618.

10 Banat, I.M., Franzetti, A., Gandolfi, I. et al. (2010). Microbial biosurfactants production, applications and future potential. *Appl. Microbiol. Biotechnol.* 87: 427–444.

11 Rahman, M.S. and Ano, T. (2009). Production characteristics of lipopeptide antibiotics in biofilm fermentation of *Bacillus subtilis*. *J. Environ. Sci.* 21: s36–s39.

12 Isoda, H., Shinmoto, H., Matsumura, M., and Nakahara, T. (1999). The neurite-initiating effect of microbial extracellular glycolipids in PC12 cells. *Cytotechnology* 31: 163–170.

13 Kitamoto, D., Yanagishita, H., Shinbo, T. et al. (1993). Surface active properties and antimicrobial activities of mannosylerythritol lipids as biosurfactant produced by *Candida antractica*. *J. Biotechnol* 29: 91–96.

14 Howe, J., Bauer, J., Andrä, J. et al. (2006). Biophysical characterization of synthetic rhamnolipids. *FEBS J.* 273: 5101–5112.

15 Sifour, M., Al-Jilawi, M.H., and Aziz, G.M. (2007). Emulsification properties of biosurfactant produced from *Pseudomonas aeruginosa* RB 28. *Pak. J. Biol. Sci.Pak J Biol Sci* 10: 1331–1335.

16 Nguyen, T.T.L., Edelen, A., Neighbors, B., and Sabatini, D.A. (2010). Biocompatible lecithin-based microemulsions with rhamnolipid and sophorolipid biosurfactants: Formulation and potential applications. *J. Colloid Interface Sci.J Colloid Interface Sci* 348: 498–504.

17 Vecino, X., Rodrıguez-Lopez, L., Ferreira, D. et al. (2018). Bioactivity of glycolipopeptide cell-bound biosurfactants against skin pathogens. *Int. J. Biol. Macromol.* 109: 971–979.

18 Franzetti, A., Gandolfi, I., Bestetti, G. et al. (2010). Production and applications of trehalose lipid biosurfactants. *Eur. J. Lipid Sci. Technol.* 112: 617–627.

19 Sánchez, M., Aranda, F.J., Teruel, J.A. et al. (2010). Permeabilization of biological and artificial membranes by a bacterial dirhamnolipid produced by *Pseudomonas aeruginosa*. *J. Colloid Interface Sci.* 341: 240–247.

20 Borzeix C, Frederique K. (2003). Use of sophorolipids comprising diacetyl lactones as agent for stimulating skin fibroblast metabolism. US Patent, 6596265.

21 Shah, V., Doncel, G.F., Seyoum, T. et al. (2005). Sophorolipids, microbial glycolipids with anti-human immunodeficiency virus and spermimmobilizing activities. *Antimicrob. Agents Chemother.* 49: 4093–4100.

22 Hagler, M., Smith-Norowitz, T.A., Chice, S. et al. (2006). Sophorolipids decrease IgE production in U266 cells by down regulation of BSAP (Pax5), TLR-2, STAT3 and IL-6. *J. Allergy Clin. Immunol.* 119: 245–249.

23 Muthusamy, K., Gopalakrishnan, S., Ravi, T.K., and Sivachidambaram, P. (2008). Biosurfactants: properties, commercial production and application. *Curr. Sci.* 94: 736–747.

24 Williams, K. (2009). Biosurfactants for cosmetic application: Overcoming production challenges. *MMG 445 Basic Biotechnol.* 5: 78–83.

25 Yoshihiko, H., Mizuyuki, R., Yuka, O. et al. (2009). Novel characteristics of sophorolipids, yeast glycolipid biosurfactants, as biodegradable low-foaming surfactants. *J. Biosci. Bioeng.* 108: 142–146.

26 Gross RA, Shah V, Doncel G. (2014). Virucidal properties of various forms of sophorolipids. Patent US 8648055:B2.

27 Mayri, A., De Rienzoa, D., Banat, I.M. et al. (2015). Sophorolipid biosurfactants: Possible uses as antibacterial and antibiofilm agent. *New Biotechnol.* 32: 720–726.

28 Yoo, D.S., Lee, B.S., and Kim, E.K. (2005). Characteristics of microbial biosurfactant as an antifungal agent against plant pathogenic fungus. *J. Microbiol. Biotechnol.* 15: 1164–1169.

29 Borsanyiova, M., Patil, A., Ruchira Mukherji, R. et al. (2016). Biological activity of sophorolipids and their possible use as antiviral agents. *Folia Microbiol.* 61: 85–89.

30 Lydon, H.L., Baccile, N., Callaghan, B. et al. (2017). Adjuvant antibiotic activity of acidic sophorolipids with potential for facilitating wound healing. *Antimicrob. Agents Chemother.* 61.

31 Sen, S., Borah, S.N., Bora, A., and Deka, S. (2017). Production, characterization, and antifungal activity of a biosurfactant produced by *Rhodotorula babjevae* YS3. *Microb. Cell Factories* 16: 95.

32 Sen, S., Borah, S.N., Kandimalla, R. et al. (2020). Sophorolipid biosurfactant can control cutaneous dermatophytosis caused by *Trichophyton mentagrophytes*. *Front. Microbiol.* 11: –329.

33 Wakamatsu, Y., Zhao, X., Jin, X.C. et al. (2001). Mannosylerythritol lipid induces characteristics of neuronal differentiation in PC12 cells through an ERK-related signal cascade. *Eur. J. Biochem.* 268: 374–383.

34 Kitamoto, D., Isoda, H., and Nakahara, T. (2002). Functional and potential application of glycerol biosurfactants. *J. Biosci. Bioeng.* 94: 187–201.

35 Lourith, N. and Kanlayavattanakul, M. (2009). Natural surfactants used in cosmetics: Glycolipids. *Int. J. Cosmet. Sci.* 31: 225–261.

36 Golubev, W.I., Kulakovskaya, T.V., and Golubeva, W. (2001). The yeast *Pseudozyma fusiformata* VKM Y-2821 producing an antifungal glycolipid. *Microbiol.* 70: 553–556.

37 Golubev, W.I., Kulakovskaya, T.V., Kulakovskaya, E.V., and Golubev, N.W. (2004). Fungicidal activity of extracellular glycolipid of *Sympodiomycopsis paphiopedili* Sugiyama et al. *Mikrobiologiia* 73: 841–845.

38 Balan, S.S., Kumar, C.G., and Jayalakshmi, S. (2019). Physicochemical, structural and biological evaluation of Cybersan (trigalactomargarate), a new glycolipid biosurfactant produced by a marine yeast, *Cyberlindnera saturnus* strain SBPN-27. *Process Biochem.* 80: 171–180.

39 Teichmann, B., Linne, U., Hewald, S. et al. (2007). A biosynthetic gene cluster for a secreted cellobiose lipid with antifungal activity from *Ustilago maydis*. *Mol. Microbiol.* 66: 525–533.

40 Reid, G., Bruce, A.W., Fraser, N. et al. (2001). Oral probiotics can resolve urogenital infections. *FEMS Immunol. Med. Microbiol.* 30: 49–52.

41 Rani, M., Weadge, J.T., and Jabaji, S. (2020). Isolation and characterization of biosurfactant-producing bacteria from oil well batteries with antimicrobial activities against food-borne and plant pathogens. *Front. Microbiol.* 27: 11–64.

42 Ohadi, M., Forootanfar, H., Dehghannoudeh, G. et al. (2020). Antimicrobial, anti-biofilm, and anti-proliferative activities of lipopeptide biosurfactant produced by *Acinetobacter junii* B6. *Microb. Pathog.* 1: 138–103806.

43 Awashti, N., Kumar, A., Makkar, R., and Cameotra, S. (1999). Enhanced biodegradation of endosulfan, a chlorinated pesticide in presence of a biosurfactant. *J. Environ. Sci. Heal. B* 34: 793–803.

44 Ahimou, F., Jacques, P., and Deleu, M. (2001). Surfactin and iturin A effects on *Bacillus subtilis* surface hydrophobicity. *Enzyme Microb. Technol.* 27: 749–754.

45 Carrillo, C., Teruel, J.A., Aranda, F.J., and Ortiz, A. (2003). Molecular mechanism of membrane permeabilization by the peptide antibiotic surfactin. *Biochim. Biophys. Acta* 1611: 91–97.

46 Hwang, M.H., Kim, M.H., Gebru, E. et al. (2008). Killing rate curve and combination effects of surfactin C produced from *Bacillus subtilis* complex BC1212 against pathogenic *Mycoplasma hyopneumoniae*. *World J. Microbiol. Biotechnol.* 24: 2277–2282.

47 Park, S.Y. and Kim, Y. (2009). Surfactin inhibits immuno stimulatory function of macrophages through blocking NK-κB, MAPK and Akt pathway. *Int. Immunopharmacol.* 9: 886–893.

48 Tang, J.S., Zhao, F., Gao, H. et al. (2010). Characterization and online detection of surfactin isomers based on HPLC-MS analyses and their inhibitory effects on the overproduction of nitric oxide and the release of TNF-α and IL-6 in LPS-induced macrophages. *Mar. Drugs* 8: 2605–2618.

49 Zhao, H., Yan, L., Xu, X. et al. (2018). Potential of *Bacillus subtilis* lipopeptides in anti-cancer I: Induction of apoptosis and paraptosis and inhibition of autophagy in K562 cells. *AMB Express* 8: 78.

50 Saini, H.S., Barragán-Huerta, B.E., Lebrón-Paler, A. et al. (2008). Efficient purification of the biosurfactant viscosin from *Pseudomonas libanensis* strain M9-3 and its physicochemical and biological properties. *J. Nat. Prod.* 71: 1011–1015.

51 Grangemard, I., Wallach, J., Maget-Dana, R., and Peypoux, F. (2001). Lichenysin: a more efficient cation chelator than surfactin. *Appl. Biochem. Biotechnol.* 90: 199–210.

52 Pradel, E., Zhang, Y., Pujol, N. et al. (2007). Detection and avoidance of a natural product from the pathogenic bacterium *Serratia marcescens* by *Caenorhabditis elegans*. *PNAS* 104: 2295–2300.

53 Chen, Y., Clardy, J., Cao, S. et al. (2012). A *Bacillus subtilis* sensor kinase involved in triggering biofilm formation on the roots of tomato plants. *Mol. Microbiol.* 85: 418–430.

54 Elad, Y. and Stewart, A. (2007). Microbial control of *Botrytis* sp. In: *Botrytis: Biology, Pathology and Control* (eds. Y. Elad, B. Williamson, P. Tudzynski and N. Delen), 223–236. Springer: The Netherlands, Dordrecht.

55 Liu, Y., Ding, S., Dietrich, R. et al. (2017). A biosurfactant-inspired heptapeptide with improved specificity to kill MRSA. *Angew. Chem. Int. Ed. Engl.* 56 (6): 1486–1490.

56 Landman, D., Georgescu, C., Martin, D.A., and Quale, J. (2008). Polymyxins revisited. *Clin. Microbiol. Rev.* 21: 449–465.

57 Karanth, N.G.K., Deo, P.G., and Veenanadig, N.K. (1999). Microbial production of biosurfactants and their importance. *Curr. Sci.* 77: 116–123.

58 Xiao, Y., Gerth, K., Müller, R., and Wall, D. (2012). Myxobacterium-produced antibiotic TA (Myxovirescin) inhibits type II signal peptidase. *Antimicrob. Agents Chemother.* 56 (4): 2014–2021.

59 Waghmode, S., Swami, S., Sarkar, D. et al. (2020). Exploring the pharmacological potentials of biosurfactant derived from *Planococcus maritimus* SAMP MCC 3013. *Curr. Microbiol.* 2: 1–8.

60 Gan, B.S., Kim, J., Reid, G. et al. (2002). *Lactobacillus fermentum* RC-14 inhibits *Staphylococcus aureus* infection of surgical implants in rats. *J. Infect. Dis.* 185: 1369–1372.

61 Huang, X., Suo, J., and Cui, Y. (2011). Optimization of antimicrobial activity of surfactin and polylysine against *Salmonella enteritidis* in milk evaluated by a response surface methodology. *Foodborne Pathog. Dis.* 8: 439–443.

62 Pirog, T.P., Konon, A.D., Sofilkanich, A.P., and Skochko, A.B. (2011). Effect of biosurfactants *Acinetobacter calcoaceticus* K-4 and *Rhodococcus erythropolis* EK-1 on some microorganisms. *Mikrobiol. Z.* 73: 14–20.

63 Javee, A., Karuppan, R., and Subramani, N. (2020). Bioactive glycolipid biosurfactant from seaweed Sargassum myriocystum associated bacteria *Streptomyces* sp. SNJASM6. *Biocatal. Agric. Biotechnol.* 1: 23–101505.

64 Huang, X., Lu, Z., Zhao, H. et al. (2006). Antiviral activity of antimicrobial lipopeptide from *Bacillus subtilis* fmbj against pseudorabies virus, porcine parvovirus, Newcastle disease virus and infectious bursal disease virus in vitro. *Int. J. Pept. Res. Ther.* 12: 373–377.

65 Remichkova, M., Galabova, D., Roeva, I. et al. (2008). Anti-herpesvirus activities of *Pseudomonas* sp. S-17 rhamnolipid and its complex with alginate. *Z. Naturforsch. C* 63: 75–81.

66 Garoff, H., Hewson, R., and Opstelten, D.J.E. (1998). Virus maturation by budding. *Microbiol. Mol. Biol. Rev.* 62: 1171–1190.

67 Khan, T.N. (2017). Cyclosporin A production from *Tolipocladium inflatum*. *Gen. Med. Open Access* 5: 4. https://doi.org/10.4172/2327-5146.1000294.

68 Smith, M.L., Gandolfi, S., Coshall, P.M., and Rahman, P.K.S.M. (2020). Biosurfactants: A Covid-19 perspective. *Front. Microbiol.* https://doi.org/10.3389/fmicb.2020.01341.

69 Velraeds-Mar-tine, M.C., Vander Mei, H.C., Reid, G., and Busscher, H.J. (1997). Inhibition of initial adhesion of uropathogenic *Enterococcus faecalis* to solid substrate by an adsorbed bio-surfactant layer from *Lactobacillus acidophilus*. *Urology* 49: 790–794.

70 Kumar, A., Ali, A., and Yerneni, L.K. (2007). Effectiveness of a mycoplasma elimination reagent on a mycoplasma-contaminated hybridoma cell line. *Hybridoma (Larchmt)* 26: 104–106.

71 Fassi, F.L., Wroblewski, H., and Blanchard, A. (2007). Activities of antimicrobial peptides and synergy with enrofloxacin against *Mycoplasma pulmonis*. *Antimicrob. Agents Chemother.* 51: 468–474.

72 Hirayama, T. and Kato, I. (1982). Novel methyl rhamnolipids from *Pseudomonas aeruginosa*. *FEBS Lett.* 139: 81–85.

73 Shibahara, M., Zhao, X.X., Wakamatsu, Y. et al. (2000). Mannosylerythritol lipid increases levels of galactoceramide in and neurite outgrowth from PC12 pheochromocytoma cells. *Cytotechnology* 33: 247–251.

74 Zhao, X. (1999). Mannosylerythritol lipid is a potent inducer of apoptosis and differentiation of mouse melanoma cells in culture. *Cancer Res.* 59: 482–486.

75 Techaoei, S., Leelapornpisid, P., Santiarwarn, D., and Lumyong, S. (2007). Preleminary screening of biosurfactant producing microorganisms isolated from hot spring and garages in Northen Thailand. *KMITL Sci. Tech. J.* 7: 38–43.

76 Thaniyavarn, J., Chongchin, A., Wanitsuksombut, N., and Thaniyavarn, S. (2006). Biosurfactant production by *Pseudomonas aeruginosa* A41using palm oil as carbon source. *J. Gen. Appl. Microbiol.* 52: 215–222.

77 Kameda, Y., Ouchira, S., Matsui Kkanatomo, S. et al. (1974). Antitumor activity of *Bacillus natto* V. isolation and characterization of surfactin in the culture medium of *Bacillus natto* KMD 2311. *Chem. Pharm. Bull.* 22: 938–944.

78 Cao, X.H., Wang, A.H., Wang, C.L. et al. (2010). Surfactin induces apoptosis in human breast cancer MCF-7 cells through a ROS/JNK-mediated mitochondrial/caspase pathway. *Chem. Biol. Interact.* 183: 357–362.

79 Chen, J., Song, X., Zhang, H., and Qu, Y. (2006). Production, structure elucidation and anticancer properties of sophorolipid from *Wickerhamiella domercqiae*. *Enzyme Microb. Technol.* 39 (3): 501–506.

80 Sudo, T., Zhao, X.X., Wakamatsu, Y. et al. (2000). Induction of the differentiation of human HL-60 promyelocytic leukemia cell line by succinoyl trehalose lipids. *Cytotechnology* 33: 259–264.

81 Chami, M., Andreau, K., Lemassu, A. et al. (2002). Priming and activation of mouse macrophages by trehalose 6, 60-dicorynomycolate vesicles from *Corynebacterium glutamicum*. *FEMS Immunol. Med. Microbiol.* 32: 141–147.

82 Das, P., Mukherjee, S., and Sen, R. (2009). Antiadhesive. action of a marine microbial surfactant. *Colloids Surf. B: Biointerfaces* 71: 183–186.

83 Rivardo, F., Turner, R.J., Allegrone, G. et al. (2009). Anti-adhesion activity of two biosurfactants produced by *Bacillus* sp. prevents biofilm formation of human bacterial pathogens. *Appl. Microbiol. Biotechnol.* 83: 541–553.

84 Mireles, J.R., Toguchi, A., and Harshey, R.M. (2001). *Salmonella enterica* serovar Typhimurium swarming mutants with altered biofilm-forming abilities: Surfactin inhibits biofilm formation. *J. Bacteriol.* 183: 5848–5854.

85 Rodrigues, L., van der Mei, H.C., Teixeira, J., and Oliveira, R. (2004). Influence of biosurfactants from probiotic bacteria on formation of biofilms on voice prostheses. *Appl. Environ. Microbiol.* 70: 4408–4410.

86 Rodrigues, L.R., Banat, I.M., van der Mei, H.C. et al. (2006a). Interference in adhesion of bacteria and yeasts isolated from explanted voice prostheses to silicone rubber by rhamnolipid biosurfactants. *J. Appl. Microbiol.* 100: 470–480.

87 Zeraik, A.E. and Nitschke, M. (2010). Biosurfactants as agents to reduce adhesion of pathogenic bacteria to polystyrene surfaces: Effect of temperature and hydrophobicity. *Curr. Microbiol.* 61: 554–559.

88 Walencka, E., Wieckowska-Szakiel, M., Rozalska, S. et al. (2007). A surface-active agent from *Saccharomyces cerevisiae* influences staphylococcal adhesion and biofilm development. *Z. Naturforsch. C* 62: 433–438.

89 Janek, T., Łukaszewicz, M., and Krasowska, A. (2012). Antiadhesive activity of the biosurfactant pseudofactin II secreted by the Arctic bacterium *Pseudomonas fluorescens* BD5. *BMC Microbiol.* https://doi.org/10.1186/1471-2180-12-24.

90 Busscher, H.J., Van de Belt-Gritter, B., Westerhof, M. et al. (1999). Microbial interference in the colonization of silicone rubber implant surfaces in the oropharynx: *Streptococcus thermophilus* against a mixed fungal/bacterial biofilm. In: *Microbial Ecology and Infectious Disease* (ed. E. Rosenberg), 66–74. Washington, DC: American Society for Microbiology.

91 Velraeds-Mar-tine, M.C., Vander Mei, H.C., Reid, G., and Busscher, H.J. (1997). Inhibition of initial adhesion of uropathogenic *Enterococcus faecalis* to solid substrate by an adsorbed biosurfactant layer from *Lactobacillus acidophilus*. *Urology* 49: 790–794.

92 Rivardo, F., Martinotti, M.G., Turner, R.J., and Ceri, H. (2011). Synergistic effect of lipopeptide biosurfactant with antibiotics against *Escherichia coli* CFT073 biofilm. *Int. J. Antimicrob. Agents* 37: 324–331.

93 de Araujo, L.V., Abreu, F., Lins, H. et al. (2011). Rhamnolipid and surfactin inhibit *Listeria monocytogenes* adhesion. *Food Res. Int.* 44: 481–488.

94 Abdollahi, S., Tofighi, Z., Babaee, T. et al. (2020). Evaluation of anti-oxidant and anti-biofilm activities of biogenic surfactants derived from *Bacillus amyloliquefaciens and Pseudomonas aeruginosa*. *Iran J. Pharm. Res.* https://doi.org/10.22037/ijpr.2020.1101033.

95 Płaza, G. and Achal, V. (2020). Biosurfactants: Eco-friendly and innovative biocides against biocorrosion. *Int. J. Mol. Sci.* 21: 2152.

96 Gharaei-Fathabad, E. (2011). Biosurfactants in pharmaceutical industry: A mini–review. *Am. J. Drug. Discov. Dev.* 1: 58–69.

97 Mittenbuhler, K., Loleit, M., Baier, W. et al. (1997). Drug specific antibodies: T-cell epitope lipopeptide conjugates are potent adjuvants for small antigens in vivo and in vitro. *Int. J. Immunopharmacol.* 19: 277–287.

98 Ishigami, Y., Zhang, Y., and Ji, F. (2000). Spiculisporic acid. Functional development of biosurfactants. *Chim. Oggi* 18: 32–34.

99 Appanna, V.D., Finn, H., and Pierre, M.S. (1995). Exocellular phosphatidylethanolamine production and multiple-metal tolerance in *Pseudomonas fluorescens*. *FEMS Microbiol. Lett.* 131: 53–56.

100 Kaplan DL, Fuhrman J, Gross RA. (2004). Emulsan adjuvant immunization formulation and use. US Patent 2004/0265340 A1.

101 Toren, A., Orr, E., Paitan, Y. et al. (2002). The active component of the bioemulsifier alasan from *Acinetobacter radioresistens* KA53 is an OmpA-like protein. *J. Bacteriol.* 184: 165–170.

102 Rosenberg, E. and Ron, E.Z. (1997). Bioemulsions: Microbial polymeric emulsifiers. *Curr. Opin. Biotechnol.* 6: 313–316.

103 Kaplan, D.L. (1998). *Biopolymers from Renewable Resources*. Berlin, Heidelberg: Springer.

104 Ron, E.Z. and Rosenberg, E. (2001). Natural roles of biosurfactants: Minireview. *Environ. Microbiol.* 3: 229–236.

105 Nerurkar, A.S., Hingurao, K.S., and Suthar, H.G. (2009). Bioemulsifiers from marine microorganisms. *J. Sci. Ind. Res.* 68: 273–277.

106 Aydaş, S.B., Ozturk, S., and Aslım, B. (2012). Phenylalanine ammonia lyase (PAL) enzyme activity and antioxidant properties of some cyanobacteria isolates. *Food Chem.* https://doi.org/10.1016/j.foodchem.

107 Maingault M. (1999). Utilization of sophorolipids as therapeutically active substances or cosmetic products, in particular for the treatment of the skin. US Patent 5,981,497.

108 Zhang, Y., Li, H., Sun, J. et al. (2010). Dc-chol/Dope cationic liposomes: A comparative study of the influence factors on plasmid pDNA and Si RNA gene delivery. *Int. J. Pharm.* 390: 198–207.

109 Maitani, Y., Igarashi, S., Sato, M., and Hattori, Y. (2007). Cationic liposome (DC-Chol/DOPE=1: 2) and a modified ethanol injection method to prepare liposomes, increased gene expression. *Int. J. Pharm.* 342: 33–39.

110 Ueno, Y., Hirashima, N., Inoh, Y. et al. (2007). Characterization of biosurfactant-containing liposomes and their efficiency for gene transfection. *Biol. Pharm. Bull.* 30: 169–172.

111 Maitani, Y., Yano, S., Hattori, Y. et al. (2006). Liposome vector containing biosurfactant-complexed DNA as herpes simplex virus thymidine kinase gene delivery system. *J. Liposome Res.* 16: 359–372.

112 Igarashi, S., Hattori, Y., and Maitani, Y. (2006). Biosurfactant MEL-A enhances cellular association and gene transfection by cationic liposome. *J. Control. Release* 112: 362–368.

113 Faivre, V. and Rosilio, V. (2010). Interest of glycolipids in drug delivery: from physicochemical properties to drug targeting. *Expert Opin. Drug Deliv.* 7: 1031–1048.

114 Nicoli, S., Eeman, M., Deleu, M. et al. (2010). Effect of lipopeptides and iontophoresis on aciclovir skin delivery. *J. Pharm. Pharmacol.* 62: 702–708.

115 Date, A.A. and Nagarsenker, M.S. (2008). Design and evaluation of microemulsions for improved parenteral delivery of Propofol. *AAPS PharmSci. Tech.* 9: 138–145.

116 Calabrese, I., Gelardi, G., Merli, M. et al. (2017). Clay-biosurfactant materials as functional drug delivery systems: Slowing down effect in the in vitro release of cinnamic acid. *Appl. Clay Sci.*: 567–574.

117 Chereshnev, V.A., Gein, S.V., Baeva, T.A. et al. (2010). Modulation of cytokine secretion and oxidative metabolism of innate immune effectors by *Rhodococcus* biosurfactant. *B Exp. Biol. Med.* 149: 734–738.

118 Gein, S.V., Kuyukina, M.S., Ivshina, I.B. et al. (2011). In vitro cytokine stimulation assay for glycolipid biosurfactant from *Rhodococcus ruber*: role of monocyte adhesion. *Cytotechnology* 63: 559–566.

119 Andrä, J., Rademann, J., Howe, J. et al. (2006). Endotoxin-like properties of a rhamnolipid exotoxin from *Burkholderia (Pseudomonas) plantarii*: Immune cell stimulation and biophysical characterization. *Biol. Chem.* 387: 301–310.

120 Hida, T., Hayashi, K., Yukishige, K. et al. (1995). Synthesis and biological activities of TAn-1511 analogues. *J. Antibiot* 48: 589–603.

121 Takizawa, M., Hida, T., Horiguchi, T. et al. (1995). Tan-1511 A, B and C, microbial lipopeptides with G-CSF and GM-CSF inducing activity. *J. Antibiot* 48: 579–588.

122 Kuyukina, M.S., Ivshina, I.B., Baeva, T.A. et al. (2015). Trehalolipid biosurfactants from nonpathogenic *Rhodococcus actinobacteria* with diverse immunomodulatory activities. *New Biotechnol.* 32: 559–568.

123 Shete, A.M., Wadhawa, G., Banat, I.M., and Chopade, B.A. (2006). Mapping of patents on bioemulsifier and biosurfactant: A review. *J. Sci. Ind. Res.* 65: 91–115.

124 Im, J.H., Yanagishita, H., Ikegami, T. et al. (2003). Mannosylerythritol lipids, yeast glycolipid biosurfactants, are potential affinity ligand materials for human immunoglobulin G. *J. Biomed. Mater. Res.* 65: 379–385.

125 Morita, Y., Tadokoro, S., Sasai, M. et al. (2011). Biosurfactant mannosyl-erythritol lipid inhibits secretion of inflammatory mediators from RBL-2H3 cells. *Biochim. Biophys. Acta* 1810: 1302–1308.

126 Byeon, S.E., Lee, Y.G., Kim, B.H. et al. (2008). Surfactin blocks NO production in lipopolysaccharide-activated macrophages by inhibiting NF-κB activation. *J. Microbiol. Biotechnol.* 18: 1984–1989.

127 Selvam, R., Maheswari, P., Kavitha, P. et al. (2009). Effect of *Bacillus subtilis* PB6, a natural probiotic on colon mucosal inflammation and plasma cytokines levels in inflammatory bowel disease. *Indian J. Biochem. Biophys.* 46: 79–85.

128 Bouffioux, O., Berquand, A., Eeman, M. et al. (2007). Molecular organization of surfactin-phospholipid monolayers: Effect of phospholipid chain length and polar head. *Biochim. Biophys. Acta Biomembr.* 1768: 1758–1768.

129 Brasseur, R., Braun, N., El Kirat, K. et al. (2007). The biologically important surfactin lipopeptide induces nanoripples in supported lipid bilayers. *Langmuir* 23: 9769–9772.

130 Francius, G., Dufour, S., Deleu, M. et al. (2008). Nanoscale membrane activity of surfactins: Influence of geometry, charge and hydrophobicity. *Biochim. Biophys. Acta* 1778: 2058–2068.

131 Deleu, M., Paquot, M., and Nylander, T. (2008). Effect of fengycin, a lipopeptide produced *by Bacillus subtilis*, on model biomembranes. *Biophys. J.* 94: 2667–2679.

132 Wang, X., Gong, L., Liang, S. et al. (2005). Algicidal activity of rhamnolipid biosurfactants produced by *Pseudomonas aeruginosa*. *Harmful Algae* 4: 433–443.

133 Aminabee, S., Prabhakar, M.C., Prasad, R.G.S.V. et al. (2012). *In vitro* pharmacological activity of biosurfactant. *IJPCBS* 2 (1): 130–133.

134 Farias, J.M., Stamford, T.C.M., Resende, A.H.M. et al. (2019). Mouthwash containing a biosurfactant and chitosan: An ecosustainable option for the control of cariogenic microorganisms. *Int. J. Biol. Macromol.* 129: 853–860.

135 da Silva IA, Bezerrac KG, Batista IJ. (2020). Evaluation of the emulsifying and antioxidant capacity of the biosurfactant produced by *Candida bombicola* URM 3718. *Chem. Engin.*, 79, 67–72. Available at: https://doi.org/10.3303/CET2079012.

136 Chami, M., Andreau, K., Lemassu, A. et al. (2002). Priming and activation of mouse macrophages by trehalose 6, 60-dicorynomycolate vesicles from *Corynebacterium glutamicum*. *FEMS Immunol. Med. Microbiol.* 32: 141–147.

137 Gein, S.V., Kuyukina, M.S., Ivshina, I.B. et al. (2011). In vitro cytokine stimulation assay for glycolipid biosurfactant from *Rhodococcus ruber*: role of monocyte adhesion. *Cytotechnology* 63: 559–566.

138 Landman, D., Georgescu, C., Martin, D.A., and Quale, J. (2008). Polymyxins revisited. *Clin. Microbiol. Rev.* 21: 449–465.

139 Mittenbuhler, K., Loleit, M., Baier, W. et al. (1997). Drug specific antibodies: T-cell epitope lipopeptide conjugates are potent adjuvants for small antigens in vivo and in vitro. *Int. J. Immunopharmacol.* 19: 277–287.

140 Park, S.Y. and Kim, Y. (2009). Surfactin inhibits immuno stimulatory function of macrophages through blocking NK-κB, MAPK and Akt pathway. *Int. Immunopharmacol.* 9: 886–893.

141 Rahman, M.S. and Ano, T. (2009). Production characteristics of lipopeptide antibiotics in biofilm fermentation of *Bacillus subtilis*. *J. Environ. Sci.* 21: s36–s39.

142 Ron, E.Z. and Rosenberg, E. (2001). Natural roles of biosurfactants. *Environ. Microbiol.* 3: 229–236.

143 Ron, E.Z. and Rosenberg, E. (2001). Natural roles of biosurfactants. *Environ. Microbiol.* 3: 229–236.

144 Sarma H, Bustamante KLT, Prasad MNV (2018) Biosurfactants for oil recovery from refinery sludge: Magnetic nanoparticles assisted purification. In: M.N.V. Prasad (ed,), *Industrial and*

Municipal Sludge, Elsevier, Cambridge, MA, USA, Butterworth-Heinemann, Osford, UK, pp. 107–132. ISBN: 9780128159071, Editor Majeti Narasimha Vara Prasad, Paulo Jorge de Campos, Favas Meththika, and Vithanage S. Venkata Mohan.

145 Vecino, X., Rodrıguez-Lopez, L., Ferreira, D. et al. (2018). Bioactivity of glycolipopeptide cell-bound biosurfactants against skin pathogens. *Int. J. Biol. Macromol.* 109: 971–979.

146 Velraeds-Mar-tine, M.C., Vander Mei, H.C., Reid, G., and Busscher, H.J. (1997). Inhibition of initial adhesion of uropathogenic *Enterococcus faecalis* to solid substrate by an adsorbed biosurfactant layer from *Lactobacillus acidophilus*. *Urology* 49: 790–794.

147 Velraeds-Mar-tine, M.C., Vander Mei, H.C., Reid, G., and Busscher, H.J. (1997). Inhibition of initial adhesion of uropathogenic *Enterococcus faecalis* to solid substrate by an adsorbed biosurfactant layer from *Lactobacillus acidophilus*. *Urology* 49: 790–794.

148 Yoshihiko, H., Mizuyuki, R., Yuka, O. et al. (2009). Novel characteristics of sophorolipids, yeast glycolipid biosurfactants, as biodegradable low-foaming surfactants. *J. Biosci. Bioeng.* 108: 142–146.

149 Zhao, X. (1999). Mannosylerythritol lipid is a potent inducer of apoptosis and differentiation of mouse melanoma cells in culture. *Cancer Res.* 59: 482–486.

14

Biosurfactant-Inspired Control of Methicillin-Resistant *Staphylococcus aureus* (MRSA)

Amy R. Nava

Department of Interdisciplinary Health Sciences, College of Health Sciences, University of Texas, El Paso, TX, USA

CHAPTER MENU

14.1 *Staphylococcus aureus*, MRSA, and Multidrug Resistance

Staphylococcus aureus is an opportunistic pathogen that is also part of the normal human flora. This organism is a gram-positive coccus that is catalase and coagulase positive. *S. aureus* is the leading cause of endocarditis [1] but is responsible for various types of infections, including food contamination, soft tissue infections, bacteremia, and biofilm formation on medical devices [2, 3]. In recent decades MRSA has started to become more prevalent in non-healthcare and community settings. Moreover, the community acquired and healthcare associated MRSA were very different from each other [2, 3]. The introduction of methicillin containing antibiotics resulted in the increasing prevalence of MRSA in the 1960s [1]. MRSA has become an epidemic in recent years as it can be acquired from both the community as well as hospital settings. The methicillin resistance is due to the penicillin binding proteins (PBPs) – specifically, the PBP2, which is coded by the mecA gene [4]. The β-lactams bind to the transpeptidase domain of the PBPs and production of peptidoglycan stops [5]. Penicillinase cleaves the β-lactam ring, which makes it useless against MRSA. While penicillinase gives partial resistance against β-lactams, the mecA is more important as it modifies PBP proteins, the target of the antibiotic [5].

Biosurfactants for a Sustainable Future: Production and Applications in the Environment and Biomedicine, First Edition. Edited by Hemen Sarma and Majeti Narasimha Vara Prasad.
© 2021 John Wiley & Sons Ltd. Published 2021 by John Wiley & Sons Ltd.

MRSA is also known for resistance against a broad spectrum of antibiotics with dissimilar properties due to various strategies of pathogenesis. Most bacteria also utilize biofilm formation as an important strategy to elude immune cells, antibiotics, and many biocides. MRSA biofilm formation on medical devices and implants has become a major healthcare concern.

For this reason, biosurfactants with different properties against the different types of infections of MRSA as well as antibiofilm properties will be discussed here. Biosurfactant properties may prove to be an effective solution to address the various types of MRSA infections without the evolution of resistance.

14.2 Biosurfactant Types Commonly Utilized Against *S. aureus* and Other Pathogens

Surfactants are amphiphilic compounds that are excreted by bacteria and fungi. Surfactants are concentrated at interphases such as between liquids, a solid and a liquid, or a gas and a liquid. The interphase is located between two substances that do not mix. At such an interphase, the hydrophobic portion of the surfactant orientates toward the surface away from the liquid while the hydrophilic portion extends into the solution. This solution is usually polarized [6, 7]. Biosurfactants are those that are made from fungi and bacteria [8, 9]. Much of the literature has concentrated on two genera of bacteria as the primary source of biosurfactants. These are *Bacillus* and *Pseudomonas* [10]. Microorganisms can be cultured from contaminated areas and with different carbon and nitrogen sources, resulting in a wide variety of chemical structures that allow for more efficiency [10].

Biosurfactants can be made from various substrates including fructose and other sugars, oils, and alkanes [10, 11]. According to Bierman et al. biosurfactants can be organized into groups based on chemical properties. These groups are phospholipids, fatty acids, lipids, glycolipids, and other compounds [12]. Very few of these compounds contain an amine giving the compound a cationic charge. The majority of these compounds have a negative or neutral charge [13]. The hydrophobic portion of the biosurfactant consists of a long fatty acid chain. The hydrophilic portion can consist of an alcohol, phosphate, or other polar compound [14]. For this reason, biosurfactants have a variety of surface properties, including low to no toxicity, pH, ionic and temperature tolerance, and antimicrobial ability [15].

The general types of biosurfactants produced by microorganisms are glycolipids, lipopeptides/lipoproteins, and surfactants [15]. Most biosurfactants are glycolipids, which are carbohydrates attached to an aliphatic acid by an ester group. Of the glycolipids there are three well-studied types. These are rhamnolipids, sophorolipids, and trehalolipids [16]. For our purposes, we will focus mostly on glycolipids and lipopeptides in this chapter.

Rhamnolipids are primarily produced by *Pseudomonas aeriginosa* and are commonly found as mono- or di-rhamnolipids. They are a rhamnose sugar molecule attached from the carboxyl end to a β-hydroxy acid [17]. They are used in a number of applications from bioremediation/oil recovery to cosmetics [18]. While the mechanism of antimicrobial ability is unclear, it is understood that biosurfactants are toxic to certain microbial pathogens due to hydrophobicity of fatty acids and lipopeptides [19].

Sophorolipids are a yeast-derived dimeric sugar sophorose, which is linked to a long chain hydroxy fatty acid. Sophorolipids are unique as they have a β-1,2 bond that is able to be acetylated at the 6′ and 6″ positions [20]. Acetylation of these bonds alters the properties of sophorolipids drastically. Acetyl groups can increase antiviral properties of the sophorolipid [20, 21]. The hydroxy fatty acid portion of the sophorolipid is usually about 16–18 carbon atoms long with saturated bonds. Similar to rhamnolipids, sophorolipids are amphiphilic and are able to reduce surface tension of water from 72.8 to 30 mN/m with a critical micelle concentration ranging from 40 to 100 mg/l to be effective [20, 22].

Sophorolipids can be in an acidic or lactonized form. For biomedical purposes, lactonic sophorolipids are considered to have a higher antimicrobial activity than those in the acidic form [20]. As with other biosurfactants, media and culture conditions strongly influence future properties and antimicrobial abilities. The majority of sophorolipids are produced by *C. bombicola, C. apicola*, and *R. bogoriensis*, but it is very likely that other yeasts from similar genre can synthesize sophorolipids. In contrast to rhamnolipids, sophorolipids are derived from non-pathogenic yeast. This makes them ideal for cosmetics and other human uses [22].

Trehalose lipids are primarily produced by bacteria that produce mycolic acid [23]. The mycolate taxon includes mycobacterium and *Rhodacoccus* [23]. Mycolic acid is found in the cell walls of these bacterium and is composed of a hydrophilic portion and a long beta hydroxy chain linked to a shorter alkyl side chain [23, 24]. This biosurfactant has been reported to reduce surface tension from 72 mN/m to approximately 19 mN/m with a critical micelle concentration (CMC) ranging from 0.7 to 37 mg/l [23]. Trehalose has gained interest due to an ability to increase emulsification or psuedosolubility of hydrophobic compounds and binding to a cellular envelope [24]. This may improve the delivery of certain hydrophobic compounds used medicinally.

Lipopeptides are composed of linear or cyclic forms of an amino acid sequence followed by a fatty acid moiety [25, 26]. The peptide portion of the lipopeptide can be positive or negatively charged. The fatty acid sequence is covalently linked to the N-terminus of the peptide. The polar head can be a lactone ring or linear in structure [25]. The mode of action of anionic lipopeptides was described by Straus and Hancock [27]. Briefly, the lipopeptide binds to the membrane of the pathogen, resulting in membrane depolarization [25]. Lipopeptides, like other biosurfactants are ideal for antimicrobial applications as resistance to them is very rare. The Asp-Gly sequence in anionic lipopeptides make them more likely to be degraded, mainly due to it being more prone to chemical reactions [27].

Lipopeptides have a broad range of applications in biomedical applications such as antimicrobial, antiviral, and anti-fungal. Biomedical applications are vast and more are being discovered. For example, a lipopeptide isolated from *Streptosporangium amethystogenes* sub sp. *fukuiense* Al-23456 was able to stimulate the proliferation of immune cells in mice [28]. This is one example of a biomedical application that is uncommon. More common applications primarily surround lichenysins, surfactins, fengycin, and iturin, which are biosurfactants produced by various *Bacillus* species [26, 29]. Very little is known about the antimicrobial mechanisms of lipopeptide. Lichenysin, for example, is a cyclic lipopeptide extracted from *Bacillus licheniformis*. Generally, it is thought that this biosurfactant was able to lyse cells. Hemolysis of erythrocyte cells by lichenysin was observed by Grangemard et al. [30] in the presence of calcium and magnesium. The study noted an unusual ability of lichenysin to chelate cations, possibly contributing to the membrane damage. Interestingly, the study also noted functional consequences of peptide sequence changes. For example, a minimal L-Glu for L-Gln increased hemolytic, surfactant, and cation chelating abilities significantly [30].

14.3 Properties of Efficient Biosurfactants Against MRSA and Bacterial Pathogens

Screening for an effective biosurfactant relies on several properties including the emulsification index, low critical micelle concentration (CMC), as well as high interfacial activity [31]. A surfactant's efficiency is measured by the CMC, which is defined as the minimal amount of surfactant needed to form a micelle [31–33]. In other words, it is the concentration of biosurfactant needed for micelle structures to form spontaneously when added to water. This is due to surfactant molecules aggregating together to reduce contact between water and hydrophobic moieties [32]. This causes a reduction of free energy of the system because it lowers the energy between the surfactant and water interphase, ultimately leading to the reduction of surface tension [33]. Addition of more biosurfactant after the CMC is reached has little effect on the system other than to create more micelles [34]. In general, the ability of a surfactant to reduce surface tension is an indication of effectiveness. The surface tension can be measured using a tensiometer, which measures the surface tension of a liquid [35]. In general, a biosurfactant is considered good if it is able to reduce the surface tension of a liquid to <40 mN/m or less [36]. Some methods of determining surface tension include the du Nouy ring method, which measures the force required to remove a ring from the surface of a liquid [32]. Another assay for determining efficiency is oil displacement, where a drop of the biosurfactant is placed over oil. If the oil disperses, the biosurfactant is considered ideal [35].

14.4 Uses for Biosurfactants

Surfactants have been used and synthesized chemically. Biosurfactants have been fairly recently discovered and are used in many industrial applications. The oil and gas industry have used surfactants as a means of petroleum recovery and to increase the solubility of oil products [37]. The use of biosurfactant in oil and petroleum is due to emulsification and degradation enhancing abilities [38]. Biosurfactants have been employed in many facets due to low environmental and eukaryote cytotoxicity. Another advantage of biosurfactants is the relatively cost-effective production using renewable substrates such as molasses, sugars, plant-derived oils/lipids, and dairy waste products [39]. The use of biosurfactants has increased in recent years and has expanded into industrial purposes such as biomedical applications, a means for treating agricultural wastes, and pharmaceutical applications [39]. For our purposes, we will discuss biomedical and pharmaceutical applications for biosurfactants successful against *S. aureus* and MRSA.

14.5 Biosurfactants Illustrating Antiadhesive Properties against MRSA Biofilms

The key to effective antibiofilm formation is the biosurfactant's ability to prevent adhesion. Central to this are the properties, as previously discussed, of the biosurfactant being utilized. These properties include CMC, oil displacement, emulsification, and the ability to reduce surface tension. While these properties are important, no exact mechanism of biofilm disruption has been clearly defined. It is difficult to state which property is most important as some biosurfactants have no antibiofilm or antimicrobial effects at all. Due to the variety of chemical structures of biofilms excreted by different species of bacteria, the efficiency of antiadhesion properties of the biosurfactant varies. For this reason, for both antimicrobial and antibiofilm properties, we will also note, where necessary, the variety of microbial sources as well as the conditions of culturing.

Lactobacillus jensenii and *Lactobacillus rhamnosis* are able to produce biosurfactant with antibiofilm, antimicrobial, and antiadhesive properties against several clinical multidrug-resistant strains of *S. aureus, Acinetobacter Baumannii, and Escherichia coli* [40]. This study by Sambanthamoorthy et al. was significant because these results were also done *in vitro*. The biosurfactant was able to kill 80–93% of *S. aureus* strains UAMS-1 and MRSA, respectively. Significant differences between the control (without biosurfactant) and biosurfactant illustrated that interference with biofilm attachment to abiotic surfaces was significantly reduced. The adherence of *S. aureus* to surfaces was interrupted with concentrations of as little as 25–50 mg/ml. The biosurfactant also showed dispersion of preformed biofilm as significantly disrupted when compared to the control with 25 mg/ml used to treat existing MRSA biofilms. After approximately 18 hours of incubation of the biofilm with the surfactant, significant reduction and disruption were observed.

The *Bacillus* genus has also demonstrated antibiofilm activity against *S. aureus*. In a 2009 study by Rivardo et al. [41], two biosurfactant strains, *Bacillus subtilis and B. licheniformis*, produced highly resilient biosurfactants that were resistant to pH and salinity. Other properties included the ability to reduce surface tension of water from 72 to 26–30 mN/m. The *B. licheniformis* produced biosurfactant V9T14 and the *B. subtilis* derived biosurfactant was V19T21. Interestingly, each biosurfactant proved to be effective against either Gram-positive (*S. aureus*) biofilms but not Gram-negative (*E. coli*) biofilms and the opposite was true for the other biosurfactant. Each biosurfactant proved effective either by coating a polystyrene surface or by adding the biosurfactant to the culture. The mechanism suggested in this study was the possibility that the biosurfactants were binding to cell surfaces modifying the hydrophobic properties. The study was of particular interest as it was able to inhibit growth of the biofilm by 90% of *S. aureus* as compared to the control.

Tahmourespour et al. [42] discussed a possible mechanism to test the efficacy of a *Lactobacillus acidophilus* biosurfactant against *Streptococcus mutans* biofilm. *S. mutans* is responsible for dental caries due to the production of biofilm that is able to adhere to the surface of teeth. The biofilm is comprised of fructans and glucans, which are encoded by the genes gtfB, gtfC, and gtfD. In this study, using real-time quantitative RT-PCR the gene expression of these *S. mutans* genes was significantly decreased in the presence of the *L. acidophilus* biosurfactant. This indicated that biosurfactants are able to down-regulate biofilm producing genes, indicating the potential of the biosurfactant as a possible signaling molecule. Since the biosurfactant was not tested on MRSA, the potential as a biofilm growth inhibitor may be likely. This mechanism can be potentially applied to biosurfactants with antimicrobial effects against *S. aureus* as both are Gram-positive cocci.

In a study by Coronel-León et al. [43], the type of medium used to culture biosurfactant-producing bacteria proved to be an important factor. The purpose of the study was to optimize response surface methodology to produce lichenysin. Lichenysin is a biosurfactant produced by *B. lichenformis*. The other purpose was to evaluate the effectiveness of lichenysin against bacterial biofilms. Lichenysin is a lipopeptide with a powerful surface activity. The lipid moiety contains β-hydroxy fatty acids with chains that average 14–16 carbons long [43]. The peptide portion usually contains glutamine as the N-terminal and a chain of specific amino acids at the N-terminal [43]. These lipid moieties are very important as they are able to interact with the cell membrane.

Different carbon sources were used to determine an optimal source as well as mineral medium and trace salts. The most optimal medium tested consisted of molasses, K_2HPO_4/KH_2PO_4, and $NaNO_3$ [43]. To analyze the method that yielded the most, lichenysin was precipitated from cell-free supernatant with acid and weighed. This medium was optimized for the purpose of extracting the most purified lichenysin. The lichenysin proved to have effective antiadhesion properties against the biofilms of *C. albicans* and MRSA, showing inhibition of 74.5 and 47.8%, respectively, in pretreatment of surfaces.

The biosurfactant was tested as a pre- and post-treatment on polystyrene surfaces. For post-treated surfaces, the biosurfactant was applied after a biofilm of various bacteria had grown. As with the pre-treated surfaces, the inhibition of biofilm was inversely proportional to the concentration of biosurfactant added. The post-treatment was also effective as a biofilm inhibitor. However, when comparing the effectiveness of pre- and post-treatments of lichenysin, the pre-treatment was better. The authors did not suggest an exact method of inhibition.

14.6 Biosurfactants with Antibiofilm and Antimicrobial Properties

It is common for biosurfactants to have both antimicrobial and antibiofilm properties but not all are both necessary and commonly may be one or the other. Having both properties allows for effective treatment against pathogenic bacteria such as MRSA, which employs multiple means of virulence. Interestingly, some biosurfactants have a specific microbial target while others have a broad range of targets both Gram positive and Gram negative. In a study by Gudiña et al. [44] a biosurfactant isolated from *L. paracasei* ssp. *paracasei* A20 was effective as an antimicrobial and antibiofilm biosurfactant and had a fairly broad range of microbial targets that were both Gram positive and Gram negative. While they are effective against a broad range of microbes, often the degree of effectiveness varied. Antimicrobial properties were also concentration dependent, which is the case for many biosurfactants. Between 25 and 50 mg/ml, complete growth inhibition was observed against *S. aureus*, *S. epidermidis*, *E. coli*, *S. pyogenes*, *and S. agalactiae*. Other observations were that the growth inhibition was high even when the concentration of biosurfactant was well below the minimum inhibitory concentration (MIC) and the minimum bactericidal concentration (MBC) levels. Some 12 of 18 organisms were inhibited by the biosurfactant. Antiadhesive properties in this study also showed some selectivity toward microbial targets. The percentage of biofilm inhibition varied over the organisms tested with *S. aureus* and *S. epidermidis* having the highest percentage of inhibition at 76.8 and 72.9%, respectively. This study was one of the first isolates from *Lactobacilli* demonstrating a broad range of targets including filamentous fungi with both antimicrobial and antibiofilm properties. The biofilm inhibition against the *Staphylococcal* species demonstrate that this biosurfactant has a great potential as a prophylactic.

Biosurfactants with both the ability to inhibit growth and biofilm formation might be a potential disinfectant. Lipopeptides have shown a promising solution for decontaminating, sterilizing surfaces, and removal and prevention of biofilm adhesion. A lipopeptide biosurfactant was successfully used as a disinfectant [45], illustrating the versatility of biosurfactants in the biomedical field. A lipopeptide was extracted from the *B. tequilensis* strain SDS21, an endo-rhizospheric bacteria isolated from the root system of *P. hysterophorus*. In this study, several derivatives of the lipopeptide were identified using liquid chromatography–mass spectrometry. The lipopeptide was reported to reduce the surface tension of water from 72 to 30 mN/m and a CMC of 40 mg/l. The study noted that the longer the fatty acid chains were on the lipid moieties, the better they were able to interact with the bacterial membrane, causing more cell damage.

The biosurfactant showed a remarkable ability to retain bactericidal properties, even after exposure to extreme pH and temperature. This is of particular interest since many disinfectants lose antimicrobial properties when out of the working pH and temperature range they have.

Karlapudi et al. isolated a biosurfactant from *A. indicus* M6 strain showed both antimicrobial and antibiofilm properties against MRSA [46]. The biosurfactant was able to reduce water surface tension from 72.0 to 39.8 mN/m with a chemical structure determined to be a glycolipoprotein. Crystal violet staining resulted in MRSA biofilm disruption with a reported value of up to 82.5% [46]. The effectiveness of the biosurfactant against biofilm formation was evident after seven days of incubation with 500 μg/ml of the biosurfactant. This study illustrates the importance of properties such as surface tension reduction and CMC in the antiadhesive mechanisms of biosurfactants.

A biosurfactant with similar properties to lichenysin was presented by Saravanakumari and Mani [47]. A xylolipid biosurfactant extracted from *L. lactis* demonstrated antimicrobial properties against *E. coli* and methicillin-resistant *S. aureus*. Using GC–MS and NMR, the glycolipid was characterized as consisting of two parts, a methyl-2-*O*-methyl-β-xylopyranoside and an octadecanoic acid. Methods for testing antimicrobial properties were done using the Kirby-Bauer method. The zone of inhibition for MRSA was 12.6 mm, in comparison to the biosurfactant isolated from *B. licheniformis*, which was 13.1. This demonstrates similar results to a well-established biosurfactant that has shown antimicrobial properties against *S. aureus* and MRSA.

The biosurfactant was fed orally to mice and the skin was injected with the biosurfactant, but no visible inflammation, allergies, or illness arose from the administration. *L. lactis* is a probiotic and an excellent source of biosurfactants that have antimicrobial properties against MRSA without the cytotoxic effects. Since the biosurfactant was also effective against *E. coli*, the biosurfactant can be used as a broad-spectrum antimicrobial compound.

14.7 Media, Microbial Source, and Culture Conditions for Antibiofilm and Antimicrobial Properties

In addition to media and culture conditions, other additives to the biosurfactants have been proven to increase effectiveness against pathogenic biofilms. Diaz de Rienzo et al. [48] discovered the synergistic effect of caprylic acid combined with a rhamnolipid biofilm derived from *P. aeruginosa* which proved to be effective against the biofilm of *S. aureus* ATCC 9144. This study illustrated the effects of rhamnolipids in mono and di forms as well as sophorolipids in the presence or absence of caprylic acid to evaluate bactericidal effects in both. The exact mechanism for this effect was suggested as the ability to intercalate into the cell membrane of *S. aureus*. The biosurfactants were tested on several bacterial species including *P. aeruginosa* PAO1, *E. coli* NCTC 10418, *B. subtilis*, and *S. aureus*. Different biosurfactants were used to evaluate which was more effective against Gram-positive or Gram-negative bacteria. The rhamnolipid was composed of a mixture of both monorhamnolipids and dirhamnolipids with a critical micelle concentration at 20 mg/l and surface tension at 28 mN/m. The sophorolipid was an equal mixture of the hydrophilic sophorose and a hydrophilic fatty acid with a surface tension at 38 mN/m and a critical micelle concentration at 40 mg/l. The caprylic acid was used with either the rhamnolipid or the sophorolipid or by itself.

Organic acids such as caprylic acid, as mentioned above, are particularly successful against *S. aureus* and *P. aeuriginosa* biofilms, as demonstrated in another study by Diaz de Rienzo et al. [49] under static and flow conditions. The combination of a rhamnolipid (0.04%) and caprylic acid (0.01%) was able to remove preformed *S. aureus* biofilm. It is important to note that in several studies the effect of antibiofilm activity works best when surfaces are pre-treated with a biosurfactant. However, in this study, the combination of the rhamnolipid and caprylic acid were effective in disaggregating the preformed biofilm.

Di-rhamnolipids derived from *P. aeruginosa* have the potential for wider applications as a protectant for metals against biofouling. Chebbi et al. [50] demonstrated the remarkable ability of a biosurfactant isolated from *P. aeruginosa* W10 strain against preformed biofilms. Antimicrobial effects were determined using a Kirby bauer to see differences in inhibition zones of biosurfactants in comparison to sodium dodecyl sulfide. The MIC for the rhamnolipid was lower than the sophorolipid, indicating that for *S. aureus*, a rhamnolipid biosurfactant might be more effective. The rhamnolipid was not as efficient against the Gram-negative bacteria. The study concluded that a synergistic effect was observed when caprylic acid was added to the media. The addition of caprylic acid to rhamnolipid at a ratio of 0.8/1% v/v inhibited growth of both *B. subtilis* and *S. aureus*. The rhamnolipid did not inhibit growth of Gram-negative bacteria. However, the combination of caprylic acid with sophorolipids at the same ratio inhibited growth in both Gram-negative and Gram-positive bacteria, including *S. aureus*. The proposed mechanism was that, since caprylic acid is a short chain fatty acid eight carbons long, it is able to intercalate fairly easily into the cell membrane.

To investigate antiadhesive properties and biofilm disruption, coverslips were incubated in varying concentrations of the biosurfactants with or without caprylic acid in bacterial culture and effects were determined using phase contrast microscopy. Biofilm disruption was determined through crystal violet staining and values were determined in percent biofilm.

Sophorolipids disrupted biofilm and proved to be an effective treatment for removing biofilm as well as prevention due to antiadhesive properties. However, the best antiadhesive effect was observed when caprylic acid was added to the rhamnolipid and sophorolipid. The study demonstrates that the use of additives such as caprylic acid are worth investigating due to their synergistic potential. The importance of understanding which biosurfactant is more effective against different organisms is crucial in treating different bacterial infections. This also includes whether the objective is biofilm prevention, biofilm disruption, or growth inhibition.

An important consideration is the microbial source for biosurfactant production. Microbes isolated from waste treatment plants, marine bacteria, and oil fields have a remarkable range of applications [51–54]. A rhamnolipid isolated from *B. amyloliquefaciens* had a significant antimicrobial effect against a broad spectrum of microbes, including on *S. aureus* and *C. albicans* [19]. The crude surfactant and rhamnolipid were thought to be effective because they came from bacteria in contaminated sources like wastewater [19]. The biosurfactant exhibited a bactericidal ability, thus illustrating the potential as an antiseptic. The authors stressed the importance of having an optimum carbon to nitrogen ratio when culturing the bacteria in MSM. Optimization of growth conditions resulted in a maximum yield of bioactive molecules. LC–MS analysis revealed similar chemical structures to surfactant-like molecules. The biosurfactant proved to be comparable to other studies as it reported similar inhibition zones for *S. aureus*. Surfactants are considered to be some of the most effective biosurfactants.

Another factor in the complexity of biosurfactant antimicrobial effectiveness is the source of microbe used as well as where in the cell the biosurfactant originates from. A 2015 study by Gudiña et al. [55] extracted a glycoprotein from the lactic acid bacteria, *Lactobacillus agillis*. Interestingly, the biosurfactant extracted was considered to be cell bound. This is the first report of an extraction from *L. agilis*. The study also reports a novel culturing process using whey from cheese. A reported 11 times increase in biosurfactant was produced by using conventional media used for biosurfactant production in lactic acid bacteria. Characterization of functional groups of the glycoprotein were determined using Fourier transform infrared spectroscopy (FTIR). The biosurfactant from different species of lactic acid bacteria were screened for the ability to lower surface tension, critical micelle concentration, and emulsification properties.

Culturing conditions prove to be crucial as Coronel-Leon et al. [56] demonstrates. As mentioned above, the study aimed to improve upon the extraction methods of biosurfactant produced by *B. licheniformis* isolated from Antarctica. It is important to note that the properties of the biosurfactants and the amount extracted were significantly influenced by temperature, carbon sources, and other culturing conditions. Moreover, microbes isolated from harsh environments like Antarctica may be able to produce biosurfactants with unique chemical properties applicable to biomedical research. It is possible that microbes isolated from Antarctica could have a higher potential to kill off pathogenic bacterium due to extreme conditions it thrives in. This was further illustrated in a study by Vollú et al. [57]. More than 80 cold-adapted spore-forming bacterial strains were taken from soil samples in the Antarctic. Spore-forming bacteria from harsh environments are able to make enzymes that have pharmaceutical and food industry preparation uses. Interestingly, of the 80 bacterial strains tested, 13.7% inhibited growth of MRSA. The antiadhesive property of the glycoprotein was effective but exclusive only to *S. aureus*. The biosurfactant was ineffective against the other microorganisms studied. This was true even at the lowest concentration of biosurfactant at 1 mg/ml. Antimicrobial properties were observed against *S. aureus*, *S. agalactiae*, and *P. aeruginosa*. The study reports the highest antiadhesive result against *S. aureus* than any other cell-bound biosurfactant extracted from a lactic acid bacterium.

The use of symbiotic lactic acid bacteria in the 2017 study by Merghni et al. [58] showed promising success against *S. aureus*. The antioxidant and antiproliferative properties of an *L. casei* extracted biosurfactant and incubating with *S. aureus* infected human epithelial cells. Two biosurfactants, BS-B1 and BS-Z9, from two *L. casei* strains were used. The study also investigated potential antimicrobial, antiadhesive, and antibiofilm properties against pathogenic *S. aureus* oral strains. It is thought that since lactic acid bacteria are symbiotic with humans, they may potentially be more able to inhibit growth of pathogenic bacteria.

Antioxidant properties are important in organic biosurfactants as they are less cytotoxic than synthetic surfactants. The antioxidant properties were investigated using a 2,2-diphenyl-1-picryl-hydrazy (DPPH) assay. Both biosurfactants demonstrated DPPH scavenging abilities at a concentration of 5.0 mg/ml as compared with ascorbic acid with BS-B1 and BS-Z9 at 74.6 and 77.3%, respectively. The antiproliferative assay demonstrated cytotoxicity with increasing concentration. At concentrations of 25, 50, and 100 mg/ml, BS-Z9 was significantly less cytotoxic than BS-L. However, at 200 mg/ml both were almost equal. Antimicrobial assays showed that BS-Z9 was more effective against the reference strain of *S. aureus* while the B-L1 biosurfactant was more potent against *S. aureus* oral strains 9P and 29P.

Antiadhesive assays were conducted to observe the ability of biosurfactants to inhibit biofilm formation. At a biosurfactant concentration of 12.5 mg/ml, there was 80–84% inhibition against the reference strain, *S. aureus* 6538. Biofilm inhibition higher than 50% was achieved by both biosurfactants against *S. aureus* reference strains 6538 and 9P. Significant differences could be seen between the two biosurfactants in the biofilm dispersal assay. Higher percentages between 80.22 and 86.21% of biofilm dispersal were evident in BS-B1 as compared to 53.38–64.42% in BS-Z9.

The antioxidant properties of both biosurfactants demonstrate the potential to be used as a protectant against oxidative stress [58]. Both biosurfactants were compared to the scavenging effect of ascorbic acid. The IC50 of both biosurfactants was below ascorbic acid. This article is of particular interest since there are few studies about the lactic acid bacteria derived biosurfactant against oral *S. aureus* strains. Moreover, the biosurfactants studied here may be used in *in vivo* therapies involving pathogenic oral strains of *S. aureus*.

Other strains of *Bacillus subtilis* have demonstrated an effective antiadhesive and antimicrobial properties against medical and food-borne pathogens. The *Bacillus* genus has proven to be beneficial to humans and are capable of producing a wide array of antibiotics. A study by Moore [59] sought to isolate several *Bacillus* species and test the biosurfactant produced by the various species against some common bacterial pathogens. The study isolated three strains of *B. subtilis* demonstrating the most effective antimicrobial abilities. These were BSB, 16 K, and 105. The study also tested the best conditions for biosurfactant production. All strains showed the highest production in starch media at 30 °C. This study also highlights the importance of how bacteria are cultured in terms of effectiveness against bacterial pathogens. The strains studied showed a high antimicrobial activity against both *Salmonella* and *Staphylococcus* cultures. The biosurfactant was effective against a broad range of species for both genera, including antibiotic-resistant clinical strains. The growth inhibition was illustrated through the high zone of growth inhibition of pre-treated surface areas with biosurfactant.

14.8 Novel Synergistic Antimicrobial and Antibiofilm Strategies Against MRSA and *S. aureus*

The application of novel strategies of biosurfactants against pathogenic bacterial biofilms seems limitless. Some already previously mentioned studies utilization of organic acids included caprylic acid for increased antimicrobial and antibiofilm activity [43, 56]. Another strategy used in a recent study utilized a combination of liposomes and a biosurfactant isolated from *L. gasseri* BC9 [60]. It is notable that in this particular experimental design, using biosurfactant-loaded liposomes (BS-LP) increased the antibiofilm activity. The liposomes provide a better delivery of the biosurfactant to the abiotic surface. This study is one of a few that target *S. aureus* specifically. The BS-LP was able to significantly disperse pre-formed biofilm of five clinically isolated MRSA strains of which three were considered multidrug resistant. The BS-LP significantly inhibited the formation of biofilm. The disruption of pre-formed biofilm and the inhibition of biofilm growth was observed at concentrations of 1.25, 2.5, and 5 mg/ml. Inhibition and disruption were observed at all three concentrations. However, efficacy varied among the different strains of *S. aureus*. The author attributed this to underlying genetic differences among the strains.

Another important factor is that cytotoxicity assays were conducted with human and murine fibroblast cells. No cytotoxicity was observed in concentrations up to 10 mg/ml. This is of particular importance in terms of dosage and administration. If very little cytotoxicity is present in a biosurfactant then it may be safe to apply it directly to the skin. The significant rise in community acquired MRSA soft tissue infections has necessitated the development of a safe method to treat various bacterial skin infections applied topically.

Synergistic applications of biosurfactants can be used to enhance antibiotics and ointments against *S. aureus* and MRSA. A 2013 study by Joshi-Navarre and Prabhune [61] demonstrates that sophorolipids can enhance the effects of the antibiotics cefaclor and tetracycline. This study is the first to investigate the effects of sophorolipids in combination with other antimicrobial compounds. Both *S. aureus* and *E. coli* were tested in this study. The antibiotics have different mechanisms of action. Cefaclor works in a similar way to beta lactam, targeting the penicillin binding proteins, which are known for crosslinking of the cell wall and forming peptidoglycan. Tetracycline inhibits the formation of peptide chains by inhibiting the attachment of the aminoacyl-tRNA to the ribosomal attachment site A [62]. The sophorolipid was characterized using high-performance liquid chromatography (HPLC) and matrix-assisted laser desorption/ionization mass spectrometry (MALDI-MS). A time-dependent assay was conducted to test the synergistic effects. Four reaction conditions were proposed to determine this. They are: the sophorolipid alone, the sophorolipid + antibiotic, no treatment control, and the antibiotic alone. Complete growth inhibition was observed when the antibiotics and the sophorolipid were left for longer periods of time. The time points used were two, four, and six hours. At six hours, complete growth inhibition was observed where the sophorolipid was used independently or with tetracycline. However, significant differences could be seen at four hours, indicating a more effective inhibition. This data is important as it demonstrates a time-dependent antimicrobial effect. The longer the sophorolipid was incubated, the more growth inhibition was observed.

The proposed mechanism is that the sophorolipid's amphipathic qualities allow antibiotics to gain access to the bacterial cell through the membrane. Scanning electron images show accumulation of damage to the cell membrane of both *S. aureus* and *E. coli*. The images show damage and structural compromisation, suggesting possible leakage. The use of the sophorolipid provides a delivery mechanism of the antibiotic to the inside of the cell, which would normally be impenetrable due to the amphipathic nature of biological membranes. The proposed mechanism states that the sophorolipid carrying the antibiotic is able to push into the cell membrane, causing it to open upon contact with the liposome and allow it to integrate into the membrane (Figure 14.1).

Another synergistic example specifically targeting MRSA was reported by Samedi et al. [63]. A rhamnolipid extracted from *P. aeruginosa* MN_1 was isolated from an oil-contaminated field and was able to inhibit all of the *S. aureus* strains regardless of susceptibility or resistance. The reported surface tension at CMC was 25 mN/m, well below the effective value as an antimicrobial. The CMC was reported at 15 mg/l and was able to enhance the effects of oxacillin by lowering the MIC value to a range of 3.12–6.25 μg/ml.

The possibilities of combining biosurfactants with other adjuvants have created a new approach to addressing antimicrobial resistance and food-borne illnesses, as illustrated by Chen et al. [64]. In this study, sophorolipids extracted from yeast could be used alone or with nisin, a polycyclic antimicrobial peptide, to inhibit growth. The authors suggested the mechanism to be an increase in membrane permeability of *S. aureus*. The 2014 study by Haba et al. [65] utilized the emulsification properties of eight rhamnolipids extracted from *P. aeruginosa*. The rhamnolipids were extracted using waste frying oil as the sole source or carbon. The rhamnolipids were combined with essential oils, tea tree, oregano, cinnamon and lavender to broaden the application of essential oils in addressing bacterial pathogens. Essential oils have natural antimicrobial properties, but the delivery of essential oils is limited by hydrophobic properties. Both the rhamnolipid and the essential oils elicited an inhibition zone

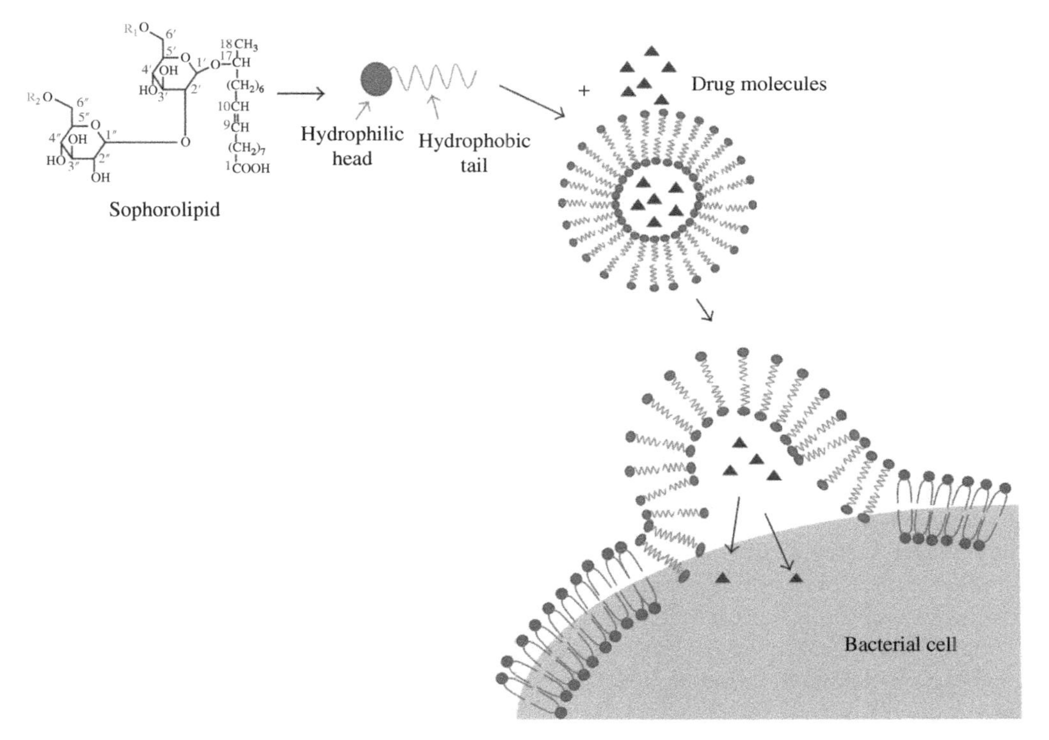

Figure 14.1 A schematic diagram from Joshi-Navarre and Prabhune [61] revealing the proposed mechanism of action. The amphipathic biosurfactant liposomes loaded with tetracycline allow for bypass of the cell membrane and access to the bacterial cytosol (See insert for color representation of the figure).

against *C. albicans* and MRSA. However, the emulsification of the two had a synergistic effect and increased the inhibition zone for both pathogens. This study illustrates the potential of biosurfactants to improve on already existing antimicrobial compounds.

14.9 Novel Potential Mechanisms of Antimicrobial and Antibiofilm Properties

The exact mechanism of how biosurfactants interfere with microbial growth and biofilm interference is not yet well established. Proposed mechanisms in several publications have been suggested thus far. Three possible mechanisms have been suggested that some polysaccharides function as signaling molecules to modulate biofilm gene expression [29, 42, 66]. Another suggestion is that the amphipathic properties of the biosurfactant alters the physical properties of bacterial cells and abiotic surfaces or the biosurfactant might block or inhibit glycoproteins or surface lectins [66].

Alterations of the properties of abiotic surfaces can happen in various methods due to the biosurfactant properties. Some marine bacteria are able to change the surface charge or alter the wettability of surfaces [66, 67]. A marine bacterium isolated from the host sponge *Dendrilla negra* produced a glycolipid under fermentative conditions in sea water [67]. The motile Gram-positive bacterium was identified as *Brevibacterium cassei* MSA 19. This study presents some novel approaches. These approaches include the method of isolation from a sponge at 12 m depth, the carbon sources and the process of submerged fermentation. These culturing conditions used to isolate the glycolipid biosurfactant may have contributed to how effective the biosurfactant was against pathogenic biofilms. The biosurfactant was also effective as an antimicrobial against a broad range of pathogens.

Another biofilm isolated from a marine source demonstrated remarkable abilities to reduce surface tension [68]. The biosurfactant differed in structure from biosurfactants isolated from *Psuedemonas aeuriginosa, Bacillus* sp., and *Psuedemonas* sp. The study noted that the use of biosurfactant extracted from marine bacteria may be more effective against biosurfactants synthesized from terrestrial sources due to molecular differences. The structure of the biosurfactant was determined to be a lipopeptide and considered to be more effective against a wider panel of pathogens than biosurfactants with rhamnolipids.

Marine bacteria have been a potential resource for the synthesis of novel enzymes and metabolites that have antimicrobial properties [51]. The isolation of a *Pseudomonas* strain obtained off the coast of India was tested for antimicrobial activity [51]. Fractionation of the secondary metabolite from this species showed inhibition effects against both Gram-positive and Gram-negative bacteria. It should be noted that the highest inhibitory effect was seen against MRSA with an inhibition zone of 22 mm. The proposed mechanism was speculated as having a few properties that could lead to cell death. The free radical activity was observed in the DPPH assay. The presence of free radicals damages the cell membrane leading to lyses.

As previously mentioned, Rivardo et al. demonstrated that pre-treatment of polystyrene surfaces gives further evidence that biosurfactants are able to change surface properties. Another example, a 2011 study by Rendueles et al. [69], demonstrates that changes in interfacial energy resulted in an increase in hydrophilicity of the surface. This study produced Ec300, which showed antimicrobial properties against Gram-positive bacteria. The polysaccharide Ec300 was an extract isolated from *E. coli* biofilm. The antiadhesive mechanism of Ec300 against *S. aureus* 15981 was determined to be due to an increase in hydrophilicity of the glass surfaces coated with Ec300. Similar to other studies, the antiadhesive properties were concentration-dependent. The ability of Ec300 to inhibit *S. aureus* biofilm formation also occurs when the two cultures are mixed at a 1 : 1 ratio. This shows that the initial secretion of Ec300 during inoculation interacted with and prevented initial *S. aureus* biofilm formation, reducing it to 5% of biofilm biomass. This was true even with co-inoculation with the impaired Ec300 producing mutant, Ec300Δ*galF-his*. This may imply that other molecular interactions are impairing pathogenic biofilm formation by commensal *E.coli*.

One model of a possible molecular mechanism of bacterial growth inhibition was a study in 2008 by Deleu et al. [70]. The study sought to shed light on some of these mechanisms using microscopy, scanning calorimetry, and ellipsometry to understand how fengycin interrupts a biological membrane. Fengycin is a lipopeptide biosurfactant produced by *B. subtilis*. Fengycin is composed of 10 amino acids of which eight are part of a cyclical structure [70]. These amino acids are linked to a β-hydroxy fatty acid. The underlying mechanism of fengycin is to make target membranes considerably more permeable, leaving them structurally vulnerable. Fengycin is a promising alternative to other biosurfactants as it has a considerably less hemolytic effect.

Deleu et al. used a dipalmitoylphosphatidylcholine (DPPC) and dioleylphosphatidylcholine (DOPC) model to simulate monolayers, bilayers, and vesicles. This model was utilized to potentially describe the first interaction between fengycin and the surface of a biological membrane. The addition of fengycin causes a rabid increase of surface pressure Π. The authors speculate that this is caused by a rapid insertion of fengycin into the membrane. The rapid increase of surface pressure is followed by a second more slow increase in surface pressure as adsorption of fengycin into the biological membrane occurs (see Figure 14.2).

The conclusion of the study was that there was a concentration-dependent two-state transition. The first state is non-perturbing and is not deeply anchored into the cell membrane while the other state is anchored and aggregated. This second state is responsible for the membrane leakage. This conclusion is in agreement across the board with other studies, indicating that antimicrobial

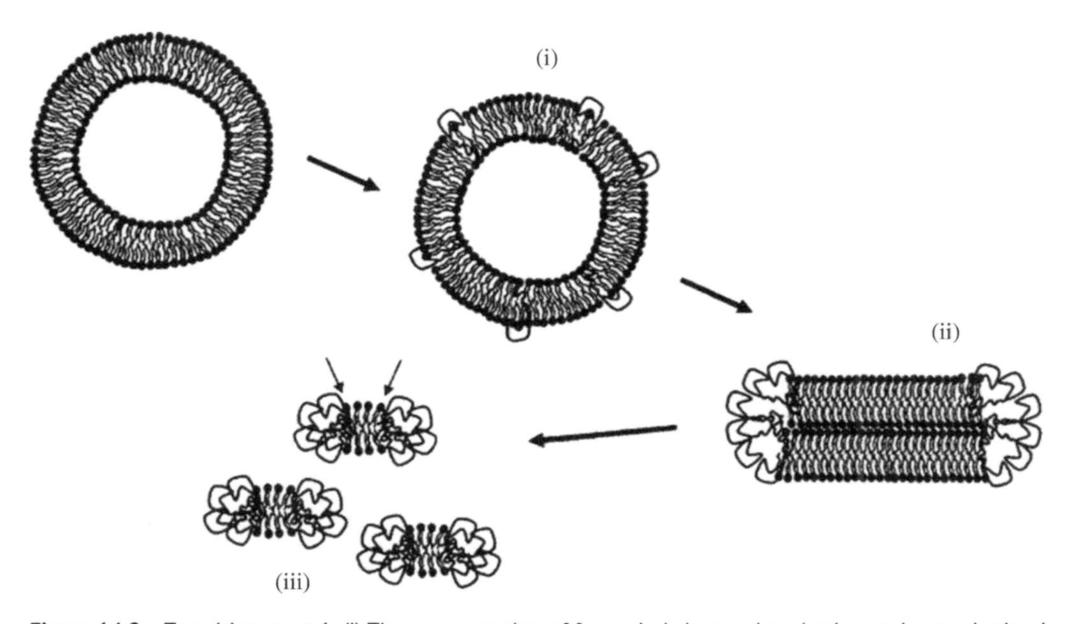

Figure 14.2 Transition state 1: (i) The concentration of fengycin is low and anchoring and perturbation is surface deep. Moderate concentration of fengycin shows membrane disruption by insertion. (ii). High levels of fengycin concentration. (iii) Total membrane disruption. *Source:* Taken from [70].

activity against *S. aureus* and MRSA are concentration-dependent. Generally, the higher the concentration of biosurfactant, the more effective it is against *S. aureus* and other pathogens.

Mechanisms of antimicrobial activity seem to be centered around compromising the membrane, as demonstrated by Liu et al. [71]. A novel biosurfactant was discovered and named bacaucin. The biosurfactant was isolated from *B. subtilis* strain CAU21. The chemical structure consisted of a cyclic heptapeptide with a fatty acid moiety. While the biosurfactant showed antimicrobial abilities against Gram-positive bacteria, it appeared ineffective against Gram-negative bacteria. Several derivatives of bacaucin were synthesized in the study. However, bacaucin-1 was the most effective against MRSA. The MIC for bacaucin-1 was at 4 μg/ml. Bacaucin-1 showed effective growth inhibition in both the stationary and exponential phase [71]. Most notably, the *S. aureus* strains did not develop a resistance against bacaucin-1. The mechanism of killing was ascertained through electron microscopy, showing a wrinkling morphology in *S. aureus* in the presence of bacaucin-1. As the concentration of bacacaucin-1 was increased the cells eventually collapsed and lysed. A reduced membrane potential was also observed, suggesting that ATP levels inside the cells were diminished. The Liu study also demonstrated that bacaucin-1 was an effective food preservative. The addition of bacaucin-1 reduced the number of *S. aureus* cells and the meat tested retained good cell structure and odor. There appeared to be no cytotoxicity or observable damage to human cells. Cytotoxicity is an important property to consider for biomedical applications. The less destructive to human cells while retaining antimicrobial properties the more ideal.

Growth inhibition was reported in both Gram-negative and Gram-positive bacteria. MIC concentrations ranged between 0.5 and 4 mg/ml. The MIC for *S. aureus* was the lowest at 0.5 mg/ml with an MBC at 1 mg/ml. These concentrations are relatively low in comparison to other biosurfactants. Inhibition of adhesion and disruption of biofilm formation on glass, steel, and polystyrene surfaces was reported to be 99%, as confirmed by confocal microscopy. Aggregation of biofilm structures is crucial to increasing bactericidal properties of the biosurfactant. The biosurfactant reported here was extremely effective in removing up to 90% of *S. aureus* biofilm and removing cells from the biofilm. Interestingly, this study was among the first to report that with 6 mg/ml of biosurfactant, 99% of the cells removed from the biofilm were inviable. The study suggested that the biosurfactant was effective against both planktonic and sessile bacteria. At lower concentrations (4 mg/ml) the cells remained viable, thus reinforcing the concentration-dependent theme that reoccurs across publications. Another interesting feature about this biosurfactant was the ability to retain properties even when challenged with extreme temperatures and pH (see Table 14.1).

14.10 Conclusion

MRSA is a highly adaptable opportunistic pathogen that causes a variety of infections from skin infections to endocarditis. Different biosurfactants presented here excel in different strategies against different *S. aureus* and MRSA infections. It is critically important to consider the application and infection being addressed. Some biosurfactants have a higher efficacy as a bactericidal against adhesion/biofilm formation and growth inhibition, while others can be applied directly to the skin. Other applications of biosurfactants against MRSA have been utilized as effective disinfectants or in the food production industry [45, 72].

The mechanism of how biosurfactants are effective against MRSA and other pathogens is not well known. Some clues to these mechanisms are given in the very properties of the biosurfactants, but to date there have been no molecular mechanisms specified and further study is needed. The

Table 14.1 Novel strategies and potential mechanisms for antimicrobial and antibiofilm properties of biosurfactants against *S. aureus* and MRSA.

Microbial source biosurfactant	Antibiofilm	Antimicrobial	Proposed mechanism	References
P. aeruginosa	Yes	Yes	Intercalation into the cell membrane	[48]
B. subtilis strain, CAU21 Baucacin-1	Yes	Yes	Cell membrane lysing/wrinkling	[71]
B. subtilis Fengycin	Not tested/ model of proposed mechanism	Not tested/model of proposed mechanism	Cell membrane destruction by insertion	[70]
L. acidophilus	Yes	Not tested	Down-regulation of biofilm genes/ possible signaling	[42]
B. subtilis and *B. licheniformis*	Yes	Yes	Binding of cell surfaces and modifies hydrophobic properties	[41]
E. coli EC300	Yes	Yes	Increases the hydrophilicity of surfaces (wettability)	[69]
P. aeruginosa	Not tested	Yes	Emulsifies essential oils with known antimicrobial properties for more efficient application	[65]
L. gasseri BC9	Yes	Not tested	Biosurfactant loaded- liposomes (BS-LP)	[60]
B. tequilensis	Yes	Yes	Hydrophobic portions of the lipopeptide imbed into the cell membrane while the hydrophilic end can compromise adhesion	[45]
P. aeruginosa W10	Yes	Yes	Synergistic use with caprylic acid which is a short chain fatty acid able to intercalate into the cell membrane	[50]
Pseudomonas genera	Not tested	Yes	A secondary metabolite had free radical activity which enabled antimicrobial properties	[51]
L. lactis	Not tested	Yes	Biosurfactant in combination with nisin causes the cell membrane to become more permeable	[64]

ability to reduce the surface tension using a small amount of biosurfactant is ideal for efficiency [48]. Generally speaking, a biosurfactant is considered efficient when it is able to lower the surface tension of water below 35 mg/ml [45, 73]. The CMC is also an important property for how effective the biosurfactant is. It is the amount of biosurfactant needed for micelles to form. Another important property is interaction with hydrophobic compounds and oil displacement. These properties tend to change abiotic surfaces, thus compromising adhesion of biofilms and interacting with bacterial cell membranes [48]. Other properties such as fatty acid length, the ability to resist extreme pH, and temperatures can determine bactericidal effects. Longer fatty acid chains have been implicated in damaging the bacterial cell membrane [45].

Several publications have suggested that the amphipathic qualities of the biosurfactant allow it to disrupt or insert into the cell membranes while others can change the properties of the abiotic surface [41, 50]. Another, commonly suggested strategy, is modification of the target cell membrane structure by reducing the membrane potential or making it more permeable [64]. Interestingly, a major commonality among biosurfactants is the concentration-dependent nature of the antibiofilm and antimicrobial properties. In general, the higher the concentration of biosurfactant the more effective it is against *S. aureus* and other pathogens. Interestingly, as in the case of fengycin, the biosurfactant does not become biologically active against membranes until there is a high concentration [43].

The consensus among publications is that the growth conditions, media, and supplements used to culture biosurfactant secreting bacterium are crucial in the formation of chemical structures and moieties that increase more desirable properties. Based on the substrates added, divalent metals and amino acids, the amphiphilic properties of the biosurfactant could be changed to manipulate the effectiveness of the antibiofilm and antimicrobial properties even with the same organism [67]. Temperature, pH, salinity, nitrogen levels, incubation time, and carbon sources are also components to consider which resulted in varying chemical structures of the biosurfactants produced [10, 68].

In addition to growth conditions, the microbe source is also important. Some sources that have been successful against MRSA have been lactic acid bacteria, yeast, marine bacteria, and some isolated from extreme environments, such as the Antarctic [55, 57, 58]. The proposed hypothesis is that bacteria from contaminated and extreme environments may produce biosurfactants with unique chemical structures that may increase the ability to interact with the bacterial membrane or change the properties of abiotic surfaces [18]. *Bacillus* strains have also illustrated both antiadhesive and antimicrobial properties against *S. aureus* [59]. Symbiotic strains such as lactic acid bacteria produce biosurfactants with less hemolytic properties while retaining antimicrobial ability. Cytotoxicity is also important in biomedical applications. The less damaging the biosurfactant is to human cells while retaining antimicrobial properties is crucial [71].

Biosurfactants have shown a very promising potential when used with adjuvants. Several studies have shown a synergistic effect when biosurfactants were used in addition to antibiotics. Increasing the delivery of tetracycline to the cell allows a bypass of the cell membrane [61]. Caprylic acid also increased antiadhesive properties, biofilm disruption, and growth inhibition when used with either a rhamnolipid or sophorolipid [56]. Biosurfactants have also demonstrated a novel means of delivering drugs to the target cells due to the hydrophobic portion of the biosurfactant. This allows for bypass of the cell membrane and delivery of the antibiotic to the cytosol and emulsification of hydrophobic medications for better application [47]. The possibilities of biomedical applications seem limitless when considering the variety of microbes, culturing conditions, and adjuvants that can be utilized to address the different infections caused by *S. aureus* and MRSA.

References

1 Abraham, J., Mansour, C., Veledar, E. et al. (2004). Staphylococcus aureus bacteremia and endocarditis: The Grady Memorial Hospital experience with methicillin-sensitive *S. aureus* and methicillin-resistant *S. aureus* bacteremia. *American Heart Journal* 147 (3): 536–539.

2 Klevens, R.M., Morrison, M.A., Nadle, J. et al. (2007). Invasive methicillin-resistant *Staphylococcus aureus* infections in the United States. *Journal of the American Medical Association* 298 (15): 1763–1771.

3 MacMorran, E., Harch, S., Athan, E. et al. (2017). The rise of methicillin resistant *Staphylococcus aureus*: Now the dominant cause of skin and soft tissue infection in Central Australia. *Epidemiology and Infection* 145 (13): 2817–2826.

4 Bush, K., and Bradford, P.A. (2016) b-Lactams and b-Lactamase inhibitors: An overview.

5 Hiramatsu, K. (1995). *Evolution of MRSA* 39 (8): 531–543.

6 Szűts, A. and Szabó-Révész, P. (2012). Sucrose esters as natural surfactants in drug delivery systems – A mini-review. *International Journal of Pharmaceutics* 433 (1–2): 1–9.

7 Zdziennicka, A., Krawczyk, J., Szymczyk, K., and Jańczuk, B. (2018). Macroscopic and microscopic properties of some surfactants and biosurfactants. *International Journal of Molecular Sciences* 19 (7).

8 Sarma, H., Bustamante, K.L.T., and Prasad, M.N.V. (2018). Biosurfactants for oil recovery from refinery sludge: Magnetic nanoparticles assisted purification. In: *Industrial and Municipal Sludge* (ed. M.N.V. Prasad). Elsevier, Massachusetts, pp. 107–132, ISBN: 9780128159071, Editors Majeti Narasimha Vara Prasad, Paulo Jorge de Campos, Favas Meththika, Vithanage S. Venkata Mohan.

9 Saikia, R.R., Deka, S., Deka, M., and Sarma, H. (2012). Optimization of environmental factors for improved production of rhamnolipid biosurfactant by *Pseudomonas aeruginosa* RS29 on glycerol. *Journal of Basic Microbiology* 52 (4): 446–457.

10 Santos, D.K.F., Rufino, R.D., Luna, J.M. et al. (2016). Biosurfactants: Multifunctional biomolecules of the 21st century. *International Journal of Molecular Sciences* 17 (3): 1–31.

11 Robert, M., Mercadé, M.E., Bosch, M.P. et al. (1989). Effect of the carbon source on biosurfactant production by *Pseudomonas aeruginosa* 44T1. *Biotechnology Letters* 11 (12): 871–874.

12 Ron, E.Z. and Rosenberg, E. (2001). Natural roles of biosurfactants. *Environmental Microbiology* 3 (4): 229–236.

13 Mulligan, C.N. (2005). Environmental applications for biosurfactants. *Environmental Pollution* 133: 183–198. https://doi.org/10.1016/J.ENVPOL.2004.06.009.

14 Lang, S. and Wagner, F. (1987). Structure and properties of biosurfactants. In: *Biosurfactants and Biotechnology*, Surfactant Science Series, vol. 25 (ed. N. Kosaric), 21–45. New York: Marcel Dekker.

15 Vijayakumar, S. and Saravanan, V. (2015). In vitro cytotoxicity and antimicrobial activity of biosurfactant produced by *Pseudomonas aeruginosa* strain PB3A. *Asian Journal of Scientific Research* 8 (4): 510–518.

16 Jarvis, F.G. and Johnson, M.J. (1949). A Glyco-lipide produced by Pseudomonas aeruginosa. *Journal of the American Chemical Society* 71 (12): 4124–4126.

17 Sekhon Randhawa, K.K. and Rahman, P.K.S.M. (2014). Rhamnolipid biosurfactants – Past, present, and future scenario of global market. *Frontiers in Microbiology* 5: 454.

18 Costa, S.G.V.A.O., Nitschke, M., Lépine, F. et al. (2010). Structure, properties and applications of rhamnolipids produced by *Pseudomonas aeruginosa* L2-1 from cassava wastewater. *Process Biochemistry* 45 (9): 1511–1516.

19 Ndlovu, T., Rautenbach, M., Vosloo, J.A. et al. (2017). Characterisation and antimicrobial activity of biosurfactant extracts produced by *Bacillus amyloliquefaciens* and *Pseudomonas aeruginosa* isolated from a wastewater treatment. *AMB Express* 7: 108. https://doi.org/10.1186/s13568-017-0363-8.

20 Van Bogaert, I.N.A., Saerens, K., De Muynck, C. et al. (2007). Microbial production and application of sophorolipids. *Applied Microbiology and Biotechnology* 76 (1): 23–34.

21 Konishi, M., Morita, T., Fukuoka, T. et al. (2017). Selective production of acid-form sophorolipids from glycerol by *Candida floricola*. *Journal of Oleo Science* 66 (12): 1365–1373.

22 Borsanyiova, M., Patil, A., Mukherji, R. et al. (2016). Biological activity of sophorolipids and their possible use as antiviral agents. *Folia Microbiologica* 61 (1): 85–89.

23 Christova, N., Lang, S., Wray, V. et al. (2015). Production, structural elucidation, and in vitro antitumor activity of trehalose lipid biosurfactant from *Nocardia farcinica* strain. *Journal of Microbiology and Biotechnology* 25 (4): 439–447.

24 Zaragoza, A., Aranda, F.J., Espuny, M.J. et al. (2009). Mechanism of membrane permeabilization by a bacterial trehalose lipid biosurfactant produced by *Rhodococcus* sp. *Langmuir* 25: 7892–7898. https://doi.org/10.1021/la900480q.

25 Biniarz, P., Łukaszewicz, M., and Janek, T. (2017). Screening concepts, characterization and structural analysis of microbial-derived bioactive lipopeptides: A review. *Critical Reviews in Biotechnology* 37 (3): 393–410.

26 Mnif, I. and Ghribi, D. (2015). Review lipopeptides biosurfactants: Mean classes and new insights for industrial, biomedical, and environmental applications. *Peptide Science* 104 (3): 129–147.

27 Straus, S.K. and Hancock, R.E.W. (2006). Mode of action of the new antibiotic for Gram-positive pathogens daptomycin: Comparison with cationic antimicrobial peptides and lipopeptides. *Biochimica et Biophysica Acta – Biomembranes* 1758 (9): 1215–1223.

28 Hida, T., Hayashi, K., Yukishige, K. et al. (1995). Synthesis and biological activities of TAN-1511 analogues. *The Journal of Antibiotics* 48 (7): 589–603.

29 Kim, P.I., Ryu, J., Kim, Y.H., and Chi, Y.T. (2010). Production of biosurfactant lipopeptides iturin A, fengycin, and surfactin A from *Bacillus subtilis* CMB32 for control of *Colletotrichum gloeosporioides*. *Journal of Microbiology and Biotechnology* 20 (1): 138–145.

30 Grangemard, I., Wallach, J., Maget-Dana, R., and Peypoux, F. (2001). Lichenysin: A more efficient cation chelator than surfactin. *Applied Biochemistry Biotechnology – Part A Enzyme Engineering Biotechnology* 90 (3): 199–210.

31 Becher, P. (1965). Nonionic surface-active compounds. X. Effect of solvent on micellar properties. *Journal of Colloid Science* 20 (7): 728–731.

32 Walter, V., Syldatk, C., and Hausmann, R. (2010). Screening concepts for the isolation of biosurfactant producing microorganisms. *Advances in Experimental Medicine and Biology* 672: 1–13. https://doi.org/10.1007/978-1-4419-5979-9_1.

33 Satpute, S.K., Kulkarni, G.R., Banpurkar, A.G. et al. (2016). Biosurfactant/s from *Lactobacilli* species: Properties, challenges and potential biomedical applications. *Journal of Basic Microbiology* 56 (11): 1140–1158.

34 Jahanbani Veshareh, M., Ganji Azad, E., Deihimi, T. et al. (2019). Isolation and screening of Bacillus subtilis MJ01 for MEOR application: biosurfactant characterization, production optimization and wetting effect on carbonate surfaces. *Journal of Petroleum Exploration and Production Technology* 9 (1): 233–245.

35 Saruni, N.H., Razak, S.A., Habib, S. et al. (2019). Comparative screening methods for the detection of biosurfactant-producing capability of Antarctic hydrocarbon-degrading. *Journal of Environmental Microbiology* 7 (1): 44–47.

36 Cooper, D.G. (1986). Biosurfactants. *Microbiological Sciences* 3 (5): 145–149.

37 Falatko, D.M. and Novak, J.T. (1992). Effects of biologically produced surfactants on the mobility and biodegradation of petroleum hydrocarbons. *Water Environment Research* 64 (2): 163–169.

38 Banat, I.M. (1995). Biosurfactants production and possible uses in microbial enhanced oil recovery and oil pollution remediation: A review. *Bioresource Technology* 51 (1): 1–12.

39 Makkar, R.S. and Cameotra, S.S. (2001). Synthesis of enhanced biosurfactant by *Bacillus subtilis* MTCC 2423 at 45 °C by foam fractionation. *Journal of Surfactants and Detergents* 4 (4): 355–357.

40 Sambanthamoorthy, K., Feng, X., Patel, R. et al. (2014). Antimicrobial and antibiofilm potential of biosurfactants isolated from *Lactobacilli* against multi-drug-resistant pathogens. *BMC Microbiology* 14 (1): 1–9.

41 Rivardo, F., Turner, R.J., Allegrone, G. et al. (2009). Anti-adhesion activity of two biosurfactants produced by *Bacillus* spp. prevents biofilm formation of human bacterial pathogens. *Applied Microbiology and Biotechnology* 83 (3): 541–553.

42 Tahmourespour, A., Salehi, R., and Kermanshahi, R.K. (2011). *Lactobacillus acidophilus*-derived biosurfactant effect on gtfB and gtfC expression level in *Streptococcus mutans* biofilm cells. *Brazilian Journal of Microbiology* 42 (1): 330–339.

43 Coronel-León, J., Marqués, A.M., Bastida, J., and Manresa, A. (2016). Optimizing the production of the biosurfactant lichenysin and its application in biofilm control. *Journal of Applied Microbiology* 120 (1): 99–111.

44 Gudiña, E.J., Rocha, V., Teixeira, J.A., and Rodrigues, L.R. (2010). Antimicrobial and antiadhesive properties of a biosurfactant isolated from *Lactobacillus paracasei* ssp. paracasei A20. *Letters in Applied Microbiology* 50 (4): 419–424.

45 Singh, A.K. and Sharma, P. (2020). Disinfectant-like activity of lipopeptide biosurfactant produced by *Bacillus tequilensis* strain SDS21. *Colloids and Surfaces B: Biointerfaces* 185 (October 2019): 110514.

46 Karlapudi, A.P., Venkateswarulu, T.C., Srirama, K. et al. (2018). Evaluation of anti-cancer, anti-microbial and anti-biofilm potential of biosurfactant extracted from an *Acinetobacter* M6 strain. *Journal of King Saud University – Science* [Internet]: 0–4. https://doi.org/10.1016/j.jksus.2018.04.007.

47 Saravanakumari, P. and Mani, K. (2010). Structural characterization of a novel xylolipid biosurfactant from *Lactococcus lactis* and analysis of antibacterial activity against multi-drug resistant pathogens. *Bioresource Technology [Internet]* 101 (22): 8851–8854. https://doi.org/10.1016/j.biortech.2010.06.104.

48 Díaz de Rienzo, M.A., Stevenson, P., Marchant, R., and Banat, I.M. (2016). Antibacterial properties of biosurfactants against selected Gram-positive and -negative bacteria. *FEMS Microbiology Letters* 363 (2).

49 Diaz de Rienzo, M.A., Stevenson, P.S., Marchant, R., and Banat, I.M. (2016). Effect of biosurfactants on *Pseudomonas aeruginosa* and *Staphylococcus aureus* biofilms in a BioFlux channel. *Applied Microbiology and Biotechnology* 100 (13): 5773–5779.

50 Chebbi, A., Elshikh, M., Haque, F. et al. (2017). Rhamnolipids from *Pseudomonas aeruginosa* strain W10; as antibiofilm/antibiofouling products for metal protection. *Journal of Basic Microbiology* 57 (5): 364–375.

51 Charyulu, E.M., Sekaran, G., Rajakumar, G.S., and Gnanamani, A. (2009). Antimicrobial activity of secondary metabolite from marine isolate, *Pseudomonas* sp. against Gram positive and negative bacteria including MRSA. *Indian Journal of Experimental Biology* 47 (12): 964–968.

52 Dobler, L., Ferraz, H.C., Araujo de Castilho, L.V. et al. (2020). Environmentally friendly rhamnolipid production for petroleum remediation. *Chemosphere* 252: 1–10.

53 Ndlovu, T., Khan, S., and Khan, W. (2016). Distribution and diversity of biosurfactant-producing bacteria in a wastewater treatment plant. *Environmental Science and Pollution Research* 23 (10): 9993–10004.

54 Geetha, S.J., Banat, I.M., and Joshi, S.J. (2018). Biosurfactants: production and potential applications in microbial enhanced oil recovery (MEOR). *Biocatalysis and Agricultural Biotechnology* 14: 23–32.

55 Gudiña, E.J., Fernandes, E.C., Teixeira, J.A., and Rodrigues, L.R. (2015). Antimicrobial and anti-adhesive activities of cell-bound biosurfactant from *Lactobacillus agilis* CCUG31450. *RSC Advances* 5 (110): 90960–90968.

56 Coronel-León, J., de Grau, G., Grau-Campistany, A. et al. (2015). Biosurfactant production by AL 1.1, a *Bacillus licheniformis* strain isolated from Antarctica: Production, chemical characterization and properties. *Annals of Microbiology* 65: 2065–2078. https://doi.org/10.1007/s13213-015-1045-x.

57 Vollú, R.E., Jurelevicius, D., Ramos, L.R. et al. (2014). Aerobic endospore-forming bacteria isolated from Antarctic soils as producers of bioactive compounds of industrial interest. *Polar Biology* 37 (8): 1121–1131.

58 Merghni, A., Dallel, I., Noumi, E. et al. (2017). Antioxidant and antiproliferative potential of biosurfactants isolated from *Lactobacillus casei* and their anti-biofilm effect in oral *Staphylococcus aureus* strains. *Microbial Pathogenesis* [Internet] 104: 84–89. https://doi.org/10.1016/j.micpath.2017.01.017.

59 Moore, T. (2013). Antagonistic activity of *Bacillus* bacteria against food-borne pathogens. *Journal of Probiotics Health* 01 (03): 1–6.

60 Giordani, B., Costantini, P.E., Fedi, S. et al. (2019). Liposomes containing biosurfactants isolated from *Lactobacillus gasseri* exert antibiofilm activity against methicillin resistant *Staphylococcus aureus* strains. *European Journal of Pharmaceutics and Biopharmaceutics* 139: 246–252.

61 Joshi-Navare, K. and Prabhune, A. (2013). A biosurfactant-sophorolipid acts in synergy with antibiotics to enhance their efficiency. *BioMed Research International* 2013: 512495. https://doi.org/10.1155/2013/512495.

62 Stepanek, J.J., Lukežič, T., Teichert, I. et al. (2016). Dual mechanism of action of the atypical tetracycline chelocardin. *Biochimica et Biophysica Acta (BBA) – Proteins and Proteomics* 1864 (6): 645–654.

63 Samadi, N., Abadian, N., Ahmadkhaniha, R. et al. (2012). Structural characterization and surface activities of biogenic rhamnolipid surfactants from *Pseudomonas aeruginosa* isolate MN1 and synergistic effects against methicillin-resistant *Staphylococcus aureus*. *Folia Microbiologica* 57 (6): 501–508.

64 Chen, J., Lü, Z., An, Z. et al. (2020). Antibacterial activities of sophorolipids and nisin and their combination against foodborne pathogen *Staphylococcus aureus*. *European Journal of Lipid Science and Technology* 122 (3): 1900333.

65 Haba, E., Bouhdid, S., Torrego-Solana, N. et al. (2014). Rhamnolipids as emulsifying agents for essential oil formulations: Antimicrobial effect against Candida albicans and methicillin-resistant *Staphylococcus aureus*. *International of Journal of Pharmaceutics [Internet]* 476 (1): 134–141. https://doi.org/10.1016/j.ijpharm.2014.09.039.

66 Rendueles, O., Kaplan, J.B., and Ghigo, J.M. (2013). Antibiofilm polysaccharides. *Environmental Microbiology* 15 (2): 334–346.

67 Kiran, G.S., Sabarathnam, B., and Selvin, J. (2010). Biofilm disruption potential of a glycolipid biosurfactant from marine *Brevibacterium casei*. *FEMS Immunology and Medical Microbiology* 59 (3): 432–438.

68 Das, P., Mukherjee, S., and Sen, R. (2008). Antimicrobial potential of a lipopeptide biosurfactant derived from a marine *Bacillus circulans*. *Journal of Applied Microbiology* 104 (6): 1675–1684.

69 Rendueles, O., Travier, L., and Latour-lambert, P. (2011). Screening of *Escherichia coli* species biodiversity reveals new biofilm. *MBio* 2 (3): 1–12.

70 Deleu, M., Paquot, M., and Nylander, T. (2008). Effect of fengycin, a lipopeptide produced by *Bacillus subtilis*, on model biomembranes. *Biophysical Journal [Internet]* 94 (7): 2667–2679. https://doi.org/10.1529/biophysj.107.114090.

71 Liu, Y., Ding, S., Dietrich, R. et al. (2017). A biosurfactant-inspired Heptapeptide with improved specificity to kill MRSA. *Angewandte Chemie, International Edition* 56 (6): 1486–1490.

72 Meylheuc, T., Renault, M., and Bellon-Fontaine, M.N. (2006). Adsorption of a biosurfactant on surfaces to enhance the disinfection of surfaces contaminated with *Listeria monocytogenes*. *International Journal of Food Microbiology* 109 (1–2): 71–78.

73 Singh, A.K. and Cameotra, S.S. (2013). Efficiency of lipopeptide biosurfactants in removal of petroleum hydrocarbons and heavy metals from contaminated soil. *Environmental Science and Pollution Research* 20 (10): 7367–7376.

74 Cameotra, S.S. and Makkar, R.S. (2004). Recent applications of biosurfactants as biological and immunological molecules. *Current Opinion in Microbiology* 7 (3): 262–266.

75 Drahansky M, Paridah M., Moradbak A, Mohamed A., Owolabi F Abdulwahab Taiwo, Asniza M, et al. 2016. Available from: https://www.intechopen.com/books/advanced-biometric-technologies/liveness-detection-in-biometrics.

76 Lin, S.C. (1996). Biosurfactants: Recent advances. *Journal of Chemical Technology and Biotechnology* 66 (2): 109–120.

77 Mandal, S.M., Barbosa, A.E.A.D., and Franco, O.L. (2013). Lipopeptides in microbial infection control: Scope and reality for industry. *Biotechnology Advances* 31 (2): 338–345.

78 Murray, C.K., Holmes, R.L., Ellis, M.W. et al. (2009). Twenty-five year epidemiology of invasive methicillin-resistant *Staphylococcus aureus* (MRSA) isolates recovered at a burn center. *Burns* 35 (8): 1112–1117.

79 Randhawa, K.K.S. and Rahman, P.K.S.M. (2014). Rhamnolipid biosurfactants-past, present, and future scenario of global market. *Frontiers in Microbiology* 5 (SEP): 1–7.

80 Jarraud, S., Mougel, C., Thioulouse, J. et al. (2002). Relationships between *Staphylococcus aureus* genetic background, virulence factors, agr groups (alleles), and human disease. *Infection and Immunity* 70 (2): 631–641. https://doi.org/10.1128/iai.70.2.631-641.2002.

81 Voyich, J.M., Otto, M., Mathema, B. et al. (2007). Is Panton-valentine Leukocidin the major virulence determinant in community-associated methicillin-resistant *Staphylococcus aureus* disease? *The Journal of Infectious Diseases* 194: 1761–1770. https://doi.org/10.1086/509506.

15

Exploiting the Significance of Biosurfactant for the Treatment of Multidrug-Resistant Pathogenic Infections

Sonam Gupta[1] *and Vikas Pruthi*[2]

[1] *Department of Biotechnology, National Institute of Technology Raipur, Raipur 492001, Chhattisgarh, India*
[2] *Department of Biotechnology, Indian Institute of Technology Roorkee, Roorkee 247667, Uttarakhand, India*

CHAPTER MENU

15.1 Introduction

Biosurfactants are multifunctional amphiphatic biomolecules that are synthesized by a variety of microorganisms, particularly residing in oil-contaminated areas such as oil sludge soil or water [1–5]. All groups of biosurfactants show amphiphilic behavior due to the presence of amphiphilic moieties, which is the key element of their physicochemical properties viz. surface-active and emulsification properties [6–9]. Due to these physicochemical properties biosurfactants offer a wide range of multifunctional biological activities. Due to their multifunctional behavior and diversified structure biosurfactants act as a "green toolbox," which can be exploited in different industrial sectors including food, pharmaceutical, dairy, bioprocess, petroleum, and agricultural [6, 10–16]. They are also important for environmental pollution control as well [6, 17, 18]. Based on biochemical properties these biomolecules can be categorized into two groups: (i) low molecular weight biosurfactants such as glycolipids, lipopeptides, polysaccharide–protein complexes, etc.; (ii) high molecular biosurfactants, which includes fatty acids, particulate biosurfactants, neutral lipids, etc. [5, 19, 20]. Studies revealed that all low molecular weight

Biosurfactants for a Sustainable Future: Production and Applications in the Environment and Biomedicine, First Edition. Edited by Hemen Sarma and Majeti Narasimha Vara Prasad.
© 2021 John Wiley & Sons Ltd. Published 2021 by John Wiley & Sons Ltd.

biosurfactants usually act as surface-active agents and alter the surface tension of a liquid system [7, 9, 110, 112–114, 116]. In contrast to this, high molecular weight biosurfactants act as emulsifiers and are often called bioemulsifiers [6, 7]. Both groups of biosurfactants are found in different structural forms and hence their applications also vary according to their structure. Studies revealed that structural diversity of these biomolecules is exclusively governed by environmental growth conditions of biosurfactant producing microorganisms and their different texas [5, 7, 21]. Notably, all groups of biosurfactants are eco-friendly in nature and thus offer several advantageous properties such as biodegradability, cytocompatibility, non/less toxicity, and stability over variable pH and temperature conditions [13, 17, 23, 96, 111]. Most conventional drugs that are in current use for different biomedical purposes are, however, associated with the severe cytotoxicity and emerging multidrug-resistant pathogens [22–25]. To address these serious health issues, recent studies have been focused on the clinical trials of bioactive compounds of both plant and microbial origins [12, 19, 25, 26]. Unlike plant bioactive compounds, microbial metabolites are easy to produce and their downstream processing is also less complicated, which enables their usage for different biomedical purposes. However, before going for commercialization, their toxicity studies are very crucial in order to avoid any health complications. In this context, biosurfactants can be used as suitable alternative candidates and several therapeutic studies showed they are very efficient and effective biomolecules [12, 19, 26, 27]. Each biosurfactant has its own specific mode of action with respect to the biomedical applications [20, 28]. However, among all of the most widely studied group of biosurfactant is that of glycolipids. This class can be further subdivided into different groups based on the difference in their carbohydrate and lipid moieties, and include rhamnolipids, sophorolipids, mannosylerythritol lipids, lipomannosyl-mannitols, lipomannas and lipoarabinomannes, diglycosyl diglycerides, monoacylglycerols, and galactosyl–diglycerides [20]. Owing to the diversified biological properties in terms of anticancerous and antifungal biosurfactants, glycolipids have been considered as a safer and green modern approach for the pharmaceutical industries. A vast range of studies and clinical trials have been performed on these amphiphatic biomolecules to be used in pharmaceutical industries that have apparently made a significant impact on the development of biomedicine and novel therapeutics. Beyond other multifunctional applications of biosurfactants, their antimicrobial, anticancerous, probiotic nature, and antiadhesive activity against pathogens have impeded a relevant contribution to their health-related applications. Their potential applications involve an antiadhesive coating for biomaterials and probiotic preparations for the treatment of urogenital tract infections and for respiratory disease therapy [12]. Apart from antimicrobial activity these surface-active agents also possess strong anticancerous and antitumor activity [12].

15.2 Microbial Pathogenesis and Biosurfactants

Clinical pathology and laboratory research suggest that microbial pathogenesis involves adherence to the intestinal mucosa, mucus secretion, biofilm formation, cytotoxic damage, and mucosal inflammation due to cytokine release [29]. However, the availability of *in vitro* tissue culture and small-animal models of infection facilitates study of complex interactions between host and pathogen during various stages of infection [30]. Among all glycolipids (rhamnolipids, trehalose lipids, sophorolipids, and mannosylerythritol lipids), rhamnolipids contribute to a wide range of biomedical applications, such as antimicrobial, wound healing and immunomodulatory activity [12, 31–38].

15.2.1 Rhamnolipids

Among these glycolipids, rhamnolipids are very important, which account for a wide range of commercial and biomedical applications [5, 31, 32, 39, 40]. Di-rhamnolipid (Rha-Rha-C_{10}-C_{10}) and mono-rhamnolipid (Rha-C_{10}-C_{10}) are the predominant rhamnolipid congeners [41, 42]. The number of rhamnose molecules in the hydrophilic moiety is a key factor responsible for the specific biological activity, which in turn depends on the environmental parameters [41]. Rhamnolipids and their nano formulations are most widely used for different nano-based drug delivery systems [43].

15.2.2 Trehalose Lipids

Trehalose lipids are mainly produced from Gram-positive eubacteria (*Mycobacterium*, *Rhodococcus*, *Nocardia*, and *Gordonia*), which fall into the order Actinomycetales. The cell wall of these organisms consists of trehalose lipids, which act as a vital component. Structurally, it is a made up of non-reducing disaccharide containing two glucose chains. All trehalose lipids including Trehalose 6,6′-dimycolate (TDM) hold different biological activities in which antitumor and immunomodulatory activities are of prime importance [44–48]. Notably, TDM was found to have an escalation effect of non-specific immunity to microbial infection, granuloma-forming activity priming of murine macrophages to produce nitric oxide [44], and enhanced cytokines production along with increased angiogenesis in experimental animals [49, 50]. However, due to pathogenicity and high toxicity of mycobacterial TDM, its therapeutic usage is restricted despite their several biological activities. In contrast to this, Sakaguchi et al. [49] reported that TDM isolated from *Rhodococcus* sp. with a shorter mycolic acid chains are less toxic under both *in vitro* and *in vivo* conditions. Similarly, another Rhodococcal TDM isolated from *Rhodococcus ruber* exhibited cytocompatibility and cell proliferative property for peripheral blood leukocytes [51], which thrive in their pharmacological importance. These glycolipid subgroups possess several biomedical properties, viz. antimicrobial, antiviral [31, 52, 53], and antitumor activities [12, 45, 54]. Moreover, they have also reported for their functioning in the niche of a cell membrane [55]. Earlier reports also revealed their effective immunomodulatory activity for a number of therapeutic applications, for example, the potential dermatological activity of trehalose lipids produced by *Rhodococcus erythropolis* 51T7 using *in vitro* assays in the human keratinocyte lines and mouse fibroblast [56]. In addition to this, they were also found to have a cytotoxic effect against human promyelocytic leukemia cell line HL60 [54, 57]. Trehalose lipids have been shown to possess antiviral and antimicrobial properties. For example, *Tsukamurella* sp. strain DSM 44370 derived trehalose lipids showed antimicrobial action toward a Gram-positive bacteria. Trehalose lipids isolated from *Rhodococcus* sp. strain 51T7 showed the presence of an efficient membrane permeable property that affected the membrane fluidity of the cell membrane, thus revealing its importance in drug delivery [58, 59].

15.2.3 Sophorolipids

Sophorolipids are another important member of glycolipids that are often produced by non-pathogenic *Candida* sp. and related microorganisms that are phylogenetically different from each other. These include *Candida bambicola*, *Yarrowia lipolytica*, *Candida apicola*, *Candida bogoriensis*, *Candida batistae*, *C. apicola*, and *Rhodotorula bogoriensis*. Sophorolipids possess biological activities similar to other glycolipids in which anticancerous and antibiofilm activities are of prime importance [10, 12, 60–65]. Haque et al. [61] demonstrated the molecular mechanism involved in the sophorolipid-induced antiadhesive and antibiofilm activities against *Candida albicans* and other pathogenic *Candida* sp. In addition, their biological activities are also investigated, such as membrane function, anti-inflammatory response, antimycoplasmal activity, and antiviral activity [66, 67]. For example, Napolitano [66] reported that these bioactive compounds can effectively reduce the cytokine production and induce macrophage activity. *In vivo* studies on the anticancerous activity of sophorolipids revealed their membrane stabilizing and destabilizing activities [67]. Studies on *in vitro* anticancerous activities showed their potential against various cancer types. They possess cytotoxic activity against human liver cancer cells (HT402), lung cancer cells, and leukemia (HL60 and K562), as reported earlier [68]. These glycolipids also induced apoptosis in esophageal cancer cells (KYSE450, KYSE109) and in pancreatic cancer cells (H7402, HPAC), as documented earlier [62, 69, 70].

15.2.4 Mannosylerythritol Lipids

Mannosylerythritol lipids offer a wide range of theranostic advantages such as antimicrobial activity, drug delivery, anticancerous activity, antitumor activity [12, 24, 26, 71–74]. The increasing interest in mannosylerythritol lipids could be attributed to their diverse pharmaceutical and biomedical applications, such as antitumor activities, cell proliferation activity, and schizophrenia or diseases caused by dopamine metabolic dysfunction.

15.3 Bio-Removal of Antibiotics Using Probiotics and Biosurfactants Bacteria

Considerable amounts of antibiotics have been consumed worldwide in different sectors, such as veterinary medicines and for humans. Among all known antibiotics, fluoroquinolones and tetracyclines are two of the most widely used classes of antibiotics having broad-spectrum therapeutic properties [75, 76]. About 36% of antibiotics have been consumed globally during 2000–2010 [77]. In Europe about 10 000 tons of these compounds are consumed every year alone [78]. However, a large fraction of these chemical compounds remains unmetabolized and excreted via urine disposal or through feces in the environment as detected in animal waste water and surface water around farms, which made antibiotics a serious environmental threat [76]. Thus, bio-removal of antibiotics has become a very important issue and various approaches have been employed to discover an effective remediation process to remove these environmental pollutants. Abiotic processes like adsorption, photocatalysis, and oxidation are being used to remediate these chemical compounds, but these methods are easily affected by environmental factors like pH and temperature [79]. Therefore, there is an urgent need to search for a sustainable and biological approach for the bio-removal of antibiotics. Probiotic biosurfactant bacteria are considered as living drugs that can reduce antibiotic consumption and increase human development [76, 80]. Lactic acid bacteria and *Bacillus* spp. are an important probiotic bacteria group that can be an effective alternative for the bio-removal of antibiotics. Recently, Liu et al. [76] showed the bio-removal efficiency of *Bacillus clausii T and Bacillus amyloliquefaciens* for tetracyclines, which account for 14% of the total antibiotic usage in medicines and animal husbandry. Norfloxacin degradation by *Bacillus subtilis* strains is able to produce biosurfactants on a bioreactor scale [81].

15.4 Antiproliferative, Antioxidant, and Antibiofilm Potential of Biosurfactant

To address the limitations of conventional antifungal drugs, different innovative strategies including novel drug formulations, effective screening, and diagnosis tools have been employed from an earlier time to develop potent, non-toxic, biocompatible, antifungal drugs as synthetic drugs possess severe health hazards [39, 82]. Therefore, clinical studies of novel phytochemical and microbial bioactive compounds can be tried to counter fungal biofilm infections [25, 83–86]. Earlier studies demonstrated that different classes of biosurfactants produced by a variety of microorganisms including bacteria and yeast such as *Pseudomonas* sp., *Bacillus* sp., *Lactobacillus* sp., and *Candida* sp. possess antiadhesive and antibiofilm activities against both fungal and bacterial pathogens [32–34, 87]. Earlier reports showed the antioxidant and antiproliferative potential of biosurfactants isolated from *Lactobacillus casei* and their antibiofilm effect in oral *Staphylococcus aureus* strains [86, 88]. On the other hand, it was found that iturin easily penetrated through the thick chitinous cell wall of these yeast cells and ruptured the cell membrane, leading to their cell death. Similarly, pseudofactin II derived from *Pseudomonas fluorescence* has been found to have effective 36–95% bacterial and 90–99% fungal antiadhesive activity [89]. In another study, biosurfactant derived from *Lactobacillus* sp. have been found to have antiadhesive activity against the *S. aureus* growth on voice prostheses and also help in the development of airflow resistance of this implant [87]. Thus, antiadhesive activities of biosurfactants prompt them to be used as a defense weapon for the removal of the post-adhesion infections associated with prolonged use of indwelling implants [90]. Biosurfactants such as lunasan, lipopeptides, and those derived from *Lactobacillus* sp. act as potent antiadhesive agents against these fungal biofilms [91–94]. Based on these studies, it can be conferred that biosurfactants exhibit three different approaches to eradicate/reduce the viability of both planktonic and biofilm growth from *C. albicans* and other emerging multidrug-resistant NACs [61, 89, 93]. Inhibition or eradication efficiency of planktonic cells or fungal biofilm can be achieved by these approaches, which include inhibition of *Candida* cell attachment to the host surface and restricting this fungal pathogen into a planktonic form of detachment of adhered *Candida* cells or by generation of reactive oxygen species to disrupt preformed mature biofilm, as depicted in Figure. 15.1.

Several techniques such as XTT (sodium 3'-[1- (phenylaminocarbonyl)- 3,4- tetrazolium]-bis (4-methoxy6-nitro) benzene sulfonic acid hydrate) assay, fluorescent microscopy, FE-SEM (field-emission scanning electron microscopic) analysis, FAC (fluorescent associated cell) sorting, and RT-PCR (reverse transcriptase polymerase chain reaction) have been exploited for the detection and analysis of antibiofilm activity of biosurfactants [61, 95–97]. Based on FACs and RT-PCR studies, Haque et al. [61] demonstrated that sophorolipids can inhibit biofilm formation and also reduce viability of pre-formed biofilm of *C. albicans* by down-regulating the associated virulent genes. Additionally, it was found that sophorolipid also inhibited and augmented the accumulation of toxic ergosterol precursors [61]. Recent studies on antibiofilm activity of other bioactive compounds involved the use of biophysical techniques such as molecular docking, which provide

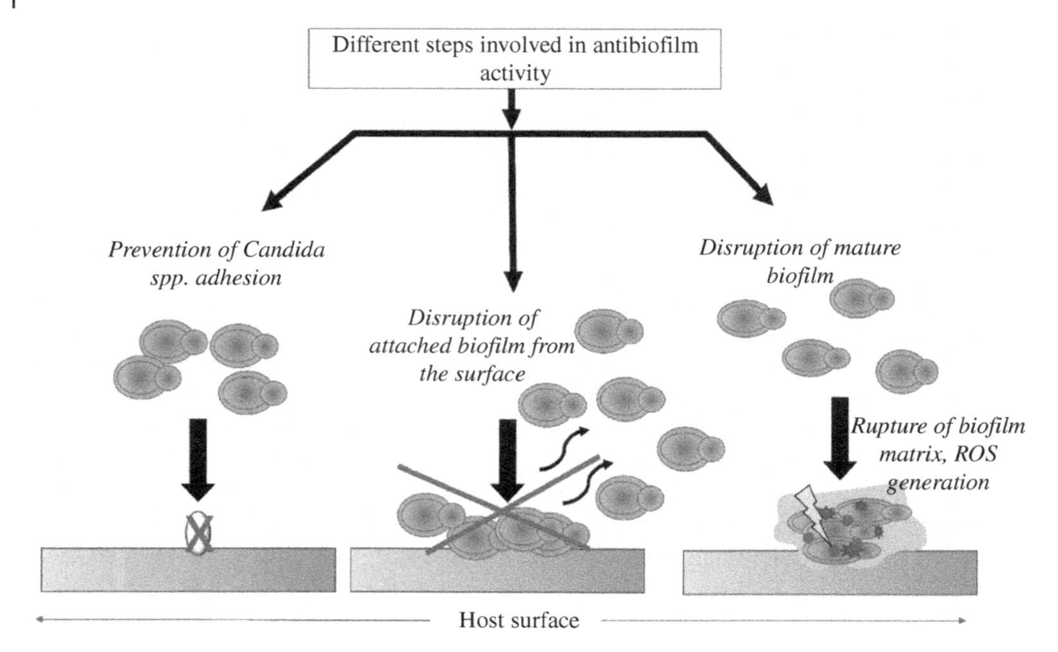

Figure 15.1 Different approaches of antibiofilm activity of biosurfactant against *Candida* biofilm.

an insight into structure and binding efficiency of biomolecules with their target site [86]. Thus, with the help of these structural based-approaches novel anticandidal and antibiofilm drug designing can be efficiently synthesized.

15.5 Wound Healing Potential of Biosurfactants

Microbial contamination at a wounded site is often associated with delayed wound healing progression and also involves the provision of a pioneer route to pathogens penetrating inside the tissues, causing acute to chronic microbial infections. The wound healing process is a dynamic and coordinated process that involves four major steps, including inflammation, migration, cellular proliferation (collagen and fibroblast synthesis), and remodeling/re-epithelization of tissues, as depicted in Figure 15.2 [30, 36, 98].

Different wound types, such as burn wounds, diabetic foot ulcers, surgical wounds, and acid wounds, are associated with different microbial colonizations as well and if not treated with proper medication become very chronic. For instance, burn wounds are a major site of *Pseudomonas* sp. colonization while *Enterococcus* sp. are often found in a diabetic foot ulcer [65]. Several strains of multi-drug resistance have been isolated from patients with burn wounds in India [99]. Biosurfactants possess strong antimicrobial activity and clinically very imperative biomolecules over their chemical counterparts. Therefore, they can also be used for wound healing ointments and drug formulations, as recently reported by various researchers [30, 36, 100–103]. *Bacillus* sp. derive lipopeptide and glycolipid and showed a tremendous *in vivo* wound healing activity on different excision wound models, such as Wistar-albino rats and Swiss–Webster mice [30, 103]. Apart from antimicrobial activity they also activate fibroblast cells and immune cells by stimulating key factors of the immune system, including neutrophils,

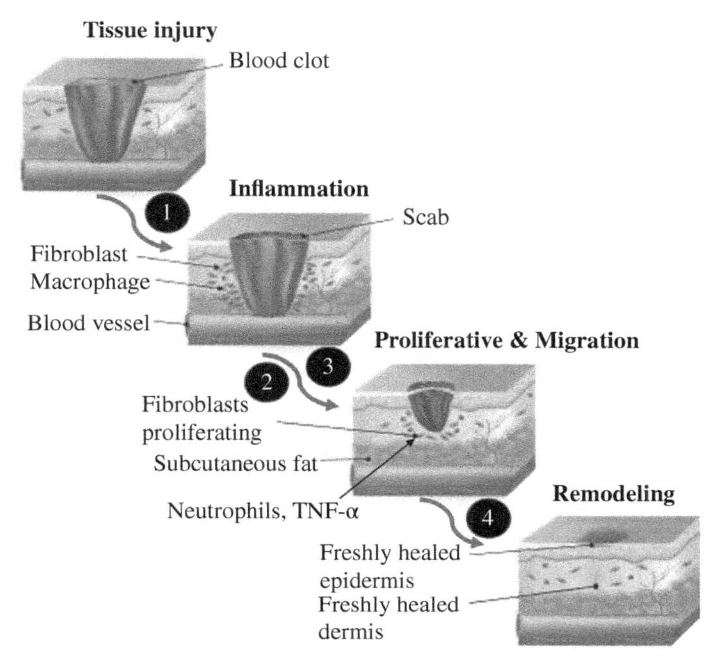

Figure 15.2 Steps involved in wound healing progression, which are inflammation, migration, proliferation and remodeling (See insert for color representation of the figure).

TNF-α, and cytokines, and hamper microbial contamination at the wounded part [30, 103, 104]. In addition, an evaluation of the *in vivo* wound healing potential involves determination to continue biochemical and histopathological studies. For instance, fibroblast proliferation at the wounded site involves increased synthesis of collagen fibers, which in turn are dependent on the hydroxyproline synthesis. Hence hydroxyproline assay is an important biochemical marker to access the wound healing potential of any wound healing agent [30]. Similarly, histopathological studies demonstrated neo-angiogenesis, re-epithelization progression, intact nuclei, and collagen synthesis, which were analyzed by different staining techniques including hematoxylin and eosin and Masson's Trichrome staining [30, 98, 103]. Further, skin irritation assay, viscosity, spreadability analysis, and tensile strength are other important parameters that have also been determined by different researchers in order to develop bioactive compounds based on topical ointment formulations [65, 105–107].

15.6 Conclusion and Future Prospects

Despite the advances made in the development of novel antibiofilm agents, devised biofilm treatment strategies are limited by their high costs and complexities, which means urgent development is required to identify cost-efficient substitutes. Nano-formulations and specifically those fabricated using bioactive compounds have been found to be effective and also economically valuable. In this context, biosurfactant-based nano-formulation synthesis with potential bactericidal and fungicidal properties have been shown to be efficient alternatives to their chemical counterparts in terms of wound care and related technical biomedical issues. In particular, these bio-based nanomaterials are worthy of serious consideration, especially in the case of biomedical devices. Nanomaterials are used as constituents of coatings, biomedical agents, and drug-delivery vehicles so implant materials and research remain active in these areas. However, key issues like nanoparticle resistance and surface interactions between nanoparticles, biofilms, and hosts need to be resolved to ensure successful clinical applications. We hope that this review of the literature persuades the reader that biosurfactants with broad spectrum therapeutic properties will be produced to replace their chemical counterparts.

References

1 Pasternak, G., Askitosari, T.D., and Rosenbaum, M.A. (2020). Biosurfactants and synthetic surfactants in bioelectrochemical systems: A mini-review. *Front. Microbiol.* 11: 358.
2 Pemmaraju, S.C., Sharma, D., Singh, N. et al. (2012). Production of microbial surfactants from oily sludge contaminated soil by *Bacillus subtilis* DSVP23. *Appl. Biochem. Biotechnol.* 167: 1119–1131.
3 Saikia, R.R., Deka, S., Deka, M., and Sarma, H. (2012). Optimization of environmental factors for improved production of rhamnolipid biosurfactant by Pseudomonas aeruginosa RS29 on glycerol. *J. Basic Microbiol.* 52 (4): 446–457.
4 Sarma, H., Bustamante, K.L.T., and Prasad, M.N.V. (2018). Biosurfactants for oil recovery from refinery sludge: Magnetic nanoparticles assisted purification. In: Industrial and Municipal Sludge (eds. M.N.V. Prasad, P.J. de Campos, F. Meththika and V.S. Venkata Mohan), 107–132. Massachusetts: Elsevier. ISBN: 9780128159071.
5 Gupta, S., Varshney, R., Jha, R.K. et al. (2017). In vitro apoptosis induction in a human prostate cancer cell line by thermotolerant glycolipid from *Bacillus licheniformis* SV1. *J. Surfactants Deterg.* 20: 1141–1151.
6 Geetha, S.J., Banat, I.M., and Joshi, S.J. (2018). Biosurfactants: production and potential applications in microbial enhanced oil recovery (MEOR). *Biocatal. Agric. Biotechnol.* 14: 23–32.
7 Satpute, S.K., Banat, I.M., Dhakephalkar, P.K. et al. (2010). Biosurfactants, bioemulsifiers and exopolysaccharides from marine microorganisms. *Biotechnol. Adv.* 28: 436–450.
8 Tokumoto, Y., Nomura, N., Uchiyama, H. et al. (2009). Structural characterization and surface-active properties of a succinoyl trehalose lipid produced by *Rhodococcus* sp. SD-74. *J. Oleo Sci.* 58: 97–102.
9 Garg, M. and Priyanka, C.,.M. (2018). Isolation, characterization and antibacterial effect of biosurfactant from *Candida parapsilosis*. *Biotechnol. Rep.* 18: e00251.
10 Callaghan, B., Lydon, H., Roelants, S.L.K.W. et al. (2016). *Lactonic sophorolipids* increase tumor burden in apcmin+/− mice. *PLos One* 11: e0156845.
11 Gudiña, E.J., Pereira, J.F.B., Costa, R. et al. (2013). Biosurfactant-producing and oil-degrading *Bacillus subtilis* strains enhance oil recovery in laboratory sand-pack columns. *J. Hazard. Mater.* 261: 106–113.
12 Gudiña, E.J., Rangarajan, V., Sen, R., and Rodrigues, L.R. (2013). Potential therapeutic applications of biosurfactants. *Trends Pharmacol. Sci.* 34: 667–675.
13 Irfan-Maqsood, M. and Seddiq-Shams, M. (2014). Rhamnolipids: Well-characterized glycolipids with potential broad applicability as biosurfactants. *Ind. Biotechnol.* 10: 285–291.
14 Lang, S. and Philp, J.C. (1998). Surface-active lipids in *Rhodococci*. *Antonie Van Leeuwenhoek* 74: 59–70.
15 Płaza, G. and Achal, V. (2020). Biosurfactants: Eco-friendly and innovative biocides against biocorrosion. *Int. J. Mol. Sci.* 21: 2152.

16 Yang, H., Yu, H., and Shen, Z. (2015). A novel high-throughput and quantitative method based on visible color shifts for screening *Bacillus subtilis* THY-15 for surfactin production. *J. Ind. Microbiol. Biotechnol.* 42: 1139–1147. https://doi.org/10.1007/s10295-015-1635-4.F.

17 Mulligan, C.N. (2005). Environmental applications for biosurfactants. *Environ. Pollut.* 133: 183–198.

18 Shekhar, S., Sundaramanickam, A., and Balasubramanian, T. (2015). Biosurfactant producing microbes and their potential applications: A review. *Crit. Rev. Environ. Sci. Technol.* 45: 1522–1554.

19 Fariq, A. and Saeed, A. (2016). Production and biomedical applications of probiotic biosurfactants. *Curr. Microbiol.* 72: 489–495.

20 Inès, M. and Dhouha, G. (2015). Glycolipid biosurfactants: Potential related biomedical and biotechnological applications. *Carbohydr. Res.* 416: 59–69.

21 Aparna, A., Srinikethan, G., and Smitha, H. (2012). Production and characterization of biosurfactant produced by a novel *Pseudomonas* sp. *Colloids Surf. B. Biointerfaces* 95: 23–29.

22 Vermitsky, J.P. and Edlind, T.D. (2004). Azole resistance in *Candida glabrata*: Coordinate upregulation of multidrug transporters and evidence for a Pdr1-like transcription factor. *Antimicrob. Agents Chemother.* 48: 3773–3781.

23 Mnif, I. and Ghribi, D. (2016). Glycolipid biosurfactants: Main properties and potential applications in agriculture and food industry. *J. Sci. Food Agric.* 96: 4310–4320.

24 Morita, T., Konishi, M., Fukuoka, T. et al. (2008). Production of glycolipid biosurfactants, mannosylerythritol lipids, by *Pseudozyma siamensis* CBS 9960 and their interfacial properties. *J. Biosci. Bioeng.* 105: 493–502.

25 Gupta, P., Gupta, S., Sharma, M. et al. (2018). Effectiveness of phyto-active molecules on transcriptional expression, biofilm matrix, and cell wall components of *Candida glabrata* and its clinical isolates. *ACS Omega* 3: 12201–12214.

26 Rodrigues, L., Banat, I.M., Teixeira, J., and Oliveira, R. (2006). Biosurfactants: Potential applications in medicine. *J. Antimicrob. Chemother.* 57: 609–618.

27 Rodrigues, L.R. and Teixeira, J.A. (2010). Biomedical and therapeutic applications of biourfactants. Biomedical and therapeutic applications of biosurfactants. *Adv. Exp. Med. Biol.* 672: 75–87.

28 Singh, P. and Cameotra, S.S. (2004). Potential applications of microbial surfactants in biomedical sciences. *Trends Biotechnol.* 22: 142–146.

29 Ellis, S.J., Crossman, L.C., McGrath, C.J. et al. (2020). Identification and characterisation of enteroaggregative *Escherichia coli* subtypes associated with human disease. *Sci. Rep.* 10: 7475.

30 Gupta, S., Raghuwanshi, N., Varshney, R. et al. (2017). Accelerated *in vivo* wound healing evaluation of microbial glycolipid containing ointment as a transdermal substitute. *Biomed. Pharmacother.* 94: 1186–1196.

31 Christova, N., Tuleva, B., Kril, A. et al. (2013). Chemical structure and *in vitro* antitumor activity of rhamnolipids from *Pseudomonas aeruginosa* BN10. *Appl. Biochem. Biotechnol.* 170: 676–689.

32 Cortés-Sánchez, A.J., Hernández-Sánchez, H., and Jaramillo-Flores, M.E. (2013). Biological activity of glycolipids produced by microorganisms: New trends and possible therapeutic alternatives. *Microbiol. Res.* 168: 22–32.

33 Desai, J.D. and Banat, I.M. (1997). Microbial production of surfactant and their commercial potential. *Microbiol. Mol. Biol. Rev.* 61: 47–64.

34 Diaz De Rienzo, M.A., Stevenson, P., Marchant, R., and Banat, I.M. (2016). Antibacterial properties of biosurfactants against selected Gram-positive and -negative bacteria. *FEMS Microbiol. Lett.* 363: 1–8.

35 Paulino, B.N., Pessôa, M.G., Mano, M.C.R. et al. (2016). Current status in biotechnological production and applications of glycolipid biosurfactants. *Appl. Microbiol. Biotechnol.* 100: 10265–10293.

36 Stipcevic, T., Piljac, A., and Piljac, G. (2006). Enhanced healing of full-thickness burn wounds using di-rhamnolipid. *Burns* 32: 24–34.

37 Abdel-Mawgoud, A.M., Lépine, F., and Déziel, E. (2010). Rhamnolipids: Diversity of structures, microbial origins and roles. *Appl. Microbiol. Biotechnol.* 86: 1323–1336.

38 De Rienzo, M.A.D. and Martin, P.J. (2016). Effect of mono and di-rhamnolipids on iofilms pre-formed by *Bacillus subtilis* BBK006. *Curr. Microbiol.* 73: 183–189. Available at: https://doi.org/10.1007/s00284-016-1046-4.

39 Banat, I.M., Franzetti, A., Gandolfi, I. et al. (2010). Microbial biosurfactants production, applications and future potential. *Appl. Microbiol. Biotechnol.* 87: 427–444.

40 Saini, H.S., Barragán-Huerta, B.E., Lebrón-Paler, A. et al. (2008). Efficient purification of the biosurfactant viscosin from *Pseudomonas libanensis* strain M9-3 and its physicochemical and biological properties. *J. Nat. Prod.* 71: 1011–1015.

41 Soberón-Chávez, G., Lépine, F., and Déziel, E. (2005). Production of rhamnolipids by *Pseudomonas aeruginosa*. *Appl. Microbiol. Biotechnol.* 68: 718–725.

42 Tiso, T., Zauter, R., Tulke, H. et al. (2017). Designer rhamnolipids by reduction of congener diversity: Production and characterization. *Microb. Cell Fac.* 16: 225.

43 Kumar, C.G., Mamidyala, S.K., Das, B. et al. (2010). Synthesis of biosurfactant-based silver nanoparticles with purified rhamnolipids isolated from *Pseudomonas aeruginosa* BS-161R. *J. Microbiol. Biotechnol.* 20: 1061–1068.

44 Chami, M., Andreau, K., Lemassu, A. et al. (2002). Priming and activation of mouse macrophages by trehalose 6,6-dicorynomycolate vesicles from *Corynebacterium glutamicum*. *FEMS Immunol. Med. Microbiol.* 32: 141–147.

45 Franzetti, A., Gandolfi, I., Bestetti, G. et al. (2010). Production and applications of trehalose lipid biosurfactants. *Eur. J. Lipid Sci. Technol.* 112: 617–627.

46 Hoq, M.M., Suzutani, T., Toyoda, T. et al. (1997). Role of γδ TCR M lymphocytes in the augmented resistance of trehalose 6,6h-dimycolate-treated mice to influenza virus infection. *J. Gen. Virol.* 78: 1597–1603.

47 Parant, M., Parant, F., Chedid, L. et al. (1977). Enhancement of nonspecific immunity to bacterial infection by cord factor (6,6′-trehalose dimycolate). *J Infect Dis* 135: 771–777.

48 Uchida, Y., Tsuchiya, R., Chino, M. et al. (1989). Extracellular accumulation of mono- and di-succinoyl trehalose lipids by a strain of *Rhodococcus erythropolis* grown on n alkanes. *Agric. Biol. Chem.* 53: 757–763.

49 Sakaguchi, I., Ikeda, N., Nakayama, M. et al. (2000). Trehalose 6,6′-dimycolate (Cord factor) enhances neovascularization through vascular endothelial growth factor production by neutrophils and macrophages. *Infect. Immun.* 68: 2043–2052.

50 Ueda, S., Fujiwara, N., Naka, T. et al. (2001). Structure–activity relationship of mycoloyl glycolipids derived from *Rhodococcus* sp. 4306. *Microb. Pathog.* 30: 91–99.

51 Kuyukina, M.S., Ivshina, I.B., Philp, J.C. et al. (2001). Recovery of *Rhodococcus* biosurfactants using methyl tertiary-butyl ether extraction. *J. Microbiol. Methods* 46: 149–156.

52 Azuma, M., Suzutani, T., Sazaki, K. et al. (1987). Role of interferon in the augmented resistance of trehalose-6,6′-dimycolate-treated mice to influenza virus infection. *J. Gen. Virol.* 68: 835–843.

53 Shao, B., Liu, Z., Zhong, H. et al. (2017). Effects of rhamnolipids on microorganism characteristics and applications in composting: A review. *Microbiol. Res.* 200: 33–44.

54 Sudo, T., Zhao, X., Wakamatsu, Y. et al. (2000). Induction of the differentiation of human HL-60 promyelocytic leukemia cell line by succinoyl trehalose lipids. *Cytotechnology* 33: 259–264.

55 Zaragoza, A., Aranda, F.J., Espuny, M.J. et al. (2009). A mechanism of membrane permeabilization by a bacterial trehalose lipid biosurfactant produced by *Rhodococcus* sp. *Langmuir* 25: 7892–7898.

56 Marqués, A.M., Pinazo, A., Farfan, M. et al. (2009). The physicochemical properties and chemical composition of trehalose lipids produced by *Rhodococcus erythropolis* 51T7. *Chem. Phys. Lipids* 158: 110–117.

57 Isoda, H., Shinmoto, H., Kitamoto, D. et al. (1997). Differentiation of human promyelocytic leukemia cell line HL60 by microbial extracellular glycolipids. *Lipids* 32: 263–271.

58 Ortiz, A., Teruel, J.A., Espuny, M.J. et al. (2008). Interactions of a *Rhodococcus* sp. biosurfactant trehalose lipid with phosphatidylethanolamine membranes. *Biochim. Biophys. Acta Mol. Basis* 1778: 2806–2813.

59 Ortiz, A., Teruel, J.A., Espuny, M.J. et al. (2009). Interactions of a bacterial biosurfactant trehalose lipid with phosphatidylserine membranes. *Chem. Phys. Lipids* 158: 46–53.

60 Elshafie, A.E., Joshi, S.J., Al-Wahaibi, Y.M. et al. (2015). Sophorolipids production by *Candida bombicola* ATCC 22214 and its potential application in microbial enhanced oil recovery. *Front. Microbiol.* 6: 1324.

61 Haque, F., Alfatah, M., Ganesan, K., and Bhattacharyya, M.S. (2016). Inhibitory effect of sophorolipid on *Candida albicans* biofilm formation and hyphal growth. *Sci. Rep.* 6: 23575.

62 Ribeiro, I.A.C., Faustino, C.M.C., Guerreiro, P.S. et al. (2015). Development of novel sophorolipids with improved cytotoxic activity toward MDA-MB-231 breast cancer cells. *J. Mol. Recognit.* 28: 155–165.

63 Spencer, J.F., Gorin, P.A., and Tulloch, A.P. (1970). *Torulopsis bombicola* sp. n. *Antonie Van Leeuwenhoek* 36: 129–133.

64 Van Bogaert, I.N.A., Saerens, K., De Muynck, C. et al. (2007). Microbial production and application of sophorolipids. *Appl. Microbiol. Biotechnol.* 76: 23–34.

65 Lydon, H.L., Baccile, N., Callaghan, B. et al. (2017). Adjuvant antibiotic activity of acidic sophorolipids with potential for facilitating wound healing. *Antimicrob. Agents Chemother.* 61: e02547–e02516.

66 Napolitano, L.M. (2006). Sophorolipids in sepsis: Antiinflammatory or antibacterial. *Crit. Care Med.* 34: 258–259.

67 Ma, X., Li, H., and Song, X. (2012). Surface and biological activity of sophorolipid molecules produced by *Wickerhamiella domercqiae* var. sophorolipid CGMCC 1576. *J. Colloid Interface Sci.* 376: 165–172.

68 Chen, J. et al. (2006). Production, structure elucidation and anticancer properties of sophorolipid from *Wickerhamiella domercqiae. Enzyme and Microbial Technology* 39 (3): 501–506.

69 Fu, S.L., Wallner, S.R., Bowne, W.B. et al. (2008). Sophorolipids and their derivatives are lethal against human pancreatic cancer cells. *J. Surg. Res.* 148: 77–82.

70 Shao, L., Song, X., Ph, D. et al. (2012). Bioactivities of sophorolipid with different structures against human esophageal cancer cells. *J. Surg. Res.* 173: 286–291.

71 Im, J.H., Nakane, T., Yanagishita, H. et al. (2001). Mannosylerythritol lipid, a yeast extracellular glycolipid, shows high binding affinity towards human immunoglobulin G. *BMC Biotechnol.* 1: 5.

72 Im, J.H., Yanagishita, H., Ikegami, T. et al. (2003). Mannosylerythritol lipids, yeast glycolipid biosurfactants are potential affinity ligand materials for human immunoglobulin G. *J. Biomed. Mater. Res.* 65: 379–385.

73 Kitamoto, D., Yanagishita, H., Shinbo, T. et al. (1993). Surface active properties and antimicrobial activities of mannosylerythritol lipids as biosurfactants produced by *Candida antarctica*. *J. Biotechnol.* 29: 91–96.

74 Fukuoka, T., Kawamura, M., Morita, T. et al. (2008). A basidiomycetous yeast, *Pseudozyma crassa*, produces novel diastereomers of conventional mannosylerythritol lipids as glycolipid biosurfactants. *Carbohydr. Res.* 343: 2947–2955.

75 Harrabi, M., Alexandrino, D.A.M., Aloulou, F. et al. (2019). Biodegradation of oxytetracycline and enrofloxacin by autochthonous microbial communities from estuarine sediments. *Sci. Total Environ.* 648: 962–972. https://doi.org/10.1016/j.scitotenv.2018.08.193.

76 Liu, C., Xu, Q., Yu, S. et al. (2020). Bio-removal of tetracycline antibiotics under the consortium with probiotics Bacillus clausii T and Bacillus amyloliquefaciens producing biosurfactants. *Sci. Total Environ.* 710: 136329. https://doi.org/10.1016/j.scitotenv.2019.136329.

77 Van Boeckel, T.P., Gandra, S., Ashok, A. et al. (2014). Global antibiotic consumption 2000 to 2010: An analysis ofnational pharmaceutical sales data. *Lancet Infect. Dis.* 14: 742–750. https://doi.org/10.1016/S1473-3099(14)70780-7.

78 Yang, L., Rybtke, M.T., Jakobsen, T.H. et al. (2009). Computer-aided identification of recognized drugs as Pseudomonas aeruginosa quorum-sensing inhibitors. *Antimicrob. Agents Chemother.* 53: 2432–2443.

79 Cao, J., Lai, L., Lai, B. et al. (2019). Degradation of tetracycline by peroxymonosulfate activated with zero-valent iron: Performance, intermediates, toxicity and mechanism. *Chem. Eng. J.* 364: 45–56. https://doi.org/10.1016/j.cej.2019.01.113.

80 Rodrigues, L., Teixeira, J., Oliveira, R., and Van Der Mei, H.C. (2006). Response surface optimization of the medium components for the production of biosurfactants by probiotic bacteria. *Process Biochem.* 41: 1–10.

81 Jałowiecki, Ł., Żur, J., Płaza, G., Kaźmierczak, B., Kutyłowska, M., Piekarska, K., TruszZdybek, A., 2017. Norfloxacin degradation by *Bacillus subtilis* strains able to produce biosurfactants on a bioreactor scale. E3S Web of Conferences, 17, 00033. doi:https://doi.org/10.1051/e3sconf/20171700033.

82 Banat, I.M., Satpute, S.K., Cameotra, S.S. et al. (2014). Cost effective technologies and renewable substrates for biosurfactants production. *Front. Microbiol.* 5: 1–18.

83 Banat, I.M., De Rienzo, M.A.D., and Quinn, G.A. (2014). Microbial biofilms: Biosurfactants as antibiofilm agents. *Appl. Microbiol. Biotechnol.* 98: 9915–9929.

84 Sadekuzzaman, M., Yang, S., Mizan, M.F.R., and Ha, S.D. (2015). Current and recent advanced strategies for combating biofilms. *Compr. Rev. Food Sci. Food Saf.* 14: 491–509.

85 Sambanthamoorthy, K., Feng, X., Patel, R. et al. (2014). Antimicrobial and antibiofilm potential of biosurfactants isolated from *Lactobacilli* against multi-drug-resistant pathogens. *BMC Microbiol.* 14: 197.

86 Merghni, A., Dallel, I., Noumi, E. et al. (2017 Mar). Antioxidant and antiproliferative potential of biosurfactants isolated from *Lactobacillus casei* and their anti-biofilm effect in oral *Staphylococcus aureus* strains. *Microb. Pathog.* 104: 84–89. doi:https://doi.org/10.1016/j.micpath.2017.01.017. Epub 2017 Jan. 11. 28087493.

87 Rodrigues, L.R., Banat, I.M., van der Mei, H.C. et al. (2004). Interference in adhesion of bacteria and yeasts isolated from explanted voice prostheses to silicone rubber by rhamnolipid biosurfactants. *Appl. Microbiol. Biotechnol.* 66: 306–311.

88 Reid, G., Bruce, A., Fraser, N. et al. (2001). Oral probiotics can resolve urogenital infections. *FEMS Immunol. Med. Microbiol.* 30: 49–52.

89 Janek, T., Radwanska, A., and Lukaszwicz, M. (2013). Lipopeptide biosurfactant pseudofactin II induced apoptosis of melanoma A 375 cells by specific interaction with the plasma membrane. *PLoS One* 8: 1–9.

90 Cochis, A., Fracchia, L., Martinotti, M.G., and Rimondini, L. (2012). Biosurfactants prevent *in-vitro C. albicans* biofilm formation on resins and silicon materials for prosthetic devices. *Oral Surg. Oral Med. Oral Pathol. Oral Radiol.* 113: 755–761.

91 Rodrigues, L.R., Banat, I.M., Van Der Mei, H.C. et al. (2006). Interference in adhesion of bacteria and yeasts isolated from explanted voice prostheses to silicone rubber by rhamnolipid biosurfactants. *J. Appl. Microbiol.* 100: 470–480.

92 Janek, T., Łukaszewicz, M., Rezanka, T., and Krasowska, A. (2010). Isolation and characterization of two new lipopeptide biosurfactants produced by *Pseudomonas fluorescens* BD5 isolated from water from the Arctic Archipelago of Svalbard. *Bioresour. Technol.* 101: 6118–6123.

93 Luna, J.M., Rufino, R.D., Sarubbo, L.A. et al. (2011). Evaluation antimicrobial and antiadhesive properties of the biosurfactant Lunasan produced by *Candida sphaerica* UCP 0995. *Curr. Microbiol.* 62: 1527–1534.

94 Gomaa, E.Z. (2013). Antimicrobial activity of a biosurfactant produced by *Bacillus licheniformis* strain M104 grown on whey. *Brazil Arch. Biol. Technol.* 56: 259–268.

95 Igarashi, S., Hattori, Y., and Maitani, Y. (2006). Biosurfactant MEL-a enhances cellular association and gene transfection by cationic liposome. *J. Control. Release* 112: 362–368.

96 Pemmaraju, S.C., Kumar, P., Pruthi, P.A. et al. (2016). Impact of oxidative and osmotic stresses on *Candida albicans* biofilm formation. *Biofouling* 32: 897–909.

97 Singh, N., Pemmaraju, S.C., Pruthi, P.A. et al. (2013). *Candida* biofilm disrupting ability of di-rhamnolipid (RL-2) produced from *Pseudomonas aeruginosa* DSVP20. *Appl. Biochem. Biotechnol.* 169: 2374–2391.

98 Vashisth, P., Srivastava, A.K., Nagar, H. et al. (2016). Drug functionalized microbial polysaccharide-based nanofibers as transdermal substitute. *Nanomedicine* 12: 1375–1385.

99 Biswal, I., Arora, B.S., and Kasana, D.N. (2014). Incidence of multidrug resistant *Pseudomonas aeruginosa* isolated from burn patients and environment of teaching institution. *J. Clin. Diagn. Res.* 8: DC26–DC29.

100 Piljac, A., Stipčevic, T., Žegarac, J.P., and Piljac, G. (2008). Successful treatment of chronic decubitus ulcer with 0.1% dirhamnolipid ointment. *J. Cutan. Med. Surg.* 12: 142–146.

101 Sana, S., Mazumder, A., Datta, S., and Biswas, D. (2017). Towards the development of an effective *in vivo* wound healing agent from *Bacillus* sp. derived biosurfactant using *Catla catla* fish fat. *RSC Adv.* 7: 13668–13677.

102 Stipcevic, T., Piljac, T., and Isseroff, R.R. (2005). Di-rhamnolipid from *Pseudomonas aeruginosa* displays differential effects on human keratinocyte and fibroblast cultures. *J. Dermatol. Sci.* 40: 141–143.

103 Zouari, R., Moalla-Rekik, D., Sahnoun, Z. et al. (2016). Evaluation of dermal wound healing and *in vitro* antioxidant efficiency of *Bacillus subtilis* SPB1 biosurfactant. *Biomed. Pharmacother.* 84: 878–891.

104 Kharazmi, A., Bibi, Z., Nielsen, H. et al. (1989). Effect of *Pseudomonas aeruginosa* rhamnolipid on human neutrophil and monocyte function. *APMIS* 97: 1068–1072.

105 Nagar, H.K., Srivastava, A.K., Srivastava, R. et al. (2016). Pharmacological investigation of the wound healing activity of *Cestrum nocturnum* (L.) ointment in Wistar albino rats. *J. Pharm.* 2016: 9249040.

106 Srivastava, A.K., Khare, P., Kumar Nagar, H. et al. (2016). Hydroxyproline: A potential biochemical marker and its role in the pathogenesis of different diseases. *Curr. Protein Pept. Sci.* 17: 596–602.

107 Raghuwanshi, N., Kumari, P., Srivastava, A.K. et al. (2017). Synergistic effects of *Woodfordia fruticosa* gold nanoparticles in preventing microbial adhesion and accelerating wound healing in Wistar albino rats *in vivo*. *Mater. Sci. Eng. C* 80: 252–262.

108 Imamura, Y., Chandra, J., Mukherjee, P.K. et al. (2008). *Fusarium* and *Candida albicans* biofilms on soft contact lenses: Model development, influence of lens type, and susceptibility to lens care solutions. *Antimicrob. Agents Chemother.* 52: 171–182.

109 Morikawa, M., Hirata, Y., and Imanaka, T. (2000). A study on the structure-function relationship of lipopeptide biosurfactants. *Biochim. Biophys. Acta* 1488: 211–218.

110 Nayarisseri, A., Singh, P., and Singh, S.K. (2018). Screening, isolation and characterization of biosurfactant producing *Bacillus subtilis* strain ANSKLAB03. *Bioinformation* 14: 304–314.

111 Pruthi, V. and Cameotra, S.S. (2000). Novel sucrose lipid produced by *Serratia marcescens* and its application in enhanced oil recovery. *J. Surfactant Deterg.* 3: 533–537.

112 Santos, D.K.F., Rufino, R.D., Luna, J.M. et al. (2016). Biosurfactants: Multifunctional biomolecules of the 21st century. *Int. J. Mol. Sci.* 17: 1–31.

113 Marchant, R. and Banat, I.M. (2012). Microbial biosurfactants: Challenges and opportunities for future exploitation. *Trends Biotechnol.* 30: 558–565.

114 Marchant, R. and Banat, I.M. (2014). Protocols for measuring biosurfactant production in microbial cultures. In: McGenity, T.J. (eds. K.N. Timmis and B. Nogales), 119–128. Hydrocarbon and Lipid Microbiology Protocols: Springer Protocols Handbooks.

115 Meenambiga, S.S. and Rajagopal, K. (2018). Antibiofilm activity and molecular docking studies of bioactive secondary metabolites from endophytic fungus *Aspergillus nidulans* on oral *Candida albicans*. *JAPS.* 8: 037–045.

116 Tiwary, M. and Dubey, A.K. (2018). Characterization of biosurfactant produced by a novel strain of *Pseudomonas aeruginosa*, isolate ADMT1. *J. Surfactants Deterg.* 21: 113–125.

117 Fu, J., Ding, Y., Wei, B. et al. (2017). Epidemiology of *Candida albicans* and non-*C. albicans* of neonatal candidemia at a tertiary care hospital in western China. *BMC Infect. Dis.* 17: 1–6.

16

Biosurfactants Against Drug-Resistant Human and Plant Pathogens

Recent Advances

Chandana Malakar[1] *and Suresh Deka*[2]

[1]*Institute of Advanced Study in Science and Technology (IASST), Garchuk, Assam, India*
[2]*Environmental Biotechnology Laboratory, Life Sciences Division, Institute of Advanced Study in Science and Technology (IASST), Guwahati, Assam, India*

CHAPTER MENU

16.1 Introduction

Medicine has been an integral part of the history of human civilization and the development of various strategies to combat infections caused by lethal microbes have been well documented across cultures and historical periods. In 1911 Erlich introduced magic bullets for the treatment of syphilis, later termed antibiotics [1]. Alexander Fleming's accidental discovery of penicillin reached the first recording of an antibiotic study. Since then a number of antibiotics, such as streptomycin, chloramphenicol, and tetracycline, were developed in the 1950s to reduce infectious diseases globally [2]. Infection-mediated mortality and morbidity have drastically decreased, marking the Golden antibiotic age. Until now, the study of antibiotic development has so far captured the

Biosurfactants for a Sustainable Future: Production and Applications in the Environment and Biomedicine, First Edition. Edited by Hemen Sarma and Majeti Narasimha Vara Prasad.
© 2021 John Wiley & Sons Ltd. Published 2021 by John Wiley & Sons Ltd.

views of many scientists globally but with continuous evolution of pathogens there is an increase in adaptation of pathogens to antibiotics. This has become a global threat as pathogens continue to resist antibiotics, causing more infections worldwide [3]. With the continuous emergence of antibiotic-resistant pathogens, there is a necessity for an antibiotic dose revision against the target pathogens. However, the revised doses of antibiotics are presumed to induce toxicity in the patient along with various effects on the environment [4]. The unrestrained use of antibiotics has not only weakened the barrier to protect human life, but also livestock and agriculture. A certain number of antibiotics escapes both the human and the animal body by excretion, which can then pollute the environment, thereby giving rise to resistomes consisting of resistant genes. Antibiotics are ubiquitous due to their intensive use and are considered to be persistent or pseudo-persistent as their rate of entry into the environment is greater than the rate at which they are eliminated, thereby causing antibiotic pollution [5, 6]. The boon of antibiotics is undeniable until now, when a number of adverse effects of antibiotic use have been reported. Decades of extensive use of antibiotics have resulted in antibiotic-resistant strains of microbes that can cause global havoc.

With the perspective to resolve the damage caused by these antibiotic-resistant microbes, worldwide research has diverted toward different green initiatives. In furthering research in the context of green alternatives to different chemical products used in agricultural and pharmaceutical prospects, initiatives have been undertaken to promote the development of naturally derived substances for the benefit of mankind and its assets. Different plant extracts and microbial metabolites are used to substitute the chemical counterpart, as the latter is likely to pose a threat to humans and the environment. Biosurfactant, a surface-active biological agent, has recently become a subject of keen interest to different scientific communities around the globe. They are the metabolite secreted under different conditions by various species of bacteria, yeast, and fungi that provide a beneficial effect against different stresses. These are amphiphilic molecules containing hydrophobic and hydrophilic moieties that can reduce the surface and interfacial tension of water or any other substrate [7]. They are widely classified on the basis of their structure as glycolipid, lipophilic, fatty acids, and polymers [8, 9]. These classes of biosurfactants are investigated from diverse perspectives for their usefulness in various fields as an alternative over their chemical counterparts. Because of their non-toxic nature, these surface-active biological agents have intensive application in remediation techniques, agriculture, pharmaceuticals, and food additives. Recently, different forms of biosurfactants have been reported for their potent antimicrobial effects against various drug-resistant pathogens. This could be a big step in the fight against antibiotic-resistant microbes causing thousands of deaths and infections worldwide. In this chapter an initiative was taken to discuss the growing concern for the increase of antibiotic resistance among the microbial groups and the prosperity of biosurfactants in dealing with the infections caused by these organisms.

16.2 Environmental Impact of Antibiotics

16.2.1 Toxicity Induced by Antibiotics

The established efficacy of numerous antibiotics includes specific forms of health-related risks that tend to be difficult to tackle from time to time. The efficacy of various classes of antibiotics such as aminoglycosides, cephalosporins, penicillins, carbapenemes, tetracyclines, macrolides, etc., used for the treatment of various infections is accompanied with certain health threats such as diarrhea, neurotoxicity, and nephrotoxicity [10, 11]. Flucloxacillin-related jaundice [12], cephalosporin-related hypersensitivity [13, 14], clindamycin-associated parenteral toxicity, acute cholestasis hepatitis, and others [15–17] are often encountered by patients. Liver-based toxicity, such as hepatitis cytotoxic, intrahepatic cholestasis, chronic active hepatitis, and microvesicular steatosis, is reported to be excised with different antibiotic exposures [18]. In addition to antibiotic-induced toxicity in patients, they are reported to evoke the immune system. Cross-allergy between antibiotics of sulfonamide and non-antibiotic sulfonamides containing drugs can induce Type I hypersensitivity, which could complicate the therapeutic effectiveness of drugs containing the sulphonamide functional group [19]. For instance, a new BAL30072 monocyclic β-lactam antibiotic was found to be effective against multi-drug – resistant Gram-negative bacteria but was reported to induce impaired hepatic mitochondrial function and glycolysis-inhibition at clinically relevant doses [20]. Decades of use of sub-therapeutic doses in animal husbandry to avoid diseases of domestic animals and to contribute to animal food processing have resulted in evolution of various antibiotic-resistant pathogens with host upgradation efficiency [21]. Tetracycline is one of the most commonly prescribed antibiotics in medicine, animal husbandry, and agriculture. The extensive use of this class of antibiotics has several environmental issues including ecological hazards and a threat to human health [22]. Although antibiotics remain an irrefutable part of human, veterinary, and farm advantages, they contribute immensely toward environmental problems as well as inducing toxicity and hypersensitivity in humans and animals.

16.2.2 Microbial Resistance to Antibiotics: A Global Concern

Resistance to antibiotics among human, animal, and plant pathogens has become a matter of concern throughout the world. Uncontrolled use of antibiotics has induced resistance in many microbial species that were previously susceptible to the antibiotic in question. A 35% increase in antibiotic consumption worldwide was reported between 2000 and 2010 [23]. The non-lethal dose of antibiotics often acts as a selective pressure for clinical, community, and farming systems that enhances the antibiotic resistance in different pathogens [24]. This led the scientific community around the globe to rethink and reinvent medicines to combat the established drug resistance in pathogens. The available therapy has been reported to be refractive to multidrug-resistant (MDR) species *Mycobacterium tuberculosis*, *Streptococcus pneumoniae*, *Shigella dysenteriae*, *Salmonella typhi*, and *Enterococcus faecium* [25]. Increased antibiotic resistance has been reported in opportunistic pathogens including *Pseudomonas aeruginosa*, *Stenotrophomonas maltophilia*, and *Acinetobacter baumannii* [26]. In 2010 Gao et al. reported the increased antibiotic resistance of *Helicobacter pylori* in 10 years against the antibiotics clarithromycin, metronidazole, and fluoroquinolone, which makes it difficult to eradicate *H. pylori* mediated infection with these antibiotics [27]. *Staphylococcus aureus* infection is a pervasive type of infection in the world among bacterial infections. The pathogen is reported to cause a wide range of human infections. The emergence of resistant variants such as methicillin-resistant *Staphylococcus aureus* (MRSA) and vancomycin-resistant *Staphylococcus aureus* (VRSA) has now become a risk to human health [28–30]. Many industrialized countries are often infected with penicillin-resistant *Streptococcus pneumonia* and vancomycin-resistant enterococci (VRE), thereby implicating a threat of antibiotic resistance among the pathogens [31]. The reckless use of antibiotics has provided an advantage to pathogens like *P. aeruginosa* to resist beta-lactams, aminoglycosides, and quinolones [32]. Usage of antibiotics for therapeutic and non-therapeutic reasons gives rise to an antibiotic-resistant gene (*arg*) that tends to protect the pathogen against the detrimental effect of antibiotics [33]. Multidrug resistant (MDR), extensive drug resistant (XDR), and pandrug resistant (PDR) strains of *Acinetobacter* exhibit resistance to polymyxins and tigecyclines, in addition to penicillines, cephalosporins, and carbapenems [34]. A pathogenic strain of yeast such as *Candida albicans* is reported to exhibit resistance against azoles, which are commonly used to treat fungal pathogenesis [35]. The development of antibiotic resistance is not obtained by a single procedure. Various genetic, cellular, as well as environmental factors can contribute toward the evolution of microbial strains with antibiotic resistance. As different antibiotics target specific cell components and pathways, the resistance to antibiotics can be achieved through the circumvention of these lethal pathways or by genetic modification. Alternating the genetic pathway to block antibiotic efficacy is not the only way to demonstrate antibiotic resistance. Specific medicines targeting the ribosomal subunits are debarred from hampering bacterial cells through cell adaptation to induce efflux, reduced influx, modification and degradation of antibiotics, as well as mutation, alteration, or over-expression of the target [36]. Table 16.1 has attempted to sum up the routes followed by different pathogens to prevent the lethal consequences of antibiotic exposure.

Some microbes tend to form biofilms to lessen the antibiotic effect on them. In most cases of pathogenesis, biofilm is reported to induce antibiotic resistance [62]. Biofilm-associated infections remain to be one of the health concerns as biofilm confers antibiotic resistance to the pathogen through limiting antibiotic penetration, nutrient limitation, slow growth, adaptive stress responses, and via the formation of persister cells [63]. Cell density-dependent regulation of gene expression is an important factor of pathogenesis in *S. aureus* biofilms [64]. Biofilm mediated recurrent and chronic infection forms 80% of infections in humans, as microbial cells within the biofilm tend to resist antibiotics about 10–1000 times compared to planktonic cells [65]. The antibiotic resistance of microbial biofilms during infection is determined by up-regulation and down-regulation of several pathways [66]. Antibiotic resistance was also observed in both bacterial and fungal phytopathogens. Phytopathogen belonging to the genera *Erwinia*, *Pectobacterium*, *Pantoea*, *Agrobacterium*, *Pseudomonas*, *Ralstonia*, *Burkholderia*, *Acidovorax*, *Xanthomonas*, *Clavibacter*, *Streptomyces*, *Xylella*, *Spiroplasma*, and *Phytoplasma* are reported to cause various biofilm-oriented diseases in fruits and vegetables, thereby causing agricultural crop loss [67]. Development of antibiotic resistance in phytopathogenic bacteria are presumed to be through horizontal gene transfer from a resistant domain of microbes. Diverse microflora that are subjected to insecticides and pesticides, often used to control agricultural losses, are reported to develop resistance that in due course develops into antibiotic-resistant strains. It has been observed that various microbes can overcome the effects of organophosphate pesticide stress in agriculture through organophosphorus hydrolase, which in turn has been found to facilitate antibiotics resistant to ampicillin, cefotaxime, chloramphenicol, streptomycin, and tetracycline [68]. Pesticides and herbicides such as Granstar herbicide, Tviks, and Alpha Super insecticides are reported to increase the number of streptomycin-resistant cells in microorganisms like *Xanthomonas translucens* and *Pseudomonas syringae* [69]. The streptomycin-resistant gene strAB in *Erwinia amylovora*, *P. syringae*, and *Xanthomonas campestris* is assumed to be acquired from non-pathogenic epiphytes under the same antibiotic stress. Various antibiotics are procured from soil-borne actinomycetes that possess the antibiotic-resistant gene (ARG). These genes in due course of time get transferred to pathogenic strains existing in plant rhizome, which ultimately result in the development of various antibiotic-resistant pathogens [70]. Knowledge regarding antibiotic resistance can be procured from the antibiotic resistance gene database (ARDB), which consists of resistance information for 13 293 genes, 377 types, 257 antibiotics, 632 genomes, 933 species, and 124 genera (see Figure 16.1 [71]).

16.3 Pathogenicity of Antibiotic-Resistant Microbes on Human and Plant Health

The increasing resistance of different infectious agents to antibiotics has led to numerous infection-mediated deaths worldwide. Owing to their rising resistance to antibiotics, traditional treatment fails to control the infection caused by these resistant pathogen species. The increased

Table 16.1 Antibiotic resistance adopted by various microbes.

Antibiotics	Pathway	Pathways adopted for antibiotic resistance	References
1. Penicillins	Binds irreversibly to the active site of transpeptidase through beta lactam rings, thereby preventing the cross-linking of peptidoglycan in cell wall	Inactivation of β-lactam ring by penicillinase via hydrolysis	[37]
2. Cephalosporin	Similar way of action as exhibited by penicillin	Hydrolytic inactivation by beta-lactamases, reduced affinity of cephalosporin to PBP, acquisition of β-lactam insensitive PBP	[38]
3. Fluoroquinolones	Target of this class of antibiotics are bacterial enzyme DNA gyrase and DNA topoisomerase thereby interfering DNA synthesis	Mutation that alters the target site of the antibiotics or reduces the drug accumulation within the cell. Qnr proteins that protect the target site from the antibiotic	[39–41]
4. Aminoglycosides	Induces inhibition in Prokaryotes by binding to 30S ribosomal subunit of Prokaryotes ribosomes	Ribosomal mutation and modification results in non-binding of AG, reduced intake of AGs, expulsion of AGs by efflux pumps, inactivation of AGs by AME	[42, 43]
5. Vancomycin	Binds with high affinity with D-ala-D-ala C terminus of the NAM pentapeptide precursor thereby inhibiting further crosslinking of peptidoglycan	Synthesis of D-ala-D-lac depsipepetide instead of D-ala-D-ala which decreases the affinity of vancomycin to the precursor pentapeptide	[44, 45]
6. Carbapenems	Binds to PBP, which eventually inhibits the inhibitor of an autolytic enzyme within the cell wall. Consequently, the glycan backbone of cell wall collapses, causing cell wall damage	Reduced influx of β-lactam, expulsion of carbapenems via efflux pump, inhibition by a class of beta-lactamases called carbapenemases	[46–48]
7. Tetracyclines	Interacts with 16S ribosomal RNA target in 30S ribosome subunit, thereby arresting translation by inhibiting docking of tRNA	Tetracycline efflux, decreased uptake, target-based mutation in rRNA, ribosomal protection by RPPs, tetracycline deactivation by tetracycline destructases	[49, 50]
8. Monobactam	Binds with PBP and thereby interferes with cell wall synthesis	Reduced influx of monobactam, hydrolysis by β-lactamases, and insensitivity of PBP toward monobactam	[51, 52]
9. Macrolides	Inhibits protein synthesis by binding stoichiometrically to 50S subunit	Target site modification or mutation that inhibits binding of macrolides to 50S subunit, drug efflux, and modification	[53, 54]

(Continued)

Table 16.1 (Continued)

Antibiotics	Pathway	Pathways adopted for antibiotic resistance	References
10. Lincosamide	Interferes with 23S of 50S subunits, thereby inhibiting protein synthesis	Similar mechanism as macrolides	[55]
11. Metronidazole	Metabolic product nitroso-free radical disrupts DNA, RNA, and protein	Increased drug efflux or decreased influx, decreased drug inactivation, increased radical scavenging, and DNA repair mechanism	[56]
12. Rifampicin	Inhibits β-subunit of RNA polymerase, thereby inhibiting RNA synthesis	Mutation in *rpo*B gene that codes for the β-subunit of RNA polymerase	[57]
13. Sulfonamides	Competitively inhibits dihydro-pteroate synthase, which is essential for folic acid	Mutations in *dhps* gene that code for DHPS, which does not bind to Sulfonamides, horizontal transfer of *dhps* gene	[58, 59]
14. Azoles	Selective inhibition of sterol P450 14αdemethylase ($P450_{14\alpha dm}$) interferes with ergosterol biosynthsis	Alteration in sterol biosynthesis, targer site, uptake, and efflux	[60, 61]

PBP: penicillin-binding proteins; AG: aminoglycosides; AME: aminoglycosides modifying enzymes; RPP: ribosomal protection proteins; DHPS: dihydro-pteroate synthase.

resistance of microbes against antibiotics has resulted in the development of different forms of beta (β)-lactamases by various multidrug-resistant (MDR) and extensive drug-resistant (XDR) strains [72]. There are reports that provide the total deaths occurring in the US each year, which is nearly 23 000 due to the havoc caused by antibiotic-resistant varieties [73]. MRSA has been reported to infect 94 000 people, killing almost 19 000 per year in the US [31]. For instance, in the US approximately 600 patients die annually from infection caused by carbapenem-resistant Enterobacteriaceae (CRE) groups of bacteria, resistant to a wide range of antibiotics. It is estimated that 93.8 million people are infected with *Salmonella*-mediated food-borne infections worldwide, killing about 155 000 patients annually. Emergence of a resistant strain of *Salmonella* against ampicillin, chloramphenicol, fluoroquinolones, and trimethoprim–sulfamethoxazole seems to be a threat to humans [74]. The hospital environment is reported to harbor an enormous number of human pathogens, which, due to the continuous use of biocides, develops resistance to various antibiotics [75]. *Candida albicans*, *Cryptococcus neoformans*, and *Aspergillus fumigatus* induced infection kill about 1.4 million people across the globe and their persistence in resisting various antifungal agents remains to be threat to mankind [76, 77]. Not only are antifungal azole agents used in human treatment for fungal infections, but they are important in the treatment of animal health, crop protection, antifouling, and preservation of timber. Increasing azole resistance of *A. fumigatus* seems to be a threat as mortality owing to aspergillosis is highest among the fungal infection-related deaths worldwide, accounting for around three million infections annually [78]. Antibiotic-resistant phytopathogen is reported to cause huge losses in agricultural production and crop yield. Fungus-mediated infection concurs to 20% loss of perennial yield globally while causing 10% loss in post-harvest crops [79]. The use of antibiotics in agriculture and animal husbandry could lead to a possible

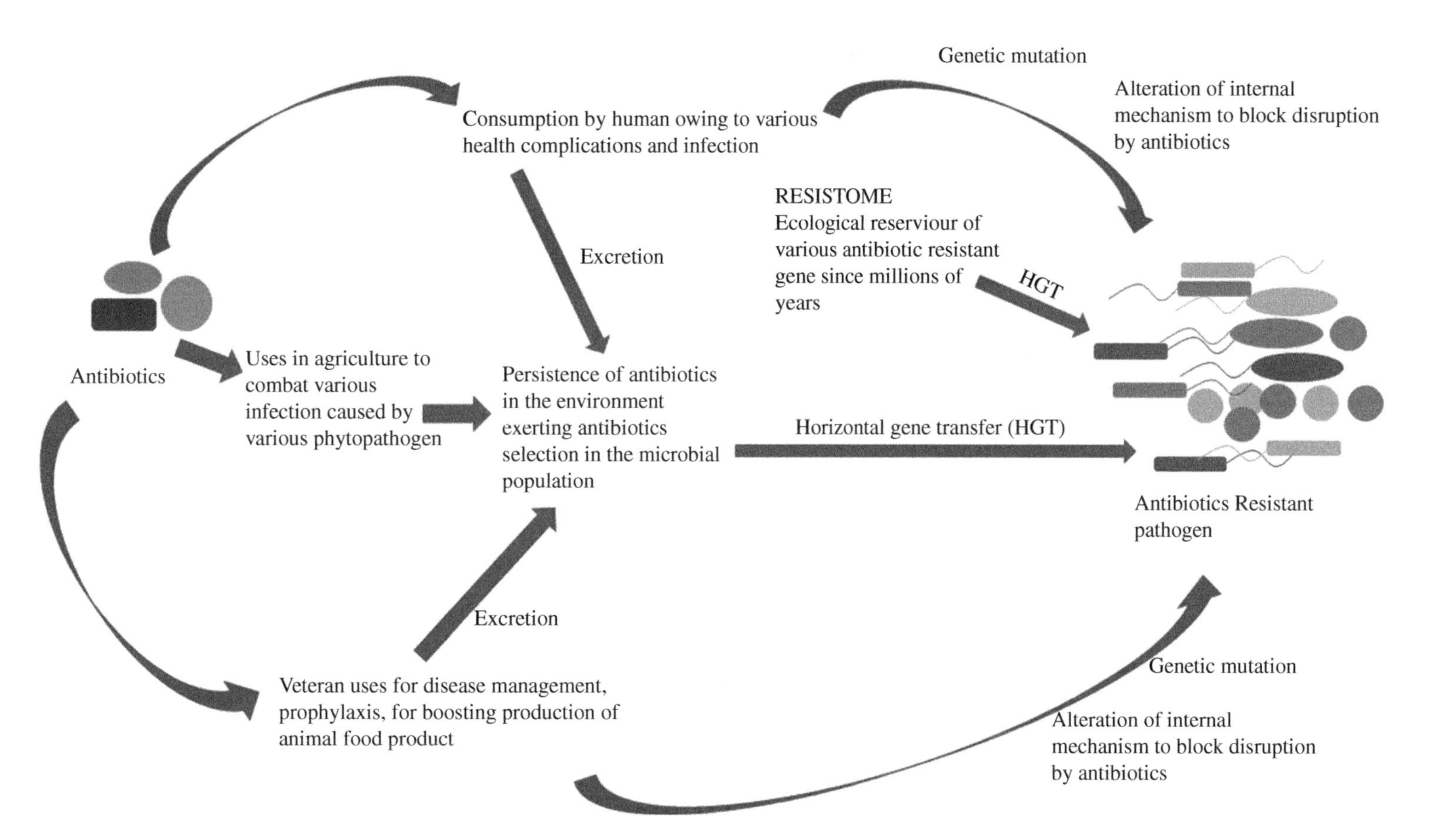

Figure 16.1 A schematic representation of how uses of antibiotics has driven the evolution of antibiotic resistant pathogenic strain [71].

stress of antibiotics resistance that induces pathogenicity in humans through (i) direct infection with resistant bacteria from an animal source, (ii) through host upgradation followed by sustained transmission of resistant strains arising in livestock into the human host, and (iii) transfer of resistance genes from agriculture into human pathogens during antibiotic stress co-evolution [80]. Antimicrobial use as stimulants and prophylactics in animal husbandry in order to increase food output has been reported to breed resistance in certain microbes causing the arrival of antimicrobial-resistant microbes [81, 82]. The discovery of antibiotics was indeed a boon to human kind. However, the co-evolution of microbes to resist this magic bullet, thereby threatening human life, has now been urging an adequate way to revert the threat of resistant strains of different pathogens. With global progress to green biotechnology, surface-active agents produced by various microbial flora can be used to process the upcoming resistant pathogens under different conditions.

16.4 Role of Biosurfactants in Combating Antibiotic Resistance: Challenges and Prospects

With rising concerns of antibiotic resistance among microorganisms there is an urgent need for an alternative to combat various infections occurring in humans, animals, and agricultural grounds effectively. Due to their promising antimicrobial property, biosurfactants have taken global attention. Although they are secreted as secondary metabolites by different microbes, these products are important in terms of medicinal properties. The biosurfactant produced by numerous microorganisms elicit countless antimicrobial properties against several pathogens including MDR pathogens. The presence of the lipophilic tail and hydrophilic head provides the detergent-like structural resemblance, enabling its insertion into the lipid bilayer of the microbial membrane causing disruption, which is responsible for its reputable antimicrobial efficacy [83–87]. The study on the use of naturally produced metabolites such as biosurfactants increased considerably with the resistance of various pathogens to conventional drugs. As biosurfactants are biodegradable products of microbial origin, their uses would pose no harm to the environment. Various glycolipid and lipopeptide biosurfactants are extensively studied for their antimicrobial efficacy against various human and plant pathogens that have given a very impressive result. Lemon, tomato, and potato coated with rhamnolipid were reported to remain free from fungal infection for 15 days at room temperature, suggesting the preventive effectiveness of biosurfactants against food spoilage [88]. Surfactins are reported to elicit efficient antiviral efficacy in an enveloped virus rather than the non-enveloped ones due to their inherent property to interact with a viral lipid envelope. The surfactin lipid chain is reported to determine its antiviral efficacy [89]. Daptomycin, a cyclic lipopeptide, was approved for the treatment of complicated skin and soft tissue infections in Europe in 2006 [90]. Biosurfactants are also reported to boost the immune system in plants to enable them to resist various phytopathogens. Lipopeptide may induce systemic resistance in the host plant at a certain concentration, which can prime the plant's resistance against phytopathogens [91, 92]. Zhang et al. reported in 2016 that the addition of rhamnolipid during aerobic composting to chicken manure may reduce the abundance of the antibiotic resistance gene, thereby limiting the horizontal gene transfer to a pathogenic strain in the environment [93]. If the abundance of antibiotic genes can be reduced, the potency of microbes to upgrade into resistant strains would therefore be minimized. Despite various research into antimicrobial efficacy of various biosurfactants, very little has been addressed as a pharmaceutical or agricultural alternative with regard to the marketing of biosurfactants. A few companies are known to produce biosurfactant with a marketing prospect. The Jeneil Biotech (Biosurfactant) Company is an agro-industrial American company known to produce large-scale rhamnolipids. Biofungicide Zonix™ is a rhamnolipid-based biofungicide marketed by the company. Sophorolipid-containing soaps are specifically marketed for acne therapy by a Korean company called MG Intobio [94].

Global concern of increasing antibiotic-resistant microbial species that have claimed millions of lives owing to their resistance against various conventional medicines needs to be addressed. The tremendous losses in food production due to the persistence of different phytopathogen in farm products could be addressed by changing traditional biocides to biosurfactants, which have reportedly generated different antimicrobial activities against a wide range of pathogens. Although less has been reported about the efficacy of biosurfactants against plant viruses, the inadequate reports yet give us an instinct about the antiviral efficacy of biosurfactants, thus showing that further commercial trials against pathogens must be carried out. Figure 16.2 indicates the prospect of a biosurfactant to act against a wide range of microbes along with its prominent antibiofilm efficacy, thus making it possible to prosper as an alternative to various antibiotics.

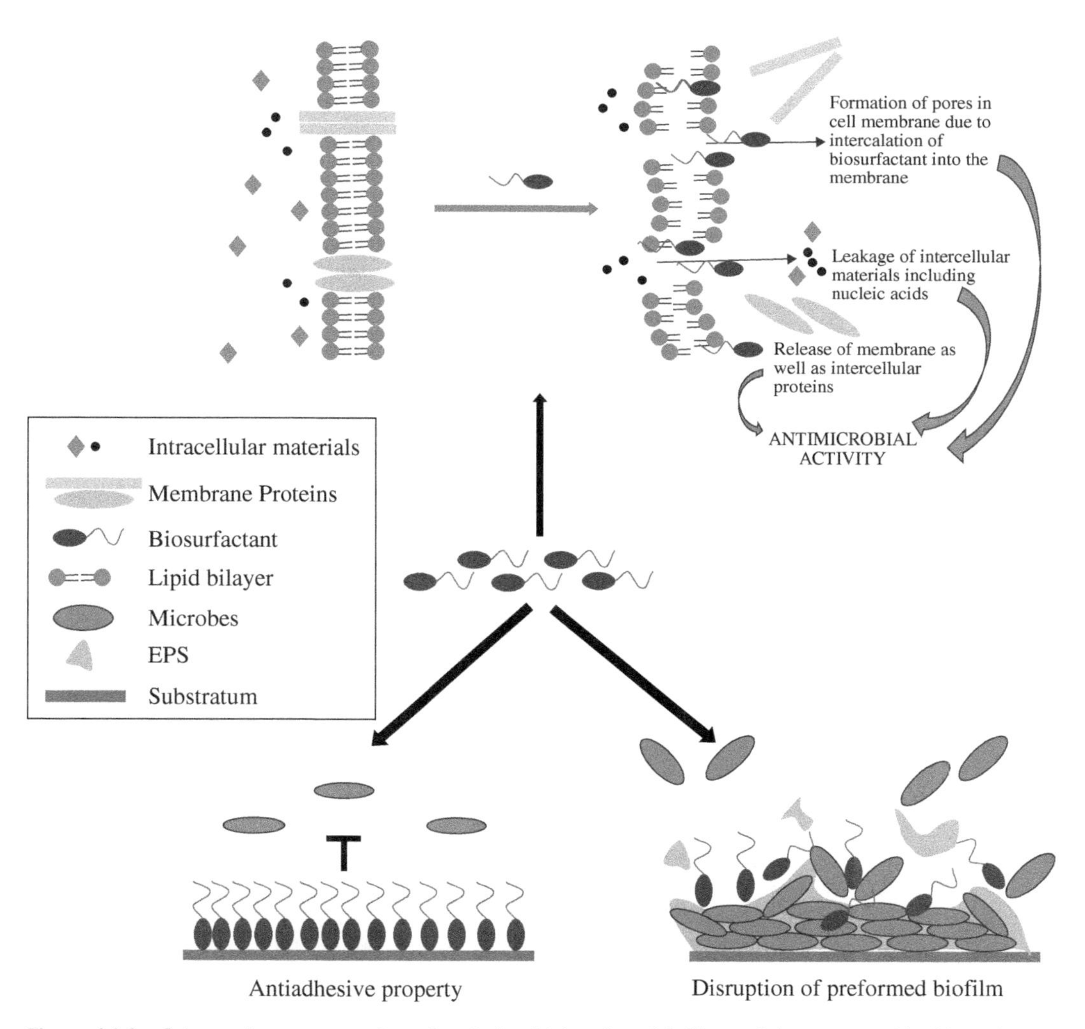

Figure 16.2 Schematic representation of antimicrobial and antibiofilm activity prompted by biosurfactant. (See insert for color representation of the figure).

16.4.1 Biosurfactants Against Pathogenic Bacteria

16.4.1.1 Human and Animal Pathogenic Bacteria

The similarity between the biosurfactant structure and a cell membrane lipid layer provides a substantial benefit in disrupting the planktonic cells of various bacterial pathogens. Although a deep study has been conducted in the susceptible variants of the pathogenic strains the same study has also recently been undertaken against the drug-resistant pathogens. Among various biosurfactants available, glycolipid and lipopeptides call for further investigation owing to their efficiency in this regard. *P. aeruginosa* MN1 produced rhamnolipids has shown antibacterial efficacy against MRSA [95]. Lipopeptide and rhamnolipid produced by *Bacillus amyloliquefaciens* and *P. aeruginosa* are reported to exhibit antibacterial potency against MRSA and drug-resistant Enteropathogenic *Escherichia coli* [83]. Lipopeptide produced by marine isolates *Bacillus circulans* and *Bacillus subtilis* were reported to exhibit antibacterial effect against the MDR strain of *E. coli* and *Klebsiella* sp., *E. faecalis* and *S. aureus* [96, 97]. Biosurfactant produced by *B. subtilis* MTCC 441 exhibited synergistic effectivity against MDR *S. aureus* when combined with various antibiotics [98]. Daptomycin, a class of lipopeptide, is also reported to exhibit the antibacterial potentiality against MRSA [99]. Apart from the major types of glycolipids and lipopeptides, other classes of biosurfactant are also reported to exhibit an antibiotic effect on drug-resistant species of bacteria. Antibacterial efficacy of a novel glycolipid named xylolipid biosurfactant has been reported against the MDR strain of *E. coli* and *S. aureus* [100]. Octapeptins, a lipopeptide produced by *B. circulans*, is reported to exhibit antibacterial efficacy against colistin-resistant pathogens [101]. Similarly, battacin produced by *Paenibacillus tianmuensis* is reported to exhibit antibacterial efficacy multidrug-resistant and extremely drug-resistant clinical isolates [102]. Emulsification of essential oil with rhamnolipid is reported to enhance the antibacterial efficacy of essential oil against MRSA [103]. A patent has been granted to a formulation that utilized a ramoplanin and rhamnolipids combination to combat vancomycin-resistant *Enterococcus*, *Clostridium difficile*, or multidrug-resistant *C. difficile* infections [104].

16.4.1.2 Phytopathogenic Bacteria

In the agricultural sector the biosurfactant-producing rhizosphere microbes are reported to antagonize various phytopathogenic bacteria [105]. However, there is little evidence of the effectiveness of the biosurfactant against drug-resistant phytopathogenic bacteria. Surfactin produced from *B. subtilis* has been reported to combat *P. syringae* infection in the *Arabidopsis* rhizosphere [106]. Rhizospheric isolates consisting of *Pseudomonas* and *Bacillus* exhibit antibacterial effects against *Pectobacterium* and *Dickeya* sp. [107]. The prospect of biosurfactant in managing drug-resistant agricultural pathogen needs attention to address loss of agricultural yield due to persistence of infection.

16.4.2 Biosurfactants Against Pathogenic Fungi

16.4.2.1 Human and Animal Pathogenic Fungi

Fungal infections form a major part of an infection that is associated with high morbidity and mortality worldwide. The reported efficacy of the biosurfactant against several dermatophytes and pathogenic yeast could be significant in controlling fungal infections. Lipopeptide produced by *B. amyloliquefaciens* was reported to exhibit anticandida efficacy against *C. albicans* [108]. The antimicrobial effect of biosurfactant produced by *Lactococcus lactis* has been reported against *C. albicans*, which is otherwise reported to be resistant to itraconazole and clotrimazole [109].

Lipopeptide biosurfactant produced by *B. subtilis* was found to be effective against planktonic as well as biofilms of *Trichosporon* sp. [110]. Similarly, *in vitro* studies of sophorolipid have revealed that they exhibit promising antifungal efficacy against *Trichophyton rubrum* [111]. Apart from the membrane disruption mode of killing, Sophorolipid is reported to initiate ROS mediated endoplasmic reticulum stress and mitochondrial dysfunction pathways in *C. albicans*, thereby exhibiting anticandida efficacy [112].

16.4.2.2 Phytopathogenic Fungi

Although antifungal efficacy of various types of biosurfactant against different phytopathogens has been established, there is limited study in regard of drug-resistant phytopathogens. There are inadequate reports wherein antifungal efficacy of biosurfactant has been studied against various phytopathogenic fungus that have acquired resistivity against various commercial fungicides. Biosurfactants are reported to exhibit sporicidal activity against phytopathogenic fungus that showed resistivity against certain commercial fungicides [105]. The efficacy of rhamnolipid biosurfactant as a fungicide has been established against *Phytophthora infestans* and *Phytophthora capsici* that were otherwise reported to be resistant to commercial fungicides [113]. Existing literature documented the efficacy of rhamnolipid in lysing the zoospore of zoosporic phytopathogenic fungus through intercalation into the plasma membrane [114, 115]. Among various antimicrobial metabolites released by *B. subtilis* and *B. amyloliquefaciens* in the rhizosphere, lipopeptide constitutes a major portion of antifungal agents [116]. For instance, there are literatures that revealed the potency of lipopeptide biosurfactant as a promising antifungal activity against phytopathogens like *Pythium ultimum*, *Botrytis cinerea* [117], *Podosphaera fusca* [118], *Gibberella zeae* [119], and *Rhizoctonia solani* [120].

16.4.3 Biosurfactants Against Pathogenic Viruses

16.4.3.1 Human and Animal Viruses

Biosurfactants have been reported to produce antiviral effects, apart from antibacterial and antifungal efficacy. Antiviral efficacy against the enveloped virus is reported to be greater than the non-enveloped virus due to its efficacy in penetrating lipid membrane of the enveloped viruses [121, 122]. Rhamnolipid produced by *P. aeruginosa* PS-17 was shown to have antiviral efficacy against herpes simplex virus type 1 and type 2 with IC_{50} at a concentration of 14.5 and 13 μg/ml [123]. *In vitro* antiviral efficacy of surfactin and fengycin produced by *B. subtilis* fmbj was reported against Pseudorabies virus (PRV), Porcine parvovirus (PPV), Newcastle Disease virus (NDV), and Infectious Bursal Disease virus (IBDV) [124]. Surfactin produced by *B. subtilis* was found to disrupt the lipid membrane of the viral coat of Semliki Forest virus (SFV), Herpes Simplex virus (HSV-1, HSV-2), Suid Herpes virus (SHV-1), Vesicular Stomatitis virus (VSV), Simian Immunodeficiency virus (SIV), Feline calicivirus (FCV), and Murine Encephalomyocarditis virus (EMCV) [121]. Similarly, sophorolipid and sophorolipid diacetate ethyl ester derivative was reported to exhibit antiviral activity against human immunodeficiency virus (HIV) [125]. Antiviral efficacy of sophorolipid has also been patented due to its promising nature against herpes virus [126].

16.4.3.2 Phytopathogenic Virus

The loss of crops due to viral infections causes a major loss of food production worldwide. Although the use of insecticides and pesticides can keep the viral vector away, these conventional biocides are ineffective against phytopathogenic virus. Although surfactin is reported to exhibit antiviral efficacy through disintegration of the viral envelope, much less has been reported regarding the bio-control efficacy of surfactin against viral infection in plants [127].

16.4.4 Biosurfactant Against Biofilms

Biofilm has always been an inevitable part of various superficial as well as invasive infections. Biofilms give pathogen antibiotic resistance by reducing the penetration of antibiotics in biofilm aggregation. Biosurfactants are reported to elicit antibiofilm efficacy either by disrupting the biofilm or by inhibiting its formation. The efficacy of these molecules lies in its efficiency in initiating physicochemical changes in biofilm that either prevents cell adhesion or cell dispersal [128]. Sophorolipid alone and in combination with amphotericin B (AmB) or fluconazole (FLZ) was reported to exhibit antibiofilm efficacy against the common while most devastating pathogen *C. albicans* [129]. Similarly, efficacy of mannosylerithritol lipid (MEL) against *S. aureus* biofilm has been reported in 2020 by Ceresa et al. [130]. At the same time, biosurfactant produced by *Lactobacillus jensenii* and *Lactobacillus rhamnosus* was reported to reduce adherence of MDR strains of *A. baumannii*, *E. coli*, and *S. aureus* (MRSA) along with biofilm disruption and inhibition of biofilm formation [131]. Lipopeptide from *Bacillus tequilensis* CH was reported to inhibit biofilm of *E. coli* and *Streptococcus mutans* at a concentration of 50 μg/ml [132]. Lichenysin, a lipopeptide produced by *Bacillus licheniformis*, was reported to exhibit antibiofilm efficacy against *C. albicans* and *MRSA* biofilm. Additionally, lipopeptide produced by marine actinobacterial strain *Nesterenkonia* sp. MSA31 was reported to inhibit biofilms formed by MRSA strains [133]. Apart from SL, MEL, and LPs, rhamnolipids play a major role in inhibiting microbial biofilms. For instance, in 2005 Irie et al. showed the effect of rhamnolipid on respiratory pathogens biofilm *Bordetella bronchiseptica* [134]. Rhamnolipid when combined with amphotericin B was observed to act synergistically on biofilm of *Trichosporon cutaneum* and *Candida parapsilosis* [135]. Although biofilm is an inevitable part of various infectious diseases that claim human lives, its involvement in phytopathogensis cannot be ignored. Biosurfactants are reported to act on various biofilms that are responsible for biofilm-mediated disease in crops, vegetables, and fruits [67]. Although biosurfactants produced by various microorganism have already been studied to a greater extent [136], details related to their production is lacking [137].

16.5 Conclusion

The beneficial impact of microbes in human life is innumerable. However, the invasion of different microbes affecting human life is also inevitable from time to time. Increased microbial resistance to various antibiotics has now forced the scientific community to look for an alternative to tackle antibiotic resistance efficiently. However, the stress of various chemically synthesized antibiotics on the physiology of plants and animals as well as on the environment cannot be ignored. With global advancement toward green technology, microbial biosurfactants can be exploited to find an effective antibiotic that can invade antibiotic-resistant microbes. While there is deep research on the multifarious activity of biosurfactants, much less is known about the effectiveness of biosurfactants against drug-resistant humans, animals, and phytopathogens. A thorough investigation is to be conducted to replace the chemically synthesized pharmaceuticals and agricultural product with microbial products such as biosurfactants.

Acknowledgements

The work was funded by the Department of Biotechnology (DBT), Government of India by a grant to Chandana Malakar as DBT-JRF. The authors thank the Director, Institute of Advanced Study in Science and Technology (IASST), Garchuk, Assam, India, for providing facilities.

References

1 Zaffiri, L., Gardner, J., and Toledo-Pereyra, L.H. (2012). History of antibiotics. From salvarsan to cephalosporins. *Journal of Investigative Surgery* 25 (2): 67–77.
2 Clardy, J., Fischbach, M.A., and Currie, C.R. (2009). The natural history of antibiotics. *Current Biology* 19 (11): R437–R441.
3 Mohr, K.I. (2016). History of antibiotics research. In: *How to Overcome the Antibiotic Crisis* (eds. M. Stadler and P. Dersch), 237–272. Cham: Springer.
4 Rolain, J.M. and Baquero, F. (2016). The refusal of the Society to accept antibiotic toxicity: missing opportunities for therapy of severe infections. *Clinical Microbiology and Infection* 22 (5): 423–427.
5 Gothwal, R. and Shashidhar, T. (2015). Antibiotic pollution in the environment: A review. *Clean: Soil, Air, Water* 43 (4): 479–489.
6 Manzetti, S. and Ghisi, R. (2014). The environmental release and fate of antibiotics. *Marine Pollution Bulletin* 79 (1–2): 7–15.
7 Karanth, N.G., Deo, P.G., and Veenanadig, N.K. (1999; 77(1)). Microbial production of biosurfactants and their importance. *Current Science*: 116–126.
8 Mulligan, C.N. and Gibbs, B.F. (2004). Types, production and applications of biosurfactants. *Proceedings-Indian National Science Academy Part B* 70 (1): 31–56.
9 de Jesus Cortes-Sanchez, A., Hernandez-Sanchez, H., and Jaramillo-Flores, M.E. (2013). Biological activity of glycolipids produced by microorganisms: New trends and possible therapeutic alternatives. *Microbiological Research* 168 (1): 22–32.
10 Grill, M.F. and Maganti, R.K. (2011). Neurotoxic effects associated with antibiotic use: management considerations. *British Journal of Clinical Pharmacology* 72 (3): 381–393.
11 Kaloyanides, G.J. (1994). Antibiotic-related nephrotoxicity. *Nephrology, Dialysis, Transplantation: Official Publication of the European Dialysis and Transplant Association-European Renal Association* 9: 130–134.
12 McNeil, J.J., Grabsch, E.A., and McDonald, M.M. (1999). Postmarketing surveillance: strengths and limitations: The flucloxacillin-dicloxacillin story. *Medical Journal of Australia* 170 (6): 270–273.
13 Campagna, J.D., Bond, M.C., Schabelman, E., and Hayes, B.D. (2012). The use of cephalosporins in penicillin-allergic patients: A literature review. *The Journal of Emergency Medicine* 42 (5): 612–620.
14 Madaan, A. and Li, J.T. (2004). Cephalosporin allergy. *Immunology and Allergy Clinics* 24 (3): 463–476.
15 Aygün, C., Kocaman, O., Gürbüz, Y. et al. (2007). Clindamycin-induced acute cholestatic hepatitis. *World Journal of Gastroenterology: WJG* 13 (40): 5408.
16 Gray, J.E., Weaver, R.N., Moran, J., and Feenstra, E.S. (1973). The parenteral toxicity of clindamycin-2-phosphate in laboratory animals. *Toxicology and Applied Pharmacology* 25: 492.
17 Wilson, A.P. (1998). Comparative safety of teicoplanin and vancomycin. *International Journal of Antimicrobial Agents* 10 (2): 143–152.
18 Westphal, J.F., Vetter, D., and Brogard, J.M. (1994). Hepatic side-effects of antibiotics. *Journal of Antimicrobial Chemotherapy* 33 (3): 387–401.

19 Brackett, C.C., Singh, H., and Block, J.H. (2004). Likelihood and mechanisms of cross-allergenicity between sulfonamide antibiotics and other drugs containing a sulfonamide functional group. *Pharmacotherapy: The Journal of Human Pharmacology and Drug Therapy* 24 (7): 856–870.

20 Paech, F., Messner, S., Spickermann, J. et al. (2017). Mechanisms of hepatotoxicity associated with the monocyclic β-lactam antibiotic BAL30072. *Archives of Toxicology* 91 (11): 3647–3662.

21 Cheng, G., Hao, H., Xie, S. et al. (2014). Antibiotic alternatives: The substitution of antibiotics in animal husbandry? *Frontiers in Microbiology* 5: 217.

22 Daghrir, R. and Drogui, P. (2013). Tetracycline antibiotics in the environment: A review. *Environmental Chemistry Letters* 11 (3): 209–227.

23 Van Boeckel, T.P., Gandra, S., Ashok, A. et al. (2014). Global antibiotic consumption 2000 to 2010: An analysis of national pharmaceutical sales data. *The Lancet Infectious Diseases* 14 (8): 742–750.

24 Andersson, D.I. and Hughes, D. (2012). Evolution of antibiotic resistance at non-lethal drug concentrations. *Drug Resistance Updates* 15 (3): 162–172.

25 Hughes, J.M. and Tenover, F.C. (1997). Approaches to limiting emergence of antimicrobial resistance in bacteria in human populations. *Clinical Infectious Diseases* 24 (Supplement 1): S131–S135.

26 Wright, G.D. (2010). Antibiotic resistance in the environment: A link to the clinic? *Current Opinion in Microbiology* 13 (5): 589–594.

27 Gao, W., Cheng, H., Hu, F. et al. (2010). The evolution of *Helicobacter pylori* antibiotics resistance over 10 years in Beijing, China. *Helicobacter* 15 (5): 460–466.

28 Sieradzki, K., Roberts, R.B., Haber, S.W., and Tomasz, A. (1999). The development of vancomycin resistance in a patient with methicillin-resistant *Staphylococcus aureus* infection. *New England Journal of Medicine* 340 (7): 517–523.

29 Gordon, R.J. and Lowy, F.D. (2008). Pathogenesis of methicillin-resistant *Staphylococcus aureus* infection. *Clinical Infectious Diseases* 46 (Supplement 5): S350–S359.

30 Huang, S.S. and Platt, R. (2003). Risk of methicillin-resistant *Staphylococcus aureus* infection after previous infection or colonization. *Clinical Infectious Diseases* 36 (3): 281–285.

31 Vashishtha, V.M. (2010). Growing antibiotics resistance and the need for new antibiotics. *Indian Pediatrics* 47 (6): 505–506.

32 Wolska, K., Kot, B., Piechota, M., and Frankowska, A. (2013). Resistance of *Pseudomonas aeruginosa* to antibiotics. *Postępy Higieny i Medycyny Doświadczalnej (Online)* 67: 1300–1311.

33 Martinez, J.L. (2009). Environmental pollution by antibiotics and by antibiotic resistance determinants. *Environmental Pollution* 157 (11): 2893–2902.

34 Manchanda, V., Sanchaita, S., and Singh, N.P. (2010). Multidrug resistant acinetobacter. *Journal of Global Infectious Diseases* 2 (3): 291.

35 Niimi, M., Firth, N.A., and Cannon, R.D. (2010). Antifungal drug resistance of oral fungi. *Odontology* 98 (1): 15–25.

36 Wilson, D.N. (2014). Ribosome-targeting antibiotics and mechanisms of bacterial resistance. *Nature Reviews Microbiology* 12 (1): 35–48.

37 Lobanovska, M. and Focus, P.G. (2017). Drug development: penicillin's discovery and antibiotic resistance: Lessons for the future? *The Yale Journal of Biology and Medicine* 90 (1): 135.

38 Livermore, D.M. (1987). Mechanisms of resistance to cephalosporin antibiotics. *Drugs* 34 (2): 64–88.

39 Jacoby, G.A. (2005). Mechanisms of resistance to quinolones. *Clinical Infectious Diseases* 41 (Supplement 2): S120–S126.

40 Hooper, D.C. and Jacoby, G.A. (2015). Mechanisms of drug resistance: Quinolone resistance. *Annals of the New York Academy of Sciences* 1354 (1): 12.
41 Redgrave, L.S., Sutton, S.B., Webber, M.A., and Piddock, L.J. (2014). Fluoroquinolone resistance: Mechanisms, impact on bacteria, and role in evolutionary success. *Trends in Microbiology* 22 (8): 438–445.
42 Garneau-Tsodikova, S. and Labby, K.J. (2016). Mechanisms of resistance to aminoglycoside antibiotics: overview and perspectives. *MedChemComm* 7 (1): 11–27.
43 Mingeot-Leclercq, M.P., Glupczynski, Y., and Tulkens, P.M. (1999). Aminoglycosides: activity and resistance. *Antimicrobial Agents and Chemotherapy* 43 (4): 727–737.
44 Courvalin, P. (2006). Vancomycin resistance in gram-positive cocci. *Clinical Infectious Diseases* 42 (Supplement 1): S25–S34.
45 Faron, M.L., Ledeboer, N.A., and Buchan, B.W. (2016). Resistance mechanisms, epidemiology, and approaches to screening for vancomycin-resistant *Enterococcus* in the health care setting. *Journal of Clinical Microbiology* 54 (10): 2436–2447.
46 Codjoe, F.S. and Donkor, E.S. (2018). Carbapenem resistance: A review. *Medical Science* 6 (1): 1.
47 Meletis, G. (2016). Carbapenem resistance: overview of the problem and future perspectives. *Therapeutic Advances in Infectious Disease* 3 (1): 15–21.
48 Elshamy, A.A. and Aboshanab, K.M. (2020). A review on bacterial resistance to carbapenems: Epidemiology, detection and treatment options. *Future Science OA* 6 (3): FSO438.
49 Grossman, T.H. (2016). Tetracycline antibiotics and resistance. *Cold Spring Harbor Perspectives in Medicine* 6 (4): a025387.
50 Speer, B.S., Shoemaker, N.B., and Salyers, A.A. (1992). Bacterial resistance to tetracycline: mechanisms, transfer, and clinical significance. *Clinical Microbiology Reviews* 5 (4): 387–399.
51 Sykes, R.B. and Bonner, D.P. (1985). Aztreonam: The first monobactam. *The American Journal of Medicine* 78 (2): 2–10.
52 Livermore, D.M. (1991). Mechanisms of resistance to lactam antibiotics. *Scandinavian Journal of Infectious Diseases* 78 (Supl): 7–16.
53 Leclercq, R. and Courvalin, P. (2002). Resistance to macrolides and related antibiotics in *Streptococcus pneumoniae*. *Antimicrobial Agents and Chemotherapy* 46 (9): 2727–2734.
54 Leclercq, R. (2002). Mechanisms of resistance to macrolides and lincosamides: nature of the resistance elements and their clinical implications. *Clinical Infectious Diseases* 34 (4): 482–492.
55 Sutcliffe, J.A. and Leclercq, R. (2002). Mechanisms of resistance to macrolides, lincosamides, and ketolides. In: *Macrolide Antibiotics* (eds. W. Schönfeld and H.A. Kirst), 281–317. Basel: Birkhäuser.
56 Dhand, A. and Snydman, D.R. (2009). Mechanism of resistance in metronidazole. In: *Antimicrobial Drug Resistance* (ed. D.L. Mayers), 223–227. Humana Press.
57 Goldstein, B.P. (2014). Resistance to rifampicin: A review. *The Journal of Antibiotics* 67 (9): 625–630.
58 Then, R.L. (1989). Resistance to sulfonamides. In: *Microbial Resistance to Drugs* (ed. L.E. Bryan), 291–312. Berlin, Heidelberg: Springer.
59 Sköld, O. (2000). Sulfonamide resistance: Mechanisms and trends. *Drug Resistance Updates* 3 (3): 155–160.
60 Joseph-Horne, T. and Hollomon, D.W. (1997). Molecular mechanisms of azole resistance in fungi. *FEMS Microbiology Letters* 149 (2): 141–149.
61 Cowen, L.E., Sanglard, D., Howard, S.J. et al. (2015). Mechanisms of antifungal drug resistance. *Cold Spring Harbor Perspectives in Medicine* 5 (7): a019752.

62 Landini, P., Antoniani, D., Burgess, J.G., and Nijland, R. (2010). Molecular mechanisms of compounds affecting bacterial biofilm formation and dispersal. *Applied Microbiology and Biotechnology* 86 (3): 813–823.

63 Stewart, P.S. (2002). Mechanisms of antibiotic resistance in bacterial biofilms. *International Journal of Medical Microbiology* 292 (2): 107–113.

64 Kong, K.F., Vuong, C., and Otto, M. (2006). *Staphylococcus quorum* sensing in biofilm formation and infection. *International Journal of Medical Microbiology* 296 (2–3): 133–139.

65 Sharma, D., Misba, L., and Khan, A.U. (2019). Antibiotics versus biofilm: An emerging battleground in microbial communities. *Antimicrobial Resistance and Infection Control* 8 (1): 76.

66 Mah, T.F. and O'Toole, G.A. (2001). Mechanisms of biofilm resistance to antimicrobial agents. *Trends in Microbiology* 9 (1): 34–39.

67 Padmavathi, A.R., Bakkiyaraj, D., and Pandian, S.K. (2015). Significance of biosurfactants as antibiofilm agents in eradicating phytopathogens. In: *Bacterial Metabolites in Sustainable Agroecosystem* (ed. D. Maheshwari), 319–336. Cham: Springer.

68 Rangasamy, K., Athiappan, M., Devarajan, N., and Parray, J.A. (2017). Emergence of multi drug resistance among soil bacteria exposing to insecticides. *Microbial Pathogenesis* 105: 153–165.

69 Buletsa, N.M., Butsenko, L.M., Pasichnyk, L.A., and Patyka, V.P. (2015). The sensitivity of phytopathogenic bacteria to streptomycin under the influence of pesticides. *Mikrobiolohichnyi zhurnal (Kiev, Ukraine: 1993)* 77 (6): 62–69.

70 Sundin, G.W. and Wang, N. (2018). Antibiotic resistance in plant-pathogenic bacteria. *Annual Review of Phytopathology* 56: 161–180.

71 Liu, B. and Pop, M. (2009). ARDB – Antibiotic resistance genes database. *Nucleic Acids Research* 37 (suppl 1): D443–D447.

72 Ansari, S., Nepal, H.P., Gautam, R. et al. (2015). Community acquired multi-drug resistant clinical isolates of *Escherichia coli* in a tertiary care center of Nepal. *Antimicrobial Resistance and Infection Control* 4 (1): 15.

73 Gross, M. Antibiotics in crisis.Current Microbiology 2013;23(24):1063–1065.

74 Eng, S.K., Pusparajah, P., Ab Mutalib, N.S. et al. (2015). *Salmonella*: A review on pathogenesis, epidemiology and antibiotic resistance. *Frontiers in Life Science* 8 (3): 284–293.

75 Smith, K. and Hunter, I.S. (2008). Efficacy of common hospital biocides with biofilms of multi-drug resistant clinical isolates. *Journal of Medical Microbiology* 57 (8): 966–973.

76 Sanglard, D. (2016). Emerging threats in antifungal-resistant fungal pathogens. *Frontiers in Medicine* 3: 11.

77 Xie, J.L., Polvi, E.J., Shekhar-Guturja, T., and Cowen, L.E. (2014). Elucidating drug resistance in human fungal pathogens. *Future Microbiology* 9 (4): 523–542.

78 Chowdhary, A., Kathuria, S., Xu, J., and Meis, J.F. (2013). Emergence of azole-resistant *Aspergillus fumigatus* strains due to agricultural azole use creates an increasing threat to human health. *PLoS Pathogens* 9 (10).

79 Fisher, M.C., Hawkins, N.J., Sanglard, D., and Gurr, S.J. (2018). Worldwide emergence of resistance to antifungal drugs challenges human health and food security. *Science* 360 (6390): 739–742.

80 Chang, Q., Wang, W., Regev-Yochay, G. et al. (2015). Antibiotics in agriculture and the risk to human health: how worried should we be? *Evolutionary Applications* 8 (3): 240–247.

81 O'Neill, J. (2015). Antimicrobials in agriculture and the environment: Reducing unnecessary use and waste. *The Review on Antimicrobial Resistance*: 1–44.

82 Nisha, A.R. (2008). Antibiotic residues-a global health hazard. *Veterinary World* 1 (12): 375.

83 Ndlovu, T., Rautenbach, M., Vosloo, J.A. et al. (2017). Characterisation and antimicrobial activity of biosurfactant extracts produced by *Bacillus amyloliquefaciens* and *Pseudomonas aeruginosa* isolated from a wastewater treatment plant. *AMB Express* 7 (1): 108.

84 Inès, M. and Dhouha, G. (2015). Glycolipid biosurfactants: Potential related biomedical and biotechnological applications. *Carbohydrate Research* 416: 59–69.

85 Chen, W.C., Juang, R.S., and Wei, Y.H. (2015). Applications of a lipopeptide biosurfactant, surfactin, produced by microorganisms. *Biochemical Engineering Journal* 103: 158–169.

86 Sen, S., Borah, S.N., Kandimalla, R. et al. (2019). Efficacy of a rhamnolipid biosurfactant to inhibit *Trichophyton rubrum* in vitro and in a mice model of dermatophytosis. *Experimental Dermatology* 28 (5): 601–608.

87 Sen, S., Borah, S.N., Kandimalla, R. et al. (2020). Sophorolipid biosurfactant can control cutaneous dermatophytosis caused by *Trichophyton mentagrophytes*. *Frontiers in Microbiology* 11: 329.

88 Sharma, V.I., Garg, M., Talukdar, D. et al. (2018). Preservation of microbial spoilage of food by biosurfactant-based coating. *Asian Journal of Pharmaceutical and Clinical Research* 11: 98–101.

89 Meena, K.R. and Kanwar, S.S. (2015). Lipopeptides as the antifungal and antibacterial agents: applications in food safety and therapeutics. *BioMed Research International* 2015.

90 Alder, J.D. (2005). Daptomycin: a new drug class for the treatment of Gram-positive infections. *Drugs of Today* 41 (2): 81–90.

91 Debois, D., Fernandez, O., Franzil, L. et al. (2015). Plant polysaccharides initiate underground crosstalk with bacilli by inducing synthesis of the immunogenic lipopeptide surfactin. *Environmental Microbiology Reports* 7 (3): 570–582.

92 Cawoy, H., Mariutto, M., Henry, G. et al. (2014). Plant defense stimulation by natural isolates of *Bacillus* depends on efficient surfactin production. *Molecular Plant-Microbe Interactions* 27 (2): 87–100.

93 Zhang, Y., Li, H., Gu, J. et al. (2016). Effects of adding different surfactants on antibiotic resistance genes and intI1 during chicken manure composting. *Bioresource Technology* 219: 545–551.

94 Hames, E.E., Vardar-Sukan, F., and Kosaric, N. (2014). 11 patents on biosurfactants and future trends. *Biosurfactants: Production and Utilization-Processes, Technologies, and Economics* 159: 165.

95 Samadi, N., Abadian, N., Ahmadkhaniha, R. et al. (2012). Structural characterization and surface activities of biogenic rhamnolipid surfactants from *Pseudomonas aeruginosa* isolate MN1 and synergistic effects against methicillin-resistant *Staphylococcus aureus*. *Folia Microbiologica* 57 (6): 501–508.

96 Das, P., Mukherjee, S., and Sen, R. (2008). Antimicrobial potential of a lipopeptide biosurfactant derived from a marine *Bacillus circulans*. *Journal of Applied Microbiology* 104 (6): 1675–1684.

97 Fernandes, P.A., Arruda, I.R., Santos, A.F. et al. (2007). Antimicrobial activity of surfactants produced by *Bacillus subtilis* R14 against multidrug-resistant bacteria. *Brazilian Journal of Microbiology* 38 (4): 704–709.

98 Irfan, M., Shahi, S.K., and Sharma, P.K. (2015). In vitro synergistic effect of biosurfactant produced by *Bacillus subtilis* MTCC 441 against drug resistant *Staphylococcus aureus*. *Journal of Applied Pharmaceutical Science.* 5 (03): 113–116.

99 Kanafani, Z.A. and Corey, G.R. (2007). Daptomycin: a rapidly bactericidal lipopeptide for the treatment of Gram-positive infections. *Expert Review of Anti-Infective Therapy* 5 (2): 177–184.

100 Saravanakumari, P. and Mani, K. (2010). Structural characterization of a novel xylolipid biosurfactant from *Lactococcus lactis* and analysis of antibacterial activity against multi-drug resistant pathogens. *Bioresource Technology* 101 (22): 8851–8854.

101 Fayad, A.A., Herrmann, J., and Müller, R. (2018). Octapeptins: Lipopeptide antibiotics against multidrug-resistant superbugs. *Cell Chemical Biology* 25 (4): 351–353.

102 Qian, C.D., Wu, X.C., Teng, Y. et al. (2012). Battacin (Octapeptin B5), a new cyclic lipopeptide antibiotic from *Paenibacillus tianmuensis* active against multidrug-resistant gram-negative bacteria. *Antimicrobial Agents and Chemotherapy* 56 (3): 1458–1465.

103 Haba, E., Bouhdid, S., Torrego-Solana, N. et al. (2014). Rhamnolipids as emulsifying agents for essential oil formulations: antimicrobial effect against *Candida albicans* and methicillin-resistant *Staphylococcus aureus*. *International Journal of Pharmaceutics* 476 (1–2): 134–141.

104 Yin X, inventor; AGAE Technologies LLC, assignee. Formulations combining ramoplanin and rhamnolipids for combating bacterial infection. US Patent Application US 14/201,633. 2014.

105 Sachdev, D.P. and Cameotra, S.S. (2013). Biosurfactants in agriculture. *Applied Microbiology and Biotechnology* 97 (3): 1005–1016.

106 Bais, H.P., Fall, R., and Vivanco, J.M. (2004). Biocontrol of Bacillus subtilis against infection of Arabidopsis roots by *Pseudomonas syringae* is facilitated by biofilm formation and surfactin production. *Plant Physiology* 134 (1): 307–319.

107 Krzyzanowska, D.M., Potrykus, M., Golanowska, M. et al. (2012). Rhizosphere bacteria as potential biocontrol agents against soft rot caused by various *Pectobacterium* and *Dickeya* spp. strains. *Journal of Plant Pathology*: 367–378.

108 Song, B., Rong, Y.J., Zhao, M.X., and Chi, Z.M. (2013). Antifungal activity of the lipopeptides produced by *Bacillus amyloliquefaciens* anti-CA against *Candida albicans* isolated from clinic. *Applied Microbiology and Biotechnology* 97 (16): 7141–7150.

109 Saravanakumari, P. and Nirosha, P. (2015). Mechanism of control of *Candida albicans* by biosurfactant purified from *Lactococcus lactis*. *International Journal of Current Microbiology and Applied Sciences* 4 (2): 529–542.

110 Cordeiro, R.D., Weslley Caracas Cedro, E., Raquel Colares Andrade, A. et al. (2018). Inhibitory effect of a lipopeptide biosurfactant produced by *Bacillus subtilis* on planktonic and sessile cells of *Trichosporon* spp. *Biofouling* 34 (3): 309–319.

111 Sen, S., Borah, S.N., Bora, A., and Deka, S. (2017). Production, characterization, and antifungal activity of a biosurfactant produced by *Rhodotorula babjevae* YS3. *Microbial Cell Factories* 16 (1): 95.

112 Haque, F., Verma, N.K., Alfatah, M. et al. (2019). Sophorolipid exhibits antifungal activity by ROS mediated endoplasmic reticulum stress and mitochondrial dysfunction pathways in *Candida albicans*. *RSC Advances* 9 (71): 41639–41648.

113 Sha, R., Jiang, L., Meng, Q. et al. (2012). Producing cell-free culture broth of rhamnolipids as a cost-effective fungicide against plant pathogens. *Journal of Basic Microbiology* 52 (4): 458–466.

114 Stanghellini, M.E. and Miller, R.M. (1997). Biosurfactants: their identity and potential efficacy in the biological control of zoosporic plant pathogens. *Plant Disease* 81 (1): 4–12.

115 Vatsa, P., Sanchez, L., Clement, C. et al. (2010). Rhamnolipid biosurfactants as new players in animal and plant defense against microbes. *International Journal of Molecular Sciences* 11 (12): 5095–5108.

116 Cawoy, H., Debois, D., Franzil, L. et al. (2015). Lipopeptides as main ingredients for inhibition of fungal phytopathogens by *Bacillus subtilis/amyloliquefaciens*. *Microbial Biotechnology* 8 (2): 281–295.

117 Ongena, M., Jacques, P., Touré, Y. et al. (2005). Involvement of fengycin-type lipopeptides in the multifaceted biocontrol potential of *Bacillus subtilis*. *Applied Microbiology and Biotechnology* 69 (1): 29.

118 Romero, D., de Vicente, A., Rakotoaly, R.H. et al. (2007). The iturin and fengycin families of lipopeptides are key factors in antagonism of *Bacillus subtilis* toward *Podosphaera fusca*. *Molecular Plant-Microbe Interactions* 20 (4): 430–440.

119 Liu, J., Liu, M., Wang, J. et al. (2005). Enhancement of the *Gibberella zeae* growth inhibitory lipopeptides from a *Bacillus subtilis* mutant by ion beam implantation. *Applied Microbiology and Biotechnology* 69 (2): 223–228.

120 Zhang, B., Dong, C., Shang, Q. et al. (2013). New insights into membrane-active action in plasma membrane of fungal hyphae by the lipopeptide antibiotic bacillomycin L. *Biochimica et Biophysica Acta (BBA) - Biomembranes* 1828 (9): 2230–2237.

121 Vollenbroich, D., Özel, M., Vater, J. et al. (1997). Mechanism of inactivation of enveloped viruses by the biosurfactant surfactin from *Bacillus subtilis*. *Biologicals* 25 (3): 289–297.

122 Banat, I.M., Franzetti, A., Gandolfi, I. et al. (2010). Microbial biosurfactants production, applications and future potential. *Applied Microbiology and Biotechnology* 87 (2): 427–444.

123 Remichkova, M., Galabova, D., Roeva, I. et al. (2008). Anti-herpesvirus activities of *Pseudomonas* sp. S-17 rhamnolipid and its complex with alginate. *Zeitschrift für Naturforschung. Section C* 63 (1–2): 75–81.

124 Huang, X., Lu, Z., Zhao, H. et al. (2006). Antiviral activity of antimicrobial lipopeptide from *Bacillus subtilis* fmbj against pseudorabies virus, porcine parvovirus, Newcastle disease virus and infectious bursal disease virus in vitro. *International Journal of Peptide Research and Therapeutics* 12 (4): 373–377.

125 Shah, V., Doncel, G.F., Seyoum, T. et al. (2005). Sophorolipids, microbial glycolipids with anti-human immunodeficiency virus and sperm-immobilizing activities. *Antimicrobial Agents and Chemotherapy* 49 (10): 4093–4100.

126 Gross RA, Shah V. Anti-herpes virus properties of various forms of sophorolipids. World patent WO2007US63701. 2007.

127 Ongena, M. and Jacques, P. (2008). *Bacillus lipopeptides*: versatile weapons for plant disease biocontrol. *Trends in Microbiology* 16 (3): 115–125.

128 Paraszkiewicz, K., Moryl, M., Płaza, G. et al. (2019). Surfactants of microbial origin as antibiofilm agents. *International Journal of Environmental Health Research* 11: 1–20.

129 Haque, F., Alfatah, M., Ganesan, K., and Bhattacharyya, M.S. (2016). Inhibitory effect of sophorolipid on *Candida albicans* biofilm formation and hyphal growth. *Scientific Reports* 6: 23575.

130 Ceresa, C., Hutton, S., Lajarin-Cuesta, M. et al. (2020). Production of Mannosylerythritol lipids (MELs) to be used as antimicrobial agents against *S. aureus* ATCC 6538. *Current Microbiology* 77: 1373–1380.

131 Sambanthamoorthy, K., Feng, X., Patel, R. et al. (2014). Antimicrobial and antibiofilm potential of biosurfactants isolated from *Lactobacilli* against multi-drug-resistant pathogens. *BMC Microbiology* 14 (1): 197.

132 Pradhan, A.K., Pradhan, N., Mall, G. et al. (2013). Application of lipopeptide biosurfactant isolated from a halophile: *Bacillus tequilensis* CH for inhibition of biofilm. *Applied Biochemistry and Biotechnology* 171 (6): 1362–1375.

133 Kiran, G.S., Priyadharsini, S., Sajayan, A. et al. (2017). Production of lipopeptide biosurfactant by a marine *Nesterenkonia* sp. and its application in food industry. *Frontiers in Microbiology* 8: 1138.

134 Irie, Y., O'toole, G.A., and Yuk, M.H. (2005 Sep 1). *Pseudomonas aeruginosa* rhamnolipids disperse *Bordetella bronchiseptica* biofilms. *FEMS Microbiology Letters* 250 (2): 237–243.

135 Maťátková, O., Kolouchová, I., Kvasničková, E. et al. (2017). Synergistic action of amphotericin B and rhamnolipid in combination on *Candida parapsilosis* and *Trichosporon cutaneum*. *Chemical Papers* 71 (8): 1471–1480.

136 Sarma H, Bustamante KLT, Prasad MNV (2019) Biosurfactants for oil recovery from refinery sludge: Magnetic nanoparticles assisted purification, In: M.N.V. Prasad (ed.), *Industrial and Municipal Sludge*, Elsevier, Butterworth-Heinemann, pp. 107–132. ISBN: 9780128159071, Editor Majeti Narasimha Vara Prasad, Paulo Jorge de Campos, Favas Meththika, Vithanage S. Venkata Mohan.

137 Saikia, R.R., Deka, S., Deka, M., and Sarma, H. (2012). Optimization of environmental factors for improved production of rhamnolipid biosurfactant by *Pseudomonas aeruginosa* RS29 on glycerol. *Journal of Basic Microbiology* 52 (4): 446–457.

17

Surfactant- and Biosurfactant-Based Therapeutics

Structure, Properties, and Recent Developments in Drug Delivery and Therapeutic Applications

Anand K. Kondapi

Laboratory for Molecular Therapeutics, Department of Biotechnology and Bioinformatics, School of Life Sciences, University of Hyderabad, Hyderabad, India

CHAPTER MENU

Biosurfactants for a Sustainable Future: Production and Applications in the Environment and Biomedicine, First Edition. Edited by Hemen Sarma and Majeti Narasimha Vara Prasad.
© 2021 John Wiley & Sons Ltd. Published 2021 by John Wiley & Sons Ltd.

17.1 Introduction

Bioactive molecules such as pharmaceuticals and biopharmaceuticals are required to be stabilized from their aggregation and non-specific interaction before they reach the intended site of action. The routes of administration frequently employed are oral, intravenous, subcutaneous, intramuscular, etc. For each route of administration, the bioactive compound encounters specific physiological environmental challenges and it is necessary that the compound should be able to overcome this before reaching the target site for conferring intended action. Thus, stabilization of a bioactive molecule through formulation would provide much needed solubility, stability, and their localization at the site of action. Various materials of distinct physicochemical properties are employed for formulating a bioactive molecule for solubilization, stabilization, and target localization for enhanced bioavailability, safety, and efficacy. Frequently used materials include polymers, carbohydrates, proteins, lipopeptides, phospholipids, metal salts, glycolipids, and glycoproteins. Since bioactive molecules are characterized by a distinct structural feature that requires an interaction by charged, polar, and non-polar groups, which will enable them to provide a solute-solvent interaction for solubilization and stabilization. An amphipathic molecule, which possesses both hydrophobic and hydrophilic groups, would provide a permissive environment for interaction with bioactive molecules. Biosurfactants possess such a unique feature, which facilitates the solubilization and stabilization of a bioactive molecule in the presence of a biosurfactant. Synthetic surfactants can be chemically prepared and are shown to be versatile in their use for formulation development. However, their use is being limited due to toxicity and non-biodegradability. The efforts to search for a surfactant with biocompatible amphipathic properties yielded an important naturally produced compound present in various microbial populations where biological surfactants with amphipathic properties could be identified. These amphipathic molecules are referred to as biosurfactants, and are natural derivatives of lipopeptides, phospholipids, glycolipids, etc. Due to the presence of natural biomolecules in their architecture, biosurfactants mimic cooperative interactions with bioactive molecules as well as cell membranes, facilitating encapsulation of a bioactive molecule and its transport through a plasma membrane. There are indeed certain limitations: (a) obtaining the same composition of the biosurfactant in each batch of respective organisms, (b) co-purification of some host-related impurities, as well as several advantages, namely (i) efficient loading of bioactive molecules, (ii) cancer-specific localization of biosurfactant, (iii) endogenous therapeutic activity of the biosurfactants, and (iv) biocompatibility and biodegradability of biosurfactants. Since the advantages outweigh the limitations, the use of biosurfactants is attracting more attention and their usage is gaining significant momentum. The technologies for improved homogeneity and quality are currently going through a phase of intense development. This chapter provides details of various surfactants and biosurfactants and their applications as therapeutic and drug delivery systems based on the studies reported by various investigators.

17.2 Determinants and Forms of Surfactants

17.2.1 Hydrophilic–Lipophilic Balance (HLB)

Hydrophilic–lipophilic balance (HLB) represents the relative contribution of the hydrophilic and lipophilic groups of the surfactant to the emulsion. A low HLB value between 3 and 6 would favor the formation of W/O microemulsions (W – water; O – oil), whereas a high HLB value in the range of 8–18 would favor O/W microemulsions. For surfactants with very high HLB values (HLBs >20), often an addition of a co-surfactant is required to reduce their effective HLB value. While the HLB value is applicable only for non-ionic molecules, it has to be determined experimentally in the case of ionic surfactants.

17.2.2 Critical Packing Parameter (CPP)

The surfactant molecule reorients its non-polar groups away from the water phase, thus reducing its contact angle of water with hydrophobic groups, such as carbon chains, thus tending to aggregate. The volume of the hydrophobic carbon chains relative to the volume of surfactant head group area at the interface will determine the type of aggregate being formed, as well as the aqueous behavior of the surfactant. This property is illustrated by a parameter known as a critical packing parameter (CPP). Based on the CPP, different surfactants pack in different orientations in multimeric aggregate forms.

17.2.3 Spontaneous Curvature (H_0)

The spontaneous curvature, H_0, is another important parameter and theoretically H_0 estimates the mechanical properties of the film as a whole. Conceptually, H_0 is similar to the CPP as the preferred curvature will also depend on the relative ratio of polar and non-polar volumes of the molecular aggregates. While the CPP is related to individual molecules, the H_0 is related to a continuum with global physical properties to explain the operational characteristics of ensemble of surfactant in the form of single or multilayer systems such as films.

17.2.4 Winsor-R Ratio

The ratio of the total interaction energies per unit area of interface of the surfactant for the oil and water phases is referred to as the Winsor-R ratio. These microemulsions may be categorized into three types:

Winsor Type I: $R < 1$, indicates that the water–surfactant interaction is stronger than the oil–surfactant interaction;

Winsor Type II: $R > 1$, indicates that the strength of the oil–surfactant interaction is stronger than the water–surfactant interaction.

Winsor Type III: $R = 1$, representing the situation in which the two (oil/water–surfactant) interactions are balanced.

17.2.5 Self-Assembly of Surfactant Molecules

In a solvent (water and polar), surfactant may self-associate and form various types of aggregates, and these can have structures of colloidal dimensions with smaller size (micelles, liposomes, and microemulsion droplets) (Figure 17.1) or larger size with connectivity in one, two, or three dimensions (liquid crystals). These assemblies are formed from blocks of all these discrete and infinite structures composed of monolayer-based or bilayer-based films.

17.2.6 Parameters Determining Self-Assembly

The following are the parameters that control the formation of self-assembly:

1) Hydrophobic effect
2) Repulsive interactions between polar head groups

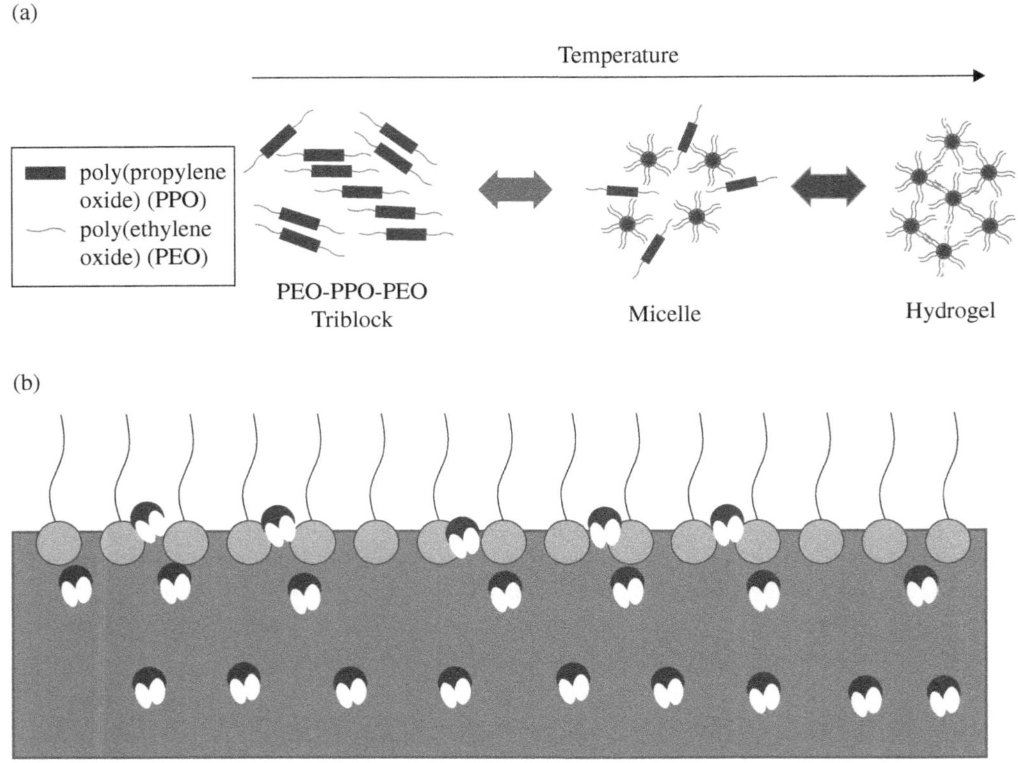

Figure 17.1 (A) Transition to micelle and hydrogel. (B) The distribution of biosurfactants between two immiscible phases leading to its accumulation at the interface by reducing surface (liquid-air) and interfacial (liquid–liquid) tension.

3) CPP
4) H_o
5) Temperature
6) Salt
7) Co-surfactants
8) Type of oil and surfactant concentration

17.2.7 Types of Self-Assembly

17.2.7.1 Bulk Self-Assembly

Based on the molecular structures of the surfactants, two types of organized structure are formed in solution.

Homogeneous (or Single-Phase) System

a) Solutions (e.g. micellar phase);
b) Liquid–crystalline phases (e.g. lamellar phase); and
c) Crystalline phases. Solutions are naturally disordered at both short- and long-range scales.

Heterogeneous (or Multiphasic) System

a) Emulsions,
b) Suspensions,
c) Foams,
d) Gels, and
e) Adsorbed films.

17.2.7.2 Self-Assembly at Interfaces

Self-assembly of surfactant not only occurs in bulk but also at interfaces such as gas–liquid and solid–liquid interfaces.

Surfactants and long-chain fatty acids, water-insoluble compounds, can orient from an organic solvent to an aqueous phase to form monomolecular films adsorbed at the gas–liquid interface, so-called monolayers. Successive deposition of monolayers yields multilayered films. Langmuir–Blodgett films are employed in biosensors.

17.3 Structural Forms of Surfactants

17.3.1 Microbial Versus Chemical Surfactants

Surfactants are a structural class of compounds composed of a lipophilic group, a long-chain hydrocarbon moiety, and a hydrophilic group, which is either a charged or an uncharged polar group. Thus, it possesses both hydrophilic and hydrophobic properties with affinity for both phases of aqueous and oil. The hydrophilic group moiety orient in the phase with stronger polarity, while the lipophilic group is placed in the phase with comparatively weaker polarity, thus reducing the Gibbs energy of the system and making the process spontaneous.

Surfactants can be classified based on the charge of their head groups into cationic (positive charge), anionic (negative charge), non-ionic (non-charged and highly hydrophilic), and zwitterionic (the presence of both negative and positive charge centers and a net charge equal to zero).

Generally, a hydrophilic group is derived from acid, peptide cations or anions, and mono-, di- or polysaccharides, while a hydrophobic group is from an unsaturated or saturated hydrocarbon chain or fatty acids.

Microbial surfactants show 10–40 times lower critical micellar concentrations than chemical surfactants and are highly stable under experimental conditions. Surfactin can be autoclaved in wide range of pH conditions.

An effective biosurfactant should significantly reduce the surface tension of water and air from 72 mN/m to less than 30 mN/m. Surfactin and rhamnolipids can reduce the surface tension of water to 27 mN/m at low concentrations, while sophorolipids from *Candida bombicola* show a reduction of the surface tension to 33 mN/m and MEL and trehalose lipids possess less than 30 mN/m [1].

Microbial surfactants are reported to partition at interfaces of fluid phases with distinct polarities and hydrogen bonding, leading to interference of adhesion of microorganisms. Surfactin induces changes in the physical membrane structure or through the disruption of protein conformations, which destabilize membranes affecting their integrity and permeability, such as transport and energy generation. Indeed, surfactin incorporated in the membrane leads to dehydration of the phospholipid polar head groups and the perturbation of lipid packaging, which strongly destabilizes the bilayer and membrane barrier properties.

When the surfactant concentration increases, glycolipid/water systems induce the formation of a range of liquid crystalline phases, leading to a spontaneous self-assemble into a variety of molecular assemblies, namely sponge (L3), cubic (V2), hexagonal (H2), or lamellar (La) configurations. Similarly, due to carboxyl groups present as side chains, rhamnolipids form micelles at pH values higher than 6.8, lipid particles at pH values between 6.2 and 6.6, lamella structures at pH 6.0–6.5, and finally vesicles sized 50–100 nm at pH 4.3–5.8 [2].

The role of. biosurfactant di-rhamnolipids in stabilizing a bilayer was shown in the formation of a stable drug-loaded lamellar structured vesicles with a combination of dioleoyl-phosphatidylethanolamine and phosphatidylethanolamine/di-rhamnolipid liposomes. These bilayer vesicles internalize into cells through endocytic pathways and release drugs in a pH-sensitive manner into the cytosol.

17.3.2 Biosurfactants

The molecular gels spontaneously assemble from small molecules that are capable of gelling in water are referred to as hydrogels. Those that can gel in organic solvents are referred to as organogels. These networks of small molecules in gels resemble a filamentous structure that connects at junctions in the form of chains, tubes, tapes, and fibers, to attain global morphology of a vesicle.

Vesicles are spherical containers formed by self-assembly of amphiphilic molecules to form a bilayer membrane structure. A combination of synthetic surfactant—co-surfactant—water systems mimics such networks of vesicular structures, wherein the closely packed vesicles resemble a solid-like phase, which swells based on the sensitivity of the chemical composition of the surfactant and co-surfactant.

Biosurfactants are surface-active molecules with an amphiphilic character, which is composed of a hydrophilic moiety and a hydrophobic group. The hydrophilic polar group is derived from mono-, oligo- or polysaccharides, peptides or proteins, and a hydrophobic non-polar moiety commonly derived from saturated, unsaturated, and hydroxylated fatty acids or fatty alcohols. They are generally classified as low molecular weight (including glycolipids and lipopeptides) and high molecular weight (polysaccharides, proteins, lipoproteins, among others) biosurfactants.

Frequently used biosurfactants can be classified as:

Glycolipids: rhamnolipids (*Pseudomonas aeruginosa*), trehalolipids (*Rhodococcus erythopolis*), sophorolipids (*Candida bombicola*) (Figure 17.2), and mannosylerythritol lipids (MELs) (*Pseudozyma yeasts*).

Lipopeptides: Surfactin, iturin, and fengicyn cyclic lipopeptides (*Bacillus* species) (Figure 17.3).

Biosurfactants are produced by microorganisms, viz. bacteria, fungi, and yeast (Table 17.1). As their origin is from biological system, they have an advantage compared to chemical surfactants in terms of lower toxicity, higher biodegradability, and higher selectivity - properties that attracted attention for their development as antiadhesive, antibacterial, antifungal, and antiviral compounds against pathogens. The process of isolation of frequently used biosurfactants is given in Table 17.2.

Lamellar aggregates are formed from a mixture of an anionic and a cationic surfactant. They are also formed from the mixing of ionic surfactants with long-chain alcohols in water or in an electrolyte solution. In the presence of a high salt-mediated increase in counter ion binding and dehydration of head groups of surfactants, the surfactant attains a lamellar structure. Further, interactions between lamellar forms lead to the formation of either unilamellar vesicles or multilayered systems. Amphiphilic molecules spontaneously self-assemble into a wide variety of structures, including spherical (surfactin) and cylindrical (rhamnolipids) micelles, depending on experimental conditions.

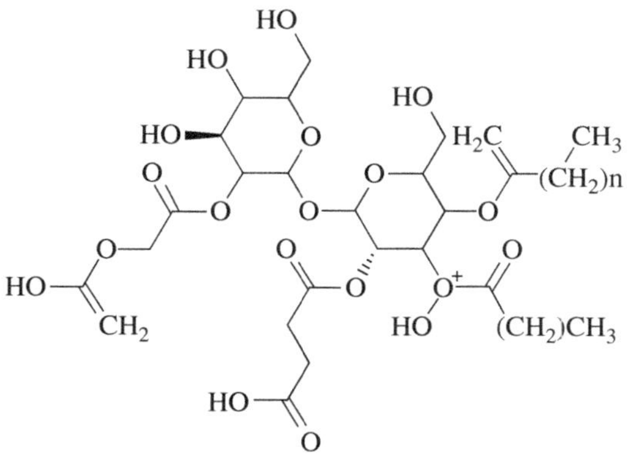

Trehalolipid

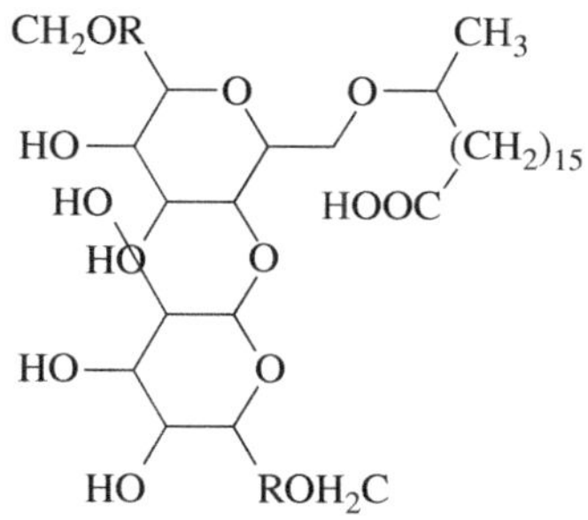

Sophorolipid

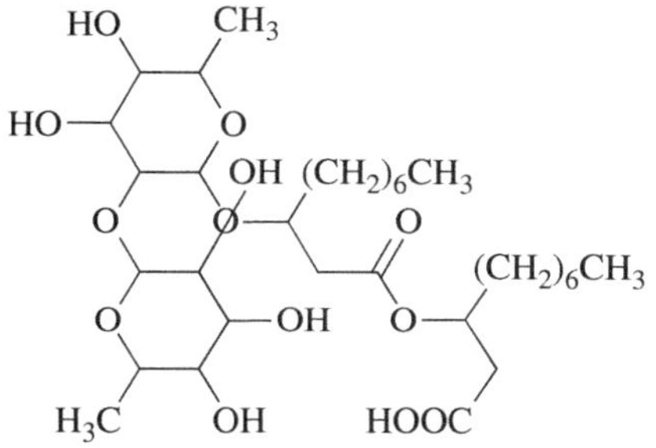

Rhamnolipid

Figure 17.2 Structure of biosurfactants.

Surfactin

Itutin

Fengicyn

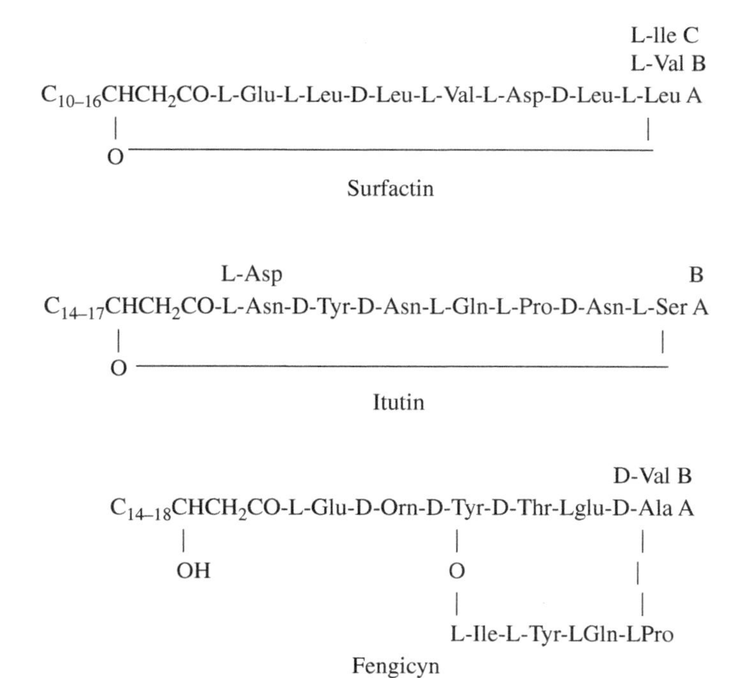

Figure 17.3 Structure of lipopeptides.

Table 17.1 Biosurfactant producing organisms.

Biosurfactant		
Group	**Class**	**Microorganism**
Fatty acids, phospholipids, and neutral lipids	Corynomycolic acid	*Corynebacterium lepus*
Glycolipids	Rhamnolipids	*Pseudomonas aeruginosa*, *Pseudomonas* sp.
	Sophorolipids	*Torulopsis apicola*, *Torulopsis bombicola*, *Torulopsis petrophilum*
	Trehalolipids	*Arthrobacter* sp., *Corynebacterium* sp. *Mycobacterium tuberculosis*, *Nocardia* sp. *Rhodococcus erythropolis*

Table 17.2 Upstream and downstream process for production of biosurfactants.

Biosurfactant type	Upstream process	Downstream process	Principles of separation
Emulsan; Biodispersan; Lipopeptides	Batch process	Ammonium sulfate precipitation	Protein-rich or polymeric biosurfactants are separated by salting-out
Surfactin		Acid precipitation	At low pH biosurfactants form insoluble precipitates Biosurfactants are soluble in organic solvents due to the hydrophobic end
Trehalolipids; Sophorolipids; Liposan		Organic solvent extraction	Due to the hydrophobic nature biosurfactants are soluble in organic solvents
Glycolipids	Continuous process	Centrifugation	Under centrifugal force insoluble biosurfactants are precipitated
Glycolipids		Ion-exchange chromatography	Based on the net charge of the biosurfactants, they bind to ion-exchange resins and thus can be eluted with salt
Glycolipids		Ultrafiltration	Above their critical micelle concentration (CMC), biosurfactants form micelles that are trapped by polymeric membranes
Rhamnolipids; Lipopeptides; Glycolipids; Mannosylerythritol Lipids (MEL)		Adsorption to wood-activated carbon	Due to hydrophobic surface, biosurfactants are adsorbed to activated carbon, can be eluted using organic solvents
Rhamnolipids; Lipopeptides; Glycolipids; MEL		Adsorption to polystyrene resins	Based on the hydrophobic, biosurfactants can be adsorbed to a suitable polystyrene resin and then eluted with organic solvents
Surfactin		Foam fractionation	Under certain surface tension conditions biosurfactants partition into foam

17.4 Drug Delivery Systems

17.4.1 Biosurfactants as Drug Delivery Agents

Drug delivery systems are derived from colloidal microemulsion comprising a water-soluble hydrophilic phase, oil-soluble hydrophobic phase, a biosurfactant, with a co-surfactant or co-solvent. The biosurfactant is the principal component of a microemulsion system, whose formulations self-aggregate to form templates of varying structures and granular sizes. The hydrophobic and hydrophilic phase facilitates the introduction of hydrophobic and hydrophilic drugs into these aggregates of different structures, which could encapsulate/solubilize a hydrophobic or hydrophilic drug in the presence of a dispersed phase (oil in the case of oil-in-water [O/W] and water in the case of water-in-oil [W/O] microemulsions) within its spherical structural core, thus partitioning the dispersed phase from the continuous phase. These structures frequently have nanometric-scaled geometries (e.g. worm-like, bi-continuous sponge-like, liquid crystalline, or hexagonal, spherical swollen micelles) involving the formation of one, two, or even three phases. Based on the required dissolution kinetics for the drug, these structures are modified and controlled. Such modifications can be regulated thermodynamically by applying temperature, pressure, drying and other techniques. Optimization of the process is very critical if any change in the parameters lead to loss of the drug as well as distortion of size and surface properties.

Historically, synthetic hydrocarbon oils such as heptane and dodecane, cyclic oils such as cyclohexane and surfactants (SUR) with 12 carbon hydrophobic chains, such as sodium dodecyl sulfate and tetra ethylene glycol monododecyl ether are studied, due to biocompatibility and toxicity issue some of them are not approved for use in pharmaceutical formulations.

Current regulatory norms permit the use of pharmaceutically acceptable excipients to design safer microemulsions. Biocompatibility has been achieved through the use of lecithins and non-ionic surfactants such as Brijs, Arlacel 186, Spans, Tweens, and AOT, which are amphiphiles and exhibit a high biocompatibility. Recent trends are focusing on replacing natural oils in place of synthetic oils to formulate non-toxic pharmaceutically acceptable microemulsion systems. Also, natural surfactants have emerged as potential alternatives for their synthetic counterparts. In particular, non-ionic surfactants such as sucrose esters, containing a hydrophilic sucrose group and fatty acid chains of varying degrees as a lipophilic group, have been widely employed in microemulsion formulations.

Biosurfactants have emerged as templates for nanoparticle synthesis. It is very difficult to predict the nature and stability of a microemulsion-based drug delivery system as results in the literature only provide a direction on the selection of the most appropriate oil/biosurfactant system. The use of rhamnolipids and surfactin is under investigation due to the complex nature of the head groups (e.g. amino acids in lipopeptide and saccharides in glycolipids) and environment-dependent structural transitions make it difficult to assess absolute structures of these compositions. An intensive effort is being made in research and development for optimization of their pharmaceutical properties.

Due to the presence of carboxylic groups, biosurfactants are anionic at high pH values and non-ionic at low pH values, along with a structure transition from micellar to lamellar in the presence of an electrolyte. Such pH- or electrolyte-dependent structural transitions promote drug release in situ based on the conditions at the site of action. Similarly, high surfactant concentrations in water frequently lead to a lyotropic lamellar liquid crystalline phase, in the same way double-tailed amphiphiles commonly form bilayer sheets, because of their most hydrated state enables the molecules to pack only in a lamellar arrangement.

Biosurfactants have been shown to be involved in several intercellular molecular recognition steps through interference with specific molecules. An interaction between two surface-active lipid molecules can lead to important modifications in the cell membrane, leading to cell death. The membrane composition of lipid determines the structure, function, and the integrity of biological membranes, and phosphatidylcholine (PC) and sphingomyelin (SM), which play a role in stabilizing the bilayer structure. The levels of PC in cancer tissue was found to be fivefold higher than that of normal tissue. Both PC and SM were found to be the major phospholipid components in cancerous and normal cervical tissues, and their compositional changes would determine fluidity of the membrane. Thus, biosurfactants can alter membrane properties, a rigid membrane with a lower surface tension can affect the penetration of drugs, while increasing membrane fluidity by biosurfactants would enhance permeability of drugs. Based on the membrane composition of normal and cancer cells, an appropriate biosurfactant could be used for selective permeation of drugs in cancer cells, thus promoting therapeutic action in a target-directed manner.

As explained earlier, a type of microemulsion formation is dependent on the magnitude of the HLB value; for example, 3–6 favor W/O microemulsions while 8–18 favor O/W microemulsions. A surfactant having HLB values (>20) requires the addition of a co-surfactant for reducing the effective HLB value for its use for a particular function. The HLB and CPP values of surfactin (10–12, 0.1435) and rhamnolipids (22–24, 0.38) indicate that these two biosurfactants are able to form O/W microemulsions, provided that alkaline conditions are maintained [3]. The presence of high surface activities suggests that only low quantities of these biosurfactants are required to formulate a microemulsion for drug encapsulation. Due to the high HLB value of rhamnolipids, a co-surfactant (alkali, alcohols) is required to be added for formation of a microemulsion. The presence of Na ions influences phase behavior of the system, minimizing electrostatic repulsions between changed head groups of rhamnolipids. Due to their strong affinity to water, Winsor type I microemulsion could be developed. Further, a phase transition could be resolved with addition of a chain length of co-surfactant alcohol, in the presence of n-butanol, the single-phase region being maximum. Also, rhamnolipids as a co-surfactant were used to modify the HLB value of a biorenewable surfactant, methyl ester ethoxylate, for formation of a microemulsion with the limonene O/W system, and oleyl alcohol may serve as a hydrophilic linker. Lipid-polymer coated hybrid nanoparticles were prepared using rhamnolipids isolated from *P. aeruginosa* biofilm for the release of an encapsulated drug in the vicinity of the *P. aeruginosa* colonies, thus improving the antibacterial effectiveness of those nanoparticles. Furthermore, the micellization and interfacial behavior was thermodynamically feasible with a mixture of surfactin and sodium dodecylbenzylsulfonate. Cationic surfactin liposomes were reported to deliver small interfering RNA (siRNA) in HeLa cells, effectively with higher biocompatibility and silencing of the gene of interest [4]. Thus, a systematic characterization of biosurfactant microemulsion systems in terms of phase-transition behavior and stability in physiological conditions to derive HLB, CPP, and Winsor R ratio would promote their use as a drug-delivery formulation.

17.4.2 Self-Emulsifying Drug-Delivery Systems

The self-emulsifying DDS could be prepared in a specific ratio of oils, surfactants, co-surfactants, and/or co-solvents. Microbial surfactants promote solubilization of drugs with reduction of interfacial tension and increases permeation of drugs along with reduced toxicity and gastric irritation. Some of the biosurfactants shown to form a self-assembled drug delivery system are formulation composed of rhamnolipid, surfactin, iturin, and pumilacidin. Indeed, a self-microemulsifying vitamin E formulation was reported to enhance pharmaceutical performance.

17.4.3 Nanoparticles

Microbial surfactants are constituents of microemulsion systems. As referred to above, the essential elements to produce a microemulsion-based colloidal DDS could be formed with a combination of an aqueous, oil, or surfactant/biosurfactant, and usually a co-surfactant or co-solvent. These structural assemblies facilitate encapsulation and/or solubilization of hydrophobic or hydrophilic drugs in the presence of a dispersed phase (oil for O/W or water for W/O microemulsions) within its structural core, thus partitioning the dispersed phase from the continuous phase. A global microemulsion system shows a wide range of thermodynamically stable structures of diverse nano-sized geometries with a high solubilization capacity and ultralow interfacial tensions of oil and water, thus suitable for development of a drug delivery systems [5].

A lipid and stearic acid–valine conjugate (biosurfactant)-based nano system (200 nm) was reported for loading of hydrophilic drug methotrexate (MTX) for sustained release using an antisolvent nanoprecipitation technique. Nanoparticles exhibited improved bioavailability with a 3.5-fold increase in cellular uptake with a significant reduction in the tumor burden [6].

Nanoparticle-based therapeutics were showing promising results in drug-delivery applications due to their ability to increase drug localization in solid tumors through enhanced permeability and retention (EPR) and MDR reversal through bypassing or inhibiting P-gp activity [7].

Sophorolipid (SL)-capped ZnO nanoparticle-mediated *C. albicans* cell death occurs via membrane bursting followed by oozing out proteins and intracellular materials. In addition to functioning as a cyclic lipopeptide, the biosurfactant (BSUR) has been found to exhibit versatile bioactive features including adjuvant for immunization and antitumor properties [8].

Based on its unique amphipathic properties, BSUR has the potential for self-assembly (under certain conditions) into nanoparticles to function as a drug carrier for loading hydrophobic drugs. Combining the anticancer activity of BSUR and the characteristics of nanoparticles such as EPR effects and multidrug resistance (MDR) reversal has been shown to improve a therapeutic potential against cancer cells. Doxorubicin (Dox) loaded BSUR nanoparticles showed significant anticancer activity against DOX-resistant human breast cancer MCF-7/ADR cells and in a nude mice model compared to free DOX [7].

Bacillus subtilis SPB1 lipopeptide (LP) showed antioxidant properties on excision wounds induced in experimental rats with entirely re-epithelized wounds with perfect epidermal regeneration, resulting in a significant increase in the percentage of wound closure compared with untreated and standard treatment groups [9].

LPs show a potential adjuvant for immunization through the oral route. They also show an increased humoral immunity against tetanus toxoid, without a decrease in serum IgG levels in a mouse model. Study on the use of LPs as adjuvants in inactivated low pathogenicity avian influenza H9N2 vaccine suggest that biosurfactant-based vaccine increased the titer of antibodies in both broiler and layer chickens and showed comparable immunogenicity to an oil-based vaccine [10].

The *Bacillus* LPs, namely Surfactin, Iturin, and Fengycin, are composed of a peptide and a fatty acid chain and were reported to exhibit antitumor activity *in vitro*. Iturin showed an inhibition of the proliferation of MDa-MB-231 cancer cells and also showed inhibition of chronic myelogenous leukemia *in vitro* via simultaneously causing paraptosis, apoptosis, and inhibition of autophagy. Fengycin can affect growth of non-small cell lung cancer cells 95D and inhibit the growth of xenografted 95D cells in nude mice [11].

17.4.4 Multilayered Nanoparticles

1) The first layer will be the encapsulation of template nanoparticles (NPs) into the vesicles and model protein drugs to be loaded into the vesicle NPs.
2) In the second process, subsequent stabilization of the vesicle NPs through incorporation between layers of the Pluronics and the lipid bilayers to form multilayer NPs.

Antidiabetic peptide, exenatide, and a glucagon-like peptide-1 (GLP-1) agonist were encapsulated in the multilayer NPs and showed a sustained activity in C57BL/6 db/db mouse models. Use of a combination of Pluronic F-127 and F-68 would modify sol–gel transformation properties under different temperatures [12].

17.5 Different Types of Biosurfactants Used for Drug Delivery

17.5.1 Glycolipids

Glycolipids, amphipathic molecules consisting of lipids with a carbohydrate attached, have been shown to be involved in growth arrest and apoptosis of mouse malignant melanoma B16 cells. The glycolipids MELs and succinyl trehalose lipids (STLs) showed growth arrest and apoptosis of tumor cells [13]. Exposure to increasing concentrations of MELs led to the accumulation of B16 cells in the sub-G0/G1 phase, which is a sign of cells undergoing apoptosis. Furthermore, a sequence of apoptotic events was observed including the condensation of chromatin and DNA fragmentation, thus confirming the apoptosis-inducing potential of MELs in these cells. The lipopeptide surfactin was reported to induce apoptosis in breast cancer cells. Results of these studies showed that regulation of the activity of protein kinase C (PKC) might be associated with apoptosis induced by MELs. Activation of PKC is the primary step for initiation of the signal transduction that leads to a cascade of cellular responses. Also, MELs have been reported to induce the differentiation of human promyelocytic leukemia HL60 cells into granulocytes lineage such as nitro blue tetrazolium reducing ability, expression of Fc receptors (Fc – a surface immunoglobulin molecule) and phagocytic activities in HL60 cells, while inhibiting PKC activity in these cells. Thus, MELs may modify membrane associated molecules and promote differentiation-inducing activity in cells. Hence, MEL biosurfactants exhibit growth inhibition, apoptosis, and induce differentiation, which may be dependent on the cell type, concentration, composition of biosurfactant, and exposure time, which requires an optimization for achieving a desired outcome. Also, STLs exhibit similar properties of MELs. Further, sophorolipids interact with plasma membrane, induce cell differentiation instead of cell proliferation, and showed inhibition of PKC activity in the HL60 human leukemia cell line [14].

Glycolipid isolated from Sphingobacterium detergents showed activity against Caco2 human colorectal cancer cells. The sophorolipid from Wickerhamiella domercqiae was reported to induce apoptosis in H7402 human liver cancer cells by blocking the cell cycle at the G1 phase, activating caspase-3, and increasing Ca^{2+} concentration in the cytoplasm. The cytotoxicity is dependent on the derivative of sopholipid, methyl ester being highly cytotoxic, and also a higher degree of acetylation is shown to enhance its cytotoxicity activity. Sophorose with a higher degree of sophorose, a low degree of unsaturated hydroxyl fatty acid, and lactonization correlated with biological activity of these compounds [15].

17.5.2 Mannosylerythritol Lipids (MELs)

A glycolipid MEL-A produced by *Pseudozyma antarctica* was analyzed for phase transitions in a ternary MEL-A/water/n-decane system and showed that a MEL-A-stabilized system could form a W/O microemulsion without addition of a co-surfactant [16]. Carbohydrate ligand systems composed of self-assembled monolayers of MELs-A served as a high-affinity, easy to handle, and low-cost ligand system for immunoglobulin G and M and lectins.

MEL-A was reported to significantly increase the efficiency of gene transfection mediated by cationic liposomes along with high activity in DNA encapsulation and membrane fusion with anionic liposomes, which are important properties for gene transfection by inducing fusion of liposomes and plasma membranes of cells. Monolayers composed of MEL-A and 1, 2-dipalmitoyl phosphatidylcholine (DPPC) had greater membrane fluidity than those containing only DPPC; indeed, unsaturated fatty acids in MEL-A had the surface pressure and packing density of a monolayer. Betulinic acid loaded with Mel A modified soy phosphatidylcholine-cholesterol liposomes affects mitochondrial membrane potential and induces apoptosis in HepG2 cells [17].

When a lipoplex of MEL-A and plasmid DNA was injected into C57BL/6J mice, the results showed significant transfection of DNA in B16/BL6 tumors [18]. Addition of MEL-A in cholesteryl-3-beta-carboxyamidoethylene-*N*-hydroxyethylamine (OH-Chol) enhanced gene expression in solid tumors by 100-fold. The higher the unsaturated fatty acid ratio were in liposomes, the more they fused well with plasma membrane, but could not deliver DNA into the nucleus. Further, MEL-A cationic liposomes can deliver SiRNA into cells with no significant immune response and cytotoxicity.

MEL-B- and MEL-C containing liposomes only increased either the encapsulation or the membrane fusion. MEL-B containing a different configuration of the erythritol moiety is reported to self-assemble into a lamellar phase over a wide range of concentrations and temperatures, namely lyotropic liquid crystalline phases. It is also showed great potential as a vesicle-forming lipid for possible application for drug and gene delivery, as well as in transdermal delivery.

EL exhibits higher emulsifying activity with soybean oil and tetradecane than polysorbate 80 along with formation W/O microemulsions without the addition of a co-surfactant or salt.

17.5.3 Lipopeptides

Microbial surfactants-based formation of a variety of liquid crystals in aqueous solutions is exploited for preparation of high-performance nanomaterials. Gold nanoparticles prepared at pH 7–9 are uniform in shape and size and stable for 2 months, while aggregates were observed at pH 5 within 24 hours. Preparation at room temperature is monodispersed and uniform compared to those prepared at 4 °C. Synthesis of cadmium sulfide nanoparticles in the presence of surfactin (*Bacillus amyloliquefaciens* KSU-109) was stable up to six months. Brushite particles (nanospheres and nanorods) were synthesized from surfactin using the reverse microemulsions method [19].

Lipopeptides and surfactin block cell proliferation by inducing proapoptotic activity and arresting the cell cycle. Also, surfactin is reported to interfere with PI3K/Akt signaling pathways involved in proapoptotic processes including cell cycle arrest [20].

Further, surfactin purified from the strain *B. subtilis* CSY191 (probiotic strain) was reported to induce apoptosis in MCF7 cells through a reactive oxygen species/c-Jun N-terminal kinase (ROS/JNK)-mediated mitochondrial/caspase pathway, resulting in mitochondrial permeability and membrane potential collapse [21]. Also, surfactin induced accumulation of the tumor suppressor p53 and cyclin kinase inhibitor p21waf1/cip1, and inhibited the activity of the G2-specific kinase, cyclin B1/p34cdc2. These findings suggest that surfactin caused the G2/M arrest of MCF7 cells through the regulation of their cell cycle factors. DNA complexed biosurfactant b-sitosterol b-D-glucoside was successfully employed in thymidine kinase gene therapy against the herpes simplex virus.

In wound healing experiments, surfactin-loaded polyvenyl alcohol showed improvement in healing. Indeed, an increase in concentration of surfactin showed a decrease in the diameter of the nanofiber in wound healing [22].

Doxorubicin encapsulated surfactin nanoparticles (Dix-Sft) were prepared through the solvent evaporation method. Doxorubicin-loaded surfactin nanoparticles showed internalization in cancer cells through macropinocytosis- and caveolin-mediated endocytosis followed by lysosomal drug release. Further, these nanoparticles showed higher localization in tumor tissue in nude mice [7]. EDC and NHS activated graphene quantum dots were conjugated with biosurfactant and folic acid. These nanoparticles showed significant localization in the cancer cells with a fluorescence imaging feature [23].

The cyclic lipopeptide Pseudofactin II (PFII) from *Pseudomonas fluorescens* BD5 showed plasma membrane incorporation in Melanoma A375 cells and NHDF cells in apoptosis pathway, while fibroblasts were not affected [24].

17.5.4 Licithin

Lecithin (i.e. phosphatidylcholine) is a zwitterionic amphiphile that is electrically neutral but has a positive charge on the choline group and a negative charge on the phosphate. Bile salts are physiological surfactants that bear unusual hydrophilic and hydrophobic faces. Mixed micelles are formed by bile salts and lecithin where a structural transition from bilayer to cylindrical and then spheroidal micelle has been reported at increasing molar ratios of bile salt to lecithin increases. These lecithin–bile salt mixed micelles in water can grow into long, flexible worm-like structures that exhibit viscoelastic behavior near an equimolar bile salt/lecithin ratio in the presence of electrolytes [25].

Also, in low polar solvents lecithin shows an interesting self-assembly structure. When mixed with bile salts, lecithin reversed spherical micelles can be induced to form worm-like micelles, leading to more than a fivefold increase in viscosity. In the presence of salts like $CaCl_2$, lecithin forms cylindrical worm-like structures of organo gels resembling fibrils [26].

A stable gel phase can be formed at low bile salt/lecithin (15% wt%) molar ratios above chain melting temperatures, with swollen multilamellar vesicles resembling solid-like hydrogel, which is sensitive to ionic concentration, where small concentrations of NaCl may induce structural transition into low turbid liquids. As the bile salt/lecithin molar ratio increases, a structural transition from multilamellar vesicles to cylindrical micelles and then to spheroidal micelles will be formed [27].

17.5.5 Rhamnolipids

The rhamnolipids form spherical nanoparticles with low polydispersity, stability in the range of between 5 and 100 nm over a broad concentration range. These are well tolerated by the human skin and thus serve as dermal delivery systems. Studies using human skin models showed that they penetrate both into the stratum, corneum, with a lesser extent into the lower epidermis. They are non-toxic to primary fibroblasts [28].

The complexes formed between the positively charged random copolymers (RCPs) of methoxy-poly-(ethylene glycol) monomethacrylate (MePEGMA) and (3-[methacryloylamino]propyl)trimethylammonium chloride (MAPTAC) with oppositely charged biosurfactants (bile salts) with less than 68 mol% of PEG content showed precipitation in water, while the complexes having co-polymer with 89 and 94 mol% of PEG content did not precipitate when the electroneutral complexes are formed, the size of these complexes being 20–30 nm, thus facilitating their use in drug delivery [29].

Metal-bound nanoparticles were synthesized and stabilized using rhamnolipids and sophorolipids. Indeed, silver nanoparticles made of rhamnolipids from *P. aeruginosa* strain BS-161R showed significant antibiotic activity against both Gram-positive and Gram-negative pathogens and *Candida albicans*. Spherical nickel oxide nanoparticles were prepared using rhamnolipids by the microemulsion technique. Rhamnolipids were used as good capping agents for preparation of ZnS nanoparticles [30].

Di-rhamnolipid inhibited growth of transformed myofibroblasts *in vitro* and reduced ear hypertrophic scars in rabbits through collagen distribution and alpha smooth muscle actin (SMA) expression in scar tissue [31].

Rhamnolipids and sophorolipids were mixed with lecithins to prepare salt and temperature tolerant, biocompatible microemulsions useful for cosmetic and drug delivery applications [32]. Further, the potential of fengycin and surfactin served as enhancers for the transdermal penetration and skin accumulation of acyclovir [33].

Similarly, a lecithin-based microemulsion in combination of rhamnolipid and sophorolipid biosurfactants showed increased stability at temperatures up to 40 °C and an electrolyte concentration of 4% (w/v).

17.5.6 Sophorolipids

Sophorolipids were produced by the yeast *Starmerella bombicola* with high titres of over 400 g/l in the presence of glucose and vegetable oil. In a Sophorolipid, a hydrophilic group b-(1,2) disaccharide sophorose is covalently linked to a hydrophobic long chain hydroxylated fatty acid.

Several congeners are present in the natural mixture, with differences in acetylation and lactonization being the most significant ones with regards to the physical behavior of the molecules, while minor variations in the nature of the fatty acid occur: length (in general 16 or 18 carbon atoms), saturation degree (0, 1, or 2 double bonds), and position of the hydroxyl group (terminal or subterminal).

Bola-sophorolipids are molecules that form an intriguing structure: a fatty acid molecule with a carbohydrate head on both sides, which behave differently from sophorolipids. The long hydrophobic spacer with hydrophilic groups at both ends renders the molecule more water soluble, yet still allows formation of micelles, vesicles, and other laminar structures. As an example, tetraether lipids present in membranes of archaea able to grow under extreme temperatures and salt concentrations. Thus, bolaform lipids strongly stabilize the membranes when exposed to extreme conditions. Further, bolaform can enhance the stability of liposomes containing unsaturated fatty acids with its hydrophilic part, sugar moiety, which would involve interactions with cell surface receptors for conjugation of target-directed ligands for drug delivery.

The ability of bolaform surfactants to interact with DNA in a compact structure allows it to use the gene delivery vehicle *in vitro* and *in vivo* [34]. Their ability in self-assembly makes them attractive in compositions to form nanomaterials. For example, bolaform-mediated synthesis of tailored zeolite nanosheets were reported for surface-active catalytic reaction at the target site. Bolaform interacts with several metals including gold and silver to form single and multiwall silver particle-coated nanotubes, gold nanocrystals, and bio-nanotubes coated with cupper nanocrystals, etc. [35].

Sophorolipid biosurfactants are comparatively more hydrophobic than sodium bis(2-ethyl) dihexyl sulfosuccinate (SBDHS), which are reported to be more hydrophobic than sodium dihexyl sulfosuccinate (SDHS) and rhamnolipid biosurfactant. Thus, the sophorolipid composition determines hydrophobicity in combination with the lecithin/rhamnolipid/sophorolipid biosurfactant formulation. This combination of biosurfactant formulation would produce Winsor Types I, II, and III microemulsions along with the corresponding ultralow IFT for limonene, decane, isopropyl myristate, and hexadecane [36].

Sophorolipids can be formulated into lipid nanostructures coated with polysorbates (Tweens) using the hot dispersion method. Formulation could be optimized for stability by varying the amount of lipid, type of surfactant, and alcohol, dilution ratio, etc. Higher entrapment of rifampicin was reported with such a carrier [37].

Lipid nanostructures covered with polysorbates (Tweens) were formulated by a hot dispersion method by optimizing the amount of lipid, type of surfactant, and alcohol, dilution ratio, etc., for stability and dispersibility for drug delivery applications.

Sophorolipids serve as good reducing and capping agents for synthesis of cobalt and silver particles. Sophorolipid-coated silver and gold nanoparticles showed anti-bacterial activity against both Gram-positive and Gram-negative bacteria [38].

17.5.7 Hydrophobins

Hydrophobins are amphiphilic proteins expressed by filamentous fungi and are characterized by a hydrophobic patch on one face, giving them an amphiphilic structure similar to surfactants or Janus particles. Intrinsically disordered proteins, caseins, form micelles through regions of hydrophobic and hydrophilic amino acids with a structural similarity to block copolymers.

Hydrophobins, derived from filamentous fungi, are a family of small (< 20 kDa), highly surface-active globular proteins with eight highly conserved cysteine residues in a specific primary sequence pattern, forming four disulfide bonds (Figure 17.4). Their amphiphilic tertiary structure is stabilized by these four disulfide bonds, which confer a surfactant-like property leading to self-assembly of hydrophobin into amphipathic layers at hydrophobic–hydrophilic interfaces.

Hydrophobins have been classified into two groups, class I and class II, based on their hydropathy, solubility characteristics, and structures formed during self-assembly. Class I hydrophobins, like SC3 from the *Schizophyllum* commune, form highly insoluble amyloid-like rodlets at interfaces, frequently proceeding through a conformational change, and are sensitive to strong acids. In contrast, class II hydrophobins, for example HFBI or HFBII from *Trichoderma reesei*, form a highly

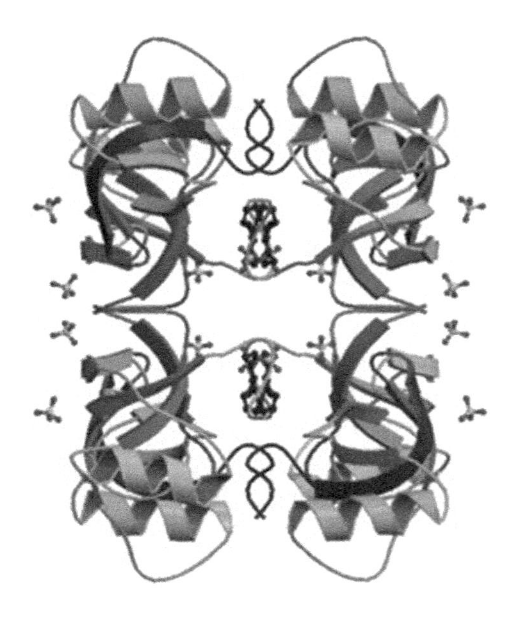

Figure 17.4 Hydrophobins. Amphiphilic nanotubes in the crystal structure of a biosurfactant protein hydrophobin HFBII (3QQT). From: https://www.rcsb.org/structure/3QQT (See insert for color representation of the figure).

ordered 2D crystalline monolayer at interfaces that can easily be dissolved with detergents, organic solvent solutions, or under high pressure.

Due to the presence of hydrophobic interaction features, hydrophobins, in combination with Tween 80 or Chremophore EL, are used for solubilization and stabilization of hydrophobic drugs under an aqueous environment [39, 40]. Due to the formation of a protective layer on hydrophobic microbial spores, hydrophobins may serve as immunosuppressors [41]. When hydrophobic drugs cyclosporine A and nifedipine solubilized with hudrophobin I, their oral bioavailability was enhanced two- and sixfold, respectively [42].

The structural and functional roles of the conserved disulfide bonds differ between the two classes, with disulfides of class I hydrophobin SC3 being involved in solubilization of protein and structurally stability, while not affecting the self-assembling ability. In the case of class II hydrophobin HFBI, disulfides are critical to both protein structure and stability as well as function at interfaces.

Structural and functional intermediates of hydrophobin were produced by genetic chimeras of class I hydrophobins EAS and DewA with class II hydrophobin NC2 These chimaras exhibit properties of both classes of hydrophobins.

Hydrophobins have a significant level of conformational plasticity, with the nature of the interfacial assemblies being highly dependent on the specific interface of the proteins with which they are interacting. Thus, engineering native surface charges on hydrophobin HFBI, viscoelastic properties of the assembled film at the air–water interface, and the ability to absorb secondary protein layers could be affected. Further, mutating the surface charges of HFBI does not affect the overall protein folding state, while specific charge mutations could be linked to inter-protein interactions at the assembled film, while other mutations were linked to protein orientation at the interface. In addition, HFBI adsorbed to the air–water interface reoriented in a pH responsive way due to changes in inter-protein interactions caused by side-chain charge states.

Class II hydrophobin-coated drug nanoparticles below 200 nm were stable for at least five hours in suspension, and for longer times after freeze-drying. Hydrophobin HFBI produced as a genetic fusion to cellulose binding domains allowed a cellulose-based nanofibrillar matrix stabilization of hydrophobin associated drug particles of around 100 nm. This formulation was stable over 10 months of storage and exhibited enhanced drug dissolution rates [43].

Functionalization of hydrophobin HFBII with thermally hydrocarbonized porous silicon nanoparticles showed improved biodistribution compared to unfunctionalized particles along with the protein adsorption profile to the particle surface.

Commercially available surfactant blend containing class I hydrophobin H star protein B could solubilize the chemotherapy drug docetaxel and the formulation was biocompatible and exhibited a high drug loading, high nanoparticle yield with small particles of narrow distribution, and delayed drug release in rats.

Class I hydrophobin SC3 was used to solubilize the hydrophobic drugs, cyclosporine A and nifedipine, showed enhanced oral bioavailability by two- and sixfold, respectively. Hydrophobins also showed as a topical drug formulation agent for nail permeation.

Coating porous silicon nano proteins with a fusion of *T. reesei* class II hydrophobins to human transferrin protein resulted in their uptake in cancer cells, conserved disulfide bonds, and promoted drug release from the particles [44, 45]. Indeed, hydrophobin-coated docetaxel nanoparticles showed improved bioavailability when administered through the intravenous route [46].

Class II hydrophobin HFBII were employed to organize and stabilize supraparticles of dodecanethiol-protected gold nanoparticles that could be loaded with hydrophobic drug and remain stable in the blood until taken up by tissues, where cytoplasmic glutathione would reduce the disulfides allowing the supraparticles to release the drug load directly in the cytoplasm, leading to a twofold enhancement of efficacy.

Lung surfactant proteins SP-B and SP-C are small alpha helical proteins, originating surface activity from the amphipathic character of their helices function for modulating the surface tension of pulmonary fluid [47].

Rsn-2 derived from the foam nest of the frog *Engystomops pustulosus* exhibits an amphiphilic character with a hydrophobic region at the N-terminus from its structure comprised of alpha helix and a four-stranded beta sheet, joined by a flexible linker region and with flexible N- and C-terminal tails [48].

17.5.8 Poloxamers

Pluronics or "Poloxamers" are composed of triblock layering of polyethylene oxide (PEO) /polypropylene oxide (PPO)/polyethylene oxide (PEO) for generating amphiphilicity (Figure 17.5). The percent composition ratios of each type of polymer determine the HLB of the gel. The cloud point can be determined at the temperature of its separation from water and is dependent on the PPO/PEO composition due to the degree of hydrogen bonding [49].

Pluronics can increase the transport of drugs across cellular barriers, namely brain endothelium, Caco-2, and polarized intestinal epithelial cells, and sensitize the multidrug-resistant (MDR) tumor cells. They alter the microviscosity of membranes and provoke an enhanced decrease in ATP levels in cells by inhibiting membrane drug efflux transporters, breast cancer resistance proteins (BCRPs), multidrug resistance proteins (MRPs), and P-glycoprotein (Pgp), leading to induction of conformational changes in the surface protein and internalization of drugs. Also, it is reported that pluronics may be incorporated in the mitochondrial membranes and stimulate the release of cytochrome C along with an increase in cytosolic reactive oxygen species (RoS) and apoptosis [50].

Poloxamer P407 was employed for delivery of insulin interleukin-2, an epidermal growth factor, bone morphogenic protein, a fibroblastic growth factor, and an endothelial cell growth factor. pH and temperature-responsive ophthalmic preparations containing poloxamer P407 and chitosan exhibit acceptable properties for the sustained release of nepafenac. Also, curcumin-loaded albumin nanoparticles were included in a hydrogel mixture of P407/P188 for treatment of diabetic retinopathy for local ocular administration [51–53].

Pharmaceutical formulations with mucoadhesive properties like poloxamers would permeate through the vagina and expose a vast network of blood vessels for localized activity as well as systemic absorption, escaping inactivation through the gastrointestinal route and hepatic first pass. A mixture of poloxamer P407 (20%) and P188 (5%) were loaded with Amphotericin B (AmB) nanosuspension to form a thermosensitive gel and addition of Carbopol would enhance mucoadhesive properties of the formulation [54]. Indeed, an expansable thermal gelling aerosol foam (ETGFA)

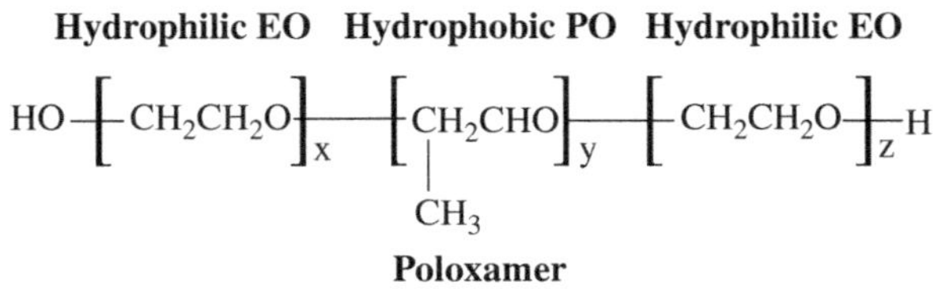

Hydrophilic EO Hydrophobic PO Hydrophilic EO

$$HO\left[CH_2CH_2O\right]_x\left[CH_2\underset{\substack{|\\CH_3}}{CH}O\right]_y\left[CH_2CH_2O\right]_zH$$

Poloxamer

Figure 17.5 Organization of poloxamer.

was developed in combination with optimized quantities of P407 (18–22% w/w) and P188 (0–5%w/w), the adhesive agents (arabic gum, sodium carboxymethyl cellulose, sodium alginate, and xanthan gum), and silver nanoparticles for localized application as a foam and gel penetration and carrier retention in the vaginal canal [52].

Pluronic P85 showed suppression of a GSH (glutathione)/GST (glutathione S-transferases) detoxification system by reducing GSH levels and deactivation of GST in different MDR cells, including human lung carcinoma epithelial cells (COR-L23/R) [55].

The micellization behavior of Pluronics in an aqueous solution in the range of 10–100 nm and their cores were loaded with various therapeutic agents and diagnostic tracers [56]. Paclitaxel (PTX) loaded in F-68 coated PEO Pluronic NP exhibits a temperature-sensitive phase transition (TIPT). These particles showed extended retention in systemic circulation, which was the prerequisite for the extended permeation and retention (EPR) effect [57].

Dox loaded Pluronics (L61 and F127 and P85) showed significant anti-cancer activity against esophageal adenocarcinoma, MDR cancers. Further, Dox was conjugated with Asp-Glu–Val-Asp (DEVD) peptide moiety for radiation-induced apoptosis-targeted chemotherapy (RIATC) to induce apoptosis and caspase-3 expression in tumor tissues. Further, a heparin/Dox/DEVD-S-Dox complex was formed via ionic interaction and further stabilization with Pluronic F-68 [58, 59].

Core/shell nanoparticles composed of a protein drug-loaded lecithin (liposome) core and a pluronic shell were prepared for encapsulation of vascular endothelial growth factor (VEGF) for induction of angiogenesis in rats with myocardial infarction and improvement of ejection fraction and cardiac output for induction of blood flow in the ischemic area through angiogenesis [60].

17.5.9 Emulsan

Emulsan, an anionic polysaccharide with fatty acid branches and an amphiphilic property, is isolated from *Acinetobacter calcoaceticus*, RAG-1 [61]. Emulsan was coated on nanoparticles, with a core filled with flax seed oil loaded with a photosensitizer, Pheophorbide (Pba). These nanoparticles showed a fast uptake in SCC7 mouse squamous cell carcinoma cells and killed the tumor cells after laser irradiation due to the photodynamic effect of Pba. The nanoparticles showed longer blood circulation when administered through a tail vein in SCC7 tumor-bearing mice, along with a 3.04-fold higher tumor accumulation in tissue compared to that of free Pba [62]. Hence, biosurfactants would aid loading of drug in the nanocore and facilitate drug loading and delivery.

17.6 Conclusions

- Biosurfactants are emerging as an important biocompatible constituent of drug delivery systems.
- A combination of biosurfactant-based phase transitions, chemical and physical sensitivity, and stability would be constructed for efficient delivery of pharmaceutically active molecules for thermanostic as well as therapeutic application in cancer and other diseases.
- Biosurfactants are versatile for preparation of various nano formulations in combination with polymers, carbon dots, metal nanocores, porous silica materials for preparation of various nanoforms nanoparticles, nanocapsules, nanofilms, nanogels, nanoimplants for drug or tracer delivery to the site of action.
- Conjugation of biosurfactant-based delivery systems with folic acid, apotransferrin, lactoferrin, and other targeted ligands would aid target delivery of the pharmaceutically active agent at the target site.

References

1. Desai, J.D. and Banat, I.M. (1997). Microbial production of surfactants and their commercial potential. *Microbiol. Mol. Biol. Rev.* 61 (1): 47–64.
2. Ozdemir, G., Peker-Basara, S., and Helvaci, S.S. (2004). Effect of pH on the surface and interfacial behavior of rhamnolipids R1 e R2. *Colloids Surf. A Physicochem. Eng. Asp.* 234 (1): 135–143.
3. De, S., Malik, S., Ghosh, A. et al. (2015). A review on natural surfactants. *RSC Adv.* 5: 65757–65767.
4. Wu, Y.-S., Ngai, S.-C., Goh, B.-H. et al. (2017). Anticancer activities of surfactin and potential application of nanotechnology assisted surfactin delivery. *Front. Pharmacol.* 8: 761.
5. Rodrigues, L.R. (2015). Microbial surfactants: Fundamentals and applicability in the formulation of nano-sized drug delivery vectors. *J. Colloid Interface Sci.* 449: 304–316.
6. Katiyar, S.S., Kushwah, V., Dora, C.P., and Jain, S. (2018). Novel biosurfactant and lipid core-shell type nanocapsular sustained release system for intravenous application of methotrexate. *Int. J. Pharm.* 2019 557: 86–96.
7. Huang, W., Lang, Y., Hakeem, A.B. et al. (2018). Surfactin-based nanoparticles loaded with doxorubicin to overcome multidrug resistance in cancers. *Int. J. Nanomedicine* 13: 1723–1736.
8. Basak, G., Das, D., and Das, N. (2014). Dual role of acidic diacetate sophorolipid as biostabilizer for ZnO nanoparticle synthesis and biofunctionalizing agent against Salmonella enterica and Candida albicans. *J. Microbiol. Biotechnol.* 24: 87–96.
9. Zouari, R., Moalla-Rekik, D., Sahnoun, Z. et al. (2016). Evaluation of dermal wound healing and in vitro antioxidant efficiency of *Bacillus subtilis* SPB1 biosurfactant. *Biomed. Pharmacother.* 84: 878–891.
10. Basit, M., Rasool, M.H., Hassan, M.F. et al. (2017). Evaluation of lipopeptide biosurfactants produced from native strains of *Bacillus cereus* as adjuvant in inactivated low pathogenicity avian influenza H9N2 vaccine. *Int. J. Agric. Biol.* 20 (6): 1419–1423.
11. Naughton, P.J., Marchant, R., Naughton, V., and Banat, I.M. (2019). Microbial biosurfactants: Current trends and applications in agricultural and biomedical industries. *J. Appl. Microbiol.* 127: 12–28.
12. Kim, J.Y., Lee, H., Oh, K.S. et al. (2013). Multilayer nanoparticles for sustained delivery of exenatide to treat type 2 diabetes mellitus. *Biomaterials* 34 (33): 8444–8449.
13. Zhao, X., Wakamatsu, Y., Shibahara, M. et al. (1999). Mannosylerythritol lipid is a potent inducer of apoptosis and differentiation of mouse melanoma cells in culture. *Cancer Res.* 59 (2): 482–486.
14. Isoda, H., Kitamoto, D., Shinmoto, H. et al. (1997). Microbial extracellular glycolipid induction of differentiation and inhibition of the protein kinase C activity of human promyelocytic leukemia cell line HL60. *Biosci. Biotechnol. Biochem.* 61 (4): 609–614.
15. Chen, J., Song, X., Zhang, H. et al. (2006). Sophorolipid produced from the new yeast strain *Wickerhamiella domercqiae* induces apoptosis in H7402 human liver cancer cell. *Appl. Microbiol. Biotechnol.* 72 (1): 52–59.
16. Worakitkanchanakul, W., Imura, T., Morita, T. et al. (2008). Formation of W/O microemulsion based on natural glycolipid biosurfactant, mannosylerythritol lipid-a. *J. Oleo Sci.* 57 (1): 55–59.
17. Shu, Q., Wu, J., and Chen, Q. (2019). Synthesis, characterization of liposomes modified with biosurfactant MEL-A loading Betulinic acid and its anticancer effect in HepG2 cell. *Molecules* 24 (3939): 1–17.
18. Inoh, Y., Furuno, T., Hirashima, N., and Nakanishi, M. (2009). Nonviral vectors with a biosurfactant MEL-A promote gene transfection into solid tumors in the mouse abdominal cavity. *Biol. Pharm. Bull.* 32 (1): 126–128.

19 Singh, B.R., Dwivedi, S., Al-Khedhairy, A.A., and Musarrat, J. (2011). Synthesis of stable cadmium sulfide nanoparticles using surfactin produced by *Bacillus amyloliquifaciens* strain KSU-109. *Colloids Surf. B: Biointerfaces* 85 (2): 207–213.
20 Kim, S.-y., Kim, J.Y., Kim, S.-H. et al. (2007). Surfactin from *Bacillus subtilis* displays anti-proliferative effect via apoptosis induction, cell cycle arrest and survival signaling suppression. *FEBS Lett.* 581 (5): 865–871.
21 Chambers, J.W. and LoGrasso, P.V. (2011). Mitochondrial c-Jun N-terminal kinase (JNK) signaling initiates physiological changes resulting in amplification of reactive oxygen species generation. *J. Biol. Chem.* 286 (18): 16052–16062.
22 Ahire, J.J., Robertson, D.D., van Reenen, A.J., and Dicks, L.M.T. (2017). Surfactin-loaded polyvinyl alcohol (PVA) nanofibers alters adhesion of Listeria monocytogenes to polystyrene. *Mater. Sci. Eng. C Mater. Biol. Appl.* 77: 27–33.
23 Bansal, S., Singh, J., Kumari, U. et al. (2019). Development of biosurfactant-based graphene quantum dot conjugate as a novel and fluorescent theranostic tool for cancer. *Int. J. Nanomedicine* 14: 809–818.
24 Janek, T., Krasowska, A., Radwanska, A., and Łukaszewicz, M. (2013). Lipopeptide biosurfactant Pseudofactin II induced apoptosis of melanoma a 375 cells by specific interaction with the plasma membrane. *PLoS One* 8 (3): e57991. https://doi.org/10.1371/journal.pone.0057991.
25 Cheng, C.Y., Oh, H., Wang, T.Y. et al. (2014). Mixtures of lecithin and bile salt can form highly viscous wormlike micellar solutions in water. *Langmuir* 30 (34): 10221–10230.
26 Tung, S.H., Huang, Y.E., and Raghavan, S.R. (2006). A new reverse wormlike micellar system: mixtures of bile salt and lecithin in organic liquids. *J. Am. Chem. Soc.* 128 (17): 5751–5756.
27 Zhang, M., Strandman, S., Waldrona, K.C., and Zhu, X.X. (2016). Supramolecular hydrogelation with bile acid derivatives: Structures, properties and applications. *J. Mater. Chem. B* 4: 7506–7520.
28 Müller, F., Hönzke, S., Luthardt, W.-O. et al. (2017). Rhamnolipids form drug-loaded nanoparticles for dermal drug delivery. *Eur. J. Pharm. Biopharm.* 116: 31–37.
29 Nisha, C.K., Manorama, S.V., Kizhakkedathu, J.N., and Maiti, S. (2004). *Langmuir* 2004 (20): 8468–8475.
30 Płaza, G.A., Chojniak, J., and Banat, I.M. (2014). Biosurfactant mediated biosynthesis of selected metallic nanoparticles. *Int. J. Mol. Sci.* 15 (8): 13720–13737.
31 Shen, C., Jiang, L., Shao, H. et al. (2016). Targeted killing of myofibroblasts by biosurfactant di-rhamnolipid suggests a therapy against scar formation. *Sci. Rep.* 6: 37553. https://doi.org/10.1038/srep37553.
32 Nguyen, T.T., Edelen, A., Neighbors, B., and Sabatini, D.A. (2010). Biocompatible lecithin-based microemulsions with rhamnolipid and sophorolipid biosurfactants: Formulation and potential applications. *J. Colloid Interface Sci.* 348 (2): 498–504.
33 Nicoli, S., Eeman, M., Deleu, M. et al. (2010). Effect of lipopeptides and iontophoresis on aciclovir skin delivery. *J. Pharm. Pharmacol.* 62 (6): 702–708.
34 Grijalvo, S., Puras, G., Zárate, J. et al. (2019). Cationic Niosomes as non-viral vehicles for nucleic acids: Challenges and opportunities in gene delivery. *Pharmaceutics* 11 (2): 1–24.
35 Liu, B., Duan, Q., Li, C. et al. (2014). Template synthesis of the hierarchically structured MFI zeolite with nanosheet frameworks and tailored structure. *New J. Chem.* 38: 4380–4387.
36 Nguyen, T.T. and Sabatini, D.A. (2011). Characterization and emulsification properties of rhamnolipid and sophorolipid biosurfactants and their applications. *Int. J. Mol. Sci.* 12 (2): 1232–1244.
37 Kanwara, R., Gradzielskib, M., Prevostc, S. et al. (2019). Experimental validation of biocompatible nanostructured lipid carriers of sophorolipid: Optimization, characterization and in-vitro evaluation. *Colloids Surf. B: Biointerfaces* 181: 845–855.

38 Kasture, M., Singh, S., Patel, P. et al. (2007). Multiutility Sophorolipids as nanoparticle capping agents: Synthesis of stable and water dispersible co nanoparticles. *Langmuir* 23: 11409–11412.

39 Paukkonen, H., Ukkonen, A., Szilvay, G. et al. (2017). Hydrophobin-nanofibrillated cellulose stabilized emulsions for encapsulation and release of BCS class II drugs. *Eur. J. Pharm. Sci.* 100: 238–248.

40 Weiszhár, Z., Czúcz, J., Révész, C. et al. (2012). Complement activation by polyethoxylated pharmaceutical surfactants: Cremophor-EL, Tween-80 and Tween-20. *Eur. J. Pharm. Sci.* 45: 492–498.

41 Aimanianda, V., Bayry, J., Bozza, S. et al. (2009). Surface hydrophobin prevents immune recognition of airborne fungal spores. *Nature* 460: 1117–1121.

42 Akanbi, M.H.J., Post, E., Meter-Arkema, A. et al. (2010). Use of hydrophobins in formulation of water insoluble drugs for oral administration. *Colloids Surf. B: Biointerfaces* 75: 526–531.

43 Valo, H., Kovalainen, M., Laaksonen, P. et al. (2011). Immobilization of proteincoated drug nanoparticles in nanofibrillar cellulose matrices – Enhanced stability and release. *J. Control. Release* 156: 390–397.

44 Maiolo, D., Pigliacelli, C., Moreno, P.S. et al. (2017). Bioreducible hydrophobin-stabilized supraparticles for selective intracellular release. *ACS Nano* 11: 9413–9423.

45 Reuter, L.J., Shahbazi, M.-A., Mäkilä, E.M. et al. (2017). Coating nanoparticles with plant-produced transferrin–hydrophobin fusion protein enhances their uptake in cancer cells. *Bioconjug. Chem.* 28: 1639–1644.

46 Fang, G., Tang, B., Liu, Z. et al. (2014). Novel hydrophobincoated docetaxel nanoparticles for intravenous delivery: in vitro characteristics and in vivo performance. *Eur. J. Pharm. Sci.* 60: 1–9.

47 Ding, J., Takamoto, D.Y., von Nahmen, A. et al. (2001). Effects of lung surfactant proteins, SP-B and SP-C, and palmitic acid on monolayer stability. *Biophys. J.* 80 (5): 2262–2272.

48 Brandani, G.B., Vance, S.J., Schor, M. et al. (2017). Adsorption of the natural protein surfactant Rsn-2 onto liquid interfaces. *Phys. Chem. Chem. Phys.* 19: 8584–8594.

49 Kim, M., Heinrich, F., Haugstad, G. et al. (2020). Spatial distribution of PEO–PPO–PEO block copolymer and PEO homopolymer in lipid bilayers. *Langmuir* 36 (13): 3393–3403.

50 Bodratti, A.M. and Alexandridis, P. (2018). Formulation of poloxamers for drug delivery. *J. Funct. Biomater.* 9 (1): 1–24.

51 Morishita, M., Barichello, J.M., Takayama, K. et al. (2001). Pluronic F-127 gels incorporating highly purified unsaturated fatty acids for buccal delivery of insulin. *Int. J. Pharm.* 212: 289–293.

52 Russo, E. and Villa, C. (2019). Poloxamer hydrogels for biomedical applications. *Pharmaceutics* 2019 (11): 1–17.

53 Varshosaz, J., Tabbakhian, M., and Salman, I.Z. (2008). Designing of a thermosensitive chitosan/ Poloxamer in situ gel for ocular delivery of ciprofloxacin. *Open Drug Deliv. J.* 2: 61–70.

54 Giuliano, E., Paolino, D., Fresta, M., and Cosco, D. (2018). Mucosal applications of Poloxamer 407-based hydrogels: An overview. *Pharmaceutics* 10 (3): 1–26.

55 Batrakova, E.V., Li, S., Alakhov, V.Y. et al. (2003). Sensitization of cells overexpressing multidrug-resistant proteins by pluronic P85. *Pharm. Res.* 20 (10): 1581–1590.

56 He, Z. and Alexandridis, P. (2018). Micellization thermodynamics of Pluronic P123 (EO20PO70EO20) amphiphilic block copolymer in aqueous ethylammonium nitrate (EAN) solutions. *Polymers* 10 (32): 1–18.

57 Tao, Y., Han, J., Ye, C. et al. (2012). Reduction-responsive gold-nanoparticle-conjugated Pluronic micelles: An effective anti-cancer drug delivery system. *J. Mater. Chem.* 22 (36): 18864–18871.

58 Batrakova, E.V., Li, S., Brynskikh, A.M. et al. (2010). Effects of pluronic and doxorubicin on drug uptake, cellular metabolism, apoptosis and tumor inhibition in animal models of MDR cancers. *J. Control. Release* 143 (3): 290–301.

59 Shiah, S.G., Chuang, S.E., and Kuo, M.L. (2001). Involvement of Asp-Glu-Val-Asp-directed, caspase-mediated mitogen-activated protein kinase, kinase 1 cleavage, c-Jun *N*-terminal kinase activation, and subsequent Bcl-2 phosphorylation for paclitaxel-induced apoptosis in HL-60 cells. *Mol. Pharmacol.* 59 (2): 254–262.

60 Khaliq, N.U., Park, D.Y., Yun, B.M. et al. (2019). Pluronics: Intelligent building units for targeted cancer therapy and molecular imaging. *Int. J. Pharm.* 556: 30–44.

61 Kaplan, N., Zosim, Z., and Rosenberg, E. (1987). Reconstitution of emulsifying activity of *Acinetobacter calcoaceticus* BD4 emulsan by using pure polysaccharide and protein. *Appl. Environ. Microbiol.* 53 (2): 440–446.

62 Yi, G., Son, J., Yoo, J. et al. (2019). Emulsan-based nanoparticles for in vivo drug delivery to tumors. *Biochem. Biophys. Res. Commun.* 508 (1): 326–331.

18

The Potential Use of Biosurfactants in Cosmetics and Dermatological Products

Current Trends and Future Prospects

Zarith Asyikin Abdul Aziz[1,2], *Siti Hamidah Mohd Setapar*[1,2,3], *Asma Khatoon*[1], *and Akil Ahmad*[1,4]

[1]Centre of Lipids Engineering and Applied Research (CLEAR), Universiti Teknologi Malaysia, Johor Bahru, Johor, Malaysia
[2]Department of Chemical Processes, Malaysia-Japan, International Institute of Technology, Universiti Teknologi Malaysia, Skudai, Johor, Malaysia
[3]SHE Empire Sdn., Jalan Pulai Ria, Bandar Baru Kangkar Pulai, Skudai, Johor, Malaysia
[4]School of Industrial Technology, Universiti Sains Malaysia, Gelugor, Penang, Malaysia

CHAPTER MENU

18.1 Introduction

Recently, increasing demand in cosmetic legislation has contributed to major concentration among cosmetic market and research in commercializing natural and organic products to develop more derma-safe and highly effective cosmetics. The awareness among consumers on the importance of safety and effectiveness in owning cosmetic products has oriented the scientific research in the cosmetic sector toward the realization of natural and organic cosmetics development [1–3]. Natural or organic cosmetic products can be defined as being products made from any raw materials from

Biosurfactants for a Sustainable Future: Production and Applications in the Environment and Biomedicine, First Edition. Edited by Hemen Sarma and Majeti Narasimha Vara Prasad.
© 2021 John Wiley & Sons Ltd. Published 2021 by John Wiley & Sons Ltd.

extracted components from natural sources such as plants, animals, minerals, or fermented substances without the involvement of any chemical compounds.

Surfactants are one of the most common chemical products that are extensively consumed on a large scale throughout the world and are also used as an important ingredient in cosmetic products, especially skin care. Specifically, in the cosmetic regime, surfactants are implemented as wetting, emulsifying, detergency, dispersing, and foaming agents that contain amphiphilic compounds, such as hydrophobic and hydrophilic segments [4, 5]. The main function of a surfactant is to reduce surface tension between two different polarities of liquids and facilitate the formation of an emulsion solution. Almost half of the surfactants produced is utilized in the washing and cleaning sectors, but in cosmetic products, sodium lauryl sulphate (SLS), ammonium lauryl sulphate (ALS), and polysorbate esters of surfactants are known to be extensively used as commercial surfactants. However, most of these surfactants are chemically synthesized from petroleum and it has been reported to cause some adverse effects on skin and significant environmental problems after long-term application [5, 6].

Biodegradability and biocompatibility are the most important characteristics of surfactants proportional to their functional performance for consumers [4, 7]. Hence, great attention has been given to the preparation of new cosmetic surfactants with safer, biocompatible, biodegradable, and natural-based properties [4]103]. A natural surfactant is defined as a surfactant that is synthesized directly from a natural renewable resource, such as plants, microbes, invertebrates or animals, and is produced by several separation processes like extraction, distillation, or precipitation without any organic synthesis involved [4, 5, 8]. Moreover, a high demand for natural surfactants is due to consumers' concern on environmental conservation and since those natural surfactants are biodegradable and non-toxic, their applications in cosmetics can be consistent [7].

Biosurfactants are also known as microbial surfactants where amphiphilic compounds with surface-active components are produced by microorganisms (yeast, fungi, or bacteria). Generally, biosurfactants are an emerging class of biomolecules that can be fully used in various applications, including cosmetics [9]. These microorganisms can be growing successfully in two conditions whether in oily and water miscible substrates or second culture broth, or remain adherent to microbial cells. Descriptions of the development of these microbial amphiphilic molecules are available for more than 225 patents [4].

Biosurfactants offer many advantages over synthetic surfactants in terms of their higher biodegradability, biocompatibility, low critical micelle concentration (CMC), low toxicity, gradual adsorption, and superior ability to form assembly and liquid crystals [10, 11]. However, there is some limitation in further applications of biosurfactants due to their high production cost and limited structural variety [12, 13].

Biosurfactants are classified in several groups broadly in terms of their chemical composition and molecular weight. According to these combinations, lipopeptides, glycolipids, fatty acids, lipoproteins, and phospholipids are groups of surfactants with low molecular weights, while polymeric biosurfactants are characterized as high molecular weight biosurfactant [14].

Glycolipids and lipopeptides are listed as microbial surfactants (biosurfactants) that are widely implemented in cosmetic formulations. Rhamnolipids, trehalolipids, sophorolipids, and mannosylerythritol lipids (MEL) are categorized as glycolipid biosurfactants [14]. The microbes involved for the synthesis of these biosurfactants include *Pseudomonas* strains [15], *Candida* species [16], *Rhodococcus* strains [17], and *Pseudozyma* yeasts [18], for rhamnolipids, sophorolipids, trehalolipids, and MEL, respectively. In addition, surfactins comprise a widely studied lipopeptide biosurfactant group that can be synthesized by *Bacillus* species [19–21].

Mannosyleryththritol lipids (MEL) are becoming the most prominent biosurfactants that are efficiently used in cosmetic formulations. Appreciable amounts of MEL were reported and produced by basidiomycetous yeast of *Pseudozyma* spp., such as *Pseudozyma aphidis*, *Pseudozyma antarctica*, *Pseudozyma parantarctica*, and *Pseudozyma rugulosa*. MEL has been applied in various cosmetic products such as eye shades, lipsticks, soap, powders, spray, body massage oils, nail care, and lipmarkers. In addition, surfactin is considered as another kind of potential biosurfactant that is applied in cosmetic products, specifically for a dermatological approach such as antiwrinkle and facial cleanser products. This is reported as an extremely potential biosurfactant and is widely used by the Japanese cosmetic industry [22].

Therefore, this review focuses on current trends of biosurfactant types of applications in dermatological and cosmetic formulations. The applications involve several current patented products with biosurfactants and research study mainly focuses on novel formulations containing biosurfactants in cosmetic and dermatological segments.

18.2 Properties of Biosurfactants

In cosmetic formulations, several parameters of biosurfactants need to be considered that are related to the biosurfactants' composition factors such as the hydrophilic–lipophilic balance (HLB) value, CMC, and ionic behavior [14]. The CMC is defined as the lowest concentration of biosurfactant to achieve maximum reduction of water surface retention, where this property has proven to be an effective factor for a surfactant. Patowary et al. [23] synthesized rhamnolipid from a *Pseudomonas aeruginosa* strain that isolated it from paneer whey waste. This study revealed a lower value of rhamnolipid at 110 mg/l. Another study demonstrated a lower CMC value of rhamnolipid of 50 mg/l, which was produced by isolation of *P. aeruginosa* strain DN1 from petroleum-contaminated soil [24].

A similar value of CMC (50 mg/l) of rhamnolipid was also reported [24] from the *P. aeruginosa* strain. This strain was isolated from low-cost substrates (corn steep liquor (CSL) and molasses), which has the ability to reduce surface tension of water up to 30 mN/m and demonstrated a high emulsifying activity of 60%. Additionally, the lowest value of rhamnolipid biosurfactant was demonstrated by Lan et al. [25]. In this study, rhamnolipid production was isolated from an innovative *Pseudomonas* strain species (*Pseumodonas* SWP-4) from waste cooking oil. The rhamnolipid CMC value was only 5 mg/l, resulting in a high yield of 13.93 g/l with a waste cooking oil utilization of around 88%. The emulsification index of rhamnolipid with n-hexadecane reached around 59%.

Other than that, some previous studies reported the synthesis of surfactin biosurfactants using *Bacillus subtilis* bacterial species with lower CMC values. Gudiña et al. [21] used CSL medium to isolate *B. subtilis* #573 for the production of surfactin biosurfactant. The result revealed that those culture mediums consisting of 10% CSL resulted in higher yields of surfactin approximately 1.3 g/l and achieved a CMC value of 160 mg/l. Hazra et al. [26] grew bacteria known as *Bacillus clausii* BS02 in basal salt medium with sunflower oil soap stock that was isolated as surfactin biosurfactant. This research demonstrated a surfactin yield of 0.6 g/l at 3% w/v of sunflower soap stock as the carbon source. The biosurfactant showed a low CMC value of 45 mg/l at 8% emulsification activity to reduce water surface tension in the range of 69.07 mN/m to 30 mN/m.

Furthermore, another type of microorganism from the lactic acid group has been reported to produce biosurfactant with a much lower range of CMC values. Madhu and Prapulla [27] successfully produced glycoprotein biosurfactant from rice-based ayurvedic fermented product and isolated the bacteria known as *Lactobacillus plantarum* CFR 2194. The study revealed that a higher yield of biosurfactant was produced at a fermentation time 72 hours under stationary conditions, with a CMC value of 6 g/l.

Another important parameter of biosurfactants to be incorporated in cosmetic formulation due to their hydrophilic lipophilic balance (HLB) value was referred to as the emulsifying capability of surfactant [8]. The functions of biosurfactants can be altered to be an emulsifier, antifoaming agent, wetting agent, and others depending on their differentiation in HLB value [8]. Low HLB values demonstrated a higher hydrophobicity of biosurfactant, while the biosurfactants are more hydrophilic with higher HLB values. Oil in water (O/W) emulsions can be developed by dispersions of higher HLB values of biosurfactants (hydrophilic); meanwhile, hydrophobic biosurfactants are suitable to be dispersed in water in oil (W/O) emulsions. The function of biosurfactants in these emulsions have a good stabilizer capacity, which is usually used in cosmetic formulations as a bioactive component carrier [8].

In dermatological applications, lipophilics (higher HLB value) of biosurfactants range between 1 to 4 and are preferred due to their skin nature, which contains lipid film and favors oil-soluble active ingredients. However, oil in water (O/W) emulsions are more favored by consumers attributable to their less greasy effect and higher absorption rate with HLB values between 8 and 16. In addition, O/W cosmetic emulsions are usually commercialized in semi-solid or liquid formulations [28]. The review on the HLB value of biosurfactants was reported in 2018 and is related to this topic and others mostly from more than the past five years.

Ohadi et al. [28] investigated physicochemical properties such as HLB, surface tension, and CMC values of lipopeptide biosurfactant produced by *Acinetobacter junii* B6, isolated from an Iranian oil excavation site. The HLB value of the biosurfactant was 10, which was suitable to be used as an O/W emulsions stabilizer. The CMC value was 300 mg/l, which was therefore able to reduce water surface tension to 36 mN/m.

Fukuoka et al. [29] produced mannosylerylthritol lipids (MELs), which are categorized under glycolipid biosurfactants and are synthesized by *P. antarctica* microorganisms, which are isolated from glucose. The biosurfactant was able to reduce water surface tension to 33.8 mN/m with a CMC value of 3.6×10^{-4} M. Additionally, the MELs biosurfactant was reported to be suitable to function as an O/W emulsion emulsifier and also as a washing detergent due to its HLB value of 12.15.

Earlier in 1998, Hillion and his research team developed a sophorolipid biosurfactant from *Turulopsis* yeasts [30]. The sophorolipids have been successfully synthesized when applied in the acid form or in the monovalent metallic salt form, and facilitate an anionic biosurfactant that forms stable emulsions. This biosurfactant was reported to have a desirable property to be implemented in cosmetic and personal care formulations due to its HLB value ranges between 13 to 15.

Additionally, the ionic behavior of biosurfactants is considered to be one of the important factors for their implementation in cosmetics. Due to their polar head charge, they can be classified as anionic, cationic, non-ionic, and amphoteric. In the aspect of a great foaming, wetting, and emulsifying performance, anionic surfactants (including biosurfactants) are reported to give successful results. Unfortunately, studies reported that anionic surfactants are the most irritating surfactants to both skin and eyes, followed by non-ionic surfactants, while it has been revealed that amphoteric surfactants give the least side effects on human skin. However, cationic surfactants have demonstrated successful antibacterial properties, as well as good emulsifier capabilities [8]. Hence this part of the review will include recent findings about ionic charge behavior in biosurfactant production.

In 2016, a research study regarding lipopeptide biosurfactant adsorption on to human hair was assessed. The biosurfactant was synthesized spontaneously from a stream of corn wet milling industry and showed that it was amphotering, including anionic and cationic segments. Moreover, when this surfactant was added on the top of human hair, the hair was able to adsorb the biosurfactant concentration close to the CMC value with a maximum capacity of 3679 μg/g [31].

18.3 Biosurfactant Classifications and Potential Use in Cosmetic Applications

18.3.1 Glycolipids

Glycolipids are compounds that consist of carbohydrate moiety linked to long-chain aliphatic or hydroxyl-aliphatic acids (fatty acids). Most glycolipids originate from bacteria and a few from fungi and yeast. In cosmetology, sophorolipids, rhamnolipids, and MEL applied in cosmetic formulations are mostly known as glycolipid biosurfactants [4].

18.3.1.1 Sophorolipids

The sophorolipids are most promising biosurfactants and promote various advantages such as low toxicity, high biodegradability, high selectivity, and ecological acceptance. According to the Food and Drug Administration (FDA), sophorolipids have no cytotoxicity effects and are approved by this administration for application in many industries [32].

Sophorolipids are mainly produced by the well-known yeast *Candida bombicola* and are classified as secondary metabolites and extracellular glycolipids, respectively [33]. Acidic and lactonic sophorolipids are the principal structures of these metabolites, and as a result of their altered biological and physicochemical properties they can beresponsible for sophorolipid applications [32].

Candida bombicola is the most implemented microorganism used to synthesize sophorolipids, which is attributable to its higher yield of production. This microorganism mainly produces sophorolipids, which consist of monounsaturated fatty acids in lactonic di-acetylated form, bonded with acidic non-acetylated forms (Figure 18.1). Therefore, as depicted by the compositions of acidic and lactonic forms, sophorolipids synthesized by *C. bombicola* are very likely to be applied in cosmetics, foods, and pharmaceuticals [34].

In the cosmetic arena, sophorolipids have been commercially implemented as active ingredients in the formulation of personal care products, mainly as foaming agents, emulsifiers, detergents, wetting agents, and stabilizers [33]. The main benefit of sophorolipids used in cosmetic formulations is due to its ability to give low cytotoxicity on human fibrolast and keratinocytes. In addition, sophorolipids promote the metabolism of fibrolast and boost collagen synthesis on skin dermis, acting as an antiaging agent in restricting repairs and toning up the skin [35].

A most recent study in 2019 has demonstrated the application of sophorolipid as a carrier for transdermal release of lignan-based transferosomal hydrogel. The lignans and sophorolipids were produced from flax seed and *C. bombicola* yeast, respectively. The transferosomal hydrogel release profile was explored using the permeation rate of the lignan major constituent, secoisolariciresinol diglucoside (SDG), on to the cellophane membrane. Before the *in vivo* study, formulation characterizations were initially conducted where those formulations with sophorolipids exhibited improved encapsulation efficiency, 75.81% of lignan-tranferosomal hydrogel compared to synthetic surfactant used (38.54%) [36].

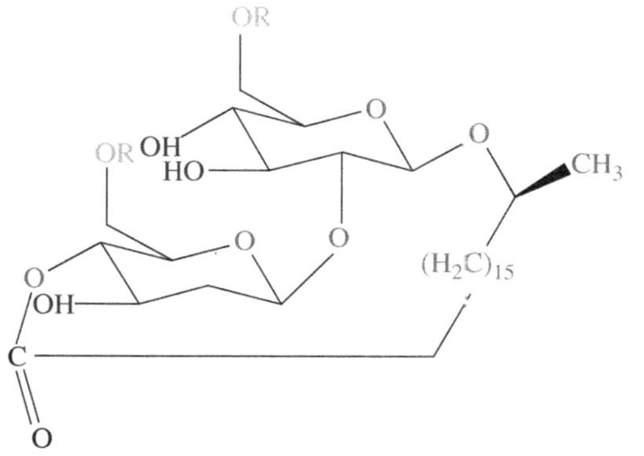

Figure 18.1 Sophorolipid chemical structure.

Then the hydrogel formulation with higher encapsulation efficiency containing sophorolipids was further tested for its release profile, which showed a sustained increase of lignan release gradually over four hours. Hence, this study concluded that this sustained release pattern made it suitable for cosmetic transdermal delivery applications [36].

Maeng et al. [37] synthesized sophorolipid from horse oil by *C. bombicola* for cosmetic application, specifically for antiwrinkle properties. An *in vitro* antiwrinkle assay was assessed using human skin fibrolast and measuring collagen-type 1 (CoI-I) mRNA expressions. The findings were compared with isolated sophorolipids, pure horse oil, hydrolysed horse oil, and vitamin C as positive controls.

Depicting the results, horse oil showed no stimulating effect on CoI-I mRNA expression, even when the concentration increased from 25 to 100 μg/ml. The positive control vitamin C was used at concentrations of 100 μg/ml, increasing the CoI-I mRNA expression by 127%. In addition, similar results were demonstrated using hydrolysed horse oil and isolated sophorolipids, which enabled stimulation of the CoI-I mRNA expression, while the sophorolipid effect was concentration-dependent [37].

18.3.1.2 Rhamnolipid

Rhamnolipids, mainly produced by the *P. aeruginosa* pathogen are the most widely studied glycolipid biosurfactants. They have significant advantageous while functioned as wetting, dispersion, decontamination, dissolution, and emulsification agents due to their low toxicity, high biodegradability, better environmental compatibility, and high performance at high pH and temperature properties [38].

The most frequently used are the mono-rhamnolipids and di-rhamnolipids, as shown in Figure 18.2. They consist of one or two rhamnose moieties, which are bonded with three hydroxyl fatty acids that contain a carbon chain from 8 to 14 [4]. The procedures involve strain production within manipulated cultivation conditions, resulting in different components and properties of rhamnolipids [39]. In one study,

Jadhav et al. [40] demonstrated that the ratio of mono- and di-rhamnolipids synthesized by *P. aeruginosa* mainly lies between 40 : 60 and 30 : 70. Another study showed that 68.35% out of a total rhamnolipids yield was reached by mono-rhamnolipids, which were produced from *P. aeruginosa* MN1 [41].

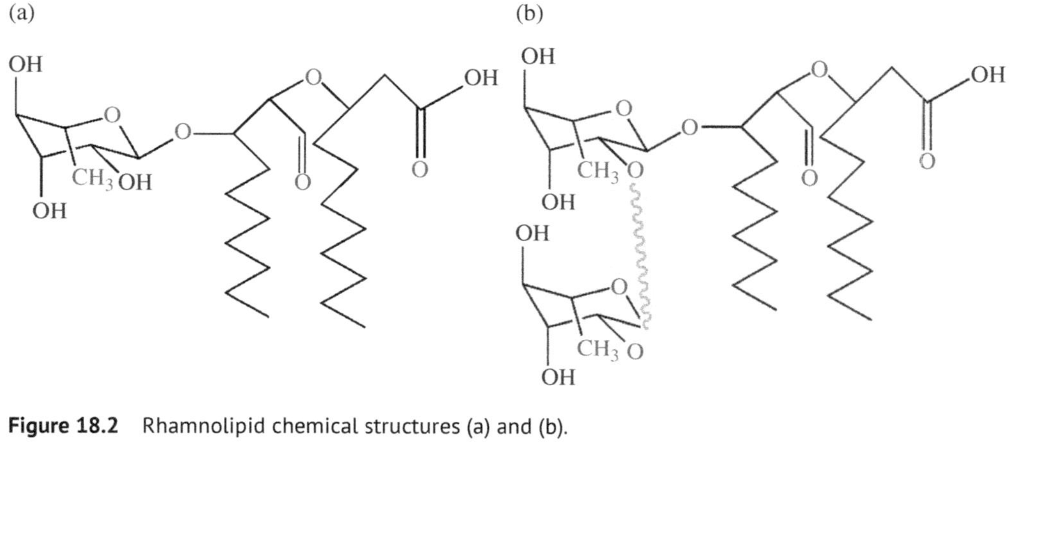

(a) (b)

Figure 18.2 Rhamnolipid chemical structures (a) and (b).

Previous research studies have been investigating the potential application of rhamnolipids in cosmetic and personal care fields, specifically as antiwrinkle, antimicrobial and antiproliferative agents. In 2007, a cosmetic formulation including rhamnolipids has been patented for its specific application as an antiwrinkles agent, where the patent concluded that rhamnolipids were demonstrated to be an effective skin re-epithelization [42].

Antimicrobial activity of rhamnolipids isolated from sunflower oil against skin microflora causes acne, as seen from *Propionibacterium acnes* reported previously. The finding showed rhamnolipids successfully inhibited the acne microbes, which resulted in a wide distinct zone of inhibition. This positive inhibitory result opens up possibilities for implementation of rhamnolipids in cosmetic and skin care products. The study also concluded that successful rhamnolipids against the skin microflora is involving a mechanism of its antimicrobial action attributable to the fact that biosurfactants may disturb the microbe membrane structure through interaction with phospholipids as well as membrane proteins. Due to rhamnolipids intrinsic properties, the surface-active compound may also interfere with cell surfaces and disrupt microbial membranes [41].

On the other hand, application of rhamnolipids as emulsifiers in oil-in-water (O/W) nanoemulsions for cosmetic formulation potential has been studied. Through the research study, rhamnolipids have been evaluated for their capability as emulsifiers to stabilize an (O/W) nanoemulsion containing a medium chain triglyceride (MCT) oil–water solution. The manipulated rhamnolipid surfactant has been used as a dependent parameter to form a smaller particle size of rhamnolipids-MCT oil droplets through a nanoemulsion solution. The result showed that smaller droplets are formed ($d < 0.15\,\mu m$) at a rhamnolipid biosurfactant-to-oil ratio (SOR) of 1 : 10. The droplets were also stable at a nanoemulsion pH value between 5 and 9, while they become unstable at an acidic pH (1 to 4) due to aggregation formation and the occurrence of a re-coaslescence phenomenon. Hence, it was concluded that rhamnolipids can be as effective as a surfactant as a natural emulsifier with a high potential to be used in certain commercial cosmetic applications [43].

18.3.1.3 Mannosyloerythritol Lipids

MELs are among biosurfactants categorized under the glycolipid group that are reported to be specifically produced by yeast strains of the genus *Pseudozyma*. The chemical structure of a MEL contains either 4-*O*-β-D-mannopyranosylerythritol or 1-*O*-β-D-mannopyranosylerythritol as the hydrophilic segment of the biosurfactant, attached with a variety of fatty acids as a hydrophobic segment [44]. MEL-A (tri-acylated MEL), MEL-B (di-acylated MEL), and MEL-C (mono-acylated MEL) are represented as different types of MEL depending on the acylation degree contained [4]. Figure 18.3 shows the MEL structure type.

Previously, several research studies reported the excellent interfacial properties demonstrated by MELs, as well as several biochemical function potentials, including differentiation induction against melanoma cells [45], rat pheochromocytoma [46], and human leukemia cells [47]. Besides, previous studies also reported that the MELs hold tremendous promise for their application in cosmetics, foods, and pharmaceuticals, due to their high biodegradability, low toxicity properties, and ease of production factors [45].

In the cosmetic arena, MEL biosurfactants have been attracting much attention as new ingredients used in cosmetic formulations attributable to their unique moisturizing and liquid-crystal foaming properties. Yamamoto et al. [48] synthesized and manipulated MEL derivatives (MEL-A, MEL-B, and MEL-C) from olive oil and evaluated their recovering effect on damaged skin cells using a three-dimensional cultured human skin model. These MELs were comparable with natural ceramide, another potential compound used as a good moisturizing agent for damaged skin. The damaged skin was treated with sodium dodecyl sulphate (SDS) and the skin's cell viability was

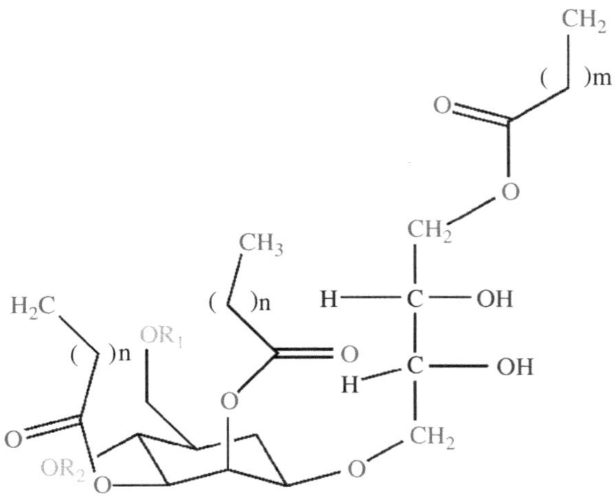

Figure 18.3 MEL chemical structure.

used as a dependent parameter to ensure capability of the MEL lipid in recovering the damaged skin.

The findings showed that all tested MEL derivatives clearly had a recovery effect on the damaged skin, with MEL-A leading with a cell viability higher than 90%, followed by MEL-B (80%), and MEL-C (<70%). All MEL results were comparable with positive control and a natural ceramide that resulted in higher than 80% of cell viability. This study reported that the higher cell viability achieved by all MEL derivatives were due to their good moisturizing activity, having the well-known skin ceramide properties. Ceramide properties of skin provide a strengthening skin structure and a reduction in water loss, which enhance the skin's epidermal water barrier and permeability [48].

Another study by Yamamoto and the research team in 2013 [49] developed MEL-B biosurfactant contained with a novel hydrophobic chain from castor oil as the carbon source to utilize different species of *Pseudozyma* yeast, named as *Pseudozyma tsukubaensis*. Depicting this result, a novel structure of MEL-B was identified as 1-*O*-β-(2'-*O*-alka(e)noyl-3'-*O*-hydroxyalka(e)noyl-6'-*O*-acetyl-D-mannopyranosyl)-D-erythritol. This "new MEL-B" was demonstrated to have an CMC value of 2.2×10^{-5} with a reduced surface tension of 28.5 mN/m. However, this study concluded that the newly identified MEL-B was likely to have a different CMC, interfacial values, and chemical structure compared with the conventional MEL-B. However, these different properties could widen the application of glycolipid biosurfactants. Moreover, a study showed that the potential of the "new MEL-B" to increase the skin water content on forearm skin and suppressed perspiration that continued for a period of at least two hours.

In a recent study, Bae et al. [50] investigated an advanced application of MEL biosurfactants in a hyperpigmentation skin treatment since depigmentation of MELs has not been evaluated. Through this study, antipigmentation activity of MELs was evaluated on three skin types, primary normal human melanocytes (NHM), α-melanocyte-stimulating hormone (MSH)–stimulated B16 cells (murine melanoma cells), and a human skin equivalent (MelanoDerm), where the results were recorded by photography. The result showed that MELs significantly reduced the melanin contents in NHMs and murine melanoma cells, and also generated a whitening effect in MelanoDerm skin by reducing the melanin content and brightening the tissue color. Additionally, this study concluded that reduction of the malignant content was due to the MEL's capability in inhibiting the ERK/CREB/MiTF signaling pathway in NHMs containing melanogenic enzymes such as tyrosinase, Tyrp-1, and Tyrp-2 [50].

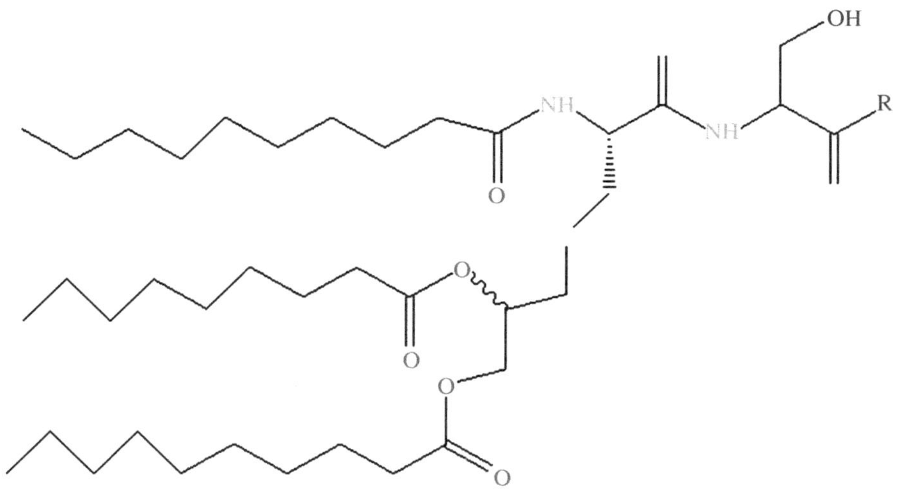

Figure 18.4 Lipopeptide chemical structure.

18.3.2 Lipopeptides

Lipopeptides are composed of 7 to 10 amino acids in amphiphilic cyclic peptides, where fengycin, iturin, and surfactin lipopeptides contain10 and 7 amino acids, respectively [51, 52]. These cyclic peptides are bonded with β-hydroxy fatty acids which are arranged in linear (n), *iso*, and *anteiso* structure forms [52]. Figure 18.4 shows the chemical structure of lipopeptides.

B. subtilis is a common microorganism used to produce lipopeptides, which were first reported in 1968 to secrete surfactin [53]. Surfactin is the predominant lipopeptide biosurfactant with excellent surface properties [54], as well as safety properties [55], so it received considerable attention and its applications has been widened to many industries, including cosmetics, foods, medicines, and pharmaceuticals [56, 57].

Implementation of lipopeptides in cosmetic formulations are mainly as emulsifiers, resulting in a low skin irritation effect that made it suitable for a transdermal application route such as transparent cosmetics with sequestering functions [58]. These biosurfactants have low CMC properties (1–240 μM) that promotes their better performance in dermatological products [59] and their safety features are confirmed by low toxicity properties, especially toward mammalian cells [60].

An *in vitro* antioxidant activity and wound healing efficacy of surfactin biosurfactant secreted from *B. subtilis* SPB1 microorganisms have been evaluated. The antioxidant assays were conducted using 2,2-diphenyl-1-picrylhydrazyl (DPPH) and a ferric reducing power. Meanwhile, the surfactin biosurfactant has been developed as a based-gel for a wound healing efficacy assessment using induction of excision wounds on experimental rats [59].

The DPPH and ferric reducing power antioxidant assay findings demonstrated a good performance, which at 1 mg/ml of surfactin the scavenging activity was 70.4%, while ferrous ion chelating activity reported was 80.32%, respectively. Furthermore, in a wound healing efficacy assessment, the percentage of wound closure on rats significantly increased over a period of 13 days of surfactin-based gel topically applied on the rats' wound site. This result was reported to be higher than that of untreated rats (negative control) and a CICAFLORA™ product-treated group (positive control). The wound healed skin using the surfactin gel was biopsied and showed a wholly re-epithelialized wound with perfect epidermal regeneration [59].

Table 18.1 Bioactive characteristics of other types of common biosurfactant categories (glycolipids and lipopeptides).

Biosurfactant categories	Biosurfactant type	Bioactive characteristics	References
Glycolipids	Sophorolipids	Antimicrobial activity	[5]
		Antioxidant activity	[62]
		Hypocholestrolemic activity	[63]
		Dermal fibrolast enhancer	[64]
		Antiaging activity	[65]
	Rhamnolipids	Antimicrobial activity	[66]
		Antiadhesive and antibiofilm activities	[67]
		Detergency agent	[68]
		Antibacterial activities	[69]
	MEL	Antioxidant activity	[70]
		Moisturizing agent	[71]
		Hair flexibility enhancer	[72]
Lipopeptides	Surfactin	Detergency agent	[73]
		Emulsifying activity	[74]
		Anticellulite agent	[75]

Most surfactin applications have been reported in patents by several materials suppliers and cosmetic companies. Surfactin has mainly been patented as an antiwrinkle agent in cosmetic formulations, which were reported to give an antiaging effect within pharmaceutically acceptable vehicles, diluents, adjuvants, and excipients [61].

Therefore, these biosurfactants showed their scientific potential in the cosmetic arena, where each review in this sub-section was among mainly biosurfactants that had been studied by previous scientific committees as natural and safer ingredients in cosmetic formulations. Table 18.1 lists a summary of biosurfactants and their potential functions in cosmetic applications.

18.4 Dermatological Approach of Biosurfactants

Sugars, proteins, and lipids secrete biosurfactants with phospholipid and protein structure similar to those structures found in human skin membranes. The phospholipid fatty acids are reported to be the compounds that provide the crucial structure of the skin, while the proteins are needed to promote skin normal cell function. According to Noughton et al. in 2019, wound healing and prebiotic activity against skin pathogens are considered as a potential use of biosurfactants in dermatological applications [12].

18.4.1 Wound Healing Application

A study was conducted by Zouari et al. [59] that evaluated the wound healing properties of lipopeptide isolated from *B. subtilis*, named as SPB1 biosurfactant. The wound healing assessment was conducted using an excision wound healing model, in which an approximately 150 mm^2 wound was made on the depilated thoracic region of rats. Then the wounded parts were topically applied by several treatments; CICAFLORA a commercial wound healing cream as the positive control, 100% glycerol as the negative control, and 5 mg/ml and 15 mg/ml of gel-based *B. subtilis* SPB1 biosurfactant. All treatments were applied every two days until complete epithelization. The wound healing evaluation parameter was based on the percentage of wound closure. The findings on wound closure of the excised area on rats found significant ($p < 0.05$) wound healing activity among animals treated with CICAFLORA and gel-based SPB1 biosurfactants.

The animals treated with 15 mg/ml of SPB1 biosurfactant gel experienced total successful wound healing properties (100% wound closure) within 13 days of treatment. Then the result was followed by a group treated with 5 mg/ml of SPB1 biosurfactant that had a significantly ($p < 0.05$) greater wound closure (97.29%) than the positive control group (95.19%). Hence, the recorded time needed to complete epithelization of the excised skin area in 15 ml/mg of SPB1 biosurfactant gel was less than those rats treated with the 5 mg/ml dose. The study suggested that successful wound healing activity of gel-based SPB1 biosurfactant was due to its antioxidant properties, reported by Jemil et al. [76]. The free radical scavenging performance of lipopeptide biosurfactant help to prevent inflammation and tissue regeneration and re-epithelization. Thus, it has been concluded the SPB1 lipopeptide biousurfactant gel has potential to use in the treatment of normal and complicated wounds as well as dermatological diseases [59].

Gupta et al. [77] found accelerated wound healing on a 6 mm^2 wounded area treated with SV1 ointment formulation containing glycolipid isolated from *Bacillus licheniformis*. In the early stage of wound healing assessment, the SV1 treatment group showed fibroblast cell proliferation and re-epithelization, with more rapid collagen deposition in the later stages. It was concluded that wound healing exhibited by the SV1 glycolipid biosurfactant may be attributed to biosurfactant-reduced oxidative stress through production of reactive oxygen prevention.

Another study reported the wound healing activity of purified glycolipid biosurfactant sophorolipids produced from *Starmerella bombicola*, where the purified sophorolipids enriched with nonacetylated acidic ($C_{18:1}$) congeners resulted tin >95% purity. Initially, the purified $C_{18:1}$ non-acetylated sophorolipids (NASLs) were evaluated for their antimicrobial activities to determine the potential application of $C_{18:1}$ NASL against wound infection and contamination that was caused by bacterial pathogens [78].

Pseudomonas aeruginosa and Enterococcus faecalis nosocomial infective agents were successfully inhibited by NASL at low concentrations of 5 mg/ml that resulted with significant reductions in colony-forming units (CFUs). In addition, *in vitro* cell viability assay was conducted to evaluate the mammalian cell toxicity effect using human cultured skin: HaCaT, human dermal microvascular endothelial cells (HDMVECs), and human umbilical vein endothelial cells (HUVECs). The cell viability of manipulated $C_{18:1}$ NASL concentrations of 0.01–500 mg/ml added to human cultured media was counted using MTT [3-(4,5-dimethyl-2-thiazolyl)-2,5-diphenyl-2H-tetrazolium] bromide assay. From the results, it was found that no adverse effects with concentrations of the acidic sophorolipids <0.5 mg/ml, which revealed that there was no effect either by keratinocytes or endothelial-derived cell lines [78].

Through the *in vivo* wound healing assay, there were wounds excised on the debilated dorsal skin of male mice. These animals were divided into several groups depending on treatments administered after wound excision. The treatments involved manipulated dosage of acidic sophorolipids at 20, 200, and 400 mg/kg and were compared with control (untreated) and vehicle groups (aqueous cream). Treatments were applied on affected areas daily for seven days and the wound healing activity was determined by measuring the wound area. The wound healing activity result demonstrated that within 16 days of study, all wounds completely closed but where animals were treated by different dosages of acidic sophorolipids the wound areas were significantly reduced, compared to the vehicle group where the wound areas increased on day 4 of the treatment. Hence, the study concluded that interesting wound healing activity exhibited by $C_{18:1}$ NASL was proportional to its consistent antimicrobial activities. It also suggested further use of the acidic sophorolipids as the antimicrobial component in dermatological cream for wound infection treatment [78].

18.4.2 Prebiotic Activity of Biosurfactants Against Skin Microflora

Several commensal microorganisms are found permanently as normal microflora of human skin, commonly known as resident microflora. Certain microbial species can also be temporarily grown on the skin for a short period of time and there are also transient microbial species where they are occasionally found on the skin [79]. The resident skin microflora consist of bacteria, fungi, and viruses that do not give any pathogenic effect. The majority of these microbes are the Gram-positive bacteria from genes of *Staphylococcus*, *Propionibacterium*, *Corynebacterium*, *Micrococcus*, and *Acinetobacter*. *Staphylococcus epidermis* are the major bacteria composition (around 90%) found as human skin microflora that function as a protection agent to prevent skin from infections and other environmental pollutants [79, 80]. However, some pathogenic microbes reported to be found on the skin, for example *Staphylococcus aureus* is a major human pathogen, known as transient microbes, can cause several skin infections such as atopic dermatitis [81, 82].

In cosmetics, prebiotic is defined as a component with the capability to promote a skin microflora re-balancing process. For example, a pathogenic microorganism of skin with acne is caused by excessive growth of *P. acnes*. This skin problem needs to be rebalanced by promoting *S. epidermis* growth and inhibiting the growth of *P. acnes*. Antimicrobial, antiviral, and antiadhesive properties of biosurfactants against several pathogens can be applied as prebiotic components in cosmetic formulations [80].

Another research study evaluated the anti-microbial activity of cell-bound biosurfactants that are produced by *Lactobacillus pentosus* (PEB), categorized as glycolipopeptide macromolecules used against several strains present in the skin microflora in claiming its potential implementation as a natural ingredient in dermatological products. PEB performance was compared with another glycolipopeptide macromolecule biosurfactant produced by *Lactobacillus paracasei* (PAB), in which both were extracted using phosphate buffer (PB) and phosphate buffer saline (PBS) [83].

Most successful antimicrobial (100% inhibition) presented by PEB went to its performance against *P. aeruginosa* (PEB extracted with PBS), *Streptococcus agalactiae* (PEB extracted for both extract), and *S. aureus* (PEB extracted with PB), followed by more than 80% inhibition against *P. aeruginosa* (85%, PEB extracted with PB), *Escherichia coli* (89% when PEB extracted with PBS) and *Streptococcus pyogenes* (about 85% for both extracts). These microbes were inhibited by the PEB biosurfactant at a concentration of 50 mg/ml. Meanwhile, biosurfactant from PAB demonstrated in general higher antimicrobial activities for all examined skin pathogens at lower concentrations, compared to PEB biosurfactants. However, this study still concluded that both biosurfactants could be used as potent natural antimicrobial agents in cosmetic and wellness formulations [83]. Later in 2018, Garg and Chatterjee [84] isolated biosurfactant from a novel bacterial source, *Candida parapsilosis*, produced by contaminated dairy products. Its antibacterial effect against skin pathogen *S. aureus* was evaluated. There are two antibacterial activity assays that have been used: agar diffusion and the microdilution method.

According to the results, the study revealed in the diffusion assay increased the biosurfactant concentrations (0.30, 0.75, and 1.50 mg/ml) and enhanced the *S. aureus* inhibition zones. The maximum zone of inhibition (2.66 cm) was demonstrated by a biosurfactant concentration value of 1.50 mg/ml. Additionally, a microdilution antibacterial assay showed 0.240 ± 0.0140 nm and 0.085 ± 0.0157 nm of *S. aureus* growth inhibited by a 5 and 10 mg/l concentration of the isolated biosurfactant, respectively. Hence, this study claimed that the successful antibacterial assay performed by the biosurfactant was due to disrupt the plasma membrane of the skin pathogen after the biosurfactant application. However, this study suggests that future studies need to be carried out to ensure that the mechanism is involved in the microbe plasma membrane disruption [84].

Demster et al. [85] developed a novel formulation containing a combination of MEL biosurfactant and common antibiotics used for skin infections (bacitracin) with the aim to reduce antibiotic concentration required in skin bacterial infection treatment. The result showed that a single application of bacitracin antibiotic at 256 and 128 μg/ml hardly showed any reduction of *S. aureus*, while MELs alone at a concentration of 0.016 mg/ml exhibited lower *S. aureus* populations. However, a combination of MELs (0.016 mg/ml) with bacitracin (128 μg/ml) demonstrated a higher bactericidal effect. Thus, this study revealed that biosurfactant and bacitracin antibiotic can be a potential combination to exhibit bacterial reduction at lower concentrations.

The most recent study was carried out by Sen et al. [86] who investigated the potential of sophorolipid (SL-YS3) produced by *Rhodotorula babjevae* YS3 against the most frequently isolated skin pathogenic fungus, *Trichophyton mentagrophytes*. Antifungal susceptibility testing was assessed using a microdilution assay to determine the optimum concentration of SL-YS3 inhibited spore germination and mycelia of the microbe. The result was compared with sophorolipid standard (SL-S) and terbinafine (TRB) as the reference standard. The results demonstrated that there was complete inhibition of spore germination by SL-YS3, SL-S, and TRB at 1.0, 1.0, and 0.031 mg/ml, respectively. However, it was also reported that a higher concentration was needed for SL-YS3 to inhibit the mycelia, where 1 mg/ml of SL-YS3 was needed to achieve 62% of inhibition. The ability of SL-YS3 against *T. mentagrophytes* spore germination and mycelia was reported due to a lactonic fatty acid compound that was widely reported for their superior antimicrobial activity.

Additionally, an *in vivo* rat assay was conducted by induction of *T. mentagrophytes* 1.5 cm dorsal portions of the animals to create a cutaneous dermatophytosis disease. The animals were divided into four groups: Group 1 – uninfected, untreated control; Group 2 – infected, untreated control; Group 3 – topical treatment with 1 mg/ml (w/v) of TRB; and Group 4 – topical treatment with 1 mg/ml (w/v) of SL-YS3. After 72 hours of the rats being infected, the treatments were topically applied on an affected area once a day. A hispathological examination was conducted as a dependent result [86]. The result showed the existence of fungal elements in the untreated skin tissues while similar therapeutic effects demonstrated by the group treated with SL-YS3 and TRB resulted in regeneration of the impaired epidermis of the infected rats. The healing of dermatophytosis on rats by SL-YS3 was due to the ability of this sophorolipid biosurfactant to regulate collagen deposition together with a proper matrix and spatial arrangement [86].

18.5 Cosmetic Formulation with Biosurfactant

Applications of biosurfactants in cosmetic formulations are categorized into patented and novel formulations prepared by inventors and scientific committees. This sub-section discusses patented and novel formulations of biosurfactant in cosmetics. Different applications of biosurfactants are summarized in Figure 18.5.

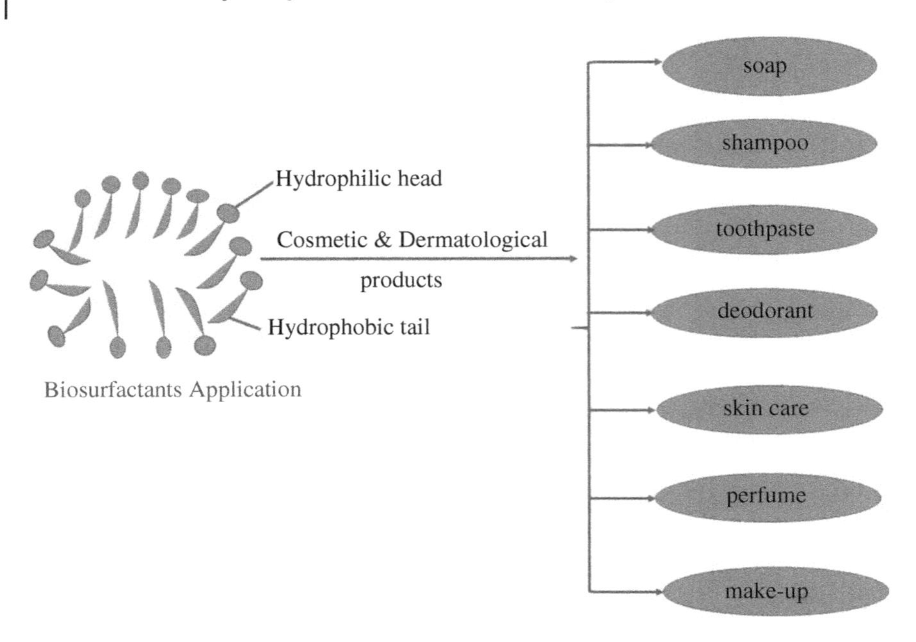

Figure 18.5 Different applications of biosurfactants in cosmetic formulation (See insert for color representation of the figure).

18.5.1 Biosurfactant Patented in a Cosmetic Product

Several biosurfactants have been patented for their specific application potential in cosmetic formulations such as antiwrinkle, whitening creams, hair products, and others. Rhamnolipids used in antiwrinkle and antiaging product formulations have been patented by Piljac and Piljac in 1999 [42]. These inventors suggested that compositions of rhamnolipids should be in the range from 0.001% up to 5% to treat signs of aging. The formulation is in the ointment phase and is recommended to be applied from 1 to 3 times per day to the affected areas.

Owen and Fan [87] patented an oligometric biosurfactant in dermatocosmetic products such as antiaging cream, facial products, and a conditioning hair mask, among others. These authors recommended that the biosurfactants should be within a range between 10 and 1000 ppm in all formulations. Other inventors proposed the usage of rhamnolipid with 2% of concentration diluted in water to be added to a shampoo product. The patent result showed that after three days of shampoo application, the hair scalp was free from odor, while maintaining a luster due to an antimicrobial effect promoted by the rhamnolipid biosurfactant [88].

Other personal care formulations such as shower gel, personal wash, and shampoo developed by a combination of sophorolipid biosurfactant and anionic surfactant have been patented. The sophorolipid biosurfactant was suggested to be added within 1–20% (w/w), while an anionic surfactant had compositions within 1–20%. Other chemicals are a foam boosting surfactant (0–10%), additional detergent additives (0–10%), an additional electrolyte (0–2%), and 40–98% consisting of water compositions [89]. Allef et al. [90] patented several cosmetic formulations: an antidandruff shampoo conditioner, a body cleanser, a moisturizing facial wash, a shower gel, and other products have at least one biosurfactant and one fatty acid. Each formulation contained rhamnolipid and sophorolipid in combination with 10% of oleic oil.

18.5.2 Novel Cosmetic Formulation Containing Biosurfactant

The most recent study was conducted by Rincón-Fontán et al. [91] who formulated a novel antidandruff substance from Zn pyrithion powder dispersed in tea tree oil emulsions stabilized by a biosurfactant obtained from CSL. This study aimed to produce a safer antidandruff shampoo formulation that previously used excessive amounts of synthetic surfactant to stabilize the low-solubility properties of Zn pyrithion powder. Zn pyrithion is widely known as a powerful antiseborrheic agent with the ability to promote a strong antibacterial effect and to improve hair appearance.

The oil/water (O/W) emulsions were composed of different surfactants (Tween 80 and biosurfactant), Zn pyrithion as the active constituent, tea tree oil as the oil phase, and deionized water. The emulsions were prepared based on a Box–Behnken factorial design, and emulsion characterizations such as particle size, stability, and solubility of Zn pyrithion in aqueous solutions were assessed. According to the factorial design, three independent and dependent variables were established, the Tween 80 concentration (X_1), the biosurfactant concentration (X_2), the tea tree oil/water ratio (X_3), the emulsions particle size (Y_1), the stability of emulsions after 30 days (Y_2), and the solubility of Zn pyrithion (Y_3), respectively. The study finally chose the most successful emulsion formulation with 5% of Tween 80 and 2.5% of biosurfactant with a tea tree oil/water ratio of 0.01, because this formulation possessed a smaller particle size (40.5 μm), good stability (91%), and the highest stability of Zn pyrithione (59%). Hence, the study recommended a combination of bio/surfactant (biosurfactant and Tween 80) to be used as a safer alternative stabilizer in an antidandruff shampoo formulation, together with another antibacterial agent, tea tree oil. Additionally, this study claimed that this was the first formulation used as a stabilizing agent in an antidandruff shampoo product [92].

Farias et al. [93] formulated a greener mouthwash with combinations of biosurfactants and microbial compounds of chitosan and peppermint essential oil (POE). Three bacteria, *Bacillus cereus* UCP 1615 (BB), *P. aeruginosa* UCP 0992 (PB), and *C. bombicola* URM 3718 (CB) were isolated for their biosurfactants. Meanwhile, the chitosan was produced from fungus biomass and Mucorales grown in yam bean broth. This study aimed to evaluate the control of first use biosurfactant combinations with a natural polymer and/or essential oil as the active constituent in a mouthwash to use against cariogenic oral microorganisms.

The antimicrobial assessment involved several substances alone, pure biosurfactants (CB, BB, and PB), chitiosan (Ch), POE, and combinations of them to be used against several microorganisms. The result showed that the yeasts of *Candida albicans* and *Lactobacillus acidophilus* were inhibited most effectively by a single POE with MIC values of 20 and 30 μg/ml, respectively. Chitosan needed higher concentrations to inhibit all microorganisms within the ranges of 200 and 300 μg/ml. For a single biosurfactant antimicrobial performance, PB demonstrated that it was the most effective with the capability to inhibit more microorganisms, such as *Streptococcus salivarius*, *E. coli*, and *S. aureus*, with MIC values of 20 μg/ml. Meanwhile, when biosurfactants combined with chitosan, the MIC values of biosurfactants were lowered for all examined microorganisms [93]. Combinations of POE with biosurfactants showed that the increase of MIC value (30 μg/ml) of the oil against *C. albicans* either combined with CB or with POE. Only those formulations with PB + POE maintained the MIC value of the oil against *L. acidophilus*, while other combinations resulted in an increased MIC value of POE. The MIC values of biosurfactants were mostly reduced after combining with POE, except formulations of CB + POE that showed a maintained MIC value [93].

The mouthwash formulations of antimicrobial activities involved several combinations of biosurfactants and POE and all bioactive compounds (biosurfactants, POE, and chitosan). As the result, those formulations contained combined bioactive compounds that exhibited maximum effectiveness against all microorganisms, which resulted in lowered MIC values, with 4 and 8 μg/ml for all biosurfactants and POE and 20 and 40 μg/ml presented by chitosan [93]. This study concluded that successful mouthwash formulations with bioactive compounds was due to antimicrobial properties of biosurfactants, chitosan, and POE, as reported previously [94, 95, 96]. Several previous studies demonstrated that lower MIC values of chitosan, POE, and biosurfactants achieved 0.5 mg/ml for Ch and POE and a higher percentage of microorganisms inhibited by biosurfactants.

Another study investigated the capability of a multifunctional biosurfactant extract (BS), produced by the corn wet-milling industry to improve the stability of vitamin C contained in cosmetic formulations with a degradation result of the vitamin. The prediction of vitamin C degradation was used in the Box–Behnken factorial design and involved independent variables, the vitamin C concentration (X_1), the biosurfactant concentration (X_2), and the storage time (X_3), whereas the dependent variable Y consisted of a percentage of vitamin C degradation. The manipulated concentrations of vitamin C and biosurfactant extract varied from 0 to 2 g/l and the storage time was between 7 and 21 days [92]. These solutions were stirred and formed a homogenous solution and a degradation percentage of vitamin C was used:

$$\text{Vitamin C degraded}(\%) = \text{Vitamin C concentration at time 0} - \text{Vitamin C concentration at specific time/Vitamin C concentration at time 0}$$

From these findings, the independent variables were standardized (coded) in dimensionless variables with variation limits from −1 to 1. The results demonstrated that those formulations with the existence of a biosurfactant resulted in a lower degradation of vitamin C. For example, in formulations 1 and 2, formulation 2 had the highest concentration of biosurfactant, 2 g/l showed a reduction of vitamin C (16.0%), compared to formulation 1, with no existence of biosurfactant promoting a higher percentage of vitamin C degradation (38.0%) [92].

Hence, the current trends of biosurfactant applications in dermatological and cosmetic formulations have been reviewed in detail. Wound healing and prebiotic applications of biosurfactants for dermatological and several novel formulations for cosmetic purposes were considered. Table 18.2 summarizes different biosurfactant potentials used in dermatological and cosmetic products.

18.6 Safety Measurement Taken for Biosurfactant Applications in Dermatology and Cosmetics

Generally, before the usage of biosurfactants for dermatological and cosmetic purposes, it was necessary to evaluate the toxicity effects of these surface-active agents on cells and/or animals. Therefore, the reviews of toxicity assessments of several biosurfactants are now given.

Sahnoun et al. [106] investigated the *in vivo* toxicity of lipopeptide biosurfactant from the *B. subtilis* SPB1 (HQ392822) strain toward male mice to ensure the capability of the examined biosurfactant to be further used in variation products, including cosmetics. The experimental study involved three groups of mice who received 47.5, 9.5, and 4.75 mg/kg intraperitoneal injections of lipopeptide biosurfactant within eight days of administration. All three dosages were fixed with serial decimal dilutions of LD50 value (475 mg/kg) that had been determined previously. The toxicity result showed no death cases, no intoxication was observed, and no unusual behavior changes were seen among the experimental rats at any dose administration. In addition, no skin irritations

Table 18.2 Dermatological and cosmetic potentials of biosurfactants.

Biosurfactants	Source	Pharmacological activity/action	Application	References
Rhamnolipids	Microbial (*Pseudomonas aeruginosa*)	Antioxidants, antibacterial, emulsifier, foaming agent	Antiaging, antidandruff, insect repellent, deodorant, toothpaste, nail treatment	[97] [43]
Surfactin	Microbial (*Bacillus* sp.)	Antibacterial, antioxidant, enhanced transdermal cosmetic formulation	Collagen stimulation, moisturizing, facial lotions, dermatological products, toothpaste	[5] [98] [99]
Saponins	Vegetables	Emulsifying, antibacterial, antioxidant, foaming	Skin care, hair products, antiaging, moisturizing cream	[100] [101] [102] [103]
Sophorolipids	Microbial (*Candida* sp.)	Antibacterial, wetting, emulsifier, solubilizer, foaming, detergent	Hair products, Acne and body odor treatment products, lipstick, antiaging, moisturizing cream	[5] [4] [104]
Mannosylerylthritol Lipids (MEL)	Microbial (*Pseudozyma* sp., *Ustilago* sp.)	Dispersant, anti-inflammatory, antioxidant, emulsifier, detergent	Skin care, hair products, antiaging, moisturizing cream	[70] [105] [44]

or dermal reactions were observed. This study concluded that the SPB1 lipopeptide has the potential to be used as an additive in cosmetics, foods, and pharmaceuticals.

Another study made a safety assessment of surfactin C biosurfactant produced by the *B. subtilis* microorganism from Korean soybean paste. This safety assessment was conducted on maternity female mice that had been given different dosages of surfactin C (15, 250, and 500) mg/kg within 18 days. On examination days 0, 6, 14, and 18 the body weights of the mice were recorded, while feed consumption was recorded on days 6, 14, and 17. On day 18, all mice were sacrificed using carbon dioxide. Next, their organ weights were measured and maternal necropsy was assessed. The wombs of the examined pregnant mice were exposed and the presence of fetuses resorption, survival, and number of implantation sites were determined. In addition, the weights of live fetuses were recorded and examined for any visceral and external malformations [107].

All of the maternal mice were found not to have any clinical signs such as vaginal bleeding, salivation, tremor, and no significant difference in maternal body weight during 17 days of manipulated surfactin C dosage administrations. Water and feed consumption among the animals also did not show statistically different results. Furthermore, surfactin C treatment also demonstrated no difference in death, late and early fetus resorption, sex ratio, and fetus body weight among the treatment groups. In addition, no fetus external and skeletal abnormalities were observed. Hence, this study suggested a 500 mg/kg of surfactin C dosage, dependent on these observed effect levels of results [107].

Recently, related industries are now looking forward to animal-free safety assessments tests, and the Organization for Economic Cooperation and Development (OECD) has evaluated a skin corrosion safety assessment test using a reconstructed human epidermis (RhE) method. The RhE is reported to have closely mimicked the histological, morphological, physiological, and biochemical properties of the upper layer of the human skin (e.g. the epidermis), which was obtained from human-derived non-transformed epidermal keratinocytes. EpiSkin™, EpiDerm™, SkinEthic™ RHE1, epiCS®, and TEST SKIN™ are among commercially available RhE and are being evaluated by the OECD using their testing assessment guidelines [108].

In 2012, a study proposed the application of EpiSkin to evaluate the safety effects of bioactive components of cosmetic formulations on skin. This method involved the formulations to be directly applied on to the RhE and their pharmaco-toxicological effects, such as skin irritation, the bioactive component absorption rate, and cell viability, are among results that need to be considered via the RhE test method. The irritation effect of examined skin can occur when toxic chemicals cross the stratum corneum of the epidermis and damage the underlying layers of skin keratinocytes and other skin cells [109].

However, the toxicity test of biosurfactants are widely studied using cell line cultures, which have been proposed as promising models for skin irritancy assessments in future formulations prior to human testing. In their study, Burgos-Díaz et al. [110] produced biosurfactants from *Sphingobacterium detergens* for a cytotoxicity assessment in different cell lines (keratinocyte and fibrolast). The biosurfactant was purified to form biosurfactant fractions A and B that were observed by thin layer chromatography (TLC). The cytotoxicity assessment of new biosurfactant fractions were conducted using neural red uptake (NRU) and MTT assays. The IC_{50} value was determined to analyze the cytotoxicity effect of fractions A and B biosurfactants via NRU and MTT assays. Depicting the IC_{50} obtained by both fractions of biosurfactants, it was observed that fraction B resulted in a less cytotoxic compound with a higher IC_{50} value. In both assays, the IC_{50} value of fraction B was determined in the NRU assays to be lower compared to MTT assays for cytotoxicity assessment on fibroblasts and keratinocytes human cell lines. It was indicated that the NRU assay is the more sensitive method for detecting deleterious effects of biosurfactants on cell viability. Thus, this study suggested that the fraction B biosurfactant was a promising alternative to other commercial synthetic surfactants in cosmetic and dermatological formulations [110].

Another research study of MEL biosurfactants proved that skin cell activation involved papilla and fibrolast cells. MEL-A was demonstrated to significantly increase the viability of fibrolast cells by more than 150% compared to control cells. Meanwhile, at only 0.001 μg.ml, MEL-A was capable of dramatically activating the papilla cells at 150% of cell viability. The activation of these two cells is a crucial point in promoting skin appendage morphogenesis and stimulating the proliferation. As an example, the study mentioned that the activation of papilla cells is a key factor for hair growth development, attributable to capability of the dermal papilla cells to induce follicle formation and hair growth by transdifferentiation of an adult epidermis [111].

Ferreira et al. [112] implemented a biosurfactant synthesized from a *L. paracasei* strain to have the potential as a cosmetic emulsifier in a novel cosmetic formulation where it was added in oil-in-water (O/W) emulsions containing essential oils and a natural antioxidant extract. This experimental formulation has been compared with those O/W emulsions combined with the synthetic surfactant, SDS. The cytotoxicity effects of the biosurfactant and emulsions containing biosurfactant were evaluated using a 3T3 mouse fibrolast cell line via a sulforhodamine B (SRB) assay.

The results demonstrated that solutions with 5.0 g/l of biosurfactant exhibited higher cell proliferation values of 97%, compared to those solutions with SDS synthetic surfactant. However, the emulsion combinations with 0.5 g/l of SDS, oil, and antioxidant compound showed low cell cytotoxicity with 83% of cell proliferation. This suggested that the antioxidant compound had a positive effect on cells, protecting them from SDS [112, 113].

18.7 Conclusion and Future Perspective

Microbial surfactants could be included in cosmetic formulations as an alternative for synthetics in order to develop more eco-friendly products. Several patented applications of biosurfactant, especially rhamnolipids and sophorolipids, are reported for use as an antiwrinkle agent for cosmetic segments. In bioactive evaluations of research studies, biosurfactants are also reported to be most successful as antimicrobial agents against skin pathogens. Studies suggested that biosurfactants could have a prebiotic role in balancing skin microflora, which would be a very good prerequisite characteristic for dermatological products. However, before further use can be made of biosurfactants, a deep study needs to be made on biosurfactant physicochemical properties and wider studies on their permeation rate through human skin are suggested. Furthermore, another future recommendation needs to be considered where additional regulations, such as the EU Cosmetic Regulation, must align with any future studies of the biosurfactant potential in cosmetic formulation.

Acknowledgement

The authors are thankful to the Department of Chemical Processes, Malaysia–Japan International Institute of Technology, Universiti Teknologi Malaysia, for providing research facilities.

References

1 Akbari, S., Abdurahman, N.H., Yunus, R.M. et al. (2018). Biosurfactants – A new frontier for social and environmental safety: A mini review. *Biotechnology Research and Innovation* 2 (1): 81–90.
2 Aleti, G., Lehner, S., Bacher, M. et al. (2016). Surfactin variants mediate species-specific biofilm formation and root colonization in Bacillus. *Environmental Microbiology* 18 (8): 2634–2645.
3 Allef P, Hartung C, Schilling M. (2014).Aqueous hair and skin cleaning compositions comprising biosurfactants. US Patent 20140349902 A1.
4 Aparajita, V. and Ravikumar, P. (2014). Liposomes as carriers in skin ageing. *International Journal of Current Pharmaceutical Research* 6 (3): 1–7.
5 Archana, K., Reddy, K.S., Parameshwar, J., and Bee, H. (2019). Isolation and characterization of sophorolipid producing yeast from fruit waste for application as antibacterial agent. *Environmental Sustainability* 2 (2): 107–115.
6 Arima, K., Kakinuma, A., and Tamura, G. (1968). Surfactin, a crystalline peptidelipid surfactant produced by *Bacillus subtilis*: Isolation, characterization and its inhibition of fibrin clot formation. *Biochem. BiophysRes. Com.* 31: 488–494.
7 Aziz, Z.A.A., Mohd-Nasir, H., Setapar, M. et al. (2019). Role of nanotechnology for design and development of cosmeceutical: Application in makeup and skin care. *Frontiers in Chemistry* 7: 739.
8 Bae, I.H., Lee, E.S., Yoo, J.W. et al. (2019). Mannosylerythritol lipids inhibit melanogenesis via suppressing ERK-CREB-MiTF-tyrosinase signalling in normal human melanocytes and a three-dimensional human skin equivalent. *Experimental Dermatology* 28 (6): 738–741.
9 Bai, L. and McClements, D.J. (2016). Formation and stabilization of nanoemulsions using biosurfactants: Rhamnolipids. *Journal of Colloid and Interface Science* 479: 71–79.
10 Ben Belgacem, Z., Bijttebier, S., Verreth, C. et al. (2015). Biosurfactant production by *Pseudomonas* strains isolated from floral nectar. *Journal of Applied Microbiology* 118 (6): 1370–1384.

11 Bockmühl, D. (2012). Biosurfactants as antimicrobial ingredients for cleaning products and cosmetics. *Tenside, Surfactants, Detergents* 49 (3): 196–198.

12 Bom, S., Jorge, J., Ribeiro, H.M., and Marto, J.A. (2019). Step forward on sustainability in the cosmetics industry: A review. *Journal of Cleaner Production* 225: 270–290.

13 Bonmatin, J.M., Laprévote, O., and Peypoux, F. (2003). Diversity among microbial cyclic lipopeptides: Iturins and surfactins. Activity–structure relationships to design new bioactive agents. *Combinatorial Chemistry & High Throughput Screening* 6 (6): 541–556.

14 Bouassida, M., Fourati, N., Krichen, F. et al. (2017). Potential application of *Bacillus subtilis* SPB1 lipopeptides in toothpaste formulation. *Journal of Advanced Research* 8 (4): 425–433.

15 Brahim, M.A.S., Fadli, M., Markouk, M. et al. (2015). Synergistic antimicrobial and antioxidant activity of saponins-rich extracts from *Paronychia argentea* and *Spergularia marginata*. *European Journal of Medicinal Plants* 7 (4): 193–204.

16 Burgos-Díaz, C., Martín-Venegas, R., Martínez, V. et al. (2013). in vitro study of the cytotoxicity and antiproliferative effects of surfactants produced by *Sphingobacterium detergens*. *International Journal of Pharmaceutics* 453 (2): 433–440.

17 Campos, J.M., Stamford, T.L.M., and Sarubbo, L.A. (2019). Characterization and application of a biosurfactant isolated from *Candida utilis* in salad dressings. *Biodegradation* 30: 313–324.

18 Cornwell, P.A. (2018). A review of shampoo surfactant technology: Consumer benefits, raw materials and recent developments. *International Journal of Cosmetic Science* 40 (1): 16–30.

19 Coronel-León, J., de Grau, G., Grau-Campistany, A. et al. (2015). Biosurfactant production by AL 1.1, a *Bacillus licheniformis* strain isolated from Antarctica: Production, chemical characterization and properties. *Annals of Microbiology* 65 (4): 2065–2078.

20 Costa, E.M., Silva, S., Pina, C. et al. (2012). Evaluation and insights into chitosan antimicrobial activity against anaerobic oral pathogens. *Anaerobe* 18 (3): 305–309.

21 De Araujo, L.V., Abreu, F., Lins, U. et al. (2011). Rhamnolipid and surfactin inhibit *Listeria* monocytogenes adhesion. *Food Research International* 44 (1): 481–488.

22 de Araujo, L.V., Guimarães, C.R., da Silva Marquita, R.L. et al. (2016). Rhamnolipid and surfactin: Anti-adhesion/antibiofilm and antimicrobial effects. *Food Control* 63: 171–178.

23 de Oliveira, M.R., Magri, A., Baldo, C. et al. (2015). Sophorolipids A promising biosurfactant and its applications. *International Journal of Advanced Biotechnology and Research* 6 (2): 161–174.

24 De Rienzo, M.A.D., Banat, I.M., Dolman, B. et al. (2015). Sophorolipid biosurfactants: possible uses as antibacterial and antibiofilm agent. *New Biotechnology* 32 (6): 720–726.

25 Dempster, C., Marchant, R. and Banat, I. (2019). Antimicrobial and antibiofilm potential of biosurfactants as novel combination therapy against bacterium that cause skin infections. In: *Microbiology Society Annual Conference*. 1 (1A). doi.org/10.1099/acmi.ac2019.po0566.

26 Desanto K. (2008).Rhamnolipid-based formulations. WO Patent 2008013899 A2.

27 Desjardins, R., Thorn, W., Schleicher, A. et al. (2013). OECD skills outlook 2013: First results from the survey of adult skills. *Journal of Applied Econometrics* 30 (7): 1144–1168.

28 Díaz, B., Gomes, A., Freitas, M. et al. (2012). Valuable polyphenolic antioxidants from wine vinasses. *Food and Bioprocess Technology* 5 (7): 2708–2716.

29 Dobler, L., de Carvalho, B.R., de Sousa Alves, W. et al. (2017). Enhanced rhamnolipid production by *Pseudomonas aeruginosa* overexpressing estA in a simple medium. *PLoS One* 12 (8): e0183857.

30 Dumas, M., Noblesse, E., Krzych, V. and Cauchard, J.H., Recherche, L. (2013). Use of an extract of common mallow as an hydrating agent, and cosmetic composition containing it. US Patent 8,455,013.

31 Elekofehinti, O.O. (2015). Saponins: Anti-diabetic principles from medicinal plants– A review. *Pathophysiology* 22 (2): 95–103.

32 Farias, J.M., Stamford, T.C.M., Resende, A.H.M. et al. (2019). Mouthwash containing a biosurfactant and chitosan: An eco-sustainable option for the control of cariogenic microorganisms. *International Journal of Biological Macromolecules* 129: 853–860.

33 Ferreira, A., Vecino, X., Ferreira, D. et al. (2017). Novel cosmetic formulations containing a biosurfactant from *Lactobacillus paracasei*. *Colloids and Surfaces B: Biointerfaces* 155: 522–529.

34 Fukuoka, T., Morita, T., Konishi, M. et al. (2007). Structural characterization and surface-active properties of a new glycolipid biosurfactant, mono-acylated mannosylerythritol lipid, produced from glucose by *Pseudozyma antarctica*. *Applied Microbiology and Biotechnology* 76 (4): 801–810.

35 Gallo, R.L. and Nakatsuji, T. (2011). Microbial symbiosis with the innate immune defense system of the skin. *Journal of Investigative Dermatology* 131 (10): 1974–1980.

36 Garg, M. and Chatterjee, M. (2018). Isolation, characterization and antibacterial effect of biosurfactant from *Candida parapsilosis*. *Biotechnology Reports* 18: e00251.

37 Grice, E.A. and Segre, J.A. (2011). The skin microbiome. *Nature Reviews Microbiology* 9 (4): 244–253.

38 Gudiña, E.J., Fernandes, E.C., Rodrigues, A.I. et al. (2015). Biosurfactant production by *Bacillus subtilis* using corn steep liquor as culture medium. *Frontiers in Microbiology* 6: 59–65.

39 Gudiña, E.J., Rodrigues, A.I., Alves, E. et al. (2015b). Bioconversion of agro-industrial by-products in rhamnolipids toward applications in enhanced oil recovery and bioremediation. *Bioresource Technology* 177: 87–93.

40 Gupta, P.L., Rajput, M., Oza, T. et al. (2019). Eminence of microbial products in cosmetic industry. *Natural Products and Bioprospecting* 9: 267–278.

41 Gupta, S., Raghuwanshi, N., Varshney, R. et al. (2017). Accelerated in vivo wound healing evaluation of microbial glycolipid containing ointment as a transdermal substitute. *Biomedicine & Pharmacotherapy* 94: 1186–1196.

42 Haba, E., Bouhdid, S., Torrego-Solana, N. et al. (2014). Rhamnolipids as emulsifying agents for essential oil formulations: antimicrobial effect against *Candida albicans* and methicillin-resistant *Staphylococcus aureus*. *International Journal of Pharmaceutics* 476 (1–2): 134–141.

43 Hanno, I., Centini, M., Anselmi, C., and Bibiani, C. (2015). Green cosmetic surfactant from rice: characterization and application. *Cosmetics* 2 (4): 322–341.

44 Hazra, C., Kundu, D., and Chaudhari, A. (2015). Lipopeptide biosurfactant from *Bacillus clausii* BS02 using sunflower oil soapstock: evaluation of high throughput screening methods, production, purification, characterization and its insecticidal activity. *RSC Advances* 5 (4): 2974–2982.

45 Hillion G, Marchal R, Stoltz C, Borzeix F. (1998). Use of a sophorolipid to provide free radical formation inhibiting activity or elastase inhibiting activity. US Patent 5756471 A.

46 Hwang, Y.H., Park, B.K., Lim, J.H. et al. (2008). Evaluation of genetic and developmental toxicity of surfactin C from *Bacillus subtilis* BC1212. *Journal of Health Science* 54 (1): 101–106.

47 Irfan-Maqsood, M. and Seddiq-Shams, M. (2014). Rhamnolipids: Well-characterized glycolipids with potential broad applicability as biosurfactants. *Industrial Biotechnology* 10 (4): 285–291.

48 Ishii, N., Kobayashi, T., Matsumiya, K. et al. (2012). Transdermal administration of lactoferrin with sophorolipid. *Biochemistry and Cell Biology* 90 (3): 504–512.

49 Jadhav, J., Dutta, S., Kale, S., and Pratap, A. (2018). Fermentative production of rhamnolipid and purification by adsorption chromatography. *Preparative Biochemistry and Biotechnology* 48 (3): 234–241.

50 Jemil, N., Ayed, H.B., Manresa, A. et al. (2017). Antioxidant properties, antimicrobial and anti-adhesive activities of DCS1 lipopeptides from *Bacillus methylotrophicus* DCS1. *BMC Microbiology* 17 (1): 144.

51 Kitamoto, D., Morita, T., Fukuoka, T. et al. (2009). Self-assembling properties of glycolipid biosurfactants and their potential applications. *Current Opinion in Colloid & Interface Science* 14 (5): 315–328.

52 Krutmann, J. (2009). Pre-and probiotics for human skin. *Journal of Dermatological Science* 54 (1): 1–5.

53 Kundu, D., Hazra, C., Dandi, N., and Chaudhari, A. (2013). Biodegradation of 4-nitrotoluene with biosurfactant production by *Rhodococcus pyridinivorans* NT2: Metabolic pathway, cell surface properties and toxicological characterization. *Biodegradation* 24 (6): 775–793.

54 Lan, G., Fan, Q., Liu, Y. et al. (2015). Rhamnolipid production from waste cooking oil using *Pseudomonas* SWP-4. *Biochemical Engineering Journal* 101: 44–54.

55 Lourith, N. and Kanlayavattanakul, M. (2009). Natural surfactants used in cosmetics: glycolipids. *International Journal of Cosmetic Science* 31 (4): 255–261.

56 Lu, J.K., Wang, H.M. and Xuan-Rui, X.U. (2016). Method for anti-aging treatment by surfactin in cosmetics via enhancing sirtuin. US Patent 9,364,413.UMO International Co., Ltd.

57 Lukic, M., Pantelic, I., and Savic, S. (2016). An overview of novel surfactants for formulation of cosmetics with certain emphasis on acidic active substances. *Tenside, Surfactants, Detergents* 53 (1): 7–19.

58 Lydon, H.L., Baccile, N., Callaghan, B. et al. (2017). Adjuvant antibiotic activity of acidic sophorolipids with potential for facilitating wound healing. *Antimicrobial Agents and Chemotherapy* 61 (5): e02547–e02516.

59 Madhu, A.N. and Prapulla, S.G. (2014). Evaluation and functional characterization of a biosurfactant produced by *Lactobacillus plantarum* CFR 2194. *Applied Biochemistry and Biotechnology* 172 (4): 1777–1789.

60 Maeng, Y., Kim, K.T., Zhou, X. et al. (2018). A novel microbial technique for producing high-quality sophorolipids from horse oil suitable for cosmetic applications. *Microbial Biotechnology* 11 (5): 917–929.

61 Makkar, R. and Cameotra (2002). An update on the use of unconventional substrates for biosurfactant production and their new applications. *Applied Microbiology and Biotechnology* 58 (4): 428–434.

62 Meena, K.R. and Kanwar, S.S. (2015). Lipopeptides as the antifungal and antibacterial agents: Applications in food safety and therapeutics. *BioMed Research International*, 2015, 1–9.

63 Minucelli, T., Ribeiro-Viana, R.M., Borsato, D. et al. (2017). Sophorolipids production by *Candida bombicola* ATCC 22214 and its potential application in soil bioremediation. *Waste and Biomass Valorization* 8 (3): 743–753.

64 Morita, T., Fukuoka, T., Imura, T., and Kitamoto, D. (2013). Production of mannosylerythritol lipids and their application in cosmetics. *Applied Microbiology and Biotechnology* 97 (11): 4691–4700.

65 Morita, T., Fukuoka, T., Imura, T., and Kitamoto, D. (2015). Mannosylerythritol lipids: Production and applications. *Journal of Oleo Science* 64 (2): 133–141.

66 Morita, T., Kitagawa, M., Yamamoto, S. et al. (2010). Glycolipid biosurfactants, mannosylerythritol lipids, repair the damaged hair. *Journal of Oleo Science* 59 (5): 267–272.

67 Muhammad, M.T. and Khan, M.N. (2018). Eco-friendly, biodegradable natural surfactant (Acacia Concinna): an alternative to the synthetic surfactants. *Journal of Cleaner Production* 188: 678–685.

68 Mukherjee, A.K. (2007). Potential application of cyclic lipopeptide biosurfactants produced by *Bacillus subtilis* strains in laundry detergent formulations. *Letters in Applied Microbiology* 45 (3): 330–335.

69 Naik, N.J., Abhyankar, I., Darne, P. et al. (2019). Sustained transdermal release of Lignans facilitated by sophorolipid based transferosomal hydrogel for cosmetic application. *International Journal of Current Microbiology and Applied Sciences* 8 (2): 1783–1791.

70 Naughton, P.J., Marchant, R., Naughton, V., and Banat, I.M. (2019). Microbial biosurfactants: current trends and applications in agricultural and biomedical industries. *Journal of Applied Microbiology* 127 (1): 12–28.

71 Nguyen, T.T., Edelen, A., Neighbors, B., and Sabatini, D.A. (2010). Biocompatible lecithin-based microemulsions with rhamnolipid and sophorolipid biosurfactants: formulation and potential applications. *Journal of Colloid and Interface Science* 348 (2): 498–504.

72 Nickzad, A. and Déziel, E. (2014). The involvement of rhamnolipids in microbial cell adhesion and biofilm development–an approach for control? *Letters in Applied Microbiology* 58 (5): 447–453.

73 Nooman, M.U., Mahmoud, M.H., Al-Kashef, A.S., and Rashad, M.M. (2017). Hypocholesterolemic impact of newly isolated sophorolipids produced by microbial conversion of safflower oil cake in rats fed high-fat and cholesterol diet. *Grasas y Aceites* 68 (3): 212.

74 Ohadi, M., Dehghannoudeh, G., Forootanfar, H. et al. (2018). Investigation of the structural, physicochemical properties, and aggregation behavior of lipopeptide biosurfactant produced by *Acinetobacter junii* B6. *International Journal of Biological Macromolecules* 112: 712–719.

75 Owen D, Fan L. (2013). Polymeric biosurfactants. Patent US 8586541 B2.

76 Pathmanathan, M.K., Uthayarasa, K., Jeyadevan, J.P., and Jeyaseelan, E.C. (2010). in vitro antibacterial activity and phytochemical analysis of some selected medicinal plants. *Int. J. Pharm. Biol. Arch.* 1 (3): 291–299.

77 Patowary, R., Patowary, K., Kalita, M.C., and Deka, S. (2016). Utilization of paneer whey waste for cost-effective production of rhamnolipid biosurfactant. *Applied Biochemistry and Biotechnology* 180 (3): 383–399.

78 Peypoux, F., Bonmatin, J.M., and Wallach, J. (1999). Recent trends in the biochemistry of surfactin. *Applied Microbiology and Biotechnology* 51: 553–563.

79 Piljac T, Piljac G. (1999). Use of rhamnolipids in wound healing, treating burn shock, atherosclerosis, organ transplants, depression, schizophrenia and cosmetics. WO Patent 1999043334 A1.

80 Rani, V.P., Mirabel, L.M., Priya, K.S. et al. (2018). Phytochemical, antioxidant and antibacterial activity of aqueous extract of *Borassus flabellifer* (L.). *Themed Section: Science and Technology* 4: 405–411.

81 Rincón-Fontán, M., Rodríguez-López, L., Vecino, X. et al. (2020). Novel multifunctional biosurfactant obtained from corn as a stabilizing agent for antidandruff formulations based on Zn pyrithione powder. *ACS Omega* 5 (11): 704–5712.

82 Rincón-Fontán, M., Rodríguez-López, L., Vecino, X. et al. (2016). Adsorption of natural surface active compounds obtained from corn on human hair. *RSC Advances* 6 (67): 63064–63070.

83 Rincón-Fontán, M., Rodríguez-López, L., Vecino, X. et al. (2020). Potential application of a multifunctional biosurfactant extract obtained from corn as stabilizing agent of vitamin C in cosmetic formulations. *Sustainable Chemistry and Pharmacy* 16: 100248.

84 Rodrigues, L.R. (2015). Microbial surfactants: fundamentals and applicability in the formulation of nano-sized drug delivery vectors. *Journal of Colloid and Interface Science* 449: 304–316.

85 Rufino, R.D., Luna, J.M., Sarubbo, L.A. et al. (2011). Antimicrobial and anti-adhesive potential of a biosurfactant Rufisan produced by *Candida lipolytica* UCP 0988. *Colloids and Surfaces B: Biointerfaces* 84 (1): 1–5.

86 Sahnoun, R., Mnif, I., Fetoui, H. et al. (2014). Evaluation of *Bacillus subtilis* SPB1 lipopeptide biosurfactant toxicity towards mice. *International Journal of Peptide Research and Therapeutics* 20 (3): 333–340.

87 Saika, A., Utashima, Y., Koike, H. et al. (2018). Biosynthesis of mono-acylated mannosylerythritol lipid in an acyltransferase gene-disrupted mutant of *Pseudozyma tsukubaensis*. *Applied Microbiology and Biotechnology* 102 (4): 1759–1767.

88 Sarma, H., Bustamante, K.L.T., and Prasad, M.N.V. (2018). Biosurfactants for oil recovery from refinery sludge: magnetic nanoparticles assisted purification. In: Industrial and Municipal Sludge (ed. M.N.V. Prasad), 107–132. Massachusetts: Elsevier. ISBN: 9780128159071, Editor Majeti Narasimha Vara Prasad, Paulo Jorge de Campos, Favas Meththika, Vithanage S. Venkata Mohan.

89 Saikia, R.R., Deka, S., Deka, M., and Sarma, H. (2012). Optimization of environmental factors for improved production of rhamnolipid biosurfactant by *Pseudomonas aeruginosa* RS29 on glycerol. *Journal of Basic Microbiology* 52 (4): 446–457.

90 Sarwar, A., Brader, G., Corretto, E. et al. (2018). Qualitative analysis of biosurfactants from *Bacillus* species exhibiting antifungal activity. *PLoS One* 13 (6): 1–15.

91 Sen, S., Borah, S.N., Kandimalla, R. et al. (2020). Sophorolipid biosurfactant can control cutaneous dermatophytosis caused by *Trichophyton mentagrophytes*. *Frontiers in Microbiology* 11: 329.

92 Setapar, M.S.H. and Nasir, M.H.M. (2018). Natural ingredients in cosmetics from Malaysian plants: aAreview. *Sains Malaysiana* 47 (5): 951–959.

93 Sil, J., Dandapat, P., and Das, S. (2017). Health care applications of different biosurfactants. *International Journal of Science and Research* 6 (10): 41–50.

94 Silveira, V.A.I., Freitas, C.A.U.Q., and Celligoi, M.A.P.C. (2018). Antimicrobial applications of sophorolipid from *Candida bombicola*: A promising alternative to conventional drugs. *Journal of Applied Biology & Biotechnology* 6 (06): 87–90.

95 Takahashi, M., Morita, T., Fukuoka, T. et al. (2012). Glycolipid biosurfactants, mannosylerythritol lipids, show antioxidant and protective effects against H_2O_2-induced oxidative stress in cultured human skin fibroblasts. *Journal of Oleo Science* 61 (8): 457–464.

96 Takahashi, T., Ohno, O., Ikeda, Y. et al. (2006). Inhibition of lipopolysaccharide activity by a bacterial cyclic lipopeptide surfactin. *The Journal of Antibiotics* 59 (1): 35–43.

97 Totté, J.E.E., Van Der Feltz, W.T., Hennekam, M. et al. (2016). Prevalence and odds of *Staphylococcus aureus* carriage in atopic dermatitis: A systematic review and meta-analysis. *British Journal of Dermatology* 175 (4): 687–695.

98 Trevor F, Crawford R, Garry L. (2013).Mild to the skin, foaming detergent composition. US Patent 8563490 B2.

99 Varvaresou, A. and Iakovou, K. (2015). Biosurfactants in cosmetics and biopharmaceuticals. *Letters in Applied Microbiology* 61 (3): 214–223.

100 Vecino, X., Cruz, J.M., Moldes, A.B., and Rodrigues, L.R. (2017). Biosurfactants in cosmetic formulations: Trends and challenges. *Critical Reviews in Biotechnology* 37 (7): 911–923.

101 Vecino, X., Rodríguez-López, L., Ferreira, D. et al. (2018). Bioactivity of glycolipopeptide cell-bound biosurfactants against skin pathogens. *International Journal of Biological Macromolecules* 109: 971–979.

102 Vollenbroich, D., Pauli, G., Ozel, M., and Vater, J. (1997). Antimycoplasma properties and application in cell culture of surfactin, a lipopeptide antibiotic from *Bacillus subtilis*. *Applied and Environmental Microbiology* 63 (1): 44–49.

103 Yamamoto, S., Fukuoka, T., Imura, T. et al. (2013). Production of a novel mannosylerythritol lipid containing a hydroxy fatty acid from castor oil by *Pseudozyma tsukubaensis*. *Journal of Oleo Science* 62 (6): 381–389.

104 Yamamoto, S., Morita, T., Fukuoka, T. et al. (2012). The moisturizing effects of glycolipid biosurfactants, mannosylerythritol lipids, on human skin. *Journal of Oleo Science* 61 (7): 407–412.

105 Yang, F., Zhao, X.H., Hu, J. et al. (2012). Preliminary studies on surface properties and antioxidant activities of sophorolipids. *Science and Technology of Food Industry* 14: 34.

106 Yoneda, T., Masatsuji, E., and Tsuzuki, T. (1999). Surfactant for use in external preparations for skin and external preparation for skin containing the same. World Patent 1999/062482, K. K. Showa Denko, Tokyo.

107 Zanotto, A.W., Valério, A., de Andrade, C.J., and Pastore, G.M. (2019). New sustainable alternatives to reduce the production costs for surfactin 50 years after the discovery. *Applied Microbiology and Biotechnology* 103 (21–22): 8647–8656.

108 Zouari, R., Moalla-Rekik, D., Sahnoun, Z. et al. (2016). Evaluation of dermal wound healing and in vitro antioxidant efficiency of *Bacillus subtilis* SPB1 biosurfactant. *Biomedicine & Pharmacotherapy* 84: 878–889.

109 Jie, Z., Xue, R., Liu, S. et al. (2019). High di-rhamnolipid production using *Pseudomonas aeruginosa* KT1115, separation of mono/di-rhamnolipids, and evaluation of their properties. *Frontiers in Bioengineering and Biotechnology* 7: 245.

110 Samadi, N., Abadian, N., Ahmadkhaniha, R. et al. (2012). Structural characterization and surface activities of biogenic rhamnolipid surfactants from *Pseudomonas aeruginosa* isolate MN1 and synergistic effects against methicillin-resistant *Staphylococcus aureus*. *Folia Microbiologica* 57 (6): 501–508.

111 Zhao, X., Murata, T., Ohno, S. et al. (2001). Protein kinase Cα plays a critical role in mannosylerythritol lipid-induced differentiation of melanoma B16 cells. *Journal of Biological Chemistry* 276 (43): 39903–39910.

112 Wakamatsu, Y., Zhao, X., Jin, C. et al. (2001). Mannosylerythritol lipid induces characteristics of neuronal differentiation in PC12 cells through an ERK-related signal cascade. *European Journal of Biochemistry* 268 (2): 374–383.

113 Isoda, H., Shinmoto, H., Kitamoto, D. et al. (1997). Differentiation of human promyelocytic leukemia cell line HL60 by microbial extracellular glycolipids. *Lipids* 32 (3): 263–271.

19

Cosmeceutical Applications of Biosurfactants

Challenges and Prospects

Káren Gercyane Oliveira Bezerra[1,2,3] *and Leonie Asfora Sarubbo*[2,3]

[1] *Northeastern Network of Biotechnology, Federal Rural University of Pernambuco, Recife, Pernambuco, Brazil*
[2] *Advanced Institute of Technology and Innovation (IATI), Recife, Pernambuco, Brazil*
[3] *Catholic University of Pernambuco, Recife, Pernambuco, Brazil*

CHAPTER MENU

19.1 Introduction

The potential for the use of biosurfactants in the most diverse fields has been proven in recent years. Particularly in the cosmeceutical/cosmetic industry, numerous studies have investigated a broad gamut of these natural tensioactive agents of a microbial or plant-based origin [1].

Biosurfactants are capable of exercising several functions in personal care and hygiene formulations and can be used to obtain detergency, cleaning, emulsification, wettability, foaming, solubilization, in addition to preservative, antimicrobial, and antioxidant actions [2]. Applications can be found at patent offices for formulations involving biosurfactants. Rhamnolipids, sophorolipids,

Biosurfactants for a Sustainable Future: Production and Applications in the Environment and Biomedicine, First Edition. Edited by Hemen Sarma and Majeti Narasimha Vara Prasad.
© 2021 John Wiley & Sons Ltd. Published 2021 by John Wiley & Sons Ltd.

surfactins, and saponins are being used for the production of cleaning agents, gel compositions, exfoliants, shampoos, skin products, toothpastes, etc. [3–5].

Another factor that strengthens the interest in the use of biosurfactants in cosmeceuticals is the worldwide trend toward the development of products with natural ingredients to replace synthetic materials in order to solve one of the major challenges for the market, which are allergies, irritations in the skin, and reactions, caused by the presence of chemical surfactants in formulations [6]. In addition, the profile of the new consumer market seeks less toxic, sustainable, and biodegradable products, thus prioritizing cosmeceuticals/cosmetics with components of natural origin in their formulations [7].

In this sense, advances in biotechnological research on biosurfactants are extremely important to contribute to the discovery and understanding of the potential of these biomolecules for use in several formulations, among them new cosmeceutical products.

In the following sections, the main properties of biosurfactants that can be used in cosmeceuticals/cosmetics will be discussed. The application perspectives and the presence of these biomolecules in this sector of the current market will also be discussed, based on scientific studies that prove the applicability of biosurfactants in various types of personal care products.

19.2 Cosmeceutical Properties of Biosurfactants

In recent years, products have emerged with functions beyond the sole purposes of cleaning and beautifying. Manufacturers have denominated these products cosmeceuticals, dermocosmetics, functional cosmetics, or even performance cosmetics [8].

These are formulations for personal use that are beneficial to the organism, causing positive, lasting changes to the health of skin, mucous membranes, and the scalp. These diverse products have numerous active ingredients as raw materials, among which biosurfactants have attracted particular interest [9].

The effects of such formulations include the control of acne, combatting aging, protection from solar rays, moisturizing, antioxidant activity, and medicinal activity [10]. Biosurfactants can benefit the body either directly or indirectly, serving as carriers and enhancers of cosmeceutical ingredients. Some of the main functions of biosurfactants that are applicable in cosmeceuticals will be addressed below, with a definition, mechanism of action, and examples from the literature.

19.2.1 Emulsifying Activity

Before discussing the cosmeceutical properties themselves, it is necessary to talk about the emulsifying capacity of biosurfactants, as it is through emulsions that many molecules and bioactive substances will generate benefits and reach the body.

Emulsions are the most widely used physical form in cosmetics to unite fat soluble, water soluble, and even insoluble ingredients into stable systems. Emulsions enable the solubilization of different active ingredients and enhance the absorption of these ingredients in skin and hair [11].

Emulsions are important vehicles for the topical release of bioactive substances. Components such as moisturizers, antioxidants, humectants, photoprotectors, antiaging agents, vitamins, and medications are added to emulsions, offering benefits in the form of creams, lotions, and conditioners [12].

Oil-in-water (O/W) emulsions are generally the most widely used for cosmetic purposes. This type of emulsion spreads more easily, does not leave an oily sensation, exerts a cooling effect on skin as the water evaporates, and comes off more easily when washing with water [13].

Numerous biosurfactants are good emulsifying agents, which is an added advantage in the preparation of "green" cosmetic products. For a biosurfactant to be considered a good emulsifier, the main criterion is the capacity to form stable emulsions at least above 50% for 24 hours or longer [14].

A biosurfactant obtained from *Lactobacillus paracasei* was used as a stabilizing agent in O/W emulsions containing essential oils and a natural antioxidant extract. In the presence of the biosurfactant, a 100% index in the volume of the emulsions was found, which is comparable to results obtained with sodium dodecyl sulphate (SDS) [9].

Emulsions containing tea saponins also exhibited an excellent performance, remaining stable even when exposed to changes in temperature (30–90 °C), pH (3–9), and salinity (0–200 mM of NaCl) [15]. Systems stabilized by hydrolysed rice glutelin and *Quillaja saponin* also maintained stability at high levels of salt and temperature [16]. Such results suggest that microbial surfactants and saponins are effective surfactants that may have applications in commercial products.

The interfacial behavior of a surfactant is one of the determinants of a stable system. This performance is evaluated using the hydrophilic–lipophilic balance (HLB), which is the value attributed to a surfactant based on its chemical composition [17]. A low HLB (3–6) indicates that the structure of the surfactant contains more non-polar groups, which is favorable to water-in-oil (W/O) emulsions. In contrast, a surfactant with a high HLB (8–18) has more polar groups in its structure and is more likely to form O/W emulsions. Surfactants with an HLB higher than 20 act as co-surfactants [18] (Figure 19.1).

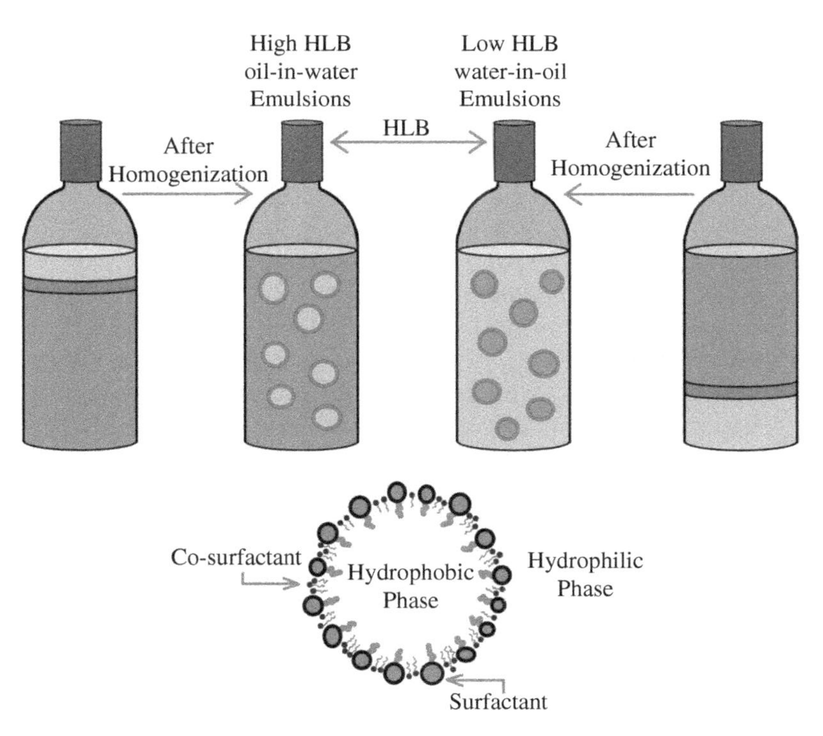

Figure 19.1 Applications of the HLB value for W/O and O/W emulsions.

Table 19.1 HLB values and possible applications.

HLB value	Application
1–2	Defoamer
3–6	W/O emulsifier agent
7–9	Wetting agent
8–16	O/W emulsifier
12–15	Detergent
15–19	Solubilizer
>20	Co-surfactant

Sources: [17, 18].

Knowing the HLB of a biosurfactant is crucial to its incorporation into a cosmetic product, as this value enables predicting its emulsification capacity as well as its possible humectant and defoaming capacity [9] (Table 19.1).

Kuyukina and Ivshina [19] calculated the HLB from the purified structure of a trehalolipid produced by *Rhodococcus ruber* IEGM 231 and obtained a value of 8. A lipopeptide produced by *Acinetobacter junii* B6 had an HLB of 10 [20]. Both biosurfactants have O/W emulsifying properties.

The HLB of saponins varies widely due to the diversity of plant species. Ginseng saponins studied by Xue et al. [21] had HBL values ranging from 1.1 to 6.3 and could be used as defoamers as well as W/O emulsifiers. Digitonin, tomatine, and glycyrrhizin are commercially available saponins with an HLB of 12.98, 12.29, and 8.9, respectively, and have been used as emulsifying agents for the formation of gels in cosmetic products.

19.2.2 Antioxidant Activity

Natural and synthetic antioxidant substances inhibit or retard damage caused by oxidants with a reduction potential through different mechanisms, such as the inhibition of free radicals and metal complexation [22].

Antioxidant compounds are widely used in "antiaging" products, as they assist in restoring the skin and preventing signs of aging by reducing lines and wrinkles [23]. On hair, antioxidants increase blood circulation, preventing hair loss and assisting in healthy hair growth [24].

One of the most widely used tests to demonstrate the antioxidant activity of a molecule is the 2,2-diphenyl-1-picrylhydrazyl (DPPH) method. DPPH is a free radical that is stable at room temperature and has a violet color in solutions with ethanol. The reduction of DPPH occurs in the presence of antioxidant molecules, turning the solution colorless [25].

The sequestering capacity of the radical cation 2,2′-azinobis-3-ethylbenzothiazoline-6-sulfonic acid (ABTS+) is another widely used method. The sample is mixed into an ABTS+ solution and absorbance is read at 734 nm. Trolox at the same concentration as the sample is used as the standard and the result is expressed using a Trolox calibration curve [26].

The ferric reducing antioxidant power (FRAP) assay is another method for testing antioxidant activity and is based on the production of the Fe II ion from the reduction of Fe III. When this reduction occurs, the tone of the reaction mixture changes from light violet to intense violet, the absorbance of which can be read at a wavelength of 595 nm. Greater absorbance or color intensity indicates greater antioxidant potential [27].

Biosurfactants are associated with antioxidant activity, as demonstrated by the DPPH, ABTS, and FRAP assays. This property means that these biomolecules can be incorporated into formulations for protection against oxidative stress [28].

Giri et al. [29] investigated the antioxidant capacity of biosurfactants produced by *Bacillus subtilis* VSG4 and *Bacillus licheniformis* VS16 using the DPPH and hydroxyl radical tests. At a concentration of 5 g/l, the VSG4 biosurfactant eliminated DPPH and the hydroxyl radical by 69.1 and 62.3%, respectively, and the same concentration of the VS16 biosurfactant achieved elimination rates of 73.5 and 68.9%, respectively. In another study, a lipopeptide biosurfactant synthesized by *Bacillus methlotrophicus* DCS1 exhibited considerable antioxidant activity, eliminating DPPH by 80.6% at a concentration of 1 mg/ml [30].

The ferric reducing activity and elimination of DPPH were also found for a biosurfactant produced by *Bacillus cereus*. In the FRAP assay, the greatest absorbance was recorded at a concentration of 2 mg/ml. The elimination of DPPH was in the range of 27–63% at concentrations of 0.5 and 2.0 mg/ml [31].

Plant extracts rich in biosurfactants, especially saponins, are also reported to have excellent antioxidant activity. The tensioactive agents in these extracts are capable of transferring hydrogen electrons to stabilize free radicals and reduce metals [26].

The extract from *Betula pendula* studied by Tmáková et al. [25] had an IC_{50} (concentration required for the inhibition of DPPH by 50%) at a concentration of 70.4 mg/ml, indicating a good free radical elimination capacity. López-Romero et al. [26] evaluated the antioxidant activity of the extract from *Agave angustifolia* (AAE) and found that AAE also exhibited a free radical elimination capacity, as demonstrated by the DPPH, ABTS, and FRAP assays (94.2, 239.1, and 148.8 μmol TE/g, respectively).

Saponins from the extract of *Abutilon indicum* leaves demonstrated an increase in the percentage of antioxidant activity, with the lowest inhibition (15.7%) at a concentration of 0.25 mg/ml and the maximum inhibition (96.17%) at a concentration of 2.5 mg/ml, using the DPPH assay [32].

All of these examples demonstrate that the use of biosurfactants as antioxidant agents in cosmeceuticals is possible.

19.2.3 Antimicrobial Activity

Cosmetic products are recognized as substrates for the survival and development of a large variety of microorganisms due to the presence of nutrients that facilitate microbial growth, such as water, lipids, polysaccharides, alcohol, proteins, amino acids, glycosides, peptides, and vitamins. The occurrence of pathogenic microorganisms in cosmetic products poses a risk to the health of consumers [33].

The Rapid Exchange of Information System (RAPEX) is a rapid alert system that offers information on cosmetic products with microbiological contamination sold in European markets. For the year 2019, RAPEX presented 18 results of contaminated products, including shampoos, liquid soaps for children, henna powder, and moistened wipes for babies. The microorganisms most frequently encountered are *Pseudomonas aeruginosa, Klebsiella oxutoca, Burkhhoderia cepacia, Staphlococcus aureus, Escherichia coli, Candida albicans, Pseudomonas putida, Pluralibacter gergoviae*. Therefore, the preservation of active ingredients in formulations is an important aspect in the cosmetic industry. As most preservatives are chemical compounds that can cause irritation, it is necessary to search for biocompatible preservatives with fewer side effects [34].

Besides preservative action, the addition of molecules with antimicrobial properties serves for the purposes of treatment. Many skin diseases are caused by bacterial and fungal infections and

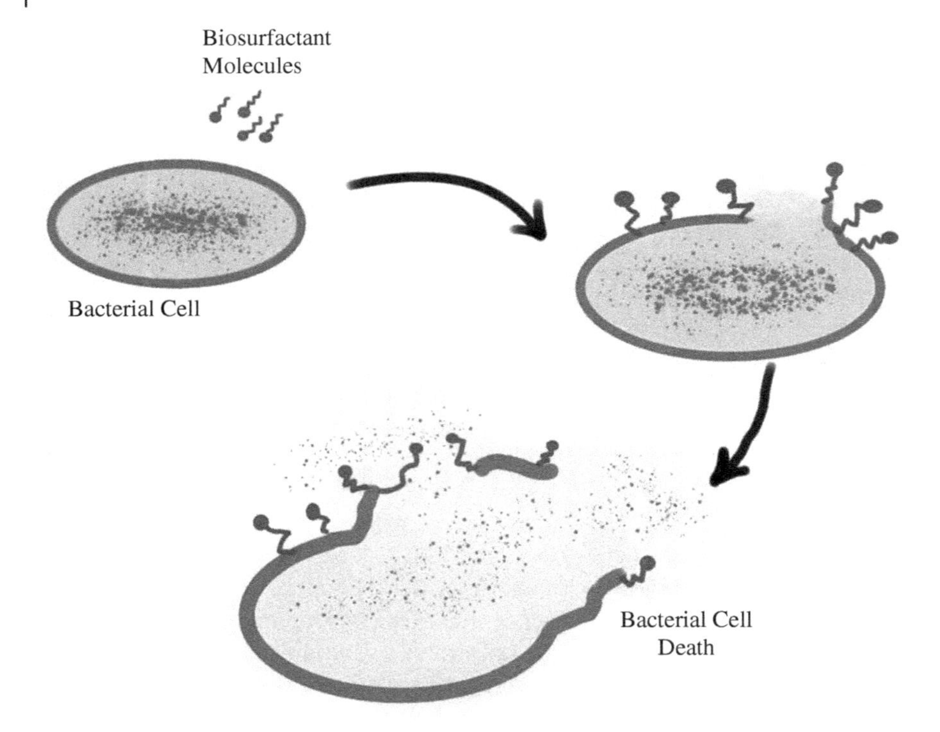

Figure 19.2 Action of biosurfactants resulting in bacterial cell death.

cosmeceuticals can be used as medication [35, 36]. *S. aureus*, *Propionibacterium acnes*, *Staphylococcus epidermidis, Streptococcus* sp., *Microsporum* sp., *C. albicans*, and *Malassezia furfur* are examples of microorganisms that cause skin and scalp infections, such as acne, folliculitis, scarlet fever, impetigo, mycosis, candidiasis, and dandruff [37].

The antibacterial capacity of biosurfactants is well known. These natural compounds cause cell death in bacteria through the expansion and rupture of the cytoplasmic membrane, causing leakage of intracellular material [38]. This action may be associated with a reduction in the hydrophobicity of the cell surface and damage to the cytoplasmic membrane. The amphipathic nature of biosurfactants enables the interaction with phospholipids, altering the permeability of the cytoplasmic membrane due to the formation of pores, with the consequent leakage of cellular components and cell death [39] (Figure 19.2).

Research has been developed for the incorporation of these biomolecules in beauty and personal care products due to their effect on harmful microorganisms. Rao and Paria [40] investigated whether the biosurfactant contained in extracts from the leaves of *Aegle marmelos* had antifungal activity against *M. furfur*, which is the dandruff-causing fungus. The aim was to obtain a formulation for a "green" dandruff shampoo based on silver nanoparticles. The silver nanoparticles in the presence of the *A. marmelos* extracts exhibited improved antifungal properties against *M. furfur*.

The antimicrobial activity of a biosurfactant produced by *Lactobacillus pentosus* was evaluated for its effect on different microorganisms found in the microflora of skin, considering its potential as a natural ingredient in cosmetics and personal care products. The results revealed 100% inhibition of *P. aeruginosa*, *S. aureus*, and *Streptococcus agalactiae* at a concentration of 50 mg/ml [41].

A glycolipid biosurfactant produced by *Staphylococcus saprophyticus* SBPS 15 exhibited potential antimicrobial activity against pathogenic bacterial and fungal lines. Among the 12 human bacterial pathogens tested, the biosurfactant exhibited antimicrobial activity against *Klebsiella pneumoniae, E. coli, P. aeruginosa, Vibrio cholerae, B. subtilis, Salmonella paratyphi*, and *S. aureus*. The biosurfactant also exhibited antifungal activity against *Cryptococcus neoformans, C. albicans*, and *Aspergillus niger* [42].

In plants, saponins serve as a chemical barrier in the defense system against attacks from bacteria and fungi, which suggests their potential as antimicrobial agents [6, 43]. A saponin purified from *Felicium decipiens* exhibited antifungal activity against *Aspergillus flavus* and *Aspergillus fumigatus* with a minimum inhibitory concentration (MIC) of 7.55 and 17.5 μg/ml, respectively [44].

Borah et al. [45] tested the antimicrobial properties of the saponin isolated from the buds of *Calamus leptospadix* Griff. The authors found an MIC of 60 μg/ml for *E. coli* and *C. albicans* and 160 μg/ml for *S. aureus*, revealing antimicrobial activity.

19.3 Other Activities

In addition to the activities presented above, biosurfactants have other functions that, although not considered cosmeceuticals, are fundamental in the development and structure of these and other formulations. The following sections will discuss some of these functions, their properties and importance.

19.3.1 Foaming Capacity

Foaming ability is a well-known activity of surfactants. It is a dynamic phenomenon where the surfactant molecules are adsorbed on the liquid film that surrounds the gaseous part. What determines the foaming, density, and elasticity of the foam is precisely the structure of the molecule [46]. The elasticity of a foam must be adequately low to allow the interface to expand and must be high enough to maintain rigidity and therefore stability.

Foaming is a desired property for various cosmetic applications, such as shampoos, soaps, and shaving cream. Depending on the stability of the foam formed by a surfactant, the intended use can be varied. In some cases, a highly stable foam is more suitable, as in bubble baths, for example, while in cleaning applications, unstable foams are more desired [47].

Generally, the foaming capacity of chemical surfactants, such as Tween 80, SDS, and Span 20, is greater when compared to that of biosurfactants. This is a negative point that needs to be reviewed and/or demystified, since the foaming capacity is not related to the cleaning capacity; however, for consumers, a personal cleanser that lacks enough foam is not well accepted [48].

Therefore, there is a search for biosurfactants with foaming capacity, in addition to other needs already known, such as low cost and high productivity, that can compete commercially with chemical surfactants [49].

An interesting solution found by a patent for a cosmetic cleaning agent containing biosurfactant, described by Schelges and Tretyakova [3], to obtain a good amount of foam was the use of a foam dispenser. According to the invention, the foam dispenser is a device having a container and a closure with an appropriate design (foam pumper) so that the liquid cosmetic cleaning agent, on passing, forms the foam. The patent also states that foam pumps are easily found on the market, making it possible to create the foam manually, without the need for foam dispensers with propellants for the formation of foam.

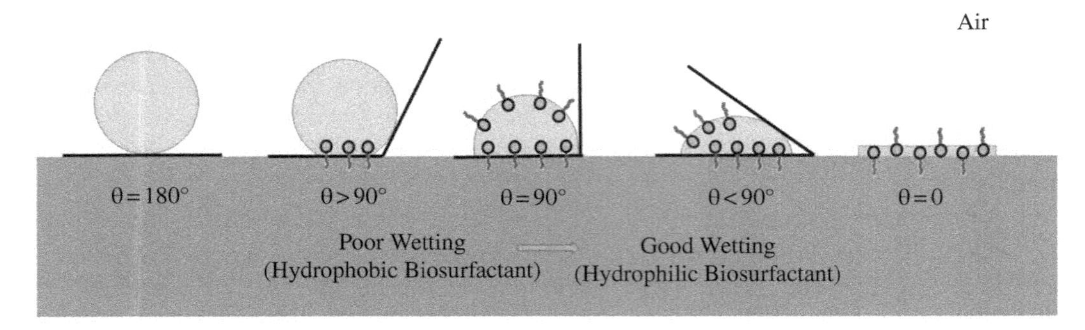

Figure 19.3 Improved wettability by surfactant adsorption at interfaces air/water-based cosmeceuticals and water-based cosmeceuticals/surface.

19.3.2 Wettability

Wettability refers to the ability of a liquid to spread on a solid surface. Sometimes this is not possible due to the hydrophobicity of the surface and the high surface tension of the water, requiring the use of surfactants. Surfactant molecules adsorb at the solid/water interface, reducing interface tension and promoting and/or improving wettability [18].

The wetting capacity of a surfactant can be analyzed by the contact angle of a drop of product with surfactant on a surface. The contact angle θ ranges from 180 to 0°, with $\theta = 180°$ when the liquid does not make contact with the surface and there is no wettability and $\theta = 0°$ when the liquid spreads completely over the surface and the wettability is complete. If $180° > \theta > 90°$, the surfactant has hydrophobic characteristics and may have little wetting capacity; on the other hand, if $0° < \theta < 90°$ the surfactant has hydrophilic characteristics and good wetting capacity, as illustrated in Figure 19.3 [18, 50].

So that a W/O cream glides smoothly and evenly over wet skin, shampoos and conditioners are able to have contact with the hair fiber, and products with active ingredients, such as bases, eyelash masks, and sunscreens, are able to penetrate the skin, a significant amount of surfactants must be incorporated in these products [51].

Many biosurfactants have good wettability and can be used for this purpose. Singh, Saha, and Padmanabhan [52] investigated the wettability of a biosurfactant produced by *Bacillus aryabhattai* SPS1001 and showed that it is capable of helping to change the wettability of a hydrophobic substrate wet in oil to wet in water. A plant biosurfactant called SUTBS also significantly reduced the contact angle from 150 to 60° in oil-wet sandstone [53].

19.3.3 Dispersion and Solubility

Pigments are extremely present and valued in the world of cosmetics and cosmeceuticals. Hair dyes, colored sunscreens, nail polishes, and make-up are manufactured in a wide range of colors and shades to suit the most varied tastes and types of consumer [54, 55].

One of the most complex issues when making cosmetics with color is the dispersion of pigments. The dispersion of a pigment incorrectly can cause variations in color, flocculation, loss of brightness, flotation, and/or fluctuation, Bénard cells, or sedimentation. To avoid this, dispersing agents are added to the formulations, which help, along with the movement machinery, to maintain the pigment particles dispersed in a satisfactory manner [56].

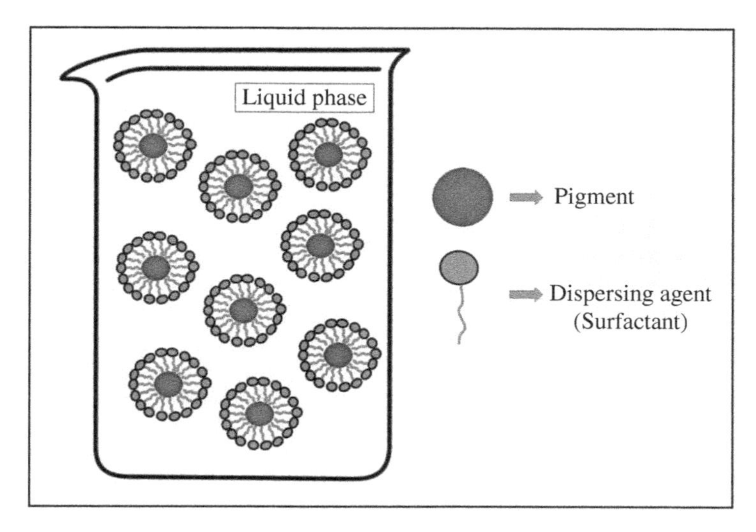

Figure 19.4 Pigment dispersion through steric impediment.

Dispersants are molecules that are adsorbed on the pigment surface, generating electrostatic repulsion and/or steric impediment, which maintains the space between the particles. The structure of the surfactants gives these compounds a dispersing capacity and they act through the steric impediment. This mechanism works as follows: the hydrophobic part of the surfactant structure will adsorb on the surface of the pigment, and the other part (hydrophilic) joins the liquid phase, favoring its "wrapping" around the pigment (Figure 19.4) [57, 58].

These surfactant additives, as they exhibit efficient surface-active properties, will not only stabilize pigment dispersion, but also function as humectants. The wettability of powdered pigments is the first step toward dispersion (Section 19.3.2). If the wettability is not good enough, formulations with a uniform and smooth touch cannot be obtained [59].

Another factor that depends on a good dispersion is the solubility, which is nothing more than particles that have dispersed and dissolved in the solvent, forming a solution. Such particles cannot be seen with the naked eye and neither can they be separated by filtration. Surfactants and biosurfactants play an important role as solubilizing agents. Relatively insoluble materials have their solubility improved by surfactants, which will interact hydrophobically with these materials through the formation of micelles, which will allow the formation of a homogeneous mixture [60].

There are several studies that show the dispersing and solubilizing potential of biosurfactants, especially studies in the environmental area with hydrophobic compounds, as is the case of a study with the raw biosurfactant of *B. cereus* UCP 1615, which showed the ability to disperse 86% of engine oil in tests carried out with sea water [61]. In another study, the addition of biosurfactants in the ternary solid dispersion system proved to be a promising excipient in the formulation of Carbamazepine in tablets for increasing the dissolution rate, reducing the dose and bioavailability [59].

In the cosmetic area, there is a study conducted by Kim and Noh [62] that described the potential of plant biosurfactants as solubilizers of bergamot oil and tocopherol acetate, both substances used in cosmetics. With the results of this research, a company has successfully developed a product for hydration, showing in practice that it is possible to use biosurfactants for this function.

19.4 Application Prospects

In the literature it is possible to find several researches that aim at the applicability of biosurfactants in cosmeceuticals and cosmetics. A brief description of the formulations that have been studied and developed recently will be explained below.

19.4.1 Shampoos

Hair has considerable economic, social, and cultural importance. It is related to self-esteem and the inclusion of individuals in a particular social group, culture, or ethnicity [6]. Thus, hair products are among the most widely used personal care items, especially shampoos, which have the function of washing the hair and scalp. Moreover, many shampoos are used for the treatment of dandruff, lice, alopecia, dermatitis, or pityriasis [63, 64].

Numerous plant extracts containing biosurfactants have been used over the years for the production of shampoos. Badi and Khan [65] used extracts from *Acacia concinna, Sapindus mukorossi, Phyllanthus emblica, Ziziphus spina-christi*, and *Citrus aurantifolia* in the preparation of a "green" shampoo, which proved to be efficient in terms of both detergency and cleansing, with a low surface tension, small bubble size, and good foam stability after five minutes.

Gour et al. [66] characterized the detergent properties of saponins from the mesocarp of the fruit of *Balanites aegyptiaca* at different concentrations for the removal of sebum and compared the results to those obtained with the chemical surfactants Tween-80, Triton X-100, and SDS. The surface tension of the mesocarp extract was similar to that of Triton X-100 and the detergency of the plant-based material was very high, as a 20% solution of the extract achieved an 88% sebum removal rate.

Bezerra et al. [6] described 13 patents for shampoo formulations containing biosurfactants in their composition, showing that in addition to cleaning functions, biosurfactants have also been incorporated for drug action, such as anti-dandruff, anti-itch and anti-inflammatory action, showing the great spectrum of action of these molecules.

19.4.2 Conditioners

Among the different hair products, conditioners moisturize hair, restoring shine, and softness and re-establishing lost/damage sebum and keratin [67]. Conditioners are composed of cationic surfactants, polymers, thickeners, emollients (oily compounds), and auxiliary emulsifiers [68]. Biosurfactants are active agents that can serve as thickeners, emulsifiers, humectants, vehicles for moisturizing products and even, as preservatives in conditioners.

The process of conditioning consists mainly in the deposition of polymers and surfactants on the damaged hair fiber, which has a negative surface charge. Fernández-Peña et al. [69] studied the adsorption of polymer–surfactant mixtures containing four types of rhamnolipids on negatively charged surfaces simulating the negative charge of hair fibers. The results showed that the amphipathic structure of the biosurfactants played a fundamental role in the control of the adsorption process, improving efficiency and moisturizing compared to conventional sulphate-based systems. The increase in the rhamnolipid concentration led to the phase separation of the system due to the formation of kinetically trapped aggregates far from the isoelectric point, playing a fundamental role in the conditioning process.

Ferreira et al. [9] discuss the potential use of a biosurfactant produced by *L. paracasei* in creams due to its good performance as an emulsifying agent in O/W systems in combination with essential oils and an antioxidant extract.

19.4.3 Skincare

When the subject is skincare, mannosylerythritol lipids (MELs), which are glycolipids produced by yeasts and fungi, are the most widely studied biosurfactants for this purpose. Moisturizing effects on dry skin, the activation of fibroblasts and cells of the papilla, antioxidant activity, and a protective effect on skin cells have been reported for this class of biosurfactant [70].

MELs also have a lightening effect on human melanocytes, exhibiting potential use in antimelanogenic cosmeceuticals, which treat hyperpigmentation in skin, such as freckles and melasma [46]. A recent study reported that saponins from *Panax notoginseng* promoted the healing of skin wounds and suppressed the formation of scars in rats, making these natural products promising candidates for medications that can accelerate the wound healing process and reduce scar formation on skin [71].

19.4.4 Toothpastes

Bouassida et al. [72] produced two toothpaste formulations using a chemical surfactant and a lipopeptide isolated from *B. subtilis* SPB1. The results showed that the biosurfactant was as efficient as the chemical surfactant with regards to spreading capacity, water activity, pH, the formation of foam, and cleansing. The biosurfactant-based toothpaste also exhibited antimicrobial activity against *Enterobacter* sp. and *Salmonella typhinirium*.

In a recent study, biosurfactants obtained from *P. aeruginosa* UCP 0992, *Bacillus metylotrophicus* UCP 1616, and *Candida bombicola* URM 3718 were used as alternatives to chemical surfactants in toothpaste formulations. The authors state that the biosurfactants tested are promising as possible replacements for synthetic products in oral hygiene products and can be used as foaming and antimicrobial agents. The formulations were effective at inhibiting biofilm formed by *Streptococcus mutans* and exhibited no toxicity [73].

19.4.5 UV Protection

Studies have demonstrated the potential of biosurfactants to provide physical protection from the harmful effects of the sun and enhance the action of other compounds designed for this purpose.

Researchers analyzed the interaction between a biosurfactant extract from corn steep liquor and mica powder to improve the solar protection factor (SPF) of a "green" formulation. The addition of the biosurfactant to the aqueous emulsion containing mica powder improved the SPF due to the capacity of the biosurfactant extract per se to protect from UV radiation and due to the formation of an emulsion with the mica powder, producing a synergic effect on the protective properties of the mica through the formation of a stronger barrier, thereby favoring physical protection [74].

Tsai and Lin [75] used the extract from *Camellia oleifera* containing saponins as ingredients in the formulation of a sunscreen formulation. The saponins had peak absorbance at wavelengths of 200 to 240 nm and could therefore absorb UV rays. The extract could be used to reduce the quantity of organic UV absorbers and inorganic agents employed in sunscreens, thereby diminishing the toxicity of the products and problems with sensitive skin.

19.5 Biosurfactants in the Market

In the previous sections, the development of promising research with different types of biosurfactants was described, which brings evidence of the great potential of these biomolecules for application in cosmeceuticals and cosmetics. In this topic, however, the real use of biosurfactants in the market will be presented, that is, companies and industries that already produce and commercialize biosurfactants for use in formulations, showing that the industrial production of microbial surfactants is a possible alternative, even with much effort [76].

Evonik, a chemical company from Germany, was the first company that in 2010 was able to use biotechnological methods to produce biosurfactants on an industrial scale. Sophorolipids were the first natural surfactants produced and purchased by the Ecover and Unilever brands for use in household products, which are already on supermarket shelves. Evonik researchers are currently developing technologies for the production of rhamnolipids on a large scale, with the aim of application as foam promoters and in cosmetic products [77].

Another company that produces biosurfactants is Jeneil Biotech, Inc., located in the United States. This company produces rhamnolipids for industries in the areas of agriculture, bioremediation, household, and personal care. The company claims that the rhamnolipids produced provide excellent properties for personal care products, such as emulsification, wetting, detergency, foam and biodegradability, and can be used in the manufacture of soaps, pastes, shampoos, creams, and cleaning products for contact lenses [78].

Kaneka, a Japanese company, has among its products a surfactin called KANEKA Surfactin. It is a biosurfactant based on cyclic lipopeptide produced by *B. subtilis*, which due to the cyclic structure works with a low concentration (3 ppm) and shows a multiplier effect to improve the performance of other surfactants, such as SDS and linear alkylbenzene sulfonate (LAS) [79]. Results of studies carried out by this company showed low human skin irritability and a high rate of biodegradability.

Another example is GlycoSurf, a chemical company started in 2013 with the goal of creating green glycolipid surfactants. GlycoSurf was founded by three researchers from the University of Arizona (United States), with technological development funding from Tech Launch Arizona at the University of Arizona. Today the company works with the production of rhamnolipids and rhamnosids of high purity (>99%), are non-toxic and biodegradable. On the company's website, purchase prices range from $10 for 10 mg of biosurfactant to $525 for 1 g of biosurfactant. The company claims that "industrial quantities" are available for order, in which case there will be a reduction in prices [80].

As shown above, there are already companies on the market that have effectively managed to develop methods for large-scale production of specific biosurfactants. A search in patent banks with the keywords biosurfactants, saponins, cosmeceuticals, and cosmetics shows a large number of results, which is another indication of how this type of market is gaining strength, and that soon it will be possible to find biosurfactants in personal care products more frequently. The results of patents on this topic, filed in the years 2019 and 2020, are reported in Table 19.2.

19.6 Challenges and Conclusion

Using completely natural raw materials for the formulation of cosmetics is a difficult task. The challenge resides in the choice of the material, which needs to have functions and accessibility comparable to those of its synthetic counterparts [73].

As explained above, biosurfactants have the potential to compete with chemical surfactants with regards to functional properties, such as surface tension, emulsifying, wetting, foaming capacity, good stability, etc. [91]. However, the big challenge, in general, remains the low concentration in production, especially with regard to microbial surfactants, making large-scale production difficult and resulting in high prices. In addition, the lack of financial resources and investments in some

Table 19.2 Patents issued on personal care formulations containing biosurfactants.

Biosurfactant	Source	Function	Product	Patent No.	Publication date
Mannosylerythritol lipid, Sophorolipid, Trealolipid, Rhamnolipid, and Surfactin	Microbial	Active ingredient carriers	Cosmetics for skin health	US20200069779A1 [81]	March 5, 2020
Saponin	Vegetable	Antibacterial and anti-inflammatory	Cosmetic composition to improve keloid scar	US20200009043A1 [82]	January 9, 2020
Saponin	Vegetable	Antibacterial and anti-inflammatory	Skin care	US10265264B2 [83]	April 23, 2019
Mannosylerythritol lipid	Microbial	Whitening	Skin lightener	US 2019 / 0231668 A1 [84]	August 1, 2019
Mannosylerythritol lipid	Microbial	Stabilizer	Skin cosmetics	US2019/0192412 A1 [85]	January 27, 2019
Surfactin	Microbial	Emulsifier	Cosmetic gel	US 2019 / 0209443 A1 [86]	July 11, 2019
Rhamnolipid/ Sophorolipid	Microbial	Solubilizer	Make-up remover	CN110151606A [87]	June 19, 2019
Glycolipid / Lecithin	Microbial	Emulsifier	Anti-wrinkle cream	CN110179730A [88]	August 30, 2019
Sophorolipid	Microbial	Cleaning	Facial cleanser	CN110123701A [89]	August 16, 2019
Tea saponin	Vegetable	Antibacterial and anti-inflammatory	Body wash	CN109464377A [90]	March 15, 2019

cases, especially in third world countries, makes it difficult for researchers to search for alternatives to achieve a high concentration in the production of biosurfactants. The companies that managed to overcome the low productivity barrier either had financial resources or received investments to pay for human and technological resources, showing that the financial valuation of research is worthwhile.

The alternatives for greater production that have been studied and developed by researchers over the years are aimed at the use of low-cost fermentation substrates, for the search for microorganisms with a better producing potential or the genetic modification of biosurfactant-producing microorganisms [92].

Regarding plant-based biosurfactants, the yield is higher and acquisition is easier compared to those of a microbial origin, as extracts rich in biotensioactive agents can be used [93, 94]. Many plant extracts are already being used by the cosmetic industry for different purposes. With regards to the search for extracts containing a significant concentration of biosurfactants, the options revolve around the advancement of extraction methods and the search for plants with a high amount of these biomolecules [95].

Finally, the efforts of scientists to obtain and study biosurfactants from all-natural sources are valid and must be supported, as the variety and potential of these biomolecules can and could offer solutions for various processes in many areas, in a biodegradable, less toxic, or nontoxic form, respecting the human being and also the environment. In addition, there is an urgent need to search for molecules from renewable sources, which in the future may and will need to replace the raw materials derived from petroleum.

References

1 Vecino, X., Cruz, M., Moldes, B., and Rodrigues, R. (2017). Critical reviews in biotechnology biosurfactants in cosmetic formulations: trends and challenges. *Critical Reviews in Biotechnology*: 1–22. https://doi.org/10.1080/07388551.2016.1269053.

2 Drakontis, C.E. and Amin, S. (2020). Biosurfactants: formulations, properties, and applications. *Current Opinion in Colloid & Interface Science* 48: 77–90. https://doi.org/10.1016/j.cocis.2020.03.013.

3 Schelges, H. and Tretyakova, M. (2017). Cleaning compositions comprising bosurfactants in a foam dispenser, 1–12. Inventors: Schelges, H. and Tretyakova, M. Application: 09 June 2016. US Patent US20170071837 A1. Antolovich, M., Prenzler, P.D., Patsalides, E., McDonald, S. and Robards, K. (2002)

4 Kaneka Corporation., (2019). Gel-state composition and production method therefor. Inventors: TSUJI, Tadao. Application: 05 December 2018. WO Patent WO2019/138739.

5 Umo International Co., Ltd., (2016). Application of surfactin in cosmetic products and thereof. Inventors: Jenn-Kan LU, Hsin-Mei and WANG, Xuan-Rui. Application: 31 Jul 2014. US Patent US2016/015536 A.

6 Bezerra, K.G.O., Rufino, R.D., Luna, J.M., and Sarubbo, L.A. (2018). Saponins and microbial biosurfactants: Potential raw materials for the formulation of cosmetics. *Biotechnology Progress* 34 (6): 1482–1493. https://doi.org/10.1002/btpr.2682.

7 Sarubbo, L.A., Rocha, R.B., Luna, J.M. et al. (2015). Some aspects of heavy metals contamination remediation and role of biosurfactants. *Chemistry and Ecology* 7540 (November): 1–17. https://doi.org/10.1080/02757540.2015.1095293.

8 Sharad, J. (2020). Cosmeceuticals, Advances in Integrative Dermatology. Chichester, UK: Wiley.

9 Ferreira, A., Vecino, X., Ferreira, D. et al. (2017). Novel cosmetic formulations containing a biosurfactant from *Lactobacillus paracasei*. *Colloids and Surfaces B: Biointerfaces* 155: 522–529. https://doi.org/10.1016/j.colsurfb.2017.04.026.

10 Draelos, Z.D. (2019). Cosmeceuticals. In: Evidence-Based Procedural Dermatology (ed. Z.D. Draelos), 479–497. Springer, Cham https://doi.org/10.1016/j.clindermatol.2007.09.005.

11 Pereira, M.J.L., Rodrigues Neto, E.M., Girão Júnior, F.J., Macedo, M.L.B.D. and Araujo, T.G. (2018) Evaluation of the behaviour of violet acid pigment 43 in formulations of shampoos and conditioners in bleached hair. doi: https://doi.org/10.5530/jyp.2018.10.63.

12 Kale, S.N. and Deore, S.L. (2017). Emulsion micro emulsion and nano emulsion: A review. *Systematic Reviews in Pharmacy* 8 (1): 39. https://doi.org/10.5530/srp.2017.1.8.

13 Dănilă, E., Moldovan, Z., Albu, G., and Ghica, V. (2019). Formulation and characterization of some oil in water cosmetic emulsions based on collagen hydrolysate and vegetable oils mixtures. *Pure and Applied Chemistry* 0 (0): 1–15. Available at: https://doi.org/10.1515/pac-2018-0911.

14 Pinto, H., Guimarães, R., and Vieira, A. (2009). Avaliação cinética da produção de biossurfactantes bacterianos. *Quimica Nova* 32 (8) https://doi.org/10.1590/S0100-40422009000800022.

15 Zhu, Z., Wen, Y., Yi, J. et al. (2019). Comparison of natural and synthetic surfactants at forming and stabilizing nanoemulsions: Tea saponin, *Quillaja saponin*, and Tween 80. *Journal of Colloid and Interface Science* 15 (536): 80–87. https://doi.org/10.1016/j.jcis.2018.10.024.

16 Xu, X., Sun, Q., and McClements, J. (2018). Enhancing the formation and stability of emulsions using mixed natural emulsifiers: Hydrolyzed rice glutelin and quillaja saponin. *Food Hydrocoll [online]* 89: 396–405. https://doi.org/10.1016/j.foodhyd.2018.11.020.

17 Costa, J., Treichel, H., Santos, L., and Martins, V. (2018). Solid-state fermentation for the production of biosurfactants and their applications. In: Current Developments in Biotechnology and Bioengineering (eds. P. Ashok, L. Christian and R. Carlos), 357–372. Amsterdam, The Netherlands: Elsevier https://doi.org/10.1016/B978-0-444-63990-5.00016-5.

18 Akbari, S., Nour, A., Yunus, R., and Farhan, A. (2018). Biosurfactants as promising multifunctional agent: A mini review. *SSRN Electronic Journal* 1: 1–6. https://doi.org/10.2139/ssrn.3323582.

19 Kuyukina, M. and Ivshina, I. (2019). Production of trehalolipid biosurfactants by *Rhodococcus*. In: Biology of Rhodococcus (ed. A. Héctor), 271–298. Perm, Russia: Espringer https://doi.org/10.1007/978-3-030-11461-9_10.

20 Ohadi, M., Dehghannoudeh, G., Forootanfar, H. et al. (2018). Investigation of the structural, physicochemical properties, and aggregation behavior of lipopeptide biosurfactant produced by *Acinetobacter junii* B6. *International Journal of Biological Macromolecules* 1 (112): 712–719. https://doi.org/10.1016/j.ijbiomac.2018.01.209.

21 Xue, P., Yang, X., Zhao, L. et al. (2020). Relationship between antimicrobial activity and amphipathic structure of ginsenosides. *Industrial Crops and Products* 1: 143. https://doi.org/10.1016/j.indcrop.2019.111929.

22 Apak, R., ÖZyürek, M., Güclų;̈.K., and Çapanoglu, E. (2016). Antioxidant activity/capacity measurement. 1. Classification, physicochemical principles, mechanisms, and Electron Transfer (ET)-based assays. *Journal of Agricultural and Food Chemistry* 64: 997–1027. https://doi.org/10.1021/acs.jafc.5b04739.

23 Afonso, C.R., Hirano, R.S., Gaspar, A.L. et al. (2019). Biodegradable antioxidant chitosan films useful as an anti-aging skin mask. *International Journal of Biological Macromolecules* 132: 1262–1273.

24 Joshi, N., Patidar, K., Solanki, R., and Mahawar, V. (2018). Growth promoting shampoo formulation. *International Journal of Green Pharmacy* 2018 (4): 835–839.

25 Tmáková, L., Sekretár, S., and Schmidt, Š. (2016). Plant-derived surfactants as an alternative to synthetic surfactants: surface and antioxidant activities. *Chemical Papers* 70 (2): 188–196. https://doi.org/10.1515/chempap-2015-0200.

26 López-Romero, J.C., Ayala-Zavala, J.F., Peña-Ramos, E.A. et al. (2018). Antioxidant and antimicrobial activity of *Agave angustifolia* extract on overall quality and shelf life of pork patties stored under refrigeration. *Journal of Food Science and Technology* 55 (11): 4413–4423. https://doi.org/10.1007/s13197-018-3351-3.

27 Antolovich, M., Prenzler, P.D., Patsalides, E. et al. (2002). Methods for testing antioxidant activity. *Analyst* 127 (1): 183–198. https://doi.org/10.1039/B009171P.

28 Mnif, I. and Ghribi, D. (2015). Review lipopeptides biosurfactants: mean classes and new insights for industrial, biomedical, and environmental applications. *Peptide Science* 104 (3): 129–147. https://doi.org/10.1002/bip.

29 Giri, S.S., Ryu, E.C., Sukumaran, V., and Park, S.C. (2019). Antioxidant, antibacterial, and anti-adhesive activities of biosurfactants isolated from *Bacillus* strains. *Microbial Pathogenesis* 132: 66–72. https://doi.org/10.1016/j.micpath.2019.04.035.

30 Jemil, N., Ayed, H.B., Manresa, A. et al. (2017). Antioxidant properties, antimicrobial and anti-adhesive activities of DCS1 lipopeptides from Bacillus methylotrophicus DCS1. *BMC Microbiology* 17 (1): 144. https://doi.org/10.1186/s12866-017-1050-2.

31 Basit, M., Rasool, M.H., Naqvi, S.A.R. et al. (2018). Biosurfactants production potential of native strains of *Bacillus cereus* and their antimicrobial, cytotoxic and antioxidant activities. *Pakistan Journal of Pharmaceutical Sciences* 31: 251–256.

32 Lokesh, R.A.V.I., Manasvi, V., and Lakshmi, B.P. (2016). Antibacterial and antioxidant activity of saponin from *Abutilon indicum* leaves. *Asian Journal of Pharmaceutical and Clinical Research* 9 (3): 344–347. https://doi.org/10.22159/ajpcr.2016.v9s3.15064.

33 Neza, E. and Centini, M. (2016). Microbiologically contaminated and over-preserved cosmetic products according Rapex 2008–2014. *Cosmetics* 3 (1): 3. https://doi.org/10.3390/cosmetics3010003.

34 Rodríguez-López, L., Rincón-Fontán, M., Vecino, X. et al. (2019). Preservative and irritant capacity of biosurfactants from different sources: A comparative study. *Journal of Pharmaceutical Sciences* 108 (7): 2296–2304. https://doi.org/10.1016/j.xphs.2019.02.010.

35 Draelos, Z.D. (2017). Cosmeceuticals for rosacea. *Clinics in Dermatology* 35 (2): 213–217. https://doi.org/10.1016/j.clindermatol.2016.10.017.

36 Nawarathne, N.W., Wijesekera, K., Wijayaratne, W.M.D.G.B. and Napagoda, M. (2019) Development of novel topical cosmeceutical formulations from *Nigella sativa* L. with antimicrobial activity against acne-causing microorganisms. *The Scientific World Journal*, 2019. doi: 10.1155/2019/5985207.

37 Byrd, A.L., Belkaid, Y., and Segre, J.A. (2018). The human skin microbiome. *Nature Reviews Microbiology* 16 (3): 143. https://doi.org/10.1038/nrmicro.2017.157.

38 Sana, S., Datta, S., Biswas, D., and Sengupta, D. (2018). Assessment of synergistic antibacterial activity of combined biosurfactants revealed by bacterial cell envelop damage. *Biochimica et Biophysica Acta (BBA) - Biomembranes* 1860 (2): 579–585. https://doi.org/10.1016/j.bbamem.2017.09.027.

39 Freitas Ferreira, J., Vieira, E.A., and Nitschke, M. (2019). The antibacterial activity of rhamnolipid biosurfactant is pH dependent. *Food Research International* 116: 737–744. https://doi.org/10.1016/j.foodres.2018.09.005.

40 Rao, K.J. and Paria, S. (2016). Anti-Malassezia furfur activity of natural surfactant mediated in situ silver nanoparticles for a better antidandruff shampoo formulation. *RSC Advances* 6 (13): 11064–11069. https://doi.org/10.1039/C5RA23174D.

41 Vecino, X., Rodríguez-López, L., Ferreira, D. et al. (2018). Bioactivity of glycolipopeptide cell-bound biosurfactants against skin pathogens. *International Journal of Biological Macromolecules* 109: 971–979. https://doi.org/10.1016/j.ijbiomac.2017.11.088.

42 Mani, P., Dineshkumar, G., Jayaseelan, T. et al. (2016). Antimicrobial activities of a promising glycolipid biosurfactant from a novel marine *Staphylococcus saprophyticus* SBPS 15. *3 Biotech* 6 (2): 163. https://doi.org/10.1007/s13205-016-0478-7.

43 Augustin, J.M., Kuzina, V., Andersen, S.B., and Bak, S. (2011). Molecular activities, biosynthesis and evolution of triterpenoid saponins. *Phytochemistry* 72 (6): 435–457. https://doi.org/10.1016/j.phytochem.2011.01.015.

44 Brandão-Costa, R.M., Nascimento, T.P., Bezerra, R.P., and Porto, A.L. (2020). FDS, a novel saponin isolated from *Felicium decipiens*: Lectin interaction and biological complementary activities. *Process Biochemistry* 88: 159–169. https://doi.org/10.1016/j.procbio.2019.10.018.

45 Borah, B., Phukon, P., Hazarika, M.P. et al. (2016). *Calamus leptospadix* Griff. A high saponin yielding plant with antimicrobial property. *Industrial Crops and Products* 82: 127–132. https://doi.org/10.1016/j.indcrop.2015.11.075.

46 Bae, I.H., Lee, E.S., Yoo, J.W. et al. (2019). Mannosylerythritol lipids inhibit melanogenesis via suppressing ERK–CREB–MiTF–tyrosinase signalling in normal human melanocytes and a three-dimensional human skin equivalent. *Experimental Dermatology* 28 (6): 738–741. https://doi.org/10.1111/exd.13836.

47 Mahmoodabadi, M., Khoshdast, H., and Shojaei, V. (2019). Efficient dye removal from aqueous solutions using rhamnolipid biosurfactants by foam flotation. *Iranian Journal of Chemistry and Chemical Engineering* 38 (4): 127–140.

48 Knoth, D., Rincón-Fontán, M., Stahr, P.L. et al. (2019). Evaluation of a biosurfactant extract obtained from corn for dermal application. *International Journal of Pharmaceutics* 564: 225–236. https://doi.org/10.1016/j.ijpharm.2019.04.048.

49 Bages-Estopa, S., White, D.A., Winterburn, J.B. et al. (2018). Production and separation of a trehalolipid biosurfactant. *Biochemical Engineering Journal* 139: 85–94. https://doi.org/10.1016/j.bej.2018.07.006.

50 Capra, P., Musitelli, G., and Perugini, P. (2017). Wetting and adhesion evaluation of cosmetic ingredients and products: correlation of in vitro–in vivo contact angle measurements. *International Journal of Cosmetic Science* 39 (4): 393–401. https://doi.org/10.1111/ics.12388.

51 Tsujii, K. (2017). Wetting and surface characterization. In: Cosmetic Science and Technology: Theoretical Principles and Applications (eds. K. Sakamoto, R. Lochhead, H. Maibach and Y. Yamashita), 373. Amsterdam, The Netherlands: Elsevier.

52 Singh, V., Saha, S., and Padmanabhan, P. (2020). Assessment of the wettability of hydrophobic solid substrate by biosurfactant produced by *Bacillus aryabhattai* SPS1001. *Current Microbiology*: 1–8. https://doi.org/10.1007/s00284-020-01985-6.

53 Simjoo, M., Rezaei, M.A., Nadri, F., Mousapour, M.S., Iravani, M. and Chahardowli, M., 2019, April. Introducing a new, low-cost biosurfactant for EOR applications: A mechanistic study. In: *IOR 2019–20th European Symposium on Improved Oil Recovery*, Vol. 2019, No. 1, pp. 1–12. European Association of Geoscientists & Engineers, Pau, France.

54 Dhakal, M., Sharma, P., Ghosh, S. et al. (2016). Preparation and evaluation of erbal lipsticks using natural pigment lycopene (*Solanum lycopersicum*). *Universal Journal of Pharmaceutical Science and Research* 2 (2): 23–29. https://doi.org/10.21276/UJPSR.2016.02.02.24.

55 Yusuf, M., Shabbir, M., and Mohammad, F. (2017). Natural colorants: Historical, processing and sustainable prospects. *Natural Products and Bioprospecting* 7 (1): 123–145. https://doi.org/10.1007/s13659-017-0119-9.

56 Brunaugh, A.D., Smyth, H.D.C., and Williams, R.O. III (eds.) (2019). Disperse systems: Emulsions. In: Essential Pharmaceutics, 111–121. Springer, Austin, TX, USA.

57 Abreu, B., Rocha, J., Fernandes, R.M. et al. (2019). Gemini surfactants as efficient dispersants of multiwalled carbon nanotubes: Interplay of molecular parameters on nanotube dispersibility and debundling. *Journal of Colloid and Interface Science* 547: 69–77. https://doi.org/10.1016/j.jcis.2019.03.082.

58 Rebello, T. (2019). Guia de produtos cosméticos. São Paulo- Brazil: Editora Senac.

59 Bolmal, U.B., Subhod, R.P., Gadad, A.P., and Patill, A.S. (2020). Formulation and evaluation of carbamazepine tablets using biosurfactant in ternary solid dispersion system. *Studies* 6: 7. https://doi.org/10.5530/ijper.54.2.35.

60 Matsuoka, K., Takahashi, N., Yada, S., and Yoshimura, T. (2019). Solubilization ability of star-shaped trimeric quaternary ammonium bromide surfactant. *Journal of Molecular Liquids* 291: 111254. https://doi.org/10.1016/j.molliq.2019.111254.

61 Durval, I.B., Resende, A., Ostendorf, T. et al. (2019). Application of *Bacillus cereus* Ucp 1615 biosurfactant for increase dispersion and removal of motor oil from contaminated seawater. *Chemical Engineering Transactions* 74: 319–324. https://doi.org/10.3303/CET1974054.

62 Kim, I.Y. and Noh, J.M. (2019). A study on synthesis of organic plant surfactant and its solubilizing action on bergamot oil. *Journal of the Korean Applied Science and Technology* 36 (4): 1208–1218. doi: 10.12925/jkocs.2019.36.4.1208.

63 Abdullah, N.A. and Kaki, R. (2017). Lindane shampoo for head lice treatment among female secondary school students in Jeddah, Saudi Arabia: An interventional study. *Journal of Infectious Diseases & Preventive Medicine* 5 (173): 2. https://doi.org/10.4172/2329-8731.1000173.

64 Gupta, A.K., Mays, R.R., Versteeg, S.G. et al. (2019). Efficacy of off-label topical treatments for the management of androgenetic *Alopecia*: A review. *Clinical Drug Investigation* 39 (3): 233–239. https://doi.org/10.1007/s40261-018-00743-8.

65 Al Badi, K. and Khan, S.A. (2014). Formulation, evaluation and comparison of the herbal shampoo with the commercial shampoos. *Beni-Suef University Journal of Basic and Applied Sciences* 3 (4): 301–305. https://doi.org/10.1016/j.bjbas.2014.11.005.

66 Gour, V.S., Sanadhya, N., Sharma, P. et al. (2015). Biosurfactant characterization and its potential to remove sebum from hair. *Industrial Crops and Products* 69: 462–465. https://doi.org/10.1016/j.indcrop.2015.03.007.

67 Barve, K. and Dighe, A. (2016). Hair conditioner. In: The Chemistry and Applications of Sustainable Natural Hair Products (eds. K. Barve and A. Dighe), 37–44. Cham: Springer.

68 D'Souza, P. and Rathi, S.K. (2015). Shampoo and conditioners: What a dermatologist should know? *Indian Journal of Dermatology* 60 (3): 248. https://doi.org/10.4103/0019-5154.156355.

69 Fernández-Peña, L., Guzmán, E., Leonforte, F. et al. (2020). Effect of molecular structure of eco-friendly glycolipid biosurfactants on the adsorption of hair-care conditioning polymers. *Colloids and Surfaces B: Biointerfaces* 185: 110578. https://doi.org/10.1016/j.colsurfb.2019.110578.

70 Morita, T., Fukuoka, T., Imura, T., and Kitamoto, D. (2013). Production of mannosylerythritol lipids and their application in cosmetics. *Applied Microbiology and Biotechnology* 97 (11): 4691–4700. https://doi.org/10.1007/s00253-013-4858-1.

71 Men, S.Y., Huo, Q.L., Shi, L. et al. (2020). Panax notoginseng saponins promotes cutaneous wound healing and suppresses scar formation in mice. *Journal of Cosmetic Dermatology* 19 (2): 529–534. https://doi.org/10.1111/jocd.13042.

72 Bouassida, M., Fourati, N., Krichen, F. et al. (2017). Potential application of *Bacillus subtilis* SPB1 lipopeptides in toothpaste formulation. *Journal of Advanced Research* 8 (4): 425–433. https://doi.org/10.1016/j.jare.2017.04.002.

73 Resende, A.H.M., Farias, J.M., Silva, D.D. et al. (2019). Application of biosurfactants and chitosan in toothpaste formulation. *Colloids and Surfaces B: Biointerfaces* 181: 77–84. https://doi.org/10.1016/j.colsurfb.2019.05.032.

74 Rincón-Fontán, M., Rodríguez-López, L., Vecino, X. et al. (2018). Design and characterization of greener sunscreen formulations based on mica powder and a biosurfactant extract. *Powder Technology* 327: 442–448. https://doi.org/10.1016/j.powtec.2017.12.093.

75 Tsai, C.E. and Lin, L.H. (2019). The liquid polyol extracts of camellia seed dregs used in sunscreen cosmetics. *Chemical Papers* 73 (2): 501–508. https://doi.org/10.1007/s11696-018-0594-4.

76 Sajna, K.V., Hofer, R., Sucumaran, R.K. et al. (2015). White biotechnology in biosurfactants. In: Industrial Biorefineries and White Biotechnology (eds. A. Pandey, R. Hofer, C. Larroche, et al.), 499–517. Amsterdam, The Netherlands: Elsevier.

77 Evonik (2020), *Evonik Biosurfactants*. Viewed: 24 May 2020, https://corporate.evonik.com/misc/micro/biosurfactants/index.en.html.

78 Jeneil Biotech, I 2020, *Biosurfactants | Jeneil Biotech*, Viewed: 24 May 2020. https://www.jeneilbiotech.com/biosurfactants.

79 Kaneka 2019, New business development biossurfactant KANEKA surfactin, Viewed: 24 May 2020. Available at: www.kaneka.co.jp/en/business/qualityoflife/nbd_002.html.

80 GlycoSurf's no date, Green Glcolipid Surfactants, Viewed: 24 May 2020 http://glycosurf.com.

81 Locus IP Company, LLC., (2020). Cosmetic compositions for skin health and methods of using same. Inventors: Sean Farmer, Ken Alibek, and Sharmistha Mazumder. Application: 30 April 2018. US Patent US2020069779A1.

82 Young Oh Park. (2020). Cosmetic composition for Keloid scar improvement. Inventors: Young Oh Park. Application: 22 March 2018. US Patent US2020009043A1.

83 Bioprocol Bioprocesos de Colombia SAS [CO]., (2019). Processes and compositions obtained from the Solanum genus of plants. Inventors: Aristizabal Alzate Carlos Esteban, Palacio Gonzalez Guillermo Leon, and Schafer Elejalde German Alfredo. Application: 1 December 2015. US Patent US10265264B2.

84 Amorepacific Corporation (2019). Skin whitening composition containing mannosylerythritol lipid. Inventors: Yoo Jae Won, Hwang Yoon Kyun, Bin Sung Ah, Kim Yong Jin, and Lee John Hwan. Application: 8 September 2016. US Patent US2019/0231668 A1.

85 Amorepacific Corporation (2019). Cosmetic composition having high dosage form stability. Inventors: Kim Yu Jung, Hyeon Chung Kim, Jae Won Yoo, Yong Jin Kim, Do Hoon Kim, and Sung Il Park. Application: 30 June 2015. US Patent US2019/0192412 A1.

86 Kaneka Corporation. Gellike composition and external-use agent for skin and cosmetic material in which said gel-like composition is used. Inventor: Tadao Tsuji. Application: 14 September 2016. US Patent US2019/0209443 A1.

87 Shanghai Meifute Biotechnology Co Ltd., (2019). Pure natural makeup removal composition and preparation method thereof. Inventors: Liu Chuanwei. Application: 19 June 2019. CN110151606A.

88 Feng Yuqin, Li Jiaqian, Zang Wei., (2019). *A kind of no line maintenance frost.* Inventors: Feng Yuqin, Li Jiaqian and Zang Wei. Appl: 19 June 2019. China Patents CN110179730A.

89 Univ Fujian Agriculture & Forestry., (2019). Konjac glucomannan facial cleanser. Inventors: Chen Siyang, Chen Xiaohan, Guo Yangyang, Jiang Jingyi, Lu Yinzhu, Pang Jie, Tong Cailing and Xu Xiaowei. Application: 15 June 2019. CN110123701A.

90 Chen Xinxin (2019). Bio-antibacterial body wash and preparation method. Inventors: Chen Xinxin. Application: 29 December 2018. CN109464377A.

91 Santos, D.K.F., Rufino, R.D., Luna, J.M. et al. (2016). Biosurfactants: Multifunctional biomolecules of the 21st century. *International Journal of Molecular Sciences* 17 (3): 401. https://doi.org/10.3390/ijms17030401.

92 Chen, C., Sun, N., Li, D. et al. (2018). Optimization and characterization of biosurfactant production from kitchen waste oil using *Pseudomonas aeruginosa*. *Environmental Science and Pollution Research* 25 (15): 14934–14943. https://doi.org/10.1007/s11356-018-1691-1.

93 Böttcher, S. and Drusch, S. (2016). Interfacial properties of saponin extracts and their impact on foam characteristics. *Food Biophysics* 11 (1): 91–100. https://doi.org/10.1007/s11483-015-9420-5.

94 Liu, Z., Li, Z., Zhong, H. et al. (2017). Recent advances in the environmental applications of biosurfactant saponins: A review. *Journal of Environmental Chemical Engineering* 5 (6): 6030–6038. https://doi.org/10.1016/j.jece.2017.11.021.

95 Gong, W., Huang, Y., Ji, A. et al. (2018). Optimisation of saponin extraction conditions with *Camellia sinensis* var. assamica seed and its application for a natural detergent. *Journal of the Science of Food and Agriculture* 98 (6): 2312–2319. https://doi.org/10.1002/jsfa.8721.

20

Biotechnologically Derived Bioactive Molecules for Skin and Hair-Care Application

Suparna Sen[1], *Siddhartha Narayan Borah*[2], *and Suresh Deka*[1]

[1]*Environmental Biotechnology Laboratory, Resource Management and Environment Section, Life Sciences Division, Institute of Advanced Study in Science and Technology, Guwahati, Assam, India*
[2]*Royal School of Biosciences, Royal Global University, Guwahati, Assam, India*

CHAPTER MENU

20.1 Introduction

Cosmetics preparations for personal care applications, which include skin, hair, and nail-care formulations, are products that are used for cleansing, beautifying, or protecting any external surface of the human body [1]. Cosmetics used on a daily basis are in the form of soaps, shampoos, deodorants, moisturizers, toothpaste, perfumes, and makeup [2]. Earlier, cosmetics were formulated from synthetic compounds, predominantly of petrochemical origin, and were

Biosurfactants for a Sustainable Future: Production and Applications in the Environment and Biomedicine, First Edition. Edited by Hemen Sarma and Majeti Narasimha Vara Prasad.
© 2021 John Wiley & Sons Ltd. Published 2021 by John Wiley & Sons Ltd.

Table 20.1 List of chemicals used in cosmetic formulations, their functions, and their adverse effects.

Sl. no.	Chemical compounds	Functions	Adverse effects
1	Parabens	Preservative	Cancer, reproductive issues
2	Phthalates	Plasticizer, fixative	Reproductive and developmental toxicity, potentially carcinogenic
3	Formaldehyde or formaldehyde releasing compounds	Nail hardener and hair relaxer, preservative	Probably carcinogenic to humans, allergic and irritant to skin
4	Polyethylene glycol (PEG)	Emulsifier, lubricant, solubilizer, and cleansing agent	Chance of presence of dioxin, a potential carcinogen
5	Propylene glycol and butylene glycol	Moisturizer and stabilizer	Cause acne and skin allergy, referred to as allergic contact dermatitis
6	Triclosan	Antimicrobial agent	Can cause allergy and endocrine disruption Can lead to the development of antibiotic-resistant microbes
7	Diethanolamine (DEA) and DEA-related compounds	Wetting agent and foaming agent	Can lead to the formation of nitrosodiethanolamine (NDEA), a potential carcinogen
8	Octocrylene and its derivatives	UV blocking agent in sunscreen lotion	Cause allergy and enhance ROS generation in skin
9	Isopropanol	Solubilizing agent or thickness modifier	Eye irritation, effects on the nervous system
10	Sodium lauryl sulfate (SLS) and sodium laureth sulfate (SLES)	Foaming and cleansing agent	Skin irritation and eye damage and chance of presence of dioxin, a potential carcinogen

Source: Adapted from [1].

primarily used for beautification purposes. Subsequently, in 1961, the concept of cosmeceuticals was presented by Raymond Reed. He defined cosmeceuticals as scientifically designed cosmetic products containing bioactive ingredients with medicinal properties and aesthetic values. The active ingredients in cosmeceutical formulations can be synthetic or natural compounds if it meets certain specific criteria, such as penetration of the skin barriers, rigorous clinical trials, etc. [1]. Chemical compounds are routinely used in cosmetics, albeit, they have several adverse effects (Table 20.1 [1]).

Considering the side effects and environmental hazards associated with the chemical compounds in cosmeceuticals, interest to replace them with biotechnologically derived active compounds like botanical extracts and several other microbially derived compounds with skin and hair-care properties have grown since the early 1990s. There are several functional aspects to be considered for any compound to be considered for cosmetic applications. Primarily, the compounds should be biocompatible with skin or hair and should not trigger any adverse immune response. The structure and physicochemical properties of the compound are critically related to its performance. Therefore, to maximize performance, appropriate downstream processing for purification of the compounds is essential to retain the structural and functional properties. A cosmetic formulation can be designed with compounds exhibiting promising skin/hair care activities after evaluation of its safety. A wide range of biotechnologically derived compounds find applications in cosmetic formulations and can be classified into several groups such as polyphenols, terpenes and carotenoids, organic acids, amino acids, and other nitrogenous compounds, vitamins and vitamin-like compounds, polysaccharides, polypeptides and proteins, essential fatty acids, sterols, and lipids derivatives.

20.2 Surfactants in Cosmetic Formulation

Surface-active agents or surfactants are amphiphiles comprising of hydrophobic and hydrophilic components that can reduce surface tension, thereby facilitating the emulsification of liquids with different polarities. They can orient themselves according to the polarities of the two opposing phases. Surfactants find usage in cosmetic formulations due to their detergency, wetting, emulsifying, solubilizing, dispersing, and foaming effects [3]. The industrial applications of surfactants are quite diverse and are utilized in virtually all industries to some extent [4]. Most surfactants are of chemical origin and derived from petrochemical sources. Surfactants of biological origin have been described only in the past few decades. Surfactants are widely used in cosmetics for creating uniform dispersing systems such as suspensions, emulsions or microemulsions, in cleansing products due to detergent properties, and as enhancers to promote drug absorption in the skin [5]. Biocompatibility is a crucial feature to consider for their use in any cosmetic formulation. Mostly, surfactants that are used in cosmetic formulations commercially are polyethylene glycol ethers. In particular, skincare formulations must meet some set standards, such as a pleasing appearance property retention during storage, and furnish long-lasting beneficial effects to the skin. Several researchers have reported that the use of chemical-based surfactants can affect the skin cells unfavorably by irritating the skin, triggering allergic reactions, and impairing the epidermal barrier function [6]. They are known to interact with proteins and solubilize lipids from the epidermal surfaces. Surfactants can also have adverse effects on the aquatic environment. Consequently, there is an urgent need for the use of surfactants that are not only efficient but are also biodegradable and biocompatible [7].

20.3 Biosurfactants in Cosmetic Formulations

Biosurfactants can act similarly to chemical surfactants. The use of surfactants from natural sources is gaining interest because of environmentally friendly properties and low-toxicity [8]. The applications of biosurfactants are multifarious and find usage in several industrial functions. Biosurfactants have gained the attention of the cosmetic industry for skin and hair-care products because of their potential use as detergents, wetting, emulsifying, foaming, solubilizing, and many other useful properties similar to chemical surfactants [9]. Biosurfactants can make excellent candidates to replace synthetic surfactants as they, owing to their natural origin, are biodegradable, display very low toxicity, have ecological acceptability, and have compatibility with human skin [10]. They find usage in personal care products such as shampoo, hair conditioners, soap, shower gel, toothpaste, creams, moisturizers, cleansers, and other skin and healthcare-related products [11].

20.3.1 Biosurfactants: Definition and Properties

The term "biosurfactants" refers to amphiphilic secondary metabolites secreted by microorganisms [12]. Biosurfactants were first discovered as extracellular amphiphilic compounds in the late

1960s [13]. Biosurfactants are produced by microorganisms like bacteria, yeasts, or fungi and are classified by their chemical composition and molecular weight. Biosurfactants exhibit a diverse range of chemical structures and are generally classified as glycolipids, lipopeptides (LPs), phospholipids, fatty acids, and polymeric compounds [14]. Despite their aqueous solubility, biosurfactants can have remarkably low levels of critical micelle concentration (CMC) when compared to structural analogs of synthetic origin. Additionally, they are considered to be good candidates for enhanced oil recovery (EOR) because of their surface activity [15, 16]. Biosurfactants have also found applications in therapeutics as antibiotic, antifungal, and antiviral compounds. Glycolipids are the best-studied microbial surfactants, with rhamnolipids (RLs), sophorolipids (SLs), and mannosylerythritol lipids (MELs) being the most popular types in the class. RLs are commercially produced by companies like Urumqi Unite Bio-Technology Co. Ltd (China) and AGAE Technologies LLC (Corvallis, OR), whereas SL biosurfactants are produced by Soliance (France) and MG Intobio Co. Ltd (South Korea) (Table 20.2). Among the LP class of biosurfactants (LPs), surfactin, produced mainly by *Bacillus* sp., is the most studied [18–21] and is marketed by Sigma Aldrich (USA). Among the different classes of biosurfactants, glycolipids and LPs are the most used biosurfactants for cosmetic applications.

Table 20.2 Biosurfactant production companies for commercial uses in cosmetic and pharmaceutical applications.

Sl. no.	Company	Location	Biosurfactant types	Application
1	TeeGeneBiotech	UK	RLs LPs	Pharmaceuticals, cosmetics, antimicrobials and anticancer ingredients
2	AGAE Technologies LLC	USA	RLs (R95)	Pharmaceutical, cosmeceutical, cosmetics, personal care,
3	Paradigm Biomedical Inc.	USA	RL	Pharmaceutical
4	Rhamnolipid Companies, Inc.	USA	RL	cosmetics, pharmaceutical
5	Fraunhofer IGB	Germany	Glycolipids	Cleansing products, shower gels, shampoos, washing-up liquids, pharmaceutical (bioactive properties)
6	Ecover Belgium	Belgium	SLs	Cosmetics, pharmaceuticals
7	Soliance	France	SLs	Cosmetics
8	MG Intobio Co. Ltd	South Korea	SLs (Sopholine – functional soap with SLs secreted by yeasts)	Personal care and bath products
9	Synthezyme LLC	USA	SLs	Cosmetics
10	Henkel	Germany	SLs, RLs, MELs	Beauty products
11	Kaneka Co.	Japan	SLs	Cosmetics, toiletries
12	Toyobo Co. Ltd	Japan	MELs	Cosmetics
13	Damy Chemicals Co. Ltd	South Korea	MELs	Beauty products, anti-cellulite formulation

Abbreviations: RL, rhamnolipid; SL, sophorolipid; MEL, mannosylerythritol lipids; LPs, lipopeptides.
Source: Adapted from [17].

20.3.2 Production of Biosurfactants

Biosurfactants can be produced from numerous substrates such as glucose, glycerol, lipidic sources such as vegetable oil, olive oil, etc. during fermentation. Despite the various benefits offered by biosurfactants, there are several concerns regarding their commercial applications. Among the different types of biosurfactants, SLs and MELs are reported to be produced in large quantities. Some of them use expensive substrates for the production process with low product yields and produce mixtures instead of single products, resulting in expensive downstream processing and purification [22]. However, strain improvement, use of renewable substrates, and process optimization can increase biosurfactant production by several folds. *Starmerella bombicola* produced more than 30 g/l of SLs in a medium containing glucose and sunflower oil [23]. During eight days of cultivation in the medium with 10% glucose, 10% sunflower oil, and 0.1% yeast extract, *S. bombicola* produced 120 g/l of SL [24], which increased to 420 g/l in a medium containing serum and rapeseed oil [25]. Some authors [26, 27] have suggested the importance of several parameters related to the composition of biosurfactants (e.g. CMCs, hydrophilic–lipophilic balance (HLB), foaming, etc.) in the cosmetic industry, since these properties will determine the type or the use of biosurfactants in cosmetic formulations.

20.3.3 Physicochemical Properties of Biosurfactants Suitable for Cosmetic Applications

20.3.3.1 Critical Micelle Concentration (CMC)

The CMC of a surfactant is an essential physiochemical parameter to consider, which denotes the minimum surfactant concentration required to achieve the lowest stable surface tension [28]. Therefore, the lower the CMC the higher will be the efficiency of a surfactant. Biosurfactants are reported to exhibit much lower CMCs in comparison to synthetic surfactants. A biosurfactant produced by *Serratia marcescens* NSK-1 exhibited a CMC of 29 mg/l and lowered the surface tension of distilled water from 72 to 38 mN/m [29]. An RL produced by *Pseudomonas aeruginosa* SWP-4 using waste cooking oil as the carbon substrate exhibited a CMC of 27 mg/l [30]. In another study, a biosurfactant produced by the genetically modified *Bacillus subtilis* BS37 strain reported a CMC of 20 mg/l [31]. In a study comparing the surface activities of surfactin and sodium lauryl sulfate (SLS) at the same concentration (0.005% w/v), the ST of surfactin was 27.9 mN/m, significantly lower than that of SLS at 56.5 mN/m [32]. However, the CMC for the commercial surfactants SLS and Triton X-100, was reported to be considerably high at 2257 and 150 mg/l, respectively [33, 34].

20.3.3.2 Emulsifying Property

HLB is a crucial factor to consider for the incorporation of biosurfactants in a cosmetic formulation as emulsifiers. The HLB represents the relative contribution of the hydrophilic and lipophilic groups of a surfactant and affects the stability of the emulsion. As a general rule, low HLB values (3–6) promote the formation of water/oil (W/O) microemulsions, whereas high HLB values (8–18) promote oil/water (O/W) microemulsions [35]. W/O emulsions (HLB values between 1 and 4) are preferable for dermatological formulations as the lipid film on the skin favors oil-soluble active ingredients. Such cosmetics exhibit a protective action along with an occlusive character. Consumers, on the other hand, tend to prefer O/W emulsions (HLB values between 8 and 16) due to their relatively grease-free feeling. Such formulations usually exhibit a semi-solid or liquid feature [36]. Fukuoka and co-workers [37] determined the HLBs of different MEL biosurfactants produced by *Pseudozyma antarctica* and observed a negative correlation to the degree of acylation. The authors reported the highest HLB values for monoacylated MELs (about 12), followed by di-acylated MELs (around 8), and tri-acylated MELs (around 6). An HLB value of 8 was recorded for a purified trehalolipid produced by *Rhodococcus ruber* IEGM 231 [37].

20.3.3.3 Foaming

Foaming is an important property of surfactants due to surface activity. However, whether foaming of surfactants is desirable or not depends on its intended use. For example, surfactants used in soaps and shampoos should possess a good foaming capacity, whereas those employed for washing purposes should create less foam. Besides the ability to form a foam, the stability of the foam is also an important criterion for surfactants, which depends on the intended application [7]. Biosurfactants like SLs are in general low foaming surfactants with high detergency [38], whereas RLs and surfactins have excellent foaming potential [39, 40].

20.3.4 Bioactive Properties of Biosurfactants for Skin and Hair-Care Applications

20.3.4.1 Moisturizing Effect

Stratum corneum (SC) is the outermost layer of skin. It forms a barrier between the outer environment and the internal body, preventing loss of body ingredients such as water, electrolyte, and the like, and simultaneously blocks the inflow of external harmful substances. Its structure is composed of anucleated corneocytes embedded in an intercellular lipid matrix. These lipids form a highly ordered lamellar structure [41]. Skin loses water balance due to various environmental, external, and internal factors, which include aging, heat, change in hormone secretion, face washes, etc. which leads to dryness and a moisture content of less than 10%. A well-hydrated skin has a moisture content of 15%, whereas a moisture content of 10% is required to maintain elasticity and flexibility of the skin. Dehydrated skin loses elasticity and flexibility, and eventually the skin protection function disappears to induce several skin conditions [42]. Frequent exposure of skin to surfactants may damage the SC barrier, resulting in skin dryness and inflammation [43]. Cleansers are usually formulated with synthetic surfactants at a concentration much higher than its CMC, wherein the majority of the surfactant molecules self-assemble into micelles [44]. A cleanser should remove unwanted exogenous lipophilic materials; however, the interaction between surfactants and skin is more complicated. Solubilization of skin components like lipids, enzymes, and natural moisturizing factors adversely affect the functioning of the skin barrier. Moreover, surfactants, even after rinsing, can remain inserted in the SC, leading to chronic surfactant exposure [45]. These surfactants disrupt the lipidic structural order of the SC, thereby causing continual degradation of the skin barrier [46]. The barrier impairment causes inflammation and oxidative stress manifested as discomfort and skin irritation [47, 48]. Ionic surfactants, like SDS, have been reported to alter the integrity of the skin barrier [49, 50]. Biosurfactants, on the other hand, do not affect the integrity of the skin barrier and form a lipidic film on the skin surface, thereby increasing hydration at deeper skin layers and providing skincare properties [51]. A biosurfactant obtained from corn steep liquor showed no adverse influence on transepidermal water loss (TEWL) and skin pH [51]. Skin hydration, pH, and TEWL are measures used to predict the impact on skin barrier functions [52]. However, the skin hydration values of the biosurfactant appeared to be lower in the study after the application of the biosurfactant, which indicated that the fatty acids formed a lipidic film on the surface, which enhanced the hydration at the deeper dermal layers [51].

It was also discovered that microbial biosurfactants could function as natural ceramides and enhance skin texture and remove ceramide deficiency in the skin [11, 53]. Ceramides are sphingolipids (glycosylceramides) that represent about 50% of intercellular lipids in the SC of skin and are also found in the cuticle of the hair. Ceramides are important for skin and hair barriers and keep them moisturized. The depletion of ceramides in SC of the skin can lead to chronic skin diseases like psoriasis, atopic dermatitis, and aged skin due to water loss and barrier dysfunction in the epidermis. Ceramides are also known to have a protective and restorative effect on hair fibers [54]. The addition of biosurfactants in cosmetics like skincare lotions and moisturizing creams can reportedly improve the quality of the product and help in reducing roughness and wrinkles [55]. Another study reported the efficacy of a microbial biosurfactant to repair damaged hair and increase the tensile strength in comparison to a synthetic ceramide, indicating its ability to recover damaged hair and improve its texture [56].

20.3.4.2 Permeation Through Skin

Delivery of actives into and across the skin plays a vital role in several topical, transdermal, dermatological, and personal care applications. While small, lipophilic molecules permeate the skin with relative ease, permeation of larger and hydrophilic molecules is hindered by the low permeability of the SC. This is particularly challenging since a significant fraction of currently marketed therapeutic drugs is hydrophilic [57]. The permeation of a compound in topical formulations can be significantly altered by the addition of permeation enhancers [58] or retarders [59], including surfactants [60]. Hence, surfactants are widely used in topical formulations; however, they might cause unwarranted effects like contact dermatitis and irritation. Therefore, synthetic surfactants are replaced with more biocompatible biosurfactants. Biosurfactants are used in pharmaceuticals or cosmeceuticals to solubilize the active ingredients and facilitate skin permeation [61]. The permeation rate of a compound is relative to the diffusion through the lipid bilayer of skin cells and depends on its lipophilicity. Hence, compounds with lipophilic properties, such as biosurfactants, cross the skin barrier with much ease as compared to the hydrophilic compounds [62]. Rodríguez-López and co-workers [63] studied the permeation capacity of 10 model drugs in the presence of a biosurfactant. They noted that compounds with a molecular mass lower than 200 increased the permeation of drugs, whereas in high molecular mass compounds permeation of the compounds decreased. This observation could be attributed to the formation of a drug-biosurfactant interaction that enhanced permeation of smaller compounds but retarded the permeation for those with a higher molecular mass.

20.3.4.3 Antioxidant Properties

Ultraviolet (UV) radiation has several harmful effects on the skin and hair, of which skin aging, due to oxidative stress, is typical. Also, skin cancer is a severe adverse effect of solar radiation and about 10 000 people die from malignancy every year. Environmental pollutants and microbial infections lead to the development of inflammatory responses that generate reactive oxygen species (ROS). The application of antioxidants could control the formation of ROS and the oxidative stress syndrome mediated by it. Cosmetics containing antioxidants as active ingredients include sunscreen lotions, skin brightening, and anti-aging products. Additionally, antioxidants stabilize and improve the shelf life of the products by preventing auto-oxidation of compounds like oil and fats. Several studies have reported the antioxidant property of biosurfactants [64–66], which might result from the fatty acid chain found in their structure [2]. The antioxidant property of biosurfactants can be exploited to reduce the damages caused by oxidative stress while providing a natural antioxidant source in cosmeceuticals [64, 66]. Rincón-Fontán and co-workers [67] studied the applicability of a biosurfactant in conjunction with mica for application as a sunscreen. They observed that the sun protection factor (SPF) of the formulation increased by more than 2000% as compared to the SPF values of only the micas in the absence of the biosurfactant. They concluded that there was a synergistic effect between the micas and the biosurfactant extract to protect against the harmful solar radiation. The observed effect of the formulation was due to the ability of the biosurfactant itself to protect against UV radiation and also due to a synergistic effect giving physical and bioactive protection. Takahashi and co-workers [68] studied the *in vitro* antioxidant property of MEL derivatives (A, B, C) and also *ex vivo* on cultured human skin fibroblasts under H_2O_2 oxidative stress. They observed that MEL-C exhibited the highest protective activity against the oxidative stress and repressed the expression of the oxidative stress marker, cyclooxygenase-2 (COX-2), confirming the antioxidant property of the biosurfactant for potential use in cosmetic formulations.

20.3.4.4 Wound Healing Properties

Wound healing is a complex procedure that involves sequential cellular events like cell proliferation, cell migration, re-epithelialization, angiogenesis, inflammatory phase, an oxidative reaction that occurs after tissue injury, or wound formation [69]. Ideally, after an injury, wound healing requires rapid wound closure with re-establishment of the damaged epidermal and dermal tissues, excluding any microbial contamination of the wound site [70]. Several studies on biosurfactant mediated wound healing have reported rapid wound closure, stimulation of the dermal fibroblast metabolism and collagen neosynthesis, macrophage activation and fibrinolytic properties, and desquamating activities, making them suitable candidates for dermatological applications [71–73]. The generation of reactive oxygen and nitrogen species at the wound site delays the healing process by lipid peroxidation, DNA damage, and inactivation of the free radical scavenger enzyme. Therefore, any wound healing agent with an antioxidant property would be desirable. The free radical scavenging property of the biosurfactants can reduce the free radicals generated at the wound site and expedite the healing process [74]. Ohadi and co-workers [75], in their study of wound healing in rats, reported that the LP produced by *Acinetobacter junii* B6 increased free-radical scavenging activities and improved histopathological remission. Microbial infections of wounds delay the healing process and create further complications. Therefore, the presence of an antimicrobial property is also used as an evaluating parameter to determine a compound's wound healing potential [73]. Interestingly, biosurfactants are well known for their antimicrobial potential against infection-causing pathogens [76, 77].

20.3.4.5 Antimicrobial Action

Antimicrobials prevent and protect cosmetic products from both bacterial and fungal contamination. Their addition functions broadly as preservatives to avoid contamination and growth of microorganisms over some threshold values that could lead to both quality deterioration and health damage of the consumers, such as skin or scalp infections [78]. The majority of the bacterial skin infections are caused by *Staphylococcus aureus* and *Streptococcus* sp. Typical viral skin infections include warts, cold sores, chickenpox, shingles, molluscum contagiosum, etc. The three main types of viruses that cause the majority of viral skin infections are the human papillomavirus (HPV), the herpes simplex virus, and the poxvirus [78]. Antifungal, antibacterial, antiviral, and antiadhesive properties of biosurfactants make them attractive candidates for use in several cosmetic formulations for skin and hair-care applications. For instance, SLs inhibit the growth of several Gram-positive microorganisms, viz. *B. subtilis*, *Staphylococcus epidermidis*, *S. aureus*, *Streptococcus faecium*, *Propionibacterium acnes*, and *Corynebacterium xerosis* in concentrations from 50 to 29 000 μg/ml [79]. In our earlier studies, we have reported the antifungal effect of an SL produced by *Rhodotorula babjevae* against the dermatophyte, *Trichophyton mentagrophytes*. The observed *in vitro* antifungal effects (MIC 1.0 mg/ml) were also corroborated *in vivo* in a mouse model of dermatophytosis. The SL also exhibited an antibiofilm effect against the dermatophyte [80]. It has been hypothesized that glycolipid biosurfactants like SLs and RLs can alter the membrane permeability of the target pathogen by irreversibly damaging its integrity, and accounts for their antifungal activity [81]. Antifungal compounds with a lipophilic character can increase the fluidity and permeability of the cell membrane of microorganisms by interfering with and destabilizing the ion transport [82]. This, in turn, leads to the inhibition of microbial growth, cell lysis, or cell death (Figure 20.1). Luna and co-workers [83] studied the antimicrobial and antiadhesive activities of a biosurfactant produced by *Candida sphaerica* UCP 0995 against different microorganisms. At a concentration of 10 g/l, the antimicrobial activity was 68% against *Streptococcus oralis* and around 58% against *Candida albicans* and *S. epidermidis*. Table 20.3 lists the different biosurfactants with reported antimicrobial effects against skin pathogens.

20.3.4.6 Hair-Care Properties

Aesthetic procedures such as hair straightening, coloring, bleaching, etc. use strong chemical treatments that leave the hair susceptible to damage. Chemical treatments disrupt hair cuticle, reduce hair elasticity, damage the texture, and leave hair unmanageable [1]. Several biotechnologically derived compounds, including biosurfactants, are now used in new generation hair treatments. The use of surfactants in the hair industry is essential, not only as cleansing agents but also as stabilizing ingredients of emulsion systems and as permeation enhancers for effective delivery of active principles. Also, in the cosmetic industry concerning hair washing and conditioning, formulations should be efficient in washing, foaming, repairing the hair fibers, and providing them with excellent sensorial properties. This can only be achieved through complex mixtures of several components, which include polyelectrolytes and surfactants bearing opposite charges [94]. Biosurfactants can substitute synthetic surfactants in hair-care formulations to fulfill the necessity for an eco-sustainable, biodegradable, and biocompatible alternative. The use of microbial biosurfactants has been reported in shampoo formulations.

The action of biosurfactants as hair cleaning agents involves the adsorption of monomers at the interface of the system that disrupts the intermolecular bonds between the hair follicle and dirt particles, thereby facilitating the release of the latter to the aqueous medium [95]. Biosurfactant extract (2.5%) was used as the stabilizing agent in anti-dandruff formulations along with Tween-80 (5%) to stabilize Zn pyrithione in O/W emulsions [96]. Another study provided insights into the conditioning process, which involves the deposition of polymer and surfactants on to damaged hair fiber with a negative surface charge [94]. The authors replaced sodium laureth sulfate (SLS) with RLs. They studied the adsorption of polymer-surfactant mixtures on to solid surfaces with a negative charge similar to that of damaged hair fibers as a model for understanding the physicochemical bases underlying the conditioning process. They found that the inclusion of RLs in the formulations increased the adsorption of polyelectrolytes in the model systems. The analysis of the degree of hydration showed that the mixtures containing mono-RL(C_{10}) could effectively replace SLS mixtures and led to an increase in surface coverage,

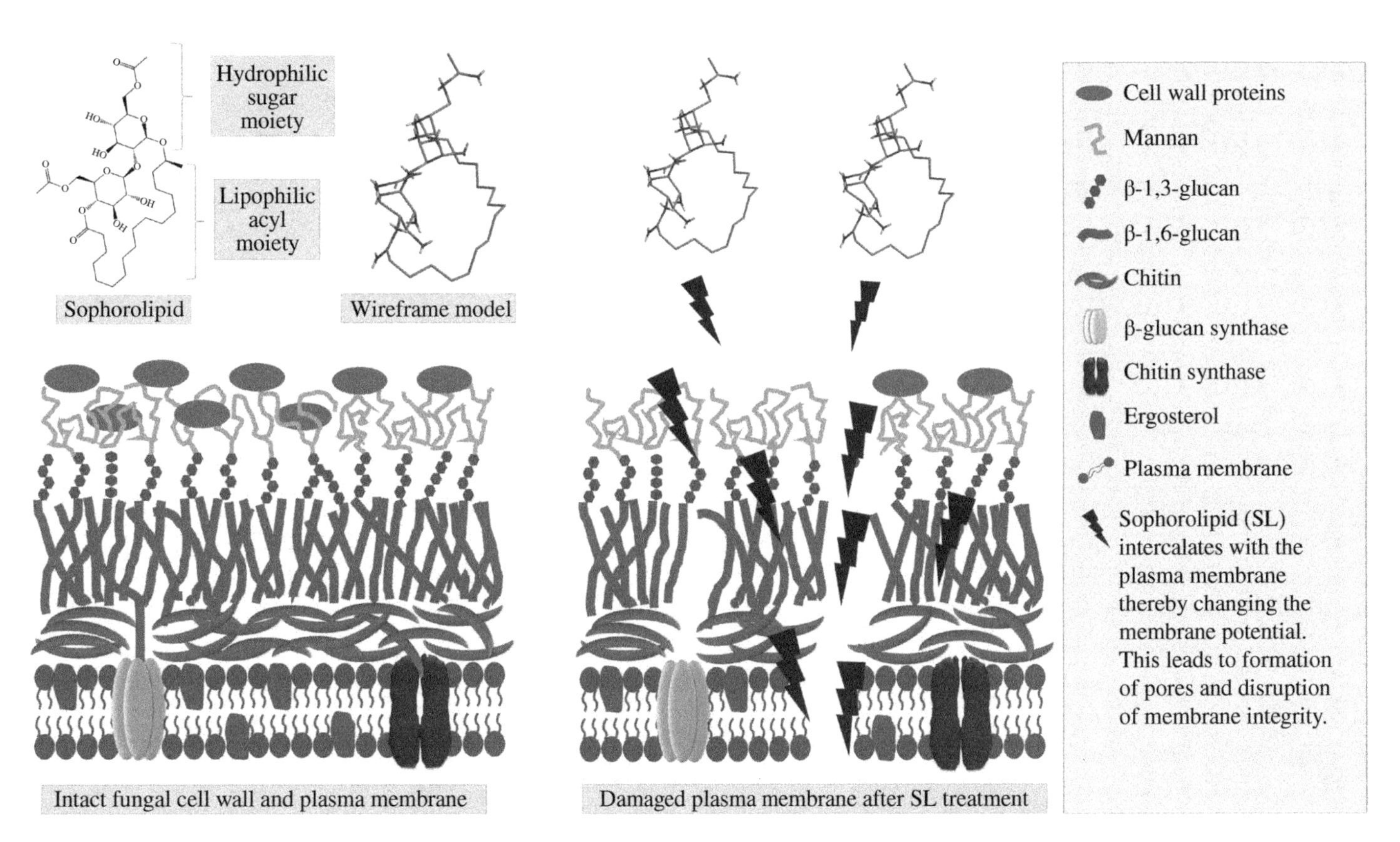

Figure 20.1 Schematic depiction of the mechanism of action of SL on the mycelial membrane of *Trichophyton mentagrophytes*. *Source:* Adapted from [80].

Table 20.3 Antimicrobial activity of biosurfactants against potential dermal pathogens.

Biosurfactant group	Biosurfactant type	Source	Pathogen	References
Lipopeptide	Pumilacidin (surfactin analog)	*Bacillus pumilus*	Herpes simplex virus-1 (HSV-1)	[84]
Glycolipid	SL	*Candida bombicola*	*Propionibacterium acnes*	[85]
	RL	*Pseudomonas aeruginosa*	*Staphylococcus aureus*	[86]
	RL	*P. aeruginosa*	*Trichophyton rubrum*	[87]
	SL	*Rhodotorula babjevae*	*T. mentagrophytes*	[80]
	RL	Jeneil Biosurfactant Co.	*P. aeruginosa*	[88]
			S. aureus	
	RL	*P. aeruginosa*	*S. epidermidis*	[89]
			S. aureus	
	SL	*C. bombicola*	*E. coli*	[90]
	RL	*P. aeruginosa*	*S. epidermidis*	[89]
			S. aureus	
	TL	*Rhodococcus fascians*	*S. epidermis*	[91]
			E. coli	
			C. albicans	
	RL	*P. aeruginosa*	*C. tropicalis*	[76]
—	—	*Lactobacillus paracasei*	*Escherichia coli*	[92]
			P. aeruginosa	
			S. aureus	
			S. epidermidis	
Glycolipopeptides	—	*L. paracasei, L. pentosus*	*E. coli*	[55]
			P. aeruginosa	
			S. aureus	
			S. epidermidis	
			Streptococcus pyogenes	
			C. albicans	
			S. agalactiae	
Cell-bound biosurfactant	—	*Lactobacillus* sp.	*S. aureus*	[93]
			Streptococcus agalactiae	
			P. aeruginosa	
			S. aureus	

Abbreviations: RL, rhamnolipid; SL, sophorolipid; TL, trehalose lipid.

enhancing the deposition and degree of hydration of the layers. The authors concluded that the RL structures with short alkyl chains could be suitable for replacement of SLS in hair washing formulations.

20.3.5 Current Trends and Other Skin/Hair-Care Applications of Biosurfactants

20.3.5.1 Glycolipids

Glycolipids are carbohydrates linked to long-chain aliphatic acids or hydroxyl-aliphatic acids. The majority of the reported glycolipids are of bacterial origin, and only a few come from yeasts and fungi. SLs, RLs, and MELs are the best-studied glycolipids with applications in cosmetics and pharmaceuticals.

Rhamnolipids (RLs) RLs are one of the most widely studied biosurfactants over the years [97]. RLs contain a hydrophilic moiety that consists of one (mono-RLs) or two (di-RLs) rhamnose molecules, and a hydrophobic part represented by one or two (or rarely more) β-hydroxy fatty acids, saturated or unsaturated, of different chain lengths (C_8–C_{24}) (Figure 20.2a). RLs have been used in antidandruff products, acne pads, and deodorants [98]. Cosmetics containing RLs have been patented and used as antiwrinkle and antiaging products [99]. Muller and co-workers [98] studied

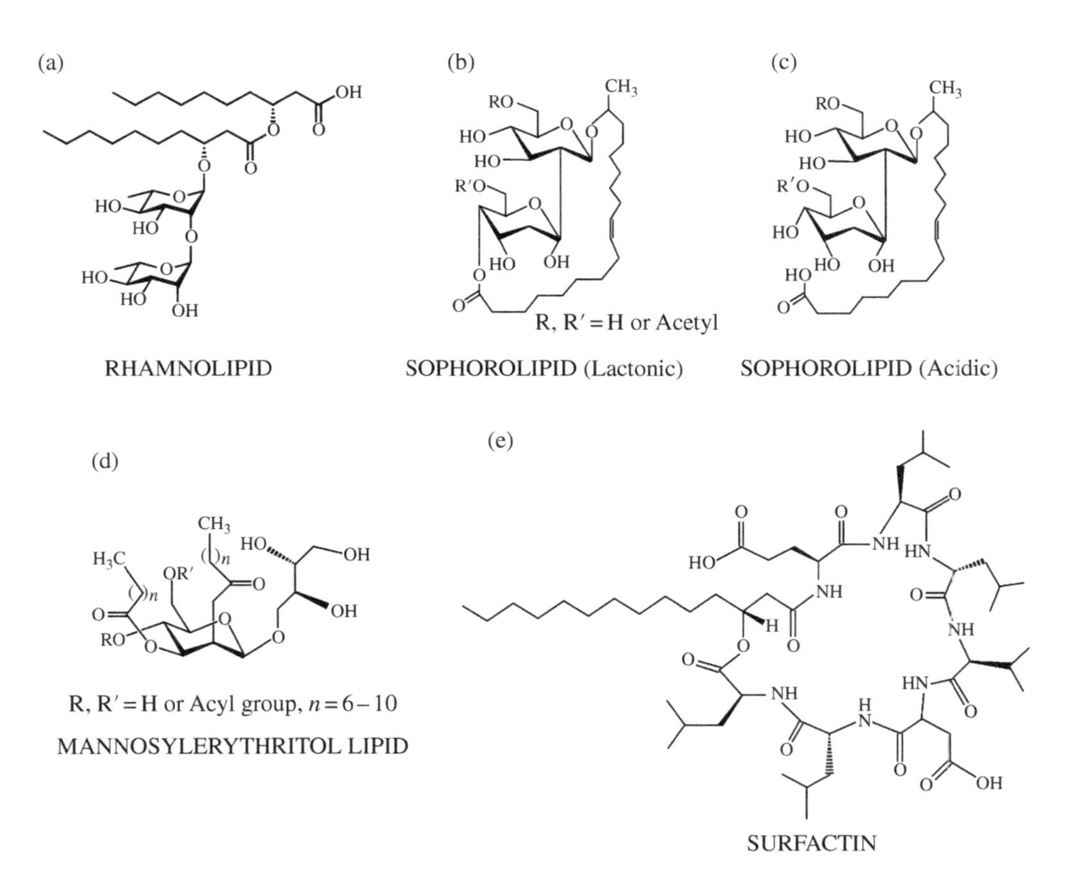

Figure 20.2 Structures of some of the most well-studied biosurfactants relevant for skin and haircare applications.

the applicability of RL nanoparticles for dermal drug delivery loaded (up to 30% drug loading) with hydrophobic drug molecules (dexamethasone and tacrolimus) used in the treatment of skin conditions like psoriasis. Based on the fluorescence microscopic results of skin samples with a fluorophore Nile red, they found efficient penetration of the model drugs into human skin, especially in the SC. Also, the RLs exhibited no toxicity toward primary human fibroblasts and were thus considered suitable for the dermal delivery of drugs. In another study, the properties of a di-RL regarding the process of cutaneous wound healing on full-thickness burn wounds were reported [100]. The therapeutic efficacy of the di-RL ointment (0.1%) was evaluated *in vivo* on Sprague–Dawley rats for 45 days. The results indicated the wound healing potential of the di-RL with accelerated closure of wounds by the end of the treatment period. Histologic analyses revealed a significantly lower collagen content in the burn wounds (47.5%) treated with di-RL as compared to the control animals. Piljac and Piljac [99] formulated a di-RL ointment (BAC-3) for wound healing (1%) and a cosmetic composition useful for the treatment of the signs of aging. DeSanto [101] patented a formulation comprising 70% RL and 30% water as a hair wash to keep scalp odorless and stimulate hair growth in bald spots. Furthermore, the patent also mentions a body wash comprising RL to treat certain dry skin conditions and also to act as an antimicrobial agent.

Sophorolipids (SLs) SLs are produced mainly by yeasts, such as *C. bombicola* (or *Torulopsis bombicola*), *C. apicola*, and *Rhodotorula bogoriensis*. SLs contain sophorose, the disaccharide comprising two glucose residues linked by a β-1,2′ bond as the hydrophilic component and a terminal or sub-terminal hydroxylated fatty acid as the hydrophobic part (Figure 20.2). The sophorose can be optionally acetylated on the 6′ and/or 6′′ position. The hydroxylated fatty acid, linked β-glycosidically to the sophorose, determines the chemical structure and nature of the SL. When the carboxylic end of the fatty acid is free, the SL exhibits an acidic structure (2c). In contrast, the esterification of the carboxylic end at the 4′′ position gives rise to the lactonic ring structure (2b). SLs can exist in the form of lactones both in monomeric or in dimeric forms [102]. SLs are produced by non-pathogenic yeasts in contrast to RLs, where the most efficient producer is the opportunistic pathogen *P. aeruginosa*. SLs are multifunctional and have several applications in the cosmetic industry. SLs are known to exhibit good compatibility with skin and hair, and excellent moisturizing properties [103]. The use of one unprocessed/acid SL or SL associated with a monovalent or divalent salt in a concentration of 0.01–30% as an antioxidant, antielastasic, and anti-inflammatory agent, in an emulsified composition for skin or hair (shampoos, moisturizers, shower gels, etc.) was described in a patent [104]. Another patent (WO2004108063A2), described a new cosmetic composition containing SLs (0.01–5%) in conjunction with a lipolytic agent (cAMP or its derivatives) as slimming agents and/or active agents stimulating the leptin synthesis through adipocytes, in the manufacture of a cosmetic composition for reducing the subcutaneous fat overload. Maingault [104] documented the use of salts of the acidic form and the ester of the deacetylated SLs in therapeutics as an activator of macrophages, a fibrinolytic agent, a healing agent, a desquamating agent, and a depigmenting agent. SLs are also present in cosmetic products like lip rouge, lip cream, eye shadow, and compressed powders [105], and are also used to treat acne [85]. Another exciting activity of SL relevant to the cosmetic and pharmaceutical industry is the enhancement of transdermal delivery by 1.3-fold to 1.7-fold of bovine lactoferrin through a model skin in a synergistic action. Lactoferrin is a multifunctional glycoprotein known to activate dermal fibroblasts, which increases cell proliferation and gene expression levels of collagen IV and hyaluronan synthases [106]. Cox and co-workers [107] prepared a mild formulation suitable for use as shower gels and shampoo using an SL (1–20% w/w) in combination with an anionic surfactant (0–10%), foam boosting surfactant, and 40–98% of water. Another study reported the application of an SL produced by fermentation of horse oil with elastase inhibition with antiaging and anti-inflammatory properties for cosmetic applications. The SL also exhibited low cytotoxicity on fibroblasts when analyzed using MTT assay [108]. Cytotoxicity of SLs on human epidermal keratinocytes was also reported to be low [38].

Mannosylerythritol Lipids (MELs) MELs are glycolipid biosurfactants containing 4-*O*-β-d-mannopyranosyl-meso-erythritol as the hydrophilic moiety and fatty acids as hydrophobic moieties (Figure 20.2e). They were recently described as one of the most promising biosurfactants with extensive applications in cosmetics due to their excellent surface activity. Kitagawa and co-workers [109] patented formulations of cosmetic pigments coated with MEL as a replacement of silicone or fluorine compounds for use in makeup products such as foundations, eyeshadow, and blushers or for basic skincare such as sunscreen products, emulsions, and creams. The coating improved the dispersibility and water-resistance while providing a hydrating effect. The moisturizing effects of MEL-A, similar to ceramide on SDS damaged skin, were reported in a study using a three-dimensional human skin model [110]. On treating the damaged cells with a 10% solution of MEL-A, the authors observed a recovery rate of 91%. MEL-B possesses the ability to suppress the perspiration, which was evident from the increase in the water content of the SC and also showed moisturizing activity by maintaining the skin barrier integrity [111]. Cosmetic cleansing agents containing (1–20% by weight) one or more biosurfactants such as MEL, RL, SL with one or more of the anionic surfactants (1–10% by weight) was patented for a prebiotic effect to treat acne [112]. Additionally, under experimental conditions, a skin whitening composition containing MEL (0.1–20% by weight) as an active ingredient was patented [113]. The composition was claimed to suppresses the formation of skin melanocytes and improve the overall skin tone. Morita and co-workers [114] investigated the hair-care properties of MEL-A and B on damaged hair. They observed that apart from the dramatic improvement of the microscopic appearance of the damaged hair, the MEL treatment also increased the tensile strength and lowered the friction coefficient indicating smoother and soft hair. The effects were comparable to synthetic ceramides and significantly better than synthetic surfactant lauryl glucoside. In another study, MEL-A at 5 μg/l increased the viability of fibroblasts by 150% and stimulated papilla cells when cultured in a medium containing MEL-A. Papilla cells are mesenchymal cells known to stimulate follicle formation and hair growth by trans-differentiation of the dermal epidermis. Stimulation of papilla cells could be effective in treating alopecia [114].

20.3.5.2 Lipopeptides (LPs)

LPs are amphiphilic compounds with hydrophilic cyclic peptides conjugated to hydrophobic fatty acids, most commonly produced by *Bacillus* sp. Based on their structures, LPs produced by *Bacillus* sp. are mainly classified into three groups: iturins, fengycins, and surfactins. Fengycin contains 10 amino acids whereas both iturin and surfactin contain seven amino acids [115] and are bonded with *β*-hydroxy fatty acids that are arranged in linear (*n*), *iso*, and *anteiso* forms [116]. Iturins are cyclopeptides composed of C_{14}–C_{17} amino fatty acids and heptapeptides. Fengycins, on the other hand, contain hydroxy fatty acids and decapeptides, which form cyclic lactone rings. Surfactins, the most well-known LP, comprise cyclic lactone rings of C_{13}–C_{16} hydroxy fatty acids and heptapeptides (Figure 20.2). LPs have found several applications in the environmental and pharmaceutical sectors. LPs are emulsifiers and a low CMC would be beneficial for usage over the synthetic counterparts [117]. LPs have been claimed to be suitable for external skin preparations, such as transparent cosmetics with sequestering properties and low skin irritation [118]. Cleansing cosmetics containing LPs (0.1–5% by mass) and polyoxyethylene glyceryl ether fatty acid ester and/or a polyoxyethylene sorbit fatty acid ester (0.1–20% by mass) show excellent washability and low skin irritation [119]. An O/W emulsion comprising of LP, xanthan gum, oil component, and water finds usage as a cosmetic compound with properties like moisture retention and stability [120]. The cosmetic, with a negligible skin-irritating property, ensures high skin comfort and stability. Some LPs have been used in whitening cosmetics to deliver a melanocyte-stimulating hormone [121]. Surfactin has also been reported to interact with lipopolysaccharides, resulting in the suppression of lipid transportation. Therefore, the incorporation of surfactin into anticellulite products could prove to be useful [122]. LPs possess antibacterial and antifungal activity [123] and, therefore, have also been used for the treatment and prevention of microbial infections and product preservation [124, 125].

20.4 Conclusion

The development of natural compounds for personal care products is gaining importance, mainly driven by the potential benefits attributed to them. Environmental and health impacts of cosmetic formulations are other vital factors that have drawn consumer interest recently. In this regard, biosurfactants have been introduced in cosmetic formulations as "greener alternatives" to their synthetic counterparts because of their biocompatibility, lack of toxicity, and dermocosmetic properties. Due to their biological (antimicrobial, antioxidant, and anti-inflammatory) activities, biosurfactants offer additional benefits beyond cleaning action and can be used as substances for self-preservation of cosmetics. However, the main concern regarding the commercial application of biosurfactants is the lack of an economically viable production and purification process on an industrial scale. In this regard, process optimization, genetic engineering of producer strains, and selection of low-cost substrates would be feasible to obtain a better yield.

References

1 Sajna, K.V., Gottumukkala, L.D., Sukumaran, R.K., and Pandey, A. (2015). White biotechnology in cosmetics. In: *Industrial Biorefineries and White Biotechnology* (eds. A. Pandey, R. Höfer, M. Taherzadeh, et al.), 607–652. Elsevier, Waltham, MA, USA.

2 Vecino, X., Cruz, J., Moldes, A., and Rodrigues, L. (2017). Biosurfactants in cosmetic formulations: trends and challenges. *Crit. Rev. Biotechnol.* 37 (7): 911–923.

3 Lourith, N. and Kanlayavattanakul, M. (2009). Natural surfactants used in cosmetics: Glycolipids. *Int. J. Cosmet. Sci.* 31 (4): 255–261.

4 Lee, S., Lee, J., Yu, H., and Lim, J. (2018). Synthesis of environment friendly biosurfactants and characterization of interfacial properties for cosmetic and household products formulations. *Colloids Surf. A Physicochem. Eng. Asp.* 536: 224–233.

5 Williams, A.C. and Barry, B.W. (2012). Penetration enhancers. *Adv. Drug Deliv. Rev.* 64: 128–137.

6 Bujak, T., Wasilewski, T., and Nizioł-Łukaszewska, Z. (2015). Role of macromolecules in the safety of use of body wash cosmetics. *Colloids Surf. B Biointerfaces* 135: 497–503.

7 Rieger, M. (2017). *Surfactants in Cosmetics*. Routledge.

8 Corley, J. (2007). All that is good: Naturals and their place in personal care. In: *Naturals and Organics in Cosmetics: From R & D to the Market Place*, 7–12. Carol Stream, IL, USA: Allured Publishing Corp.

9 Marchant, R. and Banat, I.M. (2012). Microbial biosurfactants: Challenges and opportunities for future exploitation. *Trends Biotechnol.* 30 (11): 558–565.

10 Borah, S.N., Goswami, D., Sarma, H.K. et al. (2016). Rhamnolipid biosurfactant against *Fusarium verticillioides* to control stalk and ear rot disease of maize. *Front. Microbiol.* 7: 1505.

11 Akbari, S., Abdurahman, N.H., Yunus, R.M. et al. (2018). Biosurfactants – A new frontier for social and environmental safety: A mini review. *Biotechnol. Res. Innov.* 2 (1): 81–90.

12 Henkel, M., Müller, M.M., Kügler, J.H. et al. (2012). Rhamnolipids as biosurfactants from renewable resources: Concepts for next-generation rhamnolipid production. *Process Biochem.* 47 (8): 1207–1219.

13 Varvaresou, A. and Iakovou, K. (2015). Biosurfactants in cosmetics and biopharmaceuticals. *Lett. Appl. Microbiol.* 61 (3): 214–223.

14 Kumar, A.S., Mody, K., and Jha, B. (2007). Evaluation of biosurfactant/bioemulsifier production by a marine bacterium. *Bull. Environ. Contam. Toxicol.* 79 (6): 617–621.

15 Ron, E.Z. and Rosenberg, E. (2001). Natural roles of biosurfactants: Minireview. *Environ. Microbiol.* 3 (4): 229–236.

16 Van Dyke, M.I., Gulley, S.L., Lee, H., and Trevors, J.T. (1993). Evaluation of microbial surfactants for recovery of hydrophobic pollutants from soil. *J. Ind. Microbiol.* 11 (3): 163–170.

17 Randhawa, K.K.S. and Rahman, P.K. (2014). Rhamnolipid biosurfactants – Past, present, and future scenario of global market. *Front. Microbiol.* 5: 454.

18 Vater, J., Wilde, C., and Kell, H. (2009). In situ detection of the intermediates in the biosynthesis of surfactin, a lipoheptapeptide from *Bacillus subtilis* OKB 105, by whole-cell cell matrix-assisted laser desorption/ionization time-of-flight mass spectrometry in combination with mutant analysis. *Rapid Commun. Mass Spectrom.* 23 (10): 1493–1498.

19 de Faria, A.F., Teodoro-Martinez, D.S., de Oliveira Barbosa, G.N. et al. (2011). Production and structural characterization of surfactin (C14/Leu7) produced by *Bacillus subtilis* isolate LSFM-05 grown on raw glycerol from the biodiesel industry. *Process Biochem.* 46 (10): 1951–1957.

20 Jacques, P. (2011). Surfactin and other lipopeptides from *Bacillus* spp. In: *Biosurfactants* (ed. G. Soberón-Chávez), 57–91. Springer.

21 Slivinski, C.T., Mallmann, E., de Araújo, J.M. et al. (2012). Production of surfactin by *Bacillus pumilus* UFPEDA 448 in solid-state fermentation using a medium based on okara with sugarcane bagasse as a bulking agent. *Process Biochem.* 47 (12): 1848–1855.

22 Mukherjee, S., Das, P., and Sen, R. (2006). Towards commercial production of microbial surfactants. *Trends Biotechnol.* 24 (11): 509–515.

23 Ito, S. and Inoue, S. (1982). Sophorolipids from *Torulopsis bombicola*: Possible relation to alkane uptake. *Appl. Environ. Microbiol.* 43 (6): 1278–1283.

24 Garcıa-Ochoa, F. and Casas, J. (1999). Unstructured kinetic model for sophorolipid production by *Candida bombicola*. *Enzyme Microb. Technol.* 25 (7): 613–621.

25 Daniel, H.-J., Reuss, M., and Syldatk, C. (1998). Production of sophorolipids in high concentration from deproteinized whey and rapeseed oil in a two stage fed batch process using *Candida bombicola* ATCC 22214 and *Cryptococcus curvatus* ATCC 20509. *Biotechnol. Lett.* 20 (12): 1153–1156.

26 Rodrigues, L.R. (2015). Microbial surfactants: fundamentals and applicability in the formulation of nano-sized drug delivery vectors. *J. Colloid Interface Sci.* 449: 304–316.

27 Satpute, S.K., Banpurkar, A.G., Dhakephalkar, P.K. et al. (2010). Methods for investigating biosurfactants and bioemulsifiers: A review. *Crit. Rev. Biotechnol.* 30 (2): 127–144.

28 Varjani, S.J. and Upasani, V.N. (2017). Critical review on biosurfactant analysis, purification and characterization using rhamnolipid as a model biosurfactant. *Bioresour. Technol.* 232: 389–397.

29 Anyanwu, C., Obi, S., and Okolo, B. (2011). Lipopeptide biosurfactant production by *Serratia marcescens* NSK-1 strain isolated from petroleum-contaminated soil. *J. Appl. Sci. Res.* 7: 79–87.

30 Lan, G., Fan, Q., Liu, Y. et al. (2015). Effects of the addition of waste cooking oil on heavy crude oil biodegradation and microbial enhanced oil recovery using *Pseudomonas* sp. SWP-4. *Biochem. Eng. J.* 103: 219–226.

31 Liu, Q., Lin, J., Wang, W. et al. (2015). Production of surfactin isoforms by *Bacillus subtilis* BS-37 and its applicability to enhanced oil recovery under laboratory conditions. *Biochem. Eng. J.* 93: 31–37.

32 Arima, K. (1968). Surfactin, acrystalline peptidelipid surfactant produced by *Bacillus subtilis*: Isolation, characterization and its inhibition of fibrin clot formation. *Biochem. Biophys. Res. Commun.* 31: 488–494.

33 Bade, R. and Lee, S.H. (2011). A review of studies on micellar enhanced ultrafiltration for heavy metals removal from wastewater. *J Water Sustain.* 1: 85–102.

34 Hait, S.K. and Moulik, S.P. (2001). Determination of critical micelle concentration (CMC) of nonionic surfactants by donor-acceptor interaction with lodine and correlation of CMC with hydrophile-lipophile balance and other parameters of the surfactants. *J. Surfactant Deterg.* 4 (3): 303–309.

35 Gudiña, E.J., Rangarajan, V., Sen, R., and Rodrigues, L.R. (2013). Potential therapeutic applications of biosurfactants. *Trends Pharmacol. Sci.* 34 (12): 667–675.

36 Williams, S. and Schmitt, W. (2012). *Chemistry and Technology of the Cosmetics and Toiletries Industry*. Springer Science & Business Media, New York, USA.

37 Fukuoka, T., Morita, T., Konishi, M. et al. (2007). Characterization of new glycolipid biosurfactants, tri-acylated mannosylerythritol lipids, produced by *Pseudozyma* yeasts. *Biotechnol. Lett.* 29: 1111–1118. Available at: https://doi.org/10.1007/s10529-007-9363-0.

38 Hirata, Y., Ryu, M., Oda, Y. et al. (2009). Novel characteristics of sophorolipids, yeast glycolipid biosurfactants, as biodegradable low-foaming surfactants. *J. Biosci. Bioeng.* 108 (2): 142–146.

39 Razafindralambo, H., Paquot, M., Baniel, A. et al. (1996). Foaming properties of surfactin, a lipopeptide biosurfactant from *Bacillus subtilis*. *J. Am. Oil Chem. Soc.* 73 (1): 149–151.

40 Chayabutra, C., Wu, J., and Ju, L.K. (2001). Rhamnolipid production by *Pseudomonas aeruginosa* under denitrification: Effects of limiting nutrients and carbon substrates. *Biotechnol. Bioeng.* 72 (1): 25–33.

41 Iwai, I., Han, H., Den Hollander, L. et al. (2012). The human skin barrier is organized as stacked bilayers of fully extended ceramides with cholesterol molecules associated with the ceramide sphingoid moiety. *J. Invest. Dermatol.* 132 (9): 2215–2225.

42 Park K-D, Lim C-J, Yoon S-J, Kwon S-D. Peptide analogues with an excellent moisturizing effect and use thereof. Google Patents; 2016.

43 Lemery, E., Briançon, S., Chevalier, Y. et al. (2015). Surfactants have multi-fold effects on skin barrier function. *Eur. J. Dermatol.* 25 (5): 424–435.

44 Zana, R. (2005). *Dynamics of Surfactant Self-Assemblies: Micelles, Microemulsions, Vesicles and Lyotropic Phases*. Boca Raton, FL, USA: CRC Press.

45 Downing, D., Abraham, W., Wegner, B. et al. (1993). Partition of sodium dodecyl sulfate into stratum corneum lipid liposomes. *Arch. Dermatol. Res.* 285 (3): 151–157.

46 Ghosh, S., Kim, D., So, P., and Blankschtein, D. (2008). Visualization and quantification of skin barrier perturbation induced by surfactant-humectant systems using two-photon fluorescence microscopy. *J. Cosmet. Sci.* 59 (4): 263–289.

47 Gloor, M., Senger, B., Langenauer, M., and Fluhr, J.W. (2004). On the course of the irritant reaction after irritation with sodium lauryl sulphate. *Skin Res. Technol.* 10 (3): 144–148.

48 De Jongh, C.M., Verberk, M.M., Spiekstra, S.W. et al. (2007). Cytokines at different stratum corneum levels in normal and sodium lauryl sulphate-irritated skin. *Skin Res. Technol.* 13 (4): 390–398.

49 Barba, C., Semenzato, A., Baratto, G., and Coderch, L. (2018). Action of surfactants on the mammal epidermal skin barrier. *G. Ital. Dermatol. Venereol.* 154 (4): 405–412.

50 James-Smith, M.A., Hellner, B., Annunziato, N., and Mitragotri, S. (2011). Effect of surfactant mixtures on skin structure and barrier properties. *Ann. Biomed. Eng.* 39 (4): 1215–1223.

51 Knoth, D., Rincón-Fontán, M., Stahr, P.-L. et al. (2019). Evaluation of a biosurfactant extract obtained from corn for dermal application. *Int. J. Pharm.* 564: 225–236.

52 Fujimura, T., Shimotoyodome, Y., Nishijima, T. et al. (2017). Changes in hydration of the stratum corneum are the most suitable indicator to evaluate the irritation of surfactants on the skin. *Skin Res. Technol.* 23 (1): 97–103.

53 Kitagawa M, Suzuki M, Yamamoto S, Sogabe A, Kitamoto D, Imura T, et al. Biosurfactant-containing skin care cosmetic and skin roughness-improving agent. Google Patents; 2010.

54 Van der Heyden, L. and Adachi, M. (2000). Ceramides for hair protection and conditioning. *Fragr. J.* 28 (6): 61–64.

55 Vecino, X., Rodríguez-López, L., Ferreira, D. et al. (2018). Bioactivity of glycolipopeptide cell-bound biosurfactants against skin pathogens. *Int. J. Biol. Macromol.* 109: 971–979.

56 Morita, T., Kitagawa, M., Yamamoto, S. et al. (2010). Glycolipid biosurfactants, mannosylerythritol lipids, repair the damaged hair. *J. Oleo Sci.* 59 (5): 267–272.

57 Walters, R.M., Mao, G., Gunn, E.T., and Hornby, S. (2012). Cleansing formulations that respect skin barrier integrity. *Dermatol. Res. Pract.* 2012: 495917.

58 Sinha, V. and Kaur, M.P. (2000). Permeation enhancers for transdermal drug delivery. *Drug Dev. Ind. Pharm.* 26 (11): 1131–1140.

59 Waters, L.J., Dennis, L., Bibi, A., and Mitchell, J.C. (2013). Surfactant and temperature effects on paraben transport through silicone membranes. *Colloids Surf. B Biointerfaces* 108: 23–28.

60 Sindhu, R.K., Chitkara, M., Kaur, G. et al. (2017). Skin penetration enhancer's in transdermal drug delivery systems. *Res. J. Pharm. Technol.* 10 (6): 1809–1815.

61 Alonso, C., Lucas, R., Barba, C. et al. (2015). Skin delivery of antioxidant surfactants based on gallic acid and hydroxytyrosol. *J. Pharm. Pharmacol.* 67 (7): 900–908.

62 Waters, L.J., Finch, C.V., Bhuiyan, A.M.H. et al. (2017). Effect of plasma surface treatment of poly (dimethylsiloxane) on the permeation of pharmaceutical compounds. *J. Pharm. Anal.* 7 (5): 338–342.

63 Rodríguez-López, L., Shokry, D.S., Cruz, J.M. et al. (2019). The effect of the presence of biosurfactant on the permeation of pharmaceutical compounds through silicone membrane. *Colloids Surf. B Biointerfaces* 176: 456–461.

64 Jemil, N., Ayed, H.B., Manresa, A. et al. (2017). Antioxidant properties, antimicrobial and anti-adhesive activities of DCS1 lipopeptides from *Bacillus methylotrophicus* DCS1. *BMC Microbiol.* 17 (1): 144.

65 Giri, S., Ryu, E., Sukumaran, V., and Park, S. (2019). Antioxidant, antibacterial, and anti-adhesive activities of biosurfactants isolated from *Bacillus* strains. *Microb. Pathog.* 132: 66–72.

66 Merghni, A., Dallel, I., Noumi, E. et al. (2017). Antioxidant and antiproliferative potential of biosurfactants isolated from *Lactobacillus casei* and their anti-biofilm effect in oral *Staphylococcus aureus* strains. *Microb. Pathog.* 104: 84–89.

67 Rincón-Fontán, M., Rodríguez-López, L., Vecino, X. et al. (2018). Design and characterization of greener sunscreen formulations based on mica powder and a biosurfactant extract. *Powder Technol.* 327: 442–448.

68 Takahashi, M., Morita, T., Fukuoka, T. et al. (2012). Glycolipid biosurfactants, mannosylerythritol lipids, show antioxidant and protective effects against H_2O_2-induced oxidative stress in cultured human skin fibroblasts. *J. Oleo Sci.* 61 (8): 457–464.

69 Sidhu, G.S., Mani, H., Gaddipati, J.P. et al. (1999). Curcumin enhances wound healing in streptozotocin induced diabetic rats and genetically diabetic mice. *Wound Repair Regen.* 7 (5): 362–374.

70 Naik, H.R.P., Naik, H.S.B., Naik, T.R.R. et al. (2009). Synthesis of novel benzo [h] quinolines: Wound healing, antibacterial, DNA binding and in vitro antioxidant activity. *Eur. J. Med. Chem.* 44 (3): 981–989.

71 Sana, S., Datta, S., Biswas, D. et al. (2018). Excision wound healing activity of a common biosurfactant produced by *Pseudomonas* sp. *Wound Med.* 23: 47–52.

72 Lydon, H.L., Baccile, N., Callaghan, B. et al. (2017). Adjuvant antibiotic activity of acidic sophorolipids with potential for facilitating wound healing. *Antimicrob. Agents Chemother.* 61: e02547–e02516.

73 Gupta, S., Raghuwanshi, N., Varshney, R. et al. (2017). Accelerated in vivo wound healing evaluation of microbial glycolipid containing ointment as a transdermal substitute. *Biomed. Pharmacother.* 94: 1186–1196.

74 Sana, S., Mazumder, A., Datta, S., and Biswas, D. (2017). Towards the development of an effective in vivo wound healing agent from *Bacillus* sp. derived biosurfactant using Catla catla fish fat. *RSC Adv.* 7 (22): 13668–13677.

75 Ohadi, M., Forootanfar, H., Rahimi, H.R. et al. (2017). Antioxidant potential and wound healing activity of biosurfactant produced by *Acinetobacter junii* B6. *Curr. Pharm. Biotechnol.* 18 (11): 900–908.

76 Borah, S.N., Sen, S., Goswami, L. et al. (2019). Rice based distillers dried grains with solubles as a low cost substrate for the production of a novel rhamnolipid biosurfactant having anti-biofilm activity against *Candida tropicalis*. *Colloids Surf. B Biointerfaces* 182: 110358.

77 Abalos, A., Pinazo, A., Infante, M. et al. (2001). Physicochemical and antimicrobial properties of new Rhamnolipids produced by *Pseudomonas a eruginosa* AT10 from soybean oil refinery wastes. *Langmuir* 17 (5): 1367–1371.

78 Balboa, E.M., Conde, E., Soto, M.L. et al. (2015). Cosmetics from marine sources. In: *Springer Handbook of Marine Biotechnology* (ed. S.K. Kim), 1015–1042. Springer-Verlag, Heidelberg.

79 Mulligan, C., Gibbs, B., and Kosaric, N. (1993). *Biosurfactants: Production, Properties and Applications*. New York, USA: Marcel Dekker.

80 Sen, S., Borah, S.N., Kandimalla, R. et al. (2020). Sophorolipid biosurfactant can control cutaneous Dermatophytosis caused by Trichophyton mentagrophytes. *Front. Microbiol.* 11: 329.

81 Shah, V., Doncel, G.F., Seyoum, T. et al. (2005). Sophorolipids, microbial glycolipids with anti-human immunodeficiency virus and sperm-immobilizing activities. *Antimicrob. Agents Chemother.* 49 (10): 4093–4100.

82 Di Pasqua, R., Betts, G., Hoskins, N. et al. (2007). Membrane toxicity of antimicrobial compounds from essential oils. *J. Agric. Food Chem.* 55 (12): 4863–4870.

83 Luna, J., Rufino, R., Campos, G., and Sarubbo, L. (2012). Properties of the biosurfactant produced by *Candida sphaerica* cultivated in low-cost substrates. *Chem. Eng.* 27: 67–72.

84 Naruse, N., Tenmyo, O., Kobaru, S. et al. (1990). Pumilacidin, a complex of new antiviral antibiotics. *J. Antibiot.* 43 (3): 267–280.

85 Ashby, R.D., Zerkowski, J.A., Solaiman, D.K., and Liu, L.S. (2011). Biopolymer scaffolds for use in delivering antimicrobial sophorolipids to the acne-causing bacterium *Propionibacterium acnes*. *N. Biotechnol.* 28 (1): 24–30.

86 Pierce D, Heilman TJ. Germicidal composition. Google Patents; 2001.

87 Sen, S., Borah, S.N., Kandimalla, R. et al. (2019). Efficacy of a rhamnolipid biosurfactant to inhibit *Trichophyton rubrum* in vitro and in a mice model of dermatophytosis. *Exp. Dermatol.*

88 De Rienzo, M.D., Stevenson, P., Marchant, R., and Banat, I. (2016). Effect of biosurfactants on *Pseu*domonas aeruginosa and Staphylococcus aureus biofilms in a BioFlux channel. Appl. Microbiol. Biote*chnol.* 100 (13): 5773–5779.

89 Haba, E., Pinazo, A., Jauregui, O. et al. (2003). Physicochemical characterization and antimicrobial properties of rhamnolipids produced by *Pseudomonas aeruginosa* 47T2 NCBIM 40044. *Biotechnol. Bioeng.* 81 (3): 316–322.

90 Joshi-Navare, K. and Prabhune, A. (2013). A biosurfactant-sophorolipid acts in synergy with antibiotics to enhance their efficiency. *Biomed. Res. Int.* 2013: 512495.

91 Janek, T., Krasowska, A., Czyżnikowska, Ż., and Łukaszewicz, M. (2018). Trehalose lipid biosurfactant reduces adhesion of microbial pathogens to polystyrene and silicone surfaces: An experimental and computational approach. *Front. Microbiol.* 9: 2441.

92 Gudina, E.J., Teixeira, J.A., and Rodrigues, L.R. (2010). Isolation and functional characterization of a biosurfactant produced by *Lactobacillus paracasei*. *Colloids Surf. B Biointerfaces* 76 (1): 298–304.

93 Gudiña, E.J., Fernandes, E.C., Teixeira, J.A., and Rodrigues, L.R. (2015). Antimicrobial and anti-adhesive activities of cell-bound biosurfactant from *Lactobacillus agilis* CCUG31450. *RSC Adv.* 5 (110): 90960–90968.

94 Fernández-Peña, L., Guzmán, E., Leonforte, F. et al. (2020). Effect of molecular structure of eco-friendly glycolipid biosurfactants on the adsorption of hair-care conditioning polymers. *Colloids Surf. B Biointerfaces* 185: 110578.

95 D'Souza, P. and Rathi, S.K. (2015). Shampoo and conditioners: What a dermatologist should know? *Indian J. Dermatol.* 60 (3): 248.

96 Rincón-Fontán, M., Rodríguez-López, L., Vecino, X. et al. (2020). Novel multifunctional biosurfactant obtained from corn as a stabilizing agent for antidandruff formulations based on Zn Pyrithione powder. *ACS Omega* 5 (11): 5704–5712.

97 Borah, S.N., Goswami, D., Lahkar, J. et al. (2015). Rhamnolipid produced by *Pseudomonas aeruginosa* SS14 causes complete suppression of wilt by *Fusarium oxysporum* f. sp. pisi in *Pisum sativum*. *BioControl* 60 (3): 375–385.

98 Müller, F., Hönzke, S., Luthardt, W.-O. et al. (2017). Rhamnolipids form drug-loaded nanoparticles for dermal drug delivery. *Eur. J. Pharm. Biopharm.* 116: 31–37. Available at: https://doi.org/10.1016/j.ejpb.2016.12.013.

99 Piljac T, Piljac G. Use of rhamnolipids in wound healing, treating burn shock, atherosclerosis, organ transplants, depression, schizophrenia and cosmetics. European Patent 1 889 623. Paradigm Biomedical Inc., New York. 1999.

100 Stipcevic, T., Piljac, A., and Piljac, G. (2006). Enhanced healing of full-thickness burn wounds using di-rhamnolipid. *Burns* 32 (1): 24–34.

101 DeSanto K. Rhamnolipid-based formulations. Google Patents; 2012.

102 Sen, S., Borah, S.N., Bora, A., and Deka, S. (2017). Production, characterization, and antifungal activity of a biosurfactant produced by *Rhodotorula babjevae* YS3. *Microb. Cell Fact.* 16 (1): 95.

103 Mager H, Rothlisberger R, Wagner F. Use of sophorose-lipid lactone for the treatment of dandruffs and body odor. European Patent 0 209 783. Institut Francais du Petrole, Malmaison. 1987.

104 Maingault M. Utilization of sophorolipids as therapeutically active substances or cosmetic products, in particular for the treatment of the skin. US Patent US5981497A. Institut Francais Du Petrole, Rueil-Malmaison. 1996.

105 Kawano J, Utsugi T, Inoue S, Hayashi S. Powdered compressed cosmetic material. Google Patents; 1981.

106 Ishii, N., Kobayashi, T., Matsumiya, K. et al. (2012). Transdermal administration of lactoferrin with sophorolipid. *Biochem. Cell Biol.* 90 (3): 504–512.

107 Cox TF, Crawford RJ, Gregory LG, Hosking SL, Kotsakis P. Mild to the skin, foaming detergent composition. Google Patents; 2013.

108 Maeng, Y., Kim, K.T., Zhou, X. et al. (2018). A novel microbial technique for producing high-quality sophorolipids from horse oil suitable for cosmetic applications. *J. Microbial. Biotechnol.* 11 (5): 917–929.

109 Kitagawa M, Nishimoto K, Tanaka T. Cosmetic pigments, their production method, and cosmetics containing the cosmetic pigments. Google Patents; 2015.

110 Morita, T., Konishi, M., Fukuoka, T. et al. (2006). Discovery of Pseudozyma rugulosa NBRC 10877 as a novel producer of the glycolipid biosurfactants, mannosylerythritol lipids, based on rDNA sequence. *Appl. Microbiol. Biotechnol.* 73 (2): 305.

111 Yamamoto, S., Morita, T., Fukuoka, T. et al. (2012). The moisturizing effects of glycolipid biosurfactants, mannosylerythritol lipids, on human skin. *J. Oleo Sci.* 61 (7): 407–412.

112 Schelges H, Tretyakova M, Ludwig B. Cleansing agents containing biosurfactants and having prebiotic activity. Google Patents; 2017.

113 Yoo JW, Hwang YK, Sung-Ah B, Kim YJ, Lee JH. Skin whitening composition containing mannosylerythritol lipid. Google Patents; 2019.

114 Morita, T., Kitagawa, M., Yamamoto, S. et al. (2010). Activation of fibroblast and papilla cells by glycolipid biosurfactants, mannosylerythritol lipids. *J. Oleo Sci.* 59 (8): 451–455.

115 Kanlayavattanakul, M. and Lourith, N. (2010). Lipopeptides in cosmetics. *Int. J. Cosmet. Sci.* 32 (1): 1–8.

116 Quinn, G.A., Maloy, A.P., McClean, S. et al. (2012). Lipopeptide biosurfactants from *Paenibacillus polymyxa* inhibit single and mixed species biofilms. *Biofouling* 28 (10): 1151–1166.

117 Pruthi, V. and Cameotra, S.S. (1997). Production of a biosurfactant exhibiting excellent emulsification and surface active properties by *Serratia marcescens*. *World J. Microbiol. Biotechnol.* 13 (1): 133–135.

118 Yoneda T, Masatsuji E, Tsuzuki T. Surfactant for use in external preparations for skin and external preparation for skin containing the same. World Patent WO62482 1999.

119 Yoneda T. Cosmetic composition comprising a lipopeptide. Google Patents; 2006.

120 Yoneda T, Ito N, Furuya K. Oil-in-water emulsified composition, and external preparation for skin and cosmetics ssing the composition. Google Patents; 2008.

121 Ogawa, Y., Kawahara, H., Yagi, N. et al. (1999). Synthesis of a novel lipopeptide with α-melanocyte-stimulating hormone peptide ligand and its effect on liposome stability. *Lipids* 34 (4): 387–394.

122 Takahashi, T., Ohno, O., Ikeda, Y. et al. (2006). Inhibition of lipopolysaccharide activity by a bacterial cyclic lipopeptide surfactin. *J. Antibiot.* 59 (1): 35–43.

123 Mandal, S.M., Sharma, S., Pinnaka, A.K. et al. (2013). Isolation and characterization of diverse antimicrobial lipopeptides produced by *Citrobacter* and *Enterobacter*. *BMC Microbiol.* 13 (1): 152.

124 Deleu, M., Razafindralambo, H., Popineau, Y. et al. (1999). Interfacial and emulsifying properties of lipopeptides from *Bacillus subtilis*. *Colloids Surf. A Physicochem. Eng. Asp.* 152 (1–2): 3–10.

125 Mandal, S.M., Barbosa, A.E., and Franco, O.L. (2013). Lipopeptides in microbial infection control: scope and reality for industry. *Biotechnol. Adv.* 31 (2): 338–345.

21

Biosurfactants as Biocontrol Agents Against Mycotoxigenic Fungi

Ana I. Rodrigues, Eduardo J. Gudiña, José A. Teixeira, and Lígia R. Rodrigues

CEB – Centre of Biological Engineering, University of Minho, Braga, Portugal

CHAPTER MENU

21.1 Mycotoxins

Mycotoxins are toxic secondary metabolites naturally produced by different types of filamentous fungi mainly belonging to the *Aspergillus*, *Penicillium*, and *Fusarium* species. Mycotoxins contaminate different foodstuffs and agriculture commodities either before harvest or under post-harvest conditions [1–3]. These compounds are low molecular weight molecules with a wide variety of chemical structures capable of causing adverse effects in human and animal health. Additionally, mycotoxins have a negative impact on agriculture, causing huge economic losses [4–7]. Different factors, like optimal climate conditions (moderate temperature and humidity), as well as plants and seeds vulnerability to mycotoxigenic fungi contamination increase mycotoxin production. The awareness of these compounds began in the first half of the twentieth century when human and animal diseases were associated with the ingestion of mold damaged rice and wheat in Asian countries and overwintered millet in the USSR [3]. Currently, mycotoxicosis and different chronic

Biosurfactants for a Sustainable Future: Production and Applications in the Environment and Biomedicine, First Edition. Edited by Hemen Sarma and Majeti Narasimha Vara Prasad.
© 2021 John Wiley & Sons Ltd. Published 2021 by John Wiley & Sons Ltd.

adverse effects, like carcinogenic, mutagenic, and immunosuppressive activities, are associated with the ingestion of mycotoxins present in our daily diet [7–9]. Due to their chemical stability, mycotoxins can be present in different daily products included in the diet of a large percentage of the global population, such as, for example wine, fruit juices, hazelnuts, peanuts, almonds, pistachios, maize, rice, and other processed and unprocessed cereals [10, 11]. Over the years, most reports from the Food and Agriculture Organization (FAO) estimated that 25% of the world's crop harvests may be contaminated with mycotoxins [4, 12, 13]. However, according to Eskola and co-workers [10], this value could be underestimated and actually represent 60–80% of the world's crops. This increase in the contaminated percentages is mainly due to an improvement of mycotoxin detection methods and the impact that climate change has on agriculture commodities. Weather changes like the higher frequency of cyclones, droughts, and unseasonable rains at the time of flowering during harvest and post-harvest processes are factors that contribute to the increase in mycotoxigenic fungi contamination and, consequently, the mycotoxin content [3]. A specific agriculture commodity can be contaminated with different types of mycotoxigenic fungi and some strains are capable of producing more than one type of mycotoxin. Accordingly, there is a high probability of finding different types of mycotoxins in the final food product, which increases the interaction between mycotoxins and the occurrence of adverse synergistic effects [6, 14]. Although several hundreds of different mycotoxins have been identified, the most commonly observed mycotoxins that represent the major concerns to human and animal health include aflatoxins, deoxynivalenol (DON), fumonisins, ochratoxin A (OTA), patulin, and zearalenone (ZEN) [2, 15]. Hence, several countries established limits for their presence in food and feed.

21.2 Aflatoxins

Mycotoxins gained more attention when aflatoxins were identified in the early 1960s as a result of an epidemic of "Turkey X disease," in England, where more than 100 000 turkeys became ill and died after consumption of groundnut meal from Brazil [16]. *Aspergillus* species, mainly *Aspergillus flavus* and *Aspergillus parasiticus*, are known to produce aflatoxins and of being responsible for a large contamination of different agriculture commodities around the world, particularly in temperate and damp regions, thus representing a major concern in the agriculture industry [17–19]. Aflatoxins exhibit diverse toxic effects, such as carcinogenic, teratogenic, mutagenic, neurotoxic, and immunosuppressive activities, even when low concentrations are ingested [9, 19]. Aflatoxins are very stable molecules than can maintain their integrity at temperatures as high as 150 °C [20]. Aflatoxins are polyketide-derived secondary metabolites produced by a well-defined and conserved pathway that involves more than 30 genes and at least 23 enzymatic reactions [21]. Until the present year, approximately 20 different types of aflatoxins have been identified, the most well-known ones being the aflatoxins belonging to group B (aflatoxin B_1 [AFB_1] and aflatoxin B_2 [AFB_2]), group G (aflatoxin G_1 [AFG_1] and aflatoxin G_2 [AFG_2]), and group M (aflatoxin M_1 [AFM_1]) (Figure 21.1). AFB_1 is usually the major aflatoxin produced and is considered to be the most toxic and potent carcinogenic molecule, affecting mainly the liver and kidneys, and being classified as Group I carcinogenic by the International Agency for Research on Cancer (IARC) [9, 22]. Consequently, the levels of aflatoxins present in food and feed need to be regulated at the lowest possible levels. According to the Food and Drug Administration (FDA), the maximum acceptable levels for animal feeds can go from 20 to 300 ppb for corn and peanut products depending of the final purpose, whereas for human foods the levels are lower, going up to 20 ppb for foods, Brazil nuts, peanuts and peanut products, and pistachio nuts, and in the case of milk, 0.5 ppb for AFM_1 [23]. In the European Union, the European Commission regulations 1881/2006 set lower limits for

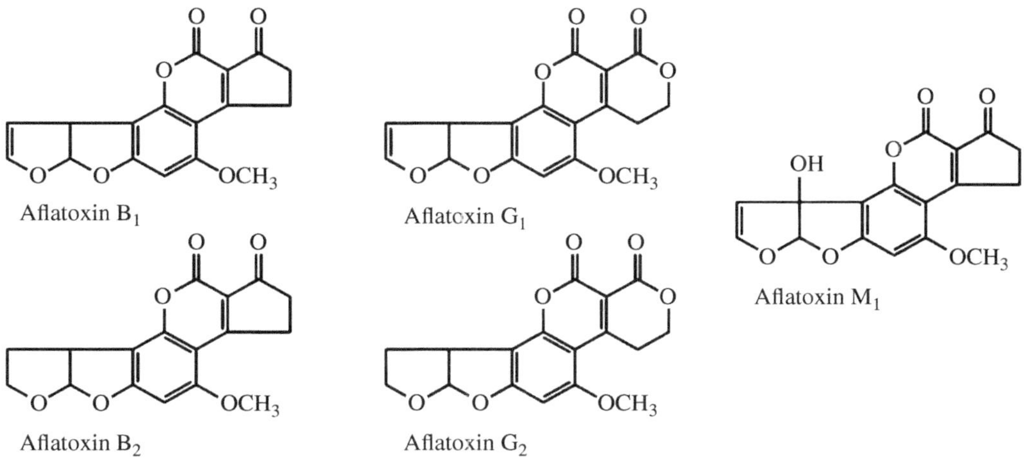

Figure 21.1 Chemical structure of aflatoxin B (AFB_1 and AFB_2), aflatoxin G (AFG_1 and AFG_2), and aflatoxin M_1.

aflatoxins when compared with the FDA. In the case of different types of cereals, the maximum levels can go from 5 to 8 ppb in the case of AFB_1 and 4 to 15 ppb in the case of the sum of AFB_1, AFB_2, AFG_1, and AFG_2. In heat-treated milk and milk-based products the limit is 0.050 ppb for AFM_1. For cereals incorporated in foods and dietary foods for special medical purposes and for babies and infants the set limits are 0.10 ppb for AFB_1 and 0.0025 ppb for AFM_1 [24].

21.3 Deoxynivalenol

Fusarium species are known to produce trichothecenes, a class of mycotoxin that can be divided into two types, A and B, according to their chemical structure (Figure 21.2). Each *Fusarium* strain usually produces only one type, and only one strain with the ability of producing both types has been identified [25]. DON, also routinely known as "vomitoxin," is a mycotoxin belonging to trichothecene B group, produced mainly by strains of *Fusarium graminearum* and *Fusarium culmorum* [17, 26, 27]. These phytopathogenic strains are plant contaminants that can cause severe diseases, like the most commonly disease known as *Fusarium* Head Blight, that affects mainly wheat, but

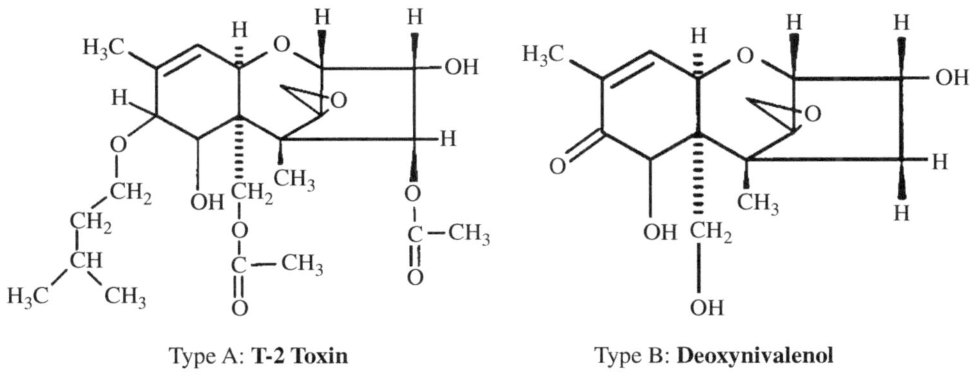

Type A: **T-2 Toxin** Type B: **Deoxynivalenol**

Figure 21.2 Chemical structure of trichothecenes of type A (T-2 Toxin) and type B (deoxynivalenol [DON]).

can also affect other cereals like corn, barley, and oats [28–30]. The eradication of this mycotoxin in food and feed products is difficult due to its high stability, remaining intact even at temperatures of 170 °C [30]. In agriculture commodities, the presence of DON-producing *Fusarium* strains is also a main issue because DON can affect the metabolic processes in germination, acting as an inhibitor of protein synthesis and causing lower yields and poor quality of the final product [26, 30]. As in the case of aflatoxins, the weather has a huge effect in the growth of these fungi and, consequently, in mycotoxin production. Overall, meteorological conditions like moderate temperature and moisture and also improper storage and process conditions lead to the reduction of the yield and crop quality [30]. The type B trichothecenes have a carbonyl group at the C8 position and DON is most commonly detected in swine, among farm animals, mostly due to their diet rich in wheat [28]. The ingestion by humans of these DON contaminated cereals could cause different symptoms, such as diarrhea, dizziness, fever, and vomiting, from where the trivial name "vomitoxin" originates [29]. The European Commission regulations 1881/2006 set the DON maximum limits for unprocessed cereals at 1250 μg/kg and 1750 μg/kg for wheat and maize, for cereals and pasta 750 μg/kg, and for bread 500 μg/kg. In the case of processed cereal-based foods for babies and infants the limit is set at 200 μg/kg [24].

21.4 Fumonisins

Interest in fumonisins increased drastically between the period of 1989 and 1990 due to numerous occurrences of mycotoxicosis in animals associated with the 1989 corn crop in the USA [31]. More than 28 fumonisins have been identified; they can be divided into four different types (A, B, C, and P) (Figure 21.3), group B being the most toxic, specially fumonisin B_1 (FB_1), which represents a major concern due to its classification as a possible carcinogenic for humans (Group 2B) by the IARC [32–34]. Animal diseases like hepatic and renal toxicity in rodents, pulmonary edema in pigs, leukoencephalomalacia in equine and nephrotoxicity and liver cancer in rats are associated with the consumption of crops, especially maize contaminated with fumonisins [35, 36]. In humans, the consumption of fumonisin can lead to esophageal cancer and neural tube defects [36–38]. Maize is frequently contaminated with fumonisins and aflatoxins [39]. Carlson and co-workers [40] and Gelderblom and co-workers [41] reported a synergistic interaction between AFB_1 and FB_1 in the development of liver cancer. *Fusarium* species are reported as fumonisin producers, mainly the strains *Fusarium verticillioides* and *Fusarium proliferatum*. Maize and rice are the most affected crops by these fungi; the disease that affects maize is called *Fusarium* kernel rot [42, 43]. The strain *Aspergillus niger* has also been reported in the literature as a fumonisin producer (the carcinogenic fumonisin B_2 [FB_2]), which implies a lot of questions and issues since this strain is used for diverse biotechnological processes, such as the production of citric acid and extracellular enzymes [44]. The fumonisin temperature stability is about 125–175 °C, and since some cooking processes do not exceed these temperatures, the probability of fumonisins being present in our daily food products is high [45]. Consequently, the European Commission regulations 1881/2006 set the limits of the sum of FB_1 and FB_2 from 200 to 2000 μg/kg for processed maize-based foods and unprocessed maize [24].

21.5 Ochratoxin A

In the 1950s, a Balkan endemic nephropathy was associated with the ingestion of OTA [46, 47]. The OTA structure consists of a pentaketide derived from the dihydrocoumarins family coupled

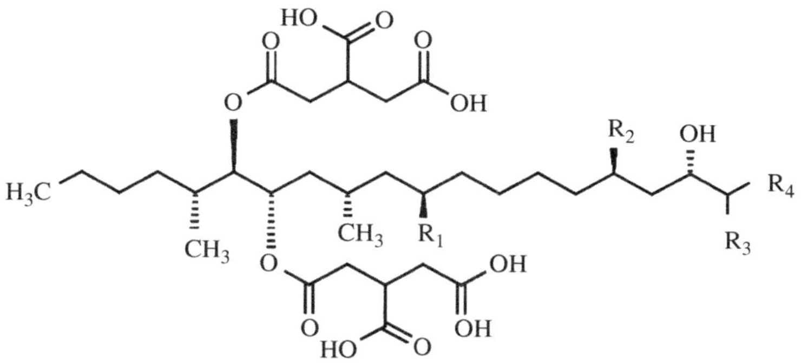

	R_1	R_2	R_3	R_4
Fumonisin A_1	OH	OH	$NHCOCH_3$	CH_3
Fumonisin A_2	H	OH	$NHCOCH_3$	CH_3
Fumonisin B_1	OH	OH	NH_2	CH_3
Fumonisin B_2	H	OH	NH_2	CH_3
Fumonisin C_1	OH	OH	NH	H
Fumonisin P_1	OH	OH	3HP	CH_3
Fumonisin P_2	H	OH	3HP	CH_3

Figure 21.3 Chemical structure of different types of fumonisins.

Figure 21.4 Chemical structure of ochratoxin A (OTA).

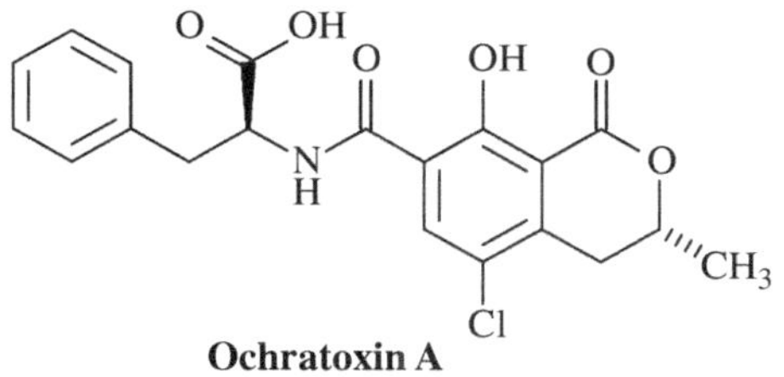

to β-phenylalanine (Figure 21.4) [48]. Similarly to the other mycotoxins previously described, OTA is stable at temperatures of 180 °C and is described as presenting hazardous adverse effects on animals and humans due to its nephrotoxic, carcinogenic, and immunosuppressive activities, being classified as possibly carcinogenic to humans (group 2B) by the IARC [17, 20, 49–51]. OTA is produced by several *Aspergillus* and *Penicillium* species, such as *Aspergillus carbonarius*, *A. niger*, *Aspergillus ochraceus*, and *Penicillium verrucosum* [17, 46]. In parallel with *A. niger* the strain *A. carbonarius* is a generally opportunistic pathogen of grapes, which can implicate the contamination with both of the mycotoxins, OTA and FB_2 [51, 52]. OTA is widely found in cereals, but has also been detected in significant levels in coffee beans, wine, dried vine fruits, spices, fish, and milk [6, 46]. In the case of OTA, the set limits regulated by the European Commission regulations 1881/2006 go from 0.5 μg/kg for dietary foods for special medical purposes and processed cereal-based foods, 2 μg/kg for wine, aromatized wine, and grape juice, 5 μg/kg for unprocessed cereals and roasted coffee beans, and 10 μg/kg for soluble coffee and dried vine fruit [24].

21.6 Patulin

Patulin is a toxic secondary metabolite produced by different genera such as *Aspergillus*, *Penicillium*, and *Byssochlamys*, with *Penicillium expansum* reported as the main producer [53, 54]. Patulin is a polyketide lactone (Figure 21.5), soluble in water with a melting point of 110 °C [54]. Significant levels of patulin can be detected in apples and apple-based food products, since *P. expansum* can infect different types of fruit, like apple, citrus fruits, pears, grapes, and peach, among others, with apples being the "preferential target" [54, 55]. Symptoms like nausea and vomiting in humans can be associated with patulin exposure, which can lead to gastrointestinal disorders including ulceration, distension, and bleeding [55, 56]. As in other mycotoxins, the levels of patulin in different food products are regulated in many countries. In the European Union the set maximum levels, according to the European Commission regulations 1881/2006, are 10 μg/kg for baby foods, apple juice and derivatives, and processed cereals-based foods for infants, 25 μg/kg for solid apple products, and 50 μg/kg for fruit juice, spirit drinks, cider, and other fermented drinks derived from apples [24].

21.7 Zearalenone

ZEN is a phenolic resorcyclic acid lactone (Figure 21.6) and is the most prevalent estrogenic mycotoxin, mainly produced by *Fusarium* species. It commonly affects farm animals, with swine being the most sensitive domestic animal [9, 57]. *Fusarium culmorum, F. cerealis.* and *F. graminearum,* among others, are described as ZEN producers and widely contaminate different cereal crops around the world in both pre- and post-harvested crops if not treated in proper conditions [57, 58].

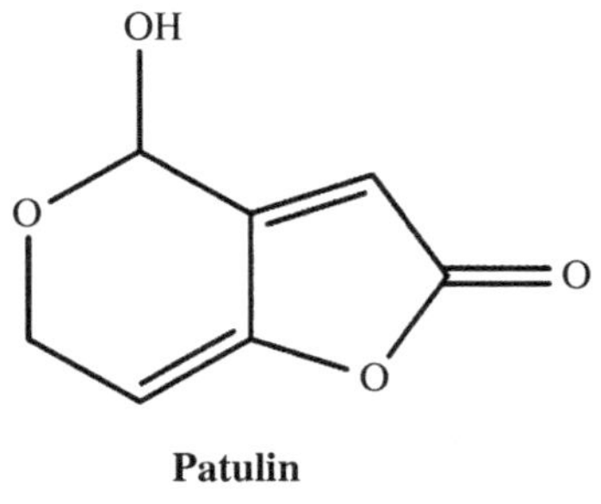

Figure 21.5 Chemical structure of patulin.

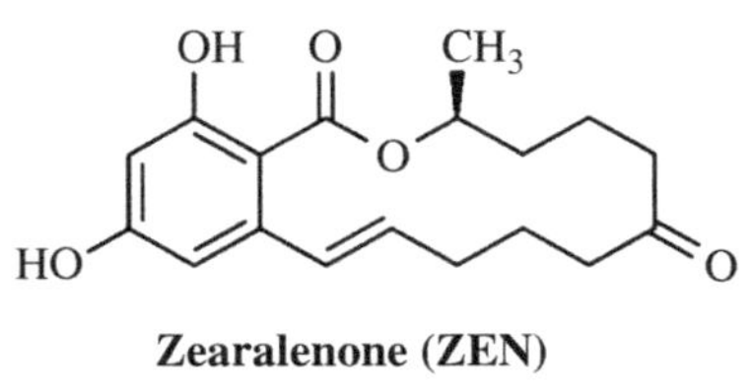

Figure 21.6 Chemical structure of zearalenone (ZEN).

The consumption of ZEN derived from contamination of these products is elevated due to their high temperature stability (around 160 °C), which makes its eradication difficult [57]. Consequently, it is necessary to regulate the presence of ZEN in different products. The European Commission regulations 1881/2006 set the limits for ZEN at 20 μg/kg for processed maize-based and cereal-based foods for infants and young children, 50 μg/kg for maize snacks, bread, pastries, or biscuits, among others, 75 μg/kg for cereals intended for direct human consumption, 100 μg/kg for unprocessed cereals other than maize, and 200 μg/kg for unprocessed maize and maize intended for direct human consumption [24].

21.8 Prevention and Control of Mycotoxins

Following the knowledge on the risk associated with the presence of mycotoxins and their fungal sources in food/feed for human and animal health, as well as the economic impact caused by the huge losses of contaminated crops, in the last few years several strategies have been pursued to reduce or eliminate mycotoxins and mycotoxigenic fungi. In developed countries, the hazard analysis and critical control point (HACCP) approach has been used to control mycotoxin contamination. Initially, the potential hazards associated with food production at all stages are identified; then the critical control points (procedures and operational steps) that can be controlled to eliminate or reduce the risks are determined; and finally, in the last stage, a monitoring system to control the critical limits for the critical control points and create a corrective system when these limits are not in accordance is established [59]. The HACCP system also foresees using good agriculture and manufacturing practices as complementary approaches. Despite the preliminary important protecting approaches against mycotoxins and mycotoxigenic fungi, such as good agriculture practices for different agriculture commodities, followed by good manufacturing practices during the subsequent phases of storage, processing, and distribution of the different crops, it is still not possible to supply quality foods free from environmental contaminants [59]. Several chemical and biological strategies, including the use of chemical fungicides (namely the formulations that require the use of surfactants) and biocontrol agents, such as using non-toxigenic strains of *A. flavus* to compete with the existing mycotoxigenic strain, have been evaluated to reduce the mycotoxin concentration [59]. Surfactants are used in most agrochemical formulations (fungicide, herbicides, and pesticides), as well as in other applications like detergents, cosmetics, and food, among others, due to their ability to reduce surface and interface tensions, increasing the solubility of hydrophobic and water-insoluble compounds, wetting ability, and foaming capacity [60–63]. The agricultural surfactant market, which includes the insecticide, herbicide, and fungicide applications, the different types of surfactants (anionic, nonionic, cationic, and amphoteric), the crop applications and geography, was evaluated at USD 1.45 billion in 2018, and by 2024, at a Compound Annual Growth Rate (CAGR) of 6.9% during the forecast period (2019–2024), is expected to rise to USD 2.19 billion [64]. It is thought that the Asia Pacific area will be the fastest growing market and North America the largest market [64]. The most widely used surfactant in fungicide formulations for a variety of agriculture commodities (tree fruits, vegetable and field crops, grapes) is the surfactant mancozeb, with sales of USD 220 million. In 2022, it is estimated that its consumption will be 250 000 tons. Other surfactants used in fungicide formulations include carbendazim and chlorothalonil, with sales of USD 200 million [60, 65]. Nonetheless, the extensive use of chemical fungicides to control plant diseases or the purposeful contamination of agriculture commodities with non-toxigenic fungal strains, raise numerous environmental and health questions since these strategies have a negative ecological impact. Other chemical strategies used to prevent fungal growth include the use of gamma irradiation or other chemical antifungal compounds like ammonium hydroxide or gaseous ammonia, concentrated ozone, heat treatment with $NaHCO_3$, H_2O_2, benomyl, fenfuram, carboxin, pyrimethanil, azoxystrobin, and oxpoconazole [66, 67]. However, some of these chemical treatments are not totally efficient against some mycotoxins and can also lead to the production of hydrolysis compounds that can be toxic for human health [66]. The chemical strategies greatly contribute to global warming, starting from their manufacturing process to their use as chemical fungicides, besides causing an ecological unbalance of the microorganisms that exist in soil and the development of resistant pathogenic strains leading to a higher risk for human and animal health [68, 69]. Despite the different strategies adopted to inhibit the mycotoxigenic fungi growth, different harvest and post-harvest measures need to be undertaken to control the contamination by mycotoxins. The timing of harvest is important to reduce the final levels of mycotoxin contaminations, for earlier harvests can lead to lower concentrations of mycotoxins. Also, the humidity level before and during storage is crucial to prevent the mycotoxigenic fungi growth and, consequently, mycotoxin production. The total water content levels required depend on the strains under evaluation, e.g. *Aspergillus* species require lower water content levels to survive than *Fusarium* sp. Temperature during storage is also a critical parameter that can affect fungal activity. The grain storage in silos requires combining cooling and drying operations, as well as ventilation systems to prevent the temperature increase in the center of the silo and, consequently, to avoid fungal activity [66]. Therefore, alternative pre- and post-harvest techniques are urgently needed to prevent the mycotoxigenic fungi contamination of plants and crops by inhibiting their growth or mycotoxins production, without altering the nutritive value of the raw product and not being harmful for human and animal health. Therefore, the development of new environmentally friendly strategies is required to replace the currently used non-advantageous and non-environmental-friendly methods. An environmentally friendly alternative comprises the use of green surfactants, so-called biosurfactants, that, due to their similarity with chemical surfactants and their antifungal activity, can be used as "natural fungicides" and anti-mycotoxin agents in different agriculture commodities.

21.9 Biosurfactants

The human population is increasing exponentially and providing quality food becomes a global threat. Therefore, the main challenge for the different countries around the world is to find the right balance between the food demand and the agriculture productivity. The protection of different agriculture commodities is the main goal for farmers to overcome the global food demand, which leads to the extensive use of fungicides to prevent fungal contamination. Surfactants used as pesticides and herbicides in agriculture commodities are mainly derived from petrochemical sources and exhibit low biodegradability and a high potential aquatic toxicity [70]. The environmental danger caused by the widespread use of surfactants, and the increasing consumer awareness regarding climate changes makes it necessary for the development of new alternatives to replace chemical fungicides with "green" fungicides. One of these approaches is the use of biosurfactants as "green" fungicides as a substitute for their chemical counterparts. Biosurfactants are a structurally diverse group of amphiphilic molecules produced by different microorganisms, including bacteria, yeasts, and filamentous fungi [71, 72]. The environmentally friendly properties of biosurfactants, like high biodegradability and low toxicity and the ability to act as a barrier to plant pathogens, make them excellent candidates to be used as "green" fungicides [73–76]. Biosurfactants are classified according to their molecular weight. Low molecular weight biosurfactants have the ability to reduce surface and interfacial tension, and include glycolipids and lipopeptides. High molecular weight biosurfactants (usually polymers) are characterized by stabilizing emulsions [77]. In the last few years, due to their similar properties with surfactants, the study of the antifungal activity of the low molecular weight biosurfactants has gained more attention. The antifungal activity of glycolipids and lipopeptides has been generally studied against different agriculture contaminant fungi, and some of the fungi tested are described in the literature as mycotoxin producers, although only a few reports evaluated the biosurfactant effect in mycotoxin production by these fungi.

21.10 Glycolipids

Glycolipid biosurfactants comprise a hydrophilic carbohydrate moiety (a mono- or oligosaccharide), like glucose, mannose, galactose, or rhamnose, linked to one or more hydrophobic chains with different lengths, that are saturated or unsaturated fatty acids, hydroxy fatty acids, or fatty alcohols [78, 79]. The most well-known glycolipid biosurfactants are rhamnolipids and sophorolipids. Rhamnolipids are the most widely studied glycolipids. They are produced by strains of *Pseudomonas* sp. and *Burkholderia* sp., mainly by *Pseudomonas aeruginosa* [62, 80]. The structure of the rhamnolipids consists in one or two rhamnoses (mono- and di-rhamnolipids, respectively) linked to one or two fatty acid chains with variable lengths between C_8 and C_{24} (Figure 21.7) [62]. The most common are the mono-rhamnolipid Rha-C_{10}-C_{10} and the di-rhamnolipid Rha-Rha-C_{10}-C_{10} [81]. Generally, *P. aeruginosa* strains produce a mixture of different congeners, which is

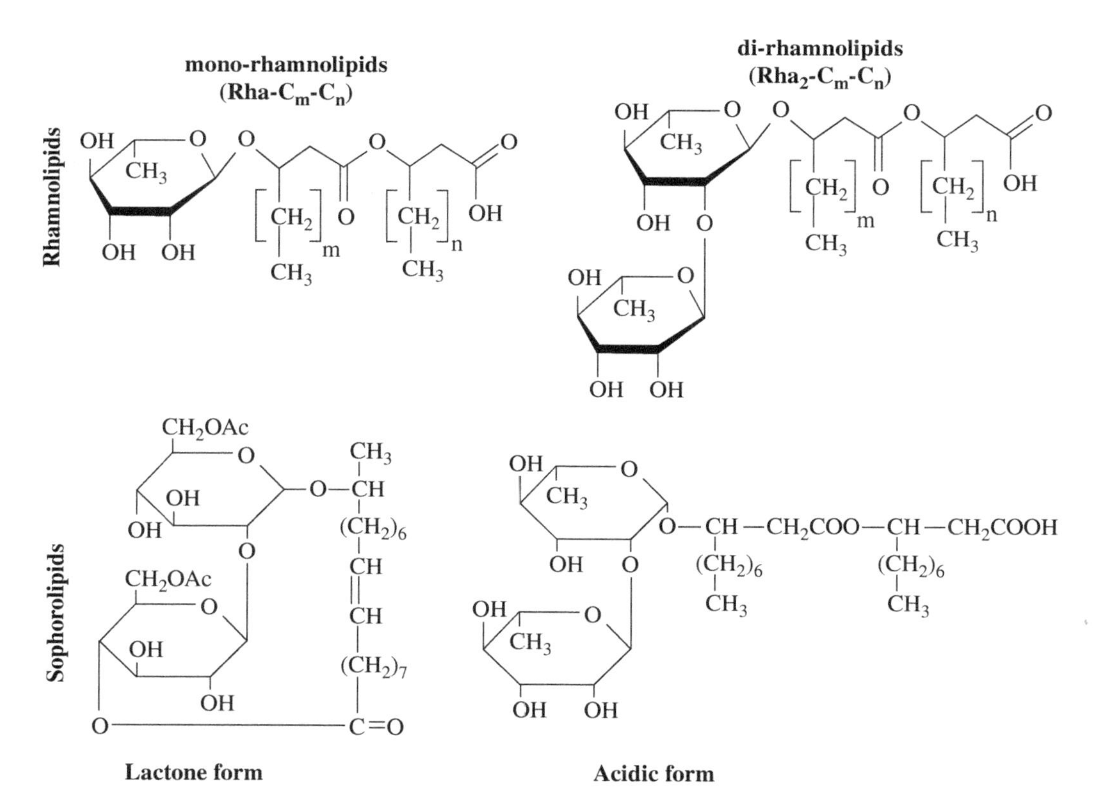

Figure 21.7 Chemical structure of the rhamnolipids congeners (m:n = 8–24) and different forms of sophorolipids.

influenced by the carbon source used in the culture medium and the fermentative conditions [62, 78, 82]. Sophorolipids are extracellular compounds mainly produced by several non-pathogenic yeasts, such as *Starmerella bombicola*, *Candida apicola*, and *Rhodotorula bogoriensis* [82–84]. Sophorolipids comprise a hydrophilic carbohydrate head (the disaccharide sophorose), β-glycosidically linked to a long hydrophobic fatty acid chain. The carboxylic end of the fatty acid can be free, originating the acidic form, or can be esterified with the sophorose head (lactone ring) forming the lactonic forms (Figure 21.7) [82, 83]. The different chemical and biological properties of sophorolipids depend on their form and composition, which in turn depend on the type of substrate used to grow the microorganisms [84]. The lactonic forms have a better capacity for reducing the surface tension and display a good antimicrobial activity, while the acidic forms have better foaming and are useful for applications in cosmetics and bioremediation [82, 83].

21.11 Lipopeptides

Lipopeptide biosurfactants include a fatty acid (hydrophobic part) linked to a linear or circularized oligopeptide with different types and number of amino acids (hydrophilic part) [79]. The most studied lipopeptides are surfactin, fengycin, and iturin. Surfactin is a cyclic lipopeptide mainly produced by *Bacillus subtilis* strains, and its structure consists of a hydrophilic moiety of a short peptide chain constituted by seven amino acids linked to a hydrophobic moiety consisting of a fatty acid of variable length (usually C12–C16) [74, 82]. Surfactin is one of the most effective biosurfactants due to a large number of biological activities, like antifungal, antitumor, and insecticidal, which can potentially be used in diverse therapeutic, industrial, and environmental applications [76]. The structure of the lipopeptide iturin differs from surfactin in the composition of the hydrophobic chain; iturin contains a β-amino fatty acid whereas surfactin contains a β-hydro fatty acid [85]. The iturin family englobes a diverse group of molecules with a good antifungal activity, such as iturin A, mycosubtilin, and bacillomycin, produced mainly by *B. subtillis* strains [86]. The lipopeptide fengycin, as the other lipopeptides are described, is mainly produced by *B. subtillis* strains. It consists of a β-hydroxy fatty acid linked to a peptide chain that comprises 10 amino acids, where eight are organized in a cyclic structure forming a lactone ring [87, 88]. The hydrophobic moiety can be saturated or unsaturated with lengths comprising 14–18 carbon atoms, and the peptide portion can exist in four isoforms [88]. Fengycin is described as an antifungal and antitumor agent and displays a hemolytic activity 40-fold lower compared with surfactin [88].

21.12 Antifungal Activity of Glycolipid Biosurfactants

Rhamnolipids are included in the biofungicide ZONIX™, commercialized by Jeneil Biosurfactant Company (USA). The inhibitory activity of rhamnolipids against different fungi has been demonstrated by a few studies. Borah and co-workers [89] demonstrated a strong antifungal activity of rhamnolipids produced by *P. aeruginosa* SS14 against the mycotoxigenic fungi *F. verticillioides*. The inhibitory activity was higher against spores (approximately 88% at the concentration of 50 mg/l) than against mycelial growth (approximately 82% at the concentration of 200 mg/l). This antifungal effect was observed in *in vitro* assays using maize seeds, where the rhamnolipids were able to reduce the infection by *F. verticillioides* without affecting the germination processes at the rhamnolipid concentrations of 50, 100, and 200 mg/l. Contrarily, the untreated maize seeds were unable to germinate. The antifungal activity of rhamnolipids produced by *P. aeruginosa* #112 was tested by Rodrigues and co-workers [90] against the agriculture contaminant species *A. niger* and *A.*

carbonarius, described as producers of FB_2 and OTA, respectively. The rhamnolipids present in the cell-free supernatant were able to completely inhibit the *A. carbonarius* growth and approximately 76% of the *A. niger* growth at the highest concentration tested, 3.0 g of rhamnolipids per liter, with a ratio of mono-rhamnolipids/di-rhamnolipids of 1.67. The authors demonstrated that di-rhamnolipid congeners were responsible for the antifungal activity and in the presence of optimal NaCl concentrations for each fungi (0.875 M for *A. niger* and 0.375 M for *A. carbonarius*) the antifungal activity of the purified rhamnolipids increased. The addition of NaCl was found to alter their aggregation behavior, thus suggesting that this led to the antifungal activity observed. Nalini and Parthasarathi [91] tested the biocontrol efficiency of the rhamnolipids produced by *Serratia rubidaea* SNAU02 against *Fusarium oxysporum*, a widely contaminant of eggplant, one of the most grown vegetables in India. The assays were performed by adding different concentrations of rhamnolipids in soil and leaves in the pot culture. The rhamnolipids mixture produced was characterized and eight congeners were identified, the congener di-rhamnolipid being the most abundant. The authors observed that 250 μg/ml of purified rhamnolipids was effective against *F. oxysporum*, completely inhibiting the adverse effects caused by the disease in eggplant. The rhamnolipid applications through both soil and foliar spray in leaves was found to be as effective as the chemical fungicide tested, namely the carbendazim (0.2%). Sen and co-workers [84] demonstrated that the sophorolipids, with both lactonic and acidic forms, produced by *Rhodotorula babjevae* YS3 and the standard lactonic sophorolipid (1–4″-sophorolactone 6′, 6″-diacete) exhibit, in *in vitro* assays, a promising antifungal activity against a broad group of mycotoxigenic fungi like *Fusarium verticilliodes* and *F. oxysporum*, known to produce fumonisin or ZEN and trichothecenes, respectively, at the minimum inhibitory concentration (MIC) of 125 μg/ml for both fungi. Despite the attractive properties of sophorolipids, the reports analyzing their antifungal activity are limited.

21.13 Antifungal and Antimycotoxigenic Activity of Lipopeptide Biosurfactants

Different studies have evaluated the antifungal activity of lipopeptide biosurfactants, *in vitro* and *in vivo* assays, against different agriculture fungal contaminants, like the mycotoxigenic fungi belonging to *Fusarium* and *Aspergillus* genera, that are general plant and cereal contaminants. The most studied lipopeptides are surfactin, iturin, and fengycin.

In *in vitro* assays, Mohammadipour and co-workers [92] demonstrated that increased purified surfactin concentrations (20–160 mg/l) produced by *B. subtilis* (BS119m) considerably reduced the *A. flavus* growth from 42 to 100%, as can be seen in Table 21.1. Sarwar and co-workers [93] also studied the antifungal activity of the purified surfactin produced by *Bacillus* (SPB) strains against rice bakanae diseases caused by *Fusarium moniliforme* and others contaminating strains that included among others the strain *F. oxysporum*. The authors performed assays *in vitro* and *in vivo*, in order to evaluate the antifungal potential from the different *Bacillus* strains and the purified surfactin produced. In *in vitro* assays, the purified surfactin produced by *B. subtilis* NH-100, at the concentration of 2000 mg/l, exhibited a significant antifungal activity against *F. moniliforme* and *F. oxysporum*, with a reduction of the fungi growth of 85%. Sun and co-workers [94] demonstrated that the lipopeptide bacillomycin D, produced by *B. subtilis* fmbj (CGMCCN0943), had an excellent antifungal activity against the mycotoxigenic fungi *F. graminearum* by damaging the hyphal and the cellular structure, while bacillomycin D at concentrations of 112.5 and 225 μg/ml completely inhibited the *F. graminearum* growth (Table 21.1). The purified bacillomycin D was also reported to completely inhibit OTA production by *A. ochraceus*, along with a complete inhibition of *A. ochraceus* growth, at a concentration of 30 μg/g [95]. The authors observed by microscopy (Scanning

Table 21.1 *In vitro* evaluation of the antifungal activity and antimycotoxigenic effect of different lipopeptides produced by different *Bacillus* strains, as compared to *Aspergillus* and *Fusarium* strains.

Bacterial strains	Biosurfactant			Fungal strain	Growth inhibition %	Spores germination inhibition %	Mycotoxin reduction %	References
B. subtilis BS119m	Surfactin	Purified	20 mg/l	*A. flavus*	42.0 ± 5.0	—	—	[92]
			40 mg/l		61.0 ± 7.0			
			80 mg/l		94.0 ± 7.0			
			160 mg/l		100 ± 0.0			
B. subtilis NH-100	Surfactin	Purified	2000 mg/l	*F. moniliform*	85.0	—	—	[93]
				F. oxysporum	85.0			
B. subtilis fmbj (CGMCCN 0943)	Bacillomycin D	Purified	18.75 μg/ml	*F. graminearum*	19.3 ± 1.5	31.7 ± 1.6	—	[94]
			37.5 μg/ml		57.9 ± 0.7	59.2 ± 0.2		
			75.0 μg/ml		94.6 ± 0.3	97.5 ± 0.8		
			112.5 μg/ml		100 ± 0.0	100 ± 0.0		
			225.0 μg/ml		100 ± 0.0	100 ± 0.0		
B. subtilis fmbj (CGMCCN 0943)	Bacillomycin D	Purified	30 μg/g	*A. ochraceus*	100.0	100.0	—	[95]
B. subtilis fmbj (CGMCCN 0943)	Bacillomycin D	Purified	200 μg/g	*A. flavus*	85.7	98.1	—	[96]
			400 μg/g		100	100		
B. subtilis UTBSP1	Surfactin Fengycin	Cell-free supernatant	10%	*A. flavus* R5	—	75.0	—	[97]
			25%			95.0		
			50%			100.0		
		Purified	500 mg/l			100.0		

Bacillus sp. P1	Iturin A Surfactin	*Bacillus* cells	10^7 cfu/ml	*A. flavus* A12	70.2 ± 1.4	—	AFB_1	99.8	[98]
				A. parasiticus 30BL	77.2 ± 0.3			—	
				A. carbonarius ITAL293	87.6 ± 0.5		OTA	97.5	
				Aspergillus sp. UCO2A	100 ± 0.0			—	
Bacillus sp. P11	Iturin A Surfactin	*Bacillus* cells	10^7 cfu/ml	*A. flavus* A12	84.2 ± 0.6	—	AFB_1	99.8	
				A. parasiticus 30BL	90.5 ± 0.7		—	—	
				A. carbonarius ITAL293	88.3 ± 0.0		OTA	97.3	
				Aspergillus sp. UCO2A	100 ± 0.0		—	—	
B. mojavensis RC1A	Surfactin Iturin A Fengycin	Cell-free supernatant	5%	*A. parasiticus* NRRRL 2999	100	—	AFB_1	100	[99]
B. velezensis Y6	Iturin (≈21%)	Purified	75 μg/ml	*F. oxysporium* f. sp. *cubense*	—	≈60.0	—	—	[100]
	Fengycin (≈67%)		125 μg/ml			≈80.0			
	Surfactin (≈13%)		150 μg/ml			100			

Electron Microscopy and Transmission Electron Microscopy) that in the presence of bacillomycin, *A. ochraceus* exhibited hyphal distortion and disruption of spores, causing apoptosis. The same damaged structure of the cell walls of the hyphae and spores in *A. flavus*, in the presence of bacillomycin D, were observed by Gong and co-workers [96]. They demonstrated that the purified bacillomycin D at 400 μg/g, in *in vivo* assays, completely inhibited the mycelial growth, spores formation, and spore germination efficiency, and at the concentration of 200 μg/g bacillomycin D was able to inhibit the mycelial growth (85.7%), spore formation (98.1%) (Table 21.1) and spore germination efficiency (96.63%) of *A. flavus*. The antifungal and antimycotoxin activity of fengycin was also studied. Hu and co-workers [101] studied the effect of fengycin produced by *B. subtilis* B-FS01 in FB_1 production by *F. verticilliodes*. They observed a 59% reduction of *F. verticilliodes* growth and through quantitative reverse-transcription PCR (RT-PCR) analysis that the transcriptional levels of the genes *FUM1* (that encodes a polyketide synthase) and *FUM8* (that encodes an aminotransferase), involved in the synthesis of FB_1, were downregulated due to the fengycin treatment (50 μg/ml), which could be related to the decrease of 28% observed in FB_1 production by unit mass mycelia. In fact, most of the studies regarding the antifungal activity of lipopeptides refer to the fact that the antifungal mechanisms are associated to the damage of the cell-wall structure of the hyphae and spores of the fungi, making them more permeable, causing multiple metabolic disorders that could influence the molecular mechanisms of the fungi, and, consequently, affecting the mycotoxin biosynthesis [102].

Different studies evaluated the synergistic effect of the different lipopeptides, produced by different *Bacillus strains*, in their antifungal and antimycotoxigenic activity. Farzaneh and co-workers [97] demonstrated that the synergistic effect of surfactin and fengycin co-produced by *B. subtilis* UTBSP1, purified and present in cell-free supernatant, were responsible for the inhibition of spore germination efficiency of *A. flavus* R5 (Table 21.1). The study performed by Veras and co-workers [98] showed that some *Bacillus* spp. strains isolated from fish intestines were able to inhibit the growth of some *Aspergillus* species, as well as AFB_1 and OTA production compared to the control. They also used a reference strain *B. subtilis* ATCC 19659 which is reported as a non-lipopeptide biosurfactants producer and did not present antifungal activity against the fungal isolates tested. The authors attributed the antifungal activity to the lipopeptide biosurfactants produced by the *Bacillus* strains isolated, which exhibited high concentrations of iturin A and surfactin. Pereyra and co-workers [99] tested the antifungal and anti-mycotoxin activity of six *Bacillus* spp. strains against the mycotoxigenic fungi *A. parasiticus*, described as an aflatoxin producer. *Bacillus mojavensis* RC1A cell-free supernatant presented a stronger inhibitory effect on the *A. parasiticus* growth, and chromatography analysis demonstrated that this strain produced surfactin, iturin A, and fengycin. The three lipopeptides were tested separately in liquid culture and each lipopeptide was able to reduce the fungal growth and aflatoxin (AFB_1) to non-detectable levels. The authors concluded that the *Bacillus* antifungal activity could be due to the action of each of the different lipopeptides produced. Cao and co-workers [100] analyzed the biocontrol action of two strains of *Bacillus velezensis* against banana *Fusarium* wilt. The authors observed that both strains produced three major types of lipopeptides at different percentages, being detected multiple isoforms of each lipopeptide: five surfactin (C12–C16), six fengycin (C12–C17), and three iturin (C14–C16). The levels of iturin and fengycin were higher in the case of *B. velezensis* Y6 (total ≈ 88%) than *B. velezensis* F7 (total ≈ 76%). The strain *B. velezensis* Y6 iturin was responsible for the antifungal activity against the *F. oxysporium* f. sp. *cubense*, in both *in vitro* and *in vivo* assays. In the *in vitro* assays, the extract from *B. velezensis* Y6 and iturin individually presented a strong inhibitory activity against *F. oxysporium* f. sp. *cubense* at concentrations between 75 and 200 μg/ml, when the fengycin did not show inhibitory activity, even at the highest concentration tested of 200 μg/ml. Kaur and co-workers [103] reported that the synergistic interaction among the lipopeptides fengycin, iturin A, and surfactin, produced by *Bacillus vallismortis* R2, significantly increased the antifungal activity against the mycotoxigenic fungi *Alternaria alternata* (zone of inhibition of 30 mm) compared with their individual performance (zone of inhibition of 0.0, 12.5, and 24 mm for surfactin, iturin, and fengycin, respectively). Moreover, the authors compared the antifungal activity of commercial surfactin and iturin (Sigma-Aldrich) against *A. alternata*, and observed a similar growth inhibition as the one obtained with the surfactin and iturin produced by *B. vallismortis* R2. They also observed that, at 200 μg/ml, fengycin displayed the highest antifungal activity (55%) followed by iturin A (23%), and surfactin did not exhibit antifungal activity. Liu and co-workers [104] reported that the synergistic effect of surfactin, fengycin, and iturin A led to a higher inhibition of spores development by *Alternaria solani*, which is described as a mycotoxigenic fungi [105]. The authors observed that the individual lipopeptide, at 50 μg/ml, presents an inhibition of spore development between 10 and 15%, while the combination of the three lipopeptides (all with equal concentrations up to a final concentration of 50 μg/ml) inhibited spore development by 30%.

In *in vivo* assays, Krishnan and co-workers [106] evaluated the potential antifungal effect of surfactin produced by *Brevibacillus brevis* KN8 against the mycotoxigenic *F. moniliforme* in maize kernels. The surfactin produced at a concentration of 50 μg/ml was more effective at inhibiting hyphae growth than the commercial fungicide carbendazim at the same concentration of 50 μg/ml. In *in vivo* assays, the surfactin completely inhibited the growth of *F. moniliforme* in maize kernels at the concentration tested of 50 μg/ml, and the seed germination rate was approximately 60% higher in the surfactin pretreated maize kernels than the surfactin untreated maize kernels. In the case of Sarwar and co-workers [93], they observed in rice basmati seeds that the consortium between the *B. subtilis* NH-100 and *Bacillus* sp. NH-217 were able to reduce approximately 78% of the *F. moniliforme* growth and decrease the FB_1 production by around 50%, when compared with untreated rice seeds. In the case of the purified surfactin produced by *B. subtilis* NH-100, at the concentration of 20 mg/ml, the *F. oxysporum* growth was reduced by approximately 99% and inhibited the FB_1 production by around 61%, compared to the control. Consequently, the purified surfactin presented a better result (*in vivo* assays) when compared to the chemical fungicide benomyl as Benlate (at the recommended dose by the manufacturing), which was able to inhibit the *F. moniliforme* growth by approximately 97% and decreased the FB_1 production by around 50%. Hazarika and co-workers [107] isolated seven bacteria from healthy sugarcane leaves to evaluate their antifungal activity against different fungi belonging to the *Alternaria*, *Saccharicola*, and *Fusarium* genera, among others, which are known to cause diseases in sugarcane leaves. *B. subitilis* SCB-1, due to the production of surfactin, in *in vitro* assays showed a strong antifungal activity against the strains *F. oxysporum* SC7.1, *F. verticillioides* SC8.1, and *Fusarium* sp. SC9.1. Consequently, the authors further tested the *B. subtilis* SCB-1 efficacy against these *Fusarium* strains on germinating mung bean seeds. The pre-treated seeds with *B. subtilis* SCB-1 were able to inhibit the fungal growth, but also presented a germination rate of approximately 65, 55, and 75% after contamination with *Fusarium oxysporum* SC7.1, *F. verticillioides* SC8.1, and *Fusarium* sp. SC9.1, respectively. Fujita and Yokota [108] grew four different types of lettuce seeds in an air-controlled greenhouse with controlled light exposure containing, separately, different concentrations of iturin A (>98% pure, purified from cultures of *B. subtilis* ATCC21556) and commercial surfactin (Wako Pure Chemical Industries, Japan) to analyze their antifungal activity against the mycotoxigenic fungi *F. oxysporum*. The purified iturin A (0.94 mg/l) and surfactin (3.75 mg/l) were able to suppress lettuce seed diseases in the four different types tested (Table 21.2). However, in the case of the lipopeptide iturin A, concentrations higher than 1.88 mg/l were unable to inhibit the *F. oxysporum* growth.

From the iturin family, the antifungal and antimycotoxigenic activity of bacillomycin D was the most studied. Sun and co-workers [94] demonstrated that bacillomycin D produced by *B. subtilis*

Table 21.2 *In vivo* evaluation of the antifungal activity and antimycotoxigenic effect of different lipopeptides produced by different *Bacillus* strains, against *Aspergillus* and *Fusarium* strains, with different seeds.

Bacterial strain	Biosurfactant produced	*In vivo* assays		Fungal strain	Growth Inhibition (%)	Mycotoxin reduction (%)		References
B. brevi KN8(2)	Surfactin	[Surfactin] 50 mg/l	Maize kernel	*F. moniliforme* ITCC 4916	100	—	—	[106]
B. subtilis NH-100	Surfactin	[Surfactin] 20 mg/ml	Basmati rice seeds	*F. moniliforme* KJ719445	≈99.0	FB_1	≈61.0	[93]
Bacillus sp. NH-100 + NH-217	Surfactin	[*Bacillus* sp. NH100 + NH217] 10^7 cfu/ml	Basmati rice seeds	*F. moniliforme* KJ719445	≈78.0	FB_1	≈50.0	
B. subtilis SCB-1	Surfactin	Bacterial suspension prepared in 0.85% sodium chloride solution, adjusted 0.6 DO_{600}	Mung bean seeds	*F. oxysporum* SC7.1	100	—	—	[107]
				F. verticillioides SC8.1	100			
				Fusarium sp. SC9.1	100			
B. subtilis	Surfactin	3.75 mg/l	Crisphead, romaine, red leaf, and green lettuce seeds	*F. oxysporum* f. sp. *lactucae*	100	—	—	[108]
B. subtilis ATCC21556	Iturin A	0.94 mg/l			100			
B. subtilis fmbj (CGCCN0943)	Bacillomycin D (BD)	75 μg_{BD}/g of wheat	Wheat	*F. graminearum*	—	DON	100	[94]
B. subtilis fmbj (CGCCN0943)	Bacillomycin D (BD)	90 μg_{BD}/g	Rice and oat	*A. ochraceus*	100	OTA	100	[95]

B. subtilis fmbj (CGMCCN 0943)	Bacillomycin D	Purified 50 μg/g	Corn	*A. flavus*	65.0	—	—	[96]
		100 μg/g			91.0			
		200 μg/g			100			
		400 μg/g			100			
B. amyloliquefaciens FZB42	Fengycin	90 μg/ml	Wheat spikes	*F. graminearium*	—	DON	≈65.0	[109]
						15-ADON	≈62.0	
						3-ADON	≈60.0	
						ZEN	≈65.0	
B. subtilis (UTBSP1)	Surfactin Fengycin	10^5 cfu/ml	Pistachio nuts	*A. flavus* R5	0.0	AFB_1	20.0	[97]
		10^6 cfu/ml			54.2		82.1	
		10^7 cfu/ml			94.1		100	
B. velezensis Y6	Iturin (≈21%) Fengycin (≈67%) Surfactin (≈13%)	10^6 cells/g of soil	Roots of the banana seedlings	*F. oxysporium* f.sp. *cubense*	48.0	—	—	[100]

fbmj (CGCCN0943) completely inhibits DON production by *F. graminearum* on wheat grain with 14% of water content at the concentration of 75 μg/g of wheat after storage for 60 days. Qian and co-workers [95] observed in rice and oat that bacillomycin D produced by the same strain, at the concentration of 400 μg/g of bacillomycin D, completely inhibited the mycelial growth, spores formation, and spores germination efficiency, and the concentration of 200 μg/g of bacillomycin D was able to inhibit the mycelial growth (85.7%), spore formation (98.1%), and spore germination efficiency (96.6%) (Table 21.2) of *A. flavus*. Gong and co-workers [96] tested the antifungal activity of the same bacillomycin D on corn at the different concentrations and observed that at the highest concentration tested (200 at 400 μg/g) *A. flavus* growth was completely inhibited (Table 21.2). However, the authors did not evaluate the production of mycotoxins. *A. flavus* is known to produce the most carcinogenic mycotoxin (AFB_1) and the observed inhibition of *A. flavus* in corn could lead to a reduction or inhibition of aflatoxin production. Therefore, further studies need to be performed.

Sporulation and germination are important steps for proper fungal growth and development. Hanif and co-workers [109] observed that fengycin produced by *Bacillus amyloliquefaciens* FZB42 could be an alternative to the chemical fungicides and could be used as a biocontrol agent against mycotoxigenic fungi in a wheat crop. In the different assays performed, fengycin exhibited a strong inhibitory effect in the formation and germination of spores from *F. gramiearum*, causing severe structural deformations in the cell wall of the fungal hyphae. Consequently, they observed that at the concentration of 90 μg/ml of fengycin, DON and ZEN production decreased significantly (≈65%) (Table 21.2). Farzaneh and co-workers [97] contaminated mature pistachio nuts with spores of *A. flavus* R5 and evaluated the effect of the addition of different concentrations of *B. subtilis* suspension in saline solution (10^5 to 10^7 cfu/ml). They observed an increasing inhibition of *A. flavus* growth at 94.1% as well as a decrease in AFB_1 production at 100% with an increasing concentration of bacterial suspension, as can be seen in Table 21.2. Cao and co-workers [100] analyzed the biocontrol action of the three lipopeptides to combine or separate (iturin A, fengycin, and surfactin), produced by *B. velezensis* Y6 against banana *Fusarium* wilt caused by *F. oxysporium* f.p. *cubense*. For that different *B. velezensis* mutants were constructed, one with a deletion of fengycin gene (*fen*C), another with iturin A gene deletion (*itu*A), and another without the fengycin and iturin gene. In *in vitro assays*, the treatment with *B. velezensis* Y6 significantly reduced the *Fusarium* wilt incidence of 48% in banana plants. The treatment with the mutant *B. velezensis* with deletion of the fengycin gene (*fen*C) was able to reduce the incidence of 51% in banana plants, instead of the mutants with iturin gene deletion (*itu*A) and in both genes (iturin A and fengycin) that were not able to suppress the *Fusarium* wilt. These results led the authors to suggest that the iturin A produced is the major contributor against *F. oxysporium* f.p. *cubense*.

21.14 Opportunities and Perspectives

Despite the antifungal and antimycotoxigenic effects of biosurfactants and their vast area of applications, their biotechnological use is still restricted to certain areas, since their production cost is not economically competitive. However, with climate change awareness by the global population, the "green" biosurfactant properties could outweigh the economic issues and be used as biofungicides. A main disadvantage that is a barrier for their large-scale commercialization is the low yields obtained at the end of the fermentation and the high production costs associated. Consequently, there are few industrial producers of biosurfactants located in Europe, North America, and Asia, such as Ecover in Belgium, Evonik and Biotensidon in Germany, Jeneil Biotech and AGAE technologies Ltd in United States of America, and Saraya in Japan. The highest industrial production corresponds to the low-molecular weight glycolipids, mainly rhamnolipids and sophorolipids. One of the efforts made in the biosurfactant field is the reduction of the cost of the culture medium, which is generally accepted that represents 30–50% of the total biosurfactant cost, using low-cost substrates rich in carbon, nitrogen, and hydrocarbons sources, such as agro-industrial wastes and byproducts [62, 110–114]. Several studies demonstrated the synergistic effect of different biosurfactant mixtures in their antifungal activity, such as surfactin and fengycin [97, 115, 116] or iturin and fengycin [86]. Therefore, another strategy that could be appealing to their commercialization is the formulation of different biosurfactants mixtures, since the synergistic effect can improve the antifungal activity and broaden the spectrum of action against different fungi, besides increasing the plant and soil viability and health. In this way farmers could reduce the contamination by mycotoxigenic fungi and improve their crop quality and in so doing could respond to the global demand for food quality. However, the formulations with different biosurfactants need to be tested since some biosurfactants at certain concentrations could induce the mycotoxin production as a stress response by the mycotoxigenic fungi (for example, a high concentration of surfactin induces fumonisin production by *F. verticillioides*) [117]. Finally, the application of biosurfactants, individual or combined, as biofungicides could possibly be a good "green" alternative to the chemical strategies currently being applied.

Acknowledgements

This study was supported by PARTEX Oil and Gas and the Portuguese Foundation for Science and Technology (FCT) under the scope of the strategic funding of the UIDB/04469/2020 unit and BioTecNorte operation (NORTE-01-0145-FEDER-000004) funded by the European Regional Development Fund under the scope of Norte 2020 – Programa Operacional Regional do Norte. The authors also acknowledge the Biomass and Bioenergy Research Infrastructure (BBRI)-LISBOA-01-0145-FEDER-022059, supported by the Operational Program for Competitiveness and Internationalization (PORTUGAL2020), by the Lisbon Portugal Regional Operational Program (Lisboa 2020), and by the North Portugal Regional Operational Program (Norte 2020) under the Portugal 2020 Partnership Agreement, through the European Regional Development Fund (ERDF).

References

1 Barrett, J.R. (2000). Mycotoxins: of molds and maladies. *Environmental Health Perspectives* 108 (1): 20–23.

2 Bennett, J.W. and Klich, M. (2003). Mycotoxins. *Clinical Microbiology Reviews* 16 (3): 497–516. https://doi.org/10.1128/CMR.16.3.497.

3 Bhat, R. V. and Miller, J. D., "Mycotoxins and food supply," 1991. http://www.fao.org/3/U3550t/u3550t0e.htm.

4 Liew, W.P.P. and Mohd-Redzwan, S. (2018). Mycotoxin: Its impact on gut health and microbiota. *Frontiers Cellular and Infection Microbiology* 8 (60): 1–17. https://doi.org/10.3389/fcimb.2018.00060.

5 Pitt, J.I. (2003). Mycotoxins, scetion 5: intoxications."Chapter 30. In: *Foodborne Infections and Intoxications, Food Science and Technology*, 4e (eds. G.J. Morris and M. Potter), 409–418. San Diego, CA: Academic Press https://doi.org/10.1016/B978-0-12-416041-5.00030-5.

6 Zain, M.E. (2011). Impact of mycotoxins on humans and animals. *Journal of Saudi Chemical Society* 15 (2): 129–144. https://doi.org/10.1016/j.jscs.2010.06.006.

7 Taheur, F.B., Fedhila, K., Chaieb, K. et al. (2017). Adsorption of aflatoxin B1, zearalenone and ochratoxin A by microorganisms isolated from Kefir grains. *International Journal of Food Microbiology* 251: 1–7. https://doi.org/10.1016/j.ijfoodmicro.2017.03.021.

8 Guimarães, A., Venancio, A., and Abrunhosa, L. (2018). Antifungal effect of organic acids from lactic acid bacteria on *Penicillium nordicum*. *Food Additives and Contaminants – Part A Chemistry, Analysis, Control, Exposure and Risk Assessment* 35 (9): 1803–1818. https://doi.org/10.1080/19440049.2018.1500718.

9 Peng, Z., Chen, L., Zhu, Y. et al. (2018). Current major degradation methods for aflatoxins: A review. *Trends Food Science and Technology* 80: 155–166. https://doi.org/10.1016/j.tifs.2018.08.009.

10 Eskola, M., Kos, G., Elliott, C.T. et al. (2019). Worldwide contamination of food-crops with mycotoxins: Validity of the widely cited 'FAO estimate' of 25%. *Critical Reviews in Food Science and Nutrition* 60: 1–17. https://doi.org/10.1080/10408398.2019.1658570.

11 Wild, C.P. and Gong, Y.Y. (2009). Mycotoxins and human disease: a largely ignored global health issue. *Carcinogenesis* 31 (1): 71–82. https://doi.org/10.1093/carcin/bgp264.

12 Kabak, B. (2009). The fate of mycotoxins during thermal food processing. *Journal of the Science of Food Agriculture* 89: 549–554. https://doi.org/10.1002/jsfa.3491.

13 Rogowska, A., Pomastowski, P., Sagandykova, G., and Buszewski, B. (2019). Zearalenone and its metabolites: Effect on human health, metabolism and neutralisation methods. *Toxicon* 162: 46–56. https://doi.org/10.1016/j.toxicon.2019.03.004.

14 Speijers, G.J.A. and Speijers, M.H.M. (2004). Combined toxic effects of mycotoxins. *Toxicology Letters* 153: 91–98. https://doi.org/10.1016/j.toxlet.2004.04.046.

15 Haschek, W.M. and Voss, K.A. (2013). Mycotoxins" Chapter 39. In: *Handbook of Toxicologic Pathology*, 3e, vol. II (eds. W.M. Haschek, C.G. Rousseaux and M.A. Wallig), 1187–1258. London, UK: Academic Press https://doi.org/10.1016/B978-0-12-415759-0.00039-X.

16 Rushing, B.R. and Selim, M.I. (2019). Aflatoxin B1: A review on metabolism, toxicity, occurrence in food, occupational exposure, and detoxification methods. *Food and Chemical Toxicology* 124: 81–100. https://doi.org/10.1016/j.fct.2018.11.047.

17 Al-Jaal, B.A., Jaganjac, M., Barcaru, A. et al. (2019). Aflatoxin, fumonisin, ochratoxin, zearalenone and deoxynivalenol biomarkers in human biological fluids: A systematic literature review, 2001–2018. *Food Chemical Toxicology* 129: 211–228. https://doi.org/10.1016/j.fct.2019.04.047.

18 Deng, J., Zhao, L., Zhang, N. et al. (2018). Aflatoxin B 1 metabolism: Regulation by phase I and II metabolizing enzymes and chemoprotective agents. *Mutation Research – Reviews in Mutation Research* 778: 79–89. https://doi.org/10.1016/j.mrrev.2018.10.002.

19 Guimarães, A., Santiago, A., Teixeira, J.A. et al. (2018). Anti-aflatoxigenic effect of organic acids produced by *Lactobacillus plantarum*. *International Journal Food Microbiology* 264: 31–38. https://doi.org/10.1016/j.ijfoodmicro.2017.10.025.

20 Raters, M. and Matissek, R. (2008). Thermal stability of aflatoxin B1 and ochratoxin A. *Mycotoxin Research* 24 (3): 130–134. https://doi.org/10.1007/BF03032339.

21 Yu, J., Chang, P., Ehrlich, K.C. et al. (2004). Clustered pathway genes in aflatoxin biosynthesis. *Applied and Environmental Microbiology* 70 (3): 1253–1262. https://doi.org/10.1128/AEM.70.3.1253.

22 IARC, Mycotoxins and human health, 2002. Available at: http://www.iarc.fr/index.php. (Accessed: 06 July 2017).

23 FDA. Guidance for industry: Action levels for poisonous or deleterious substances in human food and animal feed, 2000. Available at: https://www.fda.gov/regulatory-information/search-fda-guidance-documents/guidance-industry-action-levels-poisonous-or-deleterious-substances-human-food-and-animal-feed#afla.

24 EC (2006). Commission Regulation (EC) No 1881/2006 of 19 December 2006 – Setting maximum levels for certain contaminants in foodstuffs. *Official Journal European Union* 7 (406): 5–24.

25 Pasquali, M., Beyer, M., Logrieco, A. et al. (2016). A European database of *Fusarium graminearum* and *F. culmorum* trichothecene genotypes. *Frontiers in Microbiology* 7 (406): 1–11. https://doi.org/10.3389/fmicb.2016.00406.

26 Pestka, J.J. (2007). Deoxynivalenol: toxicity, mechanisms and animal health risks. *Animal Feed Science and Technology* 137: 283–298. https://doi.org/10.1016/j.anifeedsci.2007.06.006.

27 Pinton, P., Accensi, F., Beauchamp, E. et al. (2008). Ingestion of deoxynivalenol (DON) contaminated feed alters the pig vaccinal immune responses. *Toxicology Letters* 177: 215–222. https://doi.org/10.1016/j.toxlet.2008.01.015.

28 Döll, S. and Dänicke, S. (2011). The *Fusarium* toxins deoxynivalenol (DON) and zearalenone (ZON) in animal feeding. *Preventive Veterinary Medicine* 102: 132–145. https://doi.org/10.1016/j.prevetmed.2011.04.008.

29 FDA, "Chemical Hazards," 2019. Available: https://www.fda.gov/animal-veterinary/biological-chemical-and-physical-contaminants-animal-food/chemical-hazards.

30 Spanic, V., Marcek, T., Abicic, I., and Sarkanj, B. (2018). Effects of *Fusarium* head blight on wheat grain and malt infected by *Fusarium culmorum*. *Toxins* 10 (17): 1–12. https://doi.org/10.3390/toxins10010017.

31 Marasas, W.F.O. (1996). Occurrence of Fumonisins in foods and feeds, Chapter 1. In: *Fumonisins in Food* (eds. L. Jackson, J.W. DeVries and L.B. Bullerman), 1–39. Boston, MA, USA: Springer https://doi.org/10.1007/978-1-4899-1379-1.

32 Bordini, J.G., Ono, M.A., Garcia, G.T. et al. (2017). Impact of industrial dry-milling on fumonisin redistribution in non-transgenic corn in Brazil. *Food Chemistry* 220: 438–443. https://doi.org/10.1016/j.foodchem.2016.10.028.

33 Deepa, N. and Sreenivasa, M.Y. (2019). Molecular methods and key genes targeted for the detection of fumonisin producing *Fusarium verticillioides* – An updated review. *Food Bioscience* 32: 1–8. https://doi.org/10.1016/j.fbio.2019.100473.

34 IARC, "Fumonisin B1," 2000.

35 Marasas, W.F.O. (2001). Discovery and occurrence of the fumonisins: a historical perspective. *Environmental Health Perspectives* 109 (supplement 2): 239–243. https://doi.org/10.2307/3435014.

36 Marasas, W.F.O., Riley, R.T., Hendricks, K.A. et al. (2004). Fumonisins disrupt sphingolipid metabolism, folate transport, and neural tube development in embryo culture and in vivo: A potential risk factor for human neural tube defects among populations consuming Fumonisin-contaminated maize. *The Journal of Nutrition* 134 (4): 711–716. https://doi.org/10.1093/jn/134.4.711.

37 Sun, G., Wang, S., Hu, X. et al. (2011). Fumonisin B1 contamination of home-grown corn in high-risk areas for esophageal and liver cancer in China. *Food Additives and Contaminants* 28: 461–470. https://doi.org/10.1080/02652030601013471.

38 Missmer, S.A., Suarez, L., Felkner, M. et al. (2006). Exposure to fumonisins and the occurence of neutral tube defects along the Texas–Mexico border. *Environmental Health Perspectives* 114 (2): 237–241. https://doi.org/10.1289/ehp.8221.

39 Stasiewicz, M.J., Falade, T.D.O., Mutuma, M. et al. (2017). Multi-spectral kernel sorting to reduce aflatoxins and fumonisins in Kenyan maize. *Food Control* 78: 203–214. https://doi.org/10.1016/j.foodcont.2017.02.038.

40 Carlson, D.B., Williams, D.E., Spitsbergen, J.M. et al. (2001). Fumonisin B1 promotes aflatoxin B1 and *N*-methyl-*N*′-nitronitrosoguanidine-initiated liver tumors in rainbow trout. *Toxicolology Appllied Pharmacology* 172: 29–36. https://doi.org/10.1006/taap.2001.9129.

41 Gelderblom, W.C.A., Marasas, W.F.O., Lebepe-Mazur, S. et al. (2002). Interaction of fumonisin B1 and aflatoxin B1 in a short-term carcinogenesis model in rat liver. *Toxicology* 171: 161–173. https://doi.org/10.1016/S0300-483X(01)00573-X.

42 Kamle, M., Mahato, D.K., Devi, S. et al. (2019). Fumonisins: Impact on agriculture, food, and human health and their management strategies. *Toxins* 11 (328): 1–23. https://doi.org/10.3390/toxins11060328.

43 Wu, F. (2007). Measuring the economic impacts of *Fusarium* toxins in animal feeds. *Animal Feed Science Technology* 137: 363–374. https://doi.org/10.1016/j.anifeedsci.2007.06.010.

44 Frisvad, J.C., Smedsgaard, J., Samson, R.A. et al. (2007). Fumonisin B2 production by *Aspergillus niger*. *Journal of Agriculture and Food Chemistry* 55 (23): 9727–9732. https://doi.org/10.1021/jf0718906.

45 Bullerman, L.B., Ryu, D., and Jackson, L.S. (2002). Stability of Fumonisins in food processing. In: *Mycotoxins and Food Safety* (eds. J.W. DeVries, M.W. Trucksess and L.S. Jackson), 195–204. Boston, MA, USA: Springer https://doi.org/10.1007/978-1-4615-0629-4_20.

46 Abrunhosa, L., Santos, L., and Venâncio, A. (2006). Degradation of ochratoxin A by proteases and by a crude enzyme of *Aspergillus niger*. *Food Biotechnology* 20: 231–242. https://doi.org/10.1080/08905430600904369.

47 Varga, J., Kocinfé, S., Péteri, Z. et al. (2010). Chemical, physical and biological approaches to prevent ochratoxin induced toxicoses in humans and animals. *Toxins* 2 (7): 1718–1750. https://doi.org/10.3390/toxins2071718.

48 el Khoury, A. and Atoui, A. (2010). Ochratoxin A: General overview and actual molecular status. *Toxins* 2 (4): 461–493. https://doi.org/10.3390/toxins2040461.

49 Abrunhosa, L., Inês, A., Rodrigues, A.I. et al. (2014). Biodegradation of ochratoxin A by *Pediococcus parvulus* isolated from Douro wines. *Interntional Journal of Food Microbiology* 188: 45–52. https://doi.org/10.1016/j.ijfoodmicro.2014.07.019.

50 IARC, Ochratoxin A, 1993.

51 Lappa, I.K., Simini, E., Nychas, G.-J.E., and Panagou, E.Z. (2017). *In vitro* evaluation of essential oils against *Aspergillus carbonarius* isolates and their effects on Ochratoxin A related gene expression in synthetic grape medium. *Food Control* 73: 71–80. https://doi.org/10.1016/j.foodcont.2016.08.016.

52 Wu, T., Cheng, D., He, M. et al. (2014). Antifungal action and inhibitory mechanism of polymethoxylated flavones from *Citrus reticulata* Blanco peel against *Aspergillus niger*. *Food Control* 35 (1): 354–359. https://doi.org/10.1016/j.foodcont.2013.07.027.

53 Puel, O., Galtier, P., and Oswald, I. (2010). Biosynthesis and toxicological effects of Patulin. *Toxins* 2: 613–631. https://doi.org/10.3390/toxins2040613.

54 Saleh, I. and Goktepe, I. (2019). The characteristics, occurrence, and toxicological effects of patulin. *Food Chem. Toxicol.* 129: 301–311. https://doi.org/10.1016/j.fct.2019.04.036.

55 Ostry, V., Malir, F., Cumova, M. et al. (2018). Investigation of patulin and citrinin in grape must and wine from grapes naturally contaminated by strains of *Penicillium expansum*. *Food and Chemical Toxicology* 118: 805–811. https://doi.org/10.1016/j.fct.2018.06.022.

56 Carballo, D., Tolosa, J., Ferrer, E., and Berrada, H. (2019). Dietary exposure assessment to mycotoxins through total diet studies. A review. *Food Chemical Toxicology* 128: 8–20. https://doi.org/10.1016/j.fct.2019.03.033.

57 Zhang, G.-L., Feng, Y.-L., Song, J.-L., and Zhou, X.-S. (2018). Zearalenone: A mycotoxin with different toxic effect in domestic and laboratory animals' Granulosa cells. *Frontries in Genetics* 9 (667): 1–8. https://doi.org/10.3389/fgene.2018.00667.

58 Hueza, I.M., Raspantini, P.C.F., Raspantini, L.E.R. et al. (2014). Zearalenone, an estrogenic mycotoxin, is an immunotoxic compound. *Toxins* 6: 1080–1095. https://doi.org/10.3390/toxins6031080.

59 IARC, "Chapter 9 – Practical approaches to control mycotoxins," 2012.

60 Castro, M.J.L., Ojeda, C., and Fernández, A.C. (2013). Surfactants in agriculture, Chapter 7. In: *Green Materials for Energy, Producs and Depollution*, vol. 3 (eds. E. Lichtfouse, J. Schwarzbauer and R. Didier), 287–334. Dordrecht: Springer https://doi.org/10.1007/978-94-007-6836-9.

61 Geys, R., Soetaert, W., and Van Bogaert, I. (2014). Biotechnological opportunities in biosurfactant production. *Current Opinion in Biotechnology* 30: 66–72. https://doi.org/10.1016/j.copbio.2014.06.002.

62 Gudiña, E.J., Rodrigues, A.I., Alves, E. et al. (2015). Bioconversion of agro-industrial by-products in rhamnolipids toward applications in enhanced oil recovery and bioremediation. *Bioresource Technology* 177: 87–93. https://doi.org/10.1016/j.biortech.2014.11.069.

63 Mulligan, C.N. (2005). Environmental applications for biosurfactants. *Environmental Pollution* 133 (2): 183–198. https://doi.org/10.1016/j.envpol.2004.06.009.

64 MordorIntelligence, Agricultural surfactant market – Growth, trends, and forecast (2019–2024), 2019. Available at: https://www.mordorintelligence.com/industry-reports/agricultural-surfactant-market.

65 MarketWatch, Mancozeb market size, shared 2019 – Global market, analysis, research, business growth and forecast to 2024, 2019. Available at: https://www.marketwatch.com/press-release/mancozeb-market-size-share-2019---globally-market-analysis-research-business-growth-and-forecast-to-2024-market-reports-world-2019-11-21.

66 Jouany, J.P. (2007). Methods for preventing, decontaminating and minimizing the toxicity of mycotoxins in feeds. *Animal Feed Science and Technology* 137: 342–362. https://doi.org/10.1016/j.anifeedsci.2007.06.009.

67 Brauer, V.S., Rezende, C.P., Pessoni, A.M. et al. (2019). Antifungal agents in agriculture: Friends and foes of public health. *Biomolecules* 9 (521): 1–21. https://doi.org/10.3390/biom9100521.

68 Gerez, C.L., Carbajo, M.S., Rollán, G. et al. (2010). Inhibition of citrus fungal pathogens by using lactic acid bacteria. *Journal of Food Science* 75 (6): M354–M359. https://doi.org/10.1111/j.1750-3841.2010.01671.x.

69 Pekmezovic, M., Rajkovic, K., Barac, A. et al. (2015). Development of kinetic model for testing antifungal effect of *Thymus vulgaris* L. and *Cinnamomum cassia* L. essential oils on *Aspergillus flavus* spores and application for optimization of synergistic effect. *Biochemical Engineering Journal* 99: 131–137. https://doi.org/10.1016/j.bej.2015.03.024.

70 Deleu, M. and Paquot, M. (2004). From renewable vegetables resources to microorganisms: New trends in surfactants. *Comptes Rendus Chimie* 7: 641–646. https://doi.org/10.1016/j.crci.2004.04.002.

71 Joshi, S., Bharucha, C., and Desai, A.J. (2008). Production of biosurfactant and antifungal compound by fermented food isolate *Bacillus subtilis* 20B. *Bioresource Technology* 99 (11): 4603–4608. https://doi.org/10.1016/j.biortech.2007.07.030.

72 Luna, J.M., Rufino, R.D., Sarubbo, L.A., and Campos-Takaki, G.M. (2013). Characterisation, surface properties and biological activity of a biosurfactant produced from industrial waste by *Candida sphaerica* UCP0995 for application in the petroleum industry. *Colloids Surfaces B. Biointerfaces* 102: 202–209. https://doi.org/10.1016/j.colsurfb.2012.08.008.

73 Bharali, P., Saikia, J.P., Ray, A., and Konwar, B.K. (2013). Rhamnolipid (RL) from *Pseudomonas aeruginosa* OBP1: A novel chemotaxis and antibacterial agent. *Colloids Surfaces B Biointerfaces* 103: 502–509. https://doi.org/10.1016/j.colsurfb.2012.10.064.

74 Gudiña, E.J., Fernandes, E.C., Rodrigues, A.I. et al. (2015). Biosurfactant production by *Bacillus subtilis* using corn steep liquor as culture medium. *Frontiers in Microbiology* 6: 1–7. https://doi.org/10.3389/fmicb.2015.00059.

75 Nalini, S. and Parthasarathi, R. (2014). Production and characterization of rhamnolipids produced by *Serratia rubidaea* SNAU02 under solid-state fermentation and its application as biocontrol agent. *Bioresource Technology* 173: 231–238. https://doi.org/10.1016/j.biortech.2014.09.051.

76 Sachdev, D.P. and Cameotra, S.S. (2013). Biosurfactants in agriculture. *Applied Microbial Biotechnology* 97: 1005–1016. https://doi.org/10.1007/s00253-012-4641-8.

77 Nurfarahin, A.H., Mohamed, M.S., and Phang, L.Y. (2018). Culture medium development for microbial-derived surfactants production – An overview. *Molecules* 23 (5): 1–26. https://doi.org/10.3390/molecules23051049.

78 Müller, M.M., Kügler, J.H., Henkel, M. et al. (2012). Rhamnolipids – Next generation surfactants? *Journal of Biotechnology* 162 (4): 366–380. https://doi.org/10.1016/j.jbiotec.2012.05.022.

79 Jackson, S.A., Borchert, E., O'Gara, F., and Dobson, A.D. (2015). Metagenomics for the discovery of novel biosurfactants of environmental interest from marine ecosystems. *Current Opinion in Biotechnology* 33: 176–182. https://doi.org/10.1016/j.copbio.2015.03.004.

80 Henkel, M., Müller, M.M., Kügler, J.H. et al. (2012). Rhamnolipids as biosurfactants from renewable resources: Concepts for next-generation rhamnolipid production. *Process Biochemistry* 47 (8): 1207–1219. https://doi.org/10.1016/j.procbio.2012.04.018.

81 Sha, R., Jiang, L., Meng, Q. et al. (2012). Producing cell-free culture broth of rhamnolipids as a cost-effective fungicide against plant pathogens. *Journal of Basic Microbiology* 52 (4): 458–466. https://doi.org/10.1002/jobm.201100295.

82 Dhanarajan, G. and Sen, R. (2014). Amphiphilic molecules of microbial origin: Classification, characteristics, genetic regulations, and pathways for biosynthesis, Chapter 2. In: *Biosurfactants: Research Trends and Applications* (eds. C.N. Mulligan, S.K. Sharma and A. Mudhoo), 32–42. Boca Raton, FL, USA: CRC Press.

83 Akiyode, O., George, D., Getti, G., and Boateng, J. (2016). Systematic comparison of the functional physico-chemical characteristics and biocidal activity of microbial derived biosurfactants on blood-derived and breast cancer cells. *Journal of Colloid and Interface Science* 479: 221–233. https://doi.org/10.1016/j.jcis.2016.06.051.

84 Sen, S., Borah, S.N., Bora, A., and Deka, S. (2017). Production, characterization, and antifungal activity of a biosurfactant produced by *Rhodotorula babjevae* YS3. *Microbial Cell Factories* 16 (1): 1–14. https://doi.org/10.1186/s12934-017-0711-z.

85 Habe, H., Taira, T., Sato, Y. et al. (2019). Evaluation of yield and surface tension-lowering activity of Iturin A produced by *Bacillus subtilis* RB14. *Journal of Oleo Science* 68: 1–6. https://doi.org/10.5650/jos.ess19182.

86 Romero, D., Vicente, A.d., Rakotoaky, R.H. et al. (2007). The iturin and fengycin families of lipopeptides are key factors in antagonism of *Bacillus subtilis* toward *Podosphaera fusca*. *Molecular Plant-Microbe Interactions* 20 (4): 430–440. https://doi.org/10.1094/MPMI-20-4-0430.

87 Deleu, M., Paquot, M., and Nylander, T. (2005). Fengycin interaction with lipid monolayers at the air–aqueous interface – Implications for the effect of fengycin on biological membranes. *Journal of Colloid and Interface Science* 283: 358–365. https://doi.org/10.1016/j.jcis.2004.09.036.

88 Zakharova, A.A., Efimova, S.S., Malev, V.V., and Ostroumova, O.S. (2019). Fengycin induces ion channels in lipid bilayers mimicking target fungal cell membranes. *Scientific Reports* 9: 16034. https://doi.org/10.1038/s41598-019-52551-5.

89 Borah, S.N., Goswami, D., Sarma, H.K. et al. (2016). Rhamnolipid biosurfactant against *Fusarium verticillioides* to control stalk and ear rot disease of maize. *Frontiers in Microbiology* 7: 1–10. https://doi.org/10.3389/fmicb.2016.01505.

90 Rodrigues, A.I., Gudiña, E.J., Teixeira, J.A., and Rodrigues, L.R. (2017). Sodium chloride effect on the aggregation behaviour of rhamnolipids and their antifungal activity. *Scientific Reports* 7 (1): 12907. https://doi.org/10.1038/s41598-017-13424-x.

91 Nalini, S. and Parthasarathi, R. (2018). Optimization of rhamnolipid biosurfactant production from *Serratia rubidaea* SNAU02 under solid-state fermentation and its biocontrol efficacy against *Fusarium* wilt of eggplant. *Annals of Agrarian Science* 16: 108–115. https://doi.org/10.1016/j.aasci.2017.11.002.

92 Mohammadipour, M., Mousivand, M., Jouzani, G.S., and Abbasalizadeh, S. (2009). Molecular and biochemical characterization of Iranian surfactin-producing *Bacillus subtilis* isolates and evaluation of their biocontrol potential against *Aspergillus flavus* and *Colletotrichum gloeosporioides*. *Canadian Journal of Microbiology* 55 (4): 395–404. https://doi.org/10.1139/W08-141.

93 Sarwar, A., Hassan, M.N., Imran, M. et al. (2018). Biocontrol activity of surfactin A purified from *Bacillus* NH-100 and NH-217 against rice bakanae disease. *Microbiology Research* 209: 1–13. https://doi.org/10.1016/j.micres.2018.01.006.

94 Sun, J., Li, W., Liu, Y. et al. (2018). Growth inhibition of *Fusarium graminearum* and reduction of deoxynivalenol production in wheat grain by bacillomycin D. *Journal of Stored Products Research* 75: 21–28. https://doi.org/10.1016/j.jspr.2017.11.002.

95 Qian, S., Lu, H., Sun, J. et al. (2016). Antifungal activity mode of *Aspergillus ochraceus* by bacillomycin D and its inhibition of ochratoxin A (OTA) production in food samples. *Food Control* 60: 281–288. https://doi.org/10.1016/j.foodcont.2015.08.006.

96 Gong, Q., Zhang, C., Lu, F. et al. (2014). Identification of bacillomycin D from *Bacillus subtilis* fmbJ and its inhibition effects against *Aspergillus flavus*. *Food Control* 36: 8–14. https://doi.org/10.1016/j.foodcont.2013.07.034.

97 Farzaneh, M., Shi, Z.Q., Ahmadzadeh, M. et al. (2016). Inhibition of the *Aspergillus flavus* growth and aflatoxin B1 contamination on pistachio nut by fengycin and surfactin-producing *Bacillus subtilis* UTBSP1. *Plant Pathology Journal* 32 (3): 209–215. https://doi.org/10.5423/PPJ.OA.11.2015.0250.

98 Veras, F.F., Correa, A.P.F., Welke, J.E., and Brandelli, A. (2016). Inhibition of mycotoxin-producing fungi by *Bacillus* strains isolated from fish intestines. *International Journal of Food Microbiology* 238: 23–32. https://doi.org/10.1016/j.ijfoodmicro.2016.08.035.

99 Pereyra, M.L.G., Martínez, M.P., Petroselli, G. et al. (2018). Antifungal and aflatoxin-reducing activity of extracellular compounds produced by soil *Bacillus* strains with potential application in agriculture. *Food Control* 85: 392–399. https://doi.org/10.1016/j.foodcont.2017.10.020.

100 Cao, Y., Pi, H., Chandrangsu, P. et al. (2018). Antagonism of two plant-growth promoting *Bacillus velezensis* isolates against *Ralstonia solanacearum* and *Fusarium oxysporum*. *Scientific Reports* 8: 4360. https://doi.org/10.1038/s41598-018-22782-z.

101 Hu, L.B., Zhang, T., Yang, Z.M. et al. (2009). Inhibition of fengycins on the production of fumonisin B1 from *Fusarium verticillioides*. *Letters in Applied Microbiology* 48 (1): 84–89. https://doi.org/10.1111/j.1472-765X.2008.02493.x.

102 Deleu, M., Paquot, M., and Nylander, T. (2008). Effect of fengycin, a lipopeptide produced by *Bacillus subtilis*, on model biomembranes. *Biophysical Journal* 94: 2667–2679. https://doi.org/10.1529/biophysj.107.114090.

103 Kaur, P.K., Joshi, N., Singh, I.P., and Saini, H.S. (2016). Identification of cyclic lipopeptides produced by *Bacillus vallismortis* R2 and their antifungal activity against *Alternaria alternata*. *Journal of Applied Microbiology* 122: 139–152. https://doi.org/10.1111/jam.13303.

104 Liu, J., Hagberg, I., Novitsky, L. et al. (2014). Interaction of antimicrobial cyclic lipopeptides from *Bacillus subtilis* influences their effect on spore germination and membrane permeability in fungal plant pathogens. *Fungal Biology* 118: 855–861. https://doi.org/10.1016/j.funbio.2014.07.004.

105 Barkai-Golan, R. and Follett, P.A. (2017). Irradiation effects on Mycotoxin accumulation, Chapter 4. In: *Irradiation for Quality Improvement, Microbial Safety and Phytosanitation of Fresh Produce* (eds. R. Barkai-Golan and P. Follett), 41–46. London, UK: Academic Press.

106 Krishnan, N., Velramar, B., and Velu, R.K. (2019). Investigation of antifungal activity of surfactin against mycotoxigenic phytopathogenic fungus *Fusarium moniliforme* and its impact in seed germination and mycotoxicosis. *Pesticide Biochemistry and Physiology* 155: 101–107. https://doi.org/10.1016/j.pestbp.2019.01.010.

107 Hazarika, D.J., Goswami, G., Gautom, T. et al. (2019). Lipopeptide mediated biocontrol activity of endophytic *Bacillus subtilis* against fungal phytopathogens. *BMC Microbiology* 19: 71. https://doi.org/10.1186/s12866-019-1440-8.

108 Fujita, S. and Yokota, K. (2019). Disease suppression by the cyclic lipopeptides iturin a and surfactin from *Bacillus* spp. against *Fusarium* wilt of lettuce. *Journal of General Plant Pathology* 85: 44–48. https://doi.org/10.1007/s10327-018-0816-1.

109 Hanif, A., Zhang, F., Li, P. et al. (2019). Fengycin produced by *Bacillus amyloliquefaciens* FZB42 inhibits *Fusarium graminearum* growth and mycotoxins biosynthesis. *Toxins* 11 (5): 1–11. https://doi.org/10.3390/toxins11050295.

110 Daverey, A. and Pakshirajan, K. (2010). Sophorolipids from *Candida bombicola* using mixed hydrophilic substrates: Production, purification and characterization. *Colloids and Surfaces B: Biointerfaces* 79 (1): 246–253. https://doi.org/10.1016/j.colsurfb.2010.04.002.

111 Gudiña, E.J., Rodrigues, A.I., de Freitas, V. et al. (2016). Valorization of agro-industrial wastes towards the production of rhamnolipids. *Bioresource Technology* 212: 144–150. https://doi.org/10.1016/j.biortech.2016.04.027.

112 Kahraman, H. and Erenler, S.O. (2012). Rhamnolipid production by *Pseudomonas aeruginosa* engineered with the *Vitreoscilla* hemoglobin gene. *Applied Biochemistry and Microbiology* 48 (2): 188–193. https://doi.org/10.1134/S000368381202007X.

113 Kumar, A.P., Janardhan, A., Viswanath, B. et al. (2016). Evaluation of orange peel for biosurfactant production by *Bacillus licheniformis* and their ability to degrade naphthalene and crude oil. *3 Biotech* 6 (1): 1–10. https://doi.org/10.1007/s13205-015-0362-x.

114 Ramírez, M.I., Tsaousi, K., Rudden, M. et al. (2015). Rhamnolipid and surfactin production from olive oil mill waste as sole carbon source. *Bioresource Technology* 198: 231–236. https://doi.org/10.1016/j.biortech.2015.09.012.

115 Ongena, M., Jourdan, E., Adam, A. et al. (2007). Surfactin and fengycin lipopeptides of *Bacillus subtilis* as elicitors of induced systemic resistance in plants. *Environmental Microbiology* 9 (4): 1084–1090. https://doi.org/10.1111/j.1462-2920.2006.01202.x.

116 Sun, L., Lu, Z., Bie, X. et al. (2006). Isolation and characterization of a co-producer of fengycins and surfactins, endophytic *Bacillus amyloliquefaciens* ES-2, from *Scutellaria baicalensis* Georgi. *World Journal of Microbiology and Biotechnology* 22: 1259–1266. https://doi.org/10.1007/s11274-006-9170-0.

117 Blacutt, A.A., Mitchell, T.R., Bacon, C.W., and Gold, S.E. (2016). *Bacillus mojavensis* RRC101 lipopeptides provoke physiological and metabolic changes during antagonism against *Fusarium verticillioides*. *Molecular Plant-Microbe Interactions* 29 (9): 713–723. https://doi.org/10.1094/MPMI-05-16-0093-R.

22

Biosurfactant-Mediated Biocontrol of Pathogenic Microbes of Crop Plants

Madhurankhi Goswami[1,2] *and Suresh Deka*[1]

[1] *Environmental Biotechnology Laboratory, Resource Management and Environment Section, Life Sciences Division, Institute of Advanced Study in Science and Technology (IASST), Guwahati, Assam, India*
[2] *Life Sciences Division, Department of Molecular Biology and Biotechnology, Cotton University, Guwahati, Assam, India*

CHAPTER MENU

22.1 Introduction

India is an agrarian country with 60% of its population depending directly or indirectly on agriculture, thereby contributing 18% towards total country's gross domestic product (GDP). With an increase in human population, there is a subsequent increase in global food demand. Meeting the global food demand with limited cultivable land resource is a prodigious task. Undoubtedly, green technologies have helped the country to increase agricultural produce from 82 million tonnes in 1960–1961 to 176 million tonnes in 1991–1992 and further to 264 million tonnes in 2013–2014. During the subsequent years India will be requiring more than 450 million tonnes of food to feed the growing human population and meeting the food demand that will be a promising task since the country suffers severe yield losses every year due to the ravage of fungal and bacterial phytopathogens in many of the commercial crop-growing states across India [1]. The uncontrolled emergence of fungal phytopathogens is primarily due to the induction of mutations among species resulting in the occurrence of new species. Both environmental factors and human actions contribute toward the establishment of microbial pathogens in plants [2]. The microbial phytopathogens

Biosurfactants for a Sustainable Future: Production and Applications in the Environment and Biomedicine, First Edition. Edited by Hemen Sarma and Majeti Narasimha Vara Prasad.
© 2021 John Wiley & Sons Ltd. Published 2021 by John Wiley & Sons Ltd.

were estimated to cause ≤30% of agricultural crop diseases, resulting in crop losses ranging between 10 and 30% of crop productions. The epidemics and agricultural yield loss caused by these pathogens are of serious concern [3]. The variety of direct and indirect losses in crop production due to plant diseases includes deterioration in quality and quantity of crops produced, increase in production cost, threats to animal health and the environment, loss of natural resources, and less implementation of suitable alternatives [4]. To combat with microbial infections on crop plants and the overall loss in crop yield, farmers relentlessly use chemical fungicides, pesticides, and fertilizers in the agricultural systems. The rampant use of these chemicals in agriculture has a detrimental effect on crop quality and quantity, soil health, soil microbial diversity, and microbial biological activity [5]. Therefore, it is significantly important to understand the adverse effects of chemical fertilizers much beyond their impact on agricultural productivity in order to achieve sustainability in agricultural systems.

Sustainable agriculture is a vital factor in today's world as it offers the potential to meet the global food demand using agricultural resources while maintaining and conserving environmental quality and resources. It helps in enhancing soil biodiversity and its biological activity, thereby boosting soil fertility and soil productivity, which in turn improves crop growth while restraining pests and diseases [6]. Moreover, sustainable agriculture involves the use of efficient soil microbiota as its important component. The use of these potential soil microorganisms as soil and plant inoculants shifts microbiological equilibrium in ways that would enhance soil fertility, resulting in eco-friendly agricultural practices [7]. The soil microorganisms in association with plants produce certain growth promoting compounds and secondary metabolites like siderophores, antibiotics, and biosurfactants that improve plant growth while providing protection against a wide range of phytopathogens [8]. Among the antimicrobials, microbial biosurfactants are the most widely studied secondary metabolites. Biosurfactants are surface active, structurally diverse biomolecules produced by microorganisms with defined hydrophilic and hydrophobic moieties. These are produced either at microbial cell surfaces or excreted extra-cellularly. They exhibit unique properties like specificity, low-toxicity, biodegradability, and ease of preparation [9]. Most importantly, the phenomenal antifungal activity shown by these surface-active molecules has attracted wide interest in recent years. This chapter will focus on the antimicrobial potency of low molecular weight lipopeptidic and glycolipidic biosurfactants on fungal phytopathogens.

22.2 Biosurfactant: Properties and Types

Microbial biosurfactants are amphiphilic molecules containing a hydrophilic head and a hydrophobic tail and can reduce surface tension or interfacial tension effectively at the air–water or water–oil interfaces. Microbial surfactants have several advantages over the synthetic chemical-derived surfactants due to their low toxicity, high biodegradability and bioavailability, eco-friendly nature, low production cost, high foaming, salinity and pH tolerance, and diverse biological activities that include antifungal, antibacterial, antiviral, and insecticidal properties. Thus, they are safe and have the potential as alternatives, which draw the attention of commercial sectors like pharmaceuticals, cosmetics, food, and agriculture. Moreover, the growing environmental distress due to usage of chemical surfactants in various commercial sectors initiates a new focus on natural biosurfactants due to their harmless characteristics [10].

Biosurfactants can be classified into different classes based on their molecular weight, either low molecular weight or high molecular weight biosurfactants, ionic charges, i.e. anionic, cationic, non-ionic, and neutral, and secretion type, either intracellular, extracellular, or cell adhered

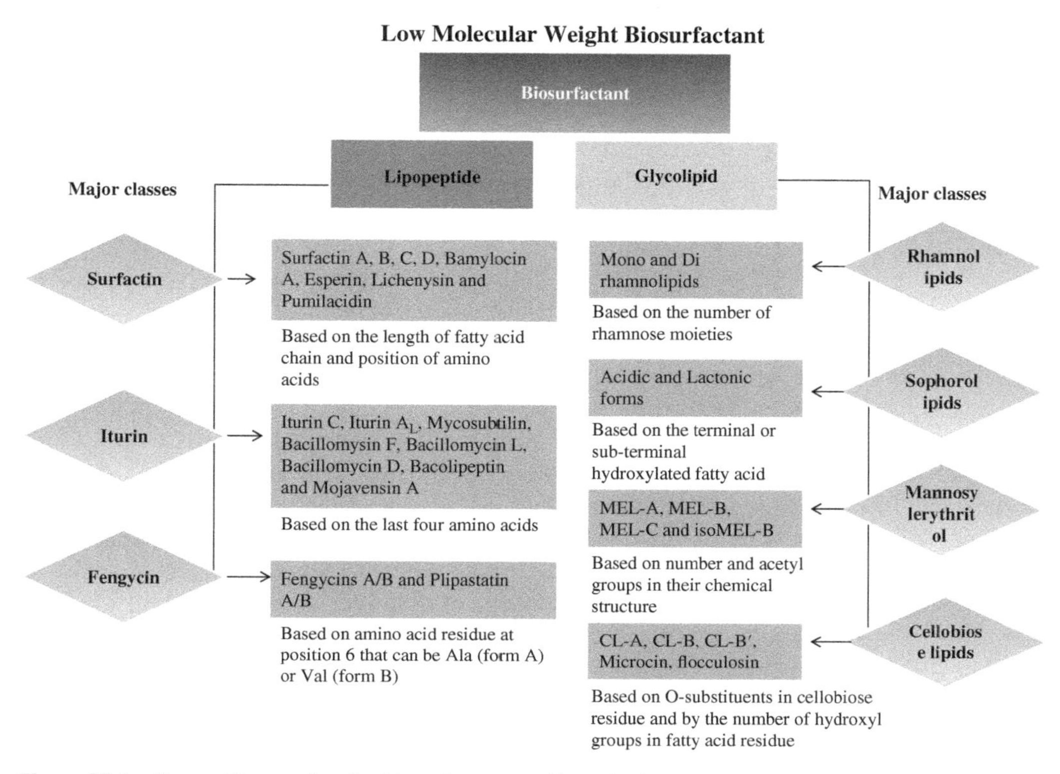

Figure 22.1 Types of low molecular biosurfactant and its sub-classes.

(Figure 22.1). However, the chemical structure presents the main criteria for classification of microbial surfactants [11]. In the present chapter we will discuss the two major low molecular weight biosurfactant class lipopeptide and glycolipids, and their role as antimicrobials in agriculture.

22.2.1 Mechanistic Insights of Biosurfactant Targeting Microbial Cells

It is always important to understand how an antimicrobial agent works to perk up the prospects for its implementation in a wide number of fields. Existing literature explains the interaction pattern between biosurfactant and microbial cells [12]. Interaction of a biosurfactant molecule with the phospholipidic bilayer of microbial cell causes incorporation or intercalation of the molecule within the lipid bilayer or adsorption on the cell surface. As a consequence, significant alterations arise in the microbial cell surface properties that include significant increase in fluidity of the phosphatidylserine acyl chains, decrease in hydration of interfacial regions of the phospholipid bilayer, changes in membrane conformation and formation of membrane pores and ion channels. Briefly, biosurfactants act against the microbial cells initially by damaging the outer cell layer followed by penetration of the cell wall or cell membrane through passive diffusion, which results in coagulation and leakage of intracellular constituents. These bioactive molecules can also target the microbial cells by affecting the interaction pattern of the microorganisms with different surfaces. They do so by altering the hydrophobicity of microbial cell surface [13].

22.2.2 Lipopeptide Biosurfactant

22.2.2.1 A Brief Overview on Lipopeptide Structure, Generation, and Its Variants

Lipopeptides (LPs) are low molecular weight cyclic amphiphilic oligopeptides produced by fungi and various bacterial genera, predominantly by *Bacillus* and *Pseudomonas* [14]. Structurally, they are constituted of fatty acid in combination with a peptide moiety corresponding to an isoform group. The isoform group differs by the composition of peptide moiety (composed of amino acids), length of the fatty acid chain, and the link between the parts [15]. The number of amino acids in the peptide moiety ranges between 7 and 25 and the length of the fatty acid chain varies from 13 to 17 carbons. They are synthesized by multienzyme complexes called non-ribosomal peptide synthetases (NRPs). These synthetases are organized on modules that permit the incorporation of each unit of a lipopeptide molecule. Initially there is incorporation of a specific amino acid, followed by condensation, termination, and cyclization of the peptide chain. The synthesized peptides include d-amino acids, β-amino acids, and hydroxyl or *N*-methylated amino acids. The complete LP molecule remains inactive until and unless the peptide moiety is coupled to a fatty acyl chain and the lipid chain fuses with the *N*-terminal residue of the peptide chain. This entire process of LP generation is further accompanied by structure modification via glycosylation or halogenation by specific enzymes associated with the NRP synthetases. LPs are highly variable and their structural analogs result from consequent changes in amino acid substitutions [16]. They are widely known for their antimicrobial, cytotoxic, immunosuppressant, antitumor, and surfactant properties. The mode of action of LPs has been represented in Figure 22.2.

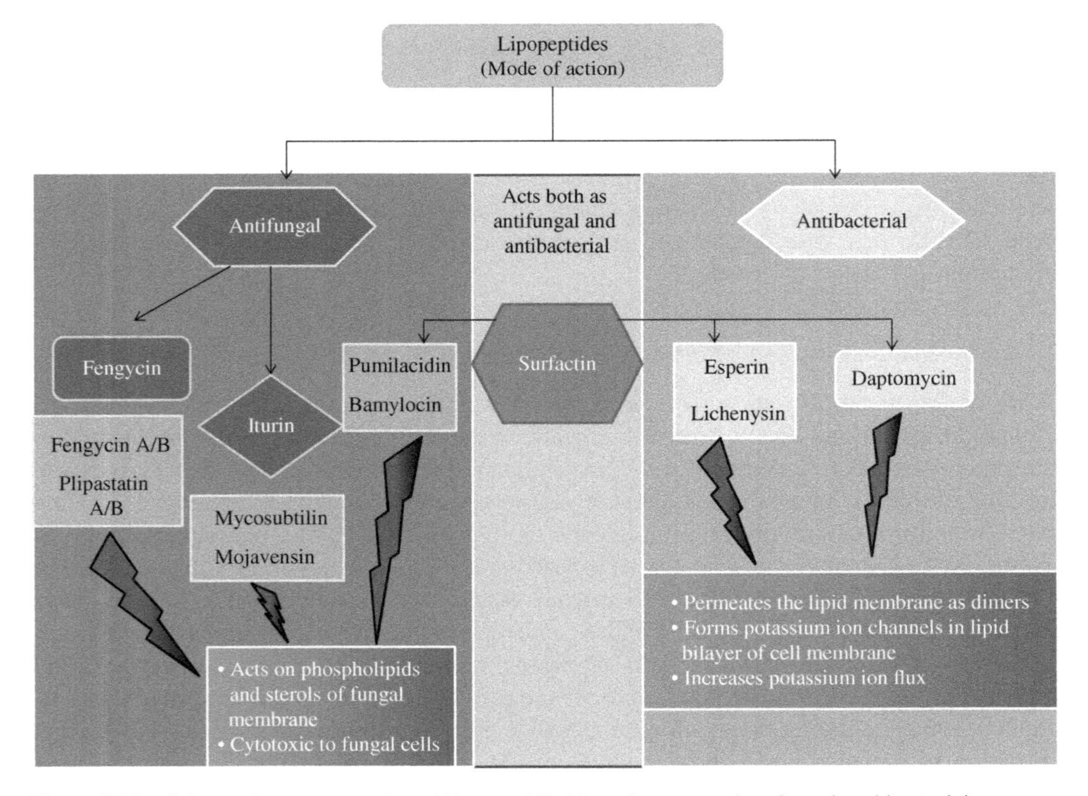

Figure 22.2 Schematic representation of lipopeptidic biosurfactant against fungal and bacterial phytopathogens.

Bacillus sp. produces a wide variety of lipopeptide biosurfactants with promising antibacterial and antifungal activity. Unlike other biosurfactants, LPs belonging to surfactin, iturin, and fengycin are formed without the participation of NRPs/or polyketidesynthases [17]. *Bacillus* specific LPs are either cyclopeptides, i.e. iturins or macrolactones that include fengycins and surfactins, characterized by the presence of D and L amino acids and variable hydrophobic tails [18]. Surfactins are a class of cyclic lipoheptapeptide containing a chiral sequence of seven D and L amino acid residues interlinked to a β-hydroxy fatty acid chain. The solubility and surface activity of surfactin depends on the arrangement of the amino acid residues. Any possible changes of amino acids at positions 2, 4, and/or 7 of surfactin (to more hydrophobic residues) will increase the surface activity of the lipopeptide molecule while decreasing its critical micelle concentration (CMC) [19]. The variation in the formation of the surfactin profile is principally due to variations in cultivation parameters that includes pH, temperature, degree of aeration, culture medium, dissolved oxygen concentration, and on genetically encoded differences [20]. The surfactin family can be further divided into Bamylocin A, Esperin, Lichenysin, and Pumilacidin based on the presence of amino acids in positions 1 and 4. Surfactin LPs are effective against Gram-positive and Gram-negative bacteria and certain fungal phytopathogens. They act on microbial cells by permeating the lipid membranes as dimers and forming ion channels in a phospholipid bilayer membrane resulting in leakage of cellular components leading to cell death.

Iturins are a class of antifungal cyclic LPs characterized by a peptide ring of seven amino acid residues including an invariable d-Tyr2 with a constant chiral sequence LDDLLDL closed by a C_{14}–C_{17} aliphatic β-amino acid. An amino acid variation in iturin molecules results in a high degree polymorphism. There are several variants of iturins like Iturin C, Iturin $A_{L,}$ Mycosubtilin, Bacillomysin F, Bacillomycin L, Bacillomycin D, Bacolipeptin, and Mojavensin A [21]. All these bioactive compounds have a similar LDDLLDL chiral sequence with a restricted number of amino residues, i.e. Asx, Glx, Pro, Ser, Thr, Tyr, a common peptide cycle, and a variable moiety containing four amino acid residues (X4 to X7) [22]. Iturins act on microbial cells by forming potassium ion-conducting channels in lipid bilayers and cause cell permeabilization that results in an overall increase in electrical conductance leading to cell death. Thus, it can be ascertained that antifungal activity of iturin is related to its interaction with the cytoplasmic membrane of the target cells, which results in increased K^+ permeability [23].

Fengycins are a class of cyclic LPs with a peptide chain of 10 amino acids linked to a β-hydroxyl tail involving C_{14} to C_{18} carbon atoms forming a cyclic lactone ring [24]. Fengycins include fengycins A/B and plipastatin A/B, which differ in their amino acid residue at position 6, i.e. Ala (form A) or Val (form B) amino acid residue [18]. Fengycins show a wide variety of bioactive properties: it is reported to inhibit fungal growth and cause perturbation, bending, and micelle formation on artificial membranes [25]. Fengycins act by creating sustainable pores in the phospholipid bilayer, thereby causing complete efflux of intracellular contents of affected cells. The antifungal activity of fengycins is mostly related to the composition of phospholipids and sterols in membranes [26].

Pseudomonas sp. are broad-based organisms found in a multitude of habitats. In recent years, *Pseudomonas* sp. were reported to produce a number of cyclic LPs of which close to 100 LPs have been individually documented to varying extents. Based on the structural characteristics, *Pseudomonas*-related LPs were classified into several groups that include viscosin, amphisin, tolaasin, syringomycin, pseudophomin, pseudomycin, nunapeptin and nunamycin, syringopeptin, arthrofactin, putisolvins I and II, orfamide, and pseudodesmins A and B [27]. They are short oligopeptides linked to a fatty acid tail. For instance, viscosin and amphisin are a class of cyclic LPs containing 9 and 11 amino acids, respectively, linked to a 3-hydroxydecanoic lipid tail. Unlike the other two, tolaasin is a class of diverse cyclic lipopeptides containing various, i.e. 19–25 unusual amino acids linked to a lipid tail (3-hydroxydecanoic or 3-hydroxyoctanoic acid). Likewise, syringomycin is another variant of *Pseudomonas*-related cyclic lipopeptide containing nine unusual amino acids and a C-terminal chlorinated threonine residue. Several novel cyclic LPs have continuously been reported in multiple habitats. These newly characterized LPs are either assigned to an existing group or are grouped into a new class of LPs. The recently identified LPs include bananamides [28] and xantholysins [29], with certain exceptions like entolysins [30], pseudofactins [31], syringopeptins and putisolvins [27], corpeptin [32], and fuscopeptins [33], which have formally not been assigned into any group (Figure 22.1). Overall, *Pseudomonas*-produced LPs have structural features either in the length of the fatty acid chain or the amino acid residues that makes them different from the four main cyclic LP group. *Pseudomonas*-related LPs were reported to possess a wide array of bioactivities that include antibacterial, antiviral, antimycoplasma, and antifungal ones. They were also reported to play an important role in virulence and motility. The bioactivities of *Pseudomonas*-produced LPs are predominantly governed by the number, nature, position, and configuration of the amino acids in the macrocyclic peptide ring as well as the type and length of the fatty acid tail [34].

22.2.2.2 Miraculous Activities of Lipopeptide in Agriculture

Lipopeptide: An Antagonistic Tool to Combat Fungal Phytopathogens The diverse structure of LP biosurfactants enables them to perform indefinite bioactivities. They are well known for disturbing membrane permeability, by inducing pores and ion channels in the lipid bilayer membrane, enabling the LP molecule to act against bacteria and fungi, thus causing cell disorganization. The advent of increased resistance towards the existing antimicrobial drugs among the pathogenic microorganisms is imperiling the value of these drugs and hence urges for comprehensive efforts to design and develop new antimicrobial agents. With this view, several studies have been undertaken to recognize *Bacillus*-related LPs and evaluate their effectiveness against different fungal phytopathogens [35]. At the same time, several *Pseudomonas* [36] and actinomycetes-related LPs were also reported to exhibit potential antifungal activity against a wide range of fungal phytopathogens. Thus, these compounds, at present, serve as potent candidates in agriculture to minimize the use of chemical fungicides and surfactants. Existing literature reports surfactin and lichenysins to be more potent as antibacterial, anti-*Mycoplasma*, antiviral, antitumor agents, and a mild antifungal agent, while iturins were reported to be stronger antifungal agents [37].

Surfactins isolated from different *Bacillus* sp. have been shown to exhibit strong antifungal activity against a wide array of fungi like *Fusarium oxysporum*, *Fusarium moniliforme*, *Fusarium solani*, *Trichoderma atroviride*, *T. reesei*, and many more. For instance, surfactin obtained from *Bacillus* NH-100 and NH-217 strains and *Bacillus thuringiensis* pak2310, were reported to show strong antifungal activity against *F. moniliforme* and *F. oxysporum* at 2000 and 50 μg/ml concentration, respectively [38, 39]. Likewise, Bamylocin A, a variant of surfactin, was reported for its potential antifungal activity against certain well-known phytopathogens, like *Botrytis cineria*, *F. oxysporum*, and *Rhizoctonia solani*, a causal agent of gray-mold disease, *Fusarium* wilt, and damping-off diseases occurring in crop plants [40]. Similarly, pumilacidin, which is an another variant of surfactin lipopeptide, was reported to exhibit potential antifungal activity against a few common fungal phytopathogens that include *R. solani*, *Pythium aphanidermatum*, and *Sclerotinia rolfsii* [41]. Iturins, a class of cyclic LPs, were known for their potent activity against a range of devastating phytopathogenic fungi like *Alternaria panax*, *Botrytis cinera*, *Colletotrichum orbiculare*, *Penicillium digitatum*, *Pyricularia grisea*, and *Sclerotinia sclerotiorum* [42]. They were believed to cause perforations in microbial cell membranes, causing spontaneous release of K^+ ions followed by the release of macromolecular components such as proteins, nucleotides, and polysaccharides, leading to cell death [43]. Among the iturins, bacillomycins are the most potent antifungal agents [44]. The two well-known group of bacillomycins, bacillomycin D and bacillomycin L, were reported to exhibit significant antifungal activity against a few common plant pathogens like *Fusarium graminearum*, *Colletotrichum gloeosporioides* (Penz.), and *Botryosphaeria dothidea* (Moug. ex. Fr). Bacillomycin treatment resulted in severe alterations and damage to fungal mycelia as obtained through ultramicroscopic studies. They are reported as being most potent due to their ability to inhibit fungal growth at a median inhibitory concentration as low as 2.162 μg/ml [29, 45, 46]. Likewise, mycosubtilin is another subclass of the iturin family and is best known for its significant activity against fungal phytopathogens. Mycosubtilin is mostly derived from different strains of *Bacillus subtilis*. Mycosubtilin (*B. subtilis* BBG100) was reported to display strong antifungal activity against *P. aphanidermatum*, the causal agent of damping-off in tomato seedlings with subsequent help in seed germination and plant growth [17]. A novel iturinic lipopeptide mojavensin A (from *Bacillus mojavensis* B0621A) was reported to exhibit dose-dependent antifungal activity against *Valsa mali* and *F. oxysporum* f. sp. *cucumerinum* at concentrations over 2 mg/ml [47]. Apart from *V. mali* and *F. oxysporum* f. sp. *cucumerinum,* mojavensin A (*B. mojavensis* RRC101) was reported to reduce disease severity of *Fusarium verticillioides* and accumulation of fumonisins when co-inoculated in maize plants [48]. Fengycins, a class of cyclic LPs, were known for their antifungal activity against a wide array of fungal phytopathogens. Fengycin homologs could effectively inhibit the growth of several filamentous fungi like *Aspergillus niger*, *S. sclerotiorum*, *Gibberella zeae*, *Mucor rouxii*, *F. graminearum*, and *Rhizopus stolonifer*. Fengycins were reported to display promising antifungal activity against the widely known *Magnaporthe grisea*, the causative agent of rice blast disease in rice cultivars. They target the cell by inducing cytoplasm content leakage and defects in cell wall integrity, resulting in cell death [49]. Literature reveals that clusters of fengycin homologs (fengycin A and fengycin B), a few novel fengycin molecules, and plipastatin A obtained from different strains of *B. subtilis* and *Bacillus amyloliquefaciens* can strongly suppress the disease incidence caused by plant pathogens like *Fusarium moliniforme* Sheldon ATCC 38932, *R. solani*, *F. oxysporum* f. sp. *cucumerinum*, *F. graminearum*, *F. oxysporum* f. sp. *cucumis melo* L, *F. oxysporum* f. sp. *Vasinfectum*, and *F. graminearum* f. sp. *zea mays* L [50–53].

A large body of research has been focused on *Pseudomonas* related LPs for its potential antifungal activity. There are a few reports that revealed the effect of viscocinamide treatment (*Pseudomonas*-related lipopeptide) on fungal phytopathogens like *Pythium ultimum* and *R. solani*. Viscocinamide treatment resulted in a prominent reduction in fungal mycelia density, oospore formation, intracellular activity, hyphal swelling, hyphal septation, esterase activity, intracellular pH, and mitochondrial organization and activity [54]. Similarly, pseudophomins (A and B) are another group of cyclic lipodepsipeptides, isolated from a range of *Pseudomonas* sp., with promising antifungal activity against several phytopathogens. As reported, pseudophomins (A and B), isolated from *Pseudomonas fluorescens* strain BRG100, showed strong inhibition activity against a few of the common phytopathogens, *Phoma lingam*, *Alternaria brassicae*, *S. sclerotiorum*, and *R. solani*. Among the two subclasses of pseudophomins, psedophomins B were reported to be more potent and effective against the phytopathogens as compared to pseudophomins A [55]. Like pseudophomins, pseudomycins are also known for their promising antifungal activity against a few of the members of sac fungi ascomycota. Pseudomycins (A, B, C, C′), predominantly isolated from *Pseudomonas syringae* strains, were known for their antifungal activity against phytopathogen *Ophiostoma* sp., causing Dutch elm disease in elm trees [56]. Syringomycins and syringopeptins are another group of *Pseudomonas*-related necrosis inducing lipopeptide phytotoxins produced mainly by plant-associated bacterium *P. syringae* pv. *syringae*. They were reported to exhibit antimicrobial activity against fungal phytopathogens *Geotrichum candidum* and *Botrytis cinerea* [57]. Among the syringomycins, syringomycin E isolated from *P. syringae* (strains ESC-10 and ESC-11) was reported to successfully inhibit the post-harvest diseases in citrus fruits caused by the fungal phytopathogen *P. digitatum* and *G. candidum* var. *citri-aurantii* at a mere concentration of 5.26–5.45 mg/ml [58]. In addition to the aforementioned LPs, nunamycins and nunapeptins are another class of LPs predominantly produced by *P. fluorescens* In5. They are well known for their promising antimicrobial activity against fungal phytopathogens like *R. solani* and *P. aphanidermatum* [59]. Overall, LPs play a major role in suppressing the havoc caused by different fungal phytopathogens in agricultural sectors resulting in severe yield loss. Due to their promising antifungal activity they can be used as alternatives to chemical substitutes. Further, a detailed study on an individual lipopeptide is essential for understanding the biological activity of each of the LPs and the way they act against various biological targets.

Mode of Action of Lipopeptide Biosurfactants LPs display a broad spectrum of activity against numerous pathogens and serve as potential candidates for plant defense. Initially the LPs interact electrostatically with the bacterial cell membrane that follows their insertion into the hydrophobic core of the cell membrane leading to the formation of non-specific ion channels/pores. These antimicrobials were believed to inhibit the synthesis of essential cell wall components that ultimately cause cell death [60]. The activity of lipopeptide molecules is enhanced by the presence of an acyl chain. The acyl chain behaves as a membrane anchor and protects the peptide sequence from proteolytic degradation. Moreover, acylation of a peptide sequence increases membrane affinity and antimicrobial activity of the lipopeptide molecule [61].

Lipopeptide as an Alternative Against Bacterial Phytopathogens for Sustainable Agriculture Plant pathogens cause severe loss in the economy as well as production in the agricultural sector. A few hundred bacterial species are responsible for causing numerous plant diseases, resulting in severe yield loss [62]. LPs from *Bacillus*, *Pseudomonas*, or *Actinomycetes* display potential antagonistic activity against a wide range of Gram-positive and Gram-negative bacteria [63]. Surfactin, the most widely studied cyclic lipopeptide, predominantly produced by various strains of *B. subtilis*, is well known for its efficacy against bacterial phytopathogens like *P. syringae* pv. *tomato* infecting *Arabidopsis* roots and *Acidovorax citrulli* and causing fruit bloth in watermelons [64, 65]. However, despite its promising biological activity, there are scanty reports that focus on their antibacterial activity against different phytopathogens and thus require further research to explore the potential of the biomolecule surfactin for wide-scale application. Similarly, iturins, a group of *Bacillus*-related LPs, were known to exhibit potential antibacterial activity against several devastating plant pathogens. There is a handful of literature explaining the promising nature of iturin-like LPs against bacterial phytopathogens like *Xanthomonas campestris* pv. *cucurbitae* and *Pectobacterium carotovorum* subsp. *carotovorum* [66]. Likewise, fengycins, although known for its potent antimicrobial activity, but its potency against bacterial phytopathogens is reported only in a handful of literature. For instance, fengycin lipopeptide derived from an autochthonous bacteria *B. amyloliquefaciens* MEP_218 was reported to show promising antibacterial activity at a minimum inhibitory concentration of 6.25 μg/ml against a phytopathogen *Xanthomonas axonopodis* pv *vesicatoria*, which causes bacterial spot disease of peppers and tomatoes. Fengycin treatment resulted in drastic alterations in bacterial surface topography and cell damage, as revealed by atomic force microscopy and potassium efflux assays. This was indicated by a decrease in bacterial cell heights and loss of intracellular components [67]. There is limited information regarding the activity of *Pseudomonas*-related LPs against bacterial phytopathogens so there is the need to develop *Pseudomonas*-related LPs as antimicrobials against fungal and bacterial phytopathogens for future consideration in agricultural sectors.

22.2.3 Glycolipid Biosurfactant

22.2.3.1 A Brief Elaboration on the Variants of Glycolipid Biosurfactant

Glycolipids are the most well known and widely studied microbial surfactants. Glycolipids are amphiphilic molecules composed of a fatty acid linked by a glycosidic bond to a carbohydrate moiety. Based on the carbohydrate moiety and lipid backbones present in the glycolipids molecule, they can be classified into a number of subclasses that includes rhamnolipids, sophorolipids, trehalolipids, cellobiose lipids (CBL), mannosylerythritol lipids (MELs), lipomannosyl-mannitols, lipomannans and lipoarabinomannanes, diglycosyl diglycerides, monoacylglycerol, and galactosyl-diglyceride [68]. Among the glycolipids, rhamnolipids produced by *Pseudomonas* sp., sophorolipid by different strains of yeast, and CBL by the members of the class ustilaginales are the most studied and commonly known glycolipids [69]. Glycolipids stand out as being the most effective antimicrobial in agriculture due to its high efficacy, low toxicity, good biodegradability, and eco-friendly nature. Most importantly, they can be easily produced in large quantities from inexpensive raw materials like industrial wastes and oily byproducts, including hydrocarbons, frying oil waste, and olive oil waste. The mode of action of glycolipids has been represented schematically in Figure 22.3.

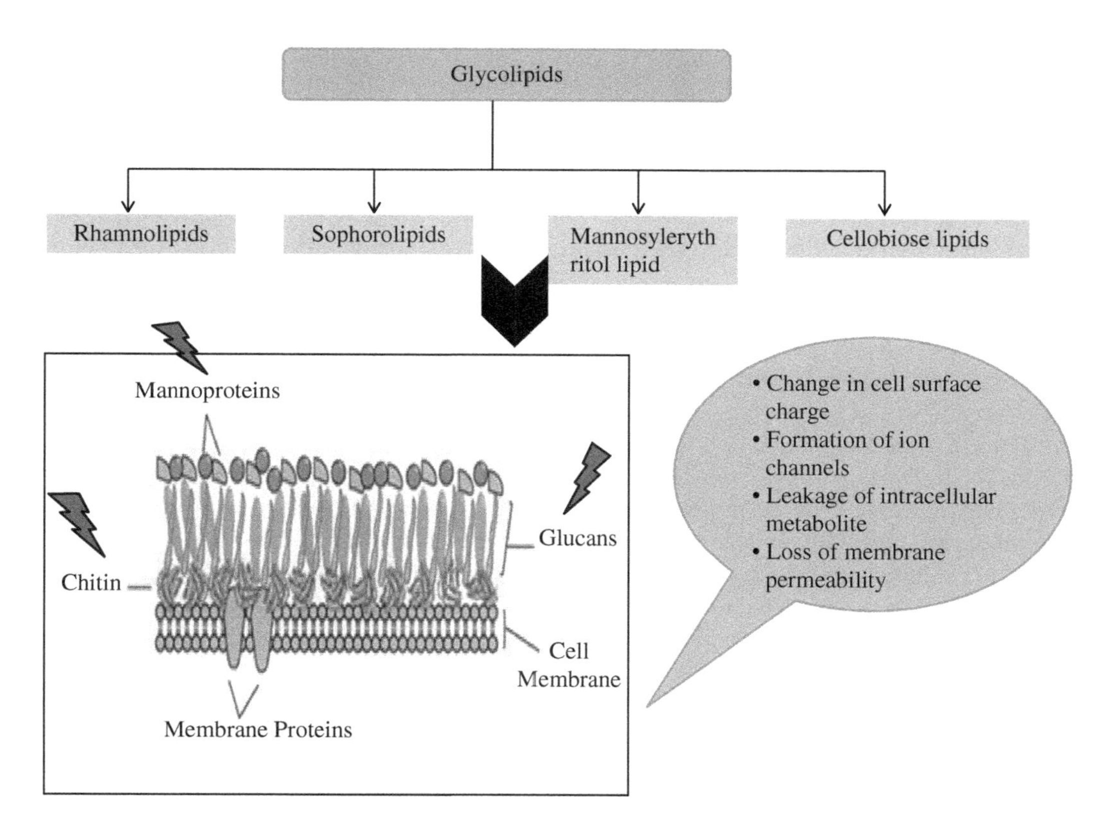

Figure 22.3 Diagrammatic representation of mode of action of glycolipids against fungal phytopathogens.

Rhamnolipids are a class of glycolipids mainly produced as an extracellular lipid predominantly by various strains of *Pseudomonas aeruginosa*. In the due course of time several rhamnolipid-producing strains were characterized that include recombinant strains like *Pseudomonas putida* [70] and *Escherichia coli* [71]. Rhamnolipids were primarily synthesized by bacterial cells in two primordial forms: monorhamnolipid (rhamnosyl-β-hydroxydecanoyl-β-hydroxydecanoate) and dirhamnolipid (rhamnosylrhamnosyl-β-hydroxydecanoyl-β-hydroxydecanoate) [72]. Genes involved in rhamnolipid synthesis are usually organized in clusters. This cluster includes two genes, rhlA and rhlB, that encode the enzyme rhamnosyltransferase that enables the transfer of rhamnose from TDP-rhamnose to β-hydroxydecanoyl-β-hydroxydecanoate and a regulatory gene that positively regulates rhamnolipid synthesis by encoding a transcriptional activator protein RhlR [73, 74]. Rhamnolipids have been known to display promising antifungal, antiviral, as well as antibacterial activity and stand as a potential alternative in pharmaceuticals, cosmetics, or agriculture over their synthetic counterparts. They are active against a wide range of Gram-positive and Gram-negative bacteria and a number of fungal phytopathogens. The rhamnolipid molecules interact with the bacterial cell and result in a reduction in lipopolysaccharide content and alterations in the outer cell membrane proteins of the bacteria [75]. Thus, they have a direct impact on the microbial cell surface by altering cell surface hydrophobicity and cell membrane permeability.

Sophorolipids are amphiphilic molecules consisting of carbohydrate sophorose as the hydrophilic part and a hydroxylated fatty acid tail of 16 or 18 carbon atoms as the hydrophobic part. There are mainly two types of sophorolipids: acidic and lactonic [76]. The hydroxyl fatty acid moiety of the acidic form has a carboxylic acid functional group while the lactonic forms of sophorolipids produce a macrocyclic lactone group with a 4″-hydroxyl group of the sophorose by intramolecular esterification [77]. They display broad spectrum biological activities by anticancer, antimycoplasma, antiviral, antiadhesive and antibiofilm methods. Additionally, they also exhibit antimicrobial activity against a number of bacteria and yeasts [78] and a few species of fungi [79]. Sophorolipids interact with the microbial cellular membrane, resulting in outpouring of cytoplasmic components resulting in disruption of cell membrane integrity and causing cell death [80].

CBLs are comprised of an O-glycosidic bond with a ω-hydroxyl group of 15, 16-dihydroxyhexadecanoic acid or 2, 15, 16 – trihydroxyhexadecanoic acid. The complex form of CBLs includes a sugar moiety with an acetyl group and a short-chain β-hydroxy fatty acid. *Ustilago maydis* was the first described producer of CBLs, referred to as ustilagic acids (UAs). Apart from *U. maydis*, there are several other species that produce a number of variants of CBL. Flocculosin, a type of cellobiose lipid secreted by *Pseudozyma flocculosa*, differs from UA in the hydroxylation pattern of the lipid chain and in the carbohydrate decoration. Likewise, *P. graminicola* and *P. fusiformata* produce CBLs that are composed of a β-hydroxylated C6 short-chain fatty acid and a α-hydroxylated long-chain tail. *Trichosporun porosum* and *Cryptococcus humicola* produce CBLs consisting of four acetyl groups attached to the cellobiose dissacharide. It is unlikely that *Sympodiomycopsis paphiopedili* produces CBLs with no decoration of the sugar head [81]. This group of glycolipids were believed to exhibit membrane damaging properties. They can intercalate into the liposomal lipid matrix, resulting in permeabilization of the lipid layer, which leads to cell death [82].

MEL comprises 4-O-β-D-mannopyranosyl-meso-erythritol as the hydrophilic group and an acetyl group as the hydrophobic moiety. They are predominantly produced by *Pseudozyma* sp. and as a minor component by *Ustilago* sp. and *Schizonella melanogramma*. MELs have been shown to reduce the surface tension of water to less than 30 mN/m. The variants of MEL are distinguished from each other based on the number of acetyl groups in their chemical structure. Recently, they have grabbed most of the attention due to their eco-friendly nature, mild production conditions, structural diversity, self-assembling properties, and versatile biological activities [83]. Although they have been used on a commercial scale for various industrial and medical applications due to their specific properties, to date these molecules have not been practically applied in agricultural sectors [84]. There are reports of MELs showing potential antifungal activity against different fungal phytopathogens but this activity has not been exploited for developing potential antimicrobials for large-scale use in agricultural sectors.

Overall, glycolipids can be a suitable alternative as antimicrobials against a range of fungi and bacteria in agricultural sectors over the chemical substitutes because of their high production yields at low cost and versatile biochemical and biological properties.

22.2.3.2 Glycolipids Biosurfactants in Agriculture

Glycolipids: A Defensive Inhibitor Molecule for Fungal Phytopathogens Among the glycolipids, rhamnolipids were the most widely studied biosurfactants owing to their significant properties. Rhamnolipids have been reported to exhibit potential antimicrobial activity against several Gram-positive and Gram-negative bacterial phytopathogens as well as phytopathogenic fungi. Several studies were undertaken to evaluate and assess the antimicrobial efficacy of rhamnolipids against different phytopathogens. Rhamnolipids, obtained from various strains of *P. aeruginosa*, were assessed against several commonly occurring phytopathogenic fungal species like *Chaetonium globosum*, *P. funiculosum*, *Gliocadium virens*, *F. solani*, *Fusarium oxysporium*, *Phytophthora nicotianae*, and *Macrophomina phaseolina*. Rhamnolipids were found to inhibit the growth of fungal mycelia of *C. globosum*, *P. funiculosum*, *G. virens*, *F. solani*, *M. phaseolina*, *F. oxysporium*, and *P. nicotianae* at a minimal inhibitory concentration of 64, 16, 32, 75, 300, 400, and 450 μg/ml, respectively [85, 86]. This inhibition was due to interaction of the rhamnolipid molecule, with the fungal cell resulting in a change in cell surface nature, forming ion channels in plasma membrane and metabolite leakage due to the loss of membrane permeability [87]. Furthermore, studies were undertaken to establish the efficacy of rhamnolipid biosurfactant as an antifungal agent under *in planta* conditions. In this context, the study undertaken by Yan et al. [88] was successful in establishing the efficacy and potency of rhamnolipid biosurfactant against the devastating pathogen *Alternaria alternata* under *in planta* conditions. A significant reduction in decay incidence of *A. alternata* on cherry tomato fruit was reported on treatment with *P. aeruginosa* derived rhamnolipid under field conditions. A combination of biocontrol yeast *Rhodotorula glutinis* with 500 μg/ml rhamnolipid biosurfactant was observed to be more effective in disease suppression of *Alternaria alternate* than mere application of rhamnolipid or *R. glutinis* individually. Combinatorial treatment stimulated peroxidase, polyphenoloxidase, and phenylalanine ammonialyase activities of cherry tomato fruit, which were stronger in inducing natural resistance of cherry tomato plants than single treatments, while enhancing colonization of yeasts on the fruit surface. There are a few more instances that reported successful implementation of rhamnolipid biosurfactant under *in planta* conditions. For instance, rhamnolipid derived from *P. aeruginosa* strains SS14, DS9, and JS29 were reported to suppress *F. oxysporum* f. sp. *pisi*, a causal agent of wilt in *Pisum sativum* L, *Fusarium sacchari*, the causal agent of pokkah boeng disease of sugarcane, and *Alternaria solani*, the causal agent of early blight of tomato under *in planta* conditions [36, 89, 90]. Likewise, rhamnolipid derived from *P. aeruginosa* ZJU211 was tested against certain members of oomycetes, such as *Phytophthora infestans*, *Phytophthora capsici*, ascomycetes like *F. oxysporum*, *B. cinerea*, and *F. graminearum*, and zygomycetes like *Mucor circinelloides* and *Mucor* sp., and it was reported to be significantly effective against all the tested phytopathogens [91]. Nalini and Parthasarthi [92] have studied the efficacy of rhamnolipid, derived from *Serratia rubidaea* SNAU02, by soil and foliar application at various concentrations against the phytopathogen *F. oxysporum* f. sp. *melongenae*. They observed that treating T9, i.e. *F. oxysporum* f. sp. *melongenae* (10^6 spores/ml), with 50 ml of 250 mg/ml biosurfactant to soil and foliar spraying of biosurfactant (250 mg/ml) was as effective as any synthetic fungicide and can be a promising agent in the biocontrol of *Fusarium* wilt of eggplant and thus help in minimizing their yield loss.

Sophorolipids biosurfactants were reported to be involved in plant protection through inhibition of phytopathogenic fungal growth. Sophorolipids show promising activity against a wide range of phytopathogens like *Fusarium* sp., *F. oxysporum*, *F. concentricum*, *Pyricularia oryzae*, *P. ultimum*, *R. solani*, *Gaeumannomyces graminis* var. *tritici*, *Alternaria kikuchiana*, and *P. infestans*. For example, the study carried out by Chen et al. [93] revealed the efficacy of sophorolipid derived from *Wickerhamiella domercqiae* Y_{2A} in inhibiting the cell growth of phytopathogenic fungi *P. infestans*, inducing morphological changes in *P. infestans* hyphae, and reducing β-1, 3-glucanase activity. Moreover, foliar application of sophorolipid at a concentration of 3 mg/ml was reported to reduce severity of late blight of potato under greenhouse conditions. There are a few other studies; for instance, Kim et al. [79] and Lee and Kim [94] have reported the potentiality of sophorolipid biosurfactant against the phytopathogen *B. cinerea*, *Phytophthora* sp., and *Pythium* sp. with overall inhibition in mycelial growth and zoospore motility with the highest zoospore lysis. Combining the results, it can therefore be ascertained that sophorolipids work potentially as antifungal agents against numerous phytopathogens. However, a further in-depth study is necessary for establishing sophorolipids as antifungal agents on a commercial scale.

CBLs derived from Basidiomycetous yeasts *C. humicola* and *P. fusiformata* were reported to exhibit fungicidal activity against different yeasts including pathogenic *Cryptococcus* and *Candida* species. *P. fusiformata* derived CBLs were reported to be more effective in inhibiting the growth of phytopathogenic fungi *S. sclerotiorum* and *Phomopsis helianthi* rather than CBLs from *C. humicola* [82]. There are very scanty literatures that have revealed the biological activity of CBLs against different fungal phytopathogens and there is a need for an in-depth study to consider cellobiose biosurfactant as an antimicrobial in agricultural sectors. MEL, a subclass of CBLs and the variants of MEL (MEL-A, MEL-B, MEL-C, and isoMEL-B) are known for their potency as antifungal agents against a broad range of fungal phytopathogens like wheat powdery mildew fungus, *Blumeria graminis* f. sp. *tritici* strain T-10, mulberry anthracnose fungus, *Colletotrichum dematium* strain S9733, strawberry anthracnose fungus, *Glomerella cingulata* strain S0709, and rice blast fungus, *M. grisea* strain Kyu89–246. MEL treatment results in inhibition of conidial germination, germ tube elongation, and suppression of appressoria formation. This is primarily due to the unique property of MEL of reducing the hydrophobicity of any solid surfaces that can behave as an inducing factor for inhibiting fungal behavior of plant surfaces leading to disease suppression [84].

The overall findings explain the efficacy of glycolipids against various phytopathogens. Although a number of books explain the role of rhamnolipids and sohorolipids against fungal phytopathogens only a handful explain their potential against bacterial phytopathogens. Furthermore, the efficacy shown by the glycolipids, CBLs and MEL, against fungal and bacterial phytopathogens are hardly reported. Thus, there is the necessity for further investigation to establish CBLs and MEL as antimicrobials for commercial use in agricultural sectors.

22.3 Biosurfactant in Agrochemical Formulations for Sustainable Agriculture

Biosurfactants exhibit certain noteworthy properties like foaming, dispersion, wetting, emulsification, penetration, and thickening that gained wide attention of researchers around the globe in recent years for use as agrochemical formulations. For instance, in the case of water-soluble insecticide acephate, the addition of rhamnolipids (a glycolipidic biosurfactant) to the insecticide formulation controls its solubility, thereby providing the insecticide at a predetermined rate. Similarly, there are instances when the solubility and translocation rate of insecticidal formulations were found to enhance by several fold on addition of rhamnolipid with water-soluble insecticide imidacloprid [95]. Apart from rhamnolipids, instances of sophorolipids used as adjuvants (substances other than water, which are non-active as an ingredient but enhance or support the effectiveness of an active ingredient) for pesticide formulations adds more evidence of using biosurfactants in agrochemical bioformulations [96]. It was reported that sophorolipids, when used as adjuvants in a formulation containing fungicide (Opus) and herbicide (Cato), both its fungicidal and herbicidal activity were enhanced by several hundred times. Biosurfactants in pesticide formulations can also increase the wettability of leaf surfaces, which helps in long-lasting residence of the pesticides on the leaf surfaces for prolonged protection against several phytopathogens. Moreover, they can be used as nutrient solubilizers in nutrient formulations for effective assimilation of nutrients by plants [97]. To date, "Serenade" (AgraQuest, Davis, CA), a fengycin-based product, is one of the well-known commercialized green antifungal products used in the agricultural sectors [98].

22.4 Biosurfactants for a Greener and Safer Environment

Biosurfactants are believed to contribute significantly toward management of resource usage that is non-renewable in nature. They act as important agents for environmental sustainability. The use of low-cost substrates for biosurfactant production confers lasting advantage over environmental sustainability as the use of industrial wastes and byproducts fosters environmental protection. Utilization of industrial wastes for mass production not only reduces its cost of production but also reduces the polluting effect caused due to disposition of these waste products either by landfill disposal or combustion. Moreover, crude biosurfactants can be suitably implemented for environmental applications due to their eco-friendly and non-toxic nature [99, 100]. Meanwhile, it reduces the use of toxic chemicals and solvents that ultimately helps in achieving sustainability [101].

22.5 Conclusion

With the rapid increase in world population, there is the need to boost global food production. Limited availability of cultivable land and an increase in microbial infection in crop plants acts as a barrier for reliable food production to meet the global demand for food supply. A high reliance on chemical fungicides put harmful impact on the environment and mankind and thus there is a need for the implementation of a biological approach. Microbial surfactants are the most promising surface-active microbial metabolites and are endowed with several advantages over synthetic chemicals, being eco-friendly, non-toxic, biodegradable, and biocompatible. They are also found to enhance the nutrient transport across membranes and affect various host–microbe interactions. Therefore, it is essential to realize various important and beneficial aspects of microbial surfactants and their application in modern agriculture. Among the microbial surfactants, LPs and glycolipids are the most popularly and widely studied biosurfactants due to their promising antimicrobial activity. Although there are several reports stating the role of biosurfactants in agriculture, the technology involving biosurfactant-mediated control of plant pathogens is nascent and requires further understanding. The success of the technology depends on further investigations and characterization of microbial biosurfactants from different microbial sources and its implementation in agriculture on a commercial scale. In order to combat various pathogenic plant infections for sustainable agriculture, the implementation of biosurfactant-based strategies provides ample opportunities that have already been identified.

References

1 Thind, T.S. (2015). Perspectives on crop protection in India. *Outlooks on Pest Management* 26 (3): 121–127.

2 Prado, S., Li, Y., and Nay, B. (2012). Diversity and ecological significance of fungal endophyte natural products. In: *Studies in Natural Products Chemistry*, vol. 36 (ed. A. Rahman), 249–296. Elsevier, The Netherlands.

3 Jain, A., Sarsaiya, S., Wu, Q. et al. (2019). A review of plant leaf fungal diseases and its environment speciation. *Bioengineered* 10 (1): 409–424.

4 Kumar, S. (2014). Plant disease management in India: Advances and challenges. *African Journal of Agricultural Research* 9 (15): 1207–1217.

5 Meena, R.S., Kumar, S., Datta, R. et al. (2020). Impact of agrochemicals on soil microbiota and management: A review. *Land* 9 (2): 34.

6 Zhang, Q.C., Shamsi, I.H., Xu, D.T. et al. (2012). Chemical fertilizer and organic manure inputs in soil exhibit a vice versa pattern of microbial community structure. *Applied Soil Ecology* 57: 1–8.

7 Singh, J.S., Pandey, V.C., and Singh, D.P. (2011). Efficient soil microorganisms: a new dimension for sustainable agriculture and environmental development. *Agriculture, Ecosystems and Environment* 140 (3–4): 339–353.

8 Shalini, D., Benson, A., Gomathi, R. et al. (2017). Isolation, characterization of glycolipid type biosurfactant from endophytic *Acinetobacter* sp. ACMS25 and evaluation of its biocontrol efficiency against *Xanthomonas oryzae*. *Biocatalysis and Agricultural Biotechnology* 11: 252–258.

9 Santos, D.K., Rufino, R.D., Luna, J.M. et al. (2016). Biosurfactants: Multifunctional biomolecules of the 21st century. *International Journal of Molecular Sciences* 17 (3): 401.

10 Akbari, S., Abdurahman, N.H., Yunus, R.M. et al. (2018). Biosurfactants – A new frontier for social and environmental safety: A mini review. *Biotechnology Research and Innovation* 2 (1): 81–90.

11 Inès, M. and Dhouha, G. (2015). Glycolipid biosurfactants: Potential related biomedical and biotechnological applications. *Carbohydrate Research* 416: 59–69.

12 Ortiz, A., Teruel, J.A., Espuny, M.J. et al. (2009). Interactions of a bacterial biosurfactant trehalose lipid with phosphatidylserine membranes. *Chemistry and Physics of Lipids* 158 (1): 46–53.

13 Elshikh, M., Marchant, R., and Banat, I.M. (2016). Biosurfactants: promising bioactive molecules for oral-related health applications. *FEMS Microbiology Letters* 363 (18): fnw213.

14 Toral, L., Rodríguez, M., Béjar, V., and Sampedro, I. (2018). Antifungal activity of lipopeptides from *Bacillus* XT1 CECT 8661 against *Botrytis cinerea*. *Frontiers in Microbiology* 9: 1315.

15 Mnif, I. and Ghribi, D. (2015). Review lipopeptides biosurfactants: Mean classes and new insights for industrial, biomedical, and environmental applications. *Peptide Science* 104 (3): 129–147.

16 Beltran-Gracia E, Macedo-Raygoza G, Villafaña-Rojas J, Martinez-Rodriguez A, Chavez-Castrillon YY, Espinosa-Escalante FM, Di Mascio P, Ogura T, Beltran-Garcia MJ. Production of lipopeptides by fermentation processes: Endophytic bacteria, fermentation strategies and easy methods for bacterial selection. In: A. F. Jozala (ed.), *Fermentation Processes* 2017 (pp.199–222). InTech, Jeneza, Croatia.

17 Leclere, V., Béchet, M., Adam, A. et al. (2005). Mycosubtilin overproduction by *Bacillus subtilis* BBG100 enhances the organism's antagonistic and biocontrol activities. *Applied and Environmental Microbiology* 71 (8): 4577–4584.

18 Jacques, P. (2011). Surfactin and other lipopeptides from *Bacillus* spp. In: *Biosurfactants* (ed. G. Soberon-Chavez), 57–91. Berlin, Heidelberg: Springer.

19 Youssef, N.H., Duncan, K.E., and McInerney, M.J. (2005). Importance of 3-hydroxy fatty acid composition of lipopeptides for biosurfactant activity. *Applied and Environmental Microbiology* 71 (12): 7690–7695.

20 Bartal, A., Vigneshwari, A., Bóka, B. et al. (2018). Effects of different cultivation parameters on the production of surfactin variants by a *Bacillus subtilis* strain. *Molecules* 23 (10): 2675.

21 Ali, S., Hameed, S., Imran, A. et al. (2014). Genetic, physiological and biochemical characterization of *Bacillus* sp. strain RMB7 exhibiting plant growth promoting and broad spectrum antifungal activities. *Microbial Cell Factories* 13 (1): 144.

22 Isogai, A., Takayama, S., Murakoshi, S., and Suzuki, A. (1982). Structure of β-amino acids in antibiotics iturin A. *Tetrahedron Letters* 23 (30): 3065–3068.

23 Ongena, M. and Jacques, P. (2008). *Bacillus* lipopeptides: Versatile weapons for plant disease biocontrol. *Trends in Microbiology* 16 (3): 115–125.

24 Cameotra, S.S. and Makkar, R.S. (2004). Recent applications of biosurfactants as biological and immunological molecules. *Current Opinion in Microbiology* 7 (3): 262–266.

25 Patel, H., Tscheka, C., Edwards, K. et al. (2011). All-or-none membrane permeabilization by fengycin-type lipopeptides from *Bacillus subtilis* QST713. *Biochimica et Biophysica Acta (BBA) – Biomembranes* 1808 (8): 2000–2008.

26 Falardeau, J., Wise, C., Novitsky, L., and Avis, T.J. (2013). Ecological and mechanistic insights into the direct and indirect antimicrobial properties of *Bacillus subtilis* lipopeptides on plant pathogens. *Journal of Chemical Ecology* 39 (7): 869–878.

27 Raaijmakers, J.M., De Bruijn, I., Nybroe, O., and Ongena, M. (2010). Natural functions of lipopeptides from *Bacillus* and *Pseudomonas*: More than surfactants and antibiotics. *FEMS Microbiology Reviews* 34 (6): 1037–1062.

28 Nguyen, D.D., Melnik, A.V., Koyama, N. et al. (2016). Indexing the *Pseudomonas* specialized metabolome enabled the discovery of poaeamide B and the bananamides. *Nature Microbiology* 2 (1): 16197.

29 Li, X., Zhang, Y., Wei, Z. et al. (2016). Antifungal activity of isolated *Bacillus amyloliquefaciens* SYBC H47 for the biocontrol of peach gummosis. *PLoS One* 11 (9): e0162125.

30 Vallet-Gely, I., Novikov, A., Augusto, L. et al. (2010). Association of hemolytic activity of *Pseudomonas entomophila*, a versatile soil bacterium, with cyclic lipopeptide production. *Applied and Environmental Microbiology* 76 (3): 910–921.

31 Janek, T., Łukaszewicz, M., Rezanka, T., and Krasowska, A. (2010). Isolation and characterization of two new lipopeptide biosurfactants produced by *Pseudomonas fluorescens* BD5 isolated from water from the Arctic Archipelago of Svalbard. *Bioresource Technology* 101 (15): 6118–6123.

32 Emanuele, M.C., Scaloni, A., Lavermicocca, P. et al. (1998). Corceptins, new bioactive lipodepsipeptides from cultures of *Pseudomonas corrugata*. *FEBS Letters* 433 (3): 317–320.

33 Ballio, A., Bossa, F., Camoni, L. et al. (1996). Structure of fuscopeptins, phytotoxic metabolites of *Pseudomonas fuscovaginae*. *FEBS Letters* 381 (3): 213–216.

34 Nybroe, O. and Sørensen, J. (2004). Production of cyclic lipopeptides by fluorescent pseudomonads. In: *Pseudomonas* (ed. J.L. Ramos), 147–172. Boston, MA, USA: Springer.

35 Goswami, M. and Deka, S. (2019). Biosurfactant production by a rhizosphere bacteria *Bacillus altitudinis* MS16 and its promising emulsification and antifungal activity. *Colloids and Surfaces B: Biointerfaces* 178: 285–296.

36 Lahkar, J., Borah, S.N., Deka, S., and Ahmed, G. (2015). Biosurfactant of *Pseudomonas aeruginosa* JS29 against *Alternaria solani*: the causal organism of early blight of tomato. *BioControl* 60 (3): 401–411.

37 Bonmatin, J.M., Laprévote, O., and Peypoux, F. (2003). Diversity among microbial cyclic lipopeptides: iturins and surfactins. Activity-structure relationships to design new bioactive agents. *Combinatorial Chemistry and High Throughput Screening* 6 (6): 541–556.

38 Sarwar, A., Hassan, M.N., Imran, M. et al. (2018). Biocontrol activity of surfactin A purified from *Bacillus* NH-100 and NH-217 against rice bakanae disease. *Microbiological Research* 209: 1–3.

39 Deepak, R. and Jayapradha, R. (2015). Lipopeptide biosurfactant from *Bacillus thuringiensis* pak2310: A potential antagonist against *Fusarium oxysporum*. *Journal de Mycologie Medicale* 25 (1): e15–e24.

40 Lee, S.C., Kim, S.H., Park, I.H. et al. (2007). Isolation and structural analysis of bamylocin A, novel lipopeptide from *Bacillus amyloliquefaciens* LP03 having antagonistic and crude oil-emulsifying activity. *Archives of Microbiology* 188 (4): 307–312.

41 Melo, F.M., Fiore, M.F., Moraes, L.A. et al. (2009). Antifungal compound produced by the cassava endophyte *Bacillus pumilus* MAIIIM4A. *Scientia Agricola* 66 (5): 583–592.

42 Ji, S.H., Paul, N.C., Deng, J.X. et al. (2013). Biocontrol activity of *Bacillus amyloliquefaciens* CNU114001 against fungal plant diseases. *Mycobiology* 41 (4): 234–242.

43 Latoud, C., Peypoux, F., and Michel, G. (1987). Action of iturin A, an antifungal antibiotic from Bacillus subtilis, on the yeast *Saccharomyces cerevisiae*: modifications of membrane permeability and lipid composition. *The Journal of Antibiotics* 40 (11): 1588–1595.

44 Maget-Dana, R. and Peypoux, F. (1994). Iturins, a special class of pore-forming lipopeptides: Biological and physicochemical properties. *Toxicology* 87 (1–3): 151–174.

45 Jin, P., Wang, H., Tan, Z. et al. (2020). Antifungal mechanism of bacillomycin D from *Bacillus velezensis* HN-2 against *Colletotrichum gloeosporioides* Penz. *Pesticide Biochemistry and Physiology* 163: 102–107.

46 Gu, Q., Yang, Y., Yuan, Q. et al. (2017). Bacillomycin D produced by *Bacillus amyloliquefaciens* is involved in the antagonistic interaction with the plant-pathogenic fungus *Fusarium graminearum*. *Applied and Environmental Microbiology* 83 (19): e01075–e01017.

47 Ma, Z., Wang, N., Hu, J., and Wang, S. (2012). Isolation and characterization of a new iturinic lipopeptide, mojavensin A produced by a marine-derived bacterium *Bacillus mojavensis* B0621A. *The Journal of Antibiotics* 65 (6): 317–322.

48 Blacutt, A.A., Mitchell, T.R., Bacon, C.W., and Gold, S.E. (2016). *Bacillus mojavensis* RRC101 lipopeptides provoke physiological and metabolic changes during antagonism against *Fusarium verticillioides*. *Molecular Plant-Microbe Interactions* 29 (9): 713–723.

49 Zhang, L. and Sun, C. (2018). Cyclic lipopeptides fengycins from marine bacterium *Bacillus subtilis* kill plant pathogenic fungus *Magnaporthe grisea* by inducing reactive oxygen species production and chromatin condensation. *Applied and Environmental Microbiology* 84: e00445–e00418.

50 Hu, L.B., Shi, Z.Q., Zhang, T., and Yang, Z.M. (2007). Fengycin antibiotics isolated from B-FS01 culture inhibit the growth of *Fusarium moniliforme* Sheldon ATCC 38932. *FEMS Microbiology Letters* 272 (1): 91–98.

51 Guo, Q., Dong, W., Li, S. et al. (2014). Fengycin produced by *Bacillus subtilis* NCD-2 plays a major role in biocontrol of cotton seedling damping-off disease. *Microbiological Research* 169 (7–8): 533–540.

52 Chen, L., Wang, N., Wang, X. et al. (2010). Characterization of two anti-fungal lipopeptides produced by *Bacillus amyloliquefaciens* SH-B10. *Bioresource Technology* 101 (22): 8822–8827.

53 Gong, A.D., Li, H.P., Yuan, Q.S. et al. (2015). Antagonistic mechanism of iturin A and plipastatin A from *Bacillus amyloliquefaciens* S76-3 from wheat spikes against *Fusarium graminearum*. *PLoS One* 10 (2): e0116871.

54 Thrane, C., Nielsen, T.H., Nielsen, M.N. et al. (2000). Viscosinamide-producing *Pseudomonas fluorescens* DR54 exerts a biocontrol effect on *Pythium ultimum* in sugar beet rhizosphere. *FEMS Microbiology Ecology* 33 (2): 139–146.

55 Pedras, M.S., Ismail, N., Quail, J.W., and Boyetchko, S.M. (2003). Structure, chemistry, and biological activity of pseudophomins A and B, new cyclic lipodepsipeptides isolated from the biocontrol bacterium *Pseudomonas fluorescens*. *Phytochemistry* 62 (7): 1105–1114.

56 Ballio, A., Bossa, F., Di Giorgio, D. et al. (1994). Novel bioactive lipodepsipeptides from *Pseudomonas syringae*: the pseudomycins. *FEBS Letters* 355 (1): 96–100.

57 Bender CL, Scholz-Schroeder BK. New insights into the biosynthesis, mode of action, and regulation of syringomycin, syringopeptin, and coronatine. In: J. L. Ramos (ed.), *Virulence and Gene Regulation* 2004 (pp. 125–158). Springer, Boston, MA, USA.

58 Bull, C.T., Wadsworth, M.L., Sorensen, K.N. et al. (1998). Syringomycin E produced by biological control agents controls green mold on lemons. *Biological Control* 12 (2): 89–95.

59 Hennessy RC, Olsson S, Stougaard P. Interaction of the psychrotroph *Pseudomonas fluorescens* In5 with phytopathogens in cold soils. In: *7th International Conference on Polar and Alpine Microbiology, Nuuk, Greenland,* 2017.

60 Mandal, S.M., Barbosa, A.E., and Franco, O.L. (2013). Lipopeptides in microbial infection control: scope and reality for industry. *Biotechnology Advances* 31 (2): 338–345.

61 Oliveras, À., Baró, A., Montesinos, L. et al. (2018). Antimicrobial activity of linear lipopeptides derived from BP100 towards plant pathogens. *PLoS One* 13 (7): e0201571.

62 Makovitzki, A., Viterbo, A., Brotman, Y. et al. (2007). Inhibition of fungal and bacterial plant pathogens in vitro and in planta with ultrashort cationic lipopeptides. *Applied and Environmental Microbiology* 73 (20): 6629–6636.

63 Geudens, N. and Martins, J.C. (2018). Cyclic lipodepsipeptides from *Pseudomonas* spp. – Biological Swiss army knives. *Frontiers in Microbiology* 9: 1867.

64 Bais, H.P., Fall, R., and Vivanco, J.M. (2004). Biocontrol of *Bacillus subtilis* against infection of *Arabidopsis* roots by *Pseudomonas syringae* is facilitated by biofilm formation and surfactin production. *Plant Physiology* 134 (1): 307–319.

65 Fan, H., Zhang, Z., Li, Y. et al. (2017). Biocontrol of bacterial fruit blotch by *Bacillus subtilis* 9407 via surfactin-mediated antibacterial activity and colonization. *Frontiers in Microbiology* 8: 1973.

66 Zeriouh, H., Romero, D., García-Gutiérrez, L. et al. (2011). The iturin-like lipopeptides are essential components in the biological control arsenal of *Bacillus subtilis* against bacterial diseases of cucurbits. *Molecular Plant-Microbe Interactions* 24 (12): 1540–1552.

67 Medeot, D.B., Fernandez, M., Morales, G.M., and Jofré, E. (2019). Fengycins from *Bacillus amyloliquefaciens* MEP218 exhibit antibacterial activity by producing alterations on the cell surface of the pathogens *Xanthomonas axonopodis* pv. *vesicatoria* and *Pseudomonas aeruginosa* PA01. *Frontiers in Microbiology* 10: 3107.

68 Kitamoto, D., Isoda, H., and Nakahara, T. (2002). Functions and potential applications of glycolipid biosurfactants. From energy-saving materials to gene delivery carriers. *Journal of Bioscience and Bioengineering* 94 (3): 187–201.

69 Shoeb, E., Akhlaq, F., Badar, U. et al. (2013). Classification and industrial applications of biosurfactants. *Academic Research International* 4 (3): 243.

70 Raza, Z.A., Khan, M.S., and Khalid, Z.M. (2007). Evaluation of distant carbon sources in biosurfactant production by a gamma ray-induced Pseudomonas putida mutant. *Process Biochemistry* 42 (4): 686–692.

71 Kulkarni, M., Chaudhari, R., and Chaudhari, A. (2007). Novel tensio-active microbial compounds for biocontrol applications. In: *General Concepts in Integrated Pest and Disease Management* (eds. A. Cianco and K.G. Mukherji), 295–304. Dordrecht: Springer.

72 Abdel-Mawgoud, A.M., Lépine, F., and Déziel, E. (2010). Rhamnolipids: diversity of structures, microbial origins and roles. *Applied Microbiology and Biotechnology* 86 (5): 1323–1336.

73 Ochsner, U.A., Fiechter, A., and Reiser, J. (1994a). Isolation, characterization, and expression in *Escherichia coli* of the *Pseudomonas aeruginosa* rhlAB genes encoding a rhamnosyltransferase involved in rhamnolipid biosurfactant synthesis. *Journal of Biological Chemistry* 269 (31): 19787–19795.

74 Ochsner, U.A., Koch, A.K., Fiechter, A., and Reiser, J. (1994b). Isolation and characterization of a regulatory gene affecting rhamnolipid biosurfactant synthesis in *Pseudomonas aeruginosa*. *Journal of Bacteriology* 176 (7): 2044–2054.

75 Vatsa, P., Sanchez, L., Clement, C. et al. (2010). Rhamnolipid biosurfactants as new players in animal and plant defense against microbes. *International Journal of Molecular Sciences* 11 (12): 5095–5108.

76 Hu, Y. and Ju, L.K. (2001). Purification of lactonic sophorolipids by crystallization. *Journal of Biotechnology* 87 (3): 263–272.

77 Lang, S., Katsiwela, E., and Wagner, F. (1989). Antimicrobial effects of biosurfactants. *Fat Science Technology* 91 (9): 363–366.

78 de Oliveira, M.R., Magri, A., Baldo, C. et al. (2015). Sophorolipids – A promising biosurfactant and its applications. *International Journal of Advanced Biotechnology and Research* 6 (2): 161–174.

79 Kim, K.J., Kim, Y.B., Lee, B.S. et al. (2002). Characteristics of sophorolipid as an antimicrobial agent. *Journal of Microbiology and Biotechnology* 12 (2): 235–241.

80 Sen, S., Borah, S.N., Kandimalla, R. et al. (2020). Sophorolipid biosurfactant can control cutaneous Dermatophytosis caused by *Trichophyton mentagrophytes*. *Frontiers in Microbiology* 11: 329.

81 Jezierska, S., Claus, S., and Van Bogaert, I. (2018). Yeast glycolipid biosurfactants. *FEBS Letters* 592 (8): 1312–1329.

82 Kulakovskaya, T., Shashkov, A., Kulakovskaya, E. et al. (2009). Extracellular cellobiose lipid from yeast and their analogues: Structures and fungicidal activities. *Journal of Oleo Science* 58 (3): 133–140.

83 Arutchelvi, J.I., Bhaduri, S., Uppara, P.V., and Doble, M. (2008). Mannosylerythritol lipids: A review. *Journal of Industrial Microbiology and Biotechnology* 35 (12): 1559–1570.

84 Yoshida, S., Koitabashi, M., Nakamura, J. et al. (2015). Effects of biosurfactants, mannosylerythritol lipids, on the hydrophobicity of solid surfaces and infection behaviours of plant pathogenic fungi. *Journal of Applied Microbiology* 119 (1): 215–224.

85 Haba, E., Pinazo, A., Jauregui, O. et al. (2003). Physicochemical characterization and antimicrobial properties of rhamnolipids produced by *Pseudomonas aeruginosa* 47T2 NCBIM 40044. *Biotechnology and Bioengineering* 81 (3): 316–322.

86 Reddy, K.S., Khan, M.Y., Archana, K. et al. (2016). Utilization of mango kernel oil for the rhamnolipid production by *Pseudomonas aeruginosa* DR1 towards its application as biocontrol agent. *Bioresource Technology* 221: 291–299.

87 Liu, Z., Zeng, Z., Zeng, G. et al. (2012). Influence of rhamnolipids and Triton X-100 on adsorption of phenol by *Penicillium simplicissimum*. *Bioresource Technology* 110: 468–473.

88 Yan, F., Xu, S., Guo, J. et al. (2015). Biocontrol of post-harvest *Alternaria alternata* decay of cherry tomatoes with rhamnolipids and possible mechanisms of action. *Journal of the Science of Food and Agriculture* 95 (7): 1469–1474.

89 Borah, S.N., Goswami, D., Lahkar, J. et al. (2015). Rhamnolipid produced by *Pseudomonas aeruginosa* SS14 causes complete suppression of wilt by *Fusarium oxysporum* f. sp. *pisi* in *Pisum sativum*. *Biocontrol* 60 (3): 375–385.

90 Goswami, D., Handique, P.J., and Deka, S. (2014). Rhamnolipid biosurfactant against *Fusarium sacchari* – The causal organism of pokkah boeng disease of sugarcane. *Journal of Basic Microbiology* 54 (6): 548–557.

91 Sha, R., Jiang, L., Meng, Q. et al. (2012). Producing cell-free culture broth of rhamnolipids as a cost-effective fungicide against plant pathogens. *Journal of Basic Microbiology* 52 (4): 458–466.

92 Nalini, S. and Parthasarathi, R. (2014). Production and characterization of rhamnolipids produced by *Serratia rubidaea* SNAU02 under solid-state fermentation and its application as biocontrol agent. *Bioresource Technology* 173: 231–238.

93 Chen, J., Liu, X., Fu, S. et al. (2020). Effects of sophorolipids on fungal and oomycete pathogens in relation to pH solubility. *Journal of Applied Microbiology* 128 (6): 1754–1763.

94 Lee, B.S. and Kim, E.K. (2005). Characteristics of microbial biosurfactant as an antifungal agent against plant pathogenic fungus. *Journal of Microbiology and Biotechnology* 15 (6): 1164–1169.

95 Awada SM, Awada MM, Spendlove RS, inventors; AGSCITECH, assignee. Method of controlling pests with biosurfactant penetrants as carriers for active agents. US Patent application US 13/181,746. 2011.

96 Giessler-Blank S, Schilling M, Thum O, Sieverding E, inventors; Evonik Degussa GmbH, assignee. Use of sophorolipids and derivatives thereof in combination with pesticides as adjuvant/additive for plant protection and the industrial non-crop field. US Patent US 9,351,485. 2016.

97 Thavasi, R., Marchant, R., and Banat, I.M. (2014). Biosurfactant applications in agriculture. In: *Biosurfactants: Production and Utilization – Processes, Technologies, and Economics* (eds. M. Kosaric and F.V. Sukan), 313–326. New York, USA: CRC Press.

98 Horn, J.N., Cravens, A., and Grossfield, A. (2013). Interactions between fengycin and model bilayers quantified by coarse-grained molecular dynamics. *Biophysical Journal* 105 (7): 1612–1623.

99 Sarma, H., Bustamante, K.L., and Prasad, M.N. (2019). Biosurfactants for oil recovery from refinery sludge: Magnetic nanoparticles assisted purification. In: *Industrial and Municipal Sludge* (eds. M.N.V. Prasad, P.J. de Campos Favas, M. Vithanage and S.V. Mohan), 107–132. Oxford, UK: Butterworth-Heinemann.

100 Saikia, R.R., Deka, S., Deka, M., and Sarma, H. (2012). Optimization of environmental factors for improved production of rhamnolipid biosurfactant by *Pseudomonas aeruginosa* RS29 on glycerol. *Journal of Basic Microbiology* 52 (4): 446–457.

101 Olasanmi, I.O. and Thring, R.W. (2018). The role of biosurfactants in the continued drive for environmental sustainability. *Sustainability* 10 (12): 4817.

Index

Biosurfactants for a Sustainable Future: Production and Applications in the Environment and Biomedicine, First Edition. Edited by Hemen Sarma and Majeti Narasimha Vara Prasad.
© 2021 John Wiley & Sons Ltd. Published 2021 by John Wiley & Sons Ltd.

g

n